Bio-Inspired Computing and Networking

Edited by Yang Xiao

CRC Press
Taylor & Francis Group
Boca Raton London New York

CRC Press is an imprint of the
Taylor & Francis Group, an **informa** business
AN AUERBACH BOOK

CRC Press
Taylor & Francis Group
6000 Broken Sound Parkway NW, Suite 300
Boca Raton, FL 33487-2742

© 2011 by Taylor and Francis Group, LLC
CRC Press is an imprint of Taylor & Francis Group, an Informa business

No claim to original U.S. Government works

Printed in the United States of America on acid-free paper
10 9 8 7 6 5 4 3 2 1

International Standard Book Number: 978-1-4200-8032-2 (Hardback)

This book is dedicated to my mother, who died of cancer just before the publication of this book.

Contents

PART III BIO-INSPIRED COMMUNICATIONS AND NETWORKS

Preface

Bio-inspired computing and communications are influenced in many ways by animal behavior, animal communication, and animal family structure as these biological systems possess similar features as those in computing and networking for the challenges faced in their natural operations. It is very important to connect computing and networking technologies to biological systems. The aim of this book is to provide state-of-the-art approaches and novel technologies in bio-inspired computing and networking, covering a range of topics so that it will be an excellent reference book for researchers in these areas.

This book investigates the fundamental aspects and applications of bio-inspired computing and networking and presents recent advances in these areas. The chapters have been contributed by several well-known researchers from around the world who are working in this field. The book contains 19 chapters, which are divided into three parts: Part I—Animal Behaviors and Animal Communications, Part II—Bio-Inspired Computing and Robots, and Part III—Bio-Inspired Communications and Networks. I hope that this book will be a good reference for researchers, practitioners, and students interested in the research, development, design, and implementation of bio-inspired computing and networking.

This book was made possible by the efforts of our contributors, who strove day and night to put together these chapters. We would also like to thank our publishers—without their encouragement and support, this book would never have seen the light of day.

For MATLAB® and Simulink® product information, please contact:

The MathWorks, Inc.
3 Apple Hill Drive
Natick, MA, 01760-2098 USA
Tel: 508-647-7000
Fax: 508-647-7001
E-mail: info@mathworks.com
Web: www.mathworks.com

Editor

Dr. Yang Xiao worked in the industry as a medium access control (MAC) architect and was involved in the IEEE 802.11 standard enhancement work before joining the Department of Computer Science at the University of Memphis in 2002. He is currently with the Department of Computer Science (with tenure) at the University of Alabama. He was a voting member of the IEEE 802.11 Working Group from 2001 to 2004. He is also a senior member of the IEEE. Dr. Xiao serves as a panelist for the U.S. National Science Foundation (NSF), the Canada Foundation for Innovation (CFI)'s Telecommunications Expert Committee, and the American Institute of Biological Sciences (AIBS). He also serves as a referee/reviewer for many national and international funding agencies. His research interests include security, communications/networks, robotics, and telemedicine. He has published more than 160 refereed journal papers and over 200 refereed conference papers and book chapters related to these areas. His research has been supported by the U.S. National Science Foundation (NSF), U.S. Army Research, the Global Environment for Network Innovations (GENI), Fleet Industrial Supply Center–San Diego (FISCSD), FIATECH, and the University of Alabama's Research Grants Committee. He currently serves as editor-in-chief for the *International Journal of Security and Networks* (*IJSN*) and the *International Journal of Sensor Networks* (*IJSNet*). He was also the founding editor-in-chief for the *International Journal of Telemedicine and Applications* (*IJTA*) (2007–2009).

Acknowledgment

This work was supported in part by the U.S. National Science Foundation (NSF) under the grant numbers CCF-0829827, CNS-0716211, and CNS-0737325.

Contributors

Neil William Adams
Department of Computer Science
University of Alabama
Tuscaloosa, Alabama

Ozgur B. Akan
Department of Electrical and
    Electronics Engineering
Koc University
Istanbul, Turkey

Quinton Alexander
Department of Computer Science
University of Alabama
Tuscaloosa, Alabama

Monica Anderson
Department of Computer Science
University of Alabama
Tuscaloosa, Alabama

S. Arakawa
Graduate School of Information
    Science and Technology
Osaka University
Osaka, Japan

Paolo Arena
Department of Electrical, Electronic
    and System Engineering
University of Catania
Catania, Italy

Panagiotis K. Artemiadis
Control Systems Laboratory
School of Mechanical Engineering
National Technical University of
    Athens
Athens, Greece

Baris Atakan
Department of Electrical and
    Electronics Engineering
Koc University
Istanbul, Turkey

Karen L. Bales
Department of Psychology
University of California
Davis, California

Calin Belta
Department of Mechanical Engineering
Division of Systems Engineering
Boston University
Boston, Massachusetts

Spring Berman
General Robotics, Automation, Sensing
    and Perception Laboratory
University of Pennsylvania
Philadelphia, Pennsylvania

Michele Bonnin
Department of Electronics
Politecnico di Torino
Turin, Italy

Iacopo Carreras
CREATE–NET
Trento, Italy

Paskorn Champrasert
Department of Computer Science
University of Massachusetts
Boston, Massachusetts

Imrich Chlamtac
CREATE–NET
Trento, Italy

Tomasz Chomiak
Transition Technologies S.A.
Warsaw, Poland

Fernando Corinto
Department of Electronics
Politecnico di Torino
Turin, Italy

Francesco De Pellegrini
CREATE–NET
Trento, Italy

S. Eum
Graduate School of Information
 Science and Technology
Osaka University
Osaka, Japan

Marco Gilli
Department of Electronics
Politecnico di Torino
Turin, Italy

Ádám Halász
General Robotics, Automation, Sensing
 and Perception Laboratory
University of Pennsylvania
Philadelphia, Pennsylvania

Heiko Hamann
Institute for Process Control and
 Robotics
Universität Karlsruhe (TH)
Karlsruhe, Germany

Stephan A. Hartmann
Corporate Technology
Siemens AG
Munich, Germany

M. Ani Hsieh
General Robotics, Automation, Sensing
 and Perception Laboratory
University of Pennsylvania
Philadelphia, Pennsylvania

Grzegorz Jarmoszewicz
Transition Technologies S.A.
Warsaw, Poland

Carolyn D. Kitzmann
Department of Psychology
University of California
Davis, California

Kostas J. Kyriakopoulos
Control Systems Laboratory
School of Mechanical Engineering
National Technical University of
 Athens
Athens, Greece

Kenji Leibnitz
National Institute of Information and
 Communications Technology
Kobe Advanced ICT Research Center
Kobe, Japan

Jian-Qin Liu
National Institute of Information and
    Communications Technology
Kobe Advanced ICT Research Center
Kobe, Japan

Andrew Markham
Department of Electrical Engineering
University of Cape Town
Rosebank, South Africa

Daniele Miorandi
CREATE–NET
Trento, Italy

Masayuki Murata
Graduate School of Information
    Science and Technology
Osaka University
Osaka, Japan

Luca Patané
Department of Electrical, Electronic
    and System Engineering
University of Catania
Catania, Italy

Pedro C. Pinto
Corporate Technology
Siemens AG
Munich, Germany

Thomas A. Runkler
Department of Mechanical Engineering
Instituto Superior Técnico
Technical University of Lisbon
Lisbon, Portugal

Thomas Schmickl
Department for Zoology
Karl-Franzens-Universität Graz
Graz, Austria

Vilmos Simon
Budapest University of Technology and
    Economics
Budapest, Hungary

João M.C. Sousa
Department of Mechanical Engineering
Instituto Superior Técnico
Technical University of Lisbon
Lisbon, Portugal

André Stauffer
Ecole Polytechnique Fédérale
Lausanne, Switzerland

Junichi Suzuki
Department of Computer Science
University of Massachusetts
Boston, Massachusetts

Konrad Swirski
Institute of Heat Engineering
Warsaw University of Technology
Warsaw, Poland

Gianluca Tempesti
University of York
York, United Kingdom

Endre Varga
Budapest University of Technology and
    Economics
Budapest, Hungary

Michal Warchol
Institute of Control and Computation
    Engineering
Warsaw University of Technology
Warsaw, Poland

Briana Wellman
Department of Computer Science
University of Alabama
Tuscaloosa, Alabama

Konrad Wojdan
Institute of Heat Engineering
Warsaw University of Technology
Warsaw, Poland

Yang Xiao
Department of Computer Science
University of Alabama
Tuscaloosa, Alabama

Tetsuya Yomo
Graduate School of Information
Science and Technology
Osaka University
Osaka, Japan

ANIMAL BEHAVIORS AND ANIMAL COMMUNICATIONS

I

Chapter 1

Animal Models for Computing and Communications: Past Approaches and Future Challenges

Karen L. Bales and Carolyn D. Kitzmann

Contents

Any model of communications that displays evolution (a selection process that acts on variation) is biologically inspired. However, here we consider models that are based explicitly on the social systems of specific animal species. Most current models focus on social insects, which display emergent behavioral properties of individuals that do not require centralized command. We discuss these previous models and present some ideas for models based on more complex social systems.

1.1 General Principles of Animal Communication

Animals communicate with conspecifics (members of their own species) and heterospecifics (members of another species) nearly every important function they perform, including foraging and predator defense. Animals must communicate these many functions in an effective way, maximizing the success of both the sender and the receiver while minimizing the cost of signaling. Just as animals themselves take on a wide variety of physical forms, the modes of communication used by animals also vary widely, including visual, acoustic, chemical, tactile, electrical, and seismic signals. These various signal modalities allow the animals using them to successfully navigate the physical and social landscapes that they inhabit. A signal must be suitable for the physical environment in which it is used,[1,2] and its form is likely to be highly adapted to perform a specific social function without imposing unnecessary costs on signalers and receivers.[3,4] In animal communication, form and function are inextricably linked because signals have been shaped by natural selection, such that animals whose signals are more adapted to their environment are more likely to survive and reproduce than competitors using less adapted signals. An important aspect of adaptation is developmental plasticity; animals that experience fluctuating social or physical environments may benefit from the ability to alter their communicative behavior to suit the changing conditions (e.g., Refs. [5–7]). Insights gained from studying animal models of communication can be informative for scientists and engineers designing communication systems for multi-agent teams.

The communication modality used by an animal is constrained both by the physical environment in which it is transmitted and by the social function of the signal. Animals that have constraints on visual contact, such as nocturnally active animals or those living in habitats with numerous visual obstructions, are likely to favor other communication modalities. An extreme example of this is the subterranean blind mole rat (subfamily Spalacinae), which spends nearly all its time underground.[8] Because visual input is severely restricted, their visual system has become markedly reduced, and they rely instead on acoustic, olfactory, and seismic communication.[8–10] Additional constraints on the signal modality stem from the social function of blind mole rat signals—that is, what information needs to be conveyed. Chemical signals can be long-lasting, remaining in a location long after the signaler moves away; for this reason, the blind mole rat uses chemical signaling to indicate territorial ownership. Acoustic communication, in contrast, transmits rapidly and then is gone. This rapid transmission, combined with physical limitations of subterranean burrows, favors the acoustic modality for immediate signaling over short distances; thus, the blind mole rat species *Spalax ehrenbergi* uses acoustic signals to communicate during courtship and other close-range situations. The burrows of blind mole rats restrict transmission of sound to short distances; as a result, mole rats also thump on burrow walls to produce seismic signals that transmit over much longer distances.[8] A final consideration regarding signal modality is that natural selection may favor the combination of multiple communication modalities within a signal, in part because one component of a signal may fail during transmission. This sort of multimodal communication is currently a highly active area of research (e.g., Refs. [11,12]), and it is thought to result in robustness in communication systems.[13]

Animal communication researchers have been investigating the factors affecting the evolution of signal form for decades, and this area continues to spark new and interesting findings. Signal form, like modality, has been shaped by natural selection to optimize transmission in the physical environment and to convey the relevant information without unnecessary costs.[4] In a now-classic study, Morton[1] showed that the structure of bird calls varies among birds living in forest and grassland habitats; specifically, forest birds produce relatively unmodulated calls within a narrow range of frequencies that are least attenuated by dense forest vegetation, while grassland birds produce rapidly modulated calls that are better transmitted through turbulent air common in open grasslands. While Morton's research shed light on the maximal distances over which sounds would propagate in different habitats, Wiley and Richards[2] proposed that there may be certain situations in which animal signals are optimized for shorter distances; when animals are more closely spaced, the use of a sound that degrades more rapidly could allow the receiver to determine the distance over which the sound traveled. Signal form is also driven by its function and the social context in which it is used. To illustrate the way that form and function are linked, we can consider auditory communication of the titi monkey as a model system. Monogamous, arboreal primate, titi monkeys have a highly complex vocal repertoire.[14,15] Titi monkeys live in nuclear family groups on small territories that they defend from neighbors and other potential intruders; the most conspicuous territory defense behavior is a loud call, produced either as a duet with both adults calling or as a solo call by the male.[16] These calls are characterized as low frequency and high amplitude, features that allow them to transmit effectively to animals on neighboring territories. In contrast, affiliative calls used within a social group consist of low-amplitude, high-frequency sounds that transmit only over short distances. An important factor in the form of signals is predation risk; signals that are conspicuous or highly localizable may draw unwanted attention to the signaler by an eavesdropping predator.[4] Thus, vocalizations intended to communicate with nearby group members, like the affiliative calls of titi monkeys, are optimized to carry only far enough and no further.

In many species, the communicative repertoire is largely innate, such that either it is intact from a very young age or it unfolds in a preprogrammed fashion in all members of a species. However, some higher vertebrates have developmental plasticity in their communication systems, enabling them to respond to fluctuations in their environment with altered responses. A problematic change for some species is increasing amounts of anthropogenic noise in their habitats, and several studies have documented ways in which animals adapt to noise. Marmosets (*Callithrix jacchus*) respond to increasing noise levels by increasing the amplitude and duration of calls, but not the number of syllables within a call sequence,[5] while many birds alter amplitude and frequency structure of vocalizations in response to noise.[7] Social change, especially changes in an individual's social group, can also induce alterations in animal signals. For example, a budgerigar (*Melopsittacus undulatus*) entering a new social group will alter its social call to match the call type used by existing group members.[17] In the greater spear-nosed bat (*Phyllostomus hastatus*), both new and resident group members alter call structure when new members join a group.[18] In both cases, altering call structure contributes to the maintenance of a common group call that signifies group membership to group members and outsiders. Furthermore, there is evidence that animals sometimes use the ability to alter call structure as an indicator of an animal's overall fitness. For example, female budgerigars prefer to mate with males that are better imitators of the female's call, and when females choose to mate with a male with impaired imitation skills, she is more likely to "cheat" on the male.[19] Assessment of the capacity for learning may be one way in which unmanned agents can avoid competing against more efficient agents.

Decades of animal communication research has uncovered wide variation in the mode and form of signals used by animals. This variation reflects adaptation of communicative systems to the physical environment as well as social context. Perhaps the most important lesson to be applied to the design of communication systems for unmanned agents is that no one solution will work for all situations, and, in fact, multiple solutions may provide superior performance in a single situation. Moreover, when an agent is expected to encounter a fluctuating physical environment (e.g., varying levels of noise) or to move in and out of changing multiagent groupings, the capacity to alter its behavior accordingly is likely to be beneficial. Given the limited range of current animal models, we argue that there is a need for a wider variety of animal models, especially models from socially complex animals such as nonhuman primates.

1.2 Current Bio-Inspired Modeling Approaches

One of the major challenges identified in bio-inspired modeling is the coordination of many separate agents, particularly ones that may be out of range of a centralized command.[20,21] This is the same problem faced by many individual animals, which may travel and forage out of sight of most of the rest of their social group. Insects, and to some extent lower vertebrates such as amphibians, have been used to model this problem.

1.2.1 Insect Models for Communication and Robotics

Most current bio-inspired models for communication and control of sensors or robots are based on social insects, in which large numbers of agents act independently and base the majority of their responses on either environmental cues or communication with a limited number of other agents (Refs. [21–24], Chapters 3, 5, and 10). Many social insects display a eusocial system in which individuals are specialized behaviorally and physiologically and have similar responses to environmental stimuli.[25] [27] Even in species without a caste system, workers may respond similarly to stimuli (although see Ref. [28]). However, the outcome of all these individual reactions is often an emergent property,[21] sometimes referred to as swarm intelligence. This has been compared to distributed, parallel computing. Various aspects of social communication such as the use of chemical trails,[29] cooperative behavior,[20] foraging, task allocation, nest construction,[22] and sensory systems[23] have been modeled in ants, bees, or wasps. Here, we particularly consider some of these models that focus on social behavior and communication.

One example of this approach, reviewed by Hirsch and Gordon,[22] is the modeling of task allocation in ants. Pacala and Gordon[30] produced two simple rules as to how ants should determine which task they undertook. As long as ants are successful in whatever task they are performing, they will keep doing so. If an ant is unsuccessful at whatever it is doing and meets a more successful individual, it will switch to the task that the other is performing. In this way, information transfer is to some extent dependent on large group size (and therefore lots of encounters between individuals). Pacala and colleagues go on to determine simple rules by which "success" at a task can be evaluated by an individual. In this and other models,[31] a worker's assessment of its own success becomes key to the success of the task-switching algorithms, at least as far as task allocation depends on local interaction with the environment rather than on social interactions. When responses to the local environment are the primary cues, group size is no longer related to sensitivity of task distribution.

Some of these insect models have been taken into simulation or even into robotics. Cazangi and colleagues[29] simulated an autonomous navigation system based on pheromone trails used by ants. Specifically, simulated robots tried to avoid collisions while capturing targets. The availability of "positive" (attractive) and "negative" (dangerous) pheromone trails, based on the trails left by ants to recruit or warn conspecifics, improved almost every measure, including increasing target captures and decreasing the mean distance traveled. This documents the utility of social information when it is available.

One study that translated insect behavior into robot behavior using distributed problem-solving rules modeled a cooperative behavior (i.e., one that requires more than one robot) based on retrieval of large prey in ants.[20] A system of five robots was constructed with the task of pushing a box too large for a single robot to move. Each individual robot was programmed to first find the box, and then to push the box. Robots were sensitive to their colleagues' behavior; they were programmed to avoid interfering with other robots, to slow down when approaching another robot and also to follow their next sensed neighbor. The main problem noted was stagnation (getting stuck around the box without pushing it). In further experiments,[32] stagnation recovery behaviors were added, based on the realignment and repositioning behaviors also shown by ants.

Another ant-inspired study focused on translating foraging behavior into robotics, using small, Khepera robots.[33] Robots received colony-level information on energy while they were at a central nest, but no information while they were out foraging. They were programmed to avoid each other to reduce competition. An intermediate group size was found to be the most efficient when food was distributed uniformly, due to negative interference at large group sizes. A second experiment found similar effects of group size with a clustered food distribution and "memory" for location of previous food sources. In a third experiment, information transfer about food source location was allowed between robots upon return to the nest; this significantly increased foraging efficiency while group size effects remained the same.

Other, non-group-living insects have also been used as models for an agent that is responsive to social stimuli.[34] Arkin used the praying mantis to model four socially oriented visuomotor behaviors: prey acquisition, predator avoidance, mating, and "chantilitaxia" or searching behavior. Though a single-robot implementation, this model included multiple simultaneous variables for the environmental stimuli (predators, mates, hiding places, pray), the internal state of the agent (fear, hunger, sex drive), and previous experience with the stimulus (i.e., learning). In this way, it incorporates both motivational or internal variables and interactions with the environment, with the robot's motion as an outcome variable. A fascinating addition to this system would be to provide robots for the other animals (mates, potential rivals) and thus more explicitly include social factors. Stoytchev and Arkin[35] produced a second robot model system, which once again included internal states, in this case, curiosity, frustration, homesickness, and anger, as well as a deliberative component, which allowed for the incorporation of high-level knowledge when available. Most interestingly, this robot was also programmed to interact with humans in several ways including getting their attention. It was tested in an office task, retrieving a fax that it had to have human help to place on its tray, and was successful on 7 out of 10 tries.

Insects (like most other animals) use multiple sensory modes, including chemical signals, that is, pheromones, to impart information. One model using dragonfly-inspired sensor node density control is presented in Ma.[23] Male dragonflies of the species *Aeschna cyanea* patrol mating sites and fight with other males they encounter. They display a density-dependent self-regulating mechanism, which could be relevant to one of the main problems in interactions between agents,[34] the avoidance of wasting resources in competition with superior agents.

1.2.2 Noninsect, Bio-Inspired Models for Communication and Robotics

Less common are models with more complex nervous systems and social hierarchies. A recently published simulation model[36] of information transfer in fish posited that individual fish would only change their behavior when they observed a large enough number (quorum) of conspecifics performing the new behavior. While this model was not advantageous for very small group sizes ($n = 4$), once groups reached $n = 8$, a quorum rule produced more "accurate" decision making for individuals. This response accurately reflected the behavior of real fish and is an example of a nonlinear, threshold effect of group size. However, while based on a more complex organism, this is still an example of a model that does not include recognition of other individuals and therefore more complex social interactions.

Weitzenfeld[37] presented a number of noninsect robot simulations. He first used frogs and toads as models for prey catching and predator avoidance, primarily modeling their visuomotor coordination; however, these models included only one agent. A multiple robot architecture was also proposed using wolf packs as a model for hunting behavior. In this model, there are two types of wolves, one alpha and several beta, and in which the betas "recognize" the alpha and operate by different rules. Alpha wolves pursue, attack, and eat prey. The betas group around the alpha, but there is no direct communication between the wolves, only local (visual) interaction with the environment. The beta wolves are provided with a number of behavioral options including wandering, formation behind the alpha, pursuit of prey, attacking, and eating. This model does include some individual recognition as well as heterogeneity of function; but it does not include a state variable (i.e., hunger) for the individuals, nor does it include any direct communication.

One unusual model (Ref. [38], Chapter 12) was intended to create a sensor communication system that would be used as a tracking system for wildlife that was less vulnerable to poachers. This system was based on the concept of a dominance hierarchy, which is a system displayed by a number of species, including Old World monkeys such as macaques and baboons.[39] As detailed in Section 1.4.1, these monkeys maintain very complex social systems in which not only is there individual recognition but also acute awareness of social status and function. This model creates dominance contests in which nodes constantly reassess their levels in the hierarchy. Fitness is assessed via levels of energy, and nodes with high social rank are more active in network tasks. This results in nodes with low amounts of energy participating less and therefore lasting longer than they would otherwise. This is a fascinating model although considerably simplified from the reality of how dominance hierarchies operate. For instance, it assumes that the largest or heaviest competitor (the one with the most energy) will be dominant. In reality, many dominance hierarchies are determined on the basis of coalitions between multiple animals[40] or on inheritance of rank from the mother. In addition, animals can differ according to dominance "style"—with some being "diplomats" rather than "bullies."[41]

Another simulation using animals with higher cognitive ability is given in Bryson's analysis of grooming behavior in chimpanzees.[42] Here, Bryson argues for embedding complexity within agents rather than within the communication process, especially when agents are heterogeneous. She suggests that the work to achieve a common framework of negotiation may be too onerous between heterogeneous agents. However, she also points out the need to test this assertion.

1.3 Challenges for Future Bio-Inspired Models

We have detailed above how insect models, which include direct communication, are strongly affected by group size. However, numbers of sensors or robots will never approach the numbers found in an insect swarm. In addition, higher vertebrates do not display the same sort of homogeneity of function and response as members of a hive. There is very little theoretical guideline for the use of heterogeneous swarms.[32] Thus, the charge is to make "smarter" models in which large numbers and homogeneity are replaced by more realistic numbers and individualism.

One of the challenges to doing so is the incorporation of more complicated and realistic variables such as internal states (motivation; hunger, sex drive, etc., as in Refs. [34,35]), as well as learning, memory, and personality, into the models. Collective group decision making (which direction a group should go, where to stop for the night) is poorly understood and could be affected by many differences at the individual level, including quality and quantity of information, and differences in experience and/or motivation.[43] A tool that might be particularly useful for this is individual-based, or agent-based, modeling.[21,44] The "Darwinian algorithms" used in this method include individual variation, assess the "fitness" of the possible solutions, and retain the fittest of these solutions for the next generation of the simulation. Experimental, real-world data are used to assess the fit of the model. For an example of the use of agent-based modeling to test models of macaque dominance hierarchies, see Bryson.[45] The model (or a modified one) can then be used to generate additional, testable predictions. This process may prove to be an important bridge between real-world behavior of complex individuals and programming of robots or sensors.

Many of the previous bio-inspired models, especially the few noninsect-based models, have included little to no communication between agents. This has often been seen as an advantage, in that environmental knowledge was gained directly and breakdowns in communication did not affect performance. However, when this information is available, it greatly improves performance. It seems that the *most* effective system would be one in which individuals could take advantage of others' knowledge when available but have rules for responding directly to the environment if necessary. In the following section, we suggest new, more complex animal models that might lead to future bio-inspiration.

1.4 Primates and Other Socially Complex Mammals as Biological Models

1.4.1 Challenge: Modeling Small vs. Large Groups and Group Heterogeneity

One challenge that has been identified for future development of bio-inspired models is the use of varied group sizes. Because primates of the New World are closely related yet live in widely varying social groupings, this family of monkeys offers much promise for investigating optimal group sizes for multiagent teams. Within the New World monkey family Cebidae, group sizes range from the small groups of monogamous monkeys (e.g., *Callicebus*) with groups as small as 2[46] to much larger groups up to 300 or more in the nonmonogamous species (e.g., *Saimiri*[47]). Moreover, in many New World primates, groups engage in fission–fusion behavior, in which larger groups break up fluidly into smaller subgroups to forage (e.g., Refs. [48,49]). Old World primates offer yet another model of large, heterogenous groups.

A great deal of variety in models can be obtained within the New World cebids. Titi monkey (*Callicebus*) group sizes are uniformly small, including some groups that consist only of a pair of adults, providing the smallest possible group size that could serve as a model for multiagent groupings. Titi monkeys have a complex signal repertoire, consisting of acoustic, visual, tactile, and olfactory signals used to communicate within the group as well as with neighboring groups. This signal repertoire is optimized to include both long- and short-range acoustic signals, for example, that accomplish the social function of each signal while minimizing the risk of eavesdropping by conspecifics or heterospecifics (e.g., predators) that are not the intended recipients. The nonmonogamous cebid monkeys live in groups that vary in composition, with the largest *Saimiri* groups including 500 or more individuals.[47] Some, but not all, of these species live in fission–fusion societies, in which group size and composition change fluidly as subgroups break off and return. Chapman[48] compared three non-monogamous cebid species, assessing the extent of subgroup formation based on ecological constraints (density and distribution of food). Spider monkeys (*Ateles geoffroyi*) and howlers (*Alouatta palliata*) both formed subgroups, and the size of these subgroups was associated with ecological constraints; capuchin monkeys (*Cebus capucinus*) did not form subgroups regardless of ecological pressures, with individuals remaining with their entire social group at all times. In the fission–fusion spider monkey society, it appears that monkeys use a single individually recognizable call, the "whinny," to maintain contact within and between subgroups. Interestingly, spider monkeys produce significantly more whinnies when another subgroup is within hearing range than they do when the nearest subgroup is out of hearing range, and whinnying appears to impact the movements of the nearby subgroup.[50] It is thought that the whinny call is crucial to the maintenance of social relationships in this fluidly changing society.

The Old World monkey subfamily Cercopithecinae, including macaques and baboons, is characterized by large multi-male, multi-female groups with obvious dominance hierarchies.[39] Female cercopithecines usually live their entire lives in the same social group into which they were born, while males reaching sexual maturity leave their natal groups and seek membership in new groups. Consequently, females maintain close relationships with female relatives, and these relationships constitute the core of cercopithecine social groups. Each group has a linear dominance hierarchy in which adult females have relatively stable dominance ranks that are passed from mother to daughter. Moreover, females tend to associate preferentially but not exclusively with matrilineal relatives,[39] although a bias toward paternal relatives may also be present.[51] These highly complex Old World monkey societies provide a model of successful communication within a heterogenous grouping that is intermediate in size between the smallest New World primate families and the very large groups that most current models use.

Current animal models of multiagent communication have not sufficiently answered how agents would behave in groups of small to intermediate size, and how the variability in group sizes might influence communication. Moreover, current models do not address agent behavior in fluidly changing group sizes and compositions, such as what is seen in many New World primate species or in heterogenous groups like the hierarchical, matrilineal cercopithecines.

1.4.2 Challenge: Incorporating Individual Differences or "Personality"

An active area of research is animal "personality" or individual differences in behavioral styles within a species.[52,53] In human personality research, there are five personality factors that are commonly described (openness, conscientiousness, extroversion, agreeableness, and neuroticism), and animal researchers have identified several of these factors in nonhuman primates, nonprimate mammals, and a few other animal species.[52] Gosling[53] argues that individual differences may be evolutionarily valuable because rare characteristics are often more advantageous than characteristics that are widespread in a population. This frequency-dependent selection is relevant to modeling autonomous agents, because it is likely that agents with a relatively rare behavior will be more successful at times than their counterparts with more typical behavior.

One species in which individual differences have been a major focus of research is the rhesus macaque (*Macaca mulatta*). John Capitanio and colleagues have built a research program in which personality assessments are done on large samples of macaques, both infants and adults, and from which two personality factors (sociability and confidence) have clearly emerged.[54,55] An interesting aspect of this research is that the researchers have linked personality traits to health outcomes,[56,57] indicating that individual differences in personality can have important implications for functioning of organisms and their long-term fitness. The inclusion of personality is one method of increasing complexity within agents, as suggested by Bryson.[42]

1.4.3 Challenge: Robustness to Damage

Failure of a signal to reach its recipient is always a risk during communication, and natural selection has shaped animal communication systems to add robustness. One way in which it is believed robustness is built into animal communication systems is via multimodality.[13] As described above, a multimodal signal is one in which multiple channels of communication are combined into one signal. Multimodal signals can be classified as "fixed," in which all components of the signal are always present, or "fluid," in which components do not always co-occur.[12] In redundant multimodal signals, the different signal components will elicit essentially the same response when produced in isolation. Redundant signals are advantageous when there is noise in the communication channel that may prevent transmission of one modality, and when it is especially important for relaying the most critical of messages. Non-redundant multimodal signals consist of more than one stream of information, with different channels providing different information. A situation in which this might be advantageous is when the signal recipient needs to pay close attention to one channel of communication; in this case, a second modality might be used to attract the recipient's attention to the signaler.[12] Multimodal signaling, although a new area of research, appears to be common among animals. It has been observed in taxa ranging from ants[58] and spiders[11] to howler monkeys.[59]

Redundant signaling does not always require the use of a single multimodal signal. An intriguing example of robustness in signaling comes from orangutans

(*Pongo pygmaeus*).[60] A recent study showed that when orangutans signal a request, their subsequent behavior is altered by whether their request was completely met, partially met, or unmet. When the request was completely met, orangutans typically ceased signaling; when it was partially met, they typically repeated their original signal. Interestingly, when the request was unmet, orangutans continued to signal, often changing the signal used. It is thought that the orangutans interpreted the "unmet" situation as a communication failure, and that they switched signaling tactics as a way to "repair" the failure. It appears, then, that an important way in which animal signals have achieved robustness is through multi-modality and other forms of redundancy.

1.4.4 Challenge: Avoiding Eavesdropping

A challenge faced by a signaler in any communication network is to ensure receipt of a signal by its intended target while minimizing eavesdropping, that is, receipt of the signal by undesired targets. Although communication is often conceived as a dyadic interaction in which a signal is passed from sender to receiver, the most common social environment in which animal communication occurs is a network in which multiple senders and receivers are present.[61] Thus, eavesdropping is widespread among animals, and signals have been shaped by natural selection to reduce the risk of detection or localizability in cases where it would be detrimental to the signaler. Animal models of eavesdropping avoidance typically involve reducing risk of detection by predators, such as the examples reviewed in the following text.

Contact calls, a class of vocalizations used to maintain contact between an individual and its groupmates, are used over very short and very long ranges. Regardless of the intended range of a contact call, its intended target includes only an animal's groupmates. Especially in the case of short-range calls, used when groupmates are nearby but need to maintain auditory contact, it is generally believed to be advantageous to minimize the risk of eavesdropping. Consistent with this notion, the structure of contact calls has been shown to vary according to whether the call is being used for short- or long-range communication. For example, pygmy marmoset varies the localizability of their trills based on the distance between sender and receiver, so that when animals are closely spaced the contact call used has relatively low localizability.[62] However, because any vocalization a marmoset produces can increase its risk to predators, all of these contact calls are produced at frequencies high enough so that predators cannot detect them.

As eluded to in the prior example, one method of eavesdropping avoidance is to produce a signal that is optimized to be detected by target receivers but not by the likely eavesdroppers. For example, in certain fish species (family Poeciliidae), males use conspicuous visual signals to attract mates, but they possess several tactics to reduce the conspicuousness of these signals to predators.[63] One tactic is that the male spots that are attractive to females use a "private wavelength," a color that is detectable to female conspecifics but not to predators. Male fish also use behavioral tactics, such as folding the spot-containing fin to obstruct the view of the spots except during courtship, when the fin is erected to display the spot for a female.

1.5 Conclusions

A variety of animal species, primarily insects, have been used as inspiration to model communications and robotics. In this chapter, we have discussed the successes and problems of previous models and suggested newer, more complex models. In particular, it would be useful for new models to incorporate complexity (personality) as well as state variables at the individual level as well as heterogeneity within groups.

Acknowledgments

This work was supported by the Good Nature Institute and NSF grant #0829828 to Yang Xiao, Xiaoyan Hong, Fei Hu, Monica Anderson, Karen Bales, and Jeffrey Schank.

References

1. E.S. Morton, Ecological sources of selection on avian sounds, *Am. Nat.*, 109, 17 (1975).
2. R.H. Wiley and D.G. Richards, Physical constraints on acoustic communication in the atmosphere: Implications for the evolution of animal vocalizations, *Behav. Ecol. Sociobiol.*, 3, 69 (1978).
3. E.S. Morton, On the occurrence and significance of motivation-structural rules in some bird and mammal sounds, *Am. Nat.*, 111, 855 (1977).
4. J.W. Bradbury and S.L. Vehrencamp, *Principles of Animal Communication* (Sinauer Associates Inc., Sunderland, MA, 1998).
5. H. Brumm, K. Voss, I. Köllmer, and D. Todt, Acoustic communication in noise: Regulation of call characteristics in a New World monkey, *J. Exp. Biol.*, 207, 443 (2004).
6. S.E.R. Egnor and M.D. Hauser, A paradox in the evolution of primate vocal learning, *Trends Neurosci.*, 27, 649 (2004).
7. G.L. Patricelli and J.L. Blickley, Avian communication in urban noise: Causes and consequences of vocal adjustment, *Auk*, 123, 639 (2006).
8. E. Nevo, G. Heth, and H. Pratt, Seismic communication in a blind subterranean mammal: A major somatosensory mechanism in adaptive evolution underground, *Proc. Natl Acad. Sci. USA*, 88, 1256 (1991).
9. B.R. Stein, Morphology of Subterranean Rodents, in *Life Underground: The Biology of Subterranean Rodents*, Eds. E.A. Lacey, J.L. Patton and G.N. Cameron (University of Chicago Press, Chicago, IL, 2000), p. 19.

10. G. Francescoli, Sensory capabilities and communication in subterranean rodents, in *Life Underground: The Biology of Subterranean Rodents*, Eds. E.A. Lacey, J.L. Patton, and G.N. Cameron (University of Chicago Press, Chicago, IL, 2000), p. 111.

11. G.W. Uetz and J.A. Roberts, Multisensory cues and multimodal communication in spiders: Insights from video/audio playback studies, *Brain, Behav. Evolut.*, 59, 222 (2002).

12. S.R. Partan and P. Marler, Issues in the classification of multimodal communication signals, *Am. Nat.*, 166, 231 (2005).

13. N. Ay, J. Flack, and D.C. Krakauer, Robustness and complexity co-constructed in multimodal signaling networks, *Phil. Trans. R. Soc. B*, 362, 441 (2007).

14. M. Moynihan, Communication in the Titi monkey, Callicebus, *J. Zool. Soc. Lond.*, 150, 77 (1966).

15. J.G. Robinson, An analysis of the organization of vocal communication in the titi monkey, Callicebus moloch, *Zeitschrift für Tierpsychologie*, 49, 381 (1979).

16. J.G. Robinson, Vocal regulation of inter- and intragroup spacing during boundary encounters in the Titi monkey, Callicebus moloch, *Primates*, 22, 161 (1981).

17. P. Bartlett and P.J.B. Slater, The effect of new recruits on the flock specific call of budgerigars (Melopsittacus undulatus), *Ethol. Ecol. Evol.*, 11, 139 (1999).

18. J.W. Boughman, Vocal learning by greater spear-nosed bats, *Proc. R. Soc. Lond. B*, 265, 227 (1998).

19. M.L. Moravec, G.F. Striedter, and N.T. Burley, Assortative pairing based on contact call similarity in budgerigars, melopsittacus undulates, *Ethol*, 112, 1108 (2006).

20. C.R. Kube and H. Zhang, Collective Robotics: From Social Insects to Robots, *Adapt. Behav.*, 2, 2 (1993).

21. K.-J. Kim and S.-B. Cho, A comprehensive overview of the applications of artificial life, *Artif. Life*, 12, 153 (2006).

22. A.E. Hirsch and D.M. Gordon, Distributed problem solving in social insects, *Ann. Math. Arti. Intell.*, 31, 199 (2001).

23. Z. Ma and A.W. Krings, Insect sensory systems inspired computing and communications, *Ad Hoc Netw.*, 7, 742 (2009).

24. F. Dressler, Benefits of bio-inspired technologies for networked embedded systems: An overview, *Dagstuhl Sem. Proc.*, Schloss Dagstuhl, Wadern, Germany, January (2006).

25. E.O. Wilson, *The Insect Societies* (Harvard University Press, Cambridge, MA, 1971).

26. A.F.G. Bourke and N.R. Franks, *Social Evolution in Ants* (Princeton University Press, Princeton, NJ, 1995).

27. B. Danforth, Bees, *Curr. Biol.*, 17, R156 (2007).

28. M.P. Schwarz, M.H. Richards, and B.N. Danforth, Changing paradigms in insect social evolution: Insights from halictine and allodapine bees, *Ann. Rev. Entomol.*, 52, 127 (2007).

29. R.R. Cazangi, F.J. Von Zuben, and M.F. Figueredo, Autonomous navigation system applied to collective robotics with ant-inspired communication, *GECCO '05*, 2005.

30. S.W. Pacala, D.M. Gordon, and H.C.J. Godfray, Effects of social group size on information transfer and task allocation, *Evol. Ecol.*, 10, 127 (1996).

31. H.M. Pereira and D.M. Gordon, A trade-off in task allocation between sensitivity to the environment and response time, *J. Theor. Biol.*, 208, 165 (2001).
32. C.R. Kube and E. Bonabeau, Cooperative transport by ants and robots, *Robot. Auton. Syst.*, 30, 85 (2000).
33. M.J.B. Krieger, J.-B. Billeter, and L. Keller, Ant-like task allocation and recruitment in cooperative robots, *Nature*, 406, 992 (2000).
34. R.C. Arkin, K. Ali, A. Weitzenfeld, and F. Cervantes-Perez, Behavioral models of the praying mantis as a basis for robotic behavior, *J. Robot. Auton. Syst.*, 32, 39 (2000).
35. A. Stoytchev and R.C. Arkin, Combining deliberation, reactivity, and motivation in the context of a behavior-based robot architecture, *IEEE* (2001).
36. A.J.W. Ward, D.J.T. Sumpter, I.D. Couzin, P.J.B. Hart, and J. Krause, Quorum decision-making facilitates information transfer in fish shoals, *PNAS*, 105, 6948 (2008).
37. A. Weitzenfeld, A prey catching and predator avoidance neural-schema architecture for single and multiple robots, *J. Intell. Robot. Syst.*, 51, 203 (2008).
38. A. Markham and A. Wilkinson, The adaptive social hierarchy: A self organizing network based on naturally occurring structures, *IEEE* (2006).
39. D.J. Melnick and M.C. Pearl, Cercopithecines in multimale groups: Genetic diversity and population structure, in *Primate Societies*, Eds. B.B. Smuts, D.L. Cheney, R.M. Seyfarth, R.W. Wrangham, and T.T. Struhsaker (University of Chicago Press, Chicago, IL, 1987), p. 121.
40. C.P. van Schaik, S.A. Pandit, and E.R. Vogel, A model for within-group coalitionary aggression among males, *Behav. Ecol. Sociobiol.*, 57, 101 (2004).
41. J.C. Ray and R.M. Sapolsky, Styles of male social behavior and their endocrine correlates among high-ranking wild baboons, *Am. J. Primatol.*, 28, 231 (1992).
42. J.J. Bryson, Where should complexity go? Cooperation in complex agents with minimal communication, *WRAC 2002*, 2564, 2003.
43. I.D. Couzin, J. Krause, N.R. Franks, and S.A. Levin, Effective leadership and decision-making in animal groups on the move, *Nature*, 433, 513 (2005).
44. J.C. Schank, Beyond reductionism: Refocusing on the individual with individual-based modeling, *Complexity*, 6, 33 (2001).
45. J.J. Bryson, Agent-based modelling as scientific method: a case study analysing primate social behavior, *Phil. Trans. R. Soc. B*, 362, 1685 (2007).
46. W.A. Mason, Social organization of the South American monkey, Callicebus moloch: A preliminary report, *Tulane Stud. Zool.*, 13, 23 (1966).
47. J.D. Baldwin and J.I. Baldwin, Squirrel monkeys (Saimiri) in natural habitats in Panama, Colombia, Brazil, and Peru, *Primates*, 12, 45 (1971).
48. C.A. Chapman, Ecological constraints on group size in three species of neotropical primates, *Folia Primat.*, 55, 1 (1990).
49. M.M. Symington, Fission-fusion social organization in Ateles and Pan, *Int. J. Primat.*, 11, 47 (1990).
50. G. Ramos-Fernández, Vocal communication in a fission-fusion society: Do spider monkeys stay in touch with close associates?, *Int. J. Primat.*, 26, 1077 (2005).

51. K. Smith, S.C. Alberts, and J. Altmann, Wild female baboons bias their social behaviour towards paternal half-sisters, *Proc. R. Soc. Lond. B*, 270, 503 (2003).

52. S.D. Gosling and O.P. John, Personality dimensions in nonhuman animals: A cross-species review, *Curr. Dir. Psych. Sci.*, 8, 69 (1999).

53. S.D. Gosling, From mice to men: What can we learn about personality from animal research?, *Psych. Bull.*, 127, 45 (2001).

54. J.P. Capitanio, Personality dimensions in adult male rhesus macaques: Prediction of behaviors across time and situation, *Am. J. Primat.*, 47, 229 (1999).

55. J.P. Capitanio and K.F. Widaman, Confirmatory factor analysis of personality structure in adult male rhesus monkeys (Macaca mulatta), *Am J. Primat.*, 65, 289 (2005).

56. J.P. Capitanio, S.P. Mendoza, and S. Baroncelli, The relationship of personality dimensions in adult male rhesus macaques to progression of simian immunodeficiency virus disease, *Brain Behav. Immun.*, 13, 138 (1999).

57. N. Maninger, J.P. Capitanio, S.P. Mendoza, and W.A. Mason, Personality influences tetanus-specific antibody response in adult male rhesus macaques after removal from natal group and housing relocation, *Am. J. Primat.*, 61, 78 (2003).

58. B. Hölldobler, Multimodal signals in ant communication, *J. Comp. Physiol. A*, 184, 129 (1999).

59. C.B. Jones and T.E. VanCantfort, Multimodal communication by male mantled howler monkeys (Alouatta palliata) in sexual contexts: A descriptive analysis, *Folia Primat.*, 78, 166 (2007).

60. E.A. Cartmill and R.W. Byrne, Orangutans Modify Their Gestural Signaling According to Their Audience's Comprehension, *Curr. Biol.*, 17, 1345 (2007).

61. P.K. McGregor and T.M. Peake, Communication networks: Social environments for receiving and signalling behavior, *Acta Ethol.*, 2, 71 (2000).

62. C.T. Snowdon and A. Hodun, Acoustic adaptations in pygmy marmoset contact calls: Locational cues vary with distances between conspecifics, *Behav. Ecol. Sociobiol.*, 9, 295 (1981).

63. J.A. Endler, Natural and sexual selection on color patterns in poeciliid fishes, *Env. Biol. Fishes*, 9, 173 (1983).

Chapter 2

Social Behaviors of the California Sea Lion, Bottlenose Dolphin, and Orca Whale

Neil William Adams and Yang Xiao

Contents

This chapter provides a survey to explore in depth the California sea lion, bottlenose dolphin, and orca whale. It covers the appearance, classification, habitat, diet, physiology, and intelligence of the three species. In order to illustrate the species' intelligence, this survey discusses social behaviors of the three species. These species have similar social abilities and functions illustrated, for example, by the common use of each in marine entertainment shows. Interaction with other species is also discussed. Important social functions of each species are breeding, birthing, pupping, and calving habits, and a large portion of this survey is dedicated to these topics.

2.1 California Sea Lion

2.1.1 Introduction to California Sea Lion

The *Merriam-Webster Online Dictionary* defines the California sea lion as "a small brown sea lion (*Zalophus californianus*) that occurs especially along the Pacific coast of North America from Vancouver Island to Baja California and the Gulf of California and in the Galapagos Islands and that is the seal most often trained to perform in circuses" [1]. California sea lions are commonly recognized for their boisterous barking, heavy socialization, and high intelligence. Animals presented as seals for a rehearsed entertainment show are usually California sea lions.

California sea lions usually have medium to dark brown fur, with males tending to be a darker shade of brown than females [2,7]. The bark of a sea lion sounds similar to that of a canine. Sea lions bark on land and while submerged under water. Barking is most prevalent during breeding season. Males establish territory with loud deep barks. Females and pups will communicate with higher-pitched barking. Sea lions show alarm by barking [7]. California sea lions have a few natural predators, including orca whales and white sharks. The diet for California sea lions primarily consists of fish and squid. Salmon, Pacific whiting, herring, and opalescent inshore squid are all common prey [7].

Breeding usually occurs in May or June. Males will establish rookeries and fast there for almost a month, living off of their blubber stores [5]. Larger males consequently have the advantage not only in territorial fights but also in fat storage, which allows them to fast for longer periods of time. Pups may be born on land or in water, grow very quickly, and learn quickly to hunt and swim. Their main breeding area is from Southern California to Central Mexico. Sea lions primarily choose to reside in coastal areas and estuaries with shallow water. Sea lions use sandy beaches, marinas, docks, jetties, and buoys as haul-out sites where large groups of sea lions (pods) will sit [5].

Sea lions use long and leathery flippers for mobility in water and on land. The flippers have nails [5]. Males weigh much more than females and have thick necks covered with a mane of hair that is noticeably longer than the rest of their fur [3].

2.1.2 Classification of the California Sea Lion

California sea lions are classified in the "Zalophus" genus, which is derived from two Greek words: "Za," which is translated as "an intensifying element," and "lophus," which is translated as "a crest" [4]. These translations refer to the large bony crest on the top of the sea lion's head.

The California sea lion is not a true seal, as it is a member of the Otariidae family and not the Phocidae family [5]. Members of the Otariidae family are commonly referred to as "eared seals," whereas members of the Phocidae family are commonly referred to as either "earless seals" or "true seals." A major distinction between otarrids and "true seals" is that sea lions use flippers to both walk and run on land rather than just to craw [5].

In one recent classification, Otarridae has seven genera: *Arctocephalus*, *Callorhinus*, *Eumetopias*, *Neophoca*, *Otaria*, *Phocarctos*, and *Zalophus* [6]. These genera among them are further divided into 16 species and 2 subspecies [6].

There are three established species of the genus *Zalophus californianus*: *Zalophus californianus californianus*, *Zalophus californianus wollebaeki*, and *Zalophus californianus japonicas*.

Z. c. californianus is found from around Vancouver Island and British Columbia to the Tres Marias Islands of Mexico [4]. Males move seasonally throughout this area, and females tend to be found around the Channel Islands of California.

Z. c. wollebaeki is found in the Galapagos Islands. Generally, they are smaller than sea lions found in California [4].

Z. c. japonica was found on the west side of Honshu Island or mainland Japan. It is now considered extinct [4].

2.1.3 Social Behaviors of Sea Lions

California sea lions live in large groups, called pods, during the fall, winter, and spring seasons. These pods are large and mostly unorganized. Sea lions are tightly packed in pods, and nonbreeding sites where sea lions gather on land are called haul-out sites. On the water's surface, sea lions float in groups called rafts.

2.1.3.1 Breeding and Perinatal Periods

With the coming of summer, sea lions divide into breeding groups of mature cows and bulls. Sea lion pups are typically born during these warm summer months. Typically, cows do not give birth to more than one pup during the 12 month gestation period [5]. For a period between 1 and 2 weeks directly after giving birth, sea lion cows will stay with the pup constantly. This period is known as the perinatal period. During this period, the mother will frequently nurse the pup, which allows the pup to grow quickly and gain energy stores soon after birth. Sea lion pups are born with open eyes and can learn to swim in as little as 2 weeks. This learning process occurs during playful interactions with other pups on the shore near the water [5]. Sea lion pups have distinctive individual calls that permit the mother to find her pup after going out on her own to hunt. Pups and mothers will also emit distinctive odors and sniff each other to identify one another after the cow returns from the hunt. In order to recognize the call and odors of the pup, the mother and pup will spend much of the perinatal period sniffing each other and recognizing each other vocally.

Because pups are raised virtually exclusively by the females, male sea lions compete in strong sexual competition. Male California sea lions tend to be polygamous and compete sexually for many females during breeding season. Successes of an individual male during a mating season depend primarily on the distribution of cow clusters. If females are spread out, a male will run out of energy stores more quickly than if the cows are closely packed [5]. Males set territories, called rookeries that extend from land into the ocean. Bulls patrol these territories and use loud barking and fighting to defend them against other bulls. Further into breeding season, fights between males will transform generally into more ritualized behaviors such as head and eye movements and threatening lunges. These territories are generally created soon after cows give birth, and copulation will occur 3–4 weeks after cows give birth to their pup [8].

2.1.3.2 Interactions between California Sea Lions and Humans

California sea lions were deployed in the Second Persian Gulf War by the United States Navy after extensive training, which allowed them to capture submerged objects and to locate enemy divers [20]. Once an enemy diver is located by the sea lion, it is trained to place a marker or restraining device on them and make a quick getaway. California sea lions were chosen for this because of their capacity to be trained by humans and their effective eyesight in near darkness [20].

2.1.3.3 Relationship between California Sea Lions and Dolphins

A study by the University of California Department of Ecology and Evolutionary Biology studied a possible relationship between California sea lions and bottlenose dolphins in inshore water hunting, and between California sea lions and bottlenose and common dolphins in offshore water [9]. In the Pacific, sea lions were found to hunt schools of fish individually or in small groups. The purpose of the study was to explore the relationship of groups of sea lions and dolphins—the species involved, the number of animals involved in each "aggregation," and the durability of these groups [9].

2.1.3.4 Methods of Study of Sea Lions and Dolphins

An association was considered formed when sea lions were less than 100 m away from dolphins for a 5 min sampling period [9]. An aggregation occurs when this relationship extended for 10 min or longer and when the sea lions and dolphins showed similar behavior. In Santa Monica Bay, sea lions were found with three dolphin species in aggregations that occurred for long periods of time—suggesting that these aggregations do not occur by chance. California sea lions associated mainly with common dolphins and then bottlenose dolphins. In more than half of the study's dolphins sightings, sea lions were also present [9]. Aggregations of sea lions and dolphins exhibited a high behavioral correlation in surface-feeding and traveling, and, to a lesser extent, socializing.

2.1.3.5 Results of Study of Sea Lions and Dolphins

In 49 sightings, sea lions synchronized surface-feeding behavior with dolphins 89% of the time [9]. Dolphins were rarely observed approaching or separating from sea lions. Forty-five sightings featured observations of one species clearly approaching the other, and 95.5% of the time sea lions made the initial approach [9]. Sea lions separated from the aggregation 97.6% of the time. Approach and separation by the sea lions were observed during traveling and feeding 80% and 73.3% of the time, respectively. Observations from the surface supported the idea that sea lions behaved similarly to dolphins with concentrated dives on encircled prey. Not once during 73 observations of sea lion/dolphin associations or aggregations were any hostile actions between the species seen [9]. There are a number of observations that indicate that sea lions "take advantage" of the dolphins' food-locating capabilities in a noncasual manner. Sea lions spend a large amount of time following dolphins during pre-feeding traveling, whereas dolphins showed no specific interest in following the sea lions during this time. Sea lions were observed to change behavior by raising their heads out of the water and by leaping to increase speed. These changes seemed to be so that they would not lose sight of the dolphins. The time sea lions spent feeding with the dolphins supports this relationship.

2.1.3.6 Discussions from Study of Sea Lions and Dolphins

In its discussions, this study concluded that aggregations primarily occurred to improve foraging for sea lions [9]. This leads the study to classify the relationship between California sea lions in Santa Monica and dolphins as social parasitism. It is suggested that dolphins may benefit somewhat from this behavior as an increased amount of predators around the fish encourage schooling behavior [9].

2.2 Bottlenose Dolphin

2.2.1 Introduction to Bottlenose Dolphin

The *Encyclopedia Britannica* describes the bottlenose dolphin as a "widely recognized species (*Tursiops truncatus*) of mammal belonging to the dolphin family, found worldwide in warm and temperate seas. Bottlenose dolphins reach an average length of 2.5–3 m and weight of 135–300 kg. Males are generally larger than females. A familiar performer at marine shows the species is characterized by a "built-in-smile" formed by the curvature of its mouth. It has also become the subject of scientific studies because of "its intelligence and its ability to communicate with its kind through sounds and ultrasonic pulses" [10]. *Merriam-Webster* defines the bottlenose dolphin as "a relatively small stout-bodied chiefly gray-toothed whale (*Tursiops truncatus*) with a prominent beak and falcate dorsal fin" [11]. The American Cetacean Society states that the bottlenose dolphin is a "relatively robust dolphin with a usually short and stubby beak—hence, the name bottlenose." It has more flexibility in its neck than other oceanic because five of the seven neck vertebrae are not fused together as in the other oceanic dolphins [12]. There are "18–26 pairs of sharp, conical teeth in each side of its jaw" [12]. National Geographic describes the bottlenose dolphin as the popular leaders of many marine animal shows. The species' intelligence and appearance are conducive to its training by humans. They are very effective swimmers and travel in groups that communicate through noises. Bottlenose dolphins hunt using clicking noises called "echolocation." The noise bounces off prey, which allows the dolphins to locate the prey [13]. Dolphins feed commonly on bottom fish, shrimp, and squid. Bottlenose dolphins are found in warm and tropical oceans around the earth. Today, there is very limited bottlenose dolphin fishing, despite the fact that they used to be hunted widely for oil and meat [13].

2.2.2 Classification of Bottlenose Dolphin

Bottlenose dolphins are of the genus *T. truncatus*, which comes from the family of oceanic whales called Delphinidae. They come from the order Cetacea, which includes whales, dolphins, and porpoises [14]. The genus *T. truncatus* divides into two species: the common bottlenose dolphin, or *T. truncatus*, and the Indo-Pacific bottlenose dolphin, or *T. aduncus*. The Indo-Pacific bottlenose dolphin is found on the east coast of the African continent to the coasts of India, China, Japan, and Australia. It lives close to the shore [14]. The Indo-Pacific bottlenose is usually smaller than the common bottlenose dolphin. The common bottlenose dolphin has a smaller rostrum (beak) than its Indo-Pacific brother [15].

2.2.3 Social Behaviors of the Bottlenose Dolphin

Bottlenose dolphins live in long-term social groups, which are called pods. Pods have many different compositions and sizes [16]. Off the eastern coast of the United States, females comprise pods of mothers and their most recent calves. Adult males were observed most often alone or in pairs and occasionally in trios. Adult males move between female groups and do not often associate with subadult males. Subadult bottlenose dolphins group together in both single-gender and mixed-gender groups. As pods are found in increasingly deeper water, they tend to be larger. Pods may sometimes join together in groups of two or more called aggregations. These aggregations sometimes include several hundred dolphins. Pods will draw closer together in cases of protection, fright, and family action. Pods will disperse somewhat during feeding, aggression, and above-average alertness [16].

Dolphins within each pod seem to establish powerful social bonds, recognize each other, and prefer certain animals over others during associations. Relationships between mothers and calves appear to be long lasting, as calves stay with their mothers for periods of 3–6 years or more. Adult male pairs also seem to share powerful, long-lasting bonds. Male pairs are often seen engaging in cooperative behaviors. Dominance between bottlenose dolphins is established commonly by biting, chasing, and smacking their teeth and tails [16]. To show aggression toward each other, dolphins most often use their teeth to scratch each other and inflict fast-healing lacerations. This has been determined by the teeth marks (parallel stripes) on the skin of dolphins. Blowing bubbles out of the blowhole is another way dolphins show aggression [16].

The point at which dolphins reach sexual maturity is somewhat more variable to the size of the dolphin than the age [17]. Females tend to reach sexual maturity after reaching 2.3 m, when they can be anywhere from 5 to 12 years old. Males tend to reach sexual maturity after reaching 2.5 m, when they can be anywhere from 10 to 12 years old. Females are observed to be most often responsible for initiating breeding behaviors and will do so throughout most of the year. Before breeding, the male may rub or nuzzle the female, and lift his head up and point his tail downward in a position called an "S-curve." Bottlenose dolphins may breed at any point throughout the year, but it varies according to location [17].

The gestation period for female bottlenose dolphins is about 12 months [18]. As mentioned earlier, breeding behaviors may be initiated throughout most of the year, but they are usually consistent according to location. For example, peak calving for dolphins on the west coast of Florida is observed to be in May, whereas peak calving on the coast of Texas is observed to be in March. A female potentially could breed every 2 years, but the most often observed interval between calf bearings is 3 years [18]. Deliveries are made in water and may be either head or tail first. Sometimes during birth, a second adult dolphin will stay close to the mother and calf during the process as an "auntie dolphin." This second adult may be either male or female [18]. Usually, this "auntie dolphin" is the only other dolphin the mother will allow close to her calf. After birth, the calf will be approximately 106–132 cm long and weigh about 20 kg. Initially, the dorsal fins of calves are flaccid, but after a few days, they become gradually stiffer. For around 6 months after birth, calves will show light lines on their side, which result from fetal folding. Calves are darker in color than adults [18].

Calves nurse from concealed abdominal mammary slits and do so close to the surface [18]. Observations show that calves begin nursing as little as 6 h after birth. In the calf's first 4–8 days of life, it may nurse up to eight times an hour and as little as three times an hour. Nursing instances are short and usually last 5–10 s. Calves may nurse for up to 18 months. Continuous nursing of the rich milk allows the rapid development of a thick layer of blubber. The mother dolphin will stay in close proximity to its calf and direct its movements. The calf will use the mother's slip stream to swim, which allows both to keep up with their pod [18].

Feeding by bottlenose dolphins peaks in the early morning and late afternoon [16]. Frequently, dolphins ride on ocean swells created naturally or by larger whales, a mother dolphin, or human vessels. Dolphins jump out of the water frequently, and they may jump as high as 16 ft from the water's surface. When the dolphin lands on its side or back, it is called a "breach." Dolphins of all ages interact with each other by chasing each other, carrying objects, tossing seaweed, and using other objects. These behaviors might be a sort of practice for feeding [16].

Dolphins have been observed in groups with other toothed whales, including pilot whales, spinner dolphins, spotter dolphins, and rough-toothed dolphins. Dolphins interact with larger whales by riding their pressure waves and swim in the waves of gray whales, humpback whales, and right whales. Often, bottlenose dolphins will force Pacific white-sided dolphins out of prime positions in these pressure waves. When coming into contact with sharks, dolphins may respond with aggression, avoidance, or sometimes even tolerance. In the wild, dolphins have been observed attacking and killing sharks. In response to humans, some dolphins try to interact by encouraging touching or feeding [16].

2.2.3.1 Bottlenose Dolphins and Behavior Imitation

Dolphins in the wild have been observed performing many activities such as swimming and surfacing in a synchronized fashion [19]. The Dolphin Institute studied the mimicking capabilities of dolphins by placing two dolphins side by side, divided by a screen that allows them to see each other, with one trainer for each. The first dolphin responds behaviorally to a request by its trainer who acts as the demonstrator. The second dolphin acts as the imitator and responds as requested by its trainer [19]. The study showed that bottlenose dolphins could reverse roles from demonstrator to imitator. The mimicking capabilities of the dolphins extend to movements by both the other dolphin and each dolphin's trainer. As a human trainer performs a pirouette, the dolphin would closely mimic the action. It was also determined that the dolphin could create anatomical analogies for mimicking human actions; when a trainer lifted his or her leg, the dolphin would respond by lifting its tail. Dolphins could mimic the movements of a human on a television screen as accurately as a live human. Also, according to the Dolphin Institute, bottlenose dolphins are very capable vocal mimickers [19].

2.2.3.2 Further Study of Interactions between Bottlenose Dolphins and Humans

In an article for National Geographic News by John Pickrell, the use of dolphins to locate sea mines in the Second Persian Gulf War is examined [20]: "Dolphins have the best sonar on this planet ... the Navy does not have any technological sonar that can find buried mines except for its dolphin system." The article quotes from Whitlow Au of the University of Hawaii's Marine Mammal Research Program, who studies marine bioacoustics. Au claims that "they can not only find objects like mines that may or may not be buried into the seabed, but they can distinguish them from clutter such as coral rock and man-made debris" [20].

2.2.3.3 Life and Social Analysis of Coastal North Carolina Bottlenose Dolphins

Duke University Doctoral student, Victoria Thayer, conducted a dissertation titled, "Life History Parameters and Social Associations of Female Bottlenose Dolphins Off North Carolina, USA." This dissertation studies the reproductive and social biology of bottlenose dolphins [21]. The coast of North Carolina was chosen because some reductions in population had occurred there. The first reduction was caused by a fishery that ran from 1797 until 1929, stopping only for the Civil War and reopening after in 1883. This fishery harvested 400–500 dolphins per year, except its peak years of 1885–1890, in which almost 2000 dolphins per year were killed. The second population reduction was an epizootic from 1987 to 1988 that is thought to have reduced the dolphin population by as much as 40%–50% [21]. Stranded dolphins at Virginia Beach, Virginia were investigated to better understand the demographics of the affected animals of the epizootic. The only disproportionately affected group of dolphins was about calves. The demography of bottlenose dolphins may also be affected by anthropogenic effects on prey. Studying the stomach contents of dolphins revealed that feeding is related to habitat type. The estuarine dolphins primarily prey on Atlantic croaker 5% and spot 26%, with coastal dolphins having a wider distribution of prey with Atlantic croaker 23% and spot 16% [21].

Sharks are the only regular predator of the bottlenose dolphin off North Carolina. Bull sharks are known to be found in estuarine waters off North Carolina. Of dolphins captured and released during study, 19% exhibited signs of violent interactions with sharks [21].

2.2.3.4 North Carolina Bottlenose Dolphin Societal Structure

Social structures of bottlenose dolphins along the Atlantic were studied by a collaborative comparison of photographs taken by studies from New Jersey to Florida [21]. This research seems to imply that at least three stocks of coastal bottlenose dolphins can be seen off North Carolina. The northern-most stock is the Northern Migratory Management Unit, which is seen off the coasts of New Jersey, Delaware, Maryland, and Virginia from May to October. This stock moves south to North Carolina from November through April. The second stock is the Northern North Carolina

Management Unit, which ranges from Virginia to Cape Lookout, North Carolina throughout the year and mixes with the Northern Migratory Unit during winter. The Southern North Carolina Management Unit is the third stock and is found from Cape Lookout, North Carolina to South Carolina year round. This stock mixes with the others off of North Carolina during the winter months [21].

2.2.3.5 Bottlenose Dolphins and Birth Factors

From 1995 to 2006, five health assessment capture operations resulted in the marking of 81 dolphins in North Carolina [21]. Interpopulation comparisons of the life history, demography, and social structure of dolphins are most likely to reveal constraints, geographical patterns, and other variations of biological traits. For example, dolphins in Sarasota Bay, Florida most often give birth from late spring to early summer. Around Shark Bay, Australia, birthing is only moderately seasonal, with the peak occurring mainly from October to December. Around Scotland and the Moray Firth, dolphins tend to birth primarily in the months of July, August, and September [21]. Sexual maturity occurs at different ages in these locations. For example, in Sarasota Bay, sexual maturity occurs between 5 and 10 years old. The first birth year for female dolphins around Australia, however, ranges from years 12 to 15 [21]. Inter-birth intervals for dolphins at Sarasota Bay and Shark Bay were similar ranging from 3 to 6 years, while inter-birth intervals for dolphins around Scotland were observed to be around 8 years [21].

Dolphins around Sarasota Bay were observed to stay with their mothers for 3–6 years. After this period, the mother would give birth to a new calf while the older calf left to join a group of mixed sex juvenile dolphins [21]. Juvenile dolphins stay together until they leave individually after becoming sexually mature. Upon reaching sexual maturity, dolphins mix more frequently with their own sex. Males form strong bonds with one or two other males. Larger groups of females are called bands. Band relationships may be based on kinship, but they are more often formed by similar reproductive states. Young calves spend some time with escort dolphins who are not their mother [21].

The objectives for the dissertation as the researcher stated them were,

> 1) to determine if seasonality of reproduction exists for North Carolina bottlenose dolphins, 2) to describe reproductive parameters, including inter-birth intervals, fecundity and mortality rates, and ages, and 3) to describe association patterns among females and determine if the strength of associations is positively correlated with similar reproductive state [21].

The dissertation studied neonatal strandings in which freshly dead or moderately decomposed dolphins were the subject. For neonatal sightings, the author would estimate the size of the observed school and then approach and position herself on a parallel course to take photographs [21]. The author defined schools as aggregations of similarly behaving dolphins within 100 m of each other. Date, location, size, composition, environmental conditions, behavior, and individual neonatal features for each school were recorded. Nineteen neonates were documented as being found stranded and had an average size of 108.2 cm. Strandings peaked in the months of April and May and were at their lowest in October, November, January, and February [21].

Birth dates for 10 neonates from 8 females were estimated. Six births occurred in May, and four occurred in June [21].

Discussion from these findings indicated a strong birth peak in the spring with low levels of strandings during the rest of the year [21]. Sightings peaked in the months of May and June. Discrepancy was accredited to the strandings being observed in estuaries and sightings occurring in coastal waters. The possibility that strandings occur because neonates are born outside the peak birthing period is also discussed [21].

The most frequent sightings of neonates occurred during the summer. Births by known females were estimated to occur primarily during May and June. Only one individual fell outside this period, which gave birth in September [21].

All of these findings lead to the conclusion that there is a spring birth peak and a smaller number of fall births. This spring peak by a population of North Carolina dolphins may reflect a mostly estuarine group during the summer [21].

The author implies that newborns have an increased survival rate in warmer waters and when there is adequate food for mothers [21]. In Virginia, for example, dolphins were spotted only in waters greater than 16°C. Dolphin births around Florida occurred primarily when waters were above 27°C [21].

This dissertation [21] also studies the female reproductive parameters of dolphins around North Carolina. Two ecotypes of Atlantic seaboard bottlenose dolphins are said to exist. The first ecotype consists of the coastal dolphins, which inhabit waters less than 25 m deep. The second type is offshore and is found typically in waters deeper than 200 m. There is an overlap between these two ecotypes in waters between 7.5 and 34 km from shore. Despite this overlap, there is a difference between the two ecotypes physiologically, ecologically, and genetically. The reproductive pattern of the coastal ecotype is the focus of a portion of the dissertation [21].

2.2.3.6 Inter-Birth Intervals for Bottlenose Dolphins

Reproductive intervals are of high interest for evaluating the status and populations of North Carolina dolphins [21]. This term is defined as the "time elapsed between successive births" [21]. The inter-birth interval for a mother was calculated by observing and calculating the years between an initial sighting of a mother and YOY (young of the year) and the subsequent year the mother was seen with another YOY [21]. If the female was not sighted for 1 year or more, then its interval was not included in the data analysis. Thirty individuals were chosen using the authors' photo-identification catalog which were sighted 15 times or more in the period between 1995 and 2004 with a YOY on three separate days. Additionally, 22 animals were chosen as known females from previous capture–release projects around North Carolina. Ninety-four YOYs were photographed that were born to 40 of the 52 total females. Of the 40 mothers that gave birth during the study period, 26 had multiple births. Fourteen female dolphins were photographed during the study period with only one YOY. Twelve females were not photographed with a YOY. Eleven mothers were photographed with two YOYs. Seven mothers were photographed with three YOYs. Six mothers were photographed with four YOYs [21]. Two females were photographed with five YOYs. Seventy-nine of the 89 total photographed dolphins were photographed the next year with their presumed mothers [21].

This data presented an average inter-birth interval of 2.9 years; however, individual females exhibited varying rates between births. Eleven females showed two different intervals, and one dolphin showed three different intervals [21]. The mode of inter-birth intervals was 2 years. In Shark Bay, inter-birth intervals tended to be between 3 and 6 years. The mode for Shark Bay dolphin inter-birth intervals was 4 years. Sarasota Bay had a much wider distribution with intervals falling usually between 2 and 10 years with a mode of 5 years [21]. Around Scotland, inter-birth intervals fall typically between 7 and 8 years. For female dolphins with surviving calves around North Carolina, the minimum inter-birth interval was 2 years. In Shark Bay, the minimum was 3 years; and in Sarasota Bay, the minimum was 2 years [21].

The author suggests several factors to explain the shorter minimum at the Atlantic coast than along Sarasota or Shark Bay. The first suggestion is that sustenance is not observed to be a limiting factor along North Carolina, leaving dolphins at their peak during pregnancy and lactation. It is also possible that dolphins exhibit a density-dependent response to changes in abundance [21].

2.2.3.7 Female Bottlenose Dolphin Social Patterns

Studies of bottlenose dolphin social structure from groups around Sarasota Bay, Shark Bay, and the Moray Firth have revealed the common tendency of fission–fusion. Fission–fusion is defined as the "association of individuals in small groups that frequently change in composition" [21]. This fluid structure, the author asserts, seems to identify the strength of individual bonds. In Sarasota Bay and Shark Bay, females were observed forming "loose groups of associates" [21]. Bands may be formed primarily through genetic relationships to the other group dolphins [21]. The relationship between individual females is measured by an estimate of time spent together on a scale ranging from 0 (never together) to 1 (always together). Females forming bands had a higher association measurement than did other females. In Shark Bay, females formed cliques or moderately stable bands that last for 5 years or more. Around the Moray Firth, Scotland, female bottlenose dolphins generally had a low association measurement. The objectives of the authors' research were "a) to describe patterns of social associations among female dolphins observed from 1995–2004 and b) to compare the association rates of females to the older calves and association rates of females with no calves" [21].

For instances in which the entire group was not photographed by observers [21], the Half-Weight Index (HWI) was used where $HWI = X/(X + 0.5)(Ya + Yb)$. X is the number of sightings of dolphin A and dolphin B in the same group, Ya is the number of sightings of dolphin A only, and Yb is the number of sightings of dolphin B only [21]. In each year, female dolphins were seen by observers, and they were classified in one of three ways [21]. The first classification was females with YOYs. The second classification was females with older calves. The third classification was females with no dependent calf. Overall, association calculations were low with a mean rate of 0.06. Association within classes was not significantly higher. These association rates are lower than those observed and calculated from different areas. Generally, estuary dolphins exhibit lower rates of seasonal movements and higher association rates, whereas offshore dolphins exhibit higher rates of seasonal movements and lower rates of association. The large number of groups around North Carolina may account for the lowest association rates to date, and a lower association rate may also be a product of the epizootic from 1987 to 1988. Bands of females were not observed around North Carolina. This is in contrast to Sarasota Bay where the survival of calves is highly related to socialization [21].

2.3 Killer Whale

2.3.1 Introduction to the Killer Whale

Merriam-Webster describes the killer whale as a relatively small-toothed whale (*Orcinus orca*). The killer whale is mostly black, with a black body and a white ventral side and patch behind the eye. It may grow to a length of 20–30 ft and is sometimes called an "orca" [22]. The *Encyclopedia Britannica* defines the killer whale as the "largest member of the dolphin family (Delphinidae). The killer whale is easy to identify by its size and striking coloration: jet-black on top and pure white below with a white patch behind each eye, another extending up each flank, and a variable 'saddle patch' just behind the dorsal fin." The killer whale is a respected carnivore, but it is not known to kill humans while in the wild. Killer whales have been captured and used in marine mammal shows [23]. Killer whales have the largest distribution amongst cetaceans. The NOAA Office of Protected Species says the killer whale "likely represents the most widely distributed mammal species in the world" [24]. Killer whales are known for their color pattern. Males are larger than females, with larger pectoral flippers, dorsal fins, tail flukes, and girths. The life expectancy in the wild is somewhat longer for females, with an average of 50 years and maximum around 90 years. Male life expectancy averages around 30 years and the maximum is 60 years. The diet of killer whales varies according to location and population. Killer whale populations in the Northeast Pacific may feed on salmonids, pinnipeds, and smaller cetaceans. Around Norway, killer whales feed primarily on herring, and populations near New Zealand feed on sharks and stingrays.

2.3.2 Social Behaviors

2.3.2.1 Basic Social Tendencies

Killer whales are extremely social creatures. They are most often observed in small social groups of 2–15 animals. These groups are generally called pods. Congregations of multiple pods sometimes occur around large populations of prey and for the purpose of mating. Echolocation is used to locate and label prey. It also serves a navigational purpose. Whales communicate with each other using whistles and pulsed calls. Calls are observed more frequently and resemble human screams and moans. Calls rapidly change tone and feature repetitive pulses [24]. Killer whales are most abundant in chilly waters, but they are also found in temperate waters. There are many distinct types of killer whales around the world. In the northeast Pacific, there are three types of killer whales: resident, transient, and offshore. Each of these categories is further divided into resident populations and then into pods. Different types of killer whales exhibit distinct diets, behaviors, habitats, and social tendencies. Interbreeding between populations or stocks is not likely, despite overlapping home ranges. Modern day threats to killer whales include ocean contaminants such as PCBs, declines in prey population, overfishing for prey, oil spills, and other human accidents such as boat collisions and accidental entanglement [24].

2.3.2.2 Coordinated Attacks on Seals and Penguins in the Antarctic

To describe the observation of seven killer whales conducting a coordinated attack on a block of ice on which a crabeater seal was hauled out, the Society for Marine Mammalogy wrote that "antarctic peninsula killer whales (*O. orca*) hunt seals and penguins on floating ice." The killer whales created a coordinated wave that forced the seal off the ice. After this observation, multiple similar events were observed and reported in the Antarctic. This paper [25] describes six occurrences, five of which involve attacks on seals and the other penguins. During these events, the number of whales involved ranged from 5 to 7, except for the attack on the penguin, which was conducted by an individual whale [25]. The attack on the Adelie Penguin did not yield a kill. Three attacks on crabeater seals yielded two definite kills and one probable kill. Two attacks on leopard seals yielded one kill and one survival by staying on the ice. An attack on a Weddell seal was observed with killer whales making five passes but having no confirmed kill. On January 15, 2006, one event was observed and filmed for about 30 min. Approximately at 2330, killer whales were seen hopping beside an ice floe in a manner that aroused suspicion regarding a possible attack. A male crabeater seal was on the ice floe. The attack group of whales consisted of five AF/SAM (adult female, small adult male), one juvenile, and one calf. Multiple attacks were coordinated by the group. The seal was dislodged and later killed [25]. At +3:12 into the filming, two whales submerged and began swimming directly under the ice floe. These whales created a large wave to tip the ice floe and then moved over the ice. This initial attack did not knock the seal off of the ice, but it did break the ice into five pieces. At +4:10, the seal left the smaller piece of broken ice and swam about 100 m to a new, larger piece of ice. From +5:37 to +14:56, the whale group created six further attack waves that reduced the size of the new ice floe. During this period in the film, the whales were also observed moving the ice to an area around which there were no other easily accessible floes. When directly moving the ice, the whales preferred using their rostrums. By +15:00, the crabeater seal was visibly and severely agitated, as evidenced by heavy breathing and completely turning on its back. During the 15th minute, four whales lined up together horizontally and swam directly under the ice floe passing around two stationary whales on the other side of the floe. This attack forced the seal into the water and the seal was observed in the mouth of a whale before being taken under water. At +20:48, the seal escaped from under water and attempted entry onto a third ice floe, but its hind flippers were grabbed by a whale and it was forced back into the ocean. At +25:32, the seal was observed dead [25]. For at least three of the five events involving seals, the seal was caught but eventually replaced on an ice floe alive. The authors suggest that this could save as training for juvenile whales or as a social-play-type behavior [25].

2.3.2.3 North Pacific Killer Wale Aggregations and Fish Hunting

The article in [29] discusses the frequent occurrence of aggregations of up to 100 North Pacific killer whales. The authors assert that the primary function of these gatherings is for reproductive purposes. It is also possible that these aggregations help establish social bonds between pods and that these aggregations aid in fish hunting, although there is no outstanding evidence toward the latter regarding more than one pod. Group living is thought to occur for three general reasons: (1) increased safety from predators, (2) increased hunting capabilities, and (3) localization of resources or suitable resting and breeding sites [26]. Since killer whales have no natural predators, that eliminates the first factor except in the possible case of protecting calves. In the North Pacific, the primary social unit of killer whales is the "matriline," which is made up of a female and several generations of her calves. For fish-eating killer whales (as opposed to mammal-eating killer whales), both sexes remain in the matriline for life. In the article, a pod is a group of whales that share a repertoire of calls and have social bonds. Fish-eating killer whale pods seem to benefit from large aggregations, as it is easier to control schools of fish and to locate dispersed groups of fish. High levels of acoustic activity were observed by large aggregations of fish-eating killer whales.

In mammal-eating killer whales, social activity increased with aggregation size. Adult males were observed swimming away apart from natal pods to interact socially and sexually with reproductive females. Sexual activity is not always reproductive, as the authors report observations of sexual interaction between adult male killer whales. These observations suggest that groups occur, in mammal-eating killer whales especially, as a sort of club to encourage social interaction [26]. The authors reported that aggregations most often were single pods, and then occurred in the order of two, three, and four pods. This led to the classification of "multi-pod" for five or more pods. The frequency of different activities by different amounts of pods was calculated pairwise by aggregation type and activity frequency. The differences for all pairwise calculations were very significant. The lowest absolute values of residuals had traveling and foraging in the single versus several pod comparison, resting in the single versus multi-pod comparison, and resting and socializing in the several versus multi-pod comparison. The highest absolute values of residuals had socializing in the single versus several pod comparison, foraging in the single versus multi-pod comparison, and foraging in the several versus multi-pod comparison [26]. The number of foraging events observed decreased as the number of pods increased, as observed mostly with one pod or subgroups of a pod. It is unlikely that food gathering plays a significant role in the formation of large aggregations of whales. During single-pod encounters, socializing was less prevalent than in several-pod or multi-pod encounters. Mating seemed to be a primary function of large aggregations of whales. Considering the fact that killer whales have relatively long lives and low birth rates, it seems likely that forming a sort of social club with which to build bonds might occur [26].

2.3.2.4 Killer Whale Interactions with Other Marine Mammals

The authors in [28] assert that while many social interactions involving the killer whale are predatory in nature, there are many types of interactions that humans are only beginning to understand. The introduction states that the feeding habits of killer whales are extremely diverse, ranging from fish and cephalopods to marine turtles, seabirds, and that they have even been found eating a deer carcass and the remains of a pig. Killer whales are most known for feeding on other marine mammals, such as the sperm whale. Ten other animals are considered marine-mammal feeders, and they include Polar Bears (*Ursus maritimus*), Steller Sea Lions (*Eumitopias jubatus*), New Zealand Sea Lions (*Phorcarctos hookeri*), Southern Sea Lions (*Otaria flavescens*), Walruses (*Odobenus rosmarus*), Leopard Seals (*Hydrurga leptonyx*), Short-finned Pilot Whales (*Globicephala macrorhynchus*), Pygmy Killer Whales (*Feresa attenuata*), False Killer Whales (*Pseudorca crassidens*), and Sperm Whales (*Physeter macrocephalus*). With the apparent exception of the killer whale, leopard seal, and polar bear, however, all of these predatory interactions with marine mammals appear to be "as a hobby" [27]. This is in contrast to the fact that some killer whales feed primarily on marine mammals. There have been a number of studies that pointed out the identified existence of two distinct types of killer whales, one feeding primarily on marine mammals and the other primarily on fish [27]. The purpose stated by the authors for this publication is to simply explore social interactions between killer whales and other marine mammals. The term "interaction" is stated by the authors to refer to the presence of two or more species in close proximity, and interaction can therefore be classified anywhere from predatory to mutual ignorance [27].

The list of cetaceans who fall prey to killer whales excludes river dolphins, but this seems to be only a function of location. The majority of reported attacks on large whales have included small groups of one to five killer whales [27]. In contrast, attacks on herds of small whales and dolphins tend to include the largest number of killer whales, usually involving 6–10 killer whales. Attacks on single dolphins or porpoises usually involve groups of one to five killer whales [27].

Pinnipeds comprise a substantial percentage of the diets of some killer whales, and killer whale attacks on pinnipeds have been reported from all over the world. The primary locations of attacks on pinnipeds are polar and subpolar climates. Monk seals appear to be the only major pinniped group with no known association as prey to killer whales. Most killer whale attacks on pinnipeds are conducted by groups of 10 or fewer. Preference or preselection of prey species is thought to occur in this social carnivore. In Argentina, killer whales have been observed gliding onto a beach and wriggling back into the ocean sometimes successfully, in an effort to capture prey on slightly sloped beaches [27].

Of other non-mentioned marine mammals, only the dugong (*Dugong dugon*) and sea otter (*Enhydra lutris*) are considered killer whale prey species [27]. The authors state that while manatees would be ideal prey for killer whales, their mostly inland distribution prevents frequent preying on the species [27].

Twenty-six species of cetaceans have been observed having non-predatory interactions with killer whales. Most of these 26 species are also known prey of killer whales. The authors point out that several species of dolphin have been observed interacting with killer whales in non-predatory functions. These specific dolphins are not considered prey for killer whales, and they include four species of Lagenorhynchus [27]. These interactions include "mixed-groups" of the two species, and observed actions are no response, concurrent feeding, flight from killer whales despite no apparent predatory actions by the killer whales, and an attraction to killer whales. For the authors, the latter interaction is the most interesting. One specific observation included a group of several humpback whales closely approaching an attack by killer whales on a steller sea lion. There have also been many reports of Dall's porpoises approaching in an apparent attempt to play [27].

At least seven individual instances have been recorded of non-predatory interactions between killer whales and pinnipeds. Most of these cases involved killer whales swimming past hauled out seals without any change in behavior for either the whale or pinniped [27].

Reports of non-predatory interactions between sea otters and killer whales appear to be more prevalent than predatory interactions. This is perhaps due to a non-preference for sea otters because they are relatively small and have no layer of blubber [27].

2.3.2.5 Evidence of Cooperative Attacks by Killer Whales

There is much evidence supporting the idea that killer whales cooperate to kill large whales, small cetaceans, and pinnipeds. The authors in [28] specify instances in which killer whales bit flukes and flippers of large whales in an apparent effort to slow their movement. Killer whales have also been known to strike pinnipeds with their bodies and to place themselves on the backs of large whales to impede movement. Killer whales have been observed surrounding groups of small marine mammals before attacking them. In addition to tipping over ice floes to trap hauled out seals and penguins, killer whales have been observed hitting a log boom to capture hauled out harbor seals [27].

In the mid-1800s, killer whales and humans cooperatively hunted humpback and right whales around Australia. After the joint kill, killer whales ate the tongue and lips of the kill, and whalers left and returned later to pick up the rest of the carcass [27].

2.3.2.6 Killer Whale Vocalization and Foraging

Mammal-eating killer whales tend to be much less vocal while foraging and feeding than fish-eating killer whales. Other marine mammals have been observed using this to their advantage by ignoring frequently vocal killer whales and tending to stay much more alert when more silent mammal-eating killer whales approach [27].

2.3.2.7 Effect of Social Affiliation on Vocal Signatures of Resident Killer Whales

The authors in [29] described the use of a "towed beam-forming array" to distinguish killer whale calls from within the same matriline. Resident (fish-eating) killer whales in the northeastern Pacific comprise very social groups called matrilines or matrilineal units, which come together to form pods. According to this study, most killer whale calls can be labeled in distinct groups using stereotyped frequency modulation. Many of these types contain "independently modulated high- and low-frequency components" [30]. Killer whale pods are already known to have different distinctive calls, and killer whale pods that share call types are linked together in units called clans. The authors compared individual whales at two distinct levels of social affiliation—inter- and intra-matriline—and used self-comparison as the control. This allowed for discussion of the influence of group membership on the individual of acoustic signals from killer whales. Recordings of whales were processed into spectograms that were linked to directograms that showed the angle of arrival of a call. Seventy-two separate recording sessions were conducted on northeastern-Pacific resident killer whales, for a total of 1508 vocalizations by 19 whales in clans R and A. Eleven individual whales from five different matrilineal units, three pods, and two clans had enough calls for the authors to analyze [30]. The authors used neural networking to identify individuals from randomly selected time–frequency contours, which indicated that shared whale calls have distinctive signature information. Calls from individuals of different matrilineal units were much more easily distinguishable. The authors did not believe that individual distinctiveness was as significant as group-specific convergence. Killer whales engage in many behaviors, such as foraging and direction changes, that might require group communication. The statistical significance of distinguishable individual signatures could be to indicate the capacity for killer whales to distinguish between calls of their matrilineal relatives [30].

2.4 Conclusions

This chapter provides a survey to explore in depth the California sea lion, bottlenose dolphin, and orca whale. It covers the appearance, classification, habitat, diet, physiology, and intelligence of the three species.

Acknowledgments

This work is supported in part by the U.S. National Science Foundation (NSF) under the grant numbers CNS-0737325, CNS-0716211, and CCF-0829827.

References

1. California sea lion (2010). In *Merriam-Webster Online Dictionary*. Retrieved March 26, 2010, from http://www.merriam-webster.com/dictionary/california sea lion
2. http://www.racerocks.com/racerock/eco/taxalab/carolinem.htm
3. California Sea Lion, http://www.marinemammalcenter.org/learning/education/pinnipeds/casealion.asp
4. Breana Delight Wheeler, The Bigeography of California Sea Lion (*Zalophus californianus*). Breana Wheeler. San Francisco State University Department of Geography. http://bss.sfsu.edu/holzman/courses/fall1%20projects/californiasealion.htm
5. Encyclopedia of Marine Mammals by William F. Perrin, Bernd Würsig, J. G. M. Thewissen. pg. 998 "Sea Lions: Overview" Daryl J. Bones.
6. ITIS, http://www.itis.gov/servlet/SingleRpt/SingleRpt?search_topic=TSN&search_value=180615

7. Sea Lion Diet, http://swfsc.noaa.gov/textblock.aspx?Division=PRD&ParentMenuId=148&id=1252
8. California, Galápagos and Japanese Sea Lions, http://www.pinnipeds.org/species/zalophus.htm
9. Bearzi, M. California sea lions use dolphins to locate food. June 2006. University of California, Los Angeles, Department of Ecology and Evolutionary Biology.
10. Bottlenose dolphin (2010). In *Encyclopædia Britannica*. Retrieved April 12, 2010, from *Encyclopædia Britannica Online*: http://www.britannica.com/EBchecked/topic/75261/bottlenose-dolphin
11. Bottlenose dolphin (2010). In *Merriam-Webster Online Dictionary*. Retrieved April 13, 2010, from http://www.merriam-webster.com/dictionary/bottlenose dolphin
12. Bottlenose Dolphin, http://www.acsonline.org/factpack/btlnose.htm
13. Bottlenose Dolphin, http://animals.nationalgeographic.com/animals/mammals/bottlenose-dolphin/
14. Iucnredlist, http://www.iucnredlist.org
15. http://www.worldscreatures.com/water-species/dolphins/indo-pacific-bottlenose-dolphin.htm
16. Behavior, http://www.seaworld.org/infobooks/bottlenose/behavdol.html
17. Reproduction, http://www.seaworld.org/infobooks/Bottlenose/reprodol.html
18. Birth and Care of Young, http://www.seaworld.org/infobooks/Bottlenose/birthdol.html
19. Behavioral Mimicry, http://www.dolphin-institute.org/our_research/dolphin_research/behavioralmimicry.htm
20. Dolphins Deployed as Undersea Agents in Iraq, http://news.nationalgeographic.com/news/2003/03/0328_030328_wardolphins.html
21. Thayer, V. Life history parameters and social associations of female bottlenose dolphins (*Tursiops truncatus*) off North Carolina, USA. PhD dissertation, Duke University, United States, North Carolina. Retrieved April 22, 2010, from Dissertations & Theses: Full Text.
22. Killer whale. (2010). In *Merriam-Webster Online Dictionary*. Retrieved April 27, 2010, from http://www.merriam-webster.com/dictionary/killer whale
23. Killer whale. (2010). In *Encyclopædia Britannica*. Retrieved April 27, 2010, from *Encyclopædia Britannica Online*: http://www.britannica.com/EBchecked/topic/317770/killer-whale
24. Killer Whale, http://www.nmfs.noaa.gov/pr/species/mammals/cetaceans/killerwhale.htm
25. Antarctic peninsula killer whales (*Orcinus orca*) hunt seals and a penguin on floating ice. Society for Marine Mammalogy. Visser, Ingrid. http://www.grupofalco.com.ar/pedefes/Visser%20et%20al%202008.%20Antarctic%20killer%20whales%20on%20ice%20-%20Marine%20Mammals%20Science.pdf
26. http://www.springerlink.com/content/3l6g66042535643l/fulltext.pdf
27. http://swfsc.noaa.gov/uploadedFiles/Divisions/PRD/Publications/Jeffersonetal.1991%288%29.pdf
28. http://rsbl.royalsocietypublishing.org/content/2/4/481.full

29. Olga A. Filatova, Ivan D. Fedutin, Tatyana V. Ivkovich, Mikhail M. Nagaylik, Alexandr M. Burdin, and Erich Hoyt. (2009). The function of multi-pod aggregations of fish-eating killer whales (*Orcinus orca*) in Kamchatka, Far East Russia, *Journal of Ethology*, 27(3): 333–341. Doi: 10.1007/s10164-008-0124-x.
30. Thomas A. Jefferson, Pam J. Stacey, and Robin W. Baird. (1991). A review of killer whale interactions with other marine mammals: Predation to co-existence, *Mammal Review*, 21(4):151–180.
31. Anna E. Nousek, Peter J.B. Slater, Chao Wang, and Patrick J.O. Miller. (2006). The influence of social affiliation on individual vocal signatures of northern resident killer whales (*Orcinus orca*), *Biological Letters*, 2(4):481–484. Doi: 10.1098/rsbl.2006.0517.

BIO-INSPIRED COMPUTING AND ROBOTS

Chapter 3

Social Insect Societies for the Optimization of Dynamic NP-Hard Problems

Stephan A. Hartmann, Pedro C. Pinto,
Thomas A. Runkler, and João M.C. Sousa

Contents

In this chapter, we discuss how social insects working in swarms manage complex situations and how their characteristics can be used in the optimization of a vast range of problems.

3.1 Introduction

In complexity theory, a problem that is $\mathcal{NP}$-hard is intractable, with exact algorithms needing, at worst, exponential time to find a solution under the assumption that $\mathcal{P} \neq \mathcal{NP}$. That is not efficient nor desirable in many practical or real-world applications where time matters and is more important than small gains in quality between a suboptimal and optimal solution. For example, the now widely popular global positioning system (GPS) car-navigating systems have to be able to compensate very quickly for deviations in the route caused by mistakes of the driver. A slightly suboptimal solution is much more desirable than having the driver stop the car while waiting for the system to finish computations. Algorithms that provide good solutions in short computation time but cannot guarantee the optimality of the solution are called approximate algorithms.

Dynamic combinatorial optimization problems add a whole new dimension to the optimization problem by allowing the change of the system parameters suddenly or over time. In practice, many applications require not the optimization of a single instance but instead the optimization of a constantly changing problem. Changes can happen either because additional information is added to the model, or due to older information becoming obsolete.

Various approximate algorithms have been proposed for solving static and dynamic combinatorial problems. The most widely spread and known algorithms of the kind are local search algorithms. Local search starts from some initial solution and iteratively tries to replace the current solution by a better solution in an appropriately defined neighborhood of the current solution. In case a better solution is found, it replaces the previous one and the cycle continues for a determined number of iterations. The danger of falling into a low suboptimal area and remaining trapped there with such a naive algorithm is clear, and, thus, several techniques were developed to circumvent this. Hence, meta-heuristics.

Meta-heuristics are a set of approximate algorithms that essentially try to combine basic heuristic methods of local search in higher level frame-works aimed at exploring the search space of a problem in an efficient way. Usually, the meta-heuristic is used to guide an underlying, more problem-specific heuristic, with the goal to increase its performance. The objective is to avoid the disadvantages of iterative improvement and, in particular, multiple descent by allowing the local search to escape from local optima. This is achieved by either allowing worsening moves or generating new starting solutions for the local search in a more "intelligent" way than just providing random initial solutions.

Meta-heuristics include algorithms such as ant colony optimization (ACO),[1] wasp swarm optimization (WSO),[2] genetic algorithms,[3] iterated local search (ILS), Simulated annealing (SA),[4] Tabu search (TS),[5] Termite optimization (TO), and honey bee optimization (HBO).[6] Of these, ACO, WSO, TO, and HBO can be classified apart, as belonging to a special class of social-insect-inspired meta-heuristics. The success of this approach is recognized by the amount of literature on the subject in late years and the number of congresses and events that include this subject, such as ICAPS 2008, AAAI 2008, ANTS 2008, and others.

This chapter begins with an introduction to optimization of dynamic $\mathcal{NP}$-hard problems and methods for optimization with emphasis on meta-heuristics, useful if the reader is not familiar with the subject. The following sections contain descriptions of each social-insect-inspired meta-heuristic, respectfully, ant, termite, wasp, and bee optimization. The application of each algorithm to a dynamic situation is presented alongside the algorithm description. The chapter ends with a conclusion on how the algorithms compare and perform and indications for the future.

3.2 Optimization of Dynamic $\mathcal{NP}$-Hard Problems

Many applications require not the optimization of a single, static situation but instead the optimization of a constantly changing problem. Changes can happen either because additional information is added to the model or because of the intrinsic changing nature in the problem. As an example of the first, there is the planning of a network that has to be prepared not only to handle the current system but also to be robust to future conditions (it is money and time-consuming to change the network configuration), and as a straightforward example of the second, there is the GPS navigation computer that has to recalculate the route when the car does not follow the proposed route.

The changed problem is often similar to the old, only differing in a few constraints or variables. While it is possible for a small change in the problem to create a significant change in the optimal solution, in most practical cases, the optimal solutions of such similar problems do not differ dramatically. Continuing the previous example, in the GPS when a car goes forward instead of turning left, the initial positions are almost the same and it may be enough to take the next turn.

A number of algorithms have been proposed for solving dynamic $\mathcal{NP}$-hard problems. They can be divided into traditional methods and methods based on meta-heuristics. While the complete methods perform better on harder static problems, the more reactive nature of the incomplete methods based on meta-heuristics makes them more suited for easier static problems and for tracking changes in dynamic problems.[7–9]

3.2.1 Approaches to Problem Optimization

There exist many techniques for problem optimization. Roughly, the spectrum of available methods can be divided into traditional methods and methods based on meta-heuristics such as evolutionary algorithms.[10] Both fields have their specific properties, advantages, and disadvantages.

Complete methods are guaranteed to find for every finite size instance of a combinatorial optimization problem an optimal solution in bounded time.[11] Yet, for problems that are $\mathcal{NP}$-hard, no polynomial time algorithm exists, assuming that $\mathcal{P} \neq \mathcal{NP}$. Therefore, complete methods might need exponential computation time in the worst case. This often leads to computation times too high for practical purposes.

If optimal solutions cannot be feasibly obtained in practice, the only solution is to trade optimality for efficiency. The guarantee of finding the optimal solution is thought as worth losing for the advantage of getting very good solutions in polynomial time. Approximate algorithms, often also called heuristic methods or *heuristics*, seek precisely that. Based on the underlying techniques that approximate algorithms use, they can be either constructive or local search methods.

In the last 20 years, a new kind of approximate algorithm has emerged, which basically tried to combine basic heuristic methods in higher level frameworks aimed at efficiently and effectively exploring a search place. These methods are nowadays commonly called meta-heuristics, a term first introduced by Glover[12] and that derives from two greek works. Heuristic derives from the verb *heuriskein*, which means "to find," while the suffix *meta* means "in a upper level." Meta-heuristics include, but are not restricted to, algorithms such as Evolutionary Computing (EC),[3] ILS,[13] SA,[4] TS,[5] and algorithms inspired in social insects.

3.2.2 Meta-Heuristics Inspired in Social Insects

Some meta-heuristic algorithms try to derive optimization procedures from the natural behavior of social insects. Social insects live in organized communities (nests or colonies) where the members depend on each other to survive. The dynamic assignment of tasks in those communities to the community members, such as foraging and brood care, and the efficient way a colony achieves goals without a conscious planning effort from any of its agents serve as inspiration source for optimization methods.

There are four basic families of social insects—ants, wasps, bees, and termites, all of which have been successfully explored as meta-heuristics. Wasp algorithms have been used to optimize scheduling and logistic processes,[14–16] and more recently dynamic constraint satisfaction problems[17,18] and clustering problems.[19] Honey-bees have been used as well in constraint optimization.[6] Heuristics based on the behavior of termites have been used in the routing of wireless networks.[20] Approaches based on ant colonies have been used to optimize a wide range of problems, such as the satisfiability problem,[21,22] the traveling salesman problem,[23] supply-chain logistics,[24,25] clustering problems,[26,27] and routing problems,[23] among others.

Now, we will consider the processes characteristic of each of the social insects and how it makes them effective optimizing tools in a selected number of examples.

3.3 Ant Colonies Keep Supply Lines

ACO[1] is a bioinspired meta-heuristic, belonging to a special group that attempts to emulate behaviors characteristic of social insects, in this case ant colonies. In ACO, the behavior of each agent in the optimization mimics the behavior of real-life ants and how they interact with each other in order to find resources and carry them to the colony efficiently. To put it briefly, during a walk, each ant deposits *pheromones*. Other ants are sensitive to these pheromones encouraging them to follow the trail, with more or less intensity depending on the concentration of the pheromone. After a period of time, the shortest path will be visited more frequently and pheromones saturate it (see Figure 3.1).

This concentration is aided by the evaporation over time of the pheromones deposited in the paths, which makes less-used and longer paths less and less attractive to the ants. This way, through involuntary collaboration, ants can find the shortest path between a food source and their colony. This type of coordination is generally known as stigmergy.[28] Besides the pheromones, ACO agents can be and are usually given some degree of "intelligence," in the form of *heuristics*, which help guiding the search.

In ACO, ant k at node i will choose one of the possible trails (i, j) connecting the actual node to one of other possible positions $j \in \{1, \ldots, n\}$, with probability

$$p_{ij}^k = f(\tau_{ij}) \tag{3.1}$$

where τ_{ij} is the pheromone concentration on the path connecting i to j. The pheromone level on a trail changes according to

$$\tau_{ij}(t + 1) = \tau_{ij}(t) \cdot \rho + \delta_{ij}^k \tag{3.2}$$

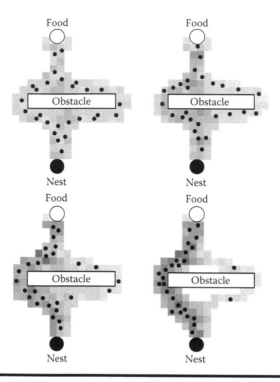

Figure 3.1   Change of pheromone concentration.

where
   δ_{ij}^{k} is the pheromone released by the ant k on the trail (i, j)
   $\rho \in [0, 1]$ is the evaporation coefficient

The system is continuous, so the time acts as the performance index, since the shortest paths will have the pheromone concentration increased in a shorter period of time.

   This is a mathematical model of a real colony of ants. However, the artificial ants that mimic this behavior can be uploaded with more characteristics, for example, memory and ability to see. The pheromone expresses the *experience* of the colony in the job of finding the shortest path. Memory and ability to see express useful *knowledge* about the problem the ants are solving. In this way, the function f in (3.1) can be defined accordingly as follows:

$$
p_{ij}^{k}(t) = \begin{cases} \dfrac{\tau_{ij}^{\alpha} \cdot \eta_{ij}^{\beta}}{\sum_{r \notin \Gamma}^{n} \tau_{ir}^{\alpha} \cdot \eta_{ir}^{\beta}} & \text{if } j \notin \Gamma \\[2ex] 0 & \text{otherwise} \end{cases} \tag{3.3}
$$

where

η_{ij} is a *heuristic function*

Γ is a *tabu list*

In this case, the heuristic expresses the capability of seeing, which is the nearest node j to travel toward the food source. Γ is a list that contains all the trails that the ant has already passed and must not be chosen again (artificial ants can go back before reaching the food source). This acts as the memory of an ant.

The combination of positive and negative reinforcement works well for static problems. There is exploration, specially in the beginning, but after a while, all solutions that are not promising are slowly cut off from the search space, since there is no reinforcement any more and the pheromones evaporate over time. However, in dynamic problems, certain solutions can be bad for a long time but suddenly become desirable after a change in the environment.[29-31] Therefore, the evaporation-reinforcement scheme of the pheromone matrix has to be adapted in dynamic problems in order to avoid the complete dismissal of those solutions over time. The best way to achieve this depends on the characteristics of the problem, but in general, one of these approaches is followed:

- *Restart-strategy*: The pheromone is restarted at times or when a change in the problem is detected.[32,33] This method is limited, since it does not take into account where the change of the problem actually occurred and loses the information about potential good solutions. On the other hand, it guarantees that good solution candidates will always be found as long as there is enough time between restarts.
- *Reactive pheromones*:[31,32] The evaporation coefficient and the pheromone update factor are variables of the rate of change in the problem. This method always tries to have a set of good candidate solutions and works best when the changes in the problem are limited in scope and location. However, it is likely to perform poorly when compared to the restart-strategy for dynamic problems where big changes occur.
- η *strategy*: This is a more locally oriented strategy that uses heuristic-based information to decide to what degree equalization is done on the pheromone values.[30]

To illustrate this algorithm, we present the dynamic variant of the well-known traveling salesman problem (TSP).[34] The TSP can be formally defined as follows: Given n cities $C = c_1, \ldots, c_n$ and a cost (distance) matrix $D = d_{ij_{n \times n}}$ where d_{ij} is the distance from c_i to c_j, find a permutation $\pi = (\pi_1, \ldots, \pi_n)$ such that

$$\sum_{i=1}^{n} d_{\pi, \pi_{i+1}} = \min \qquad (3.4)$$

where $\pi_{n1} = \pi_1$. There are several ways of making a TSP dynamic. In a typical case, C is modified over time with the removal and insertion of cities to be visited. This means, naturally, that D changes with time as well. Therefore, Equation 3.4 is modified into

$$\sum_{i=1}^{n(t)} d_{\pi,\pi_{i+1}}(t) = \min \tag{3.5}$$

There have been several publications on addressing this problem in ants, for example, in Refs [29,30]. This is the problem variant we are going to analyze here, following the procedure introduced by Guntsch et al.[30] Another useful way of turning a TSP dynamic is to consider that the cost matrix represents not the distance between cities (i.e., static) but another measure that varies inherently with time, such as traffic congestion. In that case, n is constant and (3.4) becomes

$$\sum_{i=1}^{n} d_{\pi,\pi_{i+1}}(t) = \min \tag{3.6}$$

In order to apply ACO to an optimization problem, one needs to define an appropriate representation of the problem, a problem-specific heuristic η_{ij} (where i and j are two different nodes) to help guide the ants, a way of updating the pheromone track τ_{ij}, and the probabilistic rule p_{ij} that moves the ants to the next stage.

3.3.1 Representation of the Problem

Matrix d_{ij} expresses the distance from city i to city j, and the tabu list Γ is the list of cities that the ant has already visited. Both matrices are of the size of size $N_{nodes} \times N_{nodes}$. z_k is the performance index, here representing the cost (measured in time units) of each tour.

3.3.2 Pheromone Update

Each ant deposits a pheromone δ_{ij}^k to the chosen trail:

$$\delta_{ij}^k = \tau_c \tag{3.7}$$

where τ_c is a constant. The best solution should increase even more the pheromone concentration on the shortest trail, so (3.2) is changed to

$$\tau_{ij}(t + n) = \tau_{ij}(t) \cdot \rho + \Delta\tau_{ij} \tag{3.8}$$

where $\Delta\tau_{ij}$ are pheromones deposited onto the trails (i, j),

$$\Delta\tau_{ij} = \sum_{k=1}^{q} \delta_{ij}^{k} f\left(\frac{1}{z_k}\right) \tag{3.9}$$

If the restart-strategy is followed, each city i is assigned the strategy-specific parameter $\lambda_R \in [0, 1]$ as its reset value, that is, $\gamma_i = \lambda_R$.

In the η strategy, each city i is given a value γ_i proportionate to its distance from the nearest inserted or deleted city j. This distance d_{ij}^{η} is derived from η_{ij} in a way that η_{ij} is proportional to d_{ij}^{η} and that scaling the heuristic η values has no effect:

$$d_{ij}^{\eta} = 1 - \frac{\eta_{\text{avg}}}{\lambda_E \cdot \eta_{ij}} \tag{3.10}$$

with

$$\eta_{\text{avg}} = \frac{1}{n(n+1)} \sum_{i+1}^{n} \sum_{k \neq i} \eta_{ik} \tag{3.11}$$

3.3.3 Heuristics

The heuristic information η_{ij} is defined as the inverse of the distance from city i to city j expressed in a matrix d_{ij}, $\eta_{ij} = 1/d_{ij}$.

3.3.4 Probabilistic Rule

In the TSP case, the probabilistic rule can be directly taken from Equation 3.3.

Notice that the time interval taken by the q ants to do a complete tour is $t + n$ iterations. A *tour* is a complete route between the nest and the food source, and an *iteration* is a step from i to j done by all the ants. The algorithm runs $N_{\max}$ times, where in every Nth tour, a new *ant colony* is released. The total number of iterations is $\mathcal{O}(N_{\max}(n \cdot m + q))$.

3.4 Termite Hill-Building

Similar to ants, termites use a communication concept based on stigmergy for building their mounds. In 1925, Marais presented a detailed work about termites,[35] but he was unable to answer the question of how termites are able to create grand mounds without centrally coordinating the construction. It was not until 1959 when Grassé[28] clarified the underlying mechanisms for nest construction yielding the following set of simple rules:

- First, they randomly look for appropriate building material (soil).
- Then, they perform a random walk dispensing small amounts of chewed earth and saliva on any elevated spot they encounter. Soon small soil pillars emerge.
- The concentration of saliva at those pillars stimulates nearby termites to pellet-gathering behavior and focuses their dropping activity on these patches of ground resulting in the growth of the biggest heaps into columns.

Instead of applying scents (pheromones) to paths like ant colonies, the termites' saliva plays the role of a nest-building pheromone that is deposited at the nest in order to bias the decisions of other individuals where to drop the pellets (see Figure 3.2).

The principles of swarm intelligence (positive/negative feedback, randomness, multiple interaction) also apply to the mound-building capability of termites. Positive feedback can be observed at the termites' attraction to higher pheromone

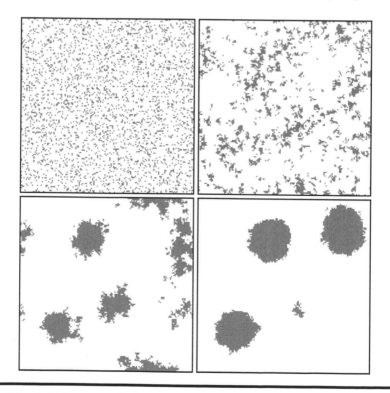

Figure 3.2 Pebble accumulation.

concentrations for dropping pellets. The bigger the pillar and the pheromone gradient the more termites are biased to move there. Evaporation of these pheromones is responsible for the negative feedback. Every move or action of a termite is determined by randomness since pheromones only bias but do not compel certain actions. Interaction of multiple individuals is necessary to obtain a critical threshold value of pheromones at a place because otherwise pheromones evaporate faster than they can be accumulated (similar to the ants).

As an example, Roth and Wicker[36] apply the principles of nest building to package routing in mobile wireless ad-hoc networks (MANET). The proposed algorithm associates a specific pheromone scent with each node in the network. As packets are dispatched from a source, each follows a bias by the pheromone gradient toward its destination while laying pheromone for their source on the traversed links and thus increases the probability of a reverse package taking the same route. The amount of pheromone deposited by a package as well as the decay rate is controlled by the pheromone accounting process.

3.4.1 Pheromone Table

Each node n maintains a table indicating the amount of pheromone for each destination when using the neighboring nodes $\mathcal{N}^n$, and it can be visualized as a matrix with one column per destination $d \leq D$ and one row per neighboring node $i \leq N$. Each entry $P_{i,d}$ represents the pheromone value for a package being routed to d via i. The maximum size of the table is given by $N \times D$, but it is also dependent on whether packets were ever set to a specific destination or a specific neighbor via the current node. If not, the respective rows and columns may not be part of the pheromone matrix. Whenever destinations are lost or added, the respective columns and/or rows are deleted or added, respectively. Destinations and neighbors are also removed for the table when all entries of the respective columns and/or rows fall below a certain value that indicates that this connection was stopped being used, due to better connections.

3.4.2 Pheromone Update

In order to determine the quality of the connections traversed so far from the source s to the current node n, each packet is carrying a utility value γ, which is used for the pheromone update upon reception at a node. Traveling packets add pheromones to the routing table of the current node upon reception using the utility of a path that is the inverse of its added costs. Since the utility—in contrast to costs—is not additive, a packet that is passed from the previous hop r to the current node n with link cost c_{rn} is updated as follows:

$$\gamma \leftarrow (\gamma^{-1} + c_{rn})^{-1} \qquad (3.12)$$

The utility γ at node r is inverted into costs, the costs c_{rn} are added, and the sum is then re-inverted into utility. Although pheromones continuously evaporate in the real-world environment, the termite algorithm only reduces the pheromone values upon package arrival. Thus, besides the evaporation rate τ, the evaporation process also has to incorporate the frequency of how often a connection is used for communication to emulate this decay behavior. The lower the frequency, the more pheromone evaporates upon arrival. Let $t^n_{r,s}$ be the time when at node n, the pheromone value to s via r was updated the last time before the current packet arrived. The complete update of the respective pheromone $P^n_{r,s}$ is defined as follows:

$$ P^n_{r,s} \longleftarrow \underbrace{P^n_{r,s} e^{-(t-t^n_{r,s})\tau}}_{\text{evaporation}} + \underbrace{\gamma}_{\text{adding}} \tag{3.13} $$

3.4.3 Routing

The routing decision at every hop is done by the forward equation that models a packet probabilistically following a pheromone gradient to its destination and repelled by gradient toward the source (*source pheromone aversion*). At first, the probability for each neighbor link is computed for both, leading to the packet's destination (Equation 3.14) as well as leading back to its source (Equation 3.15) using the respective pheromone values. Out of both for each link, the meta-probability of being chosen to forward the packet is derived (Equation 3.16).

$$ p^n_{i,d} = \frac{\left[P^n_{i,d} e^{-(t-t^n_{i,d})\tau} + K\right]^F}{\sum_{j\in\mathcal{N}^n}\left[P^n_{j,d} e^{-(t-t^n_{j,d})\tau} + K\right]^F} \tag{3.14} $$

$$ p^n_{i,s} = \frac{\left[P^n_{i,s} e^{-(t-t^n_{i,s})\tau} + K\right]^F}{\sum_{j\in\mathcal{N}^n}\left[P^n_{j,s} e^{-(t-t^n_{j,s})\tau} + K\right]^F} \tag{3.15} $$

$$ \tilde{p}^n_{i,d} = \frac{p^n_{i,d}\left(p^n_{i,s}\right)^{-A}}{\sum_{j\in\mathcal{N}^n} p^n_{j,d}\left(p^n_{j,s}\right)^{-A}} \tag{3.16} $$

The exponential expression multiplied by each pheromone accommodates the attenuation of older pheromone values as in Equation 3.13. The reason for the deemphasis is twofold: First, it copies the behavior of real pheromones that evaporate over time. Second, it incorporates the diminishing confidence in the pheromone, since the network structure and link qualities may have changed.

The constants F, K, and A are tuning parameters for the termite algorithm. $K \geq 0$ determines the sensitivity of the probabilistic routing to small pheromone values. A larger K diminishes the influence of smaller differences between small pheromone amounts. It also guarantees a minimal exploration of each connection since the probabilities are prevented from converging to zero. $F \geq 0$ influences the differences between pheromone values. $F > 1$ emphasizes the differences, while $F < 1$ attenuates them. $A \geq 0$ modulates the force by which a packet is repelled by the source pheromone.

3.4.4 Route Discovery

In case the pheromone matrix at node n does not contain any destination pheromones needed for forwarding a packet, a route discovery routine is initiated. The termite algorithm uses a traditional flooding approach, where route request packets (RREQs) are broadcasted and each recipient retransmits it if it does not have destination information for the query. At each node, only the first copy of a specific RREQ packet is forwarded all succeeding packets are dropped. In addition to adding new possible destinations at a node, RREQs also help to spread source pheromone to the network. As soon as an RREQ hits a node that is the destination or has destination pheromone in the matrix in returns a route reply packet (RREP) to the source of the RREQ packet. The routing for RREP packet is done probabilistically as with regular packets and also lays source pheromones, which are also the destination pheromones for RREQ. Data packets that initiate an RREQ are cached for a certain time and dropped if no RREP packet is received meanwhile. Traditional hello packets from a node in order to communicate its existence to other nodes are not (but can be) used in this algorithm.

3.5 Wasp Swarms and Hierarchies

Social animals live in groups in order to gain benefits such as reduced risk from predators, easier access to food, increased per capita productivity, availability of mates, or any combination of these factors. However, group living often results in occasional conflict among the elements of the group, due to the limited amount of resources generally available, may be food, shelter, or desirable mates. In many species of animals, when several unacquainted individuals are placed together in a group, they engage each other in contests for dominance. Dominance behavior has been described in hens, cows, ponies, lizards, rats, primates, and social insects, most notoriously wasps. Some of the contests are violent fights, some are fights that do not lead to any serious injury, and some are limited to the passive recognition of a dominant and a subordinate. In the initial period after placement, contests will be extremely frequent, becoming less and less frequent with time and being replaced by stable dominance–subordination relations among all group members. Once a hierarchical order is obtained, it lasts for long periods of times, only with minor modifications caused by the occasional successful attempt of a subordinate to take over. The formation of this hierarchy organizes the group in such a way that internal conflicts do not supersede the advantages of group living.

Theraulaz introduced a model for the organization characteristic of a wasp colony.[2] The model describes the nature of interactions between an individual wasp and its local environment with respect to task allocation. The colony's self-organized allocation of task is modeled using what is known as response thresholds. An individual wasp has a response threshold for each zone of the nest. Based on a wasp's threshold for a given zone and the amount of stimulus from brood located in that zone, a wasp has a certain probability to become engaged in the task of foraging for that zone. A lower response threshold for a given zone amounts to a higher likelihood of engaging in activity given a stimulus. These thresholds may remain fixed over time.[37] It was also considered that a threshold for a given task decreases during time periods when that task is performed and increases otherwise.[38] Then, the probability of wasp i bidding for a resource is given by

$$p(\eta_i) = \frac{\eta_i^\alpha}{\eta_i^\alpha + \Theta_i^\alpha}, \quad i = 1, \ldots, N \tag{3.17}$$

where
  η_i is the problem-specific heuristics related to the task performance by the wasp
  Θ_i is the bidding threshold
  N is the number of wasps
  α is the bidding exponent

Both Θ_i and α can be adjusted to diminish or increase the number of successful bidding wasps, with effects in the speed of the optimization. Notice that all wasps use the same value of α. The wasps that choose to bid will compete for the available resources.

The model was demonstrated to have a connection to market-based systems, and in applying it to a real-world vehicle paint shop problem, Campos et al.[39] succeeded in improving the results of Morley's market algorithm, a bidding mechanism for assigning trucks to paint booths at a General Motors factory.[40]

Besides the environment-to-wasp interaction, also wasp-to-wasp interactions take place within the nest.[2] When two agents of the colony encounter each other, they may with some probability interact with each other in a dominance contest. If this interaction takes place, the wasp with the higher social rank has a higher probability of winning the confrontation. Through these confrontations, wasps in the colony self-organize into a dominance hierarchy. A number of ways of modeling the probability of interaction during such an encounter exist, which range from always interacting to interacting based upon certain tendencies of the individuals.[41] This behavior is incorporated in the WSO algorithm to solve the problem of scarcity. That is, if the number of bidding wasps is too large given the amount of resources the wasps are bidding for, the winner or winners are chosen through a dominance contest where each wasp can challenge the wasp that immediately precedes it. In the basic model, wasp i wins over wasp j with a probability

$$p_{ij} = \frac{W_i^2}{W_i^2 + W_j^2}, \quad i,j = 1,\ldots,N \qquad (3.18)$$

In order to show the wasps' capabilities of handling dynamic environments, we considered a logistic process based on a real-world production line.[15,25] The optimization heuristics needs to process a large amount of information, such as the arrival and desired delivery dates, which can be extremely time-consuming. Since new orders arrive every day, the problem is inherently dynamic.

The logistic process is divided into four sequential steps: order arrival, component request, component arrival, and component assignment. In the order arrival, an order of purchase arrives for a set of different components. The order is indexed with a certain priority associated with the delivery date and the importance of the client. The arrival of the order is followed by the request and arrival of the needed components from the external suppliers. The time difference between the component request and arrival is called *supplier delay*.

The component assignment is the basic problem underneath the logistic process, since the company cannot influence the arrival rate of the orders nor the suppliers' delay. The delivery rate of the orders is thus the only variable controlled in the process. Ideally, all orders should be satisfied in time. In order to avoid large volumes of stocks and to minimize the number of delayed orders, the scheduling program assigns components to orders every day, not considering their order of arrival but their volumes, requirement, and dates of delivery. The suggested objective function[24] is shown in Equation 3.19. We impose as the most important objective that the highest number of orders is delivered at the correct date, although we consider also important to have a small tardiness variance (η_T) of the remaining orders. Thus, the objective function is given by

$$f(T) = \frac{1}{\eta_T + (1/\sum_{j=1}^{n} T_j)} \qquad (3.19)$$

The optimization starts by assigning each order to a wasp, where the strength F_i of each wasp is defined by the characteristics of the order: its urgency and its component requirements. After being assigned to an order, the wasps decide or not to bid for the available components in stock. The higher the urgency of the order, the higher the stimulus of the wasp to get the needed components to satisfy it. The probability of wasp i bidding for components in stock is given by

$$p(\eta_i) = \frac{\eta_i^\alpha}{\eta_i^\alpha + \Theta_i^\alpha} \qquad (3.20)$$

where
    η_i is the tardiness of the order
    Θ_i is the bidding threshold
    α is the bidding exponent

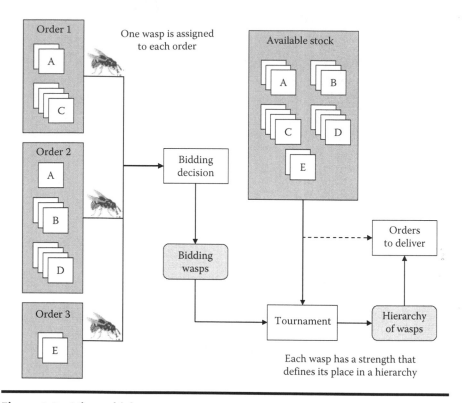

Figure 3.3 Hierarchial wasps.

Both Θ_i and α can be adjusted to diminish or increase the number of actually competing wasps, with effects in the general speed of the optimization. Notice that all wasps use the same value of α. The wasps that choose to bid will compete for the available resources, in order to satisfy their correspondent order (see Figure 3.3).

The competing algorithm is based on a formation of a hierarchy within the wasps. The position of each wasp in the hierarchy is based on its strength. Each wasp can challenge the wasp that immediately precedes it. The probability that wasp A wins the tournament against wasp B (and, thus, switches places with the defeated wasp) is defined by

$$p(A) = \frac{F_A^\beta}{F_A^\beta + F_B^\beta} \qquad (3.21)$$

where β is the tournament exponent. The force F of each wasp is defined as a combination of its order tardiness T and component availability A. Component availability is defined as the ratio between the needed components of an order and their availability in stock. If the stock is running low in the required components, the strength of the wasp diminishes.

$$F(T,A) = W_t T - W_a A \qquad (3.22)$$

W_t and W_a are constants that weigh the importance of the tardiness and the component availability, respectively. The elimination mechanism used is similar to the mechanism in the English auction, where the number of bidders decreases slowly until the number of products in stock is equal to the number of bidders still present. Here, in each iteration of the tournament, the wasp on top of the hierarchy wins the right to get the components it requires, as long as there are enough of them in stock. If not, the wasp is eliminated from the process, and the right of choice is passed to the next wasp in the hierarchy, and so on. For the next iteration, the force of each wasp is updated according to the components still in stock, there is a new tournament, and the leading wasp can again be assigned the needed components. The procedure continues until all wasps are either removed from the tournament or satisfied.

3.6 Bees

One of the rather new members of the swarm intelligence family is the Artificial Bee Colony (ABC), which tries to mimic natural behavior of honey bees in food foraging. In contrast to ants and termites that only use stigmergetic interaction, bees directly communicate by performing a *waggle dance*. By performing this dance, foragers returning from a promising flower patch share the information about both the location (distance and direction) and the amount of pollen and nectar available at that patch with their hive mates (see Figure 3.4).

Depending on the attractiveness of the single dance, other foragers are biased to follow the dancer to the patch, and, thus, productive patches can be exploited by a larger number of individuals. Previous studies[42] have shown that the waggle dance is the crucial part to successful food gathering. The duration of the performance correlates with the quality of the related food and the intensity of the waggles indicates the distance from the hive to the flower patch. The orientation of the dance on the dance floor shows the direction of the food source in relation to the sun. Bees are aware of the fact that the sun is moving during the day that can be observed at the adoption of the angle the dance is performed to the daytime. Although not all principles of the bees' behavior are completely understood,[43] there exists a set of rules that they follow.

The value of a food source is dependent on various influences such as the proximity to the hive, the richness of the energy in the nectar, and the effort for foraging the food. When the individual bees have no knowledge about any food source (yet), the foraging process begins by scout bees being sent to search for promising flower patches. They randomly move looking for patches with plentiful pollen and eventually return to the hive when they found one. Even when there are plenty of patches known to individuals of the colony, the colony keeps a percentage of the population as scout bees. The rest of the bees that do not have knowledge of the flower patches

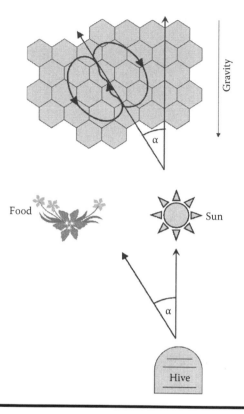

Figure 3.4   Waggle dance.

wait at the dance floor and watch returning bees that found a food source with a quality over a certain threshold performing their waggle dances. If a bee is attracted by a dance, it will start searching for food by using the knowledge from the waggle dance. When a forager is employed in food gathering (because it was a scout bee and encountered a patch or it was watching a successful waggle dance), it has to make a decision upon unloading the nectar to the food area in the hive: Communicate the patch at the dance floor, abandon the patch since the quality has dropped below a threshold, or return to the patch and continue foraging without performing a waggle dance. If it decides not to continue foraging, it attends to a waggle dance done by some other bee; if it is not attracted by any dance it then may become a scout with a certain probability.

Besides the foraging behavior of bees,[44–48] the marriage behavior[6,49–51] and the queen bee concept[52] have been studied and applied to various problems, but only the algorithm mimicking the foraging behavior has been applied to a dynamic problem yet.[53]

As an application for showing the adaptive nature of bees, we present the load-balancing problem from Nakrani and Tovey.[53] There, the bee optimization is introduced to load balancing in an internet server colony, which is an accumulation of single servers that host applications for third parties. A group of physical servers that run the same application are addressed as a virtual application server (VAS), and a load-balancing switch distributes the single requests for that specific application in the service queue to the servers contained in that VAS. Servers can be reallocated from one VAS to another for providing scalability. During this migration, the installation of the web application has to be changed, and, thus, it becomes unavailable. Since pay-per-request-served is a widespread agreement between hosting provider and service owner, the provider tries to serve as many requests as possible. The ratio of requests for different applications changes over time which is faced by adapting the number of physical servers in each VAS.

The underlying server allocation problem is mapped to the bee forager allocation to flower patches. The service queue of each hosted application corresponds to a flower patch. The number HTTP requests has a similar unpredictable dynamic pattern as the flower patch volatility, which is dependent on the daily climatic influences and other changes. Each server is represented by a foraging bee, and the service queues are thought of as nectar sources that need to be exploited efficiently. The different travel costs to food sources (and back) are equivalent to the service cost consisting of different time consumption rates for different service requests as well as the value-per-request-served. Additionally, the idle time when bees change their foraging patches can be reflected to the migration cost between two VAS.

Assume that the server colony consists of n servers grouped to M virtual application servers $V_1, \ldots, V_M$, and the respective service queues are denominated as $Q_1, \ldots, Q_M$. For each served request from service queue Q_j server $s_i \in V_j$ receives c_j cents. Each server is either a foraging server or a scout server; the dance floor is represented by an advert board whereas a waggle dance is an advert on it. Upon completion of a request for Q_j server $s_i \in V_j$ posts an advert with duration $D = c_j A$ with probability p whereas A denotes the advert scaling factor. After that, it will decide with probability r_i to select a (different) advert from the advert board and not continue serving Q_j. If it discontinues the old service, a forager or scout server will select an advert probabilistically depending on D or randomly, respectively. Probability r_i changes as a function of the servers profit rate P_i and the overall profit rate of all servers P_{colony}. P_i is defined as follows:

$$P_i = \frac{c_j R_i}{T_i} \tag{3.23}$$

whereas
    T_i is a given time interval
    R_i denotes the total number of requests served by server s_i within T_i

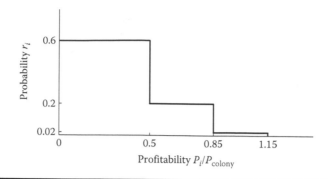

Figure 3.5 Dependency of probability r_i from s_i's profitability P_i/P_{colony}.

The overall profit rate is given by

$$P_{colony} = \frac{1}{T_{colony}} \sum_{j=1}^{M} c_j R_j \qquad (3.24)$$

where T_{colony} is the time interval where the requests R_j have been served. Depending on the *profitability* (ratio between its own profit rate and the colony's profit rate), server s_i will adapt the probability r_i according to Figure 3.5.

If s_i is serving Q_j with low profitability, it is more likely to read adverts. This is exactly the same idea as in the bee hive where bees foraging profitable patches decrease the chance of following another waggle dance (advert).

3.7 Conclusions

This chapter reviewed a branch of meta-heuristic optimization methods, which are inspired in the stigmergetic behavior of social insects. Specifically, we provided theoretical background and practical examples for each of the four different families of social insects: ants, wasps, termites, and bees. The practical applications focused on $\mathcal{NP}$-hard problems of dynamic nature, to which these types of algorithms are particularly adequate since they mimic (and improve) the adaptive nature of the insects in a changing natural environment.

All algorithms are based on how the colony agents cooperate to achieve a common goal, without a higher level supervision. With the ants, this cooperation is achieved by means of pheromones that the ants release when performing a job, thus influencing the other ants engaged in the same task in a process of reinforcement learning. Termites are also pheromone based, in contrast with the two other "flying" algorithms. However, the termites are not biased by the pheromone trails themselves, but by the pheromone gradient, that is, pheromones diffusing from accumulation points.

Wasp swarm optimization is based on how wasps compete between themselves in the swarm, forming an hierarchy based on their corresponding strengths. Thus, unlike in the pheromone-based algorithm, there is no shared memory that biases the choice of each individual of the colony. For problems that involve demand for finite resources, this kind of approach works particularly well. Like the wasps, bees do not share a collective memory. They enforce good solutions through direct communication within a group of individual members of the colony. This way the exploration of the search space is by nature very efficient and so is the escape from local minima.

This field of optimization is recent, and the growth in publications is significant each passing year. Ant-colony optimization has already been applied to a wide variety of problems since its introduction in the middle of the last decade, but is still far from being exhaustively studied. The other three meta-heuristics are much more recent. Bee optimization in particular has registered an enormous increase in literature only in the last two or three years. As the algorithms continue to find their way from the scientific community to real-world applications, the prospect for methods that have been fine tuned by millions of years of natural evolution is high.

References

1. M. Dorigo and T. Stützle, *Ant Colony Optimization*. MIT Press, Cambridge, MA (2004).
2. G. Theraulaz, S. Goss, J. Gervet, and J. L. Deneubourg, Task differentiation in polistes wasps colonies: A model for self-organizing groups of robots. In *From Animals to Animats: Proceedings of the First International Conference on Simulation of Adaptive Behavior*, pp. 346–355. MIT Press, Cambridge, MA (1991).
3. D. E. Goldberg, *The Design of Innovation (Genetic Algorithms and Evolutionary Computation)*. Kluwer Academic Publishers, Dordrecht, the Netherlands (2002).
4. E. H. L. Aarts, J. H. M. Korst, and P. J. M. V. Laarhoven, Simulated annealing. In *Local Search in Combinatorial Optimization*, pp. 91–120. Wiley-Interscience, Chichester, England (1997).
5. F. Glover and M. Laguna, *Tabu Search*. pp. 70–150 (1993).
6. H. A. Abbass, A single queen single worker honey–bees approach to 3-SAT. In eds. L. Spector, E. D. Goodman, A. Wu, W. B. Langdon, H.-M. Voigt, M. Gen, S. Sen, M. Dorigo, S. Pezeshk, M. H. Garzon, and E. Burke, *Proceedings of the Genetic and Evolutionary Computation Conference (GECCO-2001)*, pp. 807–814, San Francisco, CA (7–11, 2001). Morgan Kaufmann, San Francisco. ISBN 1-55860-774-9. URL citeseer.ist.psu.edu/abbass01single.html
7. K. Mertens, T. Holvoet, and Y. Berbers, The dynCOAA algorithm for dynamic constraint optimization problems. In eds. G. Weiss and P. Stone, *Proceedings of the Fifth International Joint Conference on Autonomous Agents and MultiAgent Systems*, pp. 1421–1423, Hakodate, Japan, May 8–12 (2006).

8. K. Mertens, T. Holvoet, and Y. Berbers, Which dynamic constraint problems can be solved by ants. In eds. G. Sutcliffe and R. Goebel, *Proceedings of the 19th International FLAIRS Conference*, pp. 439–444, Melbourne Beach, FL (2006).

9. R. Mailler, Comparing two approaches to dynamic, distributed constraint satisfaction. In *AAMAS '05: Proceedings of the Fourth International Joint Conference on Autonomous Agents and Multiagent Systems*, pp. 1049–1056, New York (2005). ACM Press, New York. ISBN 1-59593-093-0.

10. C. Blum and A. Roli, Metaheuristics in combinatorial optimization: Overview and conceptual comparison, *ACM Computational Survey.* **35**(3), 268–308 (2003). ISSN 0360-0300.

11. C. H. Papadimitriou and K. Steiglitz, *Combinatorial Optimization Algorithms and Complexity.* Dover Publications, Inc., New York (1982).

12. F. Glover, Future paths for integer programming and links to artificial intelligence, *Computers and Operations Research.* **13**, 533–549 (1986).

13. K. Smyth, H. H. Hoos, and T. Stützle, Iterated robust tabu search for MAX-SAT, *Lecture Notes in Computer Science.* Springer Verlag. **44**, 279–303 (2003).

14. V. A. Cicirello and S. F. Smith, Wasp-like agents for distributed factory coordination, *Autonomous Agents and Multi-Agent Systems.* **8**, 237–266 (2004).

15. P. Pinto, T. A. Runkler, and J. M. C. Sousa, Wasp swarm optimization of logistic systems. In *Adaptive and Natural Computing Algorithms, ICANNGA 2005, Seventh International Conference on Adaptive and Natural Computing Algorithms*, pp. 264–267 (March, 2005).

16. S. A. Hartmann and T. A. Runkler, Online optimization of a color sorting assembly buffer using ant colony optimization. In *Proceedings of the Operations Research 2007 Conference*, pp. 415–420 (2007).

17. P. Pinto, T. A. Runkler, and J. M. C. Sousa, Agent based optimization of the MAX-SAT problem using wasp swarms. In *Controlo 2006, Seventh Portuguese Conference on Automatic Control* (2006).

18. P. Pinto, T. A. Runkler, and J. M. C. Sousa, Wasp swarm algorithm for dynamic MAX-SAT problems. In *ICANNGA 2007, Eighth International Conference on Adaptive and Natural Computing Algorithms.* (2007).

19. T. A. Runkler, Wasp swarm optimization of the c-means clustering model, *International Journal of Intelligent Systems.* **23**(3), 269–285 (February, 2008).

20. R. Martin and W. Stephen, Termite: A swarm intelligent routing algorithm for mobilewireless Ad-Hoc networks. In *Studies in Computational Intelligence*, pp. 155–184. Springer, Berlin/Heidelberg (2006).

21. A. Roli, C. Blum, and M. Dorigo, ACO for maximal constraint satisfaction problems. In *MIC'2001—Metaheuristics International Conference*, pp. 187–192, Porto, Portugal (2001).

22. S. Pimont and C. Solnon, A generic ant algorithm for solving constraint satisfaction problems. In *2nd International Workshop on Ant Algorithms (ANTS 2000)*, pp. 100–108, Brussels, Belgium (2000).

23. V. A. Cicirello and S. F. Smith, Ant colony for autonomous decentralized shop floor routing. In *Proceedings of ISADS-2001, Fifth International Symposium on Autonomous Decentralized Systems* (2001).
24. C. A. Silva, T. A. Runkler, J. M. C. Sousa, and J. M. S. da Costa, Optimization of logistic processes in supply-chains using meta-heuristics. In eds. F. M. Pires and S. Abreu, *Lecture Notes on Artificial Intelligence 2902, Progress in Artificial Intelligence, 11th Portuguese Conference on Artificial Intelligence*, pp. 9–23. Springer Verlag, Beja, Portugal (December, 2003).
25. C. A. Silva, T. A. Runkler, J. M. C. Sousa, and R. Palm, Ant colonies as logistic process optimizers. In eds. M. Dorigo, G. D. Caro, and M. Sampels, *Ant Algorithms, International Workshop ANTS 2002, Brussels, Belgium, Lecture Notes in Computer Science LNCS 2463*, pp. 76–87, Heidelberg, Germany (September, 2002). Springer, New York.
26. T. A. Runkler, Ant colony optimization of clustering models, *International Journal of Intelligent Systems*. **20**(12), 1233–1261 (2005).
27. S. M. Vieira, J. M. Sousa, and T. A. Runkler, Ant colony optimization applied to feature selection in fuzzy classifiers. In *IFSA*, pp. 778–788, Cancun, Mexico (2007).
28. P.-P. Grasse, La reconstruction du nid et les coordinations inter-individuelles chez bellicositermes natalensis et cubitermes sp. la theorie de la stigmergie: Essai d'interpretation du comportement des termites constructeurs, *Insectes Sociaux*. **6**, 41–81 (1959).
29. M. Guntsch and M. Middendorf, Pheromone modification strategies for ant algorithms applied to dynamic TSP. In eds. E. J. W. Boers, S. Cagnoni, J. Gottlieb, E. Hart, P. L. Lanzi, G. Raidl, R. E. Smith, and H. Tıjınk, Applications of Evolutionary Computing. *EvoWorkshops2001: EvoCOP, EvoFlight, EvoIASP, EvoLearn, and EvoSTIM. Proceedings*, Vol. 2037, pp. 213–222, Como, Italy (18–19, 2001). Springer-Verlag, Heidelberg.
30. M. Guntsch, M. Middendorf, and H. Schmeck, An ant colony optimization approach to dynamic TSP. In eds. L. Spector, E. D. Goodman, A. Wu, W. B. Langdon, H.-M. Voigt, M. Gen, S. Sen, M. Dorigo, S. Pezeshk, M. H. Garzon, and E. Burke, *Proceedings of the Genetic and Evolutionary Computation Conference (GECCO-2001)*, pp. 860–867, San Francisco, CA (7–11, 2001). Morgan Kaufmann, San Franciso. ISBN 1-55860-774-9.
31. P. C. Pinto, T. A. Runkler, and J. M. C. Sousa, Ant colony optimization and its application to regular and dynamic MAX-SAT problems. In *Advances in Biologically Inspired Information Systems: Models, Methods, and Tools*, Springer, Berlin (2007).
32. L. Gambardella, E. Taillard, and M. D. M. Ant colonies for the quadratic assignment problem. *Journal of the Operational Research Society*. **50**, 167–176 (1999).
33. T. Stützle and H. H. Hoos, The max-min ant system and local search for the traveling salesman problem. In *Proceedings of IEEE Fourth International Conference on Evolutionary Computation (ICEC'97)*, Vol. 8, pp. 308–313. IEEE Press, Piscataway, NJ (1997).
34. D. L. Applegate, R. E. Bixby, V. Chvtal, and W. J. Cook, *The Traveling Salesman Problem: A Computational Study*. Princeton University Press, Princeton, NJ (2007).
35. E. N. Marais, *Die siel van die mier* (unknown publisher, 1925).

36. M. Roth and S. Wicker, Termite: A swarm intelligent routing algorithm for mobile wireless ad-hoc networks. In *Stigmergic Optimization*, Vol. 31. pp. 155–184. Springer, Berlin (2006).

37. E. Bonabeau, A. Sobkowski, G. Theraulaz, and J.-L. Deneubourg, Adaptive task allocation inspired by a model of division of labor in social insects. In eds. D. Lundh, B. Olsson, and A. Narayanan, *Biocomputing and Emergent Computation*, pp. 36–45. World Scientific, Singapore (1997).

38. G. Theraulaz, E. Bonabeau, and J. Deneubourg, Response threshold reinforcement and division of labour in insect societies. In *Proceedings of Royal Society of London, Biological Sciences*, Vol. 265, pp. 327–332 (1998).

39. M. Campos, E. Bonabeau, and G. T. J. Deneubourg, Dynamic scheduling and division of labor in social insects, *Adaptive Behavior.* **8**(2), 83–96 (2000).

40. D. Morley, Painting trucks at general motors: The effectiveness of a complexity-based approach. In *Embracing Complexity: Exploring the Application of Complex Adaptive Systems to Business*, pp. 53–58. The Ernst and Young Center for Business Innovation, Cambridge, MA (1996).

41. G. Theraulaz, E. Bonabeau, and J. Deneubourg, Self-organization of hierarchies in animal societies. The case of the primitively eusocial wasp polistes dominulus christ, *Journal of Theoretical Biology.* **174**, 313–323 (1995).

42. A. Baykasoğğlu, L. Özbakır, and P. Tapkan, Artificial bee colony algorithm and its application to generalized assignment problem. In *Swarm Intelligence: Focus on Ant and Particle Swarm Optimization*, pp. 113–144. Itech Education and Publishing, Vienna (2007).

43. N. Lemmens, S. de Jong, K. Tuyls, and A. Nowé, Bee behaviour in multi-agent systems. In *Adaptive Agents and Multi-Agent Systems III. Adaption and Multi-Agent Learning*, pp. 145–156. Springer, Berlin (2008).

44. P. Lucuc and D. Teodorovic, Bee system: Modeling combinatorial optimization transportation engineering problems by swarm intelligence. In *Preeprints of the TRISTAN IV Triennial Symposium on Transportation Modeling: Sao Miguel, Azores Islands*, pp. 441–445. ACM Press, 2001 (2006).

45. P. Lucic and D. B. Teodorović, Modeling Transportation Problems Using Comcepts of Swarm Intelligence and Soft Computing. PhD thesis, Virginia Polytechnic Institute and State University (2002).

46. P. Lucic and D. B. Teodorović, Vehicle routing problem with uncertain demand at nodes: The bee system and fuzzy logic approach. In *Fuzzy Sets in Optimization*, pp. 67–82. Springer, Berlin (2003).

47. D. T. Pham, E. Koç, A. Ghanbarzadeh, S. Otri, S. Rahim, and M. Zaidi, The bees algorithm—A novel tool for complex optimization problems. In *IPROMS 2006 Proceedings of the Second International Virtual Conference on Intelligent Production Machines and Systems*, pp. 454–461. Elsevier, Oxford, U.K. (2006).

48. G. Z. Marković, D. B. Teodorović, and V. S. Aćimović-Raspopović, Routing and wavelength assignment in all-optical networks based on the bee colony optimization, *AI Communications—The European Journal on Artificial Intelligence.* **20**(4), 273–285 (2007). ISSN 0921-7126.

49. H. A. Abbass, MBO: Marriage in honey bees optimization—A haplometrosis polygynous swarming approach. In *Proceedings of the 2001 Congress on Evolutionary Computation CEC2001*, pp. 207–214, COEX, World Trade Center, 159 Samseong-dong, Gangnam-gu, Seoul, Korea (27–30, 2001). IEEE Press, Seoul, Korea. ISBN 0-7803-6658-1. URL citeseer.ist.psu.edu/abbass01mbo.html

50. H. O. Bozorg and A. Afshar, Honey-bees mating optimization (HBMO) algorithm: A new heuristic approach for water resources optimization, *Water Resources Management*. **20**, 661–680 (2006).

51. A. Afshar, H. O. Bozorg, M. A. Marino, and B. J. Adams, Honey-bee mating optimization (HBMO) algorithm for optimal reservoir operation, *Journal of the Franklin Institute*. **344**, 452–462 (2007).

52. H. J. Sung, Queen-bee evolution for genetic algorithms, *Electronic Letters*. **39**, 575–576 (2003).

53. S. Nakrani and C. Tovey, On honey bees and dynamic server allocation in internet hosting centers, *Adaptive Behavior—Animals, Animats, Software Agents, Robots, Adaptive Systems*. **12**(3-4), 223–240 (2004).

Chapter 4

Bio-Inspired Locomotion Control of the Hexapod Robot Gregor III

Paolo Arena and Luca Patané

Contents

This chapter focuses on the realization of a bio-inspired hexapod robot, named *Gregor III*, a moving platform used for testing cognitive algorithms. The topic of bio-inspired locomotion control of legged robots is treated following different levels of abstraction: neurobiological basis, mathematical models, insect behavior, and bio-inspired solutions for Robotics. Gregor III was designed and developed to study cognitive systems; the final aim is to use the robot as a testbed for an insect brain architecture. The main points addressed are bio-inspired principles acquired from the insect world, dynamic simulation environments used to assess the design procedures, hardware design and realization of the control architecture based on FPGA-devices, and development of a decentralized locomotion control algorithm based on the Walknet controller. A description of all the steps followed, starting from the biological basis to the realization of the final prototype and the experimental results, is reported.

4.1 Introduction

The possibility of endowing a legged robot with a perceptual system represents a fascinating challenge, due to the intrinsic complexity of the system with respect to a wheeled vehicle. Legged robots have many advantages: in fact locomotion performances increase in terms of agility and climbing capabilities, in particular when facing with uneven terrains. Another important aspect that can be considered during the design of a legged robot is the possibility of taking inspiration from nature and in particular from the insect world. Insects, during natural evolution, were provided with a physical structure and with sensing and processing systems that make possible the adaptation to difficult environments. Several research groups are currently involved in studying different insect species from a biological point of view. These studies can be finalized to the application of specific characteristics and basic principles to realize bio-inspired robotic structures.

Among different bio-inspired hexapod robots developed in the last years, two distinct approaches were followed. The former is based on the cockroach (e.g., *Blaberus discoidalis*),[1,2] and the latter is inspired by the stick insect (e.g., *Carasius morius*).[3] The differences between the two approaches rely on several aspects: from the design of mechanical parts to the control paradigm.

Robots inspired by the *B. discoidalis* are characterized by a sprawled posture with a center of mass shifted toward the rear of the body. The structure has an asymmetric design; in fact legs are specialized, assuming different shapes and functionalities. All these characteristics increase the robot stability, speed, and climbing capabilities. The class of robots inspired by the *C. morius* includes systems with a high symmetric structure: the design of each leg is very similar. The main advantage consists in the robot agility. On the basis of these assumptions, the final aim to realize an autonomous system with high stability

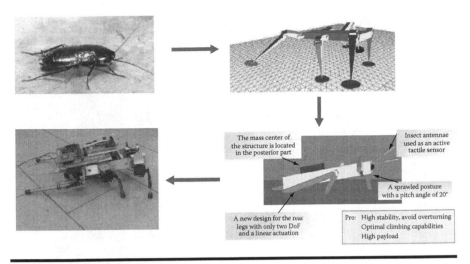

The mass center of the structure is located in the posterior part

Insect antennae used as an active tactile sensor

A sprawled posture with a pitch angle of 20°

A new design for the rear legs with only two DoF and a linear actuation

Pro: High stability, avoid overturning
Optimal climbing capabilities
High payload

Figure 4.1   Steps of the development process of a cockroach-inspired robot.

and payload capabilities directed our attention to the first class of bio-inspired robots.

Figure 4.1 shows the steps followed during the first prototype realization. The acquisition of the relevant aspects from the insect world was the first step, followed by different levels of kinematic and dynamic simulations carried out in virtual environments. The final result was a robot prototype that has been tested; the process was then iterated according to the experimental results.

From the control system point of view, we followed an incremental design process, growing up from locomotion to cognition. In this chapter, a detailed description of the locomotion control system is given, whereas more information about the cognitive architecture and algorithms is remanded to other works.[4,5]

4.2 State of the Art

The bio-robot design activities started with a deep analysis of the state of the art on bio-inspired robots. In the last decades, several research projects were carried out to develop robots taking inspiration from the animal world. In particular, we focus our attention on insect-inspired robots that present advantages in terms of adaptability to different kinds of unstructured environments.

An overview of the bio-inspired hexapod robot prototypes developed in the last years is given in Figure 4.2. A brief description of each robot is also reported, following the same presentation as that one reported in Figure 4.2.

Figure 4.2   State of the art on bio-inspired hexapod robots.

Dante II (1994) is a frame walker developed after several years of research with the aim to explore dangerous areas like volcanos.[6]

Boadicea (1992) is a pneumatically driven hexapod: the legs are modeled following the cockroach operational workspace.[7]

Hannibal (1990) is a small hexapod robot electrically powered and equipped with a complex distributed sensory system.[8]

Lobster (1998) is an underwater eight-legged robot actuated by shape memory alloy wire muscles. The sensory system equipped on it is inspired to the crustaceans: artificial air cells realized with MEMS, active antennae, accelerometers for attitude control, flux gate compass sensors, and others.[9]

RHex (2001) is an autonomous six-legged robot. Its main peculiarity is the simple leg design; in fact with a 1 degree of freedom (DoF) compliant leg, the robot is able to reach very high velocity in different terrains.[10]

Cricket (2000) micro-robot will be an autonomous robot that will fit within a 5 cm cube and will locomote by walking and jumping like real crickets.[11]

Cockroach (1996) robots are a series of biomimetic hexapods inspired by *B. discoidalis*. The last release is actuated with McKibben artificial muscles and each pair of leg is specialized as in real insects. The robot has 24 actuated DoF: 5 for the front, 4 for the middle, and 3 for the hind legs.[12]

Biobot (2000) is a six-legged, 18 DoF robot based on the features of the American cockroach, *Periplaneta americana*. It was designed with insect-like leg structure and placement, and pneumatic actuators to mimic muscles.[13]

Tarry (1998) is a 18 DoF six-legged robot actuated by servomotors. The design of the robot has been done taking inspiration from the stick insect. Moreover, the control algorithm is based on a reflex-driven strategy proposed by Prof. H. Cruse on the basis of the behavior observed in real insects.[14]

Hamlet (2000) is a six-legged robot realized to be a platform for studying different locomotion control algorithms.[15]

Sprawl (2001) is a new class of mini-legged robots developed with a shape deposition manufactured technique. Each leg has two DoF, a passive hip joint, and a pneumatic piston.[16]

Whegs (2004) robots are an innovative class of walking systems that utilize a particular tri-spoke leg trying to combine the advantages of wheels in terms of speed and payload and the advantages of legs in terms of climbing capabilities.[17]

4.3 Design Guidelines

The information acquired during the literature review was used as basis for the definition of the bio-robot design guidelines. In Figure 4.3, a flow diagram describing our legged robot prototyping activities is given. It is evident how the definition of the bio-robot design guidelines is directly related not only to the review of the state of the art but also to other important aspects.

During the definition of the robot guidelines, the following aspects have been taken into consideration:

1. **Target behaviors** identified for the robot (e.g., perceptual algorithms to be evaluated) where the requests for the sensory system are given

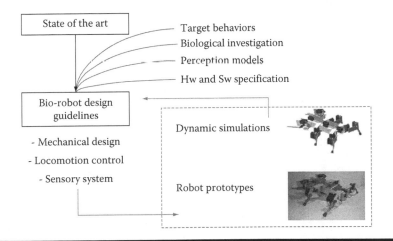

Figure 4.3   **Steps followed during the robot design and prototyping phase.**

2. **Biological investigations** that include locomotion control architecture, proprioceptive sensors, and other biological paradigms useful for bio-robot design and development
3. **Mathematical models** for perceptual systems that are important to define the sensory layer and the hardware architecture
4. The definition of the **hardware boards** and the **software framework** that constitute the control architecture

Due to the complexity of the robotic structure and control architecture, a feedforward design process has been considered unreliable, preferring a closed loop scheme. Therefore, during the design steps, dynamic simulations and preliminary prototypes have been used to improve the system capabilities.

Concerning the mechanical design, basic information was retrieved in literature[12] and the approach followed is schematized in Figure 4.4. Taking inspiration from the insect world and in particular from cockroaches, a series of hexapod robots, named Gregor, has been developed. Several experiments carried out on the first prototype, named Gregor I, underlined some faults or aspects to be improved. To make a systematic analysis of the robot performance, a high-detailed CAD model of Gregor was realized. The model was used in the software environment ADAMS[18] to perform multibody simulations. Analyzing the data acquired about speed, load distribution, requested motor torque, and the robotic structure were optimized and other prototypes developed.

The adopted dynamic simulation environment allows to perform analysis of the mechanical structure choosing different levels of details. In Figure 4.5a, the whole robot (i.e., first Gregor CAD design) and a detail of the hind leg motion mechanisms are shown.[19]

ADAMS simulations of Gregor I (see Figure 4.5b) outlined that the rotation-to-linear motion transformation mechanism preliminary used for the hind legs was not efficient. Exploiting the design phase loop (see Figure 4.3), this aspect has been optimized during the following developments, and finally, a new solution

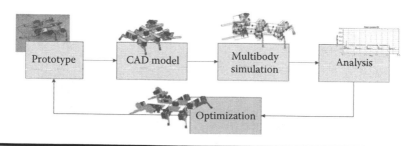

Figure 4.4   Flow diagram describing the robot design process.

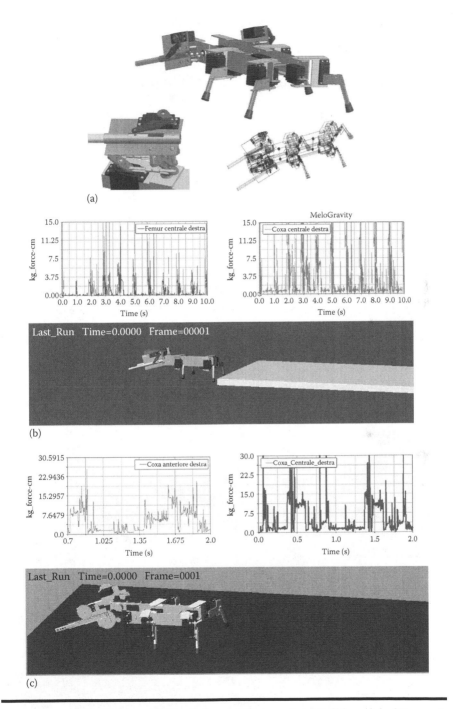

Figure 4.5 ADAMS dynamic simulation environment. (a) Solid model of Gregor I. Dynamic simulations of Gregor I (b) and Gregor III (c).

has been adopted and tested in simulation. The robot named Gregor III is shown in Figure 4.5c.

Therefore, the continuous improvement of the bio-robot characteristics led to the realization of the Gregor series:

Gregor I: Cockroach-inspired hexapod robot with 12 DoF, sprawled posture, and specialized leg functions

Gregor II: Upgrade to 16 DoF (three for front and middle legs, two for hind legs)

Gregor III: New design of the hind legs (i.e., cogwheel and rack) to increase the mechanical efficiency and the forward thrust

Both dynamical models and real prototypes of Gregors are available, as shown in Figure 4.6.

It is evident that during the testing phase of the robot capabilities, it was necessary to define a locomotion control layer. Two different approaches have been considered: Central Pattern Generator (CPG)[20] and Reflex-based control (Walknet).[21] On the basis of the results obtained from simulations and experiments, in the final system, the two control paradigms coexist. The robot is able to escape from dangerous situations using a CPG controller and is also able to slowly adapt its movements when dealing with complex terrains through the Walknet network.

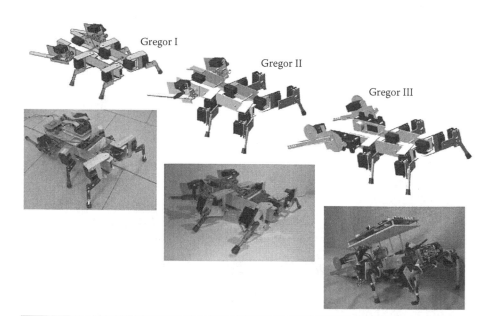

Figure 4.6 Design evolution of Gregor series.

4.4 Robotic Platform

This section is devoted to describe, in details, the mechanical aspects considered in order to develop an asymmetric hexapod robot (i.e., Gregor III), inspired by the cockroach. The characteristic of the robot was defined in order to match the following aims: development of a completely autonomous moving robot, able to face with uneven terrains, and to show perceptual skills.

Different elements must be taken into account due to the multidisciplinary aspects of the problem. The idea is to define the main blocks underlying the principal characteristics of the system dividing the work in different subtasks. In this chapter, we primarily focus on the mechanical design inspired by the biological world and on the locomotion control strategies.

4.4.1 Biological Inspiration

Gregor III design is based on biological observations in insects. In this section, we outline some of the most important results coming from insect experimental observations, with particular emphasis on the cockroach *B. discoidalis*.

Important structural features of *B. discoidalis* from an engineering viewpoint are leg structure, leg articulation, and body structure. Each cockroach leg is divided into several segments. The leg segments from the most proximal to the most distal segment are called coxa, trochanter, femur, tibia, and tarsus; the last one is indeed constituted by a series of foot joints.

The complex musculature coupled with complex mechanics confers upon the joint between body and coxa three DoF, much like that of a ball and socket joint. The joints between the coxa and trochanter, between the trochanter and femur, and between the femur and tibia are, instead, simple one DoF rotational joints. The joint between the trochanter and femur makes only a small movement and has often been referred to as fused. Each tarsal joint has several passive DoF, guaranteeing agile foot placement. Finally, a claw located on the end of the tarsus can be raised or lowered to engage the substrate during locomotion on slippery surfaces for climbing.[1]

Although the segments are reproduced in each of the three pairs of legs, their dimensions are very different in front, middle, and hind legs. Therefore, they are different in length, yielding a ratio of front:middle:hind leg lengths of 1:1.2:1.7.[13] Leg pairs with different length provide agility and adaptability. Cockroach legs articulate differently with the body, with the front legs oriented almost vertically at rest and middle and hind legs angled posteriorly of about 30° and 50°, respectively.[13] This configuration confers a *sprawled posture* able to guarantee a statically stable posture and, thus, a high margin of stability.[22,23] Finally, body is divided into three articulated segments called prothoracic, mesothoracic, and metathoracic segments. Anyway, dorsal flexion is seldom accomplished.[2]

Legs perform different functions:

- **Front legs** are mainly used to push the body of the cockroaches over obstacles. They also play an important role in turning and in decelerating.
- **Middle legs** act to push the cockroaches forward but also push the body of the cockroaches over obstacles.
- **Hind legs** generate the major part of the forward motion. They push directly toward the mass center, and the contact point is far behind to prevent the cockroaches falling on their back when climbing obstacles.

In the following, the cockroach-inspired biological principles will be applied to the robot design.

4.4.2 Robot Mechanical Design

Biological principles and past experiences on developing hexapod robots[24] guided the structure design phase. Our main concern was to replicate the cockroach features that are mainly responsible for its extreme adaptability and stability. We also took into careful consideration fundamental engineering issues like actuator selection.

The first step of the structure design is to specify what Gregor III is intended to do, since the final task deeply affects the overall design: e.g., as far as leg design is concerned, if the focus is just on horizontal walking, two DoF per leg is enough. From a locomotion point of view, the goals of Gregor I include efficiently walking on uneven terrains and climbing over obstacles whose height is at least equal to robot center of mass (CoM) height, as well as payload capability.

In the following, we outline the various steps that led to the final robot structure, from leg design to the overall structure realization. Our major concern here was on the design of rear legs (that seem to play a crucial role in obstacle climbing and payload capability), and on CoM placement.

After the identification of the biological principles, the design phase of the mechanical structure was divided into three steps:

1. Realization of the robot model on a dynamic simulator. Static analysis and preliminary dynamic simulations were carried out.
2. CAD project of the robot structure and detailed simulations of the robot model.
3. Realization of the robot prototype and experiments.

The three design stages present different levels of feedback that are fundamental to maximize the final results.

4.5 Locomotion Control

Bio-inspired solutions for locomotion control of legged robots are more and more proposed and successfully applied.[25] Two different biological paradigms are widely adopted: the CPG and the Walknet.

The CPG is a very robust mechanism that generates locomotion patterns by using oscillators controlled in a centralized way (i.e., top-down approach). On the other hand, the decentralized scheme lets the locomotion gaits emerge from the processing of sensorial stimuli coming from the tactile sensors distributed on the robotic structure (i.e., a bottom-up approach). The two paradigms have solid biological evidence and different implementations have been proposed.[21,26,27] In particular, concerning the reflex-based approach, a network named Walknet was proposed by Prof. H. Cruse. This network was designed taking into account the physiological data acquired from experiments carried out on the stick insect. The walking behavior emerges from local modules that control the movement of a single leg: Each of these local controllers switches over time between the two phases of the movement depending on its sensory inputs (stance and swing). Coordination rules—obtained in experiments on the stick insect—connect neighboring controllers and influence their behavior. In this way, the relationships are governed. The network has been successfully applied to control symmetric robotic structures inspired by the stick insect like the robot Tarry.[28]

The main focus of this research work is to investigate whether a decentralized network, based on local reflexes, can solve the problem of locomotion control of a cockroach-like hexapod robot.

The hexapod robot taken into consideration for the experiments, named Gregor III, is the last prototype developed as a suitable walking platform to study the emergence of cognitive capabilities, starting from reactive pathways, inspired by insect behavior.[4]

The main goal of the current activity[5] consists in growing from reflexive behaviors to cognitive capabilities. However, the development of a perceptual layer needs a robust locomotion layer able to supply to the upper layer the basic walking functionalities: forward walking, steering control, speed control, climbing capabilities, and others. The Walknet network has been extended to control the cockroach-inspired robot Gregor III.

4.6 Walknet and Gregor

The Walknet network is a biologically inspired structure used to control locomotion of hexapod robots.[29] In this section, the Gregor III robotic structure is described, underlying the main focuses of the system, in order to justify the application of Walknet, created to well suit the stick insect morphology and locomotion behavior, that has been modified to be adapted to the new cockroach-inspired design of the robot.

4.6.1 Gregor III

Gregor, from a locomotion point of view, should be able to efficiently walk on cluttered terrains, overcome obstacles, and carry a considerable payload. Looking in the insect world, cockroaches have been identified as an interesting source of inspiration. Our main concern was to replicate the cockroach features that are mainly responsible for its extreme agility, adaptability, and stability. However, we also took into careful consideration basic engineering issues like actuator selection, transmission mechanisms, and others. The various elements investigated that led to the final robot structure are a robust mechanical structure with specialized leg function, a sprawled posture, and powerful servomotor actuators.

The final robot configuration is characterized by 16 DoF. The active joints in the legs are actuated by servomotors Hitec HS-945MG delivering a stall torque of 11 kg cm and servomotors Hitec HSR-5995TG delivering a stall torque of 30 kg cm, placed in the most critical joints as identified during dynamical simulations. With respect to other actuators, servomotors guarantee a simple position control and low payload for power supply. The main body of the robot, parallel piped in shape, is 30 cm long, 9 cm wide, and 4 cm high; the body is simply made of two aluminum sections joined by two threaded bars. This simple structure makes room for onboard electronics and batteries. Front and middle legs are divided into three segments. All segments are made of aluminum sections. Middle legs are equal to front legs except for the length of the tibia segment that, in order to confer a sprawled posture, is a bit shorter. Gregor III has three actuators for each of these four legs. Hind legs are actuated by two motors, the former aimed to control the rotational joint, the latter employed to realize the prismatic movement (protraction/retraction) of a metal bar. To convert the rotational energy from the motor-to-linear oscillations, a transmission mechanism based on a cogwheel and a rack, to increase the mechanical efficiency and the forward thrust, has been adopted. Moreover, this kind of movement seems to support lifting the body in upward climbing in the cockroach. Due to the high payload that the hind legs have to support, a servo 5:1 gearbox has been introduced to multiply the torque supplied by the rotational joint. The drawback of speed reduction is not critical due to the limited excursion range of this joint.

4.6.2 Walknet Control on Gregor

The Walknet controller has been used in the past to control many different multi-legged robots. To adopt the Walknet model to the Gregor robot, a dynamic virtual simulation environment was considered. In the simulation environment, the robot was constructed reflecting its physical dimensions, the functionality of the joints, and the dynamical properties of its body parts. The robot was controlled by an adapted version of Walknet. Especially, a changed control scheme for the hind legs was introduced with respect to the different leg geometry.

The Walknet network is a biologically inspired control structure: it is composed of decentralized local control structures, which decide locally about the state of each leg (if it is in swing or in stance mode). Each leg state is controlled by local simple controller modules (see Figure 4.7).[30] These modules are implemented as neural networks (see Figure 4.7b).There is a network for the control of the stance mode and one for the control of the swing movement.[21] The stance movement realizes a retraction of the leg: the alpha joint moves the leg backward, while the beta joint and gamma joint are controlled by the height network: this neural net maintains the distance between body and ground. Due to the sprawled posture chosen for the robot, the height controller has been tuned to maintain this attitude during walking.

Another important block of the leg controller is the swingnet that is a neuronal implementation for the control of the joint velocities during the swing phase. It computes the trajectories of the legs. As input, it mainly uses the current joint angles and some target angles. Accordingly, these inputs are computed by two matrices: one for the target values and one for the current angular values. As output, it provides joint velocities as control signals for the joints. The swingnet is modeled as a simple feedforward neural network, which basically minimizes the difference between the current position and a predefined target position. The difference is computed through the two matrix multiplications and is used as the control signal (proportional controller). In addition, for control of a swing movement, the swingnet should not generate a linear trajectory between the PEP (posterior extreme position) as the starting point of the movement and the AEP (anterior extreme position) as the target position. It has to lift the leg from the ground by increasing the beta joint angle. For this purpose, the reference value for the beta angle equals not a target position, but it reflects the height of the swing movement. As a consequence, during the first phase of the swing movement, the leg is raised approaching this value. But in the second phase, it shall be lowered. Therefore, an additional input was introduced, which counteracted the first. This influence is delayed through a low-pass filter and became therefore dominant, influencing the beta joint at approximately half of the time of the swing movement (see Figure 4.7b). The swingnet can be extended through a sensory pathway, which leads to an avoiding movement when there is a collision with an obstacle to the front of the leg (like when running against an obstacle: in the swing phase, the leg hits the obstacle). A touch sensor is used to implement an avoidance reflex: when the animal is detecting an obstacle in front of it, it tries to climb on it and therefore has to adapt its swing movement.

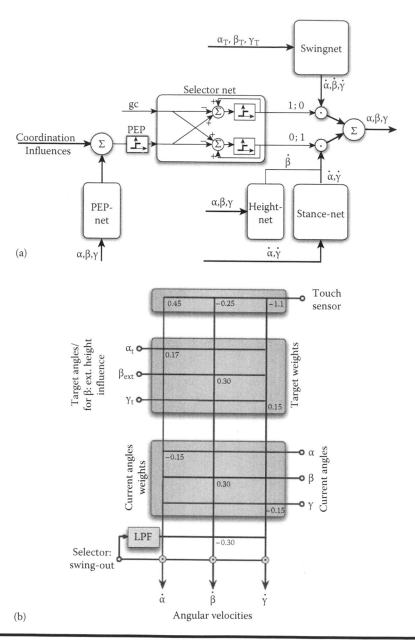

Figure 4.7 Walknet network: (a) block diagram of a single leg controller and (b) swingnet controller.

4.7 Dynamic Simulation

Virtual simulation environments are useful tools for studying locomotion control in legged robots. The simulation environment chosen to investigate the performances of the control network is named Simulator BOB.[31] It is an open source dynamic simulation environment developed with particular attention to mobile robots. The simulation core is based on the Open Dynamic Engine (ODE)[32] library that guarantees high performances in terms of speed by using optimized algorithms for collision management. The graphic interface is based on OpenGL library, and the source code of the simulator is available on the web.[31] The robot model is defined through an XML script in which the characteristics of the structure are indicated. The simulation environment guarantees a high level of details, in fact in addition to the mechanical parameters (i.e., weights, dimensions, and others), it is possible to indicate motor parameters (i.e., maximum torque, speed, and others).

As already discussed above, Walknet is a reflex-based controller that relies on distributed sensors. A feedback through the environment is important to adapt locomotion to the external conditions, generating free gaits. The sensory system introduced in the simulated robot consists of joint position sensors for the 16 actuated joints and ground contact sensors for the front, middle, and hind legs.

Exploiting the potentiality of the simulation tool, the robot model, controlled through Walknet, has been tested in different kinds of terrain configurations, as shown in Figure 4.8. A complex scenario has been created to analyze the behavior of the robot in the case of large obstacles (Figure 4.8a), slopes (Figure 4.8b), overcoming obstacles (Figure 4.8c), and uneven terrains (Figure 4.8d).

When the robot moves on flat terrains, leg controllers and coordination mechanisms produce joint reference signals that guarantee high stability to the structure. A control parameter can be used to increase the robot velocity reducing the time of the stance phase. When the speed reaches its maximum value, the free gait that emerges is similar to the tripod gait typically shown by insects during fast walking (see Figure 4.9a).

The modulation of the stance speed on one side can be used as a steering signal. The simulated robot is equipped with two distance sensors that can detect objects in a range of 120 cm. When an obstacle is detected, the robot turns as shown in Figure 4.9b, in which the left proximity sensor triggers a turning action. The minimum curvature radius is about 50 cm.

The robotic structure has been designed to easily climb over obstacles. During the swing phase, when an obstacle is detected by contact sensors equipped on the front side of the tibia of each leg, the swing net behavior is modified, and the leg after a short retracting phase increases its elevation, trying to touch down on the top of the object. In Figure 4.10a, a sequence of avoidance reflexes make the robot able to climb over an obstacle with height equal to 90% of its front body height (see environment in Figure 4.8c). Finally, as can be seen in Figure 4.10, the robot is able to adapt its coordination to maintain a stable locomotion also on uneven

Figure 4.8 **Simulated environments used to test the robot capabilities. (a) Gregor III is able to avoid large obstacles, (b) to face with slopes, (c) to climb over obstacles, and (d) to walk on uneven terrains.**

terrains (see environment in Figure 4.8d). Videos of the simulated robot are available on the web.[33]

4.8 Hardware Architecture and Robot Experiments

The realization of a completely autonomous-legged robot is a hard task under different perspectives. For instance, the problem of autonomy is not a primary target for the great part of legged systems developed until now. These robots are controlled by an external PC and/or in some cases by a network of microcontrollers that can be easily endowed in the structure. Due to the fact that we want to develop a cognitive layer for Gregor, we have chosen a hardware architecture able to handle the different concurrent tasks requested to the robot. On the basis of these constraints, we have proposed a framework based on a Field Programmable Gate Array (FPGA) that is a reprogrammable digital device that can process in parallel multiple tasks.

The characteristics that make an FPGA-based hardware an optimal solution for our purposes are the flexibility of a reprogrammable hardware and the high computational power obtained with a parallel processing. So the proposed system can independently manage several sensors with different frequencies. Dedicated channels are created for the sensors that require a high band for data transfer (visual

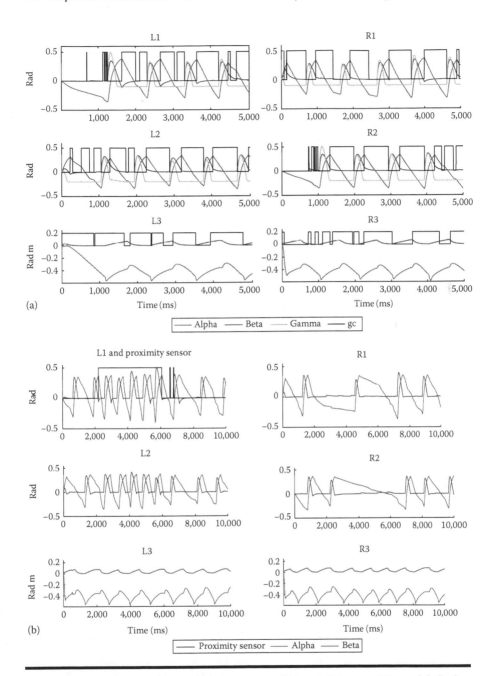

Figure 4.9 Joint positions of the 16 actuated DoF of Gregor III model during different simulation conditions. (a) Signals acquired from the robot model during walking on a flat terrain. (b) The presence of an obstacle is detected by a proximity sensor, and the robot turns to avoid it, modifying its stance velocities.

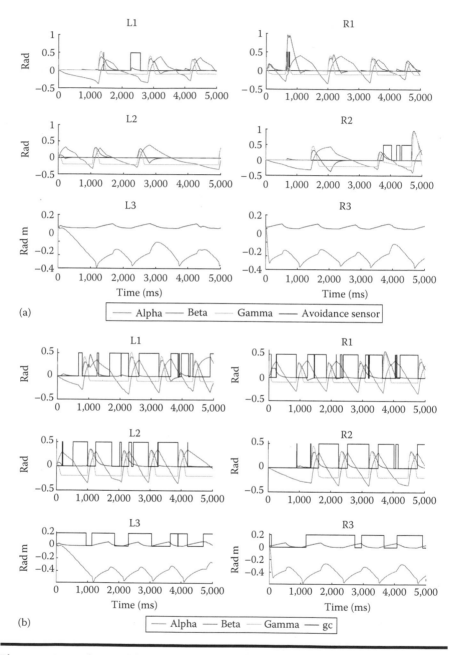

Figure 4.10 Joint positions of the 16 actuated DoF of Gregor III model during different simulation conditions. (a) Front and middle legs are equipped with contact sensors located on the tibias; when an obstacle is detected, an avoidance reflex is triggered to climb over it. (b) Adaptation of the robot movement on a cluttered environment.

system). In order to optimize the FPGA resources, a serial bus is used for the other sensors. The computational power has also been considered to guarantee the implementation of the whole cognitive architecture.

The device that we have equipped on the robot is a Stratix II EP2S60 from Altera.[34] The characteristics of the FPGA are 24.000 Adaptive Logic Modules (ALMs), a working frequency of 400 MHz, and about 1020 I/O pins. Moreover, a soft-embedded 32-bit microprocessor, the Nios II, can be synthesized in the FPGA, occupying only about 5% of the resources. The core of the hardware architecture seems to be a suitable solution for our locomotion algorithms,[4] but for a complete integration on the robot, we need to define interfaces with the sensory system and the motor system.

Insects employ a great number of distributed mechanical sensors to monitor the leg status. To acquire a complete portrait of the robot internal state variables, needed by the reflex-based control, we have equipped the robot with a set of different proprioceptive sensors: ground contact sensors, load sensors to measure the load on each leg, and potentiometers for the joint angular position.

Taking advantage of the FPGA parallelism in data management, the proposed architecture allows processing of a large number of sensors with a high degree of flexibility, speed, and accuracy. The architecture permits multisensory fusion processes. The different sensory sources are accessible by a unique interface; in fact the system can manage digital and analog sensors connected with parallel or serial buses.

Three I2C Master VHDL blocks have been included in the FPGA to interface the control board with the analog sensory board. All the I2C blocks are able to work simultaneously with different levels of priority. The first and the second I2C buses are dedicated to the devices that need analog-to-digital conversion. To make this conversion, 12-bit 4-channels ADC have been used, with a maximum of five ADC for each line. So each bus can handle up to 20 analog devices. For instance, taking into account the legged robot Gregor III, the first I2C bus processes eight infrared sensors and joint position sensors; the second I2C channel is dedicated to acquire two auditory circuit outputs and up to 18 analog inputs that are available for extra sensors. Finally, the third I2C bus is used to acquire those devices, such as the electronic compass and the accelerometer that already have a I2C interface (see Figure 4.11). The choice of the I2C bus has been made for three main reasons: it is a serial bus operating with an acceptable speed according to the desired final application; only two I/O pins are needed; it is a widespread interface in a lot of devices available on the market. From the other hand, a parallel interface has been used only for high-level sensors (e.g., connection with the visual system) that need to process huge amount of data. Finally, the digital sensors (ground contact and long distance sensors) have been connected directly to the digital input of the control board.

The servomotors that actuate the system can be controlled in position with a PWM signal at 50 Hz by means of an internal potentiometer. Slightly modifying the

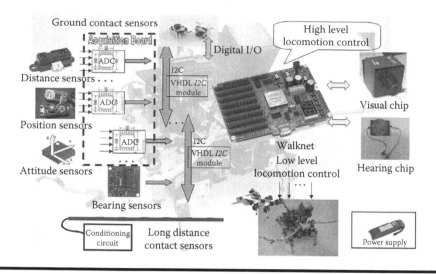

Figure 4.11   Gregor III control architecture.

motor, the position sensor signal has been acquired with an appropriate conditioning circuit and used for the Walknet network. The introduction of redundancy in the sensory system is important for fault tolerance and to improve the confidence of the measure.

The robot, as shown in Figure 4.12, is equipped with a distributed sensory system. The robot's head contains the Eye-RIS v1.2 visual processor, a cricket-inspired hearing circuit, and a pair of antennae developed using Ionic Polymer-Metal Composite (IPMC) material. A compass sensor and an accelerometer are also embedded in the robot together with four infrared distance sensors used to localize obstacles. A set of distributed tactile sensors (i.e., contact switches) is placed in each robot leg to monitor the ground contact and to detect when a leg hits with an obstacle (see Figure 4.13).

The robot is completely autonomous for the power supply. Two 11.1 V, 8 A Li Poly battery packs are used: one for the motors and the other for the electronic boards.

The robot can be used as a walking lab able to acquire and process in parallel a huge amount of different types of data. For monitoring the system status and for storing purposes, a wireless module is used to transmit the sensible information to a remote PC (see Figure 4.14).

The hexapod robot Gregor III has been designed as a moving laboratory. The robot equipped with a Li-Poly battery pack can walk autonomously for about 30 min, exchanging data with a remote workstation through a wireless communication link.

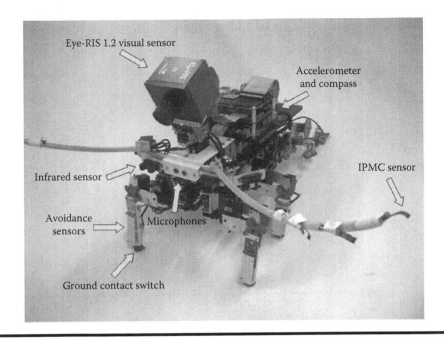

Figure 4.12 Gregor III.

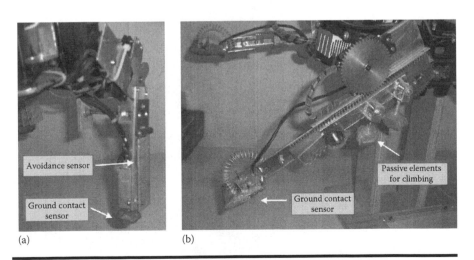

Figure 4.13 Distributed sensory system. (a) Tactile sensors distributed on front and middle legs. (b) Hind leg actuation mechanisms, sensors, and passive elements for climbing.

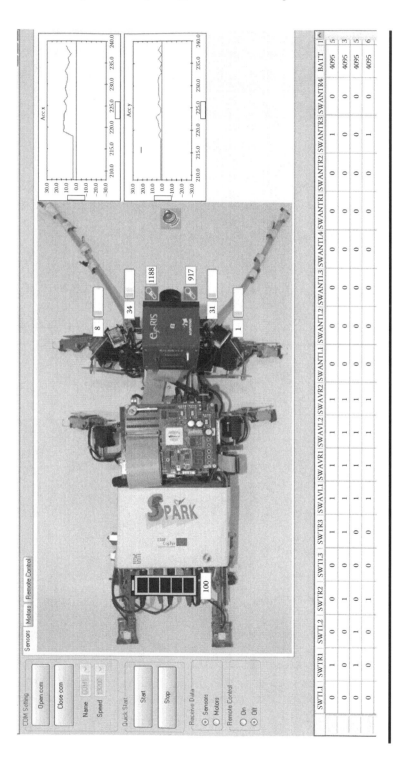

Figure 4.14 Examples of the robot interfaces created to monitor and store the status of the legged robot during experiments.

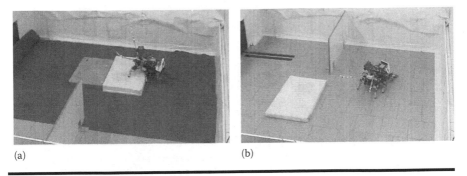

(a)   (b)

Figure 4.15 Walknet on Gregor III: (a) obstacle climbing and (b) autonomous navigation in a cluttered environment.

When the robot stability is compromised during an experiment, a recovery procedure is performed. Dangerous situations are detected using a posture controller implemented through a three-axis accelerometer equipped on board.

Figure 4.15 shows two different scenarios used to test the robot capabilities for autonomous navigation in cluttered environments and obstacle climbing. Multimedia material is available on the web.[35]

4.9 Conclusions

In this chapter, the design guidelines used to realize a bio-inspired walking robot are presented. The robot Gregor III has been presented underlying its walking capabilities obtained through the Walknet network. The Walknet controller, already developed and applied to control hexapod robots, has been adapted to the cockroach-like structure of Gregor III. The capabilities of the system were evaluated in a dynamic simulation environment. A dynamic model of Gregor has been used to evaluate the system behavior in different environments exploiting the adaptability of the Walknet network. The simulation results confirm the suitability of the approach. Moreover, the introduction of a decentralized control scheme in parallel with a CPG controller as a recovery system in case of fault of the tactile-distributed sensory network is envisaged. The hardware implementation of the control architecture is discussed, and the strategies followed to handle the distributed sensory system are analyzed. Finally, preliminary experimental results on the real robot have been proposed.

Acknowledgments

The authors acknowledge the support of the European Commission under the projects SPARK I and II *Spatial-temporal patterns for action-oriented perception in roving robots.*

References

1. R. Quinn and R. Ritzmann, Biologically based distributed control and local reflexes improve rough terrain locomotion in a hexapod robot, *Connection Science.* **10**, 239–255 (1998).
2. J. Watson, R. Ritzmann, S. Zill, and A. Pollack, Control of obstacle climbing in the cockroach, *Blaberus discoidalis.* I. Kinematics, *Journal of Comparative Physiology A.* **188**, 39–53 (2002).
3. J. Dean, T. Kindermann, J. Schmitz, M. Schumm, and H. Cruse, Control of walking in stick insect: From behavior and physiology to modeling, *Autonomous Robots.* **7**, 271–288 (1999).
4. P. Arena and L. Patané, *Spatial Temporal Patterns for Action Oriented Perception in Roving Robots.* Springer, Berlin (2008).
5. SPARK Project. Spark EU project home page. http://www.spark.diees.unict.it
6. J. Bares and D. Wettergreen, Dante II: Technical description, results, and lessons learned, *The International Journal of Robotics Research.* **18**(7), 621–649 (July 1999).
7. Home page of the Boadicea hexapod. URL http://www.ai.mit.edu/projects/boadicea/
8. Home page of the Hannibal hexapod. URL http://groups.csail.mit.edu/lbr/hannibal/
9. Biomimetic underwater robot program home page. URL http://www.neurotech nology.neu.edu/
10. Rhex the compliant hexapod robot home page. URL http://www.rhex.net/
11. M. Birch, R. Quinn, G. Hahm, S. Phillips, B. Drennan, A. Fife, H. Verma, and R. Beer, Design of a cricket microrobot. In *IEEE Conference on Robotics and Automation (ICRA '00)*, San Francisco, CA (April 2000).
12. Biorobotics lab directed by Dr. Roger Quinn. URL http://biorobots.cwru.edu/
13. J. Reichler and F. Delcomyn, Biorobotic approaches to the study of motor systems, *International Journal of Robotics Research.* **19**(1), 41–57 (January 2000).
14. Robot Tarry Home page. URL www.tarry.de/index_us.html
15. Hamlet home page. URL http://www.hamlet.canterbury.ac.nz/
16. Sprawl robots home page. URL http://www-cdr.stanford.edu/biomimetics/docu ments/sprawl/
17. T. Allen, R. Quinn, and R. R. R. Bachmann. Abstracted biological principles applied with reduced actuation improve mobility of legged vehicles. In *IROS*, Las Vegas, NV, (2003).
18. Adams software. URL http://www.mscsoftware.com/
19. P. Arena, M. Calí, L. Fortuna, C. L. Spina, S. Oliveri, L. Patané, and G. Sequenza, A cockroach-inspired hexapod robot: Performance enhancement through dynamic simulation. In *Workshop on Bio-Inspired Cooperative and Adaptive Behaviours in Robots*, Rome, Italy, (2006).
20. M. Pavone, P. Arena, and L. Patané, An innovative mechanical and control architecture for a biometic hexapod for planetary exploration, *Space Technology.* **25**(3–4), 1–12 (2006).
21. H. Cruse, T. Kindermann, M. Schumm, J. Dean, and J. Schmitz, Walknet a biologically inspired network to control six-legged walking, *Neural Networks.* **11**, 1435–1447 (1998).

22. L. Ting, R. Blickhan, and R. Full, Dynamic and static stability in hexapedal runners, *Journal of Experimental Biology.* **197**, 251–269 (August 1994).
23. J. Clark, J. Cham, S. Bailey, E. Froehlich, P. Nahata, R. Full, and M. Cutkosky. Biomimetic design and fabrication of a hexapedal running robot. In *International Conference on Robotics and Automation*, Seoul, Korea (2001).
24. Diees bio-inspired robots home page. URL http://www.scg.dees.unict.it/activities/biorobotics/movie.htm
25. J. Ayers, J. Davis, and A. Rudolph, *Neurotechnology for Biomimetic Robots.* MIT Press, Cambridge (2002).
26. F. Delcomyn, Walking robots and the central and peripheral control of locomotion in insects, *Autonomous Robots.* **7**, 259–270 (1999).
27. P. Arena, L. Fortuna, M. Frasca, and L. Patané, A cnn-based chip for robot locomotion control, *Circuits and Systems I: Regular Papers, IEEE Transactions on Circuits and Systems I.* **52**(9), 1862–1871 (2005).
28. M. Frik, M. Guddat, M. Karatas, and D. C. Losch, A novel approach to autonomous control of walking machines. In *International Conference on Climbing and Walking Robots*, pp. 333–342, Portsmouth, U.K. (September 1999).
29. M. Schilling, H. Cruse, and P. Arena, Hexapod walking: An expansion to walknet dealing with leg amputations and force oscillations, *Biological Cybernetics.* **96**(3), 323–340 (2007).
30. P. Arena, L. Patané, M. Schilling, and J. Schmitz. Walking capabilities of gregor controlled through walknet. In *Microtechnologies for the New Millennium (SPIE 07)*, Gran Canaria (SPAIN) (2007).
31. Simulator BOB 3D simulation environment for mobile robots, user and developer guide. URL http://simbob.sourceforge.net
32. ODE software. URL http://www.ode.org/
33. Robot movies web page: dynamic model of gregor III controlled through walknet. URL http://www.spark.diees.unict.it/GregorWalknet2.html
34. Altera web site. URL http://www.altera.com/products/devices/stratix2/st2-index.jsp
35. Multimedia on gregor III. URL ftp://sparkweb:sparkweb@151.97.5.245/sparkweb/Demos/pages/demo2_media.html

Chapter 5

BEECLUST: A Swarm Algorithm Derived from Honeybees

Derivation of the Algorithm, Analysis by Mathematical Models, and Implementation on a Robot Swarm

Thomas Schmickl and Heiko Hamann

Contents

We demonstrate the derivation of a powerful and simple as well as robust and flexible algorithm for a swarm robotic system derived from observations of honeybees' collective behavior. We show how such observations made in a natural system can be translated into an abstract representation of behavior (algorithm) working in the sensor-actor world of small autonomous robots. By developing several mathematical models of varying complexity, the global features of the swarm system are investigated. These models support us in interpreting the ultimate reasons of the observed collective swarm behavior, and they allow us to predict the swarm's behavior in novel environmental conditions. In turn, these predictions serve as inspiration for new experimental setups with both the natural system (honeybees and other social insects) and the robotic swarm. In this way, a deeper understanding of the complex properties of the collective algorithm, taking place in the bees and in the robots, is achieved.

5.1 Motivation

5.1.1 From Swarm Intelligence to Swarm Robotics

Swarm robotics is a relatively new field of science, which represents a conglomerate of different scientific disciplines: Robot engineers are trying to develop novel robots, which are inexpensive in the production, because for robot swarms a high number of robots is needed. Software engineers elaborate concepts from social robotics, from collective robotics in general, and from the field of distributed artificial intelligence. Their aim is to extend these concepts to a level that is suitable also for swarms consisting of hundreds or even thousands of robots. While many concepts like auction-based systems or hierarchical master–slave systems[1,2] for task allocation work with groups of several robots, these techniques are hard to apply to swarms of much larger size: Whenever a central decision-making unit is needed, or whenever a centralized communication system has to be used, solutions get harder to achieve the bigger the communication network gets, which is formed by the swarm members.

The basic challenges in developing swarm algorithms for swarm robotics are posed by the following key features, which are characteristic for a swarm robotic system:

Restrictions Posed by Robotic Hardware: A swarm robot will always be "cheaper" and less powerful in its actuator abilities, computational power, and sensory capabilities compared to a robot that is aimed to achieve the same goal alone. If an affordable swarm robot is powerful enough to achieve the goal alone sufficiently and robustly, there is no need to create a swarm system anymore. Due to this limitation, algorithms and performed behaviors in a swarm robot have to be less complex and should be applicable also in a noisy environment or on deficient platforms. Also, the available sensor data are often more noisy and less comprehensive than in a non-swarm robot.

Scaling Properties of the Swarm: All robots in a swarm form a simple and rapidly changing ad hoc network. Communication is often restricted to local neighbor-to-neighbor communication or to only few hops. Centralized communication units are in most cases unfavorable because of scaling issues.

Thus, on the one hand, a swarm algorithm has to work with poorer robots, worse sensor data, and narrow and short-ranging communication channels. On the other hand, a swarm solution for a robotic problem does also provide significant advantages: As real swarm solutions are designed to work with all that limitations and noise, they are extremely robust to errors on all organizational levels of the systems. They are often little affected by damages or breakdown of swarm members; they are often extremely tolerant against biases or errors in their sensor-actuator system. And their reaction by their nature, can be very flexible to sudden perturbations of the environment they inhabit or they can even be applied to other environments they have not been tailored for.

These features are coherent with the distinct definitions of "swarm intelligence" that were proposed by several authors. The term "swarm intelligence" was initially used by Beni and Wang[3] in the context of "cellular robotics." A strong robot-centered approach was also the source of inspiration for the algorithms presented and extended in the book "swarm intelligence":[4] "... any attempt to design algorithms or distributed problem-solving devices inspired by the collective behavior of insect colonies or other animal societies." An even wider definition was given already by Millonas[5] who defined five basic principles of a system necessary to call it a "swarm intelligent" system, which we summarize up as follows (cf. Kennedy and Eberhart:[6])

The Proximity Principle: The swarm members (and thus also the whole swarm) should be able to carry out simple calculations (in space and time).

The Quality Principle: The swarm should be able to respond to qualitative features in its environment in a well-defined way.

The Principle of Diverse response: The swarm should act in a diverse way.

The Principle of Stability: Small changes in the quality features of the environment should not alter the swarm's main mode of behavior.

The Principle of Adaptability: The swarm should be able to change its mode of behavior in response to environmental changes, whenever these changes are big enough to make it worth the computational price.

For us as one biologist and one engineer, it is important to mention that, whenever the above definitions refer to the "computational price," this should include also "other" prices as well: For biological systems, the costs of energy and the risk of death are important aspects that should also be considered in interpreting biological swarms. And for engineering approaches (like "swarm robotics"), energy efficiency (which often correlates with computational costs) and risk of damage are also significant aspects.

A swarm robotic system shares many similarities with a swarm-intelligent system. The way how solutions are approached is often very similar, although important differences exist: General swarm-intelligent algorithms reside often in a bodiless environment inside of a computer. Classical examples are the ACO algorithm,[7,8] particle-swarm optimization,[6] or multiagent simulations that incorporate swarming systems (see for an overview Refs. [2,9,10]). In contrast to such swarms formed by bodiless-simulated agents, a plethora of physical constraints apply to real robotic swarms. Thus, swarm-robotic systems have to be "embodied swarm-intelligent algorithms." This was already described by several recent reviews,[11,12] the latter one defining three basic properties being critical for a swarm-robotic system:

Robustness: Robotic swarms should operate in a way that allows continuation of operation even after the loss of swarm members or other failures.

Flexibility: The swarm behavior should consist of "modules" of solutions to distinct task sets. As the environment changes, the composition and the priorities of tasks get altered, what should be responded by the swarm by altering the collective behavior.

Scalability: A robotic swarm consists of many members. Sometimes, the swarm is formed by a huge number of robots. As in all other networks, the addition of new nodes should not impair the collective behavior of the system, for example, due to an exponential growth of communication efforts.

As can be seen by comparing the definitions of swarm robotics and of swarm intelligence, there is a high degree of overlap between those two fields of science. From our point of view, the most important difference between these two domains is the fact that swarm-robotic solutions are limited to algorithms that have the following properties:

- All actions that have to be performed by the robots are restricted to be local (limited actuator radius).
- Algorithms can utilize only a very limited local sensor range.
- In real physical embodied systems, the motion of agents is restricted to the real speed of the physical devices (robots).
- Communications are restricted to localized communication; otherwise, the scaling properties of these algorithms will be rather poor that prevents these algorithms from being used in large robotic swarms.
- Finally, as swarm robots are imperfect and use erroneous physical devices, all sensor data and all executed motion patterns are affected by a significant fraction of error and noise. Algorithms have to be sufficiently tolerant to these sources of noise.

As swarm sizes increase, the problems of interferences in communication (light pulses, radio, sound, ...) and the problems of traffic jams and local deadlocks increase in turn usually in a nonlinear manner. Thus, typical "swarm-intelligent solutions," which work perfectly in a simulated multi-agent world, often do not work as expected in a real robotic implementation.

5.1.2 From Biological Inspirations to Robotic Algorithms

The challenges and restrictions that apply for swarm robotic units do also hold for biological organisms. Also, animals have to deal with the same above-mentioned limitations: As their densities increase, problems arise in communication (limited channel width) as well as in "ordered" interaction (flocking, swarming, ...). Also for biological units, computational power is limited and reactive behavior acts fast, but not in zero time. Thus, the more interaction partners exist, the simpler the behavioral rules should be, to allow the animal to perform the required computation, decisions, and actuation. Swarms live, as most other animals, in a noisy and frequently changing environment. Sometimes, the swarm itself can add a significant amount of noise to the environment: Imagine a fish swarm and its coordinated but swarm-like motion of thousands of fish influencing local fluid motion and turbulences. These turbulences can, in turn, affect the members of the fish swarm, as they can sense them through their *myoseptum horizontale* (lateral line).

In nature, a variety of adaptations has evolved, which tackles these problems from various sides:

1. The sensors of animals developed into highly sophisticated devices. This was achieved by developing "better" sensors or by joining and combining sensors (serialization and specialization). These adaptations allow animals to deal better with a noisy environment. While such an approach can be sufficient for pure robotic engineers ("making a better robot"), scientists with more adhesion to the field of swarm intelligence might not be pleased by improving simply the single swarm robot's abilities. For them, improving the swarm's performance without improving the single robot is the main interesting goal.

2. Especially in social insects, the animals developed methods to increase the "orderedness" of their environment. Ant trails emerge, wax combs are built, and termite mounds are erected. All these adaptations of the environment help the swarm to canalize its behavior along "favorable" lines, thus allowing it to work more efficiently and to solve more complex problems. Such processes are referred to as "stigmergy" in literature, a term created by Grassé.[13,14] Its basic meaning is that the swarm's behavior changes the environment and that changed stimuli in the environment can in turn alter the swarm's behavior. This delayed feedback loop can drive the environment and the swarm into extreme conditions that can be best described as being a "highly ordered state." The "stigmergy" approach is often investigated in swarm robotics: for example, puck sorting[4] and artificial pheromone trails.[15,16]

3. In nature, the rules of interactions between the members of the swarm have also been adapted by natural selection. The simpler the rules are, the faster and better they can be executed. The more robust they are, the less they are affected by noise. The more flexible they are, the better such swarms are able to deal with environmental changes. For swarm robotics, these rules are very valuable information that can be extracted from natural swarm systems. They are very likely simple enough to be executed by cheap robots. They are robust enough to work with cheap and poor sensors. And they are flexible enough to allow a variety of fascinating and (hopefully) efficient swarm behavior.

From our perspective, the third domain of adaptation is of highest relevance for swarm robotics. The open challenge is how such rules can be identified in natural systems. Do we even notice them? How can they be extracted? If they were extracted, can they be transformed (translated) into the physical world of the target robotic swarm? Can observations done on the natural system be meaningfully compared to observations done on the artificial system? Can we look deeply into both systems, detach the specialties, and see the core elements that drive the dynamical distributed system that we call a "swarm"?

5.1.3 Modeling the Swarm

In the chapter at hand, we demonstrate how a biological inspiration taken from the example of social insects can be "translated" into an algorithm suitable for an autonomous robotic swarm. We show how behavioral experiments with honeybees can be compared to experiments performed with robotic swarms. To improve the understanding of the fundamental mechanisms that govern the swarm's mode of operation, we developed several mathematical models, which allow us to identify, characterize, and study the most important feedback loops that determine the collective behavior of the swarm.

5.2 Our Biological Inspiration

The algorithm we describe here was originally inspired by young honeybees' collective behavior of aggregation in a temperature gradient field. In a natural honeybee hive, a complex pattern of temperature fields exists: The central broodnest areas are kept on a comparably high temperature (between 32°C and 38°C), while honey comb areas and the entrance area have significantly lower temperatures. The temperature in the broodnest affects the growth and the development of larvae and, presumably, also the development of freshly emerged young honeybees. In many experiments, the self-navigation of honeybees in a temperature field was examined: It was found that young honeybees (1 day old or younger) tend to locate preferentially in a spot that is between 32°C and 38°C.[17,18] It was also found that diurnal rhythms affect the temperature preference of older bees.[19] These experiments were made with a one-dimensional device called "temperature organ" (see Figure 5.1). This device consists of one longer metal profile channel, which is rather narrow, so

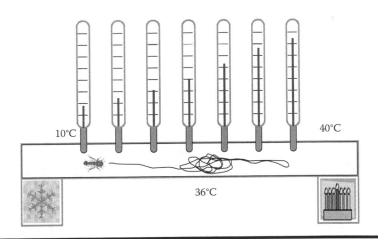

Figure 5.1 Schematic drawing of a temperature organ. Bees can move (almost) only to the left and to the right. By introducing a heating element (torch) and a cooling element (ice water), a temperature gradient is established inside of the device. Local temperature is measured with thermometers. Young bees locate themselves preferentially between 32°C and 38°C.

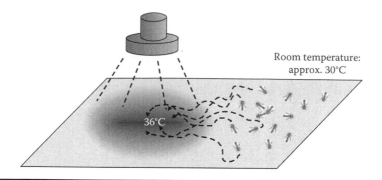

Figure 5.2 Schematic drawing of the two-dimensional aggregation arena. Bees could navigate in two dimensions and should aggregate preferentially below the spot produced by the heating lamp.

that the bees can move (almost) only to the right or to the left. By placing a heating element (or a torch) on one side and an ice-water reservoir on the other side, a rather steep temperature gradient is formed inside of the metal profile channel. The bees can be observed (and recorded) from atop, through a red filter* top cover.

This one-dimensional device was converted by ourselves to a two-dimensional temperature arena. We decided to establish the temperature gradient by using a heating lamp, which is usually used for terrariums. By implementing two feedback devices, we managed regulating the temperature maximum in the arena (hot spot in the center of the temperature gradient field), as well as keeping the ambient temperature on an as-stable-as-possible level (outer rim of the gradient field). Figure 5.2 shows the basic setup of our two-dimensional arena. Figure 5.3 shows photographs of our experimental setup.

First experiments showed that bees that were introduced in groups of 15–30 bees into the arena showed the expected behavior (as it is depicted in Figure 5.2) only if the temperature gradient was rather steep (ambient temperature around 24°C, hot spot approx. 36°C). Whenever the gradient was flat (ambient temperature around 30°C, hot spot approx. 36°C), the bees showed a different behavior: The bees started to form clusters. At the beginning of the experiments, these clusters were located randomly distributed all across the arena, and the bee clusters consisted only of 2–3 bees. As experimental time proceeded, the clusters on the right side of the arena (the cold side) disappeared, and clusters on the left side increased in size. This process continued until only one or two clusters close to the warmest spot remained. Figure 5.4 shows the course of such an experiment.

The question was as follows: How does the individual bee behave to perform this interesting collective behavior. Obviously, in a flat gradient field, it is harder for

---

* Red light is invisible for bees.

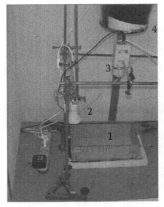

(a) Our temperature chamber   (b) Close-up of the temperature arena

Figure 5.3   Photographs of our experimental setup. (a) The temperature chamber, in which we performed our experiments. 1: The temperature arena. 2: The heating lamp. 3: IR sensitive video camera. 4: Red-light lamp for illuminating the scene for the camera. 5: The (regulated) heater (which is under the temperature arena (number 1)) for keeping the inner part of the arena at a constant temperature. (b) Wax floor, borders, and heating lamp of our temperature arena.

the bees to determine the local temperature differences, what is essential to allow a "greedy" uphill walk. We interpreted the cluster formation as a sort of "distributed search algorithm," which is performed by the bees. In addition, cluster sizes increased over time as experimental time proceeded, but this happened only in warmer areas. Clusters in colder areas break up and clusters in warmer areas grow. We concluded that these dynamics can be easily described by the following rule: "The warmer it is, the longer bees stay in a cluster."

But how did bees behave when they were not located in a cluster? Did they still walk uphill in the temperature field? Obviously, the bees stopped for some time when they met other bees; otherwise, there would be no clusters observable. But how did they behave when they hit the arena wall?

To answer these questions, additional analyses were made: First, we introduced single bees into our temperature arena and observed their motion path (trajectory) in the temperature field. We implemented an image-tracking algorithm in MATLAB®; this algorithm reported us the position and the heading of the focal bee every second. Figure 5.5 shows four exemplary trajectories of single bees in the arena. It is clearly visible that the bees did not stop below the lamp; thus, they were not able to find a suitable solution for themselves. This is in strong contrast to those experiments that were performed with bigger groups of bees. Thus, we observed a clear "swarm effect" here: Single bees frequently found no solution, but groups of bees often did. From these findings, we can derive our second rule to describe the

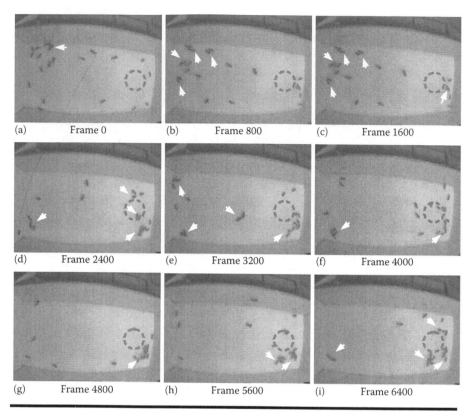

Figure 5.4 Clustering behavior of bees in a flat slope gradient. The subfigures were taken in intervals of 800 frames. White arrows indicate bee clusters that were formed. The dashed circle indicates the place of the optimal temperature spot. In total, it took the bees 8 min to reach collectively the optimal spot (except 1 distant bee cluster).

bees behavior: "A bee that is not in a cluster moves randomly (maybe with a slight bias towards warmer areas)."

In addition to randomly walking bees, other variants of bees' behaviors have also been observed: These variations range from wall following (which is rather similar to the random walks shown above, except that the center of the arena is avoided) to almost immobile bees to the seldom case that bees approached the optimum spot directly. For the development of our algorithm, the majority of tested bees, which were those that showed random (undirected) behavior, were most important, so we concentrated on those (almost) randomly moving bees.

Additional analysis of bigger groups of bees was made to determine how they behave after they encountered the arena wall. As shown in Figure 5.6, the bees stopped much more frequently after having met another bee in comparison to

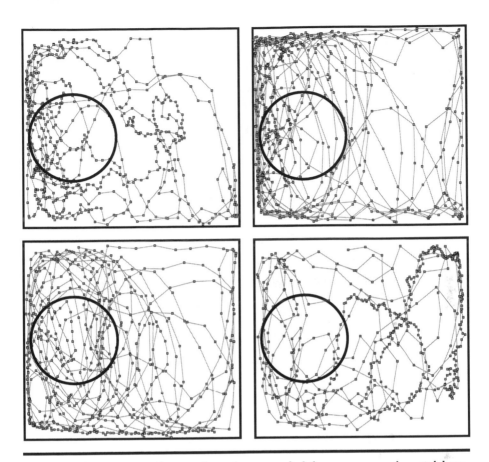

Figure 5.5 Four exemplary trajectories recorded from our experimental bees, which moved almost randomly through the arena. The lines indicate bee movements, the small squares indicate sampling points. The big circle on the left side indicates the temperature optimum. The outer rectangle indicates the arena wall. A small bias towards warmer areas can bee seen.

having encountered the arena wall. Thus, our third and last rule can be derived: "If a bee meets another bee, it is likely to stop. If a bee encounters the arena wall, it turns away and seldom stops."

We summarize the observed honeybee behavior as follows:

```
1. A bee moves randomly, in a ''correlated random walk.''
2. If the bee hits a wall, it stops with a low probability;
   otherwise, it continues immediately with step 3.
3. The bee turns away from the wall and continues with
   step 1.
```

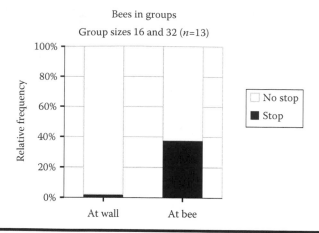

Figure 5.6 In our multi-bee experiments, the frequency of bee-to-bee encounters, as well as the frequency of bee-to-wall encounters, was measured. After a collision with another bee, the focal bee stops with a probability of $P_{stop} \approx 0.4$. After colliding with the arena wall, a bee is very unlikely to stop ($P_{stop} < 0.05$).

4. If the bee meets another bee, it stays with a certain probability. If the bee does not stay, it continues with step 1.
5. If the bee stopped, the duration of the bee's waiting time in the cluster depends on the local temperature: The warmer the local temperature at the cluster's location is, the longer the bee stands still.
6. After the waiting time is over, the bee moves on (step 1).

From the three rules we derived from the honeybees' behavior, we could construct a swarm-robotic algorithm, which is described in the following section.

5.3 Analyzing Some Basic Results of the Observed Features of the Bee's Behavior

To see whether or not the extracted rules of bees' behavior were responsible for the observed final results (formation of clusters near the optimum spot), we produced a simple multi-agent simulation of honeybees navigation in a temperature gradient field. Each bee was implemented as a single agent, which could move (in continuous space) over a grid of discreet patches. Each patch can be associated with a local temperature, to reflect the environmental conditions in our experiments. Each bee can detect obstacles (walls and other bees) if they are located in front of them.

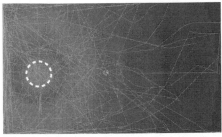 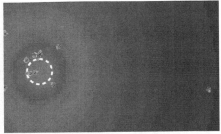

(a) A single bee in the simulation (trajectory)   (b) Twelve bees in the simulation (final positions)

Figure 5.7   Results of our multi-agent model of honeybees' navigation in a temperature gradient. (a) Motion of a single bee with a small bias toward the locally warmer temperature. (b) Final cluster position of 12 bees navigating without any bias. The warmer it is in the location of a bee-to-bee encounter, the longer the bees rest at this location. The dashed circle indicates the warmest spot in the arena.

Whenever a bee encounters a wall or another bee, it stops for an adjustable time and with an adjustable probability. The stopping time can be correlated with the local temperature, which is associated with the temperature value of the patch the bee (agent) is actually located on.

Figure 5.7 shows that with these simple rules, the basic features of the honeybees' behavior could be reproduced: Figure 5.7a shows that the simulation produces comparable trajectories for single bees, if the bees' random motion is associated with a small bias toward warmer areas: every 50th-time step, the bee turns toward the warmest local surrounding patch.

Another interesting finding using this multi-agent simulation was that the before-mentioned rules that were extracted from honeybees could be significantly reduced in complexity, without impairing the collective to settle a good collective decision: As Figure 5.7b shows a group of 12 bees could effectively aggregate at the warmest spot without having *any* bias toward the warmer area, thus performing only randomized rotations. Also, setting the stopping rate at walls to 0 did not change the final result; in contrast, it even improved the speed of the collective decision-making. The only "hard rule" that has to be followed by the bees was found to be the following: "The warmer it is at a location of a close encounter with another bee, the longer the bee stays there." Without that rule, no temperature-correlated aggregation could be found.

We can summarize the simplified bee algorithm as follows:

```
1. A bee moves randomly in a ``correlated random walk.''
2. If the bee hits a wall, it turns away from the
   wall and continues to walk (step 1).
```

3. If the bee meets another bee, it stays with this bee in an immobile cluster.
4. The warmer the local temperature at the cluster's location is, the longer the bee stands still.
5. After the waiting time is over, the bee moves on (step 1).

5.4 The Robotic Algorithm

Based on the results we obtained from the multi-agent simulations of the bees' behavior, we translated the bee-derived algorithm to a robot-focused algorithm: The world of an autonomous robot is very much different from a bee's sensory world. Our focal robot swarms consisted of either hundreds of I-SWARM robots[20] or tens of Jasmine robots.[21]

5.4.1 The Swarm Robot "Jasmine"

The Jasmine robot (see Figure 5.8 and Ref.[22]) was developed for swarm robot research: It has a small size of about $30 \times 30 \times 30$ mm^3, and it has only local communication abilities. By using six IR LEDs and six photodiodes, it can detect obstacles (walls or other robots) and it can communicate within a radius of approx. 7.5 cm. In its third generation, which we focus on in this paper, it is calculating with an Atmel Mega 168 micro-controller with 1 KB RAM and 16 KB Flash. A single LiPo

Figure 5.8 The Jasmine swarm robot.

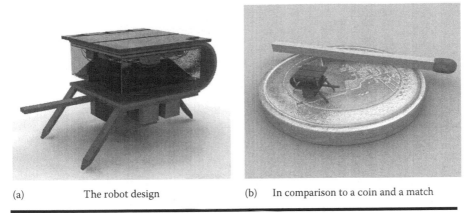

| (a) | The robot design | (b) | In comparison to a coin and a match |

Figure 5.9 The I-SWARM robot in a CAD study and a rendered photo by Paolo Corradi.

battery enables it to drive for up to 2 h, and a pair of optical encoders are used for simple odometric measurements in the mm-range.

5.4.2 The Swarm Robot "I-SWARM"

In the I-SWARM project,[20] shown in Figure 5.9, the goal was to produce a "true" (robotic) self-organizing swarm concerning the size of the swarm and the size of the individual robots. The idea was to establish a mass production for autonomous robots making, in principle, the production of numbers as high as hundreds is feasible. The robot itself has a size of just 3 mm × 3 mm × 3 mm. It is equipped with four infrared emitters and sensors, three vibrating legs, and a solar cell as an energy supply and additional sensor. The piezo-based legs allow a speed of about 1.5 mm/s at the maximum. A way of choice to build such small microsystems is to use flexible printed circuit boards. These boards are extensively used in miniature systems as consumer electronics and high-tech components. The functionality of the robot is basically focused on locomotion, an integrated tool (vibrating needle) permitting basic manipulation, a limited memory, as well as the possibility to communicate with neighboring robots via infrared light.

5.4.3 Shared and Different Properties of These Two Robots

The size difference between the robots can be compensated by changing the size of the arena the robots work in. This can create comparable swarm densities. The sensory range of the IR emitters and receptors is of comparable length in relation to the robot size. The Jasmine robot covers its environment denser, as it has more

IR units. Also, the speed of the two robots is different, even after scaling it to the respective robot size: The Jasmine robot is much faster than the I-SWARM robot.

In contrast to honeybees, both types of robots cannot sense temperature, but they both can sense information projected onto the arena from atop: The I-SWARM robot can be "fed" with simulated temperature values by an LCD projector that projects vertically onto the robot arena. For the Jasmine robot, there exists a "light sensory board," which can be mounted on top of the robot. By using a simple light source (a lamp), a light gradient can be constructed in the arena, which reflects the temperature gradient in the honeybee arena.

5.4.4 The Algorithm

The algorithm, which we called BEECLUST,* that allows the robot swarm to aggregate at the brightest (warmest) spot in the arena can now be formulated as follows:

1. Each robot moves as straight as it can until it encounters an obstacle in front.
2. If this obstacle is a wall (or any other non-robot barrier), it turns away and continues with step 1.
3. If the obstacle is another robot, the robot measures the local luminance.
4. The higher the luminance is, the longer the robot stays immobile on the place.
5. After the waiting time is over, the robot turns away from the other robot and continues with step 1.

One important aspect of the algorithm is the way how the waiting time corresponds to the local luminance. We assumed that the luminance sensor scales approximately linearly from 0 to 180 for a corresponding luminance between 0 lx and about 1500 lx that is in accordance with the technical hardware descriptions of the Jasmine robot (www.swarmrobot.org). We refer to this sensor value by e. The parameter w_{max} expresses the maximum waiting time (in seconds). The following equation can be used to map the sensor values e to waiting times of robots after encountering another robot:

$$w(e) = \frac{w_{max}e^2}{e^2 + 7000}. \tag{5.1}$$

---

* The algorithm is called BEECLUST, because we derived it from the clustering behavior of honeybees.

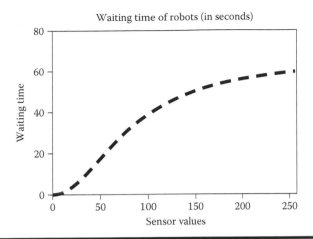

Figure 5.10 **The waiting time a robot takes after an encounter of another robot is nonlinearly correlated with the measured local luminance. Robots measure this local luminance only after encounters with other robots. This curve is defined by Equation 5.1, which is implemented in the robots' control algorithm and the (linear) behavior of the sensor.**

The resulting sigmoid function for the direct mapping of light intensities to waiting times is shown in Figure 5.10.

5.5 Swarm Experiments Using a Multi-Agent Simulation of the Robots

To verify our algorithm, we first created a bottom-up (microscopic) model of the robot swarm. We used the simulator LaRoSim,[23] which is a multi-agent simulation of both types of robots described above. The two robot types are modeled in the following way:

Jasmine Robot: Straight movements, six IR beams, high motion speed, and low number of robots

I-SWARM Robot: Fuzzy movements, four IR beams, low motion speed, and high number of robots

With both configurations, each robot was implemented as an autonomous agent. The arena is modeled as a set of patches (grid). The robots can move in 2D on this grid of patches. The local space is handled in discreet steps, and thus, the environmental cues (luminance) are assigned to the local patches. In contrast to that, the agents' (robots) motion is performed in a continuous way. The robots can detect walls and other robots by using their IR beams, which were modeled by performing "ray-tracing" of light beams that emerge from each robot in a separation of 0.1° from each other. As our algorithm does not use any higher level sensor data (e.g., distance of obstacles, rotation angle toward obstacles, . . .), the underlying sensor framework of the simulation could be held rather simple.

The simulation's main loop can be described as follows:

```
main_loop {
    read sensors(); // update the status of all sensors
                    // according to the local environment
    perform_communication(); //not used in our algorithm
    perform behavior(); // make all behavioral decisions
                        // according to our BEECLUST algorithm
    perform_movements(); // move the robots, detect physical
                         // collisions
    update_simulator_status(); // All global activities, update
                               // light conditions, count robots
                               // at target sites, ...
}
```

LaRoSim is implemented in NetLogo (http://ccl.northwestern.edu/netlogo/), which handles the quasi-parallel execution of the agents' behaviors. This is important for a multi-agent simulator to prevent artifacts, which can emerge from ordered execution stacks of agents.

5.5.1 Simulating the Swarm Robot "Jasmine"

Figure 5.11 shows a screen shot of a robot simulation using a parameterization reflecting Jasmine robots. Fifteen robots were simulated in this simulation. A bright light (>1100 lx on the brightest spot) was used in this experiment. The subfigures show the dynamics of robotic swarm configurations during the experiment. First clusters appear everywhere in the arena. The longer the experiment lasts, the more clusters are formed on the brighter side of the arena. These clusters get bigger over time. The darker side of the arena gets depleted from robots over time.

We performed this experiment again with 15 simulated Jasmine robots, but with a dimmed light (approx. 480 lx on the brightest spot). Compared to the scenario with bright light, the predicted aggregation of robots was less stable, aggregated clusters were of smaller size, and the robots tended to fill up the clusters beginning from the central (brightest) area in the light spot (see Figure 5.12).

5.5.2 Simulating the Swarm Robot "I-SWARM"

Figure 5.13 shows a scenario comparable to the previously described ones, just that I-SWARM robots were simulated in this case. In this simulation, the number

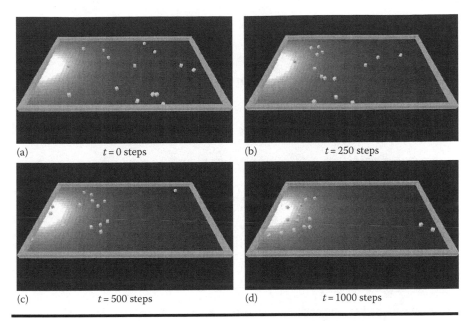

(a) $t = 0$ steps   (b) $t = 250$ steps   (c) $t = 500$ steps   (d) $t = 1000$ steps

Figure 5.11   Simulation of 15 Jasmine robots in a gradient formed by a bright lamp (>1100 lx on the brightest spot). The subfigures show stroboscopic snapshot pictures of the run of the simulation. Over time, the clusters on the bright side of the arena increase in number and size, while the clusters formed on the dark side of the arena disappear. The robots tend to aggregate more on peripheral zones of the light spot, due to blocking effects and traffic jams, and do not often fill up the brighter central area of the light spot.

of robots was significantly higher than in the simulation shown in Figure 5.11: 150 robots were simulated. The light spot (bright light condition) was shifted into the center of the arena. Again, the same collective aggregation behavior was found, but the time it took the swarm to aggregate near the brightest spot increased significantly.

Finally, we wanted to know whether or not obstacles could prevent our robotic swarm from aggregating at the light spot. As Figure 5.14 shows several variants of obstacles could not impair the swarm's ability to aggregate preferentially at the light spot. Thus, our BEECLUST algorithm can be predicted to be rather robust, concerning efficiency.

5.6 Preliminary Robotic Experiments

Using a swarm of Jasmine robots, we performed experiments depicted in Figures 5.11 and 5.12. The biggest challenge in formulating the BEECLUST algorithm in the

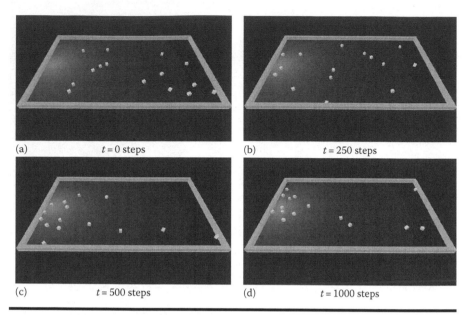

Figure 5.12 Simulation of 15 Jasmine robots in a gradient formed by a dimmed lamp (approx. 480 lx on the brightest spot). The subfigures show stroboscopic snapshot pictures of the run of the simulation. Over time, the clusters on the bright side of the arena increase slowly in number and size, while the clusters formed on the dark side of the arena disappear. The robots tend to aggregate more in the central zones of the light spot.

programming language C was to implement the robot-from-wall discrimination. This technical problem was solved as follows.

A robot that receives an IR reflection of its own emitted IR-LEDs immediately stops its motion. It starts to send longer-lasting pulses of IR emissions (to be detectable for other robots), which were altered by longer lasting pauses of any transmission. If IR pulses are sensed during these longer pauses, it cannot be reflections of own emissions, so these IR signals have to originate from other robots. The (ambient) lighting of the room, the used lamps that generate the light gradient in the arena, and even the remote control that is used to start the robots at the beginning of the experiments have to be carefully selected, so that they do not trigger the robot-to-robot detection.

As Figure 5.15 shows, the results gained from these preliminary robotic experiments correspond well with the predictions gained from the multi-agent simulations of the Jasmine robots: In bright light, more robots aggregate at the light spot and the robots aggregate in more peripheral areas, compared to the dimmed light conditions.

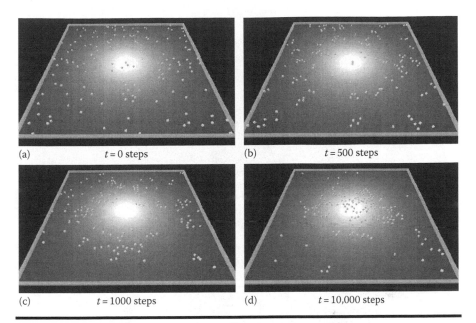

Figure 5.13 Simulation of 150 I-SWARM robots in a gradient formed by a bright lamp (>1100 lx on the brightest spot). The subfigures show stroboscopic snapshot pictures of the run of the simulation. Over time, the clusters in the central area of the arena increase in number and size, while the clusters formed on the darker peripheral areas of the arena disappear. Finally, most robots are aggregated at the light spot at the center of the arena.

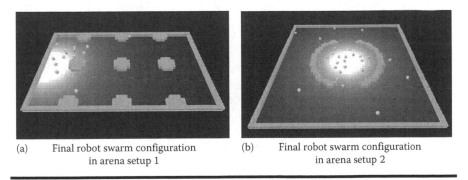

Figure 5.14 Simulation of a group of Jasmine robots in a more complex arena. The "habitat" is structured by multiple obstacles that can block the robots' movements. The robots still perform just random motion and have no controller that allows them to circumvent obstacles "deliberately." Nevertheless, the BEECLUST algorithm allows the swarm to find the light spots and to aggregate at these brighter areas.

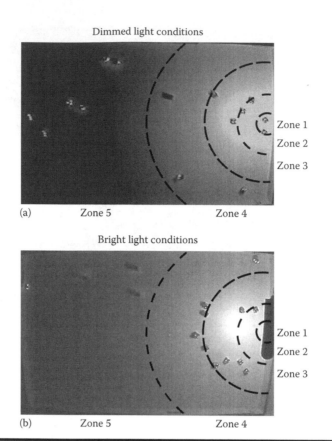

Figure 5.15 Aggregation patterns observed with 15 Jasmine robots in two light conditions. With bright light, more robots aggregate and the aggregation is located at the peripheral zones of the light spot. With dimmed light, less robots aggregate and the aggregation occurs in more central areas of the light spot. We defined four spatial zones to describe how close the robots approached the spot with the highest luminance in their final aggregated configuration. Pictures are reprinted. (From Hamann, H. et al., Spatial macroscopic models of a bio-inspired robotic swarm algorithm, In *IEEE/RSJ 2008 International Conference on Intelligent RObots and Systems (IROS'08)*, 2008. With permission.)

5.7 Macroscopic Model of the Robots' Collective Behavior

After we evaluated the swarm-robotic algorithm BEECLUST in bottom-up simulations and in real robotic systems, we wanted to generate top-down models that allow us to describe the basic feedback loops that arise within the robotic swarm in an abstract and easy-to-understand manner. Within this chapter, we demonstrate two distinct ways to achieve such top-down formulations. These approaches were described already in detail,[24] and here, we give just a short overview of the modeling steps and show the most important results that can be achieved by simulating such models.

In a first notional step towards a model, we interpret the considered swarms as agent systems, that is, these systems consist of cooperating, autonomous agents. An agent system is a microscopic description, that is, each individual is represented explicitly. Therefore, we can compute the trajectories of all agents. The many occurring agent–agent interactions in the system might lead to a global collective effect, for example, formation of patterns. However, this effect might be masked by a model based on trajectories due to an overflow of microscopic details. Therefore, we choose a macroscopic description, that is, the individual is not explicitly modeled anymore but only ensembles of many swarm configurations. A good motivation for this modeling approach in agent systems is given by Schweitzer:[25]

> To gain insight into the interplay between microscopic interactions and macroscopic features, it is important to find a level of description that, on the one hand, considers specific features of the system and is suitable for reflecting the origination of new qualities but, on the other hand, is not flooded with microscopic details. In this respect, agent models have become a very promising tool mainly for simulating complex systems. A commonly accepted theory of agent systems that also allows analytical investigations is, however, still pending because of the diversity of the various models invented for particular applications.

This necessary reduction of microscopic details can only be achieved by a probabilistic approach and through abstraction.

It was very important for us to formulate the models in a way that most model parameters can be taken directly from the implementation of the BEECLUST algorithm that we used in the real robotic experiment. Most other parameters can be directly derived from the arena geometry, from the robots' geometry, or from the hardware specifications of the robotic hardware.

The first model that we implemented was a Stock & Flow model, which is used by the school of "system dynamics."[26] To implement this rather simple model, we used the software Vensim®.[27] A Stock & Flow model depicts flow of material between compartments by using a specialized graphical notation. "Stocks" represent compartments that can hold material over time. Doubled arrows ("Flows") connect these stocks and allow material to move from one compartment to another. In an open system, specific cloud-shaped symbols depict "Sources" and "Sinks," through which material can enter or leave the system. In our model, the "material" was the robots of the robotic swarm, the "Stocks" represented the arena compartments that represented circular aggregation zones around the light spot (see Figure 5.15). The flows represented the exchange of robots among these zones, in other words: The rate at which robots leave one zone and enter one of the adjacent zones. These flows (rates) are finally expressed by ordinary differential equations (ODEs), which describe the rates of change of the corresponding stocks.

In a second modeling approach, we wanted to have a better spatial representation of the arena space. Therefore, we used partial differential equations (PDEs) to describe the drift of robots. We assumed the robots to be randomly moving units, which can be roughly compared to moving molecules in Brownian motion. In reality, the robots perform a motion pattern that can be described as "correlated random walk," as the robots change their direction only after they encountered an obstacle (wall, other robot) in front of them. However, we know that this frequently happens in the swarm of robots and that the robots themselves do not drive perfectly straight trajectories. To keep the modeling approach as simple as possible, we abstracted the motion model of the robots from a correlated random walk to a pure random walk.

We tried to reuse as much as possible when establishing the two models. This does not only save modeling time, it also keeps the models in a better comparable state. Thus, we used the same basic parameters for geometry and hardware constraints (e.g., arena size, robots size, sensory radius, robot speed), for the algorithmic expression (e.g., waiting-time calculations), as well as for modeling the light distribution in the arena.

One very important feature of both models is the correct modeling of the expected local collision frequencies. To model the number of collisions* for a number (or density) of robots, we used collision theory:[28]

First, we had to consider the likelihood (what corresponds to the frequency) of a free-driving robot, which passes through an area occupied by aggregated (standing) robots. The focal free robot f drives a distance of $v \Delta t$ within a time interval of Δt. It collides with another aggregated, thus standing, robot a if $\text{dist}(f, a) \leq r$. The radius of the emitted circle of IR pulses used for collision avoidance is denoted as $r = 0.075$ m. Based on these geometrically derived formulations, the area that is relevant for modeling the correct collision probability of a free-driving robot is

$$C_{f,a} = 2rv. \tag{5.2}$$

Please note that this expression only covers collisions of free-driving robots with aggregated (standing) robots. To calculate the actual number of collisions of free-driving robots with aggregated robots, $C_{f,a}$ has to be multiplied with the actual density of aggregated robots within the corresponding area (zone).

However, a free-driving robot can also collide with other free-driving robots. We assumed that all robots drive with the same average velocity v. The average relative speed between the focal robot and all other free-driving robots can be calculated by summing up all relative speeds over all possible uniformly distributed velocity

---

* Please note that we use the term "collision" because we used "collision theory" to model close encounters of robots. In the algorithm BEECLUST, robots do not really collide. Whenever a robot detects another robot within its sensory radius, this is accounted for a "collision."

vectors, which results in $4v/\pi$. Thus, the collision probability of a free-driving robot with other free-driving robots can be expressed by:

$$C_{f,f} = r\frac{4v}{\pi}. \qquad (5.3)$$

Again, we can calculate the actual number of collisions between pairs of free-driving robots by multiplying $C_{f,f}$ with the density of free-driving robots in the corresponding area (zone).

Both macroscopic models that are described in this chapter use the above formulated expressions for modeling the collision frequencies of robots.

5.8 The Compartment Model

The first model that we created is a compartment model,[29] which is very similar to the Stock & Flow models that can be produced with Vensim®.[27] Such compartmental models are frequently developed for describing physiologic processes, in which chemicals are built up, degraded, or exchanged among compartments like cell organelles, cells, tissues, organs, or other body compartments. In our modeling approach, the focal compartments are areas of the arena, which exchange robots. Our case is a closed system, so no robots were removed or added during runtime. Therefore, we did not have to model "sinks" or "sources" in our model. Our model could concentrate purely on the flow of robots from one compartment to another. It also has to model the changes of state of the robots, which happen in reaction, to, collision frequency and local luminance, as it is programmed in our BEECLUST algorithm.

We structured the arena space into several concentric rings, which are located around the innermost brightest spot of the light gradient field. The flow of robots among these ring-shaped zones, as well as their changes of state (free driving or aggregated still standing), can be described as follows:

Robot Diffusion: A robot can move from each ring-shaped zone only to one of the two neighboring zones. Exceptions are the outermost and the innermost zone, which have only one neighboring zone. The flow of robots from zone i to zone j is described by $\delta_{i,j}$.

Change of State (Robots Aggregate): Robots can have close encounters with other robots. The frequency of these events depends on the robot density within a zone. Robots that encounter another robot change their behavioral state: They stop moving, measure the local luminance, and begin with a waiting period, which depends on the level of local luminance. The rate at which this change of state happens is expressed by α_i.

Change of State (Robots Drive Again): The rate at which robots start to move again depends on the median local luminance in the corresponding zone and on the number of aggregated robots in that zone. This rate is expressed by β_i.

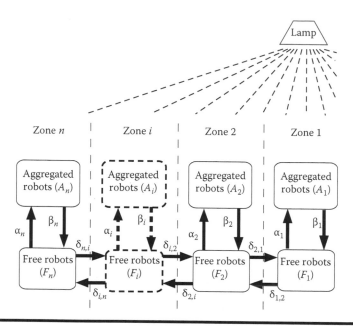

Figure 5.16 **Organization of the compartmental model: A chain of linked zones, which represent concentric ring-shaped or disc-shaped zones in the robot arena. Inside of each compartment, robots can switch from the free-driving state to the aggregated (standing) state and vice versa. Picture redrawn. (From Hamann, H. et al., Spatial macroscopic models of a bio-inspired robotic swarm algorithm, In *IEEE/RSJ 2008 International Conference on Intelligent RObots and Systems (IROS'08)*, 2008. With permission.)**

Figure 5.16 shows a schematic drawing of our modeling approach. In the model that was finally simulated, the arena was structured into five distinct zones: A central circular zone at the brightest spot with a radius of $R_1 = 11$ cm. Around this central area, there we defined three ring-shaped zones located with (outside) radii of $R_2 = 22$ cm, $R_3 = 33$ cm, and $R_4 = 66$ cm, respectively. The outermost fifth zone covered the remaining area of the arena, which had the dimension of 150 cm × 100 cm.

The variable $F_i(t)$ holds the number of free-driving robots within each zone i. In contrast to that, the variable $A_i(t)$ holds all aggregated robots within this zone. Except the innermost and the outermost zone, robots can leave a compartment with the rates $\delta_{i,i+1}$ and $\delta_{i,i-1}$. In parallel, robots enter the zone i with the rates $\delta_{i+1,i}$ and $\delta_{i-1,i}$. These exchange rates of robots can be modeled as follows:

$$\delta_{i,i-1}(t) = 0.5 v D F_i(t) \frac{R_{i-1}}{(R_i + R_{i-1})(R_i - R_{i-1})}. \tag{5.4}$$

The constant $v = 0.3$ m/s represents the average speed of robots; the constant $D = 0.5$ represents a diffusion coefficient. The border between two neighboring zones is geometrically an arc segment. To model which fraction of robots leaves zone i toward the right direction and which fraction leaves toward the left direction, the proportions of the two borders to the two neighboring zones have to be considered, as was expressed by the formulation $(R_{i-1})/(R_i + R_{i-1})$. In addition, also the width of zone i is important, as it determines how many robots leave zone i at all within a time period. The width of a zone is expressed by the term $(R_i - R_{i-1})$.

The model of robot diffusion at the innermost zone 1 is substantially simplified, as robots could enter only from zone 2 and robots can also leave only toward zone 2. The innermost zone 1 is also shaped differently: It is a half-disk, whereas all other zones are half-ring segments or more complex shapes. The following equation models the flow of robots from zone 1 to zone 2:

$$\delta_{1,2}(t) = \frac{vDF_1(t)}{R_1}. \tag{5.5}$$

Also the outermost zone 5 has only one neighboring zone (zone 4), with which robots can be exchanged. We also had to consider that this zone 5 has a special shape; it is of the form of a rectangle, from which all other zones (zones 1–4) have been subtracted. The following equation models the flow of robots from zone 5 to zone 4:

$$\delta_{5,4}(t) = \frac{vDF_5(t)}{l_{\text{arena}} - R_4}, \tag{5.6}$$

for the length of the arena $l_{\text{arena}} = 150$ cm.

Besides the diffusion of robots, the second important process in our focal robotic swarm is the aggregation of robots. As already explained in Section 5.7, the likelihood (frequency) of robot "collisions" depends on the local density (number) of robots, as well as on their behavioral state (free driving or standing): The two coefficients $C_{f,a}$ and $C_{f,f}$ express these two different likelihoods.

The parameter P_{detect} models the probability that a robot recognizes another robot and does not consider it as being an arena wall. Z_i represents the area of the focal zone i (in cm^2). Based on these parameters, we can model the rate at which robots collide, identify each other as robots, and change to the aggregated state in each zone i as

$$\alpha_i(t) = F_i(t)P_{\text{detect}}\frac{C_{f,a}A_i(t) + C_{f,f}F_i(t)}{Z_i}. \tag{5.7}$$

The third important process to model is the rate at which robots start to move again, after their waiting period has expired. This rate is denoted as β_i for each zone i. The waiting time can be approximated by $1/w_i$, where w_i is the average waiting period robots have in compartment i. The duration of w_i depends on the average luminance expected within zone i.

In our algorithm BEECLUST, the relationship of waiting time (w) to the local luminance (E) was nonlinear (see Equation 5.9): w depends nonlinearly on e, which is the sensor value reported from the luminance sensor of the robot. This sensor maps the local luminance E to the sensor output value e in a rather linear way (see Equation 5.8). As we modeled the robots within each compartment as being identical, we refer to w_i, e_i, and E_i as the average waiting time, average sensor value, and average local luminance expected for all robots within zone i, respectively.

According to the Jasmine robot's hardware specification, the luminance sensor maps luminances (E_i) between 0 and $l_{max} = 1500$ lx in a linear manner. For this range of local luminance, the sensor reports values between 0 and 180, as is expressed by the following equation:

$$e_i = \min\left(180, 256\frac{E_i}{l_{max}}\right). \qquad (5.8)$$

The average waiting time expected for a robot in zone i can be modeled based on the reported sensor value e_i as follows:

$$w_i = \max\left(1, \frac{w_{max}e_i^2}{e_i^2 + 7000}\right), \qquad (5.9)$$

whereby $w_{max} = 66\,\mathrm{s}$ refers to the maximum waiting time.

To avoid a division by 0, the waiting time w_i was restricted to values of equal to 1 or above. This necessity does also reflect empirical observations, as robots stay on the place for approx. 1 s, after they encountered a wall or misidentified another robot as being a wall.

Having calculated the average waiting time of aggregated robots in each zone i, we can model the rate at which these aggregated robots change their behavioral state and start to drive again by multiplying the number of aggregated robots $A_i(t)$ with $1/w_i$:

$$\beta_i(t) = \frac{A_i(t)}{w_i}. \qquad (5.10)$$

Finally, we had to combine the above-derived expressions that model the diffusion process, the aggregation process, and the disaggregation process into two linked ODEs, which describe the dynamics of free-driving robots (5.11) and the dynamics of aggregated robots (5.12) within each zone of our arena:

$$\frac{dF_i(t)}{dt} = \delta_{i-1,i}(t) + \delta_{i+1,i}(t) - \delta_{i,i-1}(t) - \delta_{i,i+1}(t) - \alpha_i(t) + \beta_i(t). \qquad (5.11)$$

For the rightmost compartment at the end of the chain $i = 1$, we set $\forall t$: $\delta_{i-1,i}(t) = \delta_{i,i-1}(t) = 0$. Consequently, for the leftmost compartment at the end of the chain $i_{\max}$, we set $\forall t : \delta_{i+1,i}(t) = \delta_{i,i+1}(t) = 0$ for all time. The number of aggregated robots within each zone i is modeled by

$$\frac{dA_i(t)}{dt} = \alpha_i(t) - \beta_i(t). \tag{5.12}$$

Before we discuss the issue of spatial distribution of robots in the arena in detail, we want to derive a second macroscopic model of robot motion: a model that describes spatial distribution not in four discrete compartments, but in continuous space. After the derivation of this second spatially explicit model, we will compare the predicted spatial distributions in our two macroscopic models to the observed distribution of real robots in our arena.

5.9 Macroscopic Model—Step 3

5.9.1 Macroscopic, Space-Continuous Models for Robot Swarms

In this model, we chose a full-continuous representation of space by using a PDE. This approach of modeling multi-agent or swarm-robotic systems was taken before in several studies.[25,30–33]

In our approach, we followed the concept of Brownian agents by Schweitzer,[25] which is based on Brownian motion. A Brownian agent is an active particle with an internal energy depot and self-driven motion. The most prominent feature of our approach is the analytical derivation of the macroscopic model based on the microscopic model.[30,34] This derivation is based on a Langevin equation, which is a stochastic differential equation and was used originally to describe the trajectory of a particle showing Brownian motion, that is, it is a microscopic model.[35,36]

The change of a particle's position $\mathbf{X}$ showing Brownian motion with drift can be described under certain assumptions by a Langevin equation

$$\frac{d\mathbf{X}}{dt} = -\mathbf{Q} + D\mathbf{W}, \tag{5.13}$$

for drift $\mathbf{Q}$, diffusion D, and a stochastic process $\mathbf{W}$. From the Langevin equation, it is possible to derive a Fokker–Planck or Kolmogorov forward equation, which describes the probability density of this particle, that is, it is a macroscopic model.[34,37–39] The Fokker–Planck equation corresponding to Equation 5.13 is given by

$$\frac{\partial \rho(\mathbf{x}, t)}{\partial t} = -\nabla(\mathbf{Q}\rho(\mathbf{x}, t)) + \frac{1}{2}\nabla^2(D^2\rho(\mathbf{x}, t)), \tag{5.14}$$

where $\rho(\mathbf{x}, t)dr_x dr_y$ is the probability of encountering the particle at position $\mathbf{x}$ within the rectangle defined by dx and dy at time t. The Fokker–Planck equation has a variety of applications ranging from quantum optics[40] to population genetics.[41]

In the following, we present a rather simple model based on diffusion only because here the robots' motion is simple enough to be interpreted purely stochastically. This simplifies the Fokker–Planck equation to a diffusion equation

$$\frac{\partial \rho(\mathbf{x}, t)}{\partial t} = \frac{1}{2} \nabla^2 (D^2 \rho(\mathbf{x}, t)). \tag{5.15}$$

5.9.2 Modeling the Collision-Based Adaptive Swarm Aggregation in Continuous Space

The robots were modeled as (particle) densities by F for free (moving) robots and density A for aggregated robots. According to the visualization of the compartment model (Figure 5.16), it is useful to find an interpretation based on in- and outflows. There are flows in three dimensions. The first two dimensions represent space and are due to the robot motion in the plane. As discussed, above the robot motion was modeled by a diffusion process. The mathematical description is the diffusion term $D\nabla^2 F(\mathbf{x}, t)$ for a diffusion constant D. The third dimension is time. The density of free robots $F(\mathbf{x}, t)$ is reduced by a certain amount, because they aggregate, and $F(\mathbf{x}, t)$ is also increased because they wake up and move again. This was modeled by introducing a stopping rate $s(\mathbf{x}, t)$ depending on the number of collisions and detection rates. The addition by awaking robots at spot $\mathbf{x}$ and time t is defined by the ratio of those robots that stopped before at time $t - w(\mathbf{x})$. w is the waiting time similarly defined as in Equation 5.9 but with luminance sensor values e depending on points in the plane $\mathbf{x}$. Hence, we got

$$\frac{\partial F(\mathbf{x}, t)}{\partial t} = D\nabla^2 F(\mathbf{x}, t) - s(\mathbf{x}, t)F(\mathbf{x}, t) + s(\mathbf{x}, t - w(\mathbf{x}))F(\mathbf{x}, t - w(\mathbf{x})). \tag{5.16}$$

This is a time-delay PDE. Additionally, a partial differential equation for A can be formulated although it is not necessary as the densities of aggregated robots are implicitly defined by F and w. For A we got

$$\frac{\partial A(\mathbf{x}, t)}{\partial t} = s(\mathbf{x}, t)F(\mathbf{x}, t) - s(\mathbf{x}, t - w(\mathbf{x}))F(\mathbf{x}, t - w(\mathbf{x})). \tag{5.17}$$

One way of finding an intuitive approach to these equations is to look at their simple discretization in parallel to Figure 5.16. In Figure 5.17, we show the in- and outflow of the patch at position $\mathbf{c}$ in the case of space being discretized by a grid to solve the equation numerically. The flow between neighboring patches is determined

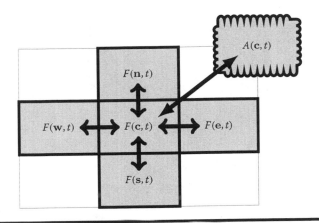

Figure 5.17 **Schematic diagram of a simple discretization of Equation 5.16 focusing on one patch in the center (density of free robots at position** c**:** $F(c, t)$**) and its neighbors to the north, east, south, and west, plus the associated patch of the aggregated robot density** $A(c, t)$**. Arrows indicate the in- and outflow. Pictures are reprinted. (From Hamann, H. et al., Spatial macroscopic models of a bio-inspired robotic swarm algorithm, In** *IEEE/RSJ 2008 International Conference on Intelligent RObots and Systems (IROS'08)*, **2008. With permission.)**

by the diffusion term $D\nabla^2 F(\mathbf{x}, t)$ in Equation 5.16. A (simplified) discretization of the Laplace operator ∇^2 using finite differences is

$$\nabla^2 F(\mathbf{c}, t) \doteq D(F(\mathbf{n}, t) + F(\mathbf{e}, t) + F(\mathbf{s}, t) + F(\mathbf{w}, t) - 4F(\mathbf{c}, t)). \qquad (5.18)$$

Thus, in our model, moving robots tend to homogenize their density in space by leaving areas of high density and by accumulating in areas of low density.

Another flow is indicated in Figure 5.17 by the diagonal arrow. These are the robots leaving patch **c**, because they detected a collision and started waiting, or entering patch **c** because their waiting time has elapsed. The patch labeled $A(\mathbf{c})$ is part of a second grid for aggregated robots. However, this grid is only optional because the aggregated robots can be administrated by using time delays and the density of moving robots F: Robots are just removed from $F(\mathbf{c})$ for their waiting time w and subsequently added.

We derive the stopping rate s by simple collision theory[28] as discussed. In the first step, the collisions per time and area $Z_{f,a}$, that is, the collision density was derived. The area of relevance here is given by Equation 5.2: $C_{f,a} = 2rv$. Area $C_{f,a}$ is populated by $C_{f,a}A$ aggregated robots. This is also the number of collisions one has to expect for a single free robot. Following classical collision theory, we multiplied by F resulting in the collision density

$$Z_{f,a}(\mathbf{x}, t) = C_{f,a}A(\mathbf{x}, t)F(\mathbf{x}, t). \qquad (5.19)$$

To derive the collision density (as defined by collision theory) of free robots colliding with free robots, we followed similar considerations: In difference to the above situation, we had to use the mean relative speed $4v/\pi$. Thus, the resulting area is $C_{f,f} = r4v/\pi$ (Equation 5.3). This result needed to be divided by two, because each collision is counted twice. We got

$$Z_{f,f}(\mathbf{x}, t) = \frac{1}{2} C_{f,f} F^2(\mathbf{x}, t). \qquad (5.20)$$

The stopping rate s determines a fraction of free robots that collide with another robot and also detect this collision. Reverting the division by 2 in the derivation of Equation 5.20 leads to $2Z_{f,f}(\mathbf{x}, t)$, which gives the number of colliding free robots instead of the number of collisions.

The probability that a robot successfully detects a collision is given by P_{detect}. This probability was multiplied to the sum of both collision densities. We got

$$s(\mathbf{x}, t) = \frac{P_{\text{detect}}}{F(\mathbf{x}, t)} (Z_{f,a}(\mathbf{x}, t) + 2Z_{f,f}(\mathbf{x}, t)). \qquad (5.21)$$

In order to limit the maximal achievable robot density, we introduced a sigmoid function $L(\mathbf{x}, t) \in [0, 1]$. It was multiplied to and, thus, incorporated in the diffusion coefficient D such that it can be interpreted as a space- and time-dependent diffusion. The flow of the robot density is slowed down as an effect. We defined

$$L(\mathbf{x}, t) = (1 + \exp(20(F(\mathbf{x}, t) + A(\mathbf{x}, t))/\rho_c - c_{\text{offset}}))^{-1} \qquad (5.22)$$

for a constant c_{offset} that shifts the sigmoid function over the density-axis, and a "critical" density ρ_c at which the robots' movements become almost impossible, which was set to $\rho_c = 1/(\pi r^2)$. The shape of L (defined by c_{offset}) and the diffusion constant D are free parameters that were fitted to the scenario. In the following, if not explicitly stated, we set $c_{\text{offset}} = 13$. The boundary conditions were set to total isolation (no robots leave or enter). The initial condition was a homogeneous distribution of robots in the dark half of the arena. We solved Equation 5.16 numerically (forward integration in time) as the time delay is increasing its complexity critically. It is numerically easy to solve though.

5.10 Results of Our Two Different Modeling Approaches

We simulated the compartmental model in three different environmental conditions: "No light," which was simulated assuming a lamp emitting light at an intensity of 0 candela; "dimmed light," which was simulated assuming that the lamp emits light at 9 candela; and finally, "bright light," which assumed that the lamp emits

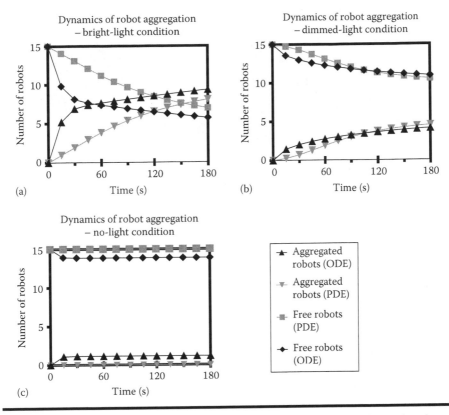

Figure 5.18 **Aggregation dynamics of our modeled robotic swarm under three different light conditions: (a) Bright light. (b) Dimmed light. (c) No light.**

27 candela. Figure 5.18 shows that in a bright-light spot, significantly more robots are predicted to aggregate than under dimmed-light conditions. With no light at all, approx 1 robot is always clustered. This can be interpreted as a modeling artifact, as one would expect no aggregation at all without any light spot. The effect arises from the fact that our minimum waiting time in our model was restricted to values above or equal to 1. Although such a result seems counterintuitive, it is still a very plausible result, because due to the high robot speed and the relatively small arena, there is approximately every 2 s a short-time encounter of two robots that will be represented by one aggregated robot per second.

The predictions of our stock and flow model about the total aggregation of robots under different light conditions (Figure 5.18) showed results that compared very well to the results we achieved with our bottom-up simulator as well as with our real robotic experiments. Of course, we also investigated the spatial distribution of the robots among the four defined arena zones to empirical results (see Figure 5.15).

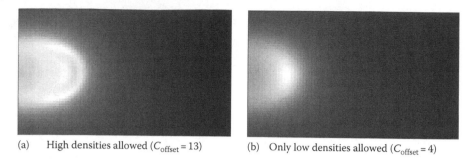

(a)   High densities allowed ($C_{offset} = 13$)   (b)   Only low densities allowed ($C_{offset} = 4$)

Figure 5.19 Expected density of aggregated robots (higher densities are brighter) for the scenario shown in Figure 5.11 but with different maximally allowed densities (different *L*-functions).

The space-continuous model indicates a slower increase of aggregated robots in Figure 5.18, especially for the bright light. This is due to the limits of the diffusion process assumption. This is discussed in the next section in more detail.

One aspect of the swarm system that cannot be described with the Stock & Flow model is the position, the spatial expansion, and the shape of the emerging clusters. At least partially, this information can be predicted by this space-continuous model. As an example, the robot density for the scenario shown in Figure 5.11 is given by Figure 5.19 for two different values of c_{offset} allowing high or only low densities, respectively. Notice that the resulting cluster shapes differ not only in quantities but also in their form.

The robot density over time for a scenario comparable to the one shown in Figure 5.13 (central light) is shown in Figure 5.20.

Besides several steps of abstraction to omit microscopic details that are unnecessary but also those that can hardly be modeled, the compartment model and the space-continuous model give good spatial predictions of the distribution of aggregated robots over the zones (Figure 5.21) compared to the empiric results. The dynamics of the aggregation process are also predicted well but with less accuracy (see Figure 5.22). This is especially true for the spatial model.

5.11 Discussion

Our process that led finally to the development and to a mathematical analysis of a robotic swarm algorithm (BEECLUST) started with a classical ethological experiment performed with real animals: We repeated some classical experiments of bee movements in a temperature gradient field, and we elaborated these experiments further by transforming these experiments from a one-dimensional setup (Figure 5.1) into a two-dimensional setup (Figure 5.2). This transformation allowed us to gain

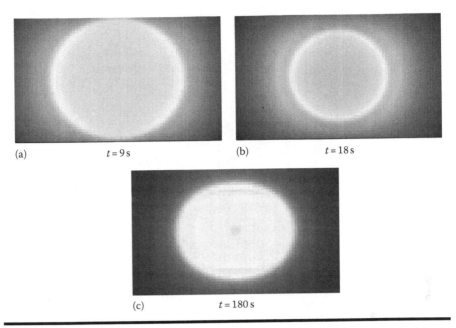

(a) $t = 9\,\text{s}$ (b) $t = 18\,\text{s}$

(c) $t = 180\,\text{s}$

Figure 5.20 **Expected density of aggregated robots (higher densities are brighter) for one light in the center (comparable to the scenario shown in Figure 5.13).**

novel insights into honeybee clustering behavior, which, from our point of view, are very interesting for the field of swarm robotics. We found that the bees show clearly emergent behavior and that they are able to solve a hard task (optimum finding) in a swarm-intelligent way (Figure 5.4), while one single bee was in most cases unable to solve the given task (Figure 5.5). We realized that the bees' behavior can be described in a small set of simple rules.

These rules were then further investigated by using an individual-based bottom-up model (Figure 5.7), which allowed us to further reformulate and reduce these rules. This way, we finally found a minimum set of rules, which was able to produce the interesting self-organized and swarm-intelligent collective behavior of a swarm of agents that search for an optimum in a noisy gradient field with very limited sensory abilities and without any channel of direct communication.

For us, it was extremely interesting that the derived algorithm was clearly a swarm-intelligent algorithm, but it uses neither direct communication nor does it use indirect communication in the form like it is described as stigmergy: No "pheromones" or other signals are spread or deposited in the environment, nor do the robots change anything else in the environment, except position themselves somewhere in this environment.

Thus, the only form of minimalistic communication that is performed is the presence or absence of robots in specific regions of the environment. Our reasoning,

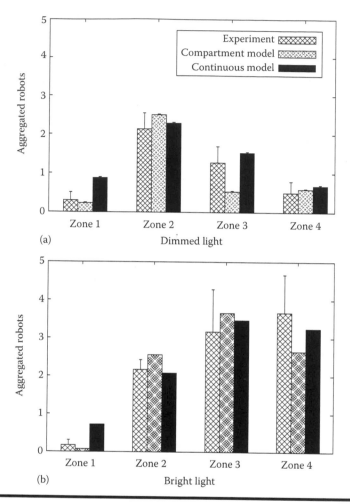

Figure 5.21 Comparing the numbers of aggregated robots in each of the four considered zones as predicted by the two models to the mean numbers of aggregated robots in each zone during the last 60 s of six robotic experiments; error bars indicate upper bound of the 95% confidence interval. Pictures are reprinted. (From Hamann, H. et al., Spatial macroscopic models of a bio-inspired robotic swarm algorithm, In *IEEE/RSJ 2008 International Conference on Intelligent RObots and Systems (IROS'08)*, 2008. With permission.)

why the swarm is able to find optimal solutions as a collective, was that the probabilities associated with the presence or the absence of robots in specific areas of the environment is associated strongly with environmental features.

First, we had to test our bio-derived robotic algorithm with bottom-up simulations of robot hardware (Figures 5.11 through 5.14). Then, we tested the algorithm

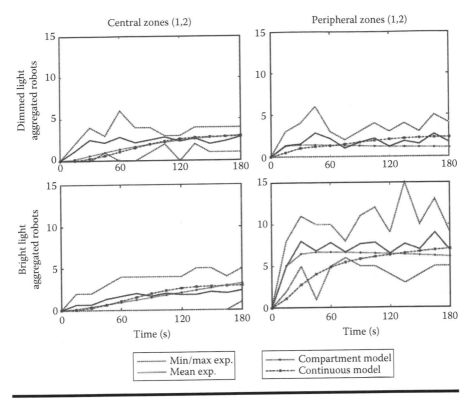

Figure 5.22 **Comparing the numbers of aggregated robots over time combining the central area (zone 1 and zone 2 pooled) and the peripheral area (zone 3 and zone 4 pooled) of the aggregation place, respectively, as predicted by the two models to mean, min., and max. of six robot experiments. Pictures are reprinted. (From Hamann, H. et al., Spatial Macroscopic Models of a bio-inspired robotic swarm algorithm. In *IEEE/RSJ 2008 International Conference on Intelligent Robots and Systems (IROS'08)*, 2008. With permission.)**

also with real robotic hardware (Figure 5.15). Finally, after we demonstrated that the simplified bio-derived robotic algorithm still shows the desired swarm behavior, we wanted to analyze the inner core of the swarm system by formulating macroscopic models, which only depict the major components of the system and which only incorporate the dominant feedback loops that reside within the components of the system. These models allowed us to further understand how this distributed and dynamic system operates and why it converges to the observed solutions.

How does the distributed system work? As in all aggregating systems, a positive feedback loop is established by the rules of our algorithm, which finally leads to the desired aggregation of robots in favorable areas: In an area occupied with many robots, it is more likely that free-driving robots meet one of standing robots, compared to an area occupied by a lower number of robots. This way, differently sized clusters of robots emerge throughout the arena. In brighter areas, clusters have a longer half-life period, as the resting time of robots is higher. All clusters are competing for free-driving robots, which are crucial for the "survival chance" of a cluster: If the frequency of free-driving robots, which join a cluster, is lower than the frequency at which robots leave the cluster, the number of robots in a cluster will decrease until the cluster is gone. Thus, it is the competition of clusters that allows the swarm to find near-optimal solutions.

Robots that join an aggregation cannot occupy the very same location that is already occupied by the previously aggregated robots. Thus, every robot indicates an own solution to the given problem. Robots within the same clusters represent nonidentical but related solutions (similar light sensor values).

If an aggregation system is driven just by positive feedback, it can be assumed that after some time there will be just one big aggregation of robots. In our focal system, we also have saturation effects, which allow the system to get dominated more and more by negative feedback, as the clusters grow in size: If a cluster is growing, robots can join in most cases only on the outer rim of the cluster, which has a lowered luminance in our gradient field. Thus, these outside robots have a lower resting time and leave the cluster earlier than the innermost robots do. In addition, the innermost robots are often blocked by the outer-rim robots; thus, they immediately reaggregate again, after their waiting period is expired. These phenomena lead to a maximum cluster size. This maximum cluster size is reached, when the positive feedback of aggregation is balanced out by the negative feedback of cluster saturation. The location of this equilibrium point depends on the shape of the used light spot, as well as on the shape of our robots' waiting time function (Figure 5.10).

In addition to the feedback loops mentioned above, there is also another effect observable in the robot swarm driven by our BEECLUST algorithm: Robots show aggregation patterns that can in most cases be described as nested strings of robots. A chain of robots showed to be a rather stable structure, as it attracts many free-driving robots, and we observed that these chains can break up only starting from their two outer edges. The whole process has a high similarity to what is known from Diffusion Limited Aggregation.[42] These effects are pure microscopic processes that emerge from the local robot-to-robot configurations; thus, these effects can be better analyzed by microscopic individual-based models than with macroscopic models.

In conclusion, we can assume that robots will distribute across several light spots, whenever the number of robots is bigger than the equilibrium number of robots for one cluster. In future, we will investigate the robot swarm also in complex environments, having different numbers of light spots, differently shaped gradients, and even in fluctuating environments, where the light spots' intensities and positions change over time.

Our main reasoning about the processes that govern the robot swarm's behavior was done by constructing macroscopic models of our swarm system. In the following, we will discuss several technical aspects of these models, as well as discrepancies and similarities of our two macroscopic models.

Besides several simplifications and assumptions that became necessary during the abstraction process, both models have the potential to give valid approximations of the observed swarm behavior. The presented models give estimates of the dynamics as well as at least some spatial features. These results were achieved without any extensive fitting of parameters although several free parameters were introduced and roughly fitted.

In our compartment modeling approach, we represented the arena space by five discrete zones. This is a rather rough approximation, especially as these five zones were of different shape and size. However, the used approach sufficed to achieve good modeling results, which compared well to our empirical observations. We did not refine the compartmental model to finer compartments, as fine resolution of the arena was achieved by the spatially resolved PDE model anyway.

With our compartmental model approach, the formulation of the aggregation process (α) and of the disaggregation process (β) was rather straightforward and easy. The biggest challenge for the compartmental model was the implementation of the diffusion process. Not only can this parameter seldom be derived from the microscopic behavioral rules of the robots, but also the complexity of the shapes of our five zones made it difficult to model this process. Besides this hard-to-determine parameter D, the robot-to-robot detection rate P_{detect} is also a crucial parameter. It is hard to determine, because the robots' internal computation cycles differ significantly from time to time; sometimes, they execute their main loop 200 times a second, and other times, it can be executed 1000 times per second. The local environment (e.g., how many nearby robots trigger collision avoidance routines) affects the frequency of execution of our detection routine significantly. Thus, P_{detect} can also vary significantly over time. We observed that (per robot-to-robot contact) the likelihood of a positive recognition of robots varies between 20% and 50%.

Concerning the space-continuous model, the correct modeling of the robot motion poses a challenge. Although the assumption of a diffusion process leads to good results and is also mathematically easy to handle. The robots in our experiments showed rather a so-called correlated random walk. They move (ideally) straight between collisions. This time-correlated behavior cannot be modeled macroscopically. The accuracy of the space-continuous model would be increased with higher numbers of robots in the experiment, that is, the robot density and, thus, the collision frequency would be increased.

In the presented scenario, the model was adapted using a high diffusion constant D. This leads to a fast "mixing" or homogenization of the density throughout the arena as observed in the experiments; but it is a counterintuitive approach because it actually would be interpreted as a higher number of collisions in the image of a diffusion process. However, this is the only way of overcoming the lack of temporal correlations and emulating the observed behavior.

By comparing both macroscopic models, we can report that both models nicely converged concerning their internal structure and results. The two models were developed at two different locations simultaneously: The compartmental model (ODEs) was developed at the University of Graz (Austria), while the space-continuous model (PDEs) was developed at the Technical University in Karlsruhe (Germany). Both model approaches focused on the same empirical study; thus, both modeling processes faced the same problems to solve. Although the two mathematical approaches differ significantly, the resulting models could be easily refined in a way, so that they reuse approximately 30%–40% of their formulation from a common stock of formulations (collision theory, light gradient distribution, ...).

5.12 Conclusion

Finally, we conclude that observing natural organisms is a valuable approach toward swarm robotics. By performing careful experimentation with these organisms under controlled laboratory conditions, the relevant behavioral patterns of natural collective behavior can be investigated, and proximate behavioral rules (or correlations) can be extracted. Computer simulations (microscopic models) can help to understand these proximate rules better and to discriminate between dominant and rather unimportant rules concerning the ultimate collective behavior. Afterward, translated, often simplified, algorithms for swarm robots can be derived this way from the biological observations, and these swarm algorithms can be tested (and tuned) in microscopic simulations of robotic swarms as well as in real robotic swarms. Finally, macroscopic modeling of the swarm system can help to understand the "reasoning" of the observed ultimate collective behavior: They remove less important microscopic details, so that a good vision of the dominant feedback rules emerges, which determine the swarm's behavior. In our future work, we will continue to investigate the focal system on all levels that have been presented here. We will investigate honeybees' behavior in more complex temperature gradient fields, as we will also investigate our robotic swarm in more complex light gradient fields. In addition to that, we will refine our macroscopic models to achieve even better predictions of the swarm's behavior and to also allow modeling of more complex and fluctuating environments.

Acknowledgments

Hamann is supported by the German Research Foundation (DFG) within the Research Training Group GRK 1194 Self-Organizing Sensor-Actuator Networks. Schmickl is supported by the following grants: EU-IST FET project "I-SWARM," no. 507006 and the FWF research grant "Temperature-induced aggregation of young honeybees," no. P19478-B16, EU-IST FET project "SYMBRION," no. 216342, EU-ICT project "REPLICATOR," no. 216240. The authors would like to thank Gerald Radspieler, Ronald Thenius, and Christoph Möslinger for providing figures and photos for proofreading of this manuscript.

References

1. G. Weiss, *Multiagent Systems: A Modern Approach to Distributed Artificial Intelligence* (MIT Press, Cambridge, MA, 1999).
2. M. Woolridge, *Introduction to Multiagent Systems* (John Wiley & Sons, Inc., New York, 2001).
3. G. Beni and J. Wang, Swarm intelligence in cellular robotic systems. In *Proceedings of the NATO Advanced Workshop on Robots and Biological Systems*, Pisa, Italy (1989).
4. E. B. M. Dorigo and G. Theraulaz, *Swarm Intelligence: From Natural to Artificial Systems* (Oxford University Press, New York, 1999).
5. M. M. Millonas. Swarms, phase transitions, and collective intelligence. In ed. C. G. Langton, *Artificial Life III* (Addison-Wesley, Reading, MA, 1994).
6. J. Kennedy and R. C. Eberhart, *Swarm Intelligence* (Morgan Kaufmann, San Francisco, CA, 2001).
7. M. Dorigo and G. D. Caro, Ant colony optimization: A new meta-heuristic. In eds. P. J. Angeline, Z. Michalewicz, M. Schoenauer, X. Yao, and A. Zalzala, *Proceedings of the 1999 Congress on Evolutionary Computation (CEC'99)*, pp. 1470–1477, Piscataway, NJ (1999). IEEE Press, Piscataway, NJ.
8. M. Dorigo and T. Stützle, *Ant Colony Optimization* (MIT Press, Cambridge, MS, London, U.K., 2004).
9. J. Ferber, *Multiagentensysteme—Eine Einführung in die Verteilte Künstliche Intelligenz* (Addison-Wesley, Paris, France, 2001).
10. F. Klügl, *Multiagentensimulationen—Konzepte, Werkzeuge, Anwendung* (Addison-Wesley, Paris, France, 2001).
11. G. Beni, From swarm intelligence to swarm robotics. In eds. E. Şahin and W. M. Spears, *Proceedings of the SAB 2004 Workshop on Swarm Robotics, Lecture Notes in Computer Science*, pp. 1–9, Santa Monica, CA (July, 2004).
12. E. Şahin, Swarm robotics: From sources of inspiration to domains of application. In eds. E. Şahin and W. M. Spears, *Swarm Robotics Workshop: State-of-The-Art Survey*, vol. 3342, Lecture Notes in Computer Science, pp. 10–20. Springer-Verlag, Berlin, Heidelberg (2005).
13. P.-P. Grassé, La reconstruction du nid et les coordinations interindividuelles chez bellicositermes natalensis et cubitermes sp. la théorie de la stigmergie:essai d'interprétation du comportement des termites constructeurs, *Insectes Sociaux.* **6**, 41–83 (1959).
14. P.-P. Grassé, Nouvelles experiences sur le termite de müller (macrotermes mülleri) et considerations sur la théorie de la stigmergie, *Insectes Sociaux.* **14**, 73–102 (1967).
15. D. Payton, M. Daily, R. Estowski, M. Howard, and C. Lee, Pheromone robotics, *Autonomous Robots.* **11**(3), 319–324 (2001).

16. K. Sugawara, T. Kazama, and T. Watanabe, Foraging behavior of interacting robots with virtual pheromone. In *Proceedings of 2004 IEEE/RSJ International Conference on Intelligent Robots and Systems* (2004).

17. H. Heran, Untersuchungen über den Temperatursinn der Honigbiene (*Apis mellifica*) unter besonderer Berücksichtigung der Wahrnehmung von Strahlungswärme, *Zeitschrift für vergleichende Physiologie.* **34**, 179–207 (1952).

18. K. Crailsheim, U. Eggenreich, R. Ressi, and M. Szolderits, Temperature preference of honeybee drones (hymenoptera: Apidae), *Entomologia Generalis.* **24**(1), 37–47 (1999).

19. P. Grodzicki and M. Caputa, Social versus individual behaviour: A comparative approach to thermal behaviour of the honeybee (*Apis mellifera* l.) and the american cockroach (*Periplaneta americana* l.), *Journal of Insect Physiology.* **51**, 315–322 (2005).

20. J. Seyfried, M. Szymanski, N. Bender, R. Estaña, M. Thiel, and H. Wörn, The I-SWARM project: Intelligent small world autonomous robots for micromanipulation. In eds. E. Şahin and W. M. Spears, *Swarm Robotics Workshop: State-of-The-Art Survey*, pp. 70–83. (Springer-Verlag, Berlin, 2005).

21. S. Kornienko, O. Kornienko, and P. Levi. R-based communication and perception in microrobotic swarms. In *Proceedings of the IEEE/RSJ International Conference on Intelligent Robots and Systems (IROS'05)*, Edmonton, Canada (2005).

22. Jasmine. Swarm robot—project website (2007). http://www.swarmrobot.org/

23. T. Schmickl and K. Crailsheim, Trophallaxis within a robotic swarm: Bio-inspired communication among robots in a swarm, *Autonomous Robots.* **25**, 171–188 (2008).

24. H. Hamann, H. Wörn, K. Crailsheim, and T. Schmickl, Spatial macroscopic models of a bio-inspired robotic swarm algorithm. In *IEEE/RSJ 2008 International Conference on Intelligent RObots and Systems (IROS'08)* (2008). Submitted.

25. F. Schweitzer, *Brownian Agents and Active Particles. On the Emergence of Complex Behavior in the Natural and Social Sciences* (Springer-Verlag, Berlin, 2003).

26. J. Forrester, *World Dynamics* (Wright-Allen, Cambridge, MA, 1971).

27. V. Systems. Vensim®. http://www.vensim.com

28. M. Trautz, Das Gesetz der Reaktionsgeschwindigkeit und der Gleichgewichte in Gasen. Bestätigung der Additivität von Cv-3/2R. Neue Bestimmung der Integrationskonstanten und der Moleküldurchmesser, *Zeitschrift für anorganische und allgemeine Chemie.* **96**(1), 1–28 (1916).

29. The NSR Physiome Project. http://www.physiome.org/model/doku.php?id=Tutorials: Compartmental#Compartmental_Modeling_Tutorial

30. H. Hamann and H. Wörn, A framework of space-time continuous models for algorithm design in swarm robotics, *Swarm Intelligence.* (2008). Submitted.

31. A. Galstyan, T. Hogg, and K. Lerman. Modeling and mathematical analysis of swarms of microscopic robots. In *Proceedings of IEEE Swarm Intelligence Symposium (SIS-2005)*, Pasadena, CA, pp. 201–208 (June, 2005).

32. T. Hogg, Coordinating microscopic robots in viscous fluids, *Autonomous Agents and Mutli-Agent Systems.* **14**(3), 271–305 (2006).

33. H. Hamann and H. Wörn. A space- and time-continuous model of self-organizing robot swarms for design support. In *First IEEE International Conference on Self-Adaptive and Self-Organizing Systems (SASO'07)*, Boston, MA, July 9–11, pp. 23–31 (2007).

34. H. Haken, *Synergetics—An Introduction* (Springer-Verlag, Berlin, 1977).

35. P. Langevin, Sur la théorie du mouvement brownien, *Comptes-rendus de l'Académie des Sciences.* **146**, 530–532 (1908).

36. D. S. Lemons and A. Gythiel, Paul langevin's 1908 paper "On the theory of Brownian motion" ["Sur la théorie du mouvement brownien," Comptes-rendus de l'Académie des Sciences (Paris) 146, 530–533 (1908)], *American Journal of Physics.* **65**(11), 1079–1081 (1997).

37. A. D. Fokker, Die mittlere Energie rotierender elektrischer Dipole im Strahlungsfeld, *Annalen der Physik.* **348**(5), 810–820 (1914).

38. M. Planck, Über einen Satz der statistischen Dynamik und seine Erweiterung in der Quantentheorie, *Sitzungsberichte der Preußischen Akademie der Wissenschaften.* **24**, 324–341 (1917).

39. A. N. Kolmogorov, Über die analytischen Methoden in der Wahrscheinlichkeitsrechnung, *Mathematische Annalen.* **104**(1), 415–458 (1931).

40. H. Risken, *The Fokker-Planck Equation* (Springer-Verlag, Berlin, 1984).

41. W. J. Ewens, *Mathematical Population Genetics* (Springer-Verlag, New York, 2004).

42. J. Tom, A. Witten, and L. M. Sander, Diffusion-limited aggregation, a kinetic critical phenomenon, *Physical Review Letters.* **19**, 1400–1403 (1981).

Chapter 6

Self-Organizing Data and Signal Cellular Systems

André Stauffer and Gianluca Tempesti

Contents

6.1 Bio-Inspired Properties

6.1.1 Cellular Architecture

An extremely simplified example, the display of the SOS acronym, is introduced in order to illustrate the basic bio-inspired properties of our self-organizing systems. The system that displays the acronym is a one-dimensional artificial organism made of three cells (Figure 6.1a). Each cell of this organism is identified by an X coordinate, ranging from 1 to 3. For coordinate values $X = 1$ and $X = 3$, the cell implements the S character; for $X = 2$, it implements the O character. Such an organism can be built with a single *totipotent cell* (Figure 6.1b) capable of displaying either the S or the O character according to the X coordinate. The totipotent cell is an arrangement of $4 \times 6 = 24$ molecules, 21 of which are invariant, one displays the S character, and two display the O character. An incrementer implementing the X coordinate calculation is embedded in the final organism.

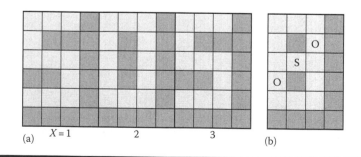

(a) $X = 1$ 2   3   (b)

Figure 6.1 SOS acronym. (a) One-dimensional organism made of three cells. (b) Totipotent cell made of 4 × 6 = 24 molecules.

6.1.2 Cloning

The *cloning* or self-replication can be implemented at the cellular level in order to build a multicellular organism and at the organismic level in order to generate a population of organisms. The cloning of the totipotent cell displayed in Figure 6.1b results thus in the SOS organism of Figure 6.1a. The cloning of the organism, defining a population of SOS acronyms (Section 6.4), rests on two assumptions: (a) there exists a sufficient number of spare cells in the array to contain at least one copy of the additional organism; and (b) the calculation of the coordinates produces a cycle $X = 1 \rightarrow 2 \rightarrow 3 \rightarrow 1$ implying $X+ = (X+1)$ mod 3. Given a sufficiently large space, the cloning of the organism can be repeated for any number of specimens in the X and/or Y axes.

6.1.3 Cicatrization

The introduction in the totipotent cell of one column of spare molecules (SM, Figure 6.2a), defined by a specific structural configuration, and the automatic detection of faulty molecules (by a built-in self-test mechanism which constantly compares two copies of the same molecule) allow *cicatrization* or self-repair at the cellular level: each faulty molecule is deactivated, isolated from the network, and replaced by the nearest right molecule, which will itself be replaced by the nearest right molecule, and so on until an SM is reached (Figure 6.2b). The number of faulty molecules handled by the cicatrization mechanism is necessarily limited: in the example of Figure 6.2a, we tolerate at most one faulty molecule per row. If more than one molecule is faulty in the same row (Figure 6.2c), cicatrization is impossible, in which case a global Kill $= 1$ is generated to activate regeneration as described thereafter.

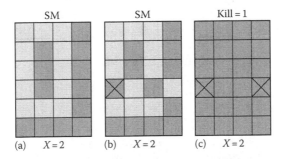

Figure 6.2   Cicatrization of the SOS organism. (a) Healthy cell displaying O. (b) Self-repaired cell with one faulty molecule. (c) Faulty cell with two faulty molecules in the same row.

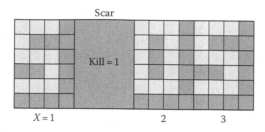

Figure 6.3   Regeneration of the SOS organism.

6.1.4 Regeneration

In order to implement *regeneration*, that is self-repair at the organismic level, we need at least one spare cell to the right of the original organism (Figure 6.1a). The existence of a fault, detected by the Kill signal generated at the cellular level (Figure 6.2c), identifies the faulty cell, and the entire column of all cells to which the faulty cell belongs is considered faulty and is deactivated (Figure 6.3; in this simple example, the column of cells is reduced to a single cell). All the functions (X coordinate and configuration) of the cells to the right of the column $X = 1$ are shifted by one column to the right. Obviously, this process requires as many spare cells to the right of the array as there are faulty cells to repair.

6.2 Functional Design

6.2.1 Structural Configuration Mechanism

The goal of the *structural configuration mechanism* is to define the boundaries of the cell as well as the living mode or spare mode of its constituting molecules. This mechanism is made up of a *structural growth process* followed by a *load process*. For a better understanding of these processes, we apply them to a minimal system, a cell made up of six molecules arranged as an array of two rows by three columns, the third column involving two SM dedicated to self-repair.

The *growth process* starts when an external *growth signal* is applied to the lower left molecule of the cell (Figure 6.4a) and this molecule selects the corresponding eastward data input (Figure 6.4b). According to the *structural configuration data* or *structural genome*, each molecule of the cell generates successively an internal *growth signal* and selects an input (Figure 6.5), in order to create a data path among the molecules of the cell (Figure 6.4b through g). When the connection path between the molecules closes, the lower left molecule delivers a *close signal* to the nearest left neighbor cell (Figure 6.4h). The structural configuration data are now moving around the data path and are ready to be transmitted to neighboring cells.

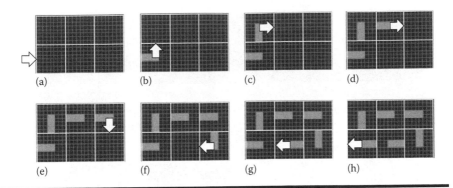

Figure 6.4 Structural growth process of a minimal system, a cell made up of six molecules. (a) External growth signal is applied to the lower left molecule. (b–g) Generation of internal growth signals to build the structural data path. (h) Closed path and close signal delivered to the nearest left neighbor cell.

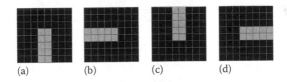

Figure 6.5 Data input selection. (a) Northward. (b) Eastward. (c) Southward. (d) Westward.

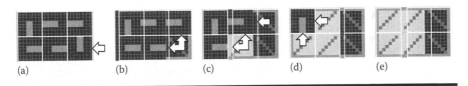

Figure 6.6 Load process. (a) External close signal applied to the lower right molecule by the nearest right neighbor cell. (b–e) Generation of internal load signals propagating westward and northward to store the molecular modes and types of the cell.

The *load process* is triggered by the *close signal* applied to the lower right molecule of the cell (Figure 6.6a). A *load signal* then propagates westward and northward through the cell (Figure 6.6b through d) and each of its molecules acquires a *molecular mode* (Figure 6.7) and a *molecular type* (Figure 6.8). We finally obtain a homogeneous tissue of molecules defining both the boundaries of the cell and the position of its *living mode* and *spare mode* molecules (Figure 6.6e). This tissue is ready for being configured by the functional configuration data.

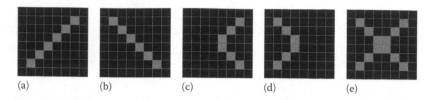

Figure 6.7 Molecular modes. (a) Living. (b) Spare. (c) Faulty. (d) Repair. (e) Dead.

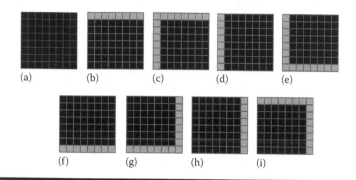

Figure 6.8 Molecular types. (a) Internal. (b) Top. (c) Top left. (d) Left. (e) Bottom left. (f) Bottom. (g) Bottom right. (h) Right. (i) Top right.

6.2.2 Functional Configuration Mechanism

The goal of the *functional configuration mechanism* is to store in the homogeneous tissue, which already contains structural data (Figure 6.6e), the functional data needed by the specifications of the current application. This mechanism is a *functional growth process*, performed only on the molecules in the *living mode* while the molecules in the *spare mode* are simply bypassed. It starts with an external *growth signal* applied to the lower left living molecule (Figure 6.9a). According to the *functional configuration data* or *functional genome*, the living molecules then successively generate an internal *growth signal*, select an input, and create a path among the living molecules of the cell (Figure 6.9b through f). The functional configuration data are now moving around the data path and are ready to be transmitted to neighboring cells.

6.2.3 Cloning Mechanism

The *cloning mechanism* or *self-replication mechanism* is implemented at the cellular level in order to build a multicellular organism and at the organismic level in order to generate a population of organisms. This mechanism supposes that there exists a sufficient number of molecules in the array to contain at least one copy of the

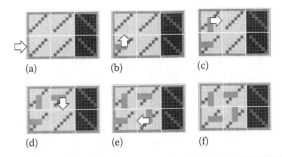

Figure 6.9 Functional configuration of the cell performed as a functional growth process applied to the living molecules. (a) External growth signal is applied to the lower left molecule. (b–e) Generation of internal growth signals in order to build the functional data path. (f) Closed functional data path.

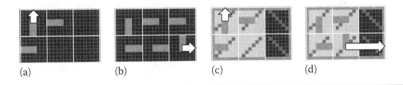

Figure 6.10 Generation of growth signals triggering the cloning mechanism. (a) Northward structural branching process. (b) Eastward structural branching process. (c) Northward functional branching process. (d) Eastward functional branching process.

additional cell or of the additional organism. It corresponds to a *branching process* which takes place when the structural and the functional configuration mechanisms deliver northward and eastward growth signals on the borders of the cell during the corresponding growth processes (Figure 6.10).

6.2.4 Cicatrization Mechanism

Figure 6.9f shows the normal behavior of a healthy minimal cell, i.e., a cell without any faulty molecule. A molecule is considered faulty, or in the *faulty mode*, if some built-in self-test detects a lethal malfunction. Starting with the normal behavior of Figure 6.9f, we suppose that two molecules will become suddenly faulty (Figure 6.11a): (1) the lower left molecule, which is in the *living mode*; and (2) the upper right molecule, which is in the *spare mode*. While there is no change for the upper right molecule, which is just no more able to play the role of an SM, the lower left one triggers a *cicatrization mechanism*. This mechanism is made up of a *repair process* involving eastward propagating *repair*

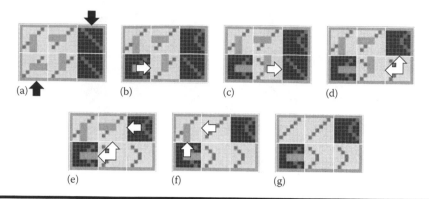

Figure 6.11 Cicatrization mechanism performed as a repair process followed by a reset process. (a) Living and spare molecules becoming faulty. (b and c) Generation of repair signals propagating eastward. (d–f) Generation of internal reset signals propagating westward and northward. (g) Cell comprising two faulty and two repair molecules ready for functional reconfiguration.

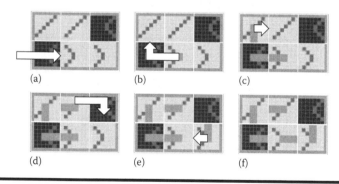

Figure 6.12 Functional reconfiguration of the living and repair molecules. (a) External growth signal bypassing the lower left faulty molecule. (b–e) Generation of internal growth signals to build a functional data path bypassing the faulty molecules. (f) Closed functional data path within the living and repair molecules.

signals (Figure 6.11b and c) followed by a *reset process* performed with westward and northward propagating internal *reset signals* (Figure 6.11d through g). This tissue, comprising now two molecules in the *faulty mode* and two molecules in the *repair mode*, is ready for being reconfigured by the functional configuration data. This implies a *functional growth process* bypassing the faulty molecules (Figure 6.12).

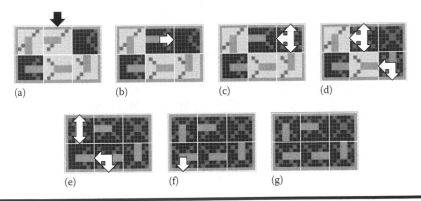

Figure 6.13 **Regeneration mechanism performed as a repair process followed by a kill process. (a) Living molecule becoming faulty. (b) Eastward repair signal. (c–f) Generation of internal and external kill signals propagating northward, westward, and southward. (g) Cell made up of six dead molecules.**

6.2.5 Regeneration Mechanism

Our minimal system comprises a single SM per row and tolerates therefore only one faulty molecule in each row. A second faulty molecule in the same row will cause the death of the whole cell and the start of a *regeneration mechanism*. Figure 6.13 illustrates the *repair process* and *kill process* involved in this mechanism. Starting with the normal behavior of the cicatrized cell (Figure 6.12f), a new molecule, the upper middle one, becomes faulty. In a first step, the new faulty molecule sends a *repair signal* eastward, in order to look for an SM that can replace it (Figure 6.13b). In a second step, the supposed SM, which is in fact a faulty one, enters the lethal *dead mode* and triggers *kill signals*, which propagate northward, westward, and southward (Figure 6.13c through f). Finally in Figure 6.13g, all the molecules of the array are dead as well as our minimal system.

6.3 Hardware Design

6.3.1 Data and Signals

In order to design self-organizing bio-inspired systems, we will split the information involved in two different types:

- The *data* constitute the information that travels through the built structure.
- The *signals* constitute the information that controls the building and maintaining of the structure.

Figure 6.14 shows the symbols and the hexadecimal codes of the *flag data* and the *structural data* that are part of the *structural genome* or structural configuration string. The boundaries of the artificial cell as well as the living or spare mode of its constituting molecules are defined according to this configuration string.

Similarly, a *functional genome* or functional configuration string, made up of flag data and functional data, is used in order to give to the cell the specifications needed by a given application.

The structural data will confer to each molecule of the artificial cell its molecular mode and its molecular type (Figure 6.15). Apart from their living or spare mode at the start, the molecules may become faulty, repair, or dead later.

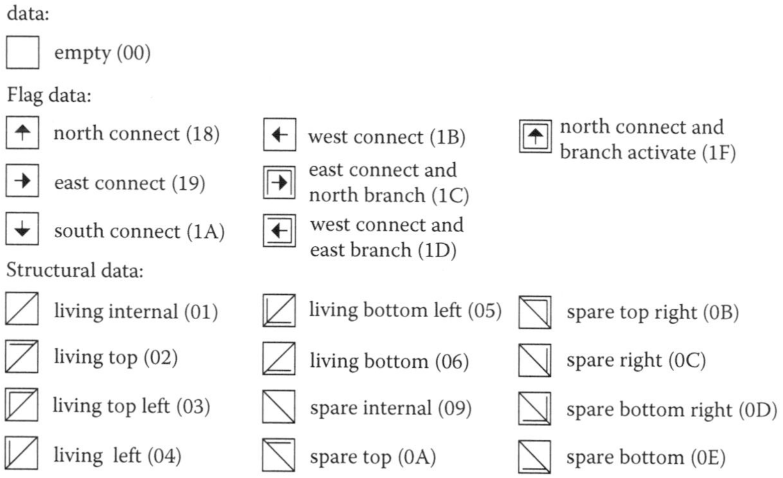

data:

empty (00)

Flag data:

north connect (18)   west connect (1B)   north connect and branch activate (1F)

east connect (19)   east connect and north branch (1C)

south connect (1A)   west connect and east branch (1D)

Structural data:

living internal (01)   living bottom left (05)   spare top right (0B)

living top (02)   living bottom (06)   spare right (0C)

living top left (03)   spare internal (09)   spare bottom right (0D)

living left (04)   spare top (0A)   spare bottom (0E)

Figure 6.14   Data symbols and hexadecimal codes.

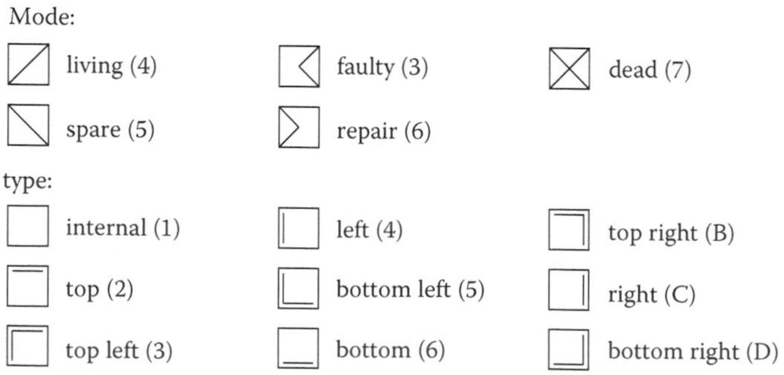

Mode:

living (4)   faulty (3)   dead (7)

spare (5)   repair (6)

type:

internal (1)   left (4)   top right (B)

top (2)   bottom left (5)   right (C)

top left (3)   bottom (6)   bottom right (D)

Figure 6.15   Molecular modes and molecular types.

Figure 6.16 shows the symbols and the hexadecimal codes of the signals. Some of them are strictly eastward or westward oriented like the close and repair signals while the others have no preferential direction.

According to the Tom Thumb algorithm,[1] the growth signals are generated for special configurations of the flag data in a $2 \times N$-level stack implemented in the molecule. The truth table of Figure 6.17 shows the configuration in the levels GN-1, PN-1, and PN producing the northward, eastward, southward, and westward growth signals, respectively.

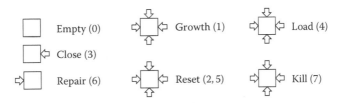

Figure 6.16   Signal symbols and hexadecimal codes.

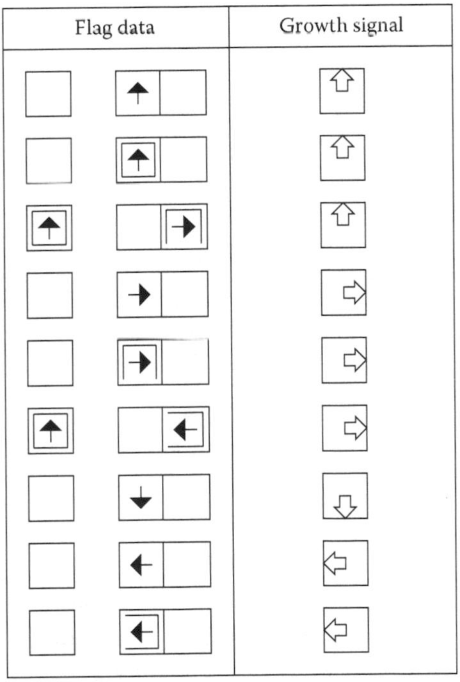

Flag data	Growth signal
↑	⇑
↑	⇑
↑ →	⇑
→	⇨
→	⇨
↑ ←	⇨
↓	⇓
←	⇦
←	⇦

Figure 6.17   Growth signal truth table.

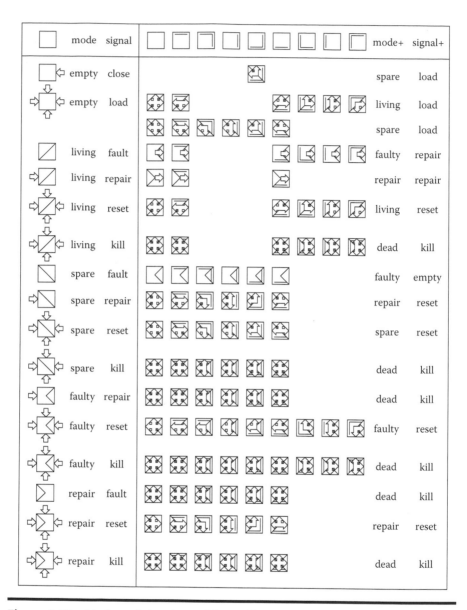

Figure 6.18 Mode and signal transition table.

Finally, Figure 6.18 represents the transition table defining both the future mode of the molecule and the future output signals generated by the molecule. According to the table, the future mode and the future output signals depend on the present mode and on the present input signals. Based on the type of the molecule, the future output signals are generated in specific directions.

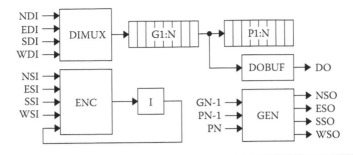

Figure 6.19   Configuration DSCA cell schematic.

6.3.2 *Configuration Level*

We will now describe the detailed architecture of our basic molecule which corresponds to a data and signals cellular automaton (DSCA) cell.[2] This DSCA cell is designed as a digital system, resulting from the interconnection of a processing unit handling the data and a control unit computing the signals.

In order to implement the propagation of the growth data, the processing unit is made up of three resources (Figure 6.19):

- An input multiplexer DIMUX, selecting one out of the four input data NDI, EDI, SDI, or WDI
- A 2 × N-level stack organized as N genotypic registers G1:N (for mobile data), and N phenotypic registers P1:N (for fixed data)
- An output buffer DOBUF producing the output data DO

The control unit consists of three resources (Figure 6.19):

- An encoder ENC for the input signals NSI, ESI, SSI, and WSI
- A transmission register I for the memorization of the input selection
- A generator GEN producing the output signals NSO, ESO, SSO, and WSO

I	DI
Zero	Empty
Seln	NDI
Sele	EDI
Sels	SDI
Selw	WDI

Figure   6.20   Input data   multiplexer DIMUX truth table.

The specifications of the input data multiplexer DIMUX correspond to the truth table of Figure 6.20. This table defines the input data *DI* selected depending on the state of the transmission register I.

The 2 × N-level stack works as a 2 × N-data shift register when its PN stage is *empty*. When the PN stage is not *empty* anymore (*nempty*), only G1:N still perform a shift operation while P1:N realize a hold operation, as described in the transition table of Figure 6.21.

PN	G1:N+	P1:N+
Empty	DI,G1:N-1	GN,P1:N-1
Nempty	DI,G1:N-1	P1:N

Figure 6.21   Data stack G1:N,P1:N transition table.

The buffer DOBUF delivers as output data DO the content of the genotypic register GN when the phenotypic register PN is *nempty* and the *empty* data otherwise (Figure 6.22).

PN	DO
Empty	Empty
Nempty	GN

Figure 6.22   Output data buffer DOBUF truth table.

The encoder ENC defines the future state of the transmission register I. According to the transition table of Figure 6.23, where *nselw* means not *selw*, this future state depends on its present one and on those of the input signals NSI, ESI, SSI, and WSI.

The generator GEN produces the output signals NSO, ESO, SSO, and WSO as described in the truth table of Figure 6.24.

6.3.3 Self-Repair Level

In order to add repair capabilities to the former DSCA cell (CONF), its control unit is made up of supplementary encoder ENC and generator GEN as well as four new resources (Figure 6.25):

- A decoder DEC defining the mode and the type of the molecule
- A signal register S
- A mode register M
- A type register T

I	NSI	ESI	SSI	WSI	I+
Zero	Empty	Empty	Empty	Empty	Zero
Seln	Empty	Empty	Empty	Empty	Seln
Sele	Empty	Empty	Empty	Empty	Sele
Sels	Empty	Empty	Empty	Empty	Sels
Nselw	Growth	Empty	Empty	Empty	Seln
Nselw	Growth	Growth	Empty	Empty	Seln
Nselw	Empty	Growth	Empty	Empty	Sele
Nselw	Empty	Empty	Growth	Empty	Sels
Nselw	Empty	Empty	Empty	Growth	Selw
Nselw	Growth	Empty	Empty	Growth	Selw
Nselw	Growth	Growth	Empty	Growth	Selw
Nselw	Empty	Growth	Empty	Growth	Selw
Selw	–	–	–	–	Selw

Figure 6.23   Input signals encoder ENC transition table.

GN-1	PN-1	PN	NSO	ESO	SSO	WSO
–	North connect	Empty	Growth	Empty	Empty	Empty
–	Branch activate	Empty	Growth	Empty	Empty	Empty
Branch activate	–	North branch	Growth	Empty	Empty	Empty
–	East connect	Empty	Empty	Growth	Empty	Empty
–	North branch	Empty	Empty	Growth	Empty	Empty
Branch activate	–	East branch	Empty	Growth	Empty	Empty
–	South connect	Empty	Empty	Empty	Growth	Empty
–	West connect	Empty	Empty	Empty	Empty	Growth
–	East branch	Empty	Empty	Empty	Empty	Growth

Figure 6.24   Output signals generator GEN truth table.

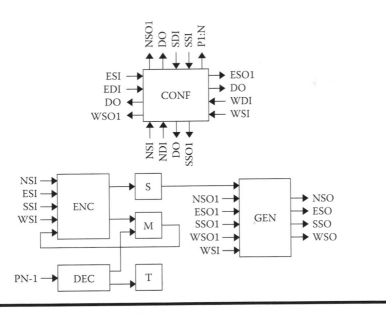

Figure 6.25   Self-repair DSCA cell schematic.

The decoder DEC extracts the molecular mode and the molecular type from the structural data stored in the phenotypic register PN-1. The specifications of this extraction correspond to the truth table of Figure 6.26.

The encoder ENC defines the future states of the mode register M and of the type register T. According to the transition table of Figure 6.27, these future states depend on the present state of the mode register M, on the global input signal GSI, and on the molecular mode provided by the decoder DEC. The global input signal GSI realizes the logical sum of the four individual input signals:

$$GSI = NSI + ESI + SSI + WSI$$

PN-1	Mode	Type
Empty	Empty	Empty
Living	Living	Internal
Living top	Living	Top
Living top left	Living	Top left
Living left	Living	Left
Living bottom left	Living	Bottom left
Living bottom	Living	Bottom
Spare	Spare	Internal
Spare top	Spare	Top
Spare top right	Spare	Top right
Spare right	Spare	Right
Spare bottom right	Spare	Bottom right
Spare bottom	Spare	Bottom

Figure 6.26 Structural data decoder DEC truth table.

M	GSI	Mode	M+	S+
Empty	Empty	–	Empty	Empty
Empty	Close	Living	Living	Load
Empty	Close	Spare	Spare	Load
Empty	Load	Living	Living	Load
Empty	Load	Spare	Spare	Load
Living	Empty	–	Living	Empty
Living	Fault	–	Faulty	Repair
Living	Repair	–	Repair	Repair
Living	Reset	–	Living	Reset
Living	Kill	–	Dead	Kill
Spare	Empty	–	Spare	Empty
Spare	Fault	–	Faulty	Empty
Spare	Repair	–	Repair	Reset
Spare	Reset	–	Spare	Reset
Spare	Kill	–	Dead	Kill
Faulty	Empty	–	Faulty	Empty
Faulty	Repair	–	Dead	Kill
Faulty	Reset	–	Faulty	Reset
Faulty	Kill	–	Dead	Kill
Repair	Empty	–	Repair	Empty
Repair	Fault	–	Dead	Kill
Repair	Reset	–	Repair	Reset
Repair	Kill	–	Dead	Kill
Dead	–	–	Dead	Empty

Figure 6.27 Input signals encoder ENC transition table.

S	NSO1	ESO1	SSO1	WSO1	WSI	NSO	ESO	SSO	WSO
Empty	Empty	Empty	Empty	Empty	Empty	Empty	Empty	Empty	Empty
Empty	Growth	Empty	Empty	Empty	Empty	Growth	Empty	Empty	Empty
Empty	Empty	Growth	Empty	Empty	Empty	Empty	Growth	Empty	Empty
Empty	Empty	Empty	Growth	Empty	Empty	Empty	Empty	Growth	Empty
Empty	Empty	Empty	Empty	Growth	Empty	Empty	Empty	Empty	Growth
Empty	Empty	Empty	Empty	Empty	Growth	Close	Close	Close	Close
Load	Empty	Empty	Empty	Empty	Empty	Load	Load	Load	Load
Repair	Empty	Empty	Empty	Empty	Empty	Repair	Repair	Repair	Repair
Reset	Empty	Empty	Empty	Empty	Empty	Reset	Reset	Reset	Reset
Kill	Empty	Empty	Empty	Empty	Empty	Kill	Kill	Kill	Kill

Figure 6.28   Output signals generator GEN truth table.

The generator GEN produces the new output signals *NSO*, *ESO*, *SSO*, and *WSO* from the ones delivered by CONF, from the present state of the signal register S, and from the input signal *WSI* (Figure 6.28).

6.3.4 Application Level

The fixed functional configuration data memorized in the phenotypic registers P1:N-1 defines the specifications of the molecule at the application level. The schematic of the molecule (Figure 6.29) interconnects therefore the self-repair DSCA cell SREP to the functional application unit APPL. In this schematic, *NAI*, *EAI*, *SAI*, and *WAI* represent the application data inputs while *NAO*, *EAO*, *SAO*, and *WAO* constitute the application data outputs.

6.3.5 Transmission Level

In order to bypass the spare, killed, or dead molecules, multiplexers and demultiplexers are added to the self-repair processing unit SREPPU (Figure 6.30), to the

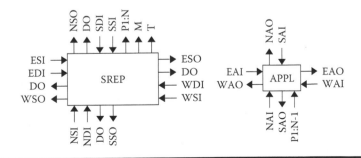

Figure 6.29   Application DSCA cell schematic.

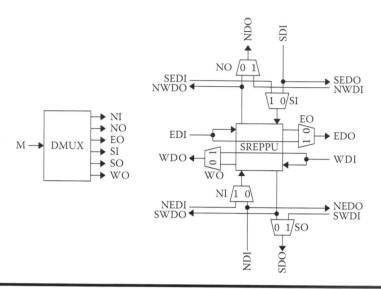

Figure 6.30   Processing unit of the transmission DSCA cell.

self-repair control unit SREPCU (Figure 6.31), and to the application unit APPL (Figure 6.32). Three more resources are needed to control them:

- A configuration data transmission unit DMUX
- A signals transmission unit SMUX
- An application data transmission unit AMUX

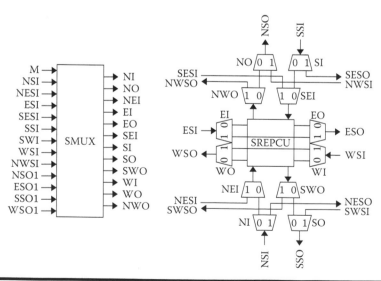

Figure 6.31   Control unit of the transmission DSCA cell.

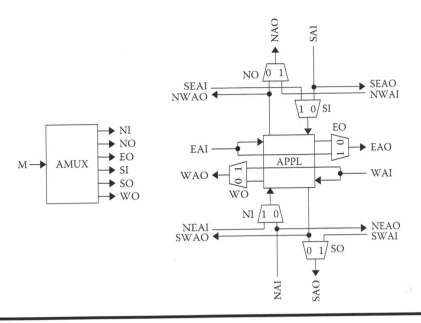

Figure 6.32 Application unit of the transmission DSCA cell.

M	NI	NO	EO	SI	SO	WO
Spare	0	0	1	0	0	1
Faulty	0	1	1	0	1	1
Repair	1	1	0	1	1	0
Dead	0	0	1	0	0	1

Figure 6.33 Configuration data multiplexers controller DMUX truth table.

M	GSI	GSO1	NI	NO	NEI	EI	EO	SEI	SI	SO	SWO	WI	WO	NWO
Spare	Growth	–	0	0	0	1	1	0	0	0	0	1	1	0
Faulty	Growth	–	1	1	0	1	1	0	1	1	0	1	1	0
Faulty	–	Growth	1	1	0	1	1	0	1	1	0	1	1	0
Repair	Growth	–	1	1	1	0	0	1	1	1	1	0	0	1
Repair	–	Growth	1	1	1	0	0	1	1	1	1	0	0	1
Dead	Growth	–	0	0	0	1	1	0	0	0	0	1	1	0

Figure 6.34 Signal multiplexers and demultiplexers controller SMUX truth table.

The control of the multiplexers operated by the configuration data transmission unit DMUX depends on the state of the mode register. The corresponding specifications are given in the truth table of Figure 6.33.

The control of the multiplexers and demultiplexers operated by the signals transmission unit SMUX depends on the state of the mode register and on the global

M	NI	NO	EO	SI	SO	WO
Spare	0	0	1	0	0	1
Faulty	0	1	1	0	1	1
Repair	1	1	0	1	1	0
Dead	0	0	1	0	0	1

Figure 6.35 **Application data multiplexers controller AMUX truth table.**

input signals as well as the global output signals of the self-repair cell. The truth table of Figure 6.34 shows the corresponding specifications.

The control of the multiplexers operated by the application data transmission unit AMUX depends on the state of the mode register. The corresponding specifications are given in the truth table of Figure 6.35.

6.3.6 Output Level

Depending on the type T of the molecule, the DSCA cell finally controls its output signals with the buffer unit SOBUF (Figure 6.36). This unit is made of buffers and

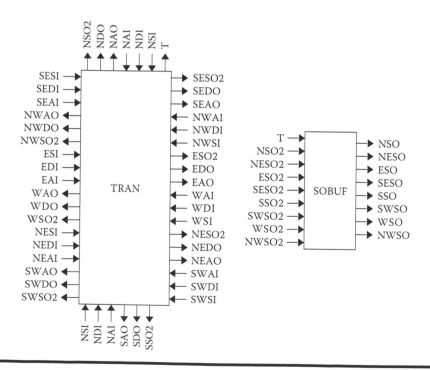

Figure 6.36 **Output DSCA cell schematic.**

Signal	Type	NO	SEO	EO	NEO	SO	SWO	WO	NWO
Empty	–	0	0	0	0	0	0	0	0
Growth	–	1	1	1	1	1	1	1	1
Load	Internal	1	0	1	0	1	0	1	0
Load	Top	0	0	1	0	1	0	1	0
Load	Top right	0	0	0	0	1	0	1	0
Load	Right	1	0	0	0	1	0	1	0
Load	Bottom right	1	0	0	0	0	0	1	0
Load	Bottom	1	0	1	0	0	0	1	0
Load	Bottom left	1	0	1	0	0	0	0	0
Load	Left	1	0	1	0	1	0	0	0
Load	Top left	0	0	1	0	1	0	0	0
Repair	–	0	0	1	0	0	0	0	0
Reset	Internal	1	0	1	0	1	0	1	0
Reset	Top	0	0	1	0	1	0	1	0
Reset	Top right	0	0	0	0	1	0	1	0
Reset	Right	1	0	0	0	1	0	1	0
Reset	Bottom right	1	0	0	0	0	0	1	0
Reset	Bottom	1	0	1	0	0	0	1	0
Reset	Bottom left	1	0	1	0	0	0	0	0
Reset	Left	1	0	1	0	1	0	0	0
Reset	Top left	0	0	1	0	1	0	0	0
Kill	Internal	1	0	1	0	1	0	1	0
Kill	Top	1	0	1	0	1	0	1	0
Kill	Top right	1	0	0	0	1	0	1	0
Kill	Right	1	0	0	0	1	0	1	0
Kill	Bottom right	1	0	0	0	1	0	1	0
Kill	Bottom	1	0	1	0	1	0	1	0
Kill	Bottom left	1	0	1	0	1	0	0	0
Kill	Left	1	0	1	0	1	0	0	0
Kill	Top left	1	0	1	0	1	0	0	0
Close	–	0	0	0	0	0	0	1	0

Figure 6.37   **Output signals buffers controller truth table.**

a controller. According to the truth table of Figure 6.37, the specifications of the controller rest also on the type of the signals generated by the transmission DSCA cell TRAN.

6.4 Hardware Simulation

6.4.1 Functional Application

The functional application introduced here is the display of the SOS acronym described in Section 6.1. The specifications given there are sufficient for designing the final architecture of the living molecules of the totipotent cell (Figure 6.38).

The living molecules of the totipotent cell are divided into six categories depending on their functionality:

1. Two busses for the horizontal transfer of the XI coordinate
2. Modulo 3 incrementation of the XI coordinate

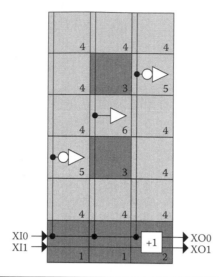

Figure 6.38 Functional specifications of the living molecules of the SOS totipotent cell.

3. One bus for the vertical distribution of the XI0 logic variable
4. Permanent display of characters S and O
5. Display of O character only (XI0 = 0 or XI0′ = 1)
6. Display of S character only (XI0 = 1)

Figure 6.39 gives the application-specific schematic of the SOS molecule. The choice of one of the six preceding categories depend on the functional configuration code *P*. This code defines also the application data connections within the unit.

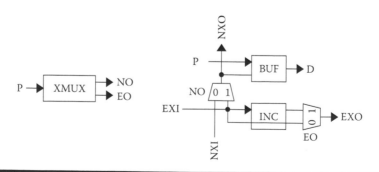

Figure 6.39 Application specific schematic of the SOS molecule.

The schematic of Figure 6.39 is made of four resources:

- Two multiplexers
- An incrementer INC
- A display buffer BUF
- A coordinate transmission unit XMUX

The specifications of the coordinate incrementer INC correspond to the truth table of Figure 6.40.

The buffer BUF drives the display D according to the truth table of Figure 6.41.

The control of the multiplexers operated by the coordinate transmission unit XMUX depends on the functional configuration code P. The truth table of Figure 6.42 shows the corresponding specifications.

For the SOS acronym application example, the number of genotypic registers as well as the number of phenotypic registers is equal to 2 in the configuration DSCA cell schematic (Figure 6.19). With one column of SM (Figure 6.2a), the corresponding structural configuration string is made up of 24 flag data and 24

XI1	XI0	XO1	XO0
0	1	1	0
1	0	1	1
1	1	0	1

Figure 6.40   Coordinate incrementer INC truth table.

P	XI	D
0,1,2,3	–	0
4	–	1
5	1,3	0
5	2	1
6	1,3	1
6	2	0

Figure 6.41   Display buffer BUF truth table.

P	NO	EO
0	–	–
1	1	0
2	1	1
3,4,5,6	0	–

Figure 6.42   Coordinate multiplexers controller XMUX truth table.

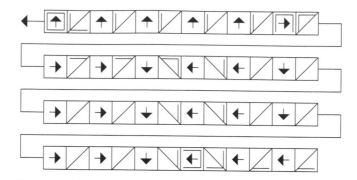

Figure 6.43 Structural configuration string of the SOS cell.

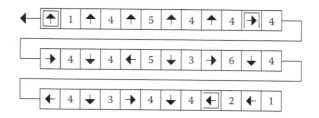

Figure 6.44 Functional configuration string of the SOS cell.

structural data (Figure 6.43). The number of flag data and functional data of the functional configuration string is reduced to 18 (Figure 6.44).

6.4.2 Multicellular Organism

Using the VHDL description language, we realized the hardware specification of the DSCA cell described in Section 6.3 including the application-specific schematic of the SOS molecule. The corresponding hardware simulation (Figure 6.45) shows how the processes of the self-organizing mechanisms grow, load, repair, reset, and kill the middle cell of the SOS organism. In this hardware simulation, the cloning mechanism is implemented at the cellular level in order to build the multicellular organism by self-replication of its totipotent cell.

6.4.3 Population of Organisms

Figure 6.46a illustrates the cloning of the multicellular organism resulting in a population of SOS acronyms. In this hardware simulation, the cloning mechanism is implemented at the organismic level. Two conditions are fulfilled to make such a self-replication possible: (1) there exist nine spare cells in the array allowing three

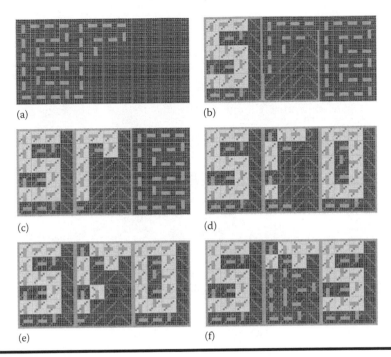

Figure 6.45 Processes performed on the middle cell. (a) Structural growth. (b) Load. (c) Functional growth. (d) Repair and reset. (e) Functional regrowth. (f) Kill.

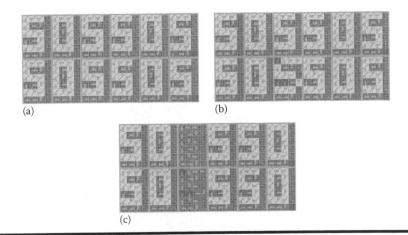

Figure 6.46 Mechanisms performed on the lower left organism. (a) Cloning. (b) Cicatrization. (c) Regeneration.

additional copies of the SOS organism; and (2) the calculation of the coordinates produces cycles respectively equal to 3 and 1 on the *X* and *Y* axes.

The graphical distortion of the S character results from the cicatrization and functional reconfiguration mechanisms applied on the lower left organism (Figure 6.46b). The scar produced by the regeneration mechanism performed on the lower left organism affects the entire column of all cells to which the faulty cell belongs (Figure 6.46c).

6.5 Hardware Implementation

6.5.1 CONFETTI Platform

CONFETTI, which stands for CONFigurable ElecTronic Tissue,[3,4] is a reconfigurable hardware platform for the implementation of complex bio-inspired architectures. The platform is built hierarchically by connecting elements of increasing complexity. The main hardware unit is the *EStack* (Figure 6.47), a stack of four layers of PCBs:

- The *ECell* boards (18 per *EStack*) represent the computational part of the system and are composed of an FPGA and some static memory. Each *ECell* is directly connected to a corresponding routing FPGA in the subjacent *ERouting* board.
- The *ERouting* board (1 per *EStack*) implements the communication layer of the system. Articulated around 18 FPGAs, the board materializes a

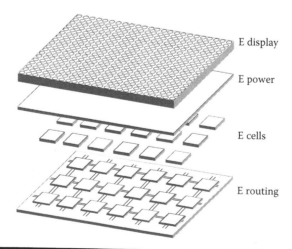

Figure 6.47 Schematic of the *EStack*.

routing network based on a regular grid topology, which provides inter-FPGA communication but also communication to other routing boards.

■ Above the routing layer lies a board called *EPower* that generates the different required power supplies.

■ The topmost layer of the *EStack*, the *EDisplay* board, consists of an RGB LED display on which a touch sensitive matrix has been added.

The complete CONFETTI platform is made up of an arbitrary number of *EStack*s seamlessly joined together (through border connectors in the *ERouting* board) in a two-dimensional array. The connection of several boards together potentially allows the creation of arbitrarily large surfaces of programmable logic. The platform used to implement the SOS acronym application example consists of six *EStack*s in a 3 × 2 array (Figure 6.48).

6.5.2 SOS Application

The 3 × 2 *EStack* platform totals a number of 108 *ECell*s organized as 6 rows of 18 units. Using one *ECell* for each molecule, the complexity of the platform thus allows the implementation of four and a half totipotent cells of the SOS acronym application.

Figure 6.49 shows the cloning of the totipotent cell in order to build a first multicellular organism SOS and sketches the cloning of this organism in order to define a population of them.

Figure 6.50 illustrates cicatrization or reparation at the cellular level as well as regeneration or reparation at the organismic level. The cicatrization of the cells having at most one faulty molecule in each of their rows causes the graphical distortion of

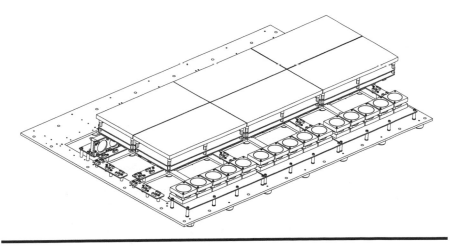

Figure 6.48   Schematic of the 3 × 2 *EStack* platform.

Figure 6.49   Cloning of the SOS acronym, totally realized at the cellular level and partially achieved at the organismic level, on a CONFETTI substrate.

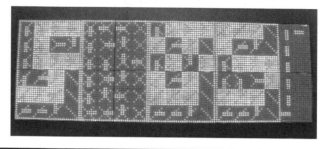

Figure 6.50   Cicatrization and regeneration of the SOS acronym on a CONFETTI substrate.

the characters S and O. The regeneration of the cell having more than one faulty molecule in one of its rows leaves a scar in the organism SOS.

References

1. D. Mange, A. Stauffer, E. Petraglio, and G. Tempesti, Self-replicating loop with universal construction, *Physica D*. **191**(1–2), 178–192 (2004).
2. A. Stauffer and M. Sipper, The data-and-signals cellular automaton and its application to growing structures, *Artificial Life*. **10**(4), 463–477 (2004).
3. F. Vannel, P.-A. Mudry, D. Mange, and G. Tempesti. An embryonic array with improved efficiency and fault tolerance. In *Proceedings of the Workshop on Evolvable and Adaptative Hardware (WEAH07)*. IEEE Computational Intelligence Society, Los Alamitos, CA (2007).
4. P.-A. Mudry, F. Vannel, G. Tempesti, and D. Mange. Confetti: A reconfigurable hardware platform for prototyping cellular architectures. In *Proceedings of the 14th Reconfigurable Architectures Workshop (RAW 2007)*. IEEE Computer Society, Los Alamitos, CA (2007).

Chapter 7

Bio-Inspired Process Control

Konrad Wojdan, Konrad Swirski, Michal Warchol,
Grzegorz Jarmoszewicz, and Tomasz Chomiak

Contents

In this chapter, a layered structure of a control system is presented. Bio-inspired methods implemented in each control layer are introduced. The utilization of an evolutionary algorithm for a proportional–integral–derivative (PID) controller tuning task is briefly discussed. Moreover, an immune-inspired single input–single output (SISO) controller is presented. The model predictive control (MPC) algorithm with an artificial neural network (ANN) is an example of a bio-inspired method applied in an advanced control layer. The design of IVY controller, which is a fuzzy MPC controller with an ANN model, is discussed in this chapter. The stochastic immune layer optimizer (SILO) system is another bio-inspired solution developed by the authors. This system is inspired by the operation of the immune system. It is used for online industrial process optimization and control. Results of IVY and SILO operation in real power plants are presented in the last section of this chapter. They confirm that bio-inspired solutions can improve control quality in real, large-scale plants.

7.1　Nature of Industrial Process Control

The problem of advanced control and steady-state optimization of industrial process has been approached in numerous research works and implementations (Morari and Lee 1997; Qin and Badgwell 2003; Qiping et al. 2003; Plamowski 2006; Plamowski and Tatjewski 2006; Tatjewski 2007). A been interest of scientists and research companies in this area is driven by significant benefits for industrial plants and the decreasing negative impact of these plants on the natural environment, resulting from successful implementation of advanced control methods.

Industrial processes, which take place in large-scale plants, with a high level of complexity, are characterized by a large number of control inputs and disturbances, long time of process response to control change, essentially nonlinear characteristics and impossible-to-omit cross transforms. Classic control systems based on PID controllers are not able to perform optimal control of such processes. The main goal of a classic control system is providing reliable plant operation and safety of the process. Optimization of industrial process should consider

- Multiplicity of (often conflicting) optimization goals
- Large number of control inputs
- Constraints of optimization task
- Measurable and nonmeasurable disturbances affecting the process
- Internal cross dependences (usually nonlinear) between inputs and outputs of the process

Thus it is necessary to introduce additional, superior control layers.

A layered structure of a control system is presented in Figure 7.1. The economical optimization layer is responsible for the calculation of the optimal operating point. Decision variables that minimize the economical quality indicator are set points for controllers in advanced and direct control layers. Control signals calculated in the superior layer are set points for PID controllers in the base control layer. Control signals calculated by controllers in the base control layer are passed directly to the controlled plant.

Process input and output vectors are shown in Figure 7.2. The following symbols are introduced:

- Vector x represents control inputs of the process.
- Vector z represents noncontrolled, measured inputs (measured disturbances).

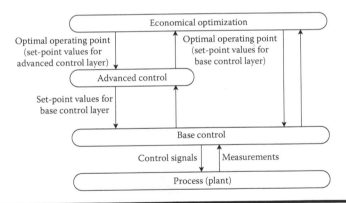

Figure 7.1 Multilayer structure of control system.

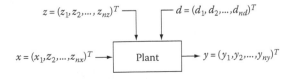

Figure 7.2 Signals description.

- Vector d represents noncontrolled inputs, which are nonmeasured, rarely measured, or omitted in the optimization task (nonmeasured disturbances).
- Vector y represents process outputs.

In a layered control system, the direct control layer is responsible for the safety of fast, dynamic processes. In real, large-scale control systems the direct control layer is implemented in the distributed control system (DCS). This system processes plant data. It is responsible for process control, visualization, and data acquisition. These control systems usually have their own integrated supervisory control and data acquisition (SCADA) systems. Characteristic features of DCS systems are

- Efficient support of large-scale plants
- Compatibility with different communication standards (measuring and control devices)
- Events and data acquisition
- Engineer and operator graphic interfaces
- Online control system modifications
- Redundancy of some control system elements, e.g., operator's station, controllers, input–output (IO) modules

For more information about DCS systems, refer to (Kim et al. 2000). PID controllers (Seaborg et al. 1989; Astrom and Hagglund 1995) dominate in the direct control layer due to their high reliability, speed, and simplicity. In the direct control layer some modifications of PID controllers are also used such as

- PID controller with correction based on disturbance signal value—feed-forward control (Ogunnaike and Ray 1994; Jones et al. 1996),
- PID controller with variable gain value—gain scheduling approach (Zhao and Tomizuka 1993).

7.2 Bio-Inspired Algorithms in Base Control Layer

There are many research works that use bio-inspired algorithms in the direct control layer. Most of them are focused on optimization of PID parameters (proportional, differential, and integral gains) using the evolutionary algorithm (Kau et al. 2002; Koza et al. 2003; Streeter et al. 2003; Chang and Yan 2004; Hassanein and Aly 2004; Ramos et al. 2005). In considered papers, bio-inspired algorithms applied to PID tuning task cause improvement of control quality in comparison with the traditional Zigler–Nichols method (Astrom and Hagglund 2004). The PID tuning procedure based on four parameters K_p, K_d, K_i, and b is presented in Astrom and Hagglund (1995). Indicator b is a weight of the set-point signal and K_p, K_i, and

K_d are gains of proportional, integral, and derivative modules respectively. The method of parameters calculation, which depends on the value of process gain K_u (ultimate plant gain) and the time constant of the process T_i (ultimate plant period), is presented in Astrom and Hagglund (1995). Improvement of control quality, achieved by the calculation of the current correction for each of the parameters K_p, K_d, K_i, and b, is performed by the genetic programming algorithm presented in Streeter et al. (2003). This correction is usually a nonlinear function of K_u and T_u parameters.

In Kim et al. (2004) the optimal tuning of PID parameters has been done based on an algorithm inspired by the operation of the immune system. Another interesting paper about the immune-inspired algorithm applied in the control loop is by Chen et al. (2006). In this chapter a mechanism of the immune system that controls the activity of B cells has been presented. Based on antigen concentration, the immune system is stimulated to create antibodies (by T_h cells) or the production of antibodies is decreased (by T_s cells). The author has introduced a nonlinear, self-adaptive single-loop controller using the following analogy: concentration of antigens represents an error (controller input) and concentration of antibodies represents a control signal.

7.3 Advanced Control Layer

The main goal of the advanced control layer (Tatjewski 2007) is the control of the process's output signals characterized by the slow rate of changes. These signals are often related with final product quality. The sampling period in the superior control layer is significantly longer than the sampling period in the direct control layer. The frequency of control changes is lower than in the direct control layer. The utilization of advanced control methods can significantly improve control quality in typical process states when the process model is accurate. Reliability and safety of a control system are provided mostly by the direct control layer. This layer is able to react to fast disturbance changes.

PID controllers from the direct control layer are not able to compute optimal control signals for industrial processes, which have nonlinear characteristics, long time of output response for control change, a large number of control inputs and disturbances, and impossible-to-omit cross transforms. Furthermore, characteristics of industrial processes are nonstationary as a result of wearing of devices, changes of chemical properties of components used in a process, or external condition changes. In the case of industrial processes that are characterized by the features mentioned above, algorithms implemented in the superior control layer and economical optimization layer are able to improve control quality in typical process states.

MPC controllers belong to a group of *advanced control* algorithms. Methods of predictive control have had a huge impact on the evolution of industrial control systems and the direction of research works in this field. MPC algorithms have achieved great success in industrial applications because of the following reasons (Tatjewski 2007):

- Consideration of constraints of control and output signals
- Consideration of plant internal dependencies between process inputs and outputs
- Simple-to-explain principle of predictive control to plant staff that is directly responsible for process control

The history of MPC algorithms goes back to the 1970s (Cutler and Ramaker 1980; Prett and Gillette 1980). Since then, algorithms of predictive control have been continuously developed and, despite the lapse of 3 decades, they are widely used in industrial applications (Qin and Badgwell 2003). Predictive control methods with receding horizon use a dynamic plant model (linear or nonlinear) to compute the control signal trajectory that minimizes a quality indicator for some time horizons. Based on

- Dynamic process model (including model of disturbances)
- Current and previous output values
- Previous values of control signals
- Future trajectory (known or assumed) of process set-points
- Assumed control goals defined in the form of quality indicator

the control vector in consecutive moments $x(k)$, $x(k+1)$, ..., $x(k+N_s-1)$ is computed in each MPC algorithm step. Parameter k represents a current time and N_s is a control horizon. In each moment k, only the first control vector $x(k)$ from the control trajectory is used. In the subsequent algorithm iteration, the whole procedure is repeated. Prediction and control time horizons are moved one step forward.

The control signal trajectory, computed by the MPC algorithm, minimizes the difference between predicted output trajectory and known or assumed set-point trajectory. Moreover, the shape of the control trajectory depends on various constraints imposed on the control signals. A typical formula of a quality indicator that is minimized by the MPC controller is shown below:

$$J_k = \sum_{s=N_d}^{N_p} \left\| y^{sp}_{k+s|k} - \hat{y}_{k+s|k} \right\|_Q^2 + \sum_{s=0}^{N_s-1} \left\| \Delta x_{k+s|k} \right\|_R^2 \tag{7.1}$$

where

N_s is the control horizon

N_p is the prediction horizon

N_d is the the shortest process delay increased by 1

$y_{k+p|k}^{sp}$ is the output set-point vector (for $k+p$ moment, calculated in k moment)

$\hat{y}_{k+p|k}$ is the prediction of process output

$\Delta x_{k+p|k}$ is the prediction of control signal changes

Q is the square, usually diagonal matrix of output signal weights

R is the square, usually diagonal matrix of control signal weights

Matrix Q represents penalty weights that are used to evaluate penalty related with deviation between predicted output trajectory and output set-point trajectory. Matrix R allows to define penalty weights that can be used to evaluate penalty related with variation of a control signal's trajectory.

The form of the process model is the main difference between linear MPC algorithms. In a DMC controller, the process model is expressed by discrete step responses. This model can be created directly based on process tests. Each test (an identification experiment) represents a dynamic process response for a single control change. The DMC algorithm is one of the first predictive control algorithms with a receding horizon. The first publication about the implementation of this algorithm dates back to 1980 (Cutler and Ramaker 1980; Prett and Gillette 1980). The DMC algorithm is very popular because of its modeling simplicity. Moreover, as one of the oldest predictive controllers, it is well known and tested. For more information about DMC, refer to Tatjewski (2007).

The generalized predictive control (GPC) algorithm (Clarke et al. 1987) is another MPC solution. It uses a process model in the form of discrete difference equations. This kind of model is more compact in comparison to the DMC algorithm. Furthermore, it allows us to consider a wider class of disturbance models (Tatjewski 2007). For more information about GPC algorithms, refer to Camacho and Bordons (1999).

There are also linear MPC algorithms that use a process model in the form of state–space equations. An example of this solution is the SMOC algorithm (Marquis and Broustail 1988). The state–space equation model allows us to simplify the theoretic analysis of MPC algorithms. However, it is not a natural approach to create a state–space equation model in industrial applications. This is the main reason why DMC and GPC algorithms are more popular.

The above mentioned MPC algorithms with a linear process model are popular solutions in the advanced control layer. However, one should remember that real plants may require nonlinear models. Utilization of the linear model in the control of essentially nonlinear plants may result in poor control quality.

Utilization of a nonlinear model is a natural approach to consider nonlinear process characteristics. In MPC-NO (MPC with nonlinear model) prediction of the output trajectory is based on the nonlinear model. The main problem related to MPC algorithms with a full nonlinear model is the selection of a suitable quality indicator minimization method. The optimization task in this case is non-square and generally non-convex.

In real applications the MPC-NSL method (MPC nonlinear with successive linearization) is a more popular approach. In the MPC-NSL algorithm a nonlinear characteristic is linearized in the neighborhood of a current process operating point (Henson 1998; Babuska et al. 1999; Megias et al. 1999; Marusak and Tatjewski 2000; Tatjewski 2007). The control trajectory is computed based on a linearized process model. One should note that the solution calculated by this method is suboptimal. However, reliability of control calculation is fulfilled, which is a very important requirement in industrial applications. The MPC-NPL method (MPC with nonlinear prediction and linearization) is an extension of the above consideration. This algorithm uses a linearized model only to solve the optimization task. Free output trajectory is estimated on the basis of a nonlinear model. For more information refer to (Gattu and Zafiriou 1992; Henson 1998; Marusak and Tatjewski 2000; Lawrynczuk and Tatjewski 2002).

In industrial applications fuzzy logic is commonly used to design nonlinear MPC controllers (Zadeh 1965; Yen and Langari 1999). An especially interesting approach is the Takagi–Sugeno (TS) fuzzy model, where conclusions are defined as linear models. A crucial element in TS fuzzy models is a set of rules (knowledge base), where each rule has a conclusion in the form of a linear function. An example of such a rule is shown below:

$$R^i: \quad \begin{aligned} &\text{IF } x_1 \text{ is } F_1^i \text{ and } x_2 \text{ is } F_2^i \text{ and } \ldots \text{ and } x_n \text{ is } F_n^i \\ &\text{THEN } y_i = f(x_1, x_2, \ldots, x_n) \end{aligned}$$

Based on the TS model, it is possible to create a fuzzy MPC controller. In this case, separate discrete differential equations, which represent a predictive control algorithm based on its own process model, are defined for each fuzzy partition. The definition of fuzzy partitions depends on the problem under consideration. Usually nonlinear signals are used to define fuzzy partitions.

There is another approach to designing fuzzy predictive controllers. In this case general equations representing the control rule are common. This approach assumes that the set of rules consists of rules in which the conclusions comprise the predictive controller's parameters and models.

7.3.1 Artificial Neural Networks in MPC Controllers (Tatjewski 2007)

Utilization of ANN (Haykin 1999) in nonlinear controllers is presented by Tatjewski and Lawrynczuk (2006). In the case of nonlinear process models efficiency and robustness of the nonlinear optimizing procedure depends mainly on the assumed model form. Utilization of ANN as a full nonlinear process model in MPC-NO and for linearization in MPC-NSL and MPC-NPL is presented by Tatjewski and Lawrynczuk (2006).

Let as assume that the output of the SISO process is defined as a nonlinear discrete-time equation

$$y(k) = g\left(x(k-d), \dots, x(k-n_x), y(k-1), \dots, y(k-n_y)\right) \qquad (7.2)$$

where g is a continuously differentiable function. Let us assume that the g function is defined as a feed-forward ANN with one hidden layer and one linear output layer.

In the MPC-NO approach an ANN is used to calculate the gradient of a quality indicator. Based on this gradient a nonlinear optimization task is solved in the online mode. The quality indicator gradient is computed based on the gradient of output trajectory with respect to the control trajectory. The output trajectory is estimated based on the ANN.

Moreover, the gradient of the output trajectory with respect to the control trajectory is used to calculate the gradient of output constraints. The optimal solution is calculated based on nonlinear optimization methods (e.g., Sequential Quadratic Programming (SQP), Bazaraa et al. 1993).

MPC-NPL is an efficient combination of nonlinear prediction of an unconstrained output trajectory and a linearized model used to optimize the future control trajectory in an online mode. The future unconstrained output trajectory is estimated based on ANN outputs, assuming that control signals will be constant in the control time horizon.

7.3.2 Hybrid Process Model

In IVY controller (a predictive controller developed by Transition Technologies) a hybrid process model is used. This model is a combination of linear model and nonlinear correction computed by an ANN. Numerous successful implementations of IVY have been done so far (Arabas et al. 1998; Domanski et al. 2000). The procedure for creating a fuzzy MPC controller with a hybrid process model that uses an ANN is presented in this section.

7.3.2.1 Step One

A MISO ARX model is created for each typical process operating point (partition)

$$\hat{y}_k^{(p)} = \sum_{i=1}^{N} a_i^{(p)} y_{k-i} + \sum_{i=1}^{K} \sum_{j=0}^{M_i} b_{(i,j)}^{(p)} u_{(i,k-j)} + c^{(p)}$$

where

 y is the output
 k is the time step
 p is the partition
 N is the time range of autoregression
 a,b is the autoregression coefficients
 u is the process inputs (control inputs and disturbances)
 K is the number of inputs
 M is the time range of inputs
 c is the constant

Fuzzy partitions are created for typical process states and their number depends on the process nonlinearity level.

7.3.2.2 Step Two

Models from individual partitions are combined by fuzzy rules

$$\hat{y}_k^{(narmax)} = \sum_{p=1}^{P} w^{(p)} \left[\sum_{i=1}^{N} a_i^{(p)} y_{k-i} + \sum_{i=1}^{K} \sum_{j=0}^{M_i} b_{(i,j)}^{(p)} u_{(i,k-j)} + c^{(p)} \right]$$

$$= \sum_{i=1}^{N} \left(\sum_{p=1}^{P} w^{(p)} a_i^{(p)} \right) y_{k-i} + \sum_{i=1}^{K} \sum_{j=0}^{M_i} \left(\sum_{p=1}^{P} w^{(p)} b_{(i,j)}^{(p)} \right) u_{(i,k-j)} + \sum_{p=1}^{P} w^{(p)} c^{(p)}$$

$$(7.3)$$

where

 w is the weights of individual partitions (a function of fuzzy inputs),
 P is the number of partitions.

As a result of this operation, a nonlinear model (NARMAX) is created. This model covers whole plant operation space. A fuzzy model ensures soft switching between operating points. The definition of fuzzy sets depends on the experience and knowledge of experts implementing the controller. In the most advanced cases it is possible to implement linear models in a fuzzy TS ANN scheme (Jang 1993; Jang and Sun 1995).

Equation 7.3 can be split into a cascade of three items:

- Delays—this part determines j in the second index of $u_{(i,k-j)}$ in Equation 7.3

$$v'_{(i,j,k)} = u_{(i,k-j)}$$

- Sum—this part represents $\sum_{i=1}^{K} \sum_{j=0}^{M_i}$ in Equation 7.3

$$v_k'' = \sum_{i=1}^{K} \sum_{j=0}^{M_i} \left(\sum_{p=1}^{P} w^{(p)} b_{(i,j)}^{(p)} \right) v_{(i,j,k)}' + \sum_{p=1}^{P} w^{(p)} c^{(p)}$$

- Autoregression—this part represents autoregression in Equation 7.3

$$\hat{y}_k^{(\text{narmax})} = v_k'' + \sum_{i=1}^{N} \left(\sum_{p=1}^{P} w^{(p)} a_i^{(p)} \right) y_{k-i}$$

Because all items are linear they can be ordered in any way. Let us invert the order of the sum and autoregression in the cascade. Because the sum contains multiple inputs, the autoregression must be applied to all its inputs:

$$\hat{v}_{(r,j,k)} = v_{(r,j,k)}' + \sum_{i=1}^{N} \left(\sum_{p=1}^{P} w^{(p)} a_i^{(p)} \right) \hat{v}_{(r,j,k-i)}, \quad r = 1, \ldots, K$$

$$\hat{y}_k^{(\text{narmax})} = \sum_{i=1}^{K} \sum_{j=0}^{M_i} \left(\sum_{p=1}^{P} w^{(p)} b_{(i,j)}^{(p)} \right) \hat{v}_{(i,j,k)} + \sum_{p=1}^{P} w^{(p)} c^{(p)}$$

The order reversion above is incorrect because of the constant $\sum_{p=1}^{P} w^{(p)} a_i^{(p)}$. This operation would be correct if the static gain of autoregression was equal to 1, or $\sum_{p=1}^{P} w^{(p)} a_i^{(p)} = 0$. Let us introduce an additional coefficient S into the sum and autoregression to make the static gain of autoregression equal to 1.

$$\hat{v}_{(r,j,k)} = Sv_{(r,j,k)}' + \sum_{i=1}^{N} \left(\sum_{p=1}^{P} w^{(p)} a_i^{(p)} \right) \hat{v}_{(r,j,k-i)}, \quad r = 1, \ldots, K$$

$$\hat{y}_k^{(\text{narmax})} = \frac{1}{S} \sum_{i=1}^{K} \sum_{j=0}^{M_i} \left(\sum_{p=1}^{P} w^{(p)} b_{(i,j)}^{(p)} \right) \hat{v}_{(i,j,k)} + \frac{1}{S} \sum_{p=1}^{P} w^{(p)} c^{(p)}$$

where

$$S = 1 - \sum_{i=1}^{N} \sum_{p=1}^{P} w^{(p)} a_i^{(p)}$$

7.3.2.3 Step Three

For a better fit to the measured data a nonlinear correction is added to the sum

$$\hat{y}_k^{(\text{narmax})} = \frac{1}{S} \sum_{i=1}^{K} \sum_{j=0}^{M_i} \left(\sum_{p=1}^{P} w^{(p)} b_{(i,j)}^{(p)} \right) \hat{v}_{(i,j,k)} + \frac{1}{S} \sum_{p=1}^{P} w^{(p)} c^{(p)} + ANN\left(\hat{V}_k \right)$$

where $\hat{V}_k$ is a subset of inputs of the sum v_k'' in a moment k. To allow the identification of ANN using training data with reasonable size, the set $\hat{V}_k$ contains only few inputs of the sum which are essentially nonlinear. The rest of the inputs of the static model are considered by the linear part only.

To identify an ANN a training set is created. This set consists of values of $\hat{v}_{(i,j,k)}$ and a vector of static deviations between NARMAX model outputs (without nonlinear correction) and historic, measured process outputs:

$$\hat{y}_k^e = \hat{y}_k^{(\text{narmax})} - y_k^{(\text{real})}$$

Then an ANN is created (e.g., multilayer perceptron). Weights of neuron connections are computed based on data from the training set. Thus an ANN is used as a nonlinear approximator of static difference between fuzzy model output and real process output

$$\hat{y}_k^e = ANN\left(\hat{V}_k \right)$$

7.3.2.4 Step Four

Nonlinear correction (neural network output) can be linearized around point before each control decision is taken:

$$ANN\left(\hat{V}_k \right) = \sum_{i=1}^{K} \sum_{j=0}^{M_i} \frac{d\hat{y}_k^e}{d\hat{v}_{(i,j,k)}} \left(\hat{V}_k^* \right) \hat{v}_{(i,j,k)} + D = \sum_{i=1}^{K} \sum_{j=0}^{M_i} F_{(i,j)} \hat{v}_{(i,j,k)} + D$$

where

$$D = ANN\left(\hat{V}_k^* \right) - \sum_{i=1}^{K} \sum_{j=0}^{M_i} \frac{d\hat{y}_k^e}{d\hat{v}_{(i,j,k)}} \left(\hat{V}_k^* \right)$$

$$F_{(i,j)} = \frac{d\hat{y}_k^e}{d\hat{v}_{(i,j,k)}} \left(\hat{V}_k^* \right)$$

Because of this linearization, the final mathematical description of the whole model can be defined as a linear equation:

$$\hat{y}_k = \sum_{i=1}^{N} \left(\sum_{p=1}^{P} w^{(p)} a_i^{(p)} \right) \hat{y}_{k-i} + \sum_{i=1}^{K} \sum_{j=0}^{M_i} \left(SF_{(i,j)} + \sum_{p=1}^{P} w^{(p)} b_{(i,j)}^{(p)} \right) u_{(i,k-j)}$$

$$+ SD + \sum_{p=1}^{P} w^{(p)} c^{(p)} \tag{7.4}$$

This form is used in the controller.

The described procedure (combination of linear model with nonlinear neural network correction and linearization in the neighborhood of process operating point) is an example of utilization of a predictive controller with a neural network in real applications. A hybrid model is characterized by good numeric resistance and accurate (in the context of control requirements) approximation of nonlinear process characteristic.

7.4 SILO—Immune-Inspired Control System

There are many successful implementations of predictive control algorithms in industrial processes control (Qin and Badgwell 1997, 2003; Desborough and Miller 2004). In spite of undisputable advantages of MPC controllers there are several serious disadvantages of this method. First of all, the implementation of an MPC controller is quite expensive. Long-lasting and labor-consuming parametric tests have to be performed to create a dynamic, mathematical model. In each test a process output response is recorded. This response is a process reaction for a single control step change. In such an experiment only one control vector element is changed. Other control signals and disturbances have to be constant. To ensure good model quality, tests should be repeated several times for each control input. Model identification is based on data recorded during parametric tests. A separate model has to be created for each fuzzy partition, when nonlinearity of process characteristics is taken into consideration and fuzzy modeling is used (TS approach). This means that more parametric tests have to be executed.

The number of fuzzy partitions is limited by the cost of creating a linear model in each partition. One should note that high cost of implementation results from high labor cost of highly qualified engineers who conduct parametric tests and identification of the dynamic process model. Moreover, long-lasting tests have to be performed under specific operating conditions, defined by the company that implements the controller. Sometimes these conditions conflict with the production schedule of industrial plant. Moreover, process is usually controlled ineffectively during tests. It causes extra financial losses. There are some plants where it is unrealistic to test all possible operating configurations. It can be caused by a large number of control inputs, failure of devices, or constraints imposed by plant staff. Economical circumstances and the necessity to ensure process safety are typical driving factors for such constraints. Honeywell's experts have calculated that the cost of creating a dynamic model varies from USD250 to USD1000 for each single dependence between the single input and single output of the controlled process (Desborough and Miller 2004). In case of a process with ten control inputs, five disturbances, and four outputs (an average dimension of the optimization task of a combustion process in power boiler), the development cost of one dynamic model, according to Honeywell experts, varies from USD15,000 to USD60,000. This amount includes test plan, process parametric tests, identification, and model validation. However, it does not include computer software and hardware price and cost of staff training. Nor does it include financial losses of the industrial plant resulting from ineffective plant operation during parametric tests.

Another disadvantage of predictive control methods is an insufficient adaptation to process characteristic changes. These changes are considered in a long-term horizon, like months or years, resulting from wearing or failure of devices, plant modernizations, changes of chemical properties of components used in a process, or external condition changes (e.g., seasonality). The easiest way to consider these changes is a manual, periodic update of the model, based on most recent tests of the process. However, this is not a satisfying solution. The implementation of an adaptation method in an MPC controller is a more desired approach. This solution, however, entails some difficulties:

- Estimation of model parameters when there is an insufficient changeability of noised signals
- Estimation of model parameters in a closed loop operation (Jung and Soderstrom 1983; Soderstom 1989)
- High computer resources usage in online operation
- Possibility of linear dependencies in an observation matrix, which consists of process measurements

Due to these problems adaptation methods implemented in MPC controllers are often insufficient.

In searching for a solution devoid of the above-mentioned drawbacks of MPC controllers, which allow the optimization of industrial processes in an online mode, one should take note of a new, dynamically developing discipline—artificial immune systems (Castro and Zuben 1999, 2000; Wierzchon 2001a). This is a new trend of solutions, inspired by the operation of immune system of live creatures. The immune system is a special system that has a memory and develops its skills by constant learning. Depending on external conditions, the system is able to develop new skills or lose old ones. From an engineering point of view, the immune system is an effective distributed data processing system. It has the ability to learn and it can adapt to variable external conditions. One should note that these properties are very useful in online control and optimization of industrial processes.

In this section a SILO (Wojdan and Swirski 2007; Wojdan et al. 2007) is presented which is an innovative solution for industrial process control and optimization. The SILO operation is based on the immune system analogy. This solution uses knowledge gathered from a static process model. Thus the model form is similar to methods used in an economical optimization layer (Tatjewski 2007) and steady-state target optimization (SSTO) algorithm (Kassmann et al. 2000; Blevins et al. 2003; Tatjewski 2007). In opposition to these methods, SILO uses a quality indicator that penalizes deviations from set-points (control form of quality indicator). The presented solution, based on knowledge gathered from immune memory, minimizes the difference between set-points and measured (or estimated) process outputs. A computed control vector (solution of optimization task) is a vector of set points for controllers in the base control layer.

SILO can be used to control processes characterized by fast but rare disturbance changes or processes where disturbances change continuously but the rate of changes is significantly slower than the dynamics of the process. In industrial plants which fulfill the above condition, SILO can be a low-cost (in an economical meaning) and effective alternative for MPC controllers because of the following differences:

- There is no need to perform parametric tests on the process. The cost for the work of a highly qualified engineer is very expensive. Moreover, process parametric tests cause extra financial losses resulting from ineffective plant operation.
- There is no need to create dynamic, mathematical model of the process based on identification experiments data. It results in a significant reduction of work time of highly qualified engineers.
- A large number of process states do not have to be considered. SILO automatically gathers knowledge about the process and adapts to current process characteristics. SILO can be forced to higher precision of nonlinear process characteristics approximation by the definition of narrower ranges of the process's signals, in which linear approximation is sufficient. In the case of MPC controllers, the number of such areas is limited by amount of labor required by parametric tests.
- The immune-inspired efficient adaptation algorithm is able to acquire knowledge about the process. In implementations of SILO that have been done so far, there was no need to perform additional model identification.

Because of the form of the quality indicator, SILO corresponds to a superior control layer in the layered control system. However, the architecture of SILO allows to define the economical quality indicator

$$J(c,y) = \sum_{k=1}^{nc} p_j^c c_j - \sum_{k=1}^{ny} p_j^y y_j$$

where

c_j is the jth element of decision variable vector in the optimization layer

p^y is the vector of output product prices

p^c is the vector of prices of input streams

The economical target function defines economical production target (usually profit). It means that SILO can be used as a static optimizer of the operating point of an industrial process. SILO usually is implemented in an advanced control layer; hence SILO with a control form of quality indicator is presented in this chapter.

7.4.1 Immune Structure of SILO

The industrial process control task is presented in this chapter in the context of immune system structure and its mechanisms. The SILO structure is compared with immune system of live creatures. Table 7.1 presents an analogy between the immune system and SILO.

7.4.1.1 Pathogen

A pathogen is recognized by the antibody thanks to epitopes, which are located on the pathogen's surface (Figure 7.3). Thus epitopes *present* the pathogen to the immune system. In SILO, a pathogen represents measured and unmeasured disturbances affecting the industrial process. A pathogen is presented to the system as a current

Table 7.1 Analogy between Immune System and SILO

Immune System	SILO
Pathogen	Measured and nonmeasured disturbances
Antibody: antigen-binding site	Process state, stored in B cell, which creates an antibody
Antibody: efector part	Optimal control vector change
B cell	Element which stores historical process state before and after control change
T_h cell	Algorithm which is responsible for selection of proper group of B cells during model creation process
Health indicator	Inverse of quality indicator value

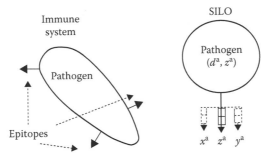

Figure 7.3   Pathogen.

process state $A = [x^a, y^a, z^a]$. In case of measured disturbances, vector z directly defines a pathogen. In case of unmeasured or rarely measured disturbances, the situation is more complicated. Values of unmeasured disturbances can be concluded from relationships between vector y and vectors x and z. Thus knowledge about the current process state is necessary to detect the presence of unmeasured disturbances.

The SILO pathogen structure is similar to the idea presented by Krishnakumar and Neidhoefer (KrishnaKumar and Neidhoefer 1995, 1997a,b; Castro and Zuben 2000). Krishnakumar and Neidhoefer have also treated disturbances as a pathogen.

7.4.1.2 B Cell

In immune system of live creatures B cells produce certain types of antibodies (Figure 7.4). B cells also take part in immune memory creation process. When an antibody located on the B cell surface binds the antigen, the pathogen is presented on the surface of the lymphocyte as an MHC class II particle. When the T_h cell recognizes the presented antigen, it activates the B cell to intensive proliferation and differentiation (clonal selection mechanism) (Solomon et al. 1993; Lydyard et al. 2000).

In the SILO system the B cell represents the process state before and after a control change. Each pair of *process state* and *process outputs response to control variables change* represents one B cell. One should note that the B cell represents such a situation only when disturbances are constant in a certain range. The formal definition of the kth B cell is as follows:

$$L_k = \left[{}^b\bar{x}^k, {}^P\bar{x}^k, {}^b\bar{y}^k, {}^P\bar{y}^k, \bar{z}^k \right]$$

where
> ${}^b\bar{x}^k$ is the average values of control signals measured before control change
> ${}^P\bar{x}^k$ is the average values of control signals measured after control change
> ${}^b\bar{y}^k$ is the average values of process outputs measured before control change

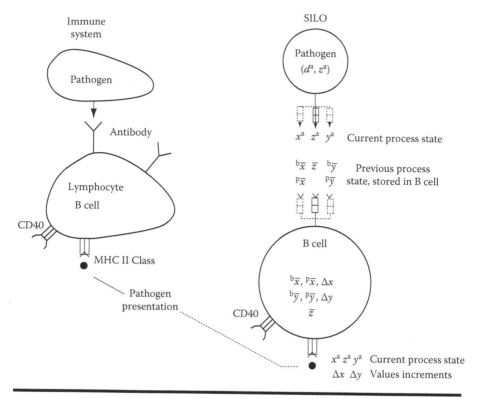

Figure 7.4 B cell.

$^{P}\bar{y}^k$ is the average values of process outputs measured after control change
$\bar{z}^k$ is the average values of measured disturbances

Increments of optimized process outputs caused by change of control inputs can be calculated based on the information stored in B cells. The static dependence between optimized output change and control signals change is used during the creation of a linear process model (refer Section 7.4.3). In later considerations the following increases of control variables and process outputs will be useful:

$$\Delta x^k = {}^{P}\bar{x}^k - {}^{b}\bar{x}^k$$
$$\Delta y^k = {}^{P}\bar{y}^k - {}^{b}\bar{y}^k$$

In a SILO system B cells are stored in immune memory, which is implemented as a database. Each B cell represents a base portion of knowledge about the optimized process. Table 7.2 shows a hypothetical record that represents the B cell for a simplified optimization task of the combustion process.

The record presented in Table 7.2 was created based on measured values of process points (elements of vectors x, y, z) in a certain time frame. The matrix that

Table 7.2   B Cell Example

Process Points	Average Value before Control Change	Average Value after Control Change	Increment
x_1: O_2 level	2.0	2.2	0.2
x_2: Left damper	10.0	10.0	0.0
x_3: Right damper	20.0	30.0	10.0
z_1: Unit load	200.0	200.0	0.0
y_1: NO_x emission	500.0	520.0	20.0
y_2: CO emission	80.0	10.0	−70.0

contains measured values of vectors x, y, z in consecutive moments is called a *time window*.

Definition   7.1   *A   time   window* with   length n is   a   matrix M $=$
$\left[[x, y, z]^t_k, \ldots, [x, y, z]^t_{k+n-1} \right]$.

A time window limited to one element from vector y and two elements of vector x is presented in Figure 7.5.

The transformation of matrix M (time window) into B cell consists of the following calculations:

- Average values of control signals before control change
- Average values of control signals after control change
- Average values of optimized process outputs before control signal change
- Average values of optimized process outputs after control signal change and when process outputs are steady
- Average values of signals representing measured disturbances (length of averaging is equal to length of time window)

B cells are being created based only on these time windows, where disturbances are constant in a whole time window.

The main task of T_h cells is the activation of suitable B cells. In SILO T_h cell function is represented by an algorithm which calculates the affinity of an antibody (that is located on B cell surface) and a pathogen. The static linear model of the process in the neighborhood of the current process operating point is created based on the information from activated B cells.

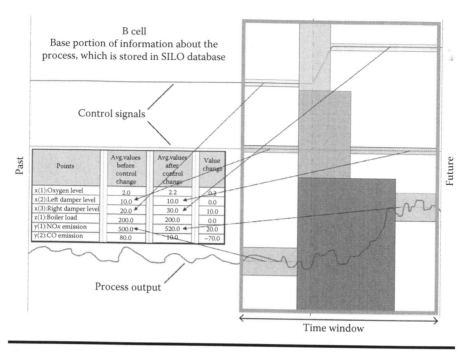

B cell
Base portion of information about the
process, which is stored in SILO database

Control signals

Past

Points	Avg.values before control change	Avg.values after control change	Value change
x(1):Oxygen level	2.0	2.2	0.2
x(2):Left damper level	10.0	10.0	0.0
x(3):Right damper level	20.0	30.0	10.0
z(1):Boiler load	200.0	200.0	0.0
y(1):NOx emission	500.0	520.0	20.0
y(2):CO emission	80.0	10.0	−70.0

Future

Process output

Time window

Figure 7.5 Time window of a B cell.

7.4.1.3 Antibody

Antibodies are produced by B cells. The task of an antibody is to bind the antigen that is located on the pathogen's surface. Each antibody is able to recognize only one sort of epitope. The antibody consists of an antigen-binding part and an effector part (Lydyard et al. 2000) (Figure 7.6).

In the SILO system the antibody effector part is an increment of the control vector calculated by the optimization algorithm, which minimizes the quality indicator (refer to Section 7.4.3). Thus setting a new control vector in the plant should compensate the negative influence of the disturbance (pathogen). This approach is similar to that proposed by Fukuda (Fukuda et al. 1998; Castro and Zuben 2000). Fukuda also treats the best solution vector as an antibody.

There are different algorithms that calculate the binding strength between antigen and antibody. One should note that the measure used for the calculation of the binding strength depends on the assumed B cell's structure (Castro and Timmis 2003). Usually in the case of an optimization task, the binding strength is calculated based on the optimized quality indicator value (Wierzchon 2001b). In a SILO system a different approach is used. It is assumed that the antibody binds the antigen when and only when the current process operating point is similar to the historical process operating point stored in the B cell that has created antibody.

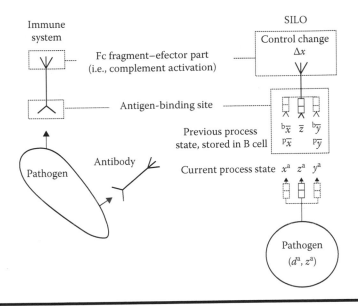

Figure 7.6 Antibody.

The set of analyzed process points can be limited merely to points representing measured disturbances. In case the characteristics of individual process points are strongly nonlinear there is a possibility of comparing current (antigen) and historical (antibody) point values from vectors x and y.

The selection of x, y, and z vector elements, which are compared, defines the process space division. Each subspace is defined by x, y, and z vector element values. There is an assumption that a linear approximation of nonlinear process characteristics is accurate enough in each such subspace. The selection of x, y, and z vector elements, which will be compared, is performed by an expert who is implementing the SILO system.

The affinity between the B cell L_k and the antigen A is defined in the following way:

$$
\mu\left(L_k, A\right) = \left(\prod_{i=1}^{nx} g_i^{x^b}\left({}^{b}\bar{x}_i^{k}, x_i^{a}\right)\right) \times \left(\prod_{i=1}^{nx} g_i^{x^p}\left({}^{p}\bar{x}_i^{k}, x_i^{a}\right)\right)\left(\prod_{i=1}^{ny} g_i^{y^b}\left({}^{b}\bar{y}_i^{k}, y_i^{a}\right)\right)
$$

$$
\times \left(\prod_{i=1}^{ny} g_i^{y^p}\left({}^{p}\bar{y}_i^{k}, y_i^{a}\right)\right) \times \left(\prod_{i=1}^{nz} g_i^{z}\left(\bar{z}_i^{k}, z_i^{a}\right)\right)
$$

where $\forall_{x_1, x_2 \in \Re} g\left(x_1, x_2\right) \in \{0, 1\}$. The antibody binds the antigen only when $\mu\left(L_k, A\right) = 1$. Examples of $g\left(x_1, x_2\right)$ functions are presented below.

Example 7.1:

$$g(x_1, x_2) = \begin{cases} 0 & \text{if} \quad |x_1 - x_2| > \varepsilon \\ 1 & \text{if} \quad |x_1 - x_2| \leq \varepsilon \end{cases}$$

Example 7.2:

$$g(x_1, x_2) = \begin{cases} 1 & \text{if} \quad x_1 < 10 \wedge x_2 < 10 \\ 1 & \text{if} \quad x_1 \geq 10 \wedge x_2 \geq 10 \\ 0 & \text{otherwise} \end{cases}$$

Example 7.3:

$$g(x_1, x_2) = 1$$

Affinity belongs to a set of $\{0, 1\}$. Zero means that the antibody does not bind the antigen, so the B cell that has created an antibody does not represent a process operating point that is similar to the current process operating point. The creation of a linear model in the neighborhood of the current process operating point, using knowledge from this B cell, is groundless. If $\mu(L_k, A) = 1$, it means that the B cell represents a process operating point that is similar to the current process operating point, and the knowledge from this B cell can be used to create a linear mathematical model in the neighborhood of a current process operating point (refer Section 7.4.3).

7.4.2 Basic Concept of SILO Operation

SILO consists of two main modules—a learning module and an optimization module (Figure 7.7). The learning module performs an online analysis of elements of x, y, and z vectors and searches for time windows, which fulfils the criteria of being a B cell. When the module finds a time window that meets those requirements, the recognized B cell is saved in a database, which represents the knowledge about the process. At least one of the control signals must change to transform a time window into a B cell. During a typical operation of a plant there are frequent changes of control inputs. Thus, the process of creating B cells is continuous. SILO updates its knowledge about the process continuously. This is one of the main features of SILO. This system updates its knowledge by recording changes of control signals and static reactions of process outputs. This feature enables SILO to gather knowledge about the process and adapt to variable operating conditions. It is especially important

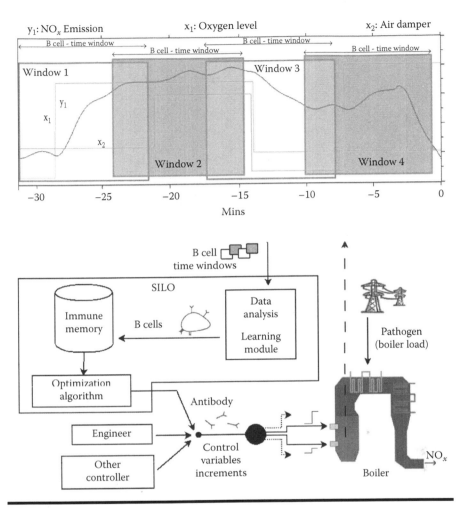

Figure 7.7   Operation of SILO.

in the case of industrial processes, where characteristics change over time as a result of wearing of devices, plant modernization, changes of chemical properties of components used in a process, or an external condition changes.

The learning module is completely independent of the optimization module. It is able to identify and store in a database (in the immune memory) B cells that were created as a result of

- Change of control signals performed by SILO
- Change of control signals performed by control operator
- Change of control signals performed by another controller
- Analysis of file with historical values of process points (batch learning)

The optimization module performs an online optimizing control of the process operating point. It computes the optimal increment of a control vector based on quality indicator coefficients and B cells from the immune memory. The quality indicator's value depends on x and y vectors. The optimization module consists of three layers. Each layer represents a different optimization algorithm. The module automatically switches between layers during normal plant operation. Utilization of three different algorithms results from

- Exploration and exploitation of solution space
- Effective response for a new, unknown pathogen (primary immune response)
- Effective usage of knowledge about the process from the immune memory for deactivation of a pathogen that is similar to a pathogen that has already tried to infect the system (secondary immune response)
- Requirement of good conditioning of model identification task

Optimization period is defined as a time range between consecutive changes of the control vector x. This time is not shorter than the time needed to reach a new steady state after a control change.

Optimization and learning modules can work independently (refer Table 7.3). The knowledge about the process comes from an observation of results of the optimization module operation. It can be said that SILO realizes the idea of dual control defined by Feldbaum (1968). This idea consists of integrated optimization and estimation.

7.4.3 Optimization Module

The optimization module is an independent module of SILO that computes the optimal increment of a control vector. The quality indicator's value depends on x and y vectors.

$$J = \sum_{k=1}^{nx} \left[\alpha_k \left(\left| x_k^a - x_k^s \right| - \tau_k^{lx} \right)_+ + \beta_k \left(\left(\left| x_k^a - x_k^s \right| - \tau_k^{sx} \right)_+ \right)^2 \right]$$
$$+ \sum_{k=1}^{ny} \left[\gamma_k \left(\left| y_k^a - y_k^s \right| - \tau_k^{ly} \right)_+ + \delta_k \left(\left(\left| y_k^a - y_k^s \right| - \tau_k^{sy} \right)_+ \right)^2 \right] \quad (7.5)$$

where
α_k is the linear penalty coefficient for kth control variable
β_k is the square penalty coefficient for kth control variable
γ_k is the linear penalty coefficient for kth optimized output
δ_k is the square penalty coefficient for kth optimized output
τ_k^{lx} is the width of insensitivity zone for linear part of penalty for kth control variable

Table 7.3   Learning Module and Optimization Module Operation

Optimization Module	Learning Module	Description
ON	ON	Normal plant operation
OFF	ON	Learning module gathers information about the process on the basis of control signal changes made by the operator or another control algorithm
ON	OFF	Learning is disabled but optimization can be still performed, e.g., calibration of measuring devices, tests of devices
OFF	OFF	Learning and optimization cannot be performed by SILO, e.g., start-up, failure of some critical devices

τ_k^{sx} is the width of insensitivity zone for square part of penalty for kth control variable

τ_k^{ly} is the width of insensitivity zone for linear part of penalty for kth optimized output

τ_k^{sy} is the width of insensitivity zone for square part of penalty for kth optimized output

$(\cdot)_+$ is the "positive part" operator $(x)_+ = \frac{1}{2}(x + |x|)$

x_k^s is the demand value for kth control variable

y_k^s is the demand value for kth optimized output

One should note that differences between set-point values and measured or estimated values of x and y vector elements are penalized. Operator $(\cdot)_+$ causes that selected elements of vectors x and y are penalized only if the absolute deviation from the set-point value is greater than some boundary value. Optimization goals can be specified more accurately thanks to linear and square modules of the quality indicator (Figure 7.8).

The optimization algorithm operates in one of three optimization layers. In each layer a different optimization algorithm is used to compute an optimal control vector increment minimizing a quality indicator. An algorithm that lets SILO switch between layers is a key algorithm of this solution (refer Figure 7.9). Activation of a particular layer depends on SILO's knowledge and the current process's operating point. It can be said that:

■ *Stochastic optimization layer* corresponds with stochastic exploration of a solution's space. It is executed when SILO has no knowledge about the process or when the knowledge about the process is insufficient. A solution found

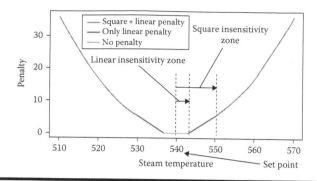

Figure 7.8   Example of penalty function for a steam temperature.

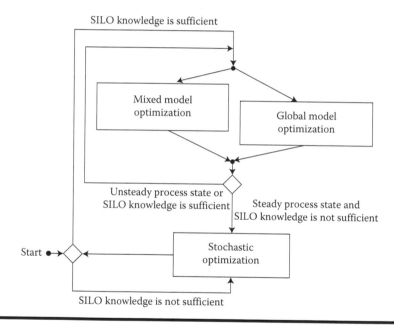

Figure 7.9   Simplified scheme of layered optimization algorithm.

in this layer is a starting point for optimization in layers that use a process model. By analogy to the immune system this layer represents the primary response of the immune system (Lydyard et al. 2000).

■ *Optimization on the global model layer*—this layer uses general knowledge about the process to compute an optimal increment of the control vector. The automatically created mathematical model used in this layer gathers knowledge about basic process dependencies. This layer is executed when SILO does not have sufficient knowledge about the process in the current operating point and thus SILO is not able to create a mathematical model that represents static process dependencies in the neighborhood of the current process operating point. The most recent portions of knowledge (B cells) from different process operating points are used to create a global model.

■ *Optimization on the mixed model layer*—this layer uses information from the immune memory to create a mathematical model that represents static process dependencies in the neighborhood of a current process operating point. The model created in this layer is based on the most recent knowledge stored in the neighborhood of the current process operating point. This model is more accurate than a global model. This layer corresponds with exploitation of solution's space. By analogy to the immune system this layer represents the secondary immune response (Lydyard et al. 2000).

In the following sections individual layers are presented in detail.

7.4.3.1 Mixed Model–Based Optimization Layer

This layer is executed when SILO has sufficient knowledge about the controlled process in the neighborhood of the current process operating point. It means that there is a sufficient number of B cells which represent similar process operating points in the immune memory. In this layer a linear, incremental mathematical model is automatically created in each optimization step. Information stored in g youngest *global* B cells and l youngest *local* B cells is used to construct this model. The set of global B cells represents all B cells stored in the immune memory. The set of local B cells represents B cells which fulfill affinity conditions. For those B cells holds $\mu(L_k, A) = 1$.

The linear, incremental process's model represents a static impact of increment of a control vector Δx on an increment of the optimized process's outputs Δy:

$$\Delta y = \Delta x K \tag{7.6}$$

where $\Delta x = [\Delta x_1, \Delta x_2, \ldots, \Delta x_{nx}]$, $\Delta y = [\Delta y_1, \Delta y_2, \ldots, \Delta y_{ny}]$ and K is a gain matrix.

Coefficients of a K matrix are estimated using information from local observation matrices ΔX_L, ΔY_L and global observation matrices ΔX_G, ΔY_G. These matrices are shown below:

$$\Delta X_L = \begin{bmatrix} \Delta x_{1,1} & \Delta x_{1,2} & \cdots & \Delta x_{1,nx} \\ \Delta x_{2,1} & \Delta x_{2,2} & \cdots & \Delta x_{2,nx} \\ \vdots & \vdots & \ddots & \vdots \\ \Delta x_{l,1} & \Delta x_{l,2} & \cdots & \Delta x_{l,nx} \end{bmatrix}$$

$$
\Delta Y_{\mathrm{L}} = \begin{bmatrix} \Delta y_{1,1} & \Delta y_{1,2} & \cdots & \Delta y_{1,ny} \\ \Delta y_{2,1} & \Delta y_{2,2} & \cdots & \Delta y_{2,ny} \\ \vdots & \vdots & \ddots & \vdots \\ \Delta y_{l,1} & \Delta y_{l,2} & \cdots & \Delta y_{l,ny} \end{bmatrix}
$$

Each of the l rows of matrix ΔX_{L} contains an increment of a control variable vector Δx stored in a local B cell (from the set of l youngest B cells). Analogically, matrix ΔX_{G} contains increments of control variables from g youngest global B cells. Each of the l rows of matrix ΔY_{L} contains an increment of an output vector Δy stored in a local B cell (from the set of l youngest B cells). Analogically, matrix ΔY_{G} contains an increment of optimized outputs from g youngest global B cells.

Coefficients of a K matrix are estimated using the least square method. The analytical solution (7.8) of the normal Equation 7.7 is presented below:

$$
WK = V \tag{7.7}
$$

$$
K = W^{-1} V \tag{7.8}
$$

where
$V = \eta \Delta X_{\mathrm{L}}^{T} \Delta Y_{\mathrm{L}} + \vartheta \Delta X_{\mathrm{G}}^{T} \Delta Y_{\mathrm{G}}$
$W = \eta \Delta X_{\mathrm{L}}^{T} \Delta X_{\mathrm{L}} + \vartheta \Delta X_{\mathrm{G}}^{T} \Delta X_{\mathrm{G}}$
η is the weight of local B cells
ϑ is the weight of global B cells

Utilization of global B cells causes SILO to be more robust for unrepresentative information stored in B cells in the immune memory. However, this could mean that the model based on global and local B cells will be less accurate than the model based on local B cells only. By changing η and ϑ weights the user can decide whether he or she wants to have better SILO robustness but worse model quality, or more accurate models where SILO will be less robust for unrepresentative data stored in the immune memory. In case of optimization of a combustion process in a power boiler, usually it is assumed that the weight of local B cells is ten times greater than the weight of global B cells. In case of strong nonlinear processes the weight of global B cells should be close to zero.

A mathematical model is created automatically in each optimization step in the mixed model layer. Every time that SILO computes control signals increments, a new model is created in the neighborhood of a current process operating point. The quality indicator (7.5) is minimized based on the model (7.6). The optimization algorithm takes into consideration constraints given by the user. There are constraints on the maximal increment of a control signal in one optimization step as well as constraints related with the range of individual elements of vector x. The optimization task is defined as follows:

$$
\min_{\Delta x} \left\{ \sum_{k=1}^{nx} \left[\alpha_k \left(\left| x_k^{\mathrm{a}} + \Delta x_k - x_k^{\mathrm{s}} \right| - \tau_k^{\mathrm{lx}} \right)_+ + \beta_k \left(\left(\left| x_k^{\mathrm{a}} + \Delta x_k - x_k^{\mathrm{s}} \right| - \tau_k^{\mathrm{sx}} \right)_+ \right)^2 \right] \right.
$$
$$
\left. + \sum_{k=1}^{ny} \left[\gamma_k \left(\left| y_k^{\mathrm{a}} + \Delta x K_k - y_k^{\mathrm{s}} \right| - \tau_k^{\mathrm{ly}} \right)_+ + \delta_k \left(\left(\left| y_k^{\mathrm{a}} + \Delta x K_k - y_k^{\mathrm{s}} \right| - \tau_k^{\mathrm{sy}} \right)_+ \right)^2 \right] \right\}
$$

$$(7.9)$$

with constraints:

$$
z_{\mathrm{low}} \leq \Delta x \leq z_{\mathrm{hi}},
$$
$$
u_{\mathrm{low}} \leq x^{\mathrm{a}} + \Delta x \leq u_{\mathrm{hi}}
$$

where K_k is a kth column of the gain matrix K.

The optimization task can be easily formulated as an LQ task after introducing additional variables

$$
\min_{\Delta x, x^{\mathrm{dlp}}, x^{\mathrm{dln}}, x^{\mathrm{ds}}, y^{\mathrm{dlp}}, y^{\mathrm{dln}}, y^{\mathrm{ds}}} \left\{ \sum_{k=1}^{nx} \left[\alpha_k \left(x_k^{\mathrm{dlp}} + x_k^{\mathrm{dln}} \right) + \beta_k \left(x_k^{\mathrm{a}} + \Delta x_k - x_k^{\mathrm{s}} - x_k^{\mathrm{ds}} \right)^2 \right] \right.
$$
$$
\left. + \sum_{k=1}^{ny} \left[\gamma_k \left(y_k^{\mathrm{dlp}} + y_k^{\mathrm{dln}} \right) + \delta_k \left(y_k^{\mathrm{a}} + \Delta x K_k - y_k^{\mathrm{s}} - y_k^{\mathrm{ds}} \right)^2 \right] \right\}
$$

$$(7.10)$$

with constraints:

$$
x_k^{\mathrm{dlp}} \geq x_k^{\mathrm{a}} + \Delta x_k - x_k^{\mathrm{s}} - \tau_k^{\mathrm{lx}}
$$
$$
x_k^{\mathrm{dlp}} \geq 0
$$
$$
x_k^{\mathrm{dln}} \geq x_k^{\mathrm{s}} - x_k^{\mathrm{a}} + \Delta x_k - \tau_k^{\mathrm{lx}}
$$
$$
x_k^{\mathrm{dln}} \geq 0
$$
$$
-\tau_k^{\mathrm{sx}} \leq x_k^{\mathrm{ds}} \leq \tau_k^{\mathrm{sx}}
$$
$$
-\tau_k^{\mathrm{sy}} \leq y_k^{\mathrm{ds}} \leq \tau_k^{\mathrm{sy}}
$$
$$
y_k^{\mathrm{dlp}} \geq y_k^{\mathrm{a}} + \Delta x K_k - y_k^{\mathrm{s}} - \tau_k^{\mathrm{ly}}
$$
$$
y_k^{\mathrm{dlp}} \geq 0
$$
$$
y_k^{\mathrm{dln}} \geq y_k^{\mathrm{s}} - y_k^{\mathrm{a}} - \Delta x K_k - \tau_k^{\mathrm{ly}}
$$
$$
y_k^{\mathrm{dln}} \geq 0
$$
$$
z_{\mathrm{low}} \leq \Delta x \leq z_{\mathrm{hi}}
$$
$$
u_{\mathrm{low}} \leq x^{\mathrm{a}} + \Delta x \leq u_{\mathrm{hi}}
$$

where

x_k^{dlp}, x_k^{dln} represent additional variables, representing distance from $\left(x_k^a + \Delta x_k\right)$ to the neighborhood of x_k^s with radius τ_k^{lx}

y_k^{dlp}, y_k^{dln} represent additional variables, representing distance from $\left(y_k^a + \Delta x K_k\right)$ to the neighborhood of y_k^s with radius τ_k^{ly}

x_k^{ds} represents additional variable, representing current part of insensitivity zone around x_k^s used in the square part of performance index

y_k^{ds} represents additional variable, representing current part of insensitivity zone around y_k^s used in the square part of performance index

One should note that if the W matrix is singular, then the K matrix may not represent real gains of the controlled process. In the mixed model optimization layer and in the global model optimization layer a special *blocking algorithm* is implemented. This algorithm tries to assure that matrix W is nonsingular and thus the Equation 7.7 is well conditioned.

The model identification task is checked for its conditioning in each optimization step on the mixed and global model layer. The conditioning level is formulated as follows:

$$cond\,(W) = \left\| W^{-1} \right\| \left\| W \right\|$$

In case of utilization of a spectral norm, the conditioning level can be defined as follows:

$$c = \left\| W^{-1} \right\|_2 \left\| W \right\|_2 = \frac{\sigma_{max}}{\sigma_{min}}$$

Conditioning of a model identification task is computed by comparing the largest σ_{max} and the smallest σ_{min} singular value of the W matrix. Coefficient c defines the conditioning level of the normal Equation 7.7. When the value of c coefficient is equal to 1, then Equation 7.7 is conditioned correctly. The greater the value of c coefficient, the worse is the conditioning of Equation 7.7. One of the optimization module's parameters is a boundary value for the c coefficient. Above this value, a special variable-blocking algorithm is activated. One should note that matrix W is symmetric and positively semi-defined. Thus the relation between singular values σ_i and eigenvalues λ_i is defined in the following way:

$$\sigma_i = \sqrt{\lambda_i},\ \lambda_i \in Spect\,(W),\ i = 1, 2, \ldots, nx$$

Before each optimization, if the c coefficient value is greater than the defined limit, some of the control variables are blocked with a certain distribution of probability. Thus the dimension of the optimization task is reduced. It causes mutation in new B cells. The main goal of this algorithm is to eliminate potential linear dependencies between columns of observation matrices (ΔX_L and ΔX_G). Elimination of these dependencies causes Equation 7.7 to be better conditioned and the quality of estimation of K matrix gains is better.

One should note that the optimization on the mixed model is similar to the secondary immune response. SILO effectively eliminates pathogens, based on knowledge stored in the immune memory, during operation in the mixed model layer.

7.4.3.2 Global Model–Based Optimization Layer

The algorithm used in the optimization on the global model layer is very similar to the algorithm used in the optimization on the mixed model layer. The only difference is that only global B cells are used to create a model in the global model layer. Local B cells' knowledge is not used. Optimization on a global model is performed when there are not enough local B cells to create a local model.

In this layer SILO does not have sufficient knowledge about static process dependencies in the neighborhood of the current process operating point. However, knowledge from other process operating points is available. This knowledge is used to create a linear process model (7.6). This is only a linear approximation of the whole nonlinear process characteristic. Thus, such a model is less accurate than a mixed model based on local B cells. However, the global model aggregates some basic process dependencies. Using this knowledge SILO can still decrease the value of a quality indicator.

7.4.3.3 Stochastic Optimization Layer

In the initial stage of SILO operation, the size of the immune memory is relatively small. By analogy to the immune system it can be said that the system is often attacked by unregistered and unknown pathogens. SILO does not have sufficient knowledge to create mathematical models and, based on these models, decrease the quality indicator value. The main goal of this layer is to gather knowledge about the process. The special heuristic used in this layer is responsible for

- Gathering knowledge about the process, by performing automatic parametric tests in the neighborhood of the current process operating point. The learning module creates new B cells based on these tests.
- Decreasing a quality indicator value in a long time horizon, assuming that disturbances do not essentially change in a long time horizon. By analogy to the immune system it can be said that the goal of this layer is an elimination of pathogen in a long time horizon, assuming that the system is attacked frequently by the same pathogen.
- Assuring that observation matrices, which are based on B cells from the immune memory (X_L and X_G matrices), are nonsingular and that the model identification task (7.7) is well conditioned.

In the heuristic used in this layer in every optimization step one control variable is randomly chosen (element of vector x) and a step signal is set for this variable.

The value of this step is user-defined. The direction of change depends on the value of the *direction table* element related to the chosen variable. The step direction is changed when the step control signal has caused an increase of the quality indicator. A change in the vector x element causes a process outputs response. The learning module creates a B cell based on this test. In the next optimization step another element of vector x is randomly chosen and the next test is performed. After several tests, the configuration of control variables related with the smallest value of the quality indicator is restored. The series of single value change tests along with the best solution restoration is called a *testing cycle*.

The heuristic mentioned above ensures that SILO can gather knowledge about a new process operating point. A change of only one control variable in one optimization step causes that:

■ Knowledge about the process is more accurate
■ Matrices X_L and X_G, used in a model identification task, are nonsingular because single changes of one element of vector x result in no linear dependencies between columns in these matrices

Moreover, the presented heuristic provides a decrease of quality indicator value in a long-term horizon. One should note that there is a clear analogy between the stochastic optimization layer and the primary immune response. In case of a primary immune response, the system fights new, unknown pathogen. An organism needs more time to overcome an unknown pathogen. After a successful defense action the knowledge about the pathogen is saved. It is used during a secondary immune response for efficient protection of the organism against similar pathogens. By analogy, in a SILO system the knowledge gathered in a stochastic optimization layer is used in other optimization layers.

7.4.3.4 Layers Switching Algorithm

Switching the optimization module between layers (presented in Figure 7.10) is a key element of the SILO algorithm. The heuristic of layer switching is based on the principle that SILO's knowledge about a current process state should be used. If this knowledge is not available, then SILO uses the knowledge about general, basic process dependencies. If optimization based on this knowledge is ineffective, then an optimal solution should be found using the stochastic optimization.

When SILO is turned on, the switching algorithm checks if there are a sufficient number of B cells in the immune memory to create a global process model. If there are not enough B cells, it means that SILO is in the initial stage of knowledge gathering. In such a case control signals are computed using the heuristic implemented in the stochastic optimization layer. As a result of tests performed in the stochastic optimization layer new B cells are created over time. If there are enough B cells to create a mathematical model, SILO switches to the optimization layer based

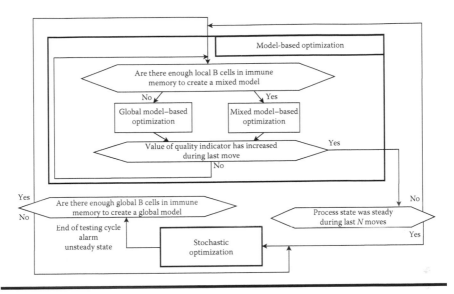

Figure 7.10 Layers switching algorithm.

on the mathematical model. If there are enough local B cells that represent a similar process operating point to the current one, then the mixed model is created automatically. Otherwise a global model is created. Optimization is performed based on a mathematical model. If the value of a quality indicator has increased after a process response to a control change, it means that the process model is not accurate enough and it is impossible to improve the solution based on this model. In such a case knowledge about the process in the neighborhood of a current operating point should be updated and SILO should try to find a better solution in a nondeterministic way in the stochastic optimization layer. Increase of the quality indicator value does not have to be related with a change of control signals. A change in the disturbance vector z can also cause the quality indicator value to increase. In this case, switching to a stochastic optimization layer is not advisable because of the following reasons:

- To perform a reliable test a process steady state is needed.
- In case of changes in the z vector the most important thing is a fast reaction for these changes. Knowledge about the process gathered in the process model is necessary for such a fast reaction.

One should note that in case of changes in the disturbances vector the best strategy is to stay in the mixed or global model optimization layer. In such a situation switching an algorithm does not allow to pass the control to the stochastic optimization layer. Thus the optimization module will stay in optimization on the mixed or global model layer. That steady-state detection task is performed by SILO's *state analyzer module.*

The stochastic optimization is interrupted when an unsteady state is detected after an optimization period, or at least one of the user-defined alarms is turned on (e.g., too high CO emission in the combustion process).

7.5 Application of Bio-Inspired Methods in Industrial Process Control

IVY as well as SILO has been designed for the control of industrial processes, in particular the combustion process in power boilers. These kinds of boilers are used in heat- and power-generating plants. In a classic, steam power plant, thermal energy from organic fuel combustion is passed to a steam generated in a power boiler. In a steam turbine, thermal energy is transformed into mechanical energy. An electrical generator transforms mechanical energy into electrical energy. For more information on power plants, refer to (Drbal and Westra 1995).

From SILO's and predictive control methods' point of view, a power boiler is a multi input multi output (MIMO) plant that has impossible-to-ignore cross transforms between its inputs and outputs. The combustion process in a power boiler is a nonlinear, dynamic process with long response time. Moreover, some disturbances have a permanent effect on this process. The main disturbances in this process are unit load and coal mill configuration. There are also some nonmeasurable or unconsidered disturbances such as quality of mills grinding or coal humidity. The characteristics of the combustion process are nonstationary. Changes of characteristics are considered in a long-term horizon and result from wearing or failure of devices, power unit modernization, changes of chemical properties of fuel, external condition changes (e.g., seasonality), changes of mills grinding quality, etc.

Typical optimization goals in the combustion process are decreasing nitric oxide (NO_x) and carbon monoxide (CO) emissions, increasing boiler efficiency, maintaining superheat and reheat steam temperature and flue gases temperature at a desired level. The goals mentioned above are often contrary. Moreover, there are some constraints of the optimization task, such as minimal and maximal value of control signals (e.g., value of oxygen level in combustion gases must be in a range 2%– 6 %), limits on increment of control signals in one optimization step (e.g., absolute increment of oxygen level in combustion gases cannot be higher than 0.15% in one optimization step). There are also constraints related with process outputs (e.g., CO emission cannot exceed 250 mg/Nm^3). These constraints can change along with changes of the combustion process operating point.

Figure 7.11 presents a power boiler in the context of a combustion process optimization.

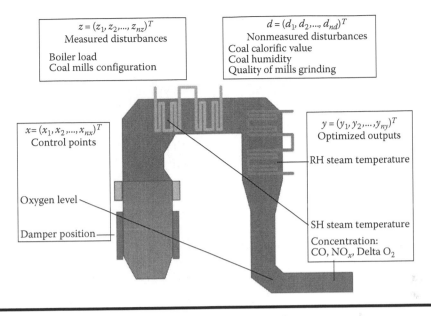

Figure 7.11 Diagram of a power boiler and its signals.

TS approach is often applied to consider the combustion process nonlinearity in MPC controllers. Model fuzzification is usually based on unit load signal. Another important information that defines the process operating point in a coal power plant is coal mill configuration and level of CO emission. However, this information is usually not used to define fuzzy partitions, because of the high cost related with a model creation process. In the SILO system, CO emission signal and coal feeder load signal are taken into consideration during computation of binding strength between antigen and antibody. Information about the binding strength is used during selection of B cells that represent the current process operating point.

7.5.1 SILO Results

SILO has been implemented in 21 units in large-scale power plants in the United States, Poland, South Korea, and Taiwan. Each implementation has given significant improvement of control quality. Results from SILO implementation in one of the Polish power plant units (maximal unit load is 200 MW) are presented in this section. The primary SILO goals were

■ To keep NO_x emission (1 h average) below 500 mg/Nm^3
■ To keep CO emission (5 min average) below 250 mg/Nm^3

The secondary SILO goals were

- To keep loss of ignition (LOI) below 5%
- To keep super heat (SH) temperature at 540°C
- To keep flue gas temperature below 140°C (flue gas desulphurization process requirement)

SILO optimized nine output signals: CO emission (left and right side), NO_x emission (left and right side), estimated SH temperature (left and right side), flue gas temperature (left and right side), and LOI signal. Six disturbance signals were chosen from the set of all process signals: unit load, coal calorific value, and status of each of four coal mills. The control vector consisted of 11 signals: O_2 level, eight secondary air dampers, and two over fire air (OFA) dampers.

Figure 7.12 presents a situation in which SILO is disabled and compares it with the situation where SILO is enabled. SILO essentially reduces NO_x emission and maintains CO emission and LOI below the defined limit. Table 7.4 presents results of a SILO operation.

A summary of the described SILO implementation is presented below:

- NO_x emission limit (500 mg/Nm³) was not exceeded even once when SILO was enabled. When SILO was turned off, 10.41% of the analyzed 1 h averages was above this limit.
- By 99.81% of SILO operation time, CO emission was kept below 250 mg/Nm³. When SILO was disabled, CO emission limit was kept below 250 mg/Nm³ by 98.85% of time.
- SH temperature was increased by 4.4°C, thus increasing process efficiency when SILO was enabled.

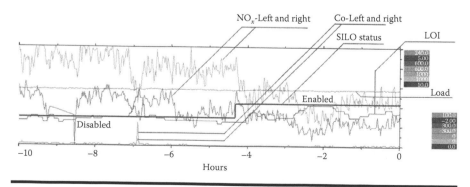

Figure 7.12   SILO influence on NO_x, CO, and LOI signals.

Table 7.4   Results of SILO Operation

	Units	SILO OFF	SILO ON
Analyzed time range	hours	822	163
Load range	MW	106.0–195.3	107.8–203.0
NO_x exceedance time	%	10.14	0.0
CO exceedance time	%	1.15	0.19
Average SH temperature	°C	532.24	536.58
LOI exceedance time	%	59.0	24.14
Average flue gas temperature	°C	117.76	120.54

■ LOI was reduced to 3.92% (from 5.16% level). SILO has kept LOI below 5% by 75.9% of its operation time. In comparison, when SILO was disabled, LOI was below the limit only by 41% of time.

■ When SILO was enabled, flue gas temperature was kept below 140°C all the time, so flue gas desulphurization process was not disturbed.

7.5.2 IVY Results

IVY controller has been implemented in more than 50 units in large-scale power plants in the United States, Europe, and Asia. Results achieved by a fuzzy predictive controller that uses a neural network are presented in this chapter.

The presented implementation has been performed in one of the Korean units (maximal unit load is 800 MW) (Figure 7.13). The IVY controller was responsible for

■ CO emission reduction without NO_x emission increase
■ Unburned carbon reduction
■ Boiler efficiency improvement
■ Flue gas temperature decrease

IVY controlled eight output signals: average CO emission, average NO_x emission, average flue gas temperature, flue gas temperature difference (left and right side), average RH temperature, average SH temperature (Figure 7.13). The unit load was a disturbance signal. The control vector consisted of the following signals: auxiliary air dampers opening, separated over fire air (SOFA) and coupled over fire air (COFA) dampers opening, forced draft (FD) and induced draft (ID) fans, positioning of SOFA tilts, pulverized outlet temperature set-point, and O_2 level set-point.

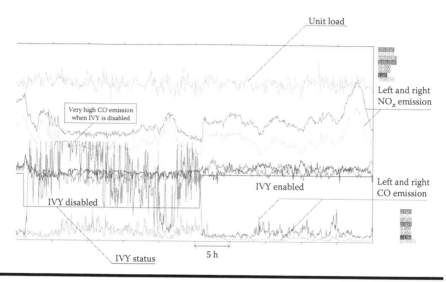

Figure 7.13   IVY influence on CO signal.

IVY results from the final acceptance test are presented below:

- CO emission was reduced from 350–500 ppm level to 50–60 ppm.
- NO_x emission was not increased.
- Unburned carbon was reduced by 10%–30% (results depend on coal and operation conditions).
- Boiler efficiency was improved by 0.17%–0.44%.
- Flue gas temperature was reduced by $1°C$–$3°C$.

7.5.3 Summary

The results presented in this chapter show that the application of bio-inspired algorithms for industrial process control allows to improve a control quality in the plant's nominal condition. In the case of optimization of a combustion process in a power boiler it is possible to significantly decrease CO and NOx emissions and increase process efficiency. Moreover, besides economical profits, it allows to decrease the environment pollution.

References

Arabas, J., Bialobrzeski, L., Domanski, P., and Swirski, K. (1998). Advanced boiler control, in *Proceedings of MMAR'98*, Vol. 5 (Miedzyzdroje, Poland), pp. 521–526.
Astrom, K. J. and Hagglund, T. (1995). *PID Control—Theory, Design and Tuning* (Instrument Society of America Research, Triangle Park, NC).

Astrom, K. J. and Hagglund, T. (2004). Revisiting the ziegler-nichols step response method for PID control, *Journal of Process Control* **14**, 635–650.

Babuska, R., Sousa, J., and Verbruggen, H. (1999). Predictive control of non-linear systems based on fuzzy and neural models, in *Proceedings of the fifth European Control Conference* (Karlsruhe, Germany), pp. CA–4:1–6.

Bazaraa, M., Roberts, P., and Griffiths, G. (1993). *Nonlinear Programming: Theory and Algorithms* (John Wiley Sons, New York).

Blevins, T., McMillan, G., Wojsznis, W., and Brown, M. (2003). *Advanced Control Unleashed* (ISA, Research Triangle Park, NC).

Camacho, E. F. and Bordons, C. (1999). *Model Predictive Control* (Springer Verlag, London).

Castro, L. N. D. and Timmis, J. I. (2003). Artificial immune systems as a novel soft computing paradigm, *Soft Computing* **7**, 8, 526–544, http://www.cs.kent.ac.uk/pubs/2003/1508

Castro, L. N. D. and Zuben, F. V. (1999). Artificial immune systems: Part I—Basic theory and applications, Technical Report RT DCA 01/99, Department of Computer Engineering and Industrial Automation, School of Electrical and Computer Engineering, State University of Campinas, Campinas, SP, Brazil.

Castro, L. N. D. and Zuben, F. V. (2000). Artificial immune systems: Part II—A survey of applications, Technical Report RT DCA 02/00, Department of Computer Engineering and Industrial Automation, School of Electrical and Computer Engineering, State University of Campinas, Campinas, SP, Brazil, ftp://ftp.dca.fee.unicamp.br/pub/docs/techrep/2000/DCA00-002.pdf

Chang, W. and Yan, J. (2004). Optimum setting of PID controllers based on using evolutionary programming algorithm, *Journal—Chinese Institute of Engineers* **27**(3), 439–442.

Chen, W., Zhou, J., and Wei, H. (2006). Compensatory controller based on artificial immune system, in *Proceedings of the 2006 IEEE International Conference on Mechatronics and Automation* (Luoyang, Henan), pp. 1608–1613.

Clarke, D. W., Mohtadi, C., and Tuffs, P. S. (1987). Generalised predictive control-parts I and II, *Autimatica* **23**, 137–160.

Cutler, C. R. and Ramaker, B. L. (1980). Dynamic matrix control—A computer control algorithm, in *Proceedings of Joint Automatic Control Conference* (San Fransisco), pp. WP5–B.

Desborough, L. and Miller, R. (2004). Increasing customer value of industrial control performance monitoringÚhoneywell's experience, in *IEEE Proceedings of American Control Conference*, Vol. 6 (San Fransisco), pp. 5046–5051.

Domanski, P., Swirski, K., and Williams, J. (2000). Application of advanced control technologies to the emission control and optimization, in *Proceedings of Conference on Power Plant Emission Control and Monitoring Technologies* (London).

Drbal, L., Westra, K., and Boston, P. (1995). *Power Plant Engineering* (Springer, New York).

Feldbaum, A. (1968). *Foundations of Optimal Control Systems Theory* (PWN, Warsaw), in Polish.

Fukuda, T., Mori, K., and Tsukiama, M. (1998). Parallel search for multi-modal function optimization with diversity and learning of immune algorithm, in D. Dasgupta (ed.), *Artificial Immune Systems and Their Applications* (Springer, Berlin), pp. 210–220.

Gattu, G. and Zafiriou, E. (1992). Nonlinear quadratic dynamic matrix control with state estimation, *Ind. Eng. Chem.* **31**(4), 1096–1104.

Hassanein, O. and Aly, A. (2004). Genetic—PID control for a fire tube boiler, in *Proceedings of Second IEEE International Conference on Computational Cybernetics ICCC 2004* (Vienna University of Technology, Austria), pp. 19–24.

Haykin, S. (1999). *Neural Networks—A Comprehensive Foundation* (Prentice Hall, Englewood Cliffs, NJ).

Henson, M. (1998). Nonlinear model predictive control: Current status and future directions, *Computers and Chemical Engineering* **23**, 187–202.

Jang, J. (1993). Anfis: Adaptive-network-based fuzzy inference system, *IEEE Transactions on Systems, Man and Cybernetics* **23**, 665–685.

Jang, J.-S. R. and Sun, C.-T. (1995). Neuro-fuzzy modeling and control, in *Proceedings of IEEE*, **83**(3), 378–406.

Jones, A., Ajlouni, N., Lin, Y.-C., Kenway, S., and Uzam, M. (1996). Genetic design of robust PID plus feedforward controllers, in *Proceedings of IEEE Conference on Emerging Technologies and Factory Automation EFTA'96*, Vol. 1 (Kauai), pp. 267–271.

Jung, L. and Soderstrom, T. (1983). *Theory and Practice of Recursive Identification* (MIT Press, Cambridge, MA).

Kassmann, D. E., Badgwell, T. A., and Hawkins, R. B. (2000). Robust steady-state target calculation for model predictive control, *AIChE Journal* **46**(5), 1007–1024.

Kau, G., Mukherjee, S., Loo, C., and Kwek, L. (2002). Evolutionary PID control of non-minimum phase plants, in *Proceedings of Seventh International Conference on Control, Automation, Robotics and Vision ICARCV 2002*, Vol. 3, pp. 1487–1492.

Kim, D. H., Jo, J. H., and Lee, H. (2004). Robust power plant control using clonal selection of immune algorithm based multiobjective, in *Proceedings of the Fourth International Conference on Hybrid Intelligent Systems (HIS'04)* (Japan), pp. 450–455.

Kim, H. S., Lee, J. M., Park, T. R., and Kwon, W. H. (2000). Design of networks for distributed digital control systems in nuclear power plants, in *Proceedings of 2000 International Topical Meeting on NPIC and HMIT*, Washington DC, pp. 629–633.

Koza, J., Keane, M., Streeter, M., Matthew, J., Mydlowec, W., Yu, J., and Lanza, G. (2003). *Genetic Programming IV: Routine Human-Competitive Machine Intelligence* (Kluwer Academic Publishers, Dordrecht, the Netherlands).

KrishnaKumar, K. and Neidhoefer, J. (1995). An immune system framework for integrating computational intelligence paradigms with applications to adaptive control, in M. Palaniswami, Y. Attikiouzel, R. J. I. Marks, D. Fogel, and T. Fukuda (eds.), *Computational Intelligence: A Dynamic System Perspective* (IEEE Press, Piscataway, NJ), pp. 32–45.

KrishnaKumar, K. and Neidhoefer, J. (1997a). Immunized adaptive critics for level 2 intelligent control, in *1997 IEEE International Conference on Systems, Man, and Cybernetics*, Vol. 1 (IEEE, Orlando, FL), pp. 856–861.

KrishnaKumar, K. and Neidhoefer, J. (1997b). Immunized neurocontrol, *Expert Systems with Applications* **13**, 3, 201–214.

Lawrynczuk, M. and Tatjewski, P. (2002). A computationally efficient nonlinear predictive control algorithm based on neural models, in *Proceedings of the Eighth IEEE Conference on Methods and Models in Automation and Robotics MMAR'02* (Szczecin), pp. 781–786.

Lydyard, P. M., Whelan, A., and Fanger, M. W. (2000). *Instant Notes in Immunology*, Instant Notes Series (BIOS Scientific Publishers Limited, Oxford, U.K.).

Marquis, P. and Broustail, J. (1988). Smoc, a bridge between state space and model predictive controllers: Application to the automation of a hydrotreating unit, in *Proceedings of the 1988 IFAC Workshop on Model Based Process Control* (Atlanta, GA), pp. 37–43.

Marusak, P. and Tatjewski, P. (2000). Fuzzy dynamic matrix control algorithms for nonlinear plants, in *Proceedings of the Sixth International Conference on Methods and Models in Automation and Robotics MMAR'00* (Miedzyzdroje), pp. 749–754.

Megias, D., Serrano, J., and Ghoumari, M. (1999). Extended linearized predictive control: practical control algorithms for non-linear systems, in *Proceedings of the Fifth European Control Conference* (Karlsruhe, Germany).

Morari, M. and Lee, J. H. (1997). Model predictive control: Past, present and future, in *Proceedings of Joint 6th International Symposium on Process Systems Engineering (PSE'97) and 30th European Symposium on Computer Aided Process Engineering (ESCAPE-7)* (Trondheim).

Ogunnaike, B. A. and Ray, W. H. (1994). *Process Dynamics, Modeling, and Control* (Oxford University Press, Melbourne, Australia).

Plamowski, S. (2006). *Safe implementation of advanced control in a diagnostic-based switching structure*, PhD thesis (Warsaw University of Technology, Warsaw), in polish.

Plamowski, S. and Tatjewski, P. (2006). Safe implementation of advanced control in a diagnostic-based switching structure, in *IFAC – Safeprocess Proceedings*, Beijing, China.

Prett, D. M. and Gillette, R. D. (1980). Optimization and constrained multivariable control of catalytic cracking unit, in *Proceedings of Joint Automatic Control Conference* (San Fransisco).

Qin, S. J. and Badgwell, T. A. (1997). An overview of industrial model predictive control technology, in *Proceedings of Fifth International on Chemical Process Control*, Tahoe City, California, pp. 232–256.

Qin, S. J. and Badgwell, T. A. (2003). A survey of industrial model predictive control technology, *Control Engineering Practice* **11**, 733–764.

Qiping, Z., Xiangyu, G. J. W., and Youhua, W. (2003). An industrial application of APC technique in fluid catalytic cracking process control, in *SICE 2003 Annual Conference*, Vol. 1, pp. 530–534.

Ramos, M., Morales, L. M., Juan, L., and Bazan, G. (2005). Genetic rules to tune proportional + derivative controllers for integrative processes with time delays, in *Proceedings of 15th International Conference on Electronics, Communications and Computers CONIELECOMP 2005*, pp. 143–148.

Seaborg, D., Edgar, T., and Mellichamp, D. (1989). *Process Dynamics and Control* (John Wiley Sons, New York).

Solomon, E. P., Berg, L. R., Martin, D. W., and Villee, C. A. (1993). *Biology*, 3rd edn. (Saunders College Publishing, Fort Worth, TX).

Streeter, M. J., Keane, M. A., and Koza, J. R. (2003). Automatic synthesis using genetic programming of improved PID tuning rules, in A. E. Ruano (ed.), *Preprints of the 2003 Intelligent Control Systems and Signal Processing Conference* (Faro, Portugal), pp. 494–499, http://www.genetic-programming.com/jkpdf/icons2003.pdf

Soderstom, T. and Stoica, P. (1989). *System Identification* (Prentice Hall, Englewood Cliffs, NJ).

Tatjewski, P. (2007). *Advanced Control of Industrial Processes: Structures and Aglorithms* (Springer Verlag, London).

Tatjewski, P. and Lawrynczuk, M. (2006). Soft computing in model-based predictive control, *International Journal of Applied Mathematics and Computer Science* **16**(1), 7–26.

Wierzchon, S. (2001a). *Artificial Immune Systems—Theory and Applications* (Exit, Warsaw), in polish.

Wierzchon, S. T. (2001b). Immune algorithms in action: Optimization of nonstationary functions, in *SzI–16'2001: XII Oglnopolskie Konwersatorium nt. Sztuczna Inteligencja— nowe wyzwania (badania—zastosowania—rozwj)* (Siedlce–Warsaw, Poland), pp. 97–106, http://www.ipipan.eu/staff/s.wierzchon/ais/index.html, in Polish.

Wojdan, K. and Swirski, K. (2007). Immune inspired system for chemical process optimization on the example of combustion process in power boiler, in *Proceedings of the 20th International Conference on Industrial, Engineering and Other Applications of Applied Intelligent Systems* (Kyoto, Japan).

Wojdan, K., Swirski, K., Warchol, M., and Chomiak, T. (2007). New improvements of immune inspired optimizer SILO, in *Proceedings of the 19th IEEE International Conference on Tools with Artificial Intelligence* (Patras, Greece).

Yen, J. and Langari, R. (1999). *Fuzzy Logic—Intelligence, Control and Information* (Prentice Hall, Upper Saddle River, NJ).

Zadeh, L. (1965). Fuzzy sets, *Information and Control* **8**, 338–353.

Zhao, Z. Y. and Tomizuka, M. (1993). Fuzzy gain scheduling of PID controllers, *IEEE Transactions on Systems, Man, and Cybernetics* **23**, 1392–1398.

Chapter 8

Multirobot Search Using Bio-Inspired Cooperation and Communication Paradigms

Briana Wellman, Quinton Alexander, and Monica Anderson

Contents

Animal systems are recognized for their ability to successfully forage and hunt in unknown and changing conditions. If the mechanisms of this adaptability and flexibility can be determined and implemented in multi-robot systems, autonomous systems could cost-effectively assist with homeland security and military applications by providing situational awareness.

In this chapter, cooperation and communication within teams are studied for evidence of salient mechanisms. Specifically, we find that primates make use of three paradigms: decentralized asynchronous action selection, limited communications modalities, and transient role selection. Although there has been some use of these features in recent research, their implementation tends to occur in isolation. An example is presented that compares a traditional coverage approach to a bio-inspired coverage approach that uses decentralized, asynchronous decision making.

8.1 Introduction

Multi-robot systems have been widely studied in the areas of exploration, search, and coverage. Possible applications in homeland security, search and rescue, military and surveillance have increased as researchers have demonstrated the benefits of using multi-robot systems rather than single-robot systems.

There are several advantages in using a multi-robot system over a single-robot system. Having multiple robots can improve fault tolerance, provide faster performance, and make use of less expensive robots. Fault tolerance is essential when implementing a time-critical system. The issue of reliability is especially important in search and rescue where time is of the essence. Systems that use multiple robots reduce sensitivity to single system failures. Multi-robot systems can also provide faster performance. By allowing a task to occur in parallel, task completion can be faster. Balch and Arkin[20] conducted experiments with robot teams to show how additional resources can result in speed-up or faster completion of foraging, gazing, and consumption tasks.

A final benefit to implementing a multi-robot system is the possibility of using less complex, less expensive robots. Rather than having a single complex robot with high-end sensors, several less expensive robots with lower-end sensors can be used. However, to realize these benefits, an appropriate cooperation paradigm must be used. Studying cooperation in nature can provide cues to successful strategies.

8.1.1 Successes in Nature

There have been several attempts in defining cooperation in animals. Chase[1] describes it as "when an animal invests resources in a common interest shared with other group members." Together, animals breed, raise their young, and forage. In fact, there are many examples of animals cooperatively hunting.

Meerkats exhibit cooperation while foraging. As meerkats hunt together, one group member acts as a guard by looking out for predators while the rest of the group forage.[2] The guards announce their duty vocally. They also make alarm calls whenever a predator is spotted, allowing the rest of the group to focus on hunting rather than predators.

Cooperation in lions has also been analyzed. Lionesses take roles of "wingers" and "centers."[3] In hunts, "wingers" stalk prey by circling them and then attacking from the front or side. "Centers" move short distances to prey and wait for them to come near while avoiding another lioness. Such behavior decreases the chances of prey escaping.

Harris' hawks initiate a search by splitting into groups on perch sites such as trees or polls.[4] To search for prey, they use a "leapfrog" movement in which they split and rejoin in groups while moving in one direction. They continue this until a prey is captured. To capture a prey, the most common tactic is several hawks attacking all at once.

While hunting, primates change roles depending on the state of the prey and other members.[5] For example, if at first a prey is on the ground and then runs up a tree, the roles of primate team members change. In addition, they adapt to their environment by changing the way they hunt and communicate. For instance, chimpanzees living in forests coordinate and plan hunts more than those living in grasslands. In the grasslands where trees are short, chimpanzees are more opportunistic when they hunt.[6]

Because of their highly cooperative and coordinated hunts, primates can be used as an inspiration to ideas for cooperation in a multi-robot system. Primate cooperation and communication give insight into challenges faced in multi-robot systems.

8.1.2 Challenges and Problems

Although multi-robot systems are advantageous in many applications, they also introduce challenges. These issues provide active research topics, especially in cooperative multi-robot systems. Challenges include limited communications range and bandwidth, unknown and/or dynamic environments, and restricted sensor range.

In order for a team of robots to coordinate and cooperate, members must communicate. However, communication is constrained to limited distances and limited users. Cooperation that relies upon digital messages can degrade as messages do not reach their intended targets. Cooperation paradigms must be developed that can operate within these constraints.

Unknown or dynamic area configuration is another problem. Robots must be able to make reasonable motion choices. However, uncertainty about the environment can obscure relevant environmental information. As they search, they have to adapt their activities according to the state of the environment to enhance search performance.

Sensor ranges are restricted meaning that they have only a certain field of view or range of detection which limits the amount of information that can be obtained from the environment. Information that is obtained is tainted by noise, further complicating the process of deciding appropriate actions for the current environment and system state.

These challenges are not unique to multi-robot systems. Animal systems contend and even thrive where communication is limited, environments are changing, and sensor information is ambiguous. Primates show patterns of cooperation that can be applied to multi-robot systems. The goal of this chapter is to show that bio-inspired paradigms gleaned from primates provide cues to fault-tolerant, robust cooperative coverage.

Section 8.2 provides an introduction to three primate teams and evaluates the way they cooperatively hunt and travel. Role selection, communication and navigation, and the impact of the environment on cooperation and communication are examined in each primate. Section 8.3 discusses key features of primate systems that can be applied to multi-robot systems. Section 8.4 presents a proof of concept system that takes advantage of primate features. Finally, Section 8.5 presents the chapter summary and conclusion.

8.2 Study of Primates

Imagine living and researching in a tropical rainforest with a group of people for a week. Attempting to live the full rainforest experience, you decide to adjust to your natural surroundings by searching for food. While searching together, you try to cover all possible places that food may be found. There is no map so you have to rely on each other or someone who is somewhat familiar with the environment. As you search for food, someone suggests that you head west and then later someone else implies that you navigate to the northwest because it is less dense. Electronic devices such as radios and cell phones are not present; thus, you have to remain close enough to communicate either visually or verbally. Visual and verbal communication is sometimes limited depending on the situation. If you are searching for moving prey you have to avoid scaring it off so you are quiet and rely heavily on hand signals or other body movements. At the same time, if you are under tall thick trees or bushes, there is limited visibility and low light so verbal communication is more important. Throughout your search, the role of the navigator changes as well as the way you communicate and travel because of your unpredictable environment.

The above situation describes many important aspects of cooperation among animal and robot teams. Cooperation is not successful without some form of communication. Even if it is implicit communication, teams have to communicate to accomplish tasks as a team and in a certain time frame. In addition, communication and navigation are affected by the environment.

In this section, we examine three primates including Tai and Gombe chimpanzees, lion tamarins, and capuchin monkeys. First, there is an introduction on each primate by presenting basic information such as social structure, habitat, and location. Then we examine the way they cooperate and communicate as they forage and travel. We discuss their role selections, communication while navigating, and then the impact of the environment on their travel and conquest for food.

8.2.1 Introduction to Primates

Chimpanzees can be found in rainforests, woodlands, and grasslands but for this chapter the chimpanzee populations of the Tai Forest and Gombe National Park are examined. Chimpanzees of the Tai Forest live in dense rainforests while chimpanzees of the Gombe National Park live in savanna woodlands, or in other words, grasslands with scattered trees.[7,8] Although they are located in different places and environments, they have similar social structures.

The social structures of chimpanzees have been greatly studied. Chimpanzees live in fission–fusion societies.[9] Individuals form parties that change significantly in size and composition throughout the day. There is one community that breaks up into subgroups or parties depending on what is going on at the time. A chimpanzee may be found to hunt alone or with a group. Unlike the next two primates presented, the chimpanzee societies are very flexible and unpredictable.

Golden lion tamarins are an endangered species native to Brazil's Atlantic coastal forests. Lion tamarins are recognized as New World Monkeys, or monkeys living in South America. They are slightly smaller than squirrels and have reddish orange fur on their entire body except on their face. They live in family groups of up to 14 and are led by the breeding parents. Their family units are very close in which both parents care for the young. As they are omnivorous, lion taramins forage cohesively for both plants and animals.

Capuchin monkeys are also considered New World Monkeys. They can be found in the forests of South America and Central America. Their fur color varies but they commonly have black hair with tan fur around their face, shoulders, and neck. There are also omnivorous, territorial, and are known to live in groups containing between 6 and 40 members.

Although primates have different family and social structures and live in different places, like many other animals they cooperate to survive. The following sections discuss role selections, communication and navigation, and the impact of the environment on cooperation and communication.

8.2.2 Role Selection

The roles of primates during a search are dynamic. It is assumed that roles aren't explicitly decided on ahead of time and change depending on the state of the search. Observation of roles and state drives cooperation among primates. They decide on what roles to take as they search. In this section, each primate's eating habits are presented because they impact the way they hunt. Then, their hunting behaviors are presented. Finally and most importantly, their role selection process is discussed.

The main prey of both Tai and Gombe chimpanzees are colobus monkeys. Unlike Gombe chimpanzees, half of the Tai chimpanzees' prey population is adult. Gombe chimpanzees are afraid of adult colobus monkeys so they usually attack juvenile or female colobus monkeys. Although both Tai and Gombe chimpanzees hunt for colobus monkeys, Tai chimpanzees hunt in coordinated groups more than Gombe chimpanzees.[6] As illustrated in Figure 8.1, the role selections of Tai chimpanzee group hunts can be described as implicit cooperation with coordinated group hunts. They implicitly cooperate, meaning that roles are decided on simultaneously depending on the state of the hunt and on other team members. Gombe chimpanzee groups also cooperate implicitly but in comparison to Tai chimpanzees they coordinate less in group hunts.[5]

In Tai chimpanzees, roles have been described in terms of bystander and hunter and later in terms of driver, blocker, chaser, and ambusher.[5,6,10] In observations of Tai chimpanzee hunts, it was found that members usually performed different roles in the same hunt. As the hunt progressed, roles changed depending on what was going on with the prey during the hunt.

Gombe Chimpanzees usually hunt solitarily but simultaneously.[6,10] They are less likely to hunt in groups. When they do hunt in groups, there aren't any specific

Figure 8.1 Tai chimpanzees often hunt in coordinated groups while Gombe chimpanzees hunt more solitarily.

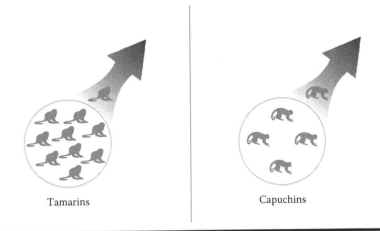

Figure 8.2 Tamarins travel cohesively while capuchins travel in a more dispersed pattern.

defined roles that are used to capture a prey. Gombe chimpanzees usually don't coordinate hunts and, like Tai chimpanzees, Gombe chimpanzees don't implicitly agree on division of labor before hunting.

Both tamarins and capuchins are omnivorous. Although their diets mostly consist of fruits and insects, they occasionally eat bird eggs and vertebrates such as lizards, birds, and squirrels. Both are very social animals and they share food with each other. They search for food in teams but roles are not as distinct as in chimpanzees. Instead of roles, positions are described of members in a traveling troop. For example, while capuchins travel in troops, the two positions, vanguard and rearguard, are described.[11,12] Similar to Figure 8.2, the vanguard is located at the starting position of the troop. Rearguards are all positioned behind the vanguard. The positions of vanguards and rearguards change when the direction of the troop changes. A group member vocally expresses an intention to change direction.

8.2.3 Communication and Navigation

From studies of chimpanzees searching for prey, we can assume that they search together by observing one another. They hunt very quietly while walking close together one after another in a straight line in the same direction.[6] To communicate a change of direction, they use auditory signals such as drumming or pant-hooring. Instead of detecting prey visually they have to rely on acoustics because of low visibility in the dense forest.

Figure 8.2 illustrates travel in tamarins and capuchins. Tamarins travel cohesively while capuchins travel in a more dispersed pattern.[11] Generally, foraging less than

15 m apart, it is suggested that taramins travel cohesively in close proximity. Because they forage cohesively, several individuals are found to arrive and feed on the same food at the same time. In a study of the communication of taramins, the vocal calls of wah-wahs, tsicks, whines, clucks, and trills were observed.[12] The wah-wah call was used to initiate travel of a stationary group in a specific direction. Whines were more frequent when the nearest neighbor was more than 5 m away compared to when they were less than 5 m away. It is suggested that these calls are used for group cohesion and provide information about the caller's forthcoming activities.

Capuchins often forage dispersed with members 50–100 m apart.[11] Because they are more dispersed than tamarins, a few capuchins arrive at a feeding site at the same time. In contrast, tamarins arrive all at once. They also make calls to navigate the troop.[13,14]

8.2.4 Impact of Environment on Cooperation and Communication

The environment impacts foraging and traveling by affecting cooperation and communication. Although animals are closely related and in the same species, they cooperate differently depending on their environment. For example, Tai chimpanzees live in the rainforest at Tai National Forest along the Ivory Coast. In this forest, trees are over 40–50 m high. Because the forest trees are high, prey have a greater chance of escaping than if the forest trees were lower. Thus, Tai chimpanzees find it more effective to hunt in groups. They hunt quietly because they observe one another. Tai Chimpanzees have to rely on acoustics because of low visibility in the dense Tai forest. Gombe chimpanzees live at the Gombe Stream National Park in Tanzania. In this forest, trees are about 15 m high. It is suggested that their hunting techniques are less organized than that of Tai chimpanzees.[6] Their prey have less escape routes than if the forest were higher. While Gombe chimpanzees are considered opportunist hunters, Tai chimpanzees' hunts are more intentional and more organized.

Tamarins and capuchins also have an impact of environment on cooperation and communication. They are so close that the animals can rely on visual coordination.[12–14] Both of these New World Monkeys cohesively travel and vocalize to initiate travel and to change directions during travel. In extreme danger, tamarins occasionally travel without any vocalization. It is suggested that because they are small and have many predators, they remain close to each other.

8.3 Bio-Inspired Multi-Robot Systems

To gain the benefits of primate-based coverage systems, the paradigms that have made primates successful while being flexible and fault-tolerant should be analyzed for applicability to multi-robot systems. Primate cooperative coverage is marked by three interconnected features: decentralized, asynchronous decision making; limited communications modalities; and transient role selection.

8.3.1 Decentralized, Asynchronous Decision Making

Cooperative search requires that all members disperse to cover the environment. Making good movement decisions is predicated on each robot having enough information about the current state of the environment to choose actions that compliment the team and enhance coverage performance. However, the framework that is used to communicate information and make decisions can both help and constrain the cooperative process. For example, in early multi-robot systems a centralized computer received information about the environment state from all robots from which it calculated optimal actions for the team. Robots were stationary between the time that sensor information was sent and the new commands were received. This synchronous paradigm not only caused the robots to move sporadically while they waited for the calculation of the next instruction but relied heavily upon a two-way communications channel between each robot and the central computer (Figure 8.3). Most current research focuses on decentralized cooperation where team members are peers. Robots make action decisions based on collective state information made available to all team members. Some approaches[15,16] use an explicit cooperation paradigm where action selection results from a team-based negotiation. Although near optimal selections can result, it requires that members synchronously engage in reasoning about the best approach.

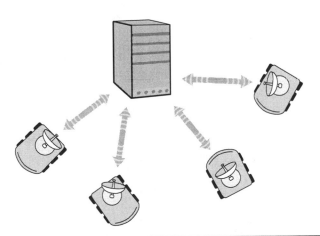

Figure 8.3 Robot communication with a central arbitrator that compiles sensor data, calculates global state, and determines next actions for each robot.

There is no evidence that primates negotiate action selection. Cooperative coverage within primate groups can be explained by primarily individual behavior that is influenced by the actions of nearby teammates. This translates into greedy coverage by individual robots based on completely or partially shared views of the world.

Greedy coverage approaches are central to work by Yamauchi,[17] where robots select next actions from a global list of search areas. However, such algorithms are susceptible to duplication of coverage and interference among team members. Newer approaches[18] allow each robot to keep individual action lists, rather than work from a shared list. Completed actions are removed from the lists of other members but actions are only added via local discovery. This shift reduces duplication and interference and results in faster, more consistent coverage performance due to the large differences that exist in task lists.

Although the greedy approach relates to primates in terms of action selection being individual and asynchronous, there are differences in the scope of knowledge. In robot teams, each robot is expected to have a complete list of searched areas, usually communicated digitally. This is the mechanism for reducing duplication as each member removes already searched areas from the list. However, primates do not appear to communicate this information. It is unclear how or if primates reduce duplication of coverage. Perhaps, observation of the current actions of others precipitates the prediction of future actions.

Observation-based intent information and action prediction have not been studied as components of multi-robot search. Given primate-based cooperation paradigms, these mechanisms may provide the ability to reduce duplication without communicating and sharing a global map.

8.3.2 Limited Communications Modalities

The greedy approach to cooperative coverage displayed by primates has most likely evolved to deal with the limitations of their communications system. Unlike many multi-robot systems that rely upon digital communications for cooperation, primate systems function in the absence of heavily connected networks. Although communications modalities vary within each primate group, the distinct signal types and propagation distance of messages are limited. Communications networks in primates use an opportunistic model where information is shared when possible. The connections mirror Figure 8.4 rather than the multi-robot model of complete connections (Figure 8.5).

Not all multi-robot systems use complete peer-to-peer digital communication. Other research has focused on motivating cooperative paradigms that can function in less complex communications structures. Rybski et al.[19] demonstrated a multi-robot foraging system that used observation of lights to communicate the presence of targets. The low-dimensional communications mechanism did not enhance the task completion time over no communication but did reduce the variance in

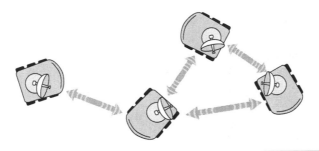

Figure 8.4 **Robots share information only with other robots within their line of sight or close proximity. Cooperation is based on only a partial view of the environment.**

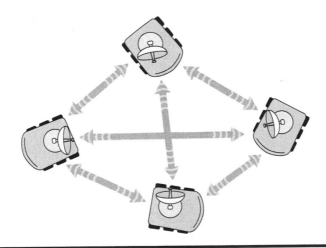

Figure 8.5 **Robots cooperate via a complete connection communications mechanism that facilitates data sharing.**

performance. These results mirror reports from Balch and Arkin[20] that more complex communications frameworks did not guarantee quicker coverage.

Observation-based cooperation can also be found in social potential field methods. These methods rely upon the recognition of team members in the search field to facilitate dispersion. Implemented via range sensors,[21–23] robots assume sensed obstacles are other robots. Robots calculate an action vector that maximizes distance from neighbors. These methods have proven useful for local dispersion but not for dispersion in large areas where once robots move away from neighbors, there is no additional heuristic available.

Pheromone-based methods use indirect observation for communication. Dispersion is accomplished on a global level by moving away from the trails left by other robots. Analogous physical implementations use Radio Frequency Identification (RFID)[24] and beacons[25] to accomplish indirect communication for the purposes of cooperative coverage.

Both utterances and observation appear to serve as the basis for information sharing. Multi-robot systems that rely upon both observation and opportunistic information sharing are still evolving. The challenges are how competing information should be used to influence action. The ability to make use of joint information may hold the key to reasonable cooperation under realistic communications constraints.

8.3.3 Transient Role Selection

Some primate groups use roles throughout coverage to enhance cooperation. Unlike the traditional roles used in cooperative paradigms, primates do not hold roles for the complete duration of the search but only for as long as the role is needed. In addition, the choice to assume a role is decided individually, not by group negotiation or assignment and only when it is perceived that the role is beneficial to task completion.

Transient role selection requires an understanding of the task that decomposes the possible synergies into specific actions. Illustrated by the roles used in hunting Tai chimpanzees, roles dictate spatial placement within the group and relative movement choices. Although this example shows the importance of roles in pursuing prey, the prerequisite act, finding prey, also seems to benefit from using roles to facilitate useful relative placement and coordinated movement, especially in more complex environments with limited visibility (movement on the ground vs. movement in the trees).

Current research on role-based cooperation focuses on assignment via ability or playbooks. Chaimowicz et al.[26] present role assignment as a function of utility. Each individual robot changes or exchanges roles based upon the higher utility of assuming another role.

Much of the role assignment research has been motivated by RoboCup, a robotic soccer tournament. In soccer, roles suggest spatial placement around the soccer ball, proximity to the goal, and passing logistics. In Vail and Veloso,[27] three roles—primary attacker, offensive supporter, and defensive supporter—are determined in parallel by each robot based on shared information. This contrasts with the primate example where roles are observed and chosen asynchronously.

8.4 Toward Bio-Inspired Coverage: A Case Study in Decentralized Action Selection

Work done by Anderson and Papanikolopoulos[18] examines a decentralized, asynchronous approach to cooperative search. A team of robots search an unknown area by recursively discovering, queuing, and scouting the boundaries between open and unknown spaces. The search algorithm used by this technique is based on the reactive, layered approach. It is composed of three behaviors: stall recovery, obstacle avoidance, and search. The robots maintain a worldview that is comprised of a grid of cells.[28] Each cell can either be marked empty or seen. Cells that have been marked as seen are further described as being occupied or unoccupied. An occupied cell can be later labeled as unoccupied but an unoccupied status cannot be changed. In this way static features of the environment are easily identified. For this coverage system each robot keeps a queue of target cells and moves to consume each target until the queue is empty. When there are no targets available, the search employs a random walk.

Lightweight cooperation is achieved by allowing the robots to share information about previously searched areas. Team members remove already searched areas from their queues to reduce duplication of coverage. By restricting the kind of information being shared between team members, this system exhibits decentralized, asynchronous action selection. The individual robots use a greedy approach to search. Members select the next search action that is best for itself in the absence of coordination from other team members. In fact, robots maintain different lists of areas to search since only local sensor information is used to determine what areas to search. The implications are that no one robot has a global picture of all the known areas to search. However, since the future intentions of team members is never shared or considered, the removal of a robot does not cause an area to go uncovered.

Both simulation and physical experiments using K-Tem Koala robots (Figure 8.6) were performed to analyze the implications of using an individual greedy approach

Figure 8.6 K-Team Koala robots equipped with LED laser range finder and 500 MHz PC-104 for control. Stargazers provide ground truth positioning for performance analysis.

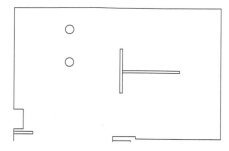

Figure 8.7 **Robot experiments were conducted using up to three robots searching in a 50 m² area for 20 min each.**

instead of a more coordinated approach (robots share a global view of progress and known areas to search). Figure 8.7 describes the experimental setup. It was observed that the average coverage percentage for the individual greedy approach was higher. The individual greedy search also performed more consistently, having a lower variance between trials. When robots have identical search task lists, they tend to display more path overlap and a higher number of collisions.

These results support the efficiency of using more loosely coordinated teams of robots modeled after primate groups. Although this research only implemented one of the three paradigms, the results are promising. Future research may discover ways to implement more primate behaviors effectively in multi-robot teams.

8.5 Summary and Conclusions

Paradigms from primate-based coverage systems provide cues to fault-tolerant, robust cooperative coverage in multi-robot systems. During a search, primates demonstrate decentralized asynchronous action selection, limited communications modalities, and transient role selection. Together, these features should be analyzed for applicability in multi-robot systems.

In this chapter, we have presented three primates and examined how they cooperatively travel and forage. Primate-based coverage topics include role selection, communication and navigation, and the impact of the environment on cooperation and communication. The benefits of primate-based coverage systems have been discussed. An example of a traditional approach to coverage compared to a primate-inspired approach that uses decentralized asynchronous action selection, limited communications modalities, and transient role selection has been presented. A direction for future work is to apply these features to multi-robot systems.

References

1. I. D. Chase, Cooperative and noncooperative behavior in animals, *The American Naturalist*. **115**(6), 827–857 (1980).
2. M. B. Manser, Response of foraging group members to sentinel calls in suricates, suricata suricatta, *Proceedings: Biological Sciences*. **266**(1423), 1013–1019 (1999).
3. P. E. Stander, Cooperative hunting in lions: The role of the individual, *Behavioral Ecology and Sociobiology*. **29**(6), 445–454 (1992).
4. J. C. Bednarz, Cooperative hunting harris' hawks (parabuteo unicinctus), *Science*. **239**(4847), 1525–1527 (1988).
5. C. Boesch, Cooperative hunting roles among Taï chimpanzees, *Human Nature*. **13**, 27–46 (2002).
6. C. Boesch and H. Boesch, Hunting behavior of wild chimpanzees in the Tai national park, *American Journal of Physical Anthropology*. **78**, 547–573 (1989).
7. C. Boesch and H. Boesch-Achermann, *The Chimpanzees of the Tai Forest: Behavioural Ecology and Evolution* (Oxford University Press, Oxford, 2000).
8. J. Goodall, *The Chimpanzees of Gombe* (Replica Books, Bridgewater, CT, 1997).
9. J. Lehmann and C. Boesch, To fission or to fusion: Effects of community size on wild chimpanzee (pan troglodytes verus) social organisation, *Behavioral Ecology and Sociobiology*. **56**, 207–216 (2004).
10. C. Boesch, Hunting strategies of Gombe and Tai chimpanzees, *Chimpanzee Cultures*. pp. 77–91 (Howard University Press, Cambridge, MA, 1994).
11. S. Boinski, P. A. Garber, and S. Boinski, *On the Move: How and Why Animals Travel in Groups* (University of Chicago Press, Chicago, IL, 2000).
12. S. Boinski, E. Moraes, D. G. Kleiman, J. M. Dietz, and A. J. Baker, Intra-group vocal behaviour in wild golden lion tamarins, leontopithecus rosalia: Honest communication of individual activity, *Behaviour*. **130**, 53–75 (1994).
13. S. Boinski, Vocal coordination of troop movement among white-faced capuchin monkeys, cebus capucinus, *American Journal of Primatology*. **30**, 85–100 (1993).
14. S. Boinski and A. F. Campbell, Use of trill vocalizations to coordinate troop movement among white-faced capuchins: A second field test, *Behaviour*. **132**, 875–901 (1995).
15. A. Gage and R. R. Murphy, Affective recruitment of distributed heterogeneous agents. In *Proceedings of the Nineteenth National Conference on Artificial Intelligence*, San Jose, CA, pp. 14–19 (2004).
16. R. Zlot, A. Stentz, M. B. Dias, and S. Thayer, Multi-robot exploration controlled by a market economy. In *Proceedings of the IEEE/RSJ International Conference on Robotic Automation (ICRA)*, Washington, DC, Vol. 3 (2002).
17. B. Yamauchi, Frontier-based exploration using multiple robots. In *International Conference on Autonomous Agents*, Minneapolis MN (1998).
18. M. Anderson and N. Papanikolopoulos, Implicit cooperation strategies for multi-robot search of unknown areas, *Journal of Intelligent and Robotic Systems*. **53**(4), 381–397, (2008).
19. P. E. Rybski, S. A. Stoeter, M. Gini, D. F. Hougen, and N. Papanikolopoulos, Performance of a distributed robotic system using shared communications channels, *Transactions on Robotics Automation*. **22**(5), 713–727 (2002).

20. T. Balch and R. C. Arkin, Communication in reactive multiagent robotic systems, *Autonomous Robots*. **1**(1), 27–52 (1994).

21. G. Fang, G. Dissanayake, and H. Lau, A behaviour-based optimisation strategy for multi-robot exploration. In *IEEE Conference on Robotics, Automation and Mechatronics*, Singapore (2004).

22. H. Lau, Behavioural approach for multi-robot exploration. In *Proceedings of 2003 Australasian Conference on Robotics and Automation*, Brisbane, Australia, December (2003).

23. M. A. Batalin and G. S. Sukhatme, Spreading out: A local approach to multi-robot coverage. In *Proceedings of 6th International Symposium on Distributed Robotic Systems*, pp. 373–382 (2002).

24. A. Kleiner, J. Prediger, and B. Nebel, RFID technology-based exploration and SLAM for research and rescue. In *Proceedings of the IEEE/RSJ International Conference on Intelligent Robots and Systems (IROS)*, Beijing (2006).

25. M. A. Batalin and G. S. Sukhatme, Efficient exploration without localization. In *Proceedings of the IEEE International Conference on Robotics and Automation (ICRA)*. Vol. 2 (2003).

26. L. Chaimowicz, M. Campos, and V. Kumar, Dynamic role assignment for cooperative robots. In *Proceedings of IEEE International Conference on Robotics and Automation (ICRA'02)*, Vol. 1 (2002).

27. D. Vail and M. Veloso, Multi-robot dynamic role assignment and coordination through shared potential fields, *Multi-Robot Systems* (Kluwer, Dordrecht, 2003).

28. A. Elfes, Using occupancy grids for mobile robot perception and navigation, *Computer*. **22**(6), 46–57 (1989).

Chapter 9

Abstractions for Planning and Control of Robotic Swarms

Calin Belta

Contents

In this chapter, we outline a hierarchical framework for planning and control of arbitrarily large groups (swarms) of robots. At the first level of hierarchy, we aggregate the high dimensional control system of the swarm into a small dimensional control system capturing its essential features, such as position of the center, shape, orientation, and size. At the second level, we reduce the problem of controlling the essential features of the swarm to a model checking problem. In the obtained hierarchical framework, high-level specifications given in natural language such as linear temporal logic formulas over linear predicates in the essential features are automatically mapped to provably correct robot control laws. We present simulation results for the particular case of a continuous abstraction based on centroid and variance of a planar swarm made of fully actuated robots with polyhedral control constraints.

As a result of tremendous advances in computation, communication, sensor, and actuator technology, it is now possible to build teams of hundreds and even thousands of small and inexpensive ground, air, and underwater robots. They are light and easy to transport and deploy and can fit into small places. Such *swarms of autonomous agents* provide increased robustness to individual failures, the possibility to cover wide regions, and improved computational power through parallelism. However, planning and controlling such large teams of agents with limited communication and computation capabilities is a hard problem that received a lot of attention in the past decade.

It is currently believed that inspiration should be taken from the behavior of biological systems such as flocks of birds, schools of fish, swarms of bees, or crowds of people. Over the past 2 decades, researchers from several areas, including ecology and evolutionary biology, social sciences, statistical physics, and computer graphics, have tried to understand how local interaction rules in distributed multiagent systems can produce emergent global behavior. Over the past few years, researchers in control theory and robotics have drawn inspiration from the above areas to develop distributed control architectures for multiagent mobile systems. Even though, in some cases, it was observed or proved that local interaction rules in distributed natural or engineered multiagent systems produce global behavior, the fundamental questions still remain to be answered. What are the essential features of a large group? How do we specify its behavior? How can we generate control laws for each agent so that a desired group behavior is achieved? In this chapter, we outline some ideas and results that could partially answer the above questions. The approach exploits the interplay between the "continuous" world of geometric nonlinear control and the traditionally "discrete" world of automata, languages, and temporal logics.

9.1 Specification-Induced Hierarchical Abstractions

The starting point for the approach presented here is the observation that tasks for large groups evolving in complex environments are "qualitatively" specified. This notion has a dual meaning. First, a swarm is naturally described in terms of a small set of "features," such as shape, size, and position of the region in a plane or space occupied by the robots, while the exact position or trajectory of each robot is not of interest. Second, the accomplishment of a swarming mission usually does not require exact values for swarm features, but rather their inclusion in certain sets. For example, in the planar case, if the robots are constrained to stay inside an ellipse, there is a whole set of values for the pose and semi-axes of the ellipse that guarantees that the swarm will not collide with an obstacle of given geometry. Moreover, specifications for mobile robots are often temporal, even though time is not necessarily captured explicitly. For example, a swarm might be required to reach a certain position and shape "eventually" or maintain a size smaller than a specified value "until" a final desired value is achieved. Collision avoidance among robots, obstacle avoidance, and cohesion are required "always." In a surveillance mission, a certain area should be visited "infinitely often."

Motivated by the above ideas, in this chapter we briefly describe an approach for planning and control of robotic swarms based on *abstractions*. Our framework is hierarchical. At the first level, we construct a *continuous abstraction* by extracting a small set of features of interest of the swarm. The result of the continuous abstraction will be a small dimensional continuous control system, whose dimension does not scale with the number of robots, and with constraints induced by the individual constraints of the robots. Intuitively, its state can be seen as the coordinates of a geometrical object spanning the area or volume occupied by the robots. An illustration is given in Figure 9.1 for a planar case, where the "spanning" object is an ellipse. In this example, if we can control the pose and semiaxes of the ellipse with the guarantee that the robots stay inside and their control and communication constraints are satisfied, then in principle we reduce a very high dimensional control problem to a five-dimensional one (two coordinates for position of center, one for rotation, and one for each of the two semiaxes).

Figure 9.1 Cooperative tasks for robotic swarms can be specified in terms of low dimensional abstractions capturing the pose and shape of geometrical objects describing the area (volume) occupied by the robots, such as spanning ellipses.

At the second level of the hierarchy, we use *discrete abstractions* to generate control strategies for the continuous abstraction from high-level, human-like task specifications given as temporal and logic statements about the reachability of regions of interest by the swarm. Simply put, through "smart" partitioning, discrete abstractions reduce the problem of controlling the continuous abstraction to a control problem for an automaton. In the discrete and finite world of automata, such human-like specifications translate to temporal logic formulas, and the control problem resembles model checking, which is just a more sophisticated search on a graph.

When combined, the two approaches lead to the *hierarchical abstraction architecture* shown in Figure 9.2, where the *continuous abstraction reduces the dimension of the problem by focusing on swarm cohesion and robot dynamics and constraints*, and the *discrete abstraction captures the complexity of the environment*. The geometry of the obstacles and the specifications of the swarming task determine a partition of the state

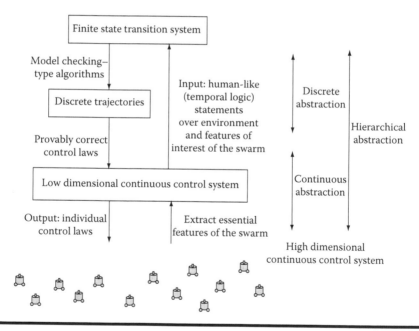

Figure 9.2 Hierarchical abstraction architecture for planning and control of robotic swarms: high-level specifications given in human-like language such as temporal logic formulas are used to construct a discrete and finite description of the problem. Tools resembling model checking are involved in finding a solution for this problem, which is then implemented as a hybrid control strategy for the continuous abstraction. Individual robot control laws are then generated through projection. (From Kloetzer, M. and Belta, C., *IEEE Trans. Robotics*, 23(2), 320, 2007.)

space of the continuous abstraction produced at the first level. Discrete abstractions will then be used to check the existence and construct admissible control laws for the continuous abstraction, which captures the individual robot constraints. Therefore, in this hierarchical abstraction framework, "qualitatively" specified swarming tasks, such as temporal and logical statements about the reachability of regions of interest, are in principle reduced to analyses of finite state automata.

9.2 Continuous Abstractions: Extracting the Essential Features of a Swarm

Coordinates of geometrical objects describing the volume (or area) occupied by the robots can serve as abstract descriptions of the swarm. For example, for a planar task, controlling the pose (position and orientation) and semiaxes of a spanning ellipse, while making sure that the robots stay inside it, can be used for a large class of cooperative tasks. A pictorial illustration of this idea is given in Figure 9.1, where the ellipse can be shrunk, reshaped, and/or reoriented for avoiding obstacles.

Inspired by this idea, one can think of abstractions as having a product structure of a Lie *group* capturing the pose of the team in the world frame and a *shape*, which is an intrinsic description of the team, invariant to the world coordinate frame. In our example from Figure 9.1, the group would be the position of the center and the orientation of the ellipse in the world frame, while the shape would correspond to its two semiaxes. For such a group-shape representation of the swarm, it makes sense to require that the group part be controlled separately from the shape. Indeed, the idea of having a coordinate free definition of the shape together with shape control law decoupled from the group is fundamental in formation control. Without knowing the coordinates of the robot in the world frame, only shape variables can be measured and controlled using on-board sensors of the robots.[1,2] In addition, a description of the swarm should be invariant to robot permutations. This would lead to robustness against individual failures for the planning and control architecture to be developed.

To introduce the main ideas, let us assume for simplicity that the swarm is composed of N identical planar fully actuated point-like robots with polyhedral control constraints. In other words, each robot is described by a control system of the form

$$\dot{r}_i = u_i, \quad u_i \in U, \quad i = 1, \ldots, N, \tag{9.1}$$

where
  $r_i \in \mathbb{R}^2$ is the position vector of robot i in a world frame
  u_i is its velocity, which can be directly controlled
  $U \subseteq \mathbb{R}^2$ is a set capturing the control constraints

We collect all the robot states in $r = \left[r_1^T, \ldots, r_N^T\right]^T \in \mathbb{R}^{2N}$ and the robot controls in $u = \left[u_1^T, \ldots, u_N^T\right]^T \in \mathbb{R}^{2N}$. We denote by Q the set of all swarm configurations, which equals $\mathbb{R}^{2N}$ if no obstacles and environment limits are imposed.

To construct a continuous abstraction with a product structure of group and shape, we build a smooth surjective map

$$h : Q \to A, \quad h(r) = a, \tag{9.2}$$

where

    h is called the (continuous) *abstraction*, or *aggregation*, or *quotient* map
    a is denoted as the *abstract state* of the swarm

We require A to have a product structure of a group G and a shape S, i.e.,

$$A = G \times S, \quad a = (g,s), \quad a \in A, \quad g \in G, \quad s \in S. \tag{9.3}$$

$g \in G$ and $s \in S$ are the group and the shape part of $a \in A$, respectively. The invariance to robot identifications translates to invariance of h to permutations of r_i. The invariance to the world frame translates to left invariance for h, which simply means that, if a change occurs in the world frame, which can be modelled as an action by an element $\bar{g} \in G$, then, in the image of the swarm configuration through the map h, the group part will be left translated by $\bar{g}$, while the shape part will not change.

The notions of invariance with respect to world frames and permutations of robots, which are fundamental for swarms, are properly approached in shape theory in the well-known "n-body problem," traditionally studied in theoretical physics,[3] and more recently applied to robotics.[1] However, these works focus on maximal shape spaces (i.e., largest subspaces of configuration spaces invariant to permutations and world frames) and are therefore restricted to very small teams of robots. To use such ideas for robotic swarms, one can start, as in the works enumerated above, from Jacobi vectors.[4] However, instead of computing maximal Jacobi spaces, one can try "democratic" sums over dot products and triple products of Jacobi vectors, as suggested by Zhang et al.[1] Jacobi vectors are already reduced by translation, the dot products are invariant to rotations, and permutations of particles are easy to characterize as actions of the "democracy" group $O(N-1)$. Moreover, such sums can give useful estimates of the area occupied by the robots in the planar case and of the volume of a spanning region in the spatial case.

Alternatively, one can investigate the eigenstructure of the *inertia tensor* of the system of particles r_i, $i = 1, \ldots, N$. Properly normed and ordered eigenvectors can give the rotational part of the group, while the eigenvalues (and physically significant combinations of them) are good candidates for shape variables. Related to this, one can think of a probabilistic approach, starting from the assumption that the vectors r_i are normally distributed. Then their sample mean and covariance converge to the parameters of a normal distribution when N is large enough. The corresponding *concentration ellipsoid* is guaranteed to enclose an arbitrary percentage of the robots inside it, and can therefore be used as a spanning region. If the distribution is not normal, higher central moments can be useful in defining abstractions as well. For example, the third moment (*skewness*) is a measure of how far the distribution is from being symmetric, while the fourth central moment (*kurtosis*) can be used as a measure of the coverage by the robots of the area of the ellipsoid.

In addition to providing a description of the swarm position, size, and shape that is invariant to permutations and world frames, we will require h to perform a *correct aggregation* of the large dimensional state space Q of the swarm. Let $u(r)$ and $w(a)$ be the vector fields giving the full dynamics of the swarm and of the continuous abstraction, respectively, with $a = h(r)$. In our view, a correct aggregation has three requirements. First, the flow $\dot{r} = u(r)$ and the quotient map (9.2) have to be *consistent*, which intuitively means that swarm configurations that are equivalent (indistinguishable) with respect to h are treated in an equivalent way by the flow of the swarm. Second, a correct aggregation should allow for any motion on the abstract space A, which we call the *actuation* property. Third, we do not allow the swarm to spend energy in motions which are not "seen" in the abstract space A (*detectability*). Kloetzer and Belta,[5] using a geometrical approach, provide necessary and sufficient conditions for correct aggregation. In short, the consistency of the flow $\dot{r} = u(r)$ and of the quotient map (9.2) is equivalent to the "matching condition" $dh(r)u(r) = dh(r')u(r')$, for all r, r' with $h(r) = h(r')$, where $dh(r)$ denote the differential (tangent) map of h at point r. The actuation property is equivalent to requiring h to be a submersion, while the detectability requirement is equivalent to restricting $u(r)$ to the range of $dh^T(r)$. Note that all these characterizations are all very simple conditions that can be easily checked computationally. In conclusion, if all the above conditions are satisfied, then arbitrary "abstract" vector fields $w(a)$ ($a = h(r)$) in A can be produced by "swarm" vector fields $u(r)$.

9.2.1 Examples of Continuous Abstractions

Belta and Kumar[6] defined a five-dimensional abstraction (9.2) that can be used for swarms of fully actuated planar robots. The abstraction satisfies all the requirements stated above. An extension to unicycles, based on a simple input–output regulation intermediate step, is described by Belta et al.[7] In the abstraction $a = (g, s)$, the three-dimensional group part is $g = (\mu, \theta) \in SE(2)$, where μ is the centroid of the swarm and $\theta \in S^1$ parameterizes the rotation that diagonalizes the inertia tensor of the system of particles. The two-dimensional shape $s \in \mathbb{R}^2$ is given by the eigenvalues of the inertia tensor. In this interpretation, the abstraction captures the pose and side lengths of a rectangle with the guarantee that the robots are always inside. Equivalently, if the robots are assumed to be normally distributed, μ can be interpreted as a sample mean, θ as rotation diagonalizing the sample covariance matrix, while the shape variables s are the eigenvalues of the covariance matrix. In this second interpretation, the abstraction parameterizes an ellipse, with the guarantee that an arbitrary percent of the

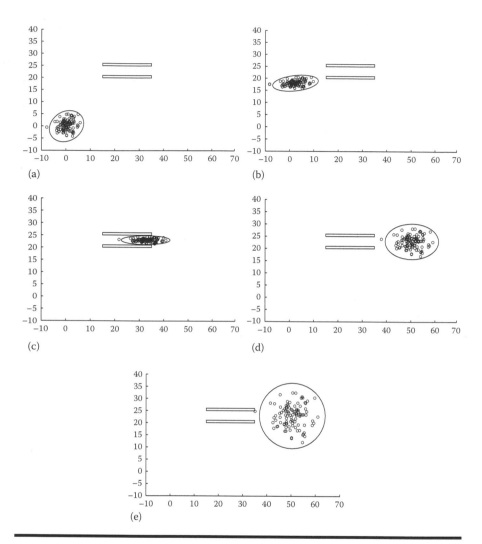

Figure 9.3 Ninety nine of *N* = 100 normally distributed planar robots are driven through a tunnel by designing five-dimensional controls for the corresponding equiprobability ellipse: the robots start from some initial configuration (a), gather in front if the tunnel (b), pass the tunnel (c), and then expand at the end of the tunnel to cover a large area (d), (e). (From Belta, C. and Kumar, V., *IEEE Trans. Robotics*, 20(5), 865, 2004.)

robots stay inside. A simulation result is shown in Figure 9.3, where a swarm of *N* = 100 robots (evolving on a 200-dimensional space!) is driven through a tunnel by designing controls on a five-dimensional space parameterizing a spanning ellipse.

When orientation is not relevant for a certain application, we can define a simpler three-dimensional abstraction, by restricting the group *g* to the position

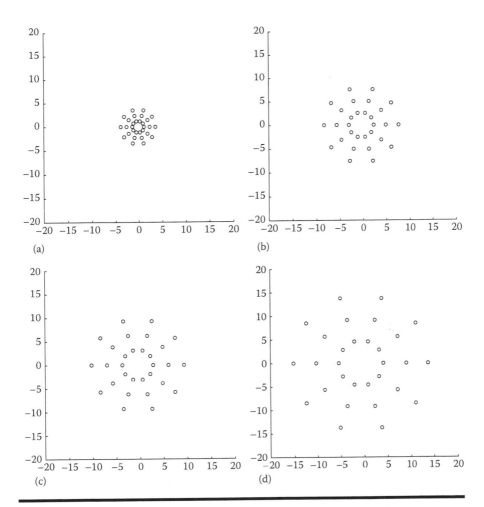

Figure 9.4 $N = 30$ robots experiencing an expansion using control laws based on three-dimensional abstraction: ((a)–(d) are consecutive snapshots). The centroid is kept fixed. Orientation, parallelism, angles, and ratios of lengths are preserved. (From Belta, C. and Kumar, V., *IEEE Trans. Robotics*, 20(5), 865, 2004.)

of the centroid μ, and by collapsing the shape s to the sum of the two shape variables from above. In this case, the N robots described by a three-dimensional abstract variable $a = (\mu, s)$ are enclosed in a circle centered at μ and with radius proportional to s. In Figure 9.4, we show a simulation for controlling $N = 30$ robots using this simpler abstraction. Initially, the robots are distributed on three concentric circles.

A generalization of these ideas to three-dimensional environments is possible. Michael et al.[8] constructed a nine-dimensional abstraction ($a = (g, s)$) for a swarm in three dimension, where the group part $g = (\mu, R) \in SE(3)$ consists of centroid

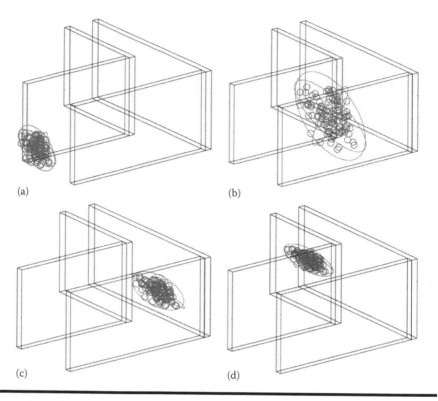

(a)   (b)

(c)   (d)

Figure 9.5 $N = 100$ **robots pass a corridor by controlling the nine-dimensional abstraction parameterizing a concentration ellipsoid: the robots start from some initial configuration (a), gather in front if the tunnel (b), pass the tunnel (c), and then expand at the end of the tunnel to cover a large area (d), (e). (From Michael, N. et al., Controlling three dimensional swarms of robots, In *IEEE International Conference on Robotics and Automation*, Orlando, FL, 2006.)**

$\mu \in \mathbb{R}^3$ and rotation $R \in SO(3)$ that diagonalizes the inertia tensor. As before, the shape $s \in \mathbb{R}^3$ captures the eigenvalues of the inertia tensor. In Figure 9.5, we show a simulation of controlling a swarm of $N = 100$ robots in space based on this nine-dimensional abstraction, which can be seen as parameterizing an ellipsoid spanning the volume occupied by the swarm.

9.3 Discrete Abstractions: Accommodating Rich Specifications

Once the large dimensional state of the swarm is correctly aggregated by properly choosing the aggregation map and the robot control laws, we have the freedom to assign arbitrary vector fields $w(a)$ in the abstract space A. These vector fields will be constructed from high-level temporal and logical specifications over the continuous abstraction a. For example, assume that $a = (\mu, s) \in \mathbb{R}^3$, where $\mu \in \mathbb{R}^2$ gives the centroid of a swarm and $s \in \mathbb{R}$ is its size (e.g., area). If it is desired that the swarm converges to a configuration in which its centroid belongs to a polygon P^d and with a size smaller than s^d, this can be written more formally as "eventually always ($\mu \in P^d$ and $s < s^d$)," with the obvious interpretation that it will eventually happen that $\mu \in P^d$ and $s < s^d$ and this will remain true for all future times. If during the convergence to the final desired configuration it is necessary that the swarm visits a position $\bar{\mu}$ with a size $\bar{s}$, then the specification becomes "eventually (($\mu = \bar{\mu}$ and $s = \bar{s}$) and (eventually always ($\mu \in P^d$ and $s < s^d$)))." If in addition it is required that the size s is smaller than $\bar{s}$ for all times before $\bar{s}$ is reached, the specification changes to "$s < \bar{s}$ until (($\mu = \bar{\mu}$ and $s = \bar{s}$) and (eventually always ($\mu \in P^d$ and $s < s^d$)))."

The starting point in the development of the discrete abstraction is the observation that such specifications translate to formulas of temporal logics over linear predicates in the abstract variables. Informally, linear temporal logic (LTL) formulas are made of temporal operators, Boolean operators, and atomic propositions connected in any "sensible way."[9] Examples of temporal operators include **X** ("next time"), **F** ("eventually" or "in the future"), **G** ("always" or "globally"), and **U** ("until"). The Boolean operators are the usual $\neg$ (negation), $\vee$ (disjunction), $\wedge$ (conjunction), $\Rightarrow$ (implication), and $\Leftrightarrow$ (equivalence). The atomic propositions are properties of interest about a system, such as the linear inequalities in μ and s enumerated above. For example, the last specification in the above paragraph corresponds to the LTL formula $(s < \bar{s})\mathbf{U}(\mu = \bar{\mu} \wedge s = \bar{s}) \wedge (\mathbf{FG}(\mu \in P^d \wedge s < s^d))$.

Kloetzer and Belta[10] developed a computational framework for the control of arbitrary linear systems (i.e., $\dot{x} = Ax + b + Bu$) from specifications given as LTL formulas in arbitrary linear predicates over the state x. To illustrate the method, consider the two-dimensional linear system

$$\dot{x} = \begin{bmatrix} 0.2 & -0.3 \\ 0.5 & -0.5 \end{bmatrix} x + \begin{bmatrix} 1 & 0 \\ 0 & 1 \end{bmatrix} u + \begin{bmatrix} 0.5 \\ 0.5 \end{bmatrix}, \quad x \in P, \quad u \in U, \qquad (9.4)$$

where P is the bounding polytope shown in Figure 9.6, and the control constraint set is $U = [-2, 2] \times [-2, 2]$. Assume the task is to visit a sequence of three regions r_1, r_2, and r_3, in this order, infinitely often, while avoiding regions o_1, o_2, and o_3 for all times (see Figure 9.6). Using the temporal and logic operators defined above, this specification translates to the LTL formula

$$\mathbf{G}(\mathbf{F}(r_1 \wedge \mathbf{F}(r_2 \wedge \mathbf{F}r_3)) \wedge \neg(o_1 \vee o_2 \vee o_3)). \qquad (9.5)$$

The solution is given as the set of initial states from which the formula can be satisfied (see Figure 9.6) and a corresponding set of feedback control strategies.

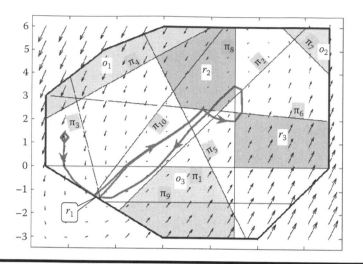

Figure 9.6 A polygonal two-dimensional state space, the target polygonal regions r_1, r_2, r_3 (shown in dark grey) and the regions to be avoided o_1, o_2, o_3 (shown in grey), which can be expressed as conjunctions of the linear predicates $\pi_1, \pi_2, \ldots, \pi_{10}$. The arrows indicate the drift vector field of system (9.4). There exist control strategies such that all trajectories of the closed loop system (9.4) satisfy formula (9.5) for any initial state in *P*, except for the light regions. A sample trajectory originating in the allowed set of initial conditions is shown. (From Kloetzer, M. and Belta, C., *IEEE Trans. Automatic Control*, 53(1), 287, 2008.)

9.4 Hierarchical Abstractions: Automatic Deployment of Swarms from Human-Like Specifications

We are now ready to present a computational framework allowing for automatic deployment of arbitrarily large swarms of fully actuated robots from specifications given as arbitrary temporal and logic statements about the satisfaction of linear predicates by the swarm's mean and variance. These types of specifications seem to be enough for a fairly large class of tasks. In addition, the method allows for automatic containment inside the environment, obstacle avoidance, cohesion, and inter-robot collision avoidance for all times.

To extract the mean and variance of the swarm, we define the following three-dimensional abstraction map:

$$h(r) = a, \quad a = (\mu, \sigma), \quad \mu = \frac{1}{N} \sum_{i=1}^{N} r_i, \quad \sigma = \sqrt{\frac{1}{N} \sum_{i=1}^{N} (r_i - \mu)^T (r_i - \mu)}.$$

$$(9.6)$$

Kloetzer and Belta[5] show that the robot controllers

$$u_i(r_i, a) = \left[I_2 \quad \frac{r_i - \mu}{\sigma} \right] w(a), \quad i = 1, \dots, N, \tag{9.7}$$

where $w(a)$ is an arbitrary vector field in the abstract space $\mathbb{R}^3$, provide a correct aggregation with respect to the abstraction map (9.6). Under the assumption that the environment and the obstacles are polyhedral, we also show that containment inside the environment, obstacle avoidance, cohesion, and inter-robot collision avoidance can be expressed as LTL formulas over linear predicates in the abstract variables (μ, σ). In addition, arbitrary polyhedral control constraints can be accommodated. The proofs are based on the following two main facts: (a) under the control laws (9.7), the swarm undergoes an affine transformation, which preserves the vertices of its convex hull, and the pairs of robots giving the maximum and minimum pairwise distances and (b) quantifier elimination in the logic of the reals with addition and comparison is decidable.

To illustrate, consider a swarm consisting of $N = 30$ robots moving in a rectangular environment P with two obstacles O_1 and O_2 as shown in Figure 9.7a. The initial configuration of the swarm is described by mean $\mu(0) = [-3.5, \ 4.5]^T$ and variance $\sigma(0) = 0.903$. The convex hull of the swarm is initially the square of center $\mu(0)$ and side 2 shown in the top left corner of Figure 9.7a.

Consider the following swarming task given in normal language: *Always stay inside the environment P, avoid the obstacles O_1 and O_2, and maintain maximum and minimum inter-robot distances of 3.5 and 0.01, respectively. In addition, the centroid μ must eventually visit region R_1. Until then, the minimum pairwise distance must be greater than 0.03. After R_1 is visited, the swarm must reach such a configuration that its centroid is in region R_2 and the spanned area is greater than the initial one, and remain in this configuration forever.* Regions R_1 and R_2 are the two squares shown in Figure 9.7a.

This task translates to the following LTL formula over linear predicates in μ and σ:

$$\mathbf{G}(P_{co} \wedge P_d) \wedge \{(\sigma > 0.54)\mathbf{U}[(\mu \in R_1) \wedge \mathbf{FG}((\mu \in R_2) \wedge (\sigma > \sigma(0)))]\}, \tag{9.8}$$

where P_{co} guarantees containment and obstacle avoidance and is a propositional logic formula consisting of 27 occurrences of 19 different linear predicates in μ and σ, P_d gives upper and lower bounds for σ that guarantee cohesion and inter-robot collision avoidance, and $\sigma > 0.54$ corresponds to pairwise distance greater than 0.03. A solution is shown in Figure 9.7b as the trace of the convex hull of the swarm. By close examination, it can be seen that the specified task is accomplished.

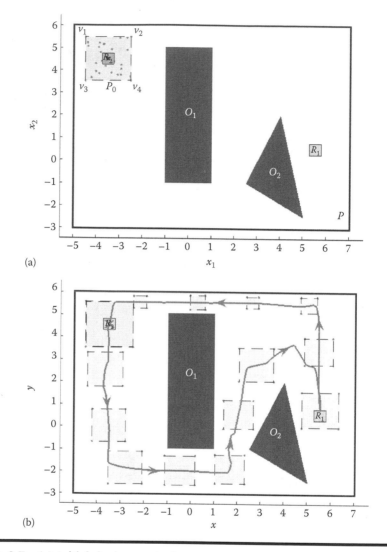

(a)

(b)

Figure 9.7 **(a) Initial deployment of a swarm consisting of 30 robots in a rect-angular environment *P* with two obstacles *O*₁ and *O*₂. The regions *R*₁ and *R*₂ are used in the task specification. (b) Trace of the convex hull of the swarm and trajectory of centroid μ. (From Kloetzer, M. and Belta, C., *IEEE Trans. Robotics*, 23(2), 320, 2007.)**

9.5 Limitations of the Approach and Directions for Future Work

While providing a fully automated framework for deployment of arbitrarily large swarms from rich specifications given in human-like language, the approach presented in the previous section has several limitations. First, it is computationally expensive, since, in addition to polyhedral operations, it involves quantifier elimination and a graph search resembling LTL model checking. Second, in order to guarantee that the swarm does not collide with an obstacle, we impose that the whole convex hull of the swarm has empty intersection with the obstacle. One can imagine that there might exist motions of the swarm where an obstacle enters the convex hull without hitting any of the robots. Third, we restrict our attention to a very small set of essential features, which only allow for translation and scaling of the convex hull. For example, if rotation was allowed, motions where the spanning polytope would rotate to avoid obstacles rather than just unnecessarily shrinking would have been possible. Fourth, our approach to the control of the essential features of the swarm based on discrete abstractions is conservative, as detailed by Kloetzer and Belta.[10]

Fifth, in its current form, the approach assumes that the environment is static and known. Any change in the boundaries of the environment or in the obstacles should be followed by a full re-computation. This being said, for a static and known environment, our framework is robust with respect to small errors in knowledge about the environment. This results from the fact that the discrete abstraction procedure for control of the continuous abstraction from temporal logic specifications over linear predicates (developed in our previous work[10]) is robust against small changes in these predicates. Sixth, throughout this work, we assume that the robots are fully actuated point masses. However, to accommodate robots of non-negligible sizes, one can enlarge the obstacles and shrink the environment boundaries.[11] To accommodate more complex dynamics, one can add another level in the hierarchy. For example, using input–output regulation, controlling a unicycle can be reduced to controlling the velocity of an off-axis reference point.[12] Alternatively, one might try to capture robot under-actuation constraints by properly constructing the continuous abstraction, as suggested by Belta and Kumar.[6]

Last but not least, the resulting communication architecture is centralized. In fact, for all the continuous abstractions reviewed in this article, the control law for each robot depends on the current state of the robot and the state of the continuous abstraction (see Equation 9.7). However, this does not seem very restrictive due to two main reasons. First, in a practical application, this communication architecture can be easily implemented if a swarm of ground vehicles is deployed together with an unmanned aerial vehicle (UAV), such as a blimp. The UAV can be equipped with a camera, and each time it can localize the robots, compute the abstract variable a, and then disseminate it back to the robots. Note that the bandwidth of the broadcast variable a is low (since a is low dimensional) and does not depend on the size of the swarm. Second, recent results[13] in consensus algorithms show that global variables, such that the centroid μ, can be estimated using local information, under some mild assumptions of the topology of the communication graph. Starting from this idea, as suggested by Yang et al.,[14] one can build an "abstract state estimator" based only on local information, and feed the estimated values into the framework described above.

9.6 Conclusion

We reviewed some basic ideas and preliminary results allowing for fully automated deployment of arbitrarily large swarms from high-level and rich specifications. Our approach is hierarchical. In the first level of the hierarchy, we aggregate the large dimensional state space of the swarm into a small dimensional continuous abstract space which captures essential features of the swarm. In the second level, we control the continuous abstraction so that specifications given in linear temporal logic over linear predicates in the essential features are satisfied. For planar robots with polyhedral control constraints moving in polygonal environments with polygonal obstacles, and a three-dimensional continuous abstraction consisting of mean and variance, we show that a large class of specifications are captured.

Acknowledgment

This work is partially supported by NSF CAREER 0447721 at Boston University.

References

1. F. Zhang, M. Goldgeier, and P. S. Krishnaprasad, Control of small formations using shape coordinates. In *Proceedings of IEEE ICRA*, Taipei, Taiwan (2003).
2. E. Justh and P. Krishnaprasad. Steering laws and continuum models for planar formations. In *Proceedings of IEEE Conference on Decision and Control*, Maui, Hawaii, pp. 3609–3614 (2003).
3. R. G. Littlejohn and M. Reinsch, Internal or shape coordinates in the n-body problem, *Phys. Rev. A.* **52**(3), 2035–2051 (1995).
4. C. G. J. Jacobi, *Vorlesungen uber Dynamik* (Reimer, Berlin, 1866).
5. M. Kloetzer and C. Belta, Temporal logic planning and control of robotic swarms by hierarchical abstractions, *IEEE Trans. Robotics.* **23**(2), 320–331 (2007).
6. C. Belta and V. Kumar, Abstraction and control for groups of robots, *IEEE Trans. Robotics.* **20**(5), 865–875 (2004).
7. C. Belta, G. Pereira, and V. Kumar, Control of a team of car-like robots using abstractions. In *42nd IEEE Conference on Decision and Control*, Maui, Hawaii (2003).
8. N. Michael, C. Belta, and V. Kumar, Controlling three dimensional swarms of robots. In *IEEE International Conference on Robotics and Automation*, Orlando, FL (2006).
9. E. A. Emerson, Temporal and modal logic, In ed. J. van Leeuwen, *Handbook of Theoretical Computer Science: Formal Models and Semantics*, Vol. B, pp. 995–1072 (North-Holland Pub. Co./MIT Press, Cambridge, MA, 1990).

10. M. Kloetzer and C. Belta, A fully automated framework for control of linear systems from LTL specifications. In eds. J. Hespanha and A. Tiwari, *Hybrid Systems: Computation and Control: 9th International Workshop*, Vol. 3927, *Lecture Notes in Computer Science*, pp. 333–347 (Springer, Berlin/Heidelberg, 2006).

11. J. Latombe, *Robot Motion Planning* (Kluwer Academic Pub., Boston, 1991).

12. J. Desai, J. Ostrowski, and V. Kumar, Modeling and control of formations of nonholonomic mobile robots, *IEEE Trans. Robotics Automation.* **17**(6), 905–908, (2001).

13. R. Olfati-Saber and R. M. Murray, Consensus problems in networks of agents with switching topology and time-delays, *IEEE Trans. Automatic Control.* **49**(9), 1520–1533.

14. P. Yang, R. A. Freeman, and K. M. Lynch, Multi-agent coordination by decentralized estimation and control, *IEEE Trans. Automatic Control.* **53**(11), 2480–2496, (2003).

Chapter 10

Ant-Inspired Allocation: Top-Down Controller Design for Distributing a Robot Swarm among Multiple Tasks

Spring Berman, Ádám Halász, and M. Ani Hsieh

Contents

We present a decentralized, scalable, communication-less approach to the dynamic allocation of a swarm of homogeneous robots to multiple sites in specified fractions. This strategy has applications in surveillance, search-and-rescue, environmental monitoring, and other task allocation problems. Our work is inspired by an experimentally based model of ant house-hunting, a decentralized process in which a colony attempts to emigrate to the best nest site among several alternatives. In our approach, we design a continuous model of the swarm that satisfies the global objectives and use this model to define stochastic control policies for individual robots that produce the desired collective behavior. We define control policies that are derived from a linear continuous model, a model that includes navigation delays, and a model that incorporates switching behaviors based on quorum sensing. The stability and convergence properties of these models are analyzed. We present methods of optimizing a linear model for fast redistribution subject to a constraint on inter-site traffic at equilibrium, both with and without knowledge of the initial distribution. We use simulations of multi-site swarm deployments to compare the control policies and observe the performance gains obtained through model optimization, quorum-activated behaviors, and accounting for time delays.

10.1 Introduction

Advances in embedded processor and sensor technology that have made individual robots smaller, more capable, and less expensive have also enabled the development and deployment of teams of robotic agents. While multi-robot systems may seem like a recent paradigm shift, distributed robotics research had its genesis in the early 1980s, when the initial focus was primarily on the control and coordination of multiple robotic arms.[1] The unrelenting progression of Moore's law in the 1990s with improvements in sensor and actuation technology, coupled with ubiquitous wireless communication, made it possible to create and deploy teams of robots numbered in the tens and potentially hundreds and thousands. However, as team size increases, it becomes difficult, if not impossible, to efficiently manage or control the team through centralized algorithms or tele-operation. Accordingly, it makes sense to develop strategies in which robots can be programmed with simple but identical behaviors that can be realized with limited on-board computational, communication, and sensing resources.

In nature, the emergence of complex group behaviors from simple agent behaviors is often seen in the dynamics of bee[2] and ant[3] colonies, bird flocks,[4] and fish schools.[5] These systems generally consist of large numbers of organisms that individually lack either the communication or computational capabilities required for centralized control. As such, when considering the deployment of large robot teams, it makes sense to consider such "swarming paradigms" where agents have the capability to operate asynchronously and determine their trajectories based on local sensing and/or communication. One of the earliest works to take inspiration from biological swarms for motion generation was presented in 1987 by Reynolds,[6] who proposed a method for generating visually satisfying computer animations of bird flocks, often referred to as *boids*. Almost a decade later, Vicsek et al. showed through simulations that a team of autonomous agents moving in the plane with the same speeds but different headings converges to the same heading using nearest neighbor update rules.[7] The theoretical explanation for this observed phenomenon was provided by Jadbabaie et al.[8] and Tanner et al. extended these results to provide detailed analyses of the stability and robustness of such flocking behaviors.[9] These works show that teams of autonomous agents can stably achieve consensus through local interactions alone, i.e., without centralized coordination, and have attracted much attention in the multi-robot community.

We are interested in the deployment of a swarm of homogeneous robots to various distinct locations for parallel task execution at each locale. The ability to autonomously distribute a swarm of robots to physical sites is relevant to many applications such as the surveillance of multiple buildings, large-scale environmental monitoring, and the provision of aerial coverage for ground units. In these applications, robots must have the ability to distribute themselves among many locations/sites as well as to autonomously redistribute to ensure task completion at each site, which may be affected by robot failures or changes in the environment.

This problem of (re)distribution is similar to the task/resource allocation problem, in which the objective is to determine the optimal assignment of robots to tasks. Such combinatorial optimization problems are known to scale poorly as the numbers of agents and tasks increase. Furthermore, in the multi-robot domain, existing methods often reduce to market-based approaches[10–12] in which robots must execute complex bidding schemes to determine the appropriate allocation based on the various perceived costs and utilities. While market-based approaches have gained much success in various multi-robot applications,[13–16] the computation and communication requirements often scale poorly in terms of team size and number of tasks.[17,18] Therefore, as the number of agents increases, it is often impractical, if not unrealistic, to expect small, resource-constrained agents to always have the ability to explicitly communicate with other team members, especially those physically located at other sites, e.g., in mining and search-and-rescue applications. For this reason, we are interested in developing a decentralized strategy to the allocation problem that can lead to optimal outcomes for the population in a manner that is efficient, robust, and uses *minimal* communication.

In this paper, we present a bio-inspired approach to the deployment of a robot swarm to multiple sites that is decentralized, robust to changes in team size, and requires no explicit inter-agent wireless communication. Our work draws inspiration from the process by which an ant colony selects a new home from several sites using simple stochastic behaviors that arise from local sensing and physical contact.[3,19] While there are many existing bio-inspired swarm coordination strategies,[8,9,20–23] group behaviors are often obtained from the synthesis of a collection of individual agent behaviors. Rather than follow these bottom-up approaches to group behavior synthesis, we propose a methodology that enables the design of group behaviors from global specifications of the swarm which can then be realized on individual robots. In other words, we provide a top-down approach to group behavior synthesis such that the resulting agent closed-loop control laws will lead the population to behave in the prescribed manner.

10.2 Background

10.2.1 Related Work

Social insect colonies have inspired much research on the development of self-organized task allocation strategies for multi-robot systems. In these decentralized strategies, robots switch between action states based on environmental stimuli and, in some cases, interactions with other robots. A study on the effects of group size on task allocation in social animal groups concluded, using deterministic and stochastic swarm models, that larger groups tend to be more efficient because of higher rates of information transfer.[24] A similar analysis on robotic swarm systems used the implementation of ant-inspired algorithms to demonstrate the effect of group size and recruitment on collective foraging efficiency.[25] Another study on foraging showed how division of labor and greater efficiency can be achieved in a robot swarm via simple adaptation of a transition probability based on local information.[26]

A common technique in these types of allocation approaches is to employ a threshold-based response, in which a robot becomes engaged in a task, deterministically or probabilistically, when a stimulus or demand exceeds its response threshold. A comparison of market-based and threshold methods for task allocation has shown that threshold methods perform more efficiently when information is inaccurate.[27] The efficiency and robustness of three threshold-based allocation algorithms have been analyzed and compared for an object aggregation scenario.[28] One task allocation strategy, based on a model of division of labor in a wasp colony, minimizes global demand by modeling agents' preferences for particular tasks as stochastic, threshold-based transition rates that are refined through learning.[29]

Other recent work on distributed task allocation focuses on deriving a continuous model of the system with a high degree of predictive power by defining individual robot controllers and averaging their performance. In these "bottom-up" methods, the main challenge is to derive an appropriate mathematical form for the task transition rates in the model.[30] The rates are computed from physical robot parameters, sensor measurements, and geometrical considerations, under the assumption that robots and their stimuli are uniformly spatially distributed. The resulting continuous model is validated by comparing steady-state variables and other quantities of interest to the results of embodied simulations and experiments. The applications that have been modeled include collaborative stick-pulling,[20] object clustering,[21] and adaptive multi-foraging.[22] In the last application, which uses a swarming paradigm similar to the one we consider, the task is modeled as a stochastic process and it involves no explicit communication or global knowledge. However, the only way to control robot task reallocation is to modify the task distribution in the environment.

In the task allocation methods discussed, it is necessary to perform simulations of the system under many different conditions to investigate the effect of changing parameters. In contrast, our top-down synthesis approach allows us to a priori design the set of transition rates that will guarantee a desired system outcome by carrying out the design at the population level, which can then be used to synthesize individual robot controllers. This work draws inspiration from the process by which a *Temnothorax albipennis* ant colony selects a new home from several sites using simple behaviors that arise from local sensing and physical contact with neighbors. We describe this process in the following section.

10.2.2 Ant House-Hunting

T. albipennis ants engage in a process of collective decision making when they are faced with the task of choosing between two new nest sites upon the destruction of their old nest.[19] This selection process involves two phases. Initially, undecided ants choose one of the sites quasi-independently, often after visiting both. A recruiter ant, i.e., one who has chosen one of the two candidate sites, returns to the old nest site and recruits another ant to its chosen site through a *tandem run*. The "naive" ant learns the path to the new site by following the recruiter ant. Once the path is learned, the naive ant is highly likely to become a recruiter for the particular candidate site; in essence, showing a preference for this site. In such a fashion, ants with preferences for different sites recruit naive ants from the old nest to their preferred site. Once a critical population level or *quorum* at a candidate site has been attained, the convergence rate to a single choice is then boosted by recruiters who, instead of leading naive ants in tandem runs, simply transport them from the old nest to the new one. While it has been shown that this quorum-activated recruitment significantly speeds up convergence of the colony to the higher-quality site,[3,31,32] the exact motivations for the quorum mechanism by the ants are not well understood. However, the process itself is completely decentralized and results in the formation of a consensus by the colony on which of the two candidate sites to select as its new home.

This "house-hunting" process has been modeled from experimental observations for the case of two available nest sites of differing quality.[3,19] During the selection process, ants transition spontaneously but at well-defined and experimentally measurable rates between behaviors. The pattern of transition rates, which determines the average propensity of individual ants to switch behaviors, ensures that the higher-quality nest is chosen and that no ants are stranded in the worse nest. This outcome has been observed to be robust to both environmental noise and to changes in the colony population.

In our initial studies on applying the ant-inspired swarm paradigm to multi-robot task allocation, we synthesized robot controllers to produce antlike house-hunting activity that resulted in convergence to the better of two sites.[31] We then explored the idea of controlling the degree of distribution; we altered the house-hunting model with realistic ant behaviors such that the synthesized controllers cause the swarm to split between the two sites in a predetermined ratio.[32] In this chapter, we extend this concept to the problem of distributing robots among many sites in predefined fractions. We abandon the ant roles in favor of a simpler set of tasks defined as site occupation. Furthermore, we address the problem of performance specification and compare the performance of the various proposed models.

In the remainder of this chapter, we provide a detailed description and analysis of the various models that can be used to represent the swarm at the population level. We formulate the problem in Section 10.3 and discuss the stability of our derived controllers in Section 10.4. Sections 10.5 and 10.6 summarize our design methodology and the derivation of agent-level controllers from our proposed population models. We present our simulation results in Section 10.7, which is followed by a brief discussion in Section 10.8. We conclude with some final thoughts in Section 10.9.

10.3 Problem Statement

We are interested in the distribution of a large population of N homogeneous agents among M sites. We begin with a brief summary of relevant definitions and a detailed discussion of the various models/abstractions used to represent the swarm of agents.

10.3.1 Definitions

We denote the number of agents at site $i \in \{1, \ldots, M\}$ at time t by $n_i(t)$ and the desired number of agents at site i by $\bar{n}_i$. We specify that $\bar{n}_i > 0$ at each site. The population fraction at each site at time t is defined as $x_i(t) = n_i(t)/N$, and the system state vector is given by $\mathbf{x}(t) = [x_1(t), \ldots, x_M(t)]^T$. We denote the desired final distribution of the swarm as

$$\bar{x}_i = \frac{\bar{n}_i}{N}, \quad \forall\, i = 1, \ldots, M. \tag{10.1}$$

A specification in terms of fractions rather than absolute agent numbers is practical for scaling as well as for potential applications where losses of agents to attrition and breakdown are common. The task is to redeploy the swarm of robots from an initial distribution $\mathbf{x}^0$ to the desired configuration $\bar{\mathbf{x}}$ while minimizing inter-agent communication.

To model the interconnection topology of the M sites, we use a directed graph, $\mathcal{G} = (\mathcal{V}, \mathcal{E})$, such that the set of vertices, $\mathcal{V}$, represents sites $\{1, \ldots, M\}$ and the set of edges, $\mathcal{E}$, represents physical routes between sites. Two sites $i, j \in \{1, \ldots, M\}$ are *adjacent*, $i \sim j$, if a route exists for agents to travel directly from i to j. We represent this relation by (i, j) such that the edge set is defined as $\mathcal{E} = \{(i, j) \in \mathcal{V} \times \mathcal{V} | i \sim j\}$. We assume that the graph $\mathcal{G}$ is *strongly connected*, i.e., a path exists for any $i, j \in \mathcal{V}$. Here, a *path* from site i to site j is defined as a sequence of vertices $\{v_0, v_1, \ldots, v_p\} \in \mathcal{V}$ such that $v_0 = i$, $v_p = j$, and $(v_{k-1}, v_k) \in \mathcal{E}$ where $k = 1, \ldots, p$. An example of such a graph is shown in Figure 10.3.

We consider $\mathbf{x}(t)$ to represent the distribution of the state of a Markov process on $\mathcal{G}$, for which $\mathcal{V}$ is the state space and $\mathcal{E}$ is the set of possible transitions. We assign to every edge in $\mathcal{E}$ a constant *transition rate*, $k_{ij} > 0$, where k_{ij} defines the transition probability per unit time for one agent at site i to go to site j. Here k_{ij} is essentially a stochastic transition rule and in general $k_{ij} \neq k_{ji}$. In addition, we assume there is a $k_{ij}^{\max}$ associated with every edge $(i, j) \in \mathcal{E}$ which represents the maximum capacity for the given edge (i.e., road). Lastly, we define the flux from site i to site j, denoted by ϕ_{ij}, as the fraction of agents per unit time moving from i to j and denote the time required to travel from site i to site j as τ_{ij}.

For a graph that is strongly connected with bidirectional edges $((i, j) \in \mathcal{E}$ if and only if $(j, i) \in \mathcal{E})$ but not necessarily fully connected, we consider the case of having a *reversible* Markov process on the graph. A reversible Markov process with respect to $\bar{\mathbf{x}}$ satisfies the *detailed balance equations*

$$k_{ij}\bar{x}_i = k_{ji}\bar{x}_j, \quad (i, j) \in \mathcal{E}. \tag{10.2}$$

In this case, we can define a strongly connected, *undirected* graph $\mathcal{G}_u = (\mathcal{V}, \mathcal{E}_u)$ that corresponds to $\mathcal{G}$. $\mathcal{E}_u$ is the set of unordered pairs (i, j) such that the ordered pair (i, j) (and thus also (j, i)) is in the edge set $\mathcal{E}$ of graph $\mathcal{G}$. Each edge $(i, j) \in \mathcal{E}_u$ is associated with a weight w_{ij}, defined as

$$w_{ij} = k_{ij}\bar{x}_i = k_{ji}\bar{x}_j, \quad (i,j) \in \mathcal{E}_u. \tag{10.3}$$

We assume that every agent has complete knowledge of $\mathcal{G}$ as well as all the transition rates k_{ij} and their corresponding $k_{ij}^{\max}$. This is equivalent to providing a map of the environment to every agent. Finally, we assume that agents have the ability to localize themselves within the given environment.

10.3.2 Linear Model

Our baseline strategy[33] endows each agent with a small set of instructions based on the transition rates k_{ij} and achieves (re)deployment of the swarm among the sites using no explicit wireless communication. Rather than model the system as a collection of individual agents, this strategy models the swarm as a continuum. In the limit of large N, the time evolution of the population fraction at site i is given by a linear equation, the difference between the total influx and total outflux at the site:

$$\dot{x}_i(t) = \sum_{\forall j|(j,i)\in\mathcal{E}} k_{ji}x_j(t) - \sum_{\forall j|(i,j)\in\mathcal{E}} k_{ij}x_i(t). \tag{10.4}$$

The system of equations for the M sites is given by the linear model

$$\dot{\mathbf{x}} = \mathbf{Kx} \tag{10.5}$$

where $\mathbf{K} \in \mathbb{R}^{M \times M}$. The entries of $\mathbf{K}$ are defined as $\mathbf{K}_{ij} = k_{ji}$, $j \neq i$, if $(j,i) \in \mathcal{E}$ and 0 otherwise, and $\mathbf{K}_{ii} = -\sum_{(i,j)\in\mathcal{E}} k_{ij}$. We note that the columns of $\mathbf{K}$ sum to 0. Additionally, since the number of agents is conserved, the system is subject to the conservation constraint

$$\sum_{i=1}^{M} x_i(t) = 1. \tag{10.6}$$

In general, for some desired final distribution $\bar{\mathbf{x}}$, the entries of $\mathbf{K}$ are chosen such that in the limit $t \to \infty$, $\mathbf{x}(t) \to \bar{\mathbf{x}}$. This, in turn, results in agent-level closed-loop controllers given by $\mathbf{K}$ that encode the set of agent-level instructions necessary for (re)deployment of the swarm to the various sites.

10.3.3 Time-Delayed Model

The linear model assumes that agents instantaneously switch from one site to another. In practice, it takes a finite time τ_{ij} to travel between sites i and j (or to switch between two tasks). We note that the loss of agents at a site due to transfers to other sites is immediate, while the gain due to incoming agents from other sites is delayed. The linear model can be extended to take into consideration the time needed to travel between sites by converting (10.4) into a delay differential equation (DDE),

$$\dot{x}_i(t) = \sum_{\forall j|(j,i)\in\mathcal{E}} k_{ji}x_j(t-\tau_{ji}) - \sum_{\forall j|(i,j)\in\mathcal{E}} k_{ij}x_i(t) \tag{10.7}$$

for $i = 1, \ldots, M$.

Unlike (10.4), the time delays in (10.7) will always result in a finite number of agents *en route* between sites at equilibrium. Furthermore, the fraction of agents en route versus the fraction at sites increases as the time delays increase. Let $n_{ij}(t)$ be the number of robots traveling from site i to site j at time t and $y_{ij}(t) = n_{ij}(t)/N$. Then the conservation equation for this system is

$$\sum_{i=1}^{M} x_i(t) + \sum_{i=1}^{M} \sum_{\forall j|(i,j)\in\mathcal{E}} y_{ij}(t) = 1. \tag{10.8}$$

10.3.4 Quorum Model

In the linear and time-delayed models, agents will move between sites even at equilibrium, when the net flux at each site is zero. This is because these models force a trade-off between maximizing the link capacities for fast equilibration and achieving long-term efficiency at equilibrium. In this section, we propose an extension to model (10.5) to enable fast convergence to the desired state using switching behaviors based on quorum sensing. We define the *quorum* as a threshold occupancy at a site; when a site is above quorum, the agents there will travel to adjacent sites at an increased rate. In addition to assuming that agents have a map of the environment and know $\mathcal{G}$, $\mathbf{K}$, and each $k_{ij}^{\max}$, we assume that they can detect a quorum at a site based on their encounter rate with other agents at the site.[34]

Each site i is then characterized by a quorum q_i, a threshold number of agents which we specify as a fraction of the design occupancy $\bar{x}_i$. If site i is above quorum, the transition rate from i to an adjacent site j can be automatically set to either a multiple of the existing transition rate, αk_{ij}, with $\alpha > 0$ chosen to satisfy max $\alpha k_{ij} <$ min $k_{ij}^{\max}$, or simply the maximum transition rate, $k_{ij}^{\max}$. We refer to such an edge as *activated*, and activation is maintained until x_i drops below q_i. We assume that the site graph $\mathcal{G}$ has bidirectional edges and that there is a reversible Markov process on the graph, so that Equations 10.2 hold.

The differential equation model with quorum is

$$\dot{x}_i(t) = \sum_{\forall j|(j,i)\in\mathcal{E}} k_{ji}x_j(t) - \sum_{\forall i|(i,j)\in\mathcal{E}} \phi_{ij}(t), \tag{10.9}$$

where the fluxes ϕ_{ij} are defined as

$$\phi_{ij}(t) = k_{ij}x_i(t) + \sigma_i(x_i, q_i)(\alpha - 1)k_{ij}x_i(t) \qquad (10.10)$$

or

$$\phi_{ij}(t) = k_{ij}x_i(t) + \sigma_i(x_i, q_i)\left(k_{ij}^{\max} - k_{ij}\right)x_i(t) \qquad (10.11)$$

depending on the choice of edge activation. The function $\sigma_i \in [0, 1]$ is an analytic switch given by

$$\sigma_i(x_i, q_i) = \left(1 + e^{\gamma(q_i - x_i/\bar{x}_i)}\right)^{-1}, \qquad (10.12)$$

where $\gamma \gg 1$ is a constant. We note that as $(q_i - x_i/\bar{x}_i) \to -\infty$, $\sigma_i \to 1$ with $\sigma_i \to 0$ otherwise. The constant γ is chosen such that $\sigma_i \approx 1$ when $x_i/\bar{x}_i = q_i + \epsilon$, where $\epsilon > 0$ is small. This is similar to threshold methods described by Bonabeau et al.[29] and Agassounon and Martinoli.[28]

To incorporate time delays due to nonzero quorum estimation time and nonzero travel time between sites, we can formulate (10.9) as a delay differential equation:

$$\dot{x}_i(t) = \sum_{\forall j|(j,i)\in\mathcal{E}} k_{ji}x_j(t - \tau_{ji} - \tau_{E_j}) - \sum_{\forall j|(i,j)\in\mathcal{E}} \phi_{ij}(t - \tau_{E_i}), \qquad (10.13)$$

where τ_{E_i} denotes the time required to estimate the quorum at site i. The fluxes ϕ_{ij} are given by Equation 10.10 or 10.11 depending on the choice of edge activation.

10.4 Analysis

In this section we consider the uniqueness and stability properties of each model's equilibrium point, which is the desired final distribution.

10.4.1 Linear Model

We state our first theorem for the linear model and provide a brief sketch of the proof.

THEOREM 10.1[33] For a strongly connected graph $\mathcal{G}$, the system (10.5) subject to (10.6) has a unique, stable equilibrium point.

Proof The proof of this theorem can be easily constructed in two parts. The first part builds on the fact that the rank of $\mathbf{K}$ is $(M - 1)$ since the columns of the matrix

K sum to 0. Furthermore, the vector **1** exists in the nullspace of $\mathbf{K}^T$. From here we can conclude that the system $\mathbf{Kx} = \mathbf{0}$ subject to (10.6) has a unique equilibrium point. Next, to show that the equilibrium point is in fact a stable one, consider the matrix given by $\mathbf{S} = (1/s)(s\mathbf{I} + \mathbf{K}^T)$, where $s > 0$ and **I** is the $M \times M$ identity matrix. We note that for large enough s, **S** is a Markov matrix with nonnegative entries which then allows us to prove that $\mathbf{K}^T$ is negative semi-definite. We refer the interested reader to Halász et al.[33] for more detailed exposition. ■

10.4.2 Time-Delayed Model

We can arrive at a similar conclusion for the time-delayed model (10.7) if we view this model as an abstraction of a more realistic one in which the time delays τ_{ij} are random variables from some distribution. If we approximate the probability density function of these delays as an Erlang distribution, then the DDE model given by (10.7) can be transformed into an ordinary differential equation (ODE) model of the form (10.4).

To achieve this, we replace each edge (i, j) with a directed path composed of a sequence of dummy sites, $u = 1, \ldots, D_{ij}$. Assume that the dummy sites are equally spaced; then τ_{ij}/D_{ij} is the deterministic time to travel from dummy site $u \in \{1, \ldots, D_{ij}\}$ to its adjacent site. The rate at which an agent transitions from one dummy site to the next is defined as the inverse of this time which we denote by $\lambda_{ij} = D_{ij}/\tau_{ij}$. The number of transitions between two adjacent dummy sites in a time interval of length t has a Poisson distribution with parameter $\lambda_{ij}t$. If we assume that the numbers of transitions in non-overlapping intervals are independent, then the probability density function of the travel time T_u between dummy sites u and $u + 1$ is given by

$$f(t) = \lambda_{ij}e^{-\lambda_{ij}t}. \tag{10.14}$$

Let $T_{ij} = \sum_{u=1}^{D_{ij}} T_u$ be the total time to travel from site i to site j. Since $T_1, \ldots, T_{D_{ij}}$ are independent random variables drawn from the common density function (10.14), their sum T_{ij} follows the Erlang density function[35]

$$g(t) = \frac{\lambda_{ij}^{D_{ij}} t^{D_{ij}-1}}{(D_{ij} - 1)!}e^{-\lambda_{ij}t} \tag{10.15}$$

with expected value $E(T_{ij}) = \tau_{ij}$ and variance $Var(T_{ij}) = \tau_{ij}^2/D_{ij}$. Thus, in an equivalent ODE model, the average travel time from site i to site j is always the delay τ_{ij} from the DDE model, and the number of dummy sites D_{ij} can be chosen to reflect the travel time variance. Lastly, note that $Var(T_{ij}) \to 0$ as $D_{ij} \to \infty$.

Therefore, the equivalent ODE model for the system (10.7) with Erlang-distributed time delays can be obtained by replacing each $\dot{x}_i(t)$ with the following set of equations:

$$\dot{x}_i(t) = \sum_{j|(j,i)\in\mathcal{E}} \lambda_{ji} y_{ji}^{(D_{ji})}(t) - \sum_{j|(i,j)\in\mathcal{E}} k_{ij} x_i(t) \,,$$

$$\dot{y}_{ij}^{(1)}(t) = k_{ij} x_i(t) - \lambda_{ij} y_{ij}^{(1)}(t) \,,$$

$$\dot{y}_{ij}^{(m)}(t) = \lambda_{ij} \left(y_{ij}^{(m-1)}(t) - y_{ij}^{(m)}(t) \right),$$

$$m = 2, \ldots, D_{ij}, \tag{10.16}$$

where
$y_{ij}^{(l)}(t)$ denotes the fraction of the population that is at dummy site $l \in \{1, \ldots, D_{ij}\}$
λ_{ij} denotes the transition rates between the dummy sites, and $(i,j) \in \mathcal{E}$

Figure 10.1 illustrates how an edge from model (10.4) is expanded with dummy states $y_{ij}^{(l)}$ in (10.16).

From here, it is easy to see that the equivalent ODE model (10.16) is in fact an expanded version of the linear model (10.4) whose interconnection topology can also be modeled as a directed graph, $\mathcal{G}' = (\mathcal{V}', \mathcal{E}')$, where $\mathcal{V}' = \{1, \ldots, M'\}$ and $\mathcal{E}' = \{(i,j) \in \mathcal{V}' \times \mathcal{V}' \mid i \sim j\}$ with $M' = M + \sum_{i \sim j} D_{ij}$. Since $\mathcal{G}$ is strongly connected, so is $\mathcal{G}'$. Furthermore, the system (10.16) is subject to a similar conservation constraint as (10.6), where the total number of agents is conserved across all real and dummy sites. We refer to this system as the *chain model*, since it incorporates a chain of dummy sites between each pair of physical sites. We will refer to the corresponding linear model (10.5) (the system without dummy sites) as the *switching model*, since it describes a system in which agents switch instantaneously between sites. This approach is similar to the "linear chain trick" used by MacDonald[36] to transform a system of integro-differential equations with Erlang-distributed delays into an equivalent system of ODEs.

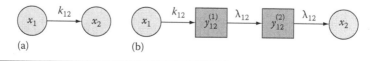

(a) (b)

Figure 10.1 A labeled edge $(i,j) = (1,2)$ that consists of (a) the physical sites, corresponding to model (10.4), and (b) both physical and dummy sites (for $D_{12} = 2$), corresponding to (10.16). (From S. Berman, Á. Halász, M. A. Hsieh, and V. Kumar, Navigation-Based Optimization of Stochastic Strategies for Allocating a Robot Swarm among Multiple Sites. In *Proceedings of the 2008 IEEE Conference on Decision and Control*, Cancun, Mexico, pp. 4316–4381. With permission.)

THEOREM 10.2[37]   If the graph $\mathcal{G}$ is strongly connected, then the chain model has a unique, stable equilibrium.

Proof   Since the system can be represented in the same form as (10.5) subject to (10.6), Theorem 10.1 can be applied to show that there is a unique, stable equilibrium.   ■

10.4.3 Quorum Model

In this section we state our final theorem, which concerns stability of the quorum model given by (10.9). We begin with the edge activation scheme given by (10.10) and assume that $q_i = q$ for all i. Consider the following candidate Lyapunov function given by

$$V = \sum_{i=1}^{M} \frac{x_i^2}{2\bar{x}_i}. \tag{10.17}$$

THEOREM 10.3[38]   The system defined by Equation 10.9 for $i = 1, \ldots, M$ subject to condition (10.2) and the conservation constraint (10.6) converges asymptotically to $\bar{\mathbf{x}} = [\bar{x}_1, \ldots, \bar{x}_M]^T$, defined by the specification (10.1).

Proof   We first show that the system is stable. We note that V is a radially unbounded function of $\|\mathbf{x}\|$. The net flux from site i to site j is defined as $\Phi_{ij} = -\phi_{ij} + k_{ji}x_j$. Note that $\Phi_{ij} = -\Phi_{ji}$. By design, $\Phi_{ij} = -\phi_{max} + k_{ji}x_j < 0$ if $x_i/\bar{x}_i > q$ and $x_j/\bar{x}_j < q$ and $\Phi_{ij} = -k_{ij}x_i + \phi_{max} > 0$ if $x_i/\bar{x}_i < q$ and $x_j/\bar{x}_j > q$. If both sites are above quorum, then Φ_{ij} simplifies to

$$\Phi_{ij} = \alpha(-k_{ij}x_i + k_{ji}x_j).$$

Using (10.2), the above equation can be rewritten as

$$\Phi_{ij} = \alpha\left(-k_{ij}x_i + \frac{\bar{x}_i}{\bar{x}_j}k_{ij}x_j\right) = \alpha k_{ij}\bar{x}_i\left(-\frac{x_i}{\bar{x}_i} + \frac{x_j}{\bar{x}_j}\right).$$

The above relationship holds when both sites are below quorum except when $\alpha = 1$. Consider the time derivative of the Lyapunov function (10.17):

$$\frac{dV}{dt} = \sum_{i=1}^{M} \frac{x_i}{\bar{x}_i}\frac{dx_i}{dt} = \sum_{\forall i \mid (i,j) \in \mathcal{E}} \frac{1}{2}\left(\frac{x_i}{\bar{x}_i} - \frac{x_j}{\bar{x}_j}\right)\Phi_{ij}. \tag{10.18}$$

By design, if $\Phi_{ij} < 0$, then $x_i/\bar{x}_i > x_j/\bar{x}_j$ and similarly, if $\Phi_{ij} > 0$, then $x_i/\bar{x}_i < x_j/\bar{x}_j$. In the event that sites i and j are both above quorum, Φ_{ij} will be opposite in sign to $(x_i/\bar{x}_i - x_j/\bar{x}_j)$. Thus, by (10.18), the time derivative of the Lyapunov function is always negative and so the system is stable. To show that the equilibrium point is given by (10.1), consider the set of equilibrium states $\mathbf{x}^e$ satisfying (10.6), such that $dV/dt = 0$. The time derivative of the Lyapunov function evaluates to zero when all $\Phi_{ij} = 0$ or when $x_i = \bar{x}_i$ for all i. By design, $\Phi_{ij} \neq 0$ for all $(i,j) \in \mathcal{E}$ whenever $x_i/\bar{x}_i \neq x_j/\bar{x}_j$, so $((x_i/\bar{x}_i) - (x_j/\bar{x}_j))\Phi_{ij} < 0$ for all i,j. Thus, the only stable equilibrium is $\mathbf{x}^e = \bar{\mathbf{x}}$, so the system converges asymptotically to (10.1). ■

Similarly, we can show that the quorum model (10.9) with edge activation scheme (10.11) is also stable. However, rather than view the system as a single system described by (10.9), we treat it as a hybrid system with one mode described by (10.9) and the other described by (10.4). The system is in the *quorum mode* when $x_i > q_i$ for some i, i.e., some sites are above quorum, and in the *linear mode* when $x_i < q_i$ $\forall i$, where all sites are below quorum. For simplicity we assume that $q_i = q$ for all i and $k_{ij}^{\max} = k^{\max}$ for all $(i,j) \in \mathcal{E}$. Next, consider the following function:

$$W_q = \sum_{i=1}^{M} \max\{x_i - q\bar{x}_i, 0\}, \qquad (10.19)$$

where W_q denotes the fraction of the population that is operating in the quorum mode. We note that agents who transition between above-quorum sites have no net effect on W_q, while the flux between sites above quorum and sites below quorum does produce an effect. Using non-smooth analysis, one can show that the time rate of change of W_q is always negative by design. This means that the number of agents operating in the quorum mode is always decreasing. Additionally, let W_l denote the scaled occupancy of the most populated site in the linear mode,

$$W_l = \max_{i}\left\{\frac{x_i}{q}\right\}. \qquad (10.20)$$

Similarly, one can show that the time rate of change of W_l is always decreasing since $\max_i\{x_i/q_i\} \geq (x_j/q_j)$ for all j. This means that even in the linear mode, the occupancy at the most populated site always decreases. Therefore, as the system switches between the quorum and linear modes, the system is always exiting each mode at a lower energy state than when the system first entered it. Furthermore, once the system enters the linear mode, it will not return to the quorum mode and thus the quorum model (10.9) with edge activation scheme (10.11) is stable.

10.5 Design of Transition Rate Matrix K

As mentioned previously, while these models accomplish the multi-site deployment task, the solutions can be relatively inefficient since the rate of convergence of the system to the desired configuration depends on the magnitudes of the transition rates k_{ij}. While large transition rates ensure fast convergence, they result in many idle trips once the design configuration is achieved. In actual robotic systems, the extraneous traffic resulting from the movement between sites at equilibrium comes at a significant cost. In light of this trade-off, we define an *optimal deployment strategy* as a choice of K that maximizes convergence toward the desired distribution while bounding the number of idle trips at equilibrium. In other words, an *optimal transition rate matrix* is one that can balance short-term gains, i.e., fast convergence, against long-term losses, i.e., inter-site traffic at equilibrium.

10.5.1 Linear Model

In general, determining an optimal transition rate matrix that satisfies both the short- and long-term restrictions is not trivial. In addition, given the same set of short- and long-term requirements, one can either determine a transition rate matrix that is optimal for the entire domain of initial distributions or one that is optimal for the given initial distribution, x^0. In this section, we describe our methodology for obtaining a general optimal transition rate matrix K_* that is suitable for a large range of initial distributions and an optimal transition rate matrix $K_*(x^0)$ that is specifically tailored for a given initial configuration.

While obtaining K_* may seem computationally expensive, this matrix can in fact be calculated with limited assumptions using convex optimization. Since system (10.5) is linear, the rate of convergence of x to $\bar{x}$ is governed by the real parts of the eigenvalues of K, of which one is zero and the rest are negative by Theorem 10.1. The smallest nonzero eigenvalue of K, denoted by $\lambda_2(K)$, governs the asymptotic rate of convergence of (10.5) to $\bar{x}$. Thus, when designing K, one way to produce fast convergence to $\bar{x}$ is to maximize $Re(\lambda_2(-K))$ (note that this quantity is *positive*) subject to the following constraint on the total flux (traffic) at equilibrium,

$$\sum_{(i,j)\in\mathcal{E}} k_{ij}\bar{x}_i \leq c_{tot}, \tag{10.21}$$

where c_{tot} is a positive constant.

To achieve this, we assume that $\mathcal{G}$ has bidirectional edges and that there is a reversible Markov process on the graph. The site topology can be modeled by a corresponding undirected graph $\mathcal{G}_u = (\mathcal{V}, \mathcal{E}_u)$, as described in Section 10.3.1. The problem of maximizing $Re(\lambda_2(-K))$ subject to (10.21) can then be posed as

$$\min_{\mathbf{w}} \sum_{(i,j)\in\mathcal{E}_u} 2w_{ij}$$

$$\text{subject to} \quad \Pi^{-1/2}\mathbf{L}\Pi^{-1/2} \succeq \mathbf{I} - \mathbf{v}\mathbf{v}^T$$

$$\mathbf{w} \geq 0 \tag{10.22}$$

where w_{ij} is the set of undirected edge weights, $\mathbf{v} = [(\bar{x}_1)^{1/2} \cdots (\bar{x}_M)^{1/2}]^T$, and $\Pi = diag(\bar{\mathbf{x}})$. $\mathbf{L}$ is the $M \times M$ weighted Laplacian of $\mathcal{G}_u$; its (i,j) entries are defined as $\mathbf{L}_{ij} = -w_{ij}$ if $(i,j) \in \mathcal{E}_u$, $\mathbf{L}_{ii} = \sum_{(i,j)\in\mathcal{E}_u} w_{ij}$, and $\mathbf{L}_{ij} = 0$ otherwise. Here, the optimization variable is $\mathbf{w}$ and the transition rates k_{ij} are derived from $\mathbf{w}$ according to (10.3). This program is similar to program (11) by Sun et al.[39] for finding the fastest mixing reversible Markov process on a graph. The optimization problem can be further extended to more general strongly connected graphs.[40]

To find the optimal $\mathbf{K}$ for a given initial configuration $\mathbf{x}^0$, $\mathbf{K}_*(\mathbf{x}^0)$, Metropolis optimization[41] is used with the entries of $\mathbf{K}$ as the optimization variables. The objective is to minimize the convergence time subject to upper bounds on the number of idle trips at equilibrium according to (10.21) and possibly on the transition rates, $k_{ij} \leq k_{ij}^{\max}$. We use the linear model (10.5) to calculate the convergence time to a set fraction of misplaced agents $\Delta(\mathbf{x},\bar{\mathbf{x}})$, the total disparity between the actual and desired population fractions at all sites, in closed form. This is achieved by decomposing $\mathbf{K}$ into its normalized eigenvectors and eigenvalues and mapping (10.5) into the space spanned by its normalized eigenvectors. Then, given an initial state $\mathbf{x}^0$, we can apply the appropriate transformation to compute the new state $\mathbf{x}(t)$ using the matrix exponential of the diagonal matrix of eigenvalues of $\mathbf{K}$ multiplied by time. We use $\mathbf{x}(t)$ to calculate $\Delta(\mathbf{x},\bar{\mathbf{x}})$. Since (10.5) is stable by Theorem 10.1, $\Delta(\mathbf{x},\bar{\mathbf{x}})$ always decreases monotonically with time, so a Newton scheme can be used to calculate the exact time when $\Delta(\mathbf{x},\bar{\mathbf{x}})$ is reduced to 10% of its initial value. Figure 10.2 plots the convergence time for a sample stochastic optimization of the transition rates.

10.5.2 *Time-Delayed Model*

The optimization methods described above can be extended to the time-delayed model (10.7) when it is represented as an equivalent ODE model as outlined in Section 10.4.2. We employ our multilevel swarm representation to empirically determine the travel time distributions for all pairs of sites. These distributions can be affected by road congestion, population levels at the target sites, and time spent on collision avoidance. Once an Erlang distribution of the form (10.15) is fitted to the distribution for each edge, the resulting parameters D_{ij} and λ_{ij} give the number of dummy sites to be added and the transition rates between these dummy sites, respectively. This allows us to construct the chain model, which in turn allows for the optimization of the transition rates following the procedures outlined in the previous section.

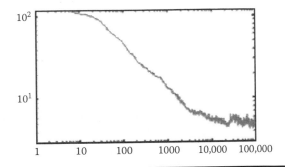

Figure 10.2 **Convergence time for stochastic optimization of the transition rates. The random walk in the space of all K configurations is biased with the convergence time so that lower times are eventually found. The horizontal axis is the number of the respective configuration. (From M. A. Hsieh, Á. Halász, S. Berman, and V. Kumar, Biologically inspired redistribution of a swarm of robots among multiple sites. In *Swarm Intelligence*, 2(2), 121–141, 2008. With permission.)**

10.6 Simulation Methodology

As in recent work on modeling and analyzing swarm robotic systems[20–22,30] we employ a multi-level representation of swarm activity. At the physical level, the swarm consists of a collection of individual robots whose dynamics are governed by their control laws. In our case, the control policies, i.e., $\mathbf{K}_*$ and $\mathbf{K}_*(\mathbf{x}^0)$, are derived from the continuous linear model (10.5), which assumes an infinitely large number of agents in a swarm. In practice, while the population size of a swarm is not infinitely large, it is often large enough to render most agent-based simulations costly. As such, it makes sense to develop an equivalent intermediate level of description, termed macro-discrete, as opposed to the micro-continuous (agent level) and macro-continuous (ODE) models, that will allow for relatively inexpensive simulations and retain some of the features of an agent-based simulation.

The correspondence between a given ODE model and a set of individual stochastic transition rules is straightforward if we choose to implement the latter as Poisson transitions controlled by fixed transition rates. Consider two states, which can either be behaviors or correspond to physically separate locations or sites. Assume that initially all agents are at site i and they all follow a stochastic transition rule by which they move to site j at rate $k = 1.0 \times 10^{-4}$ per second. At every iteration (assume 1 second per iteration for simplicity), each agent runs a random process with two possible outcomes, 0 or 1, such that the probability of 1 is given by $k\Delta t = 1.0 \times 10^{-4}$. If the outcome is 1, the agent moves to site j; otherwise it stays at i. It is clear that, given a large population size, the number of agents remaining at site i after time t is well approximated by $n_i(t) = n_i(0)e^{-kt}$. Alternatively, instead of generating a random number each time, an agent could generate a random number T distributed according to the Poisson distribution

$$f(t) = \frac{1}{k} e^{-kt} \tag{10.23}$$

and perform its transition at time $t = T$. The two methods described above are mathematically equivalent in the limit of very short sampling times Δt. If the random number generators used by the agents are independent, then the individual transition times will be distributed according to the Poisson law (10.23) and the time dependence of the number of agents remaining at site i will approximate the continuous formula.

The idea for the macro-discrete model is as follows. We begin with a population of $n_i(0)$ agents at site i. The transition probability per unit time for *each* agent is k, so that the individual probability of transition between 0 and (an infinitesimally small) Δt is $\Delta p = k\Delta t$. The probability for *any one* of the $n_i(0)$ agents to transition in the same time interval is $n_i(0)k\Delta t$. Thus, the distribution of the time of the first transition is similar to that for a single agent, only with probability rate (or *propensity*) $n_i(0)k$:

$$f(t, n_i(0)) = \frac{1}{n_i(0)k} e^{-n_i(0)kt}. \tag{10.24}$$

We can then simulate the consecutive transitions in a single program, where we only follow the *number* of agents at each site and generate transitions according to (10.24). Of course, once the first transition takes place, the number $n_i(0)$ is decreased by one agent and the next transition is generated using propensity $(n_i(0) - 1)k$.

This is an illustration of the more general Gillespie algorithm,[42] in which the system is described in terms of the number of agents at each site and transition times are generated consecutively using properly updated propensities. We stress that this method is *mathematically equivalent* to an agent-based simulation in which individual agents follow the respective Poisson transition rules. This method has the advantage of much faster execution compared to an agent-based simulation. We refer the interested reader to Berman et al.[31] for further discussion of this framework.

10.7 Results

We implemented simulations of multi-site deployment scenarios to compare the performance of different choices of $\mathbf{K}$ in terms of convergence to the desired distribution and the number of idle trips at equilibrium. We begin with the comparison of optimal and non-optimal $\mathbf{K}$ matrices and demonstrate the benefit of using quorum-activated transition rates.[38] Next, we show how a chain model, which takes into account the time delays due to navigation, performs better than the corresponding switching model.[37]

10.7.1 Linear Model vs. Quorum Model

We consider the deployment of 20,000 planar homogeneous agents to 42 sites, each executing controllers derived from the model (10.5) with a non-optimal choice $\mathbf{K}_u$ for $\mathbf{K}$, an optimal matrix $\mathbf{K}_*$ that is independent of the initial configuration, and an optimal matrix $\mathbf{K}_*(\mathbf{x}^0)$ that is specifically chosen for a particular initial configuration $\mathbf{x}^0$. We then compare the performance of the same system with agents executing controllers derived from the quorum model with the edge activation scheme given by (10.11).

For our simulations, $\mathbf{K}_*$ and $\mathbf{K}_*(\mathbf{x}^0)$ were obtained following the methodology described in Section 10.5.1. Both $\mathbf{K}_*$ and $\mathbf{K}_*(\mathbf{x}^0)$ were computed assuming the same upper bounds on the number of idle trips at equilibrium and the transition rates. The transitions between sites were simulated according to the methodology in Section 10.6. The interconnection topology of the sites is shown in Figure 10.3, in which each arrow represents a "one-way road" connecting two sites.

Agents are initially scattered at sites configured to form the number 0, and the task is to redistribute the swarm to another set of sites that form the number 8. While our focus is on the global design and properties of the swarm, our simulation methodology takes into account the exact number of agents assigned to each site as well as the travel initiation and termination times for each individual traveler. Snapshots of the simulation are shown in Figure 10.4, in which the red circles

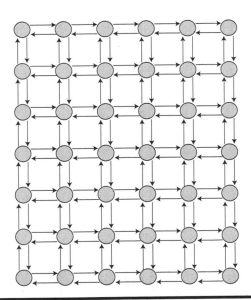

Figure 10.3 **The network of the 42 sites used in our simulations. (From M. A. Hsieh, Á. Halász, S. Berman, and V. Kumar, Biologically inspired redistribution of a swarm of robots among multiple sites. In *Swarm Intelligence*, 2(2), 121–141, 2008. With permission.)**

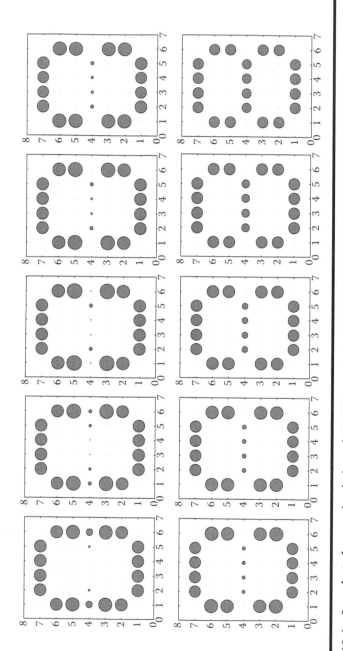

Figure 10.4 Snapshots from a simulation of 20,000 agents, sequenced from top left to bottom right. The initial configuration forms the number 0, and the design specification is the number 8. This simulation is based on the linear model (10.5) with a non-optimal choice for K. (From M. A. Hsieh, Á. Halász, S. Berman, and V. Kumar, Biologically inspired redistribution of a swarm of robots among multiple sites. In *Swarm Intelligence*, 2(2), 121–141, 2008. With permission.)

represent the number of the agents at each site. The larger the circle, the higher the agent population.

For the first set of simulations, the transition rate matrix $\mathbf{K}$ was set to $\mathbf{K}_u$, $\mathbf{K}_*$, and $\mathbf{K}_*(\mathbf{x}^0)$. The agents switch between sites with controllers derived from (10.5), i.e., no quorum activation is used. In all of these simulations, the equilibrium initiation rate for a transition from site i to site j was set to 1. This is equivalent to bounding the total number of idle trips at equilibrium. Also, $k^{\mathrm{max}} = 12$ in all simulations.

Figure 10.5 shows the fraction of misplaced agents over time for the three different choices of $\mathbf{K}$. It is not surprising that both $\mathbf{K}_*$ and $\mathbf{K}_*(\mathbf{x}^0)$ outperform $\mathbf{K}_u$ in terms of convergence speed for the same bound on idle trips at equilibrium. Note that the stochastic runs fluctuate around the corresponding ODE simulations, which verifies that the transition rates designed using the continuous model produce similar system performance when run on individual robots. The properties of each $\mathbf{K}$ are summarized in the first table.

Choice of $\mathbf{K}$	Time Units to 2/3 of Initial Deviation	Idle Trip Initiation Rate at Equilibrium	Maximum k_{ij}
$\mathbf{K}_u$	21.81	1	1.35
$\mathbf{K}_*$	4.51	1	11.88
$\mathbf{K}_*(\mathbf{x}^0)$	1.84	1	6.94

Source: M. A. Hsieh, Á. Halász, S. Berman, and V. Kumar, Biologically inspired redistribution of a swarm of robots among multiple sites. In *Swarm Intelligence*, 2(2), 121–141, 2008. With permission.

Figure 10.6 shows the fraction of misplaced agents over time for the linear model (10.5) with $\mathbf{K}_u$ and two other choices of $\mathbf{K}$. The first, $\mathbf{K}_{\mathrm{max}}(\mathbf{x}^0)$, is the optimal transition rate matrix given an initial configuration $\mathbf{x}^0$ subject solely to constraints $k_{ij} \leq k^{\mathrm{max}}$, with no constraints on idle trips at equilibrium. This means that $\mathbf{K}_{\mathrm{max}}(\mathbf{x}^0)$ is the optimal transition rate matrix with respect to convergence speed. The second choice of $\mathbf{K}$ is the quorum model (10.9) with below-quorum transition rates chosen from $\mathbf{K}_u$ and the edge activation scheme (10.11). In these simulations, we see that the quorum model allows us to maximize transient transfer rates between sites without sacrificing the limit on the number of idle trips at equilibrium. The second table summarizes the different properties of the three systems shown in Figure 10.6.

10.7.2 Linear Model vs. Time-Delayed Model

To investigate the utility of the chain model in optimizing the transition rates, we simulated a surveillance task with transition rates from a linear model (10.5)

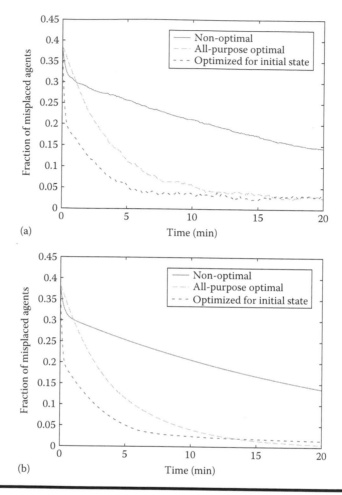

Figure 10.5 Fraction of misplaced agents over time for system (10.5) with three choices of K: K_u (solid lines), K_* (light dashed lines), and $K_*(x^0)$ (dark dashed lines). (a) Stochastic simulation. (b) Differential equation simulation. (From M. A. Hsieh, Á. Halász, S. Berman, and V. Kumar, Biologically inspired redistribution of a swarm of robots among multiple sites. In *Swarm Intelligence*, 2(2), 121–141, 2008. With permission.)

(a "switching model"; see Section 10.4.2) and a chain model computed from travel time distributions. The workspace was chosen to be a four-block region on the University of Pennsylvania campus, shown in Figure 10.7. The task was for a collection of 200 robots to perform perimeter surveillance of four campus buildings, highlighted in light dashed lines in Figure 10.7. We used a graph $\mathcal{G}$ for these four sites with the structure in Figure 10.8. The robots are initially distributed equally between sites 3 and 4, and they are required to redistribute in equal fractions among

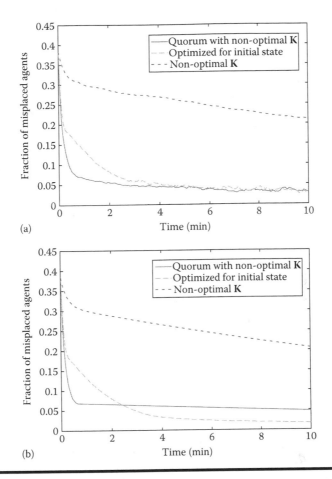

(a)

(b)

Figure 10.6 **Fraction of misplaced agents over time for system (10.5) with three choices of K: the quorum model (10.9) with K_u and edge activation scheme (10.11) (solid lines), $K_{max}(x^0)$ (light dashed lines), and K_u (dark dashed lines). (a) Stochastic simulation. (b) Differential equation simulation. (From M. A. Hsieh, Á. Halász, S. Berman, and V. Kumar, Biologically inspired redistribution of a swarm of robots among multiple sites. In *Swarm Intelligence*, 2(2), 121–141, 2008. With permission.)**

all sites. The transitions between sites are simulated according to the methodology in Section 10.6.

Agents exhibit two types of motion: perimeter tracking and site-to-site navigation. An agent that is monitoring a building circulates around its perimeter, slowing down if the agent in front of it enters its sensing radius. This results in an approximately uniform distribution of agents around the perimeter. This motion can be easily achieved with feedback controllers of the form given by Hsieh et al.[43]; in this simulation the agents simply aligned their velocities with the straight-line perimeters.

Choice of **K**	Time Units to 2/3 of Initial Deviation	Idle Trip Initiation Rate at Equilibrium	Maximum k_{ij}
K$_u$	179.3	1	0.396
K$_{max}(\mathbf{x}^0)$	14.18	12.64	12
Quorum with **K**$_u$	19	1	12

Source: M. A. Hsieh, Á. Halász, S. Berman, and V. Kumar, Biologically inspired redistribution of a swarm of robots among multiple sites. In *Swarm Intelligence*, 2(2), 121–141, 2008. With permission.

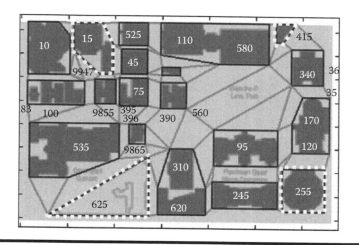

Figure 10.7 Workspace with cell decomposition of the free space used for navigation.

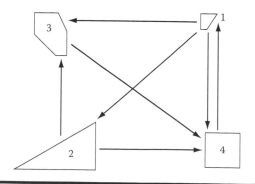

Figure 10.8 Numbering and connectivity of surveyed buildings, which are highlighted in Figure 10.7.

To implement inter-site navigation, we first decomposed the free space into a tessellation of convex cells, shown in Figure 10.7. This results in a discrete roadmap on which shortest-path computations between cells can be easily obtained using any standard graph search algorithm. Each edge $(i,j) \in \mathcal{E}$ was defined as a path from building i to building j that begins at a distinct exit point on the perimeter of i and ends at an entry point on j. The exit and entry point for (i,j) were associated with the adjacent cells, and Dijkstra's algorithm was used to determine the shortest path between these cells, which consisted of a sequence of cells to be traversed by each agent moving from i to j. Navigation between cells was achieved by composing local potential functions such that the resulting control policy ensured arrival at the desired goal position.[44] These navigation controllers were then composed with repulsive potential functions to achieve inter-agent collision avoidance within each cell. Inter-site navigation was achieved by providing each agent a priori with the sequence of cells corresponding to each edge (i,j), and at each time step the agent computed the feedback controller to move from one cell to the next based on its current position and the set of agents within its sensing range.

We first used the Metropolis optimization of Section 10.5.1 to obtain an optimal switching model $\mathbf{K}_*^{sw}(\mathbf{x}^0)$ (ignoring travel times). From the simulation using this $\mathbf{K}$, we collected 750–850 travel times τ_{ij} per edge. We then fit an Erlang distribution to the resulting histograms to obtain D_{ij} and λ_{ij} for each edge (Section 10.4.2); a sample fitting is shown in Figure 10.9. The average equilibrium traveler fraction

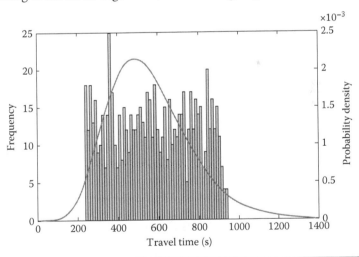

Figure 10.9 Histogram of the travel times from site 1 to site 4 (758 data points) and the approximate Erlang distribution. Based on these data, $D_{14} = 9$ and $\lambda_{14} = 0.0162$. (From S. Berman, Á. Halász, M. A. Hsieh, and V. Kumar, Navigation-based optimization of stochastic deployment strategies for a robot swarm to multiple sites. In *Proceedings of the 2008 IEEE Conference on Decision and Control*, Cancun, Mexico, pp. 4376–4381. With permission.)

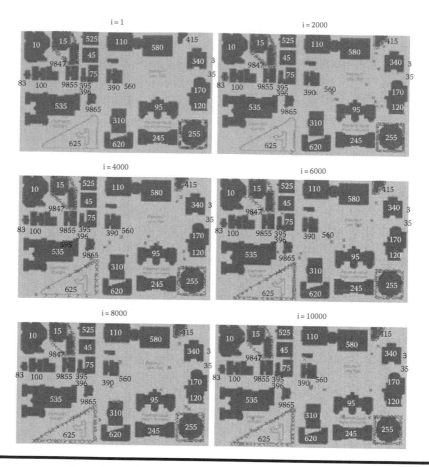

Figure 10.10 Snapshots of the chain model simulation at intervals of 2000 time steps.

from ∼30,000 data points was approximately 0.27. This fraction was used as a constraint in the optimization of the chain model $\mathbf{K}_*^{ch}(\mathbf{x}^0)$ according to Section 10.5.2. The simulation was then run with the transition rates from $\mathbf{K}_*^{ch}(\mathbf{x}^0)$.

The snapshots in Figure 10.10 illustrate the redistribution of the agents in equal fractions among the four sites for the chain model. The agents represented by stars have committed to traveling between two buildings; the agents represented by *x*'s are not engaged in a transition. Figure 10.11 shows that the traveler fraction for both models oscillates close to the mean switching model fraction, 0.27. Both models therefore have approximately the same equilibrium inter-site traffic. Figure 10.12 shows that the misplaced agent fraction of the chain model converges to 10% of its original value faster than the switching model. This provides evidence that the chain model is a better approximation of the simulated surveillance system.

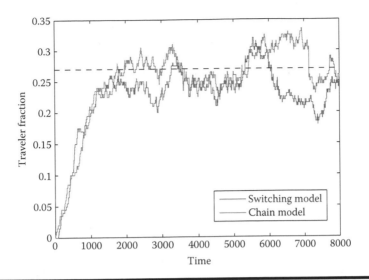

Figure 10.11   Fraction of travelers vs. time for the switching and chain models.

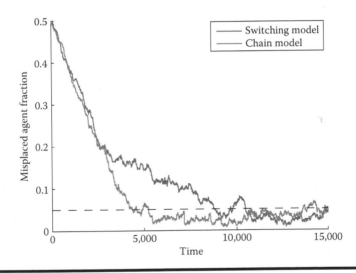

Figure 10.12   Fraction of misplaced agents vs. time for the switching and chain models.

10.8 Discussion

In biology, complex group behaviors often arise from the numerous interactions of individual agents that switch between a small set of "standard" behaviors. This is clearly seen in the house-hunting process of the *T. albipennis* ants. Additionally, it has been shown how the differential equations describing the global evolution of the swarm stem from simple, stochastic switching rules executed by the individuals.[31,32] These rules are equivalent to the transition rates k_{ij} encoded in $\mathbf{K}$. In this work, we built on the deployment scheme presented by Halász et al.[33] and independently considered the effects of a quorum-based strategy and the effects of travel time delays on the (re)distribution problem.

Our baseline model is a set of states that can be interpreted as physical sites or internal states/behaviors. As we demonstrated, this model forces a trade-off between fast convergence to the desired distribution and number of idle trips at equilibrium. Fortunately, under certain circumstances, the optimization of the transition rates can be posed as a convex optimization problem. Given an initial configuration of the swarm, Metropolis optimization can be used to obtain the optimal transition rates for the desired final configuration. Hence, the multi-site linear model can be analyzed with some very powerful tools: a well-developed theoretical framework,[39] closed-form solutions of the corresponding ODEs, and an efficient macro-discrete simulation method that easily scales up to tens of thousands of agents.

In the quorum model, we endowed agents with the ability to choose between two sets of transition rates depending on whether the current site is above or below a certain threshold occupancy. As discussed earlier, the exact motivations for these quorum strategies in ants are not clearly understood; however, one explanation is that the quorum mechanism is used to secure the choice of a candidate new nest site for ants by speeding up convergence. This can be seen in our simulations, which indicate that the quorum strategy, used with a nonoptimal choice for $\mathbf{K}$, consistently outperforms the linear models regardless of whether $\mathbf{K}$ is chosen to be $\mathbf{K}_*(\mathbf{x}^0)$ or $\mathbf{K}_*$. While we have only presented results for one pair of initial and final configurations, our exploration of a varied set of initial and final conditions supports this finding. The quorum model is an example of a biologically inspired heuristic that provides a large benefit under limited sensing capability. We used our simulation framework to quantify the gain in efficiency, allowing for a future cost–benefit analysis when the cost of the necessary sensing can be properly defined.

Lastly, in a step toward more real-world applications, the final contribution is a method that can account for realistic distributions of travel times without leaving the linear ODE framework. We achieve this by augmenting the network of sites with specifically constructed sets of virtual sites that represent the progress of agents along the paths connecting the physical sites. This approach relies on two important points: in practice, travel times are highly variable due to factors such as collision avoidance and errors in localization; and the travel time over a chain of D_{ij} sites which are connected through Poisson transition rules converges to an Erlang distribution whose relative standard deviation is proportional to $D_{ij}^{-1/2}$. Thus, the mean and standard deviation of an experimentally derived distribution of travel times can be matched by a properly chosen linear chain of sites. Our results illustrate the possibility of building larger linear ODE models that provide good approximations of variable agent travel times. The predictive value of such models is of course dependent on how well the distribution of the travel times is characterized. As our results show, the additional insight allows for better design of stochastic transition rate systems.

10.9 Conclusions

We have presented a bio-inspired approach to the (re)distribution of a swarm of robots among a set of available sites. Our methodology is built on top of our baseline strategy, which models the swarm as a continuum via a system of deterministic linear ordinary differential equations. We extended our linear model to a hybrid system in which agents switch between maximum transfer rates and constant transition rates dictated by the model. We also considered the effects of travel time delays and proposed a methodology to synthesize optimal transition rates given Erlang-distributed travel times. Although we model our system as a continuum, our methodology enables us to synthesize decentralized controllers that can be implemented at the agent level with little to no explicit wireless communication.

In physical systems, inaccuracies in navigation are common due to noisy sensors and actuators. However, these inaccuracies can be easily captured via our delay differential equation model. As discussed in Section 10.4.2, this model is in fact an abstraction of a more realistic model in which the delays are represented as random variables that follow a distribution which allows the time-delayed model to be converted into an equivalent linear model with no delays. This equivalence can be exploited as long as the inaccuracies introduced by the physical system can be mapped to stochastic time delays. Thus, the effects of crowding, localization errors, collision avoidance, and quorum estimation can be readily incorporated into our framework if we are able to model the distribution of the resulting delays. This is a topic of great interest for future work.

While stochastic controllers seem appealing because they involve significantly less communication, sensing, and planning compared to deterministic approaches, the purposeful introduction of stochasticity into an engineered system raises a number of questions. One issue is the development of simulation tools for the assessment of the performance of a mesoscopic swarm. In this work, we developed a mathematical framework, using methods borrowed from chemistry, to bridge the gap between the agent-based description (necessary for the implementation and testing of individual controllers) and the top-level description, which is traditionally continuous. Our long-term goal is to investigate the utility and limitations of this approach when used either by itself or in conjunction with more traditional models of swarm behavior synthesis.

Acknowledgments

We gratefully acknowledge the comments and support from Professor Vijay Kumar from the University of Pennsylvania and the support of NSF grants

CCR02-05336 and IIS-0427313, and ARO Grants W911NF-05-1-0219 and W911NF-04-1-0148.

References

1. G. Bekey, *Autonomous Robots: From Biological Inspiration to Implementation and Control* (MIT Press, Cambridge, MA, 2005).
2. N. F. Britton, N. R. Franks, S. C. Pratt, and T. D. Seeley, Deciding on a new home: How do honeybees agree? *Proceedings of the Royal Society of London B.* **269** (2002).
3. S. Pratt, E. B. Mallon, D. J. T. Sumpter, and N. R. Franks, Quorum sensing, recruitment, and collective decision-making during colony emigration by the ant *Leptothorax albipennis, Behavioral Ecology and Sociobiology.* **52**, 117–127 (2002).
4. I. D. Couzin, J. Krause, R. James, G. D. Ruxton, and N. R. Franks, Collective memory and spatial sorting in animal groups, *Journal of Theoretical Biology.* **218** (2002).
5. J. K. Parrish, S. V. Viscido, and D. Grunbaum, Self-organized fish schools: An examination of emergent properties, *Biological Bulletin.* **202**, 296–305 (2002).
6. C. W. Reynolds, Flocks, herds and schools: A distributed behavioral model. In *Proceedings of the 14th Annual Conference on Computer Graphics (SIGGRAPH'87)*, pp. 25–34. ACM Press, Anaheim, California (1987).
7. T. Vicsek, A. Czirok, E. Ben-Jacob, I. Cohen, and O. Shochet, Novel type of phase transition in a system of self-driven particles, *Physical Review Letters.* **75**(6), 1226–1229 (1995).
8. A. Jadbabaie, J. Lin, and A. Morse, Coordination of groups of mobile autonomous agents using nearest neighbor rules, *IEEE Transactions on Automatic Control* **48**(6), 988–1001 (June, 2003).
9. H. G. Tanner, A. Jadbabaie, and G. J. Pappas, Flocking in fixed and switching networks. In *IEEE Transactions on Automatic Control,* **52**(5), 863–868 (2007).
10. L. Lin and Z. Zheng, Combinatorial bids based multi-robot task allocation method. In *Proceedings of the 2005 IEEE International Conference on Robotics and Automation (ICRA'05)*, Barcelona, Spain (2005).
11. J. Guerrero and G. Oliver, Multi-robot task allocation strategies using auction-like mechanisms, In *Artificial Intelligence Research and Development,* edited by Aguiló et al., IOS Press, pp. 111–122 (2003).
12. E. G. Jones, B. Browning, M. B. Dias, B. Argall, M. Veloso, and A. T. Stentz, Dynamically formed heterogeneous robot teams performing tightly-coordinated tasks. In *International Conference on Robotics and Automation*, Orlando, FL, pp. 570–575 (May, 2006).
13. M. B. Dias, R. M. Zlot, N. Kalra, and A. T. Stentz, Market-based multirobot coordination: A survey and analysis, *Proceedings of the IEEE.* **94**(7), 1257–1270 (2006).
14. D. Vail and M. Veloso, Multi-robot dynamic role assignment and coordination through shared potential fields. In eds. A. Schultz, L. Parker, and F. Schneider, *Multi-Robot Systems* (Kluwer, Dordrecht, 2003).

15. B. P. Gerkey and M. J. Mataric, Sold!: Auction methods for multi-robot control, *IEEE Transactions on Robotics & Automation: Special Issue on Multi-robot Systems.* **18**(5), 758–768 (2002).

16. E. G. Jones, M. B. Dias, and A. Stentz, Learning-enhanced market-based task allocation for oversubscribed domains. In *Proceedings of the Conference on Intelligent Robot Systems (IROS'07)*, San Diego, CA (2007).

17. M. B. Dias, *TraderBots: A New Paradigm for Robust and Efficient Multirobot Coordination in Dynamic Environments.* PhD thesis, Robotics Institute, Carnegie Mellon Univ., Pittsburgh, PA (January, 2004).

18. M. Golfarelli, D. Maio, and S. Rizzi, Multi-agent path planning based on task-swap negotiation. In *Proceedings of 16th UK Planning and Scheduling SIG Workshop*, Durham University, UK (1997).

19. N. Franks, S. C. Pratt, N. F. Britton, E. B. Mallon, and D. T. Sumpter, Information flow, opinion-polling and collective intelligence in house-hunting social insects, *Philosophical Transactions of Biological Sciences.* (2002).

20. A. Martinoli, K. Easton, and W. Agassounon, Modeling of swarm robotic systems: A case study in collaborative distributed manipulation, *International Journal of Robotics Research: Special Issue on Experimental Robotics.* **23**(4–5), 415–436 (2004).

21. W. Agassounon, A. Martinoli, and K. Easton, Macroscopic modeling of aggregation experiments using embodied agents in teams of constant and time-varying sizes, *Autonomous Robots.* **17**(2–3), 163–192 (2004).

22. K. Lerman, C. V. Jones, A. Galstyan, and M. J. Mataric, Analysis of dynamic task allocation in multi-robot systems, *International Journal of Robotics Research.* **25**(4), 225–242 (2006).

23. M. A. Hsieh, L. Chaimowicz, and V. Kumar, Decentralized controllers for shape generation with robotic swarms, Accepted to *Robotica.* **26**(5), 691–701 (September, 2008).

24. S. W. Pacala, D. M. Gordon, and H. C. J. Godfray, Effects of social group size on information transfer and task allocation, *Evolutionary Ecology.* **10**(2), 127–165 (March, 1996).

25. M. J. B. Krieger, J.-B. Billeter, and L. Keller, Ant–like task allocation and recruitment in cooperative robots, *Nature.* **406**, 992–995 (2006).

26. T. H. Labella, M. Dorigo, and J.-L. Deneubourg, Division of labor in a group of robots inspired by ants' foraging behavior, *ACM Transactions on Autonomous and Adaptive Systems.* **1**(1), 4–25 (2006). ISSN 1556-4665. http://doi.acm.org/10.1145/1152934.1152936.

27. N. Kalra and A. Martinoli, A comparative study of market-based and threshold-based task allocation. In *Proceedings of the 8th International Symposium on Distributed Autonomous Robotic Systems (DARS)*, Minneapolis/St. Paul, MN (July, 2006).

28. W. Agassounon and A. Martinoli, Efficiency and robustness of threshold-based distributed allocation algorithms in multi-agent systems. In *Proceedings of the First International Joint Conference on Autonomous Agents and Multi-Agent Systems AAMAS-02*, Bologna, Italy, pp. 1090–1097 (2002).

29. E. Bonabeau, A. Sobkowski, G. Theraulaz, and J.-L. Deneubourg, Adaptive task allocation inspired by a model of division of labor in social insects. In eds. D. Lundh, B. Olsson, and A. Narayanan, *Biocomputing and Emergent Computation*, pp. 36–45 (World Scientific, Singapore, 1997).

30. K. Lerman, A. Martinoli, and A. Galstyan, A review of probabilistic macroscopic models for swarm robotic systems. In *Swarm Robotics, Workshop: State-of-the-art Survey*, edited by E. Sahin and W. Spears, LNCS 3342, pp. 143–152, Springer-Verlag, Berlin (2004).

31. S. Berman, Á. Halász, V. Kumar, and S. Pratt, Algorithms for the analysis and synthesis of a bio-inspired swarm robotic system. In eds. E. Sahin, W. M. Spears, and A. F. T. Winfield, *Swarm Robotics*, Vol. 4433, *Lecture Notes in Computer Science*, pp. 56–70. Springer, Heildelberg (2006).

32. S. Berman, Á. Halász, V. Kumar, and S. Pratt, Bio-inspired group behaviors for the deployment of a swarm of robots to multiple destinations. In *Proceedings of the International Conference on Robotics and Automation (ICRA)*, pp. 2318–2323 (2007).

33. Á. Halász, M. A. Hsieh, S. Berman, and V. Kumar, Dynamic redistribution of a swarm of robots among multiple sites. In *Proceedings of the Conference on Intelligent Robot Systems (IROS)*, pp. 2320–2325 (2007).

34. S. C. Pratt, Quorum sensing by encounter rates in the ant Temnothorax albipennis, *Behavioral Ecology*. **16**(2) (2005).

35. B. Harris, *Theory of Probability* (Addison-Wesley, Reading, MA, 1966).

36. N. MacDonald, *Time-Lags in Biological Models*. Vol. 27, *Lecture Notes in Biomathematics* (Springer, Berlin, 1978).

37. S. Berman, Á. Halász, M. A. Hsieh, and V. Kumar, Navigation-Based Optimization of Stochastic Strategies for Allocating a Robot Swarm among Multiple Sites. In *Proceedings of the 2008 IEEE Conference on Decision and Control*, Cancun, Mexico. pp. 4376–4381.

38. M. A. Hsieh, Á. Halász, S. Berman, and V. Kumar, Biologically inspired redistribution of a swarm of robots among multiple sites. In *Swarm Intelligence*, 2(2), 121–141 (2008).

39. J. Sun, S. Boyd, L. Xiao, and P. Diaconis, The fastest mixing Markov process on a graph and a connection to the maximum variance unfolding problem, *SIAM Review*. **48**(4) (2006).

40. S. Berman, Á. Halász, M. A. Hsieh, and V. Kumar, Optimized Stochastic Policies for Task Allocation in Swarms of Robots. In *IEEE Transactions on Robotics*. **25**(4), 927–937.

41. D. P. Landau and K. Binder, *A Guide to Monte-Carlo Simulations in Statistical Physics* (Cambridge University Press, Cambridge, U.K., 2000).

42. D. Gillespie, Stochastic simulation of chemical kinetics, *Annual Review of Physical Chemistry*. **58**, 35–55 (2007).

43. M. A. Hsieh, S. Loizou, and V. Kumar, Stabilization of multiple robots on stable orbits via local sensing. In *Proceedings of the International Conference on Robotics and Automation (ICRA) 2007*, Rome, Italy (April, 2007).

44. D. C. Conner, A. Rizzi, and H. Choset, Composition of local potential functions for global robot control and navigation. In *Proceedings of 2003 IEEE/RSJ International Conference on Intelligent Robots and Systems (IROS 2003)*, Vol. 4, pp. 3546–3551. IEEE, Las Vegas, NV (October, 2003).

Chapter 11

Human Peripheral Nervous System Controlling Robots

Panagiotis K. Artemiadis
and Kostas J. Kyriakopoulos

Contents

With the introduction of robots in everyday life, especially in developing services for people with special needs (i.e., elderly or impaired persons), there is a strong necessity of simple and natural control interfaces for robots. In particular, an interface that would allow the user to control a robot in a continuous way by performing natural motions with his/her arm would be very effective. In this chapter, electromyographic (EMG) signals from muscles of the human upper limb are used as the control interface between the user and a robot. EMG signals are recorded using surface EMG electrodes placed on the user's skin, letting the user's upper limb free of bulky interface sensors or machinery usually found in conventional teleoperation systems. The muscle synergies during motion are extracted using a dimensionality reduction technique. Moreover, the arm motion is represented into a low-dimensional manifold, from which motion primitives are extracted. Then using a decoding method, EMG signals are transformed into a continuous representation of arm motions. The accuracy of the method is assessed through real-time experiments including arm motions in three-dimensional (3D) space with various hand speeds. The method is also compared with other decoding algorithms, while its efficiency in controlling a robot arm is compared to that of a magnetic position tracking system.

11.1 Introduction

Despite the fact that robots came to light approximately 50 years ago, the way humans control them is still an important issue. In particular, since the use of robots is increasingly widening to everyday life tasks (e.g., service robots, robots for clinical applications), the human–robot interface plays a role of utmost significance. A large number of interfaces have been proposed in previous works, where some examples can be found.[1-3] Most of the previous works propose complex mechanisms or systems of sensors, while in most of the cases the user should be trained to map his/her action (i.e., 3D motion of a joystick or a haptic device) to the resulting motion of the robot. In this chapter a new means of control interface is proposed, according to which the user performs natural motions with his/her upper limb. Surface electrodes recording the electromyographic (EMG) activity of the muscles of the upper limb are placed on the user's skin. The recorded muscle activity is transformed to kinematic variables that are used to control the robot arm. Since in this study, an anthropomorphic robot arm is used, the user doesn't need to be acquainted with the interface mapping, since natural arm motions essentially suffice to directly control the robot arm.

EMG signals have often been used as control interfaces for robotic devices. However, in most cases, only discrete control has been realized, focusing only, for example, on the directional control of robotic wrists[4] or on the control of multi-fingered robot hands to a limited number of discrete postures.[5–7] Quite recently Thakor et al.[8] achieved in identifying 12 individuated flexion and extension movements of the fingers using EMG signals from muscles of the forearm of an able-bodied subject. However, controlling a robot by using only finite postures can cause many problems regarding smoothness of motion, especially in cases where the robot performs everyday life tasks. Moreover, from a teleoperation point of view, a small number of finite commands or postures can critically limit the areas of application. Therefore, effectively interfacing a robot arm with a human one entails the necessity of continuous and smooth control.

Continuous models have been built in the past in order to decode arm motion from EMG signals. The Hill-based muscle model,[9] whose mathematical formulation can be found in the study by Zajac,[10] is more frequently used in the literature.[11,12] However, only a few degrees of freedom (DoFs) were analyzed (i.e., 1 or 2), since the nonlinearity of the model equations and the large numbers of the unknown parameters for each muscle make the analysis rather difficult. Similarly, musculoskeletal models have been analyzed in the past,[13–15] focusing on a small number of muscles and actuated DoFs. A neural network model was used for extracting continuous arm motion in the past using EMG signals[16]; however, the movements analyzed were restricted to single-joint, isometric motions.

A lot of different methodologies have been proposed for the decoding of EMG signals so that they may be used as control variables for robotic devices. A statistical log-linearized Gaussian mixture neural network has been proposed by Fukuda et al.[4] to discriminate EMG patterns for wrist motions. Neural networks are widely used in many EMG-based control applications, especially in the field of robotic hands. A standard back propagation network was used by Hiraiwa et al.[17] for estimating five-finger motion while the same kind of network was used by Huang and Chen[18] for classifying eight motions based on the signal features extracted (e.g., zero-crossing and variance). Farry et al.[19] used the frequency spectrum of EMG signals to classify motions of the human hand in order to remotely operate a robot hand, while support vector machines were used for the classification of hand postures using EMG signals by Bitzer and van der Smagt.[6] A Hill-based muscle model[9] was used to estimate human joint torque for driving an exoskeleton by Cavallaro et al.[11] The authors have used an EMG-based auto-regressive moving average with exogenous output (ARMAX) model for estimating elbow motion.[20]

Study on the human motor control system brings up several challenges. Billions of control units (neurons) stimulate large numbers of motor units (muscle fibers) actuating the numerous DoFs in the human body. Narrowing our interest down to the human upper limb, approximately 30 individual muscles actuate the human arm during motion in 3D space. Moreover, the human upper limb can be modeled as a seven-DoF arm (excluding finger motions). Considering these numbers, one can conclude that it is a highly complex, over-actuated mechanism that essentially provides humans with incredible dexterity. Dealing with such complexity from an engineering point of view is a rather demanding task, especially when we desire to decode muscle activity to arm motion. Recent work in the field of biomechanics, though, proposes that muscles are activated collectively, forming the so-called time-varying muscle synergies.[21] This finding suggests that muscle activations can be represented into a low-dimensional space, where these synergies can be represented instead of individual muscle activations. Several studies in human motor control have also suggested that a low-dimensional representation is feasible at the arm kinematic level (i.e., joint angles) too.[22] Identifying those underlying low-dimensional representations of muscle activations and movements performed, one could come up with a more robust way of decoding EMG signals to motion. A robust decoder can then lead to a continuous representation of the human arm motion from multiple muscle recordings, to be used as the control algorithm for an EMG-based robot control scenario.

In this chapter, a methodology for controlling an anthropomorphic robot arm using EMG signals from the muscles of the upper limb is proposed. Surface EMG electrodes are used to record from muscles of the shoulder and the elbow. The user performs random movements in 3D space. The EMG signals are decoded to a continuous profile of motion that is utilized accordingly to control an anthropomorphic robot arm. The user has visual contact with the robot arm that replicates his/her movements in space, and thus is able to teleoperate it in real time.

The chapter is organized as follows: the proposed system architecture is analyzed in Section 11.2, the experiments are reported in Section 11.3, while Section 11.4 concludes the chapter.

11.2 EMG-Based Control

In this section, the experimental setup used will be analyzed. Specifically, the user-analyzed motion will be kinematically defined, while the system of sensors used for recording the motion of the arm will be introduced. Then, the processing of the recorded EMG signals will be presented, in order to consequently define the motion-decoding methodology proposed. Finally, the EMG-based motion estimates will be used in a control law for actuating an anthropomorphic robot arm.

11.2.1 System Overview

Eleven bipolar surface EMG electrodes record the muscular activity of the equal number of muscles acting on the shoulder and the elbow joints. The system architecture is divided into two phases: the training and the real-time operation. During the training phase, the user is instructed to move his/her arm in random patterns with variable speed in the 3D space. A position-tracking system is used to record the arm motion during reaching. The procedure lasts for 1 min. Four DoFs are analyzed (i.e., two for the shoulder and two for the elbow). To tackle the dimensionality problem, the activation of the 11 muscles recorded and the 4 joint angle profiles are represented into two low-dimensional spaces via the appropriate technique. The mapping between these two low-dimensional spaces is realized through a linear model whose parameters are identified using the previously collected data. As soon as the linear model is trained, the real-time operation phase commences. During this phase, the trained model outputs the decoded motion using only the EMG recordings. A control law that utilizes these motion estimates is applied to the robot arm actuators. In this phase, the user can teleoperate the robot arm in real time, while he/she can correct any possible deviations since he/she has visual contact with the robot. The efficacy of the proposed method is assessed through a large number of experiments, during which the user controls the robot arm in performing random movements in the 3D space.

11.2.2 Background and Problem Definition

The motion of the upper limb in the 3D space will be analyzed, though not including the wrist joint for simplicity. Therefore the shoulder and the elbow joints are of interest. Since the method proposed here will be used for the control of a robot arm, equipped with two rotational DoFs at each of the shoulder and the elbow joints, we will model the human shoulder and the elbow as having two DoFs too, without any loss of generality. In fact, it can be proved from the kinematic equations of a simplified model of the upper limb, that the motion of the human shoulder can be addressed by using two rotational DoFs, with perpendicularly intersecting axes of rotation. The elbow is modeled with a similar pair of DoFs corresponding to the flexion–extension and pronation–supination of this joint. Hence, four DoFs will be analyzed from a kinematic point of view.

11.2.3 Recording Arm Motion

For the training of the proposed system, the motion of the upper limb should be recorded and joint trajectories should be extracted. Therefore, in order to record the motion and then extract the joint angles of the four modeled DoFs, a magnetic position tracking system was used. The system is equipped with two position trackers and a reference system, with respect to which the 3D position of the trackers is provided. In order to compute the four joint angles, one position tracker is placed at the user's elbow joint and the other one at the wrist joint. The reference system is placed on the user's shoulder. The setup as well as the four modeled DoFs are shown in Figure 11.1. Let $\mathbf{T_1} = \begin{bmatrix} x_1 \ y_1 \ z_1 \end{bmatrix}^T$, $\mathbf{T_2} = \begin{bmatrix} x_2 \ y_2 \ z_2 \end{bmatrix}^T$ be the position of the trackers with respect to the tracker reference system. Let q_1, q_2, q_3, q_4 be the four joint angles modeled as shown in Figure 11.1. Finally, by solving the inverse kinematic equations (see Appendix 11.A.1) the joint angles are given by

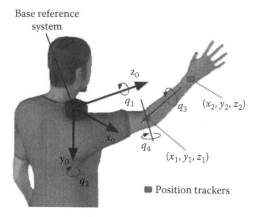

Base reference
system

Figure 11.1 The user moves his arm in the 3D space. Two position tracker measurements are used for computing the four joint angles. The tracker base reference system is placed on the shoulder.

$$q_1 = \arctan 2 \left(\pm y_1, x_1 \right)$$
$$q_2 = \arctan 2 \left(\pm \sqrt{x_1^2 + y_1^2}, z_1 \right)$$
$$q_3 = \arctan 2 \left(\pm B_3, B_1 \right)$$
$$q_4 = \arctan 2 \left(\pm \sqrt{B_1^2 + B_3^2}, -B_2 - L_1 \right)$$

(11.1)

where

$$B_1 = x_2 \cos \left(q_1 \right) \cos \left(q_2 \right) + y_2 \sin \left(q_1 \right) \cos \left(q_2 \right) - z_2 \sin \left(q_2 \right)$$
$$B_2 = -x_2 \cos \left(q_1 \right) \sin \left(q_2 \right) - y_2 \sin \left(q_1 \right) \sin \left(q_2 \right) - z_2 \cos \left(q_2 \right)$$
$$B_3 = -x_2 \sin \left(q_1 \right) + y_2 \cos \left(q_1 \right)$$

(11.2)

where L_1 is the length of the upper arm. The length of the upper arm can be computed from the distance of the first position tracker from the base reference system:

$$L_1 = \| \mathbf{T_1} \| = \sqrt{x_1^2 + y_1^2 + z_1^2}$$

(11.3)

Likewise, the length of the forearm L_2 can be computed from the distance between the two position trackers, i.e.,

$$L_2 = \sqrt{(x_2 - x_1)^2 + (y_2 - y_1)^2 + (z_2 - z_1)^2}$$

(11.4)

It must be noted that since the position trackers are placed on the skin and not in the center of the modeled joints, the lengths L_1, L_2 may vary as the user moves the arm. However, it was found that the variance during a 4 min experiment was less than 1 cm (i.e., approximately 3% of the mean values for the lengths L_1, L_2). Therefore, the mean values of L_1, L_2 for a 4 min experiment were used for the following analysis.

The position-tracking system provides the position vectors $\mathbf{T_1}$, $\mathbf{T_2}$ at the frequency of 30 Hz. Using an anti-aliasing FIR filter these measurements are resampled at a frequency of 1 kHz, to be consistent with the muscle activations sampling frequency.

11.2.4 Recording Muscles Activation

Based on the biomechanics literature,[23] a group of 11 muscles mainly responsible for the studied motion is recorded: deltoid (anterior), deltoid (posterior), deltoid (middle), pectoralis major, teres major, pectoralis major (clavicular head), trapezius, biceps brachii, brachialis, brachioradialis, and triceps brachii. A smaller number of muscles could have been recorded (e.g., focusing on one pair of agonist–antagonist muscles for each joint). However, in order to investigate a wider arm motion variability, where less significant muscles could play an important role in specific arm configurations, a group of 11 muscles were selected. Surface bipolar EMG electrodes used for recording are placed on the user's skin following the directions given by Cram and Kasman.[23] Recording sites and muscles are shown in Figure 11.2. Raw EMG signals after amplification are digitized at a sampling frequency of 1 kHz. Then a full wave rectification takes place, and then the signals are low-pass filtered using a fourth-order Butterworth filter. Finally the signals from each muscle are normalized to their maximum voluntary isometric contraction value.[10] As mentioned earlier, the system initiates with a training phase, where EMG recordings and motion data are collected for model training. Concerning EMG recordings, after data collection we have the muscle activations $u_{kT}^{(i)}$ of each muscle i at time kT, where T is the sampling period and $k = 1, 2, \ldots, 11$. Regarding motion data, using the position tracker readings and (11.1), the corresponding joint angles $q_{1k}, q_{2k}, q_{3k},$ and q_{4k} are collected. The dimension of these sets is quite large (i.e., 11 variables for muscle activations and 4 for joint angles), making the mapping between them excessively hard. In order to deal with this dimensionality issue, a dimension reduction technique is applied.

11.2.5 Dimensionality Reduction

The problem of dimension reduction is introduced as an efficient way to overcome the curse of the dimensionality when dealing with vector data in high-dimensional spaces and as a modeling tool for such data. It is generally defined as the search for a low-dimensional manifold that embeds the high-dimensional data. The goal of

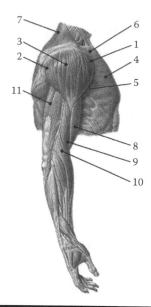

Figure 11.2 Muscles recorded: 1, deltoid (anterior); 2, deltoid (posterior); 3, deltoid (middle); 4, pectoralis major; 5, teres major; 6, pectoralis major (clavicular head); 7, trapezius; 8, biceps brachii; 9, brachialis; 10, brachioradialis; 11, triceps brachii. (From Putz, R. and Pabst, R., *Sobotta Atlas of Human Anatomy*. Lippincott Williams and Wilkins, Philadelphia, PA, 2001.)

dimension reduction is to find a representation of that manifold (i.e., a coordinate system) that will allow to project the original data vectors on it and obtain a low-dimensional, compact representation of them. In our case, muscle activations and joint angles are the high-dimensional data, which will be embedded into two manifolds of lower dimension. This should extract muscle synergies and motion primitives, which could be represented into the two new low-dimensional manifolds using their new coordinate systems. Afterwards, having decreased the dimensionality of both the muscle activations and joint angles, the mapping between these two sets will be achievable.

The most widely used dimension reduction technique is principal component analysis (PCA). It is widely used due to its conceptual simplicity and the fact that relatively efficient algorithms exist for its computation. The central idea of PCA, since it is a dimension reduction method, is to reduce the dimensionality of a data set consisting of a large number of interrelated variables, while retaining as much as possible of the variation present in the data set. This is achieved by transforming to a new set of variables the principal components (PCs), which are uncorrelated, and which are ordered so that the first few retain most of the variation present in all of the original variables. In this study, the PCA algorithm will be implemented twice: once for finding the new representation of the muscle activation data, and then again for the representation of joint angles. For details about the method, the reader should refer to the literature.[25,26]

In order to collect data during the training period, the user is instructed to move the arm in 3D space, as shown in Figure 11.1. Muscle activation and joint angles are available after the pre-processing described before. Let

$$\mathbf{U} = \begin{bmatrix} \mathbf{u}^{(1)} & \mathbf{u}^{(2)} & \cdots & \mathbf{u}^{(11)} \end{bmatrix}^T \tag{11.5}$$

be an $11 \times m$ matrix containing the m samples of the muscle activations from each of the 11 muscles recorded, i.e.,

$$\mathbf{u}^{(i)} = \begin{bmatrix} u_1^{(i)} & u_2^{(i)} & \cdots & u_m^{(i)} \end{bmatrix}^T, \quad i = 1, \ldots, 11, \quad m \in \mathbb{N} \tag{11.6}$$

Likewise, let

$$\mathbf{Y} = \begin{bmatrix} \mathbf{y}^{(1)} & \mathbf{y}^{(2)} & \mathbf{y}^{(3)} & \mathbf{y}^{(4)} \end{bmatrix}^T \tag{11.7}$$

be a $4 \times m$ matrix containing m samples of the four joint angles, i.e.,

$$\begin{aligned}
\mathbf{y}^{(1)} &= \begin{bmatrix} q_{1_1} & q_{1_2} & \cdots & q_{1_m} \end{bmatrix}^T \\
\mathbf{y}^{(2)} &= \begin{bmatrix} q_{2_1} & q_{2_2} & \cdots & q_{2_m} \end{bmatrix}^T \\
\mathbf{y}^{(3)} &= \begin{bmatrix} q_{3_1} & q_{3_2} & \cdots & q_{3_m} \end{bmatrix}^T \\
\mathbf{y}^{(4)} &= \begin{bmatrix} q_{4_1} & q_{4_2} & \cdots & q_{4_m} \end{bmatrix}^T
\end{aligned} \tag{11.8}$$

where $q_{1_k}, q_{2_k}, q_{3_k}, q_{4_k}, k = 1, \ldots, m$ denote the kth measurement of the joint angle q_1, q_2, q_3, q_4 respectively.

The computation of the PCA method entails the singular value decomposition (SVD) of the zero-meaned data covariance matrix. Hence, regarding muscle activations, mean values of each row of the matrix $\mathbf{U}$ in (11.5) are subtracted from the corresponding row. Then, the covariance matrix $\mathbf{F}_{11 \times 11}$ is computed. The singular value decomposition of the latter results in

$$\mathbf{F} = \mathbf{G}\mathbf{J}\mathbf{G}^T \tag{11.9}$$

where
    $\mathbf{J}$ is the 11×11 diagonal matrix with the eigenvalues of $\mathbf{F}$
    $\mathbf{G}$ is the 11×11 matrix with columns that are the eigenvectors of $\mathbf{F}$

The principal component transformation of the muscle activation data can then be defined as

$$\mathbf{M} = \mathbf{G}^T \mathbf{K} \tag{11.10}$$

where **K** is the $11 \times m$ matrix computed from **U** by subtracting the mean value of each muscle across the m measurements. The principal component transformation of the joint angles can be computed likewise. If Λ is the $4 \times m$ matrix computed from **Y** by subtracting the mean values of each joint angle across the m measurements, and **H** is the 4×4 matrix with columns that are the eigenvectors of the covariance matrix, then the principal component transformation of the joint angles is defined as

$$\mathbf{Q} = \mathbf{H}^T \Lambda \tag{11.11}$$

The PCA algorithm results in a representation of the original data to a new coordinate system. The axes of that system are the eigenvectors computed as analyzed before. However, the data do not appear to have the same variance across those axes. In fact, in most cases only a small set of eigenvectors is enough to describe most of the original data variance. Here is where the dimension reduction comes into use. Let p be the total number of variables in the original data, i.e., the number of the eigenvectors calculated by the PCA algorithm. It can be proved that only the *first* r, $r \ll p$, eigenvectors can describe most of the original data variance. The *first* eigenvectors are the first eigenvectors we have if they are ranked in a descending order with respect to their eigenvalues. Therefore, the eigenvectors with the highest eigenvalues describe most of the data variance.[26] Describing the p original variables with fewer dimensions r is finally the goal of the proposed method. Many criteria have been proposed for choosing the right number of principal components to be kept, in order to retain most of the original data variance. The reader should refer to the literature[26] for a complete review of those methods. Perhaps the most obvious criterion for choosing r is to select a cumulative percentage of total variation which one desires that the selected PCs contribute, i.e., 80% or 90%. The required number of the PCs is then the smallest value of r, for which this chosen percentage is exceeded. Since the eigenvalue of each eigenvector coincides with its contribution to the total variance, the smallest value of r, r_- can be found by using the following inequality:

$$\frac{\sum_{i=1}^{r_-} \lambda_i}{\sum_{n=1}^{p} \lambda_n} \geq P \%, \quad r_-, p \in \mathbb{N} \tag{11.12}$$

where

$P \%$ is the total variance which one desires that the selected PCs contribute, i.e., 90 %

r_- is the minimum integer number for which (11.12) is satisfied

Using the above criterion, we can select the minimum number of eigenvectors of muscle activation and joint angles capable of retaining most of the original data variance. In particular, for muscle recordings, the first two principal components were capable of describing 95% of the total variance. Regarding joint angles, the first two principal components described 93% of the total variance too. Therefore the low-dimensional representation of the activation of 11 muscles during 3D motion of the arm is defined by

$$\Xi = \mathbf{V}^T \mathbf{K} \tag{11.13}$$

where $\mathbf{V}$ is an 11×2 matrix, whose columns are the two first eigenvectors of $\mathbf{G}$ in (11.10), resulting in the size of $2 \times m$ for matrix Ξ. Likewise, the low-dimensional representation of joint angles during 3D motion of the arm is defined by

$$\Phi = \mathbf{W}^T \Lambda \tag{11.14}$$

where $\mathbf{W}$ is a 4×2 matrix, whose columns are the two first eigenvectors of $\mathbf{H}$ in (11.11), resulting in the size of $2 \times m$ for matrix Φ.

Using the above dimension reduction technique, the high-dimensional data of muscle activations and corresponding joint angles were represented into two manifolds of fewer dimensions. In particular for joint angles, using two instead of four variables to describe arm movement suggests motor primitives, which is a general conception that has been extensively analyzed in the literature.[21,27] For analysis reasons, it is interesting to see what those two variables describe in the high-dimensional space. In other words, how the variation in the two axes of the low-dimensional manifold can be represented back into the high-dimensional space of the four joint angles, and consequently the arm movement. This is shown in Figure 11.3. In the top figure, the arm motion depicted corresponds to the variation in the first axis of the low-dimensional manifold, i.e., along the first eigenvector extracted from the PCA on the arm kinematics. It is evident that the first principal component describes the motion of the arm on a plane parallel to the coronal plane.[28] Considering the

Figure 11.3 Motion primitives computed from the first two principal components. Above: the motion described along the first eigenvector. Below: the motion described along the second eigenvector.

second principal component, it can be regarded as describing motion in the transverse plane. The authors do not claim that the human motor control system uses these two motor primitives to perform any 3D motion in general. As noted before, the proof of the presence of internal coordination mechanisms of the human motor control is way beyond the scope of this study. On the contrary, this study focuses on extracting task-specific motor primitives and, by using the proper mathematical formulation, employs them for controlling robots. The former was achieved through the PCA method on kinematic data collected during the training phase, while the resulting motor primitives that were extracted are depicted in Figure 11.3 as proof of concept. It must be noted that representing a 4D space into two dimensions does not affect the available workspace or the analyzed DoFs. The number of DoFs analyzed remains the same (i.e., 4), and the available arm workspace is always the 3D space. Using the dimension reduction technique, the same variability is represented into another space, taking advantage of any underlying covariance, without losing any dimension of the original data. Until this point, muscle activations and joint angles have been represented into low-dimensional manifolds. Next we show how this representation facilitates the mapping between the EMG signals and the corresponding arm motion.

11.2.6 Motion Decoding Model

Raw EMG signals from 11 muscles are pre-processed and represented into a low-dimensional space resulting in only two variables (i.e., low-dimensional EMG embeddings). In addition, four joint angles are embedded into a 2D space. Having those variables, we can define the problem of decoding as follows: find a function f that can map muscle activations to arm motion in real time, in a way that it may be identified using training data. Generally, we can define it by

$$\mathbb{Y} = f(\mathbb{U}) \tag{11.15}$$

where
 $\mathbb{Y}$ denotes human arm kinematic embeddings
 $\mathbb{U}$ represents muscle activation embeddings

As noted in the introduction the scope of this study, unlike related past works, is to obtain a continuous representation of motion using EMG signals. From a physiological point of view, a model that would describe the function of skeletal muscles actuating the human joints would be generally a complex one. This would entail highly nonlinear musculoskeletal models with a great number of parameters that should be identified.[12] For this reason, we can adopt a more flexible decoding model in which we introduce hidden or latent variables we call **x**. These hidden variables can model the unobserved, intrinsic system states and thus facilitate the correlation between the observed muscle activation $\mathbb{U}$ and arm kinematics $\mathbb{Y}$. Therefore, (11.15) can be rewritten as shown below:

$$\mathbb{Y} = f(\mathbb{X}, \mathbb{U}) \tag{11.16}$$

where $\mathbb{X}$ denotes the set of the "hidden" states. Regarding function f that describes the relation between the input and the hidden states with the output of the model, a selection of a linear one can be made in order to facilitate the use of well-known algorithms for training. The latter selection results in the following state space model:

$$\mathbf{x}_{k+1} = \mathbf{A}\mathbf{x}_k + \mathbf{B}\boldsymbol{\xi}_k + \mathbf{w}_k$$
$$\mathbf{z}_k = \mathbf{C}\mathbf{x}_k + \mathbf{v}_k \tag{11.17}$$

where

$\mathbf{x}_k \in \mathbb{R}^d$ is the hidden state vector at time instance kT, $k = 1, 2, \ldots$
T is the sampling period
d is the dimension of this vector
$\boldsymbol{\xi} \in \mathbb{R}^2$ is the vector of the low-dimensional muscle activations
$\mathbf{z}_k \in \mathbb{R}^2$ is the vector of the low-dimensional joint kinematics

The matrix $\mathbf{A}$ determines the dynamic behavior of the hidden state vector $\mathbf{x}$, $\mathbf{B}$ is the matrix that relates muscle activations $\boldsymbol{\xi}$ to the state vector $\mathbf{x}$, while $\mathbf{C}$ is the matrix that represents the relationship between the joint kinematics $\mathbf{z}_k$ and the state vector $\mathbf{x}$. $\mathbf{w}_k$ and $\mathbf{v}_k$ represent zero-mean Gaussian noise in the process and observation equations respectively, i.e., $\mathbf{w}_k \sim N(\mathbf{0}, \boldsymbol{\Psi})$, $\mathbf{v}_k \sim N(\mathbf{0}, \boldsymbol{\Gamma})$, where $\boldsymbol{\Psi} \in \mathbb{R}^{d \times d}$, $\boldsymbol{\Gamma} \in \mathbb{R}^{2 \times 2}$ are the covariance matrices of $\mathbf{w}_k$ and $\mathbf{v}_k$, respectively.

Model training entails the estimation of the matrices $\mathbf{A}$, $\mathbf{B}$, $\mathbf{C}$, $\boldsymbol{\Psi}$, and $\boldsymbol{\Gamma}$. Given a training set of length m, including the low-dimensional embeddings of the muscle activations and joint angles, the model parameters can be found using an iterative prediction-error minimization (i.e., maximum like-lihood) algorithm.[29] The dimension d of the state vector should also be selected for model fitting. This is done in parallel with the fitting procedure, by deciding the number of states, nothing that any additional states do not contribute more to the model input–output behavior. The way to arrive at this conclusion is by observing the singular values of the covariance matrix of the system output.[29]

Having trained the model, the real-time operation phase commences. Raw EMG signals are collected, pre-processed, and then represented into the low-dimensional manifolds using (11.13), where the eigenvectors included in $\mathbf{V}$ were computed during training. Then, the fitted model (11.17) is used. The model outputs the low-dimensional arm kinematics vector $\mathbf{z}_k$ at each time instance kT. This vector is transformed back to the 4D space, representing the estimates for the four joint angles of the upper limb. This is done by using (11.14) and solving for the high-dimensional vector:

$$\mathbf{q}_{\mathrm{H}_k} = \left(\mathbf{W}\mathbf{W}^T\right)^{-1}\mathbf{W}\mathbf{z}_k \tag{11.18}$$

where $\mathbf{q_{H}}_{k}$ is the 4×1 vector with the four estimates for the arm joint angles at time instance kT, i.e., the high-dimensional representation of arm kinematics. Having computed the estimated joints angles

$$\mathbf{q_{H}}_{k} = \begin{bmatrix} \hat{q}_{1_k} & \hat{q}_{2_k} & \hat{q}_{3_k} & \hat{q}_{4_k} \end{bmatrix}^{T} \qquad (11.19)$$

we can then command the robot arm. However, since the robot's and the user's links have different lengths, the direct control in joint space would lead the robot end-effector to a different position in space than that desired by the user. Consequently, the user's hand position should be computed by using the estimated joint angles, and then the robot commanded to position its end-effector at this point in space. This is realized by using the forward kinematics of the human arm to compute the user's hand position and then solving the inverse kinematics for the robot arm to drive its end-effector to the same position in 3D space. Hence, the final command to the robot arm is in joint space. Therefore, the subsequent robot controller analysis assumes that a final vector $\mathbf{q_d} = \begin{bmatrix} q_{1d} & q_{2d} & q_{3d} & q_{4d} \end{bmatrix}^{T}$ containing the four desired robot joint angles is provided, where these joint angles are computed through the robot inverse kinematics as described earlier.

11.2.7 Robot Control

A seven-DoF anthropomorphic robot arm (PA-10, Mitsubishi Heavy Industries) is used. Only four DoFs of the robot are actuated (joints of the shoulder and elbow) while the others are kept fixed at zero position via electromechanical brakes. The arm is horizontally mounted to mimic the human arm. The robot arm along with the actuated DoFs is depicted in Figure 11.4. The robot motors are controlled in torque. In order to control the robot arm using the desired joint angle vector $\mathbf{q}_d$, an inverse dynamic controller is used, defined by

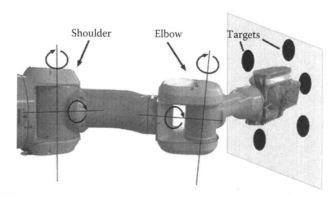

Figure 11.4 The robot arm used for teleoperation is shown. Four DoFs are teleoperated, two of the shoulder and two of the elbow. Targets are placed in the robot workspace in an identical manner to those presented to the user.

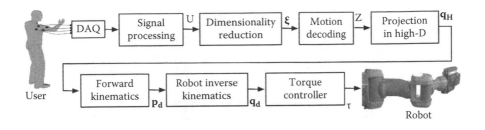

Figure 11.5 **Proposed system architecture. p_d denotes the human hand position vector computed from the estimated joint angles through the human forward kinematics.**

$$\tau = I\left(q_r\right)\left(\ddot{q}_d + K_v \dot{e} + K_P e\right) + G\left(q_r\right) + C\left(q_r, \dot{q}_r\right)\dot{q}_r + F_{fr}\left(\dot{q}_r\right) \qquad (11.20)$$

where

$\tau = \begin{bmatrix} \tau_1 & \tau_2 & \tau_3 & \tau_4 \end{bmatrix}^T$ is the vector of robot joint torques

$q_r = \begin{bmatrix} q_{1r} & q_{2r} & q_{3r} & q_{4r} \end{bmatrix}^T$ are the robot joint angles

K_v and K_p are gain matrices

e is the error vector between the desired and the robot joint angles

i.e.,

$$e = \begin{bmatrix} q_{1d} - q_{1r} & q_{2d} - q_{2r} & q_{3d} - q_{3r} & q_{4d} - q_{4r} \end{bmatrix}^T \qquad (11.21)$$

I, G, C, and F_{fr} are the inertia tensor, the gravity vector, the Coriolis-centrifugal matrix, and the joint friction vector of the four actuated robot links and joints respectively, identified by Mpompos et al.[30] The vector $\ddot{q}_d$ corresponds to the desired angular acceleration vector that is computed through simple differentiation of the desired joint angle vector q_d using a necessary low-pass filter to cut off high frequencies. A block diagram depicting the total architecture proposed for decoding EMG signals to motion and controlling the robot arm is depicted in Figure 11.5.

11.3 Experimental Results

11.3.1 Hardware and Experiment Design

The proposed architecture is assessed through remote teleoperation of the robot arm using only EMG signals from the 11 muscles as analyzed earlier. Two personal computers are used, running Linux operating system. One of the personal computers communicates with the robot servo-controller through the ARCNET protocol in the frequency of 500 Hz, while the other acquires the EMG signals and the position tracker measurements (during training). The two personal computers are connected through serial communication (RS-232) interface for synchronization purposes. EMG signals are acquired using a signal acquisition board (NI-DAQ 6036E, National Instruments) connected to an EMG system (Bagnoli-16, Delsys Inc.). Single differential surface EMG electrodes (DE-2.1, Delsys Inc.) are used. The position tracking system (Isotrak II, Polhemus Inc.) used during the training phase is connected with the personal computer through serial communication interface (RS-232). The size of the position sensors is 2.83(W), 2.29(L), 1.51(H) cm.

The user is initially instructed to move his/her arm toward targets placed randomly to random positions in the 3D space, with variable speed, in order to cover a wide range of the arm workspace, as shown in Figure 11.1. During this phase EMG signals and position tracker measurements are collected for not more than 3 min. These data are enough to train the model analyzed earlier. The model computation time is less than 1 min. As soon as the model is estimated, the real-time operation phase takes place. The user is instructed to move the arm in 3D space, having visual contact with the robot arm. The position tracker measurements are not used during this phase. Using the fitted model, estimations about the human motion are computed using only the recorded EMG signals from the 11 muscles mentioned earlier. Despite the fact that the estimation is done in joint space, the human hand position in 3D space is of interest in most cases. Thus, the human joint angles are transformed to hand positions in Cartesian space through the kinematics. The robot end-effector is driven to the same position (with respect to its base reference system) using the robot inverse kinematics. Finally, the position trackers are kept in place (i.e., on the human arm) for offline validation reasons. The estimated 3D trajectory by the user's hand along with the ground truth is depicted in Figure 11.6. As can be seen the method could estimate the hand trajectory with high accuracy. The system was tested by three able-bodied persons, who found it convenient and accurate, while they were easily acquainted with its operation.

11.3.2 Method Assessment

Two criteria will be used for assessing the accuracy of the reconstruction of human motion using the proposed methodology. These are the root-mean-squared error (RMSE) and the correlation coefficient (CC). The latter describes essentially the similarity between the reconstructed and the true motion profiles and constitutes the most common means of reconstruction assessment for decoding purposes. If $\hat{\mathbf{q}} = \begin{bmatrix} \hat{q}_1 & \hat{q}_2 & \hat{q}_3 & \hat{q}_4 \end{bmatrix}^T$ are the estimated joint angles and $\mathbf{q_t} = \begin{bmatrix} q_{1_t} & q_{2_t} & q_{3_t} & q_{4_t} \end{bmatrix}^T$ the corresponding true values, then the RMSE and CC are defined as follows:

$$RMSE_i = \sqrt{\frac{1}{n} \sum_{k=1}^{n} \left(q_{i_t k} - \hat{q}_{ik}\right)^2}, \quad i = 1, 2, 3, 4 \qquad (11.22)$$

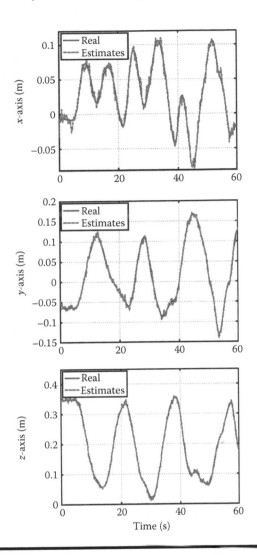

Figure 11.6   Real and estimated hand trajectory along the *x*-, *y*-, and *z*-axes, for a 1 min period.

$$CC_i = \frac{\sum_{k=1}^{n} \left(q_{i_t k} - \bar{q}_{i_t}\right)\left(\hat{q}_{ik} - \bar{\hat{q}}_{i}\right)}{\sqrt{\sum_{k=1}^{n} \left(q_{i_t k} - \bar{q}_{i_t}\right)^2 \sum_{k=1}^{n} \left(\hat{q}_{ik} - \bar{\hat{q}}_{i}\right)^2}}, \quad i = 1, 2, 3, 4 \qquad (11.23)$$

where $\bar{q}_i$ represents the mean of the *i*th joint angle across *n* testing samples. Perfect matching between the estimated and the true angles corresponds to $CC = 1$.

Real and estimated motion data were recorded for 30 s during the real-time operation phase. The above-mentioned criteria concerning joint angles were computed and the values are reported in Table 11.1. Using the hand kinematics, the corresponding values in Cartesian space were also computed and are listed in Table 11.2.

A characteristic of the method worth assessing is the use of the low-dimensional representation of the muscle activation and human kinematic variables. In order to conclude if this approach finally facilitated the decoding method, we tried to estimate a model given by (11.17) using the high-dimensional data for muscle activations and human joint angles; i.e., the input vector ξ had length equal to 11 (11 muscles recorded), while the output vector **z** had length equal to 4 (4 joint angles). The same training and testing data were used as previously, so that the comparison could be meaningful. The results are shown in Tables 11.1 and 11.2. One can see that the decoding method using the high-dimensional data resulted in a model of order 10, requiring 18 min to be trained, while the results were worse than those of the proposed decoding in the low-dimensional space. This result proves the efficiency of the proposed methodology in the task of the EMG-based dexterous control of robots.

Another parameter worth assessing is the type of the model used (i.e., linear model with hidden states). The authors feel that a comparison with a well-known algorithm of similar complexity is rational. Thus, a comparison with the linear filter method was made. The linear filter method is a widely used method for decoding arm motions (especially when using neural signals) which has achieved exciting results so far.[31] Briefly, if $\mathcal{Q}_k$ are the kinematics variables decoded (i.e., joint angles) at time $t_k = kT$ and $\mathcal{E}_{i,k-j}$ is the muscle activation of muscle i at time t_{k-j}, the computation of the linear filter entails finding a set of coefficients $\pi = \begin{bmatrix} a\, f_{1,1} \cdots f_{i,j} \end{bmatrix}^T$ so that

$$\mathcal{Q}_k = a + \sum_{i=1}^{\upsilon} \sum_{j=0}^{N} f_{i,j}\mathcal{E}_{i,k-j} \tag{11.24}$$

where
   a is a constant offset
   $f_{i,j}$ are the filter coefficients
   υ is the number of muscles recorded
   the parameter N specifies the number of time bins used

A typical value of the latter is 100 ms; thus $N = 100$ for a sampling period of 1 ms.[31] The coefficients can be learned from training data using simple least-squares regression. In our case, for the sake of comparison, the same training data were used for both the state space model and the linear filter, and after training, both models were tested using the same testing data as previously. Values for RMSE and CC for these two methods are reported in Tables 11.1 and 11.2. It must be noted that

Table 11.1 Method Efficiency Comparison between High- and Low-Dimensional Representation of Data and the Linear Filter Method, in Joint Space

Decoding Model	Training Time (s)	Order	CC_1	CC_2	CC_3	CC_4	$RMSE_1$ (°)	$RMSE_2$ (°)	$RMSE_3$ (°)	$RMSE_4$ (°)
Low-D state space	10	3	0.97	0.99	0.98	0.95	1.98	2.46	1.12	4.23
High-D state space	1076	10	0.85	0.81	0.83	0.78	9.89	4.86	13.22	14.91
Linear filter	25	—	0.84	0.83	0.83	0.81	7.19	3.45	5.32	7.91

Table 11.2 Method Efficiency Comparison between High- and Low-Dimensional Representation of Data and the Linear Filter in Cartesian Space

Decoding Model	CC_x	CC_y	CC_z	$RMSE_x(cm)$	$RMSE_y(cm)$	$RMSE_z(cm)$
Low-D state space	0.98	0.97	0.99	2.21	2.16	1.95
High-D state space	0.86	0.79	0.81	8.82	11.01	14.14
Linear filter	0.87	0.78	0.79	9.45	14.14	12.34

the low-dimensional representation for muscle activations and kinematics was used since the linear filter behaved better using those kind of data rather than using the high-dimensional data.

In general, the proposed methodology was used very efficiently for controlling the robot arm. The users were able to teleoperate the robot arm in real time, having visual contact with it. In terms of motion prediction the proposed method was proved to be quite successful. The proposed decoding model outperformed the mostly used one (i.e., linear filter), while the complexity of the method and the time of training were negligible. All the three users who tested the system found it quite comfortable and efficient.

11.3.3 EMG-Based Control vs. Motion-Tracking Systems

From the previous analysis of the results it turns out that the method could be used for the real-time control of robots using EMG signals, for long periods, compensating for the EMG changes due to muscle fatigue. However, a comparison of the proposed architecture with a position-tracking system is worth analyzing. In other words, why is the proposed EMG-based method for controlling robots more efficient than a position-tracking system that can track the performed user's arm motion?

One of the main reasons for which an EMG-based system is more efficient is the high frequency and the smoothness of the motion estimates that it provides, which is very crucial when a robot arm is to be controlled. The EMG-based system provides motion estimates at the frequency of 1 kHz (i.e., equal to the acquisition frequency of the EMG signal, which is usually high), compared to the 30 Hz frequency of the most often used position-tracking systems. Moreover, the proposed methodology guarantees smoothness in the computed estimates because of the linear dynamics of the hidden states used in the model. In order to depict the necessity for smooth and high-frequency estimates from the robot control point of view, a robot arm teleoperation test was carried out using the EMG-based proposed architecture and a position-tracking system providing motion data at the frequency of 30 Hz. The resulted robot arm motion in the 3D space is shown in Figure 11.7. As can be seen,

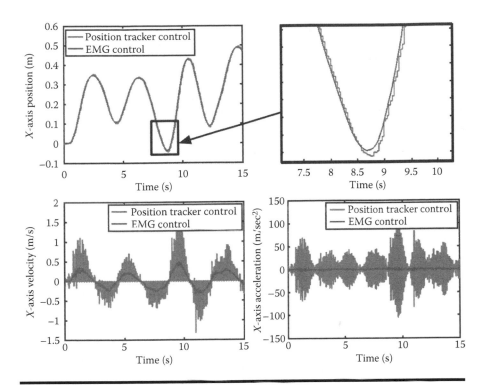

Figure 11.7 Comparison of robot performance using EMG-based control and a position tracker. Top left figure: The robot end-effector position at the X robot axis. Top right figure: a detailed view of the robot end-effector position. Control using a position tracker proved to be very jerky, resulting in high robot velocity (bottom left figure) and high acceleration (bottom right figure), in comparison with EMG-based control that resulted in smooth robot trajectory. Robot end-effector along the X-axis only is depicted, since similar behavior was noticed at the other axes.

the resulted robot motion was very smooth when EMG-based estimates were used, while it was very jerky when the position tracker measurements were used instead. It must be noted that filtering the position tracker measurements with an averaging filter was also attempted; however, the delays that the filter imposed on the desired motion made it difficult for the user to teleoperate the robot arm.

Another reason for which an EMG-based system is more efficient than a conventional position-tracking system can be found if one realizes that EMG signals represent muscle activity, which is not only responsible for motion but also for force. Therefore, EMG signals can be used for the estimation of exerted force, when the user wants to teleoperate a robot in exerting force to the environment, or for the control of a human–robot coupled system, i.e., arm exoskeletons. The authors have used EMG signals to decode arm-exerted force and control human–robot coupled systems in the past.[32,33] Since conventional motion-tracking systems cannot provide information about exerted force, EMG signals constitute a more convenient, inexpensive, and efficient interface for the control of robotic systems.

11.4 Conclusions

A novel human–robot interface for 3D robot teleoperation was introduced. EMG signals recorded from muscles of the upper limb were used for extracting kinematic variables (i.e., joint angles) in order to control an anthropomorphic robot arm in real time. Activations from 11 different muscles were used in order to estimate the motion of four DoFs of the user's arm during 3D tasks. Muscle activations were transformed into a low-dimensional manifold utilizing signal correlation (i.e., muscle synergies). Moreover, arm motion was represented into a low-dimensional space, revealing motion primitives in the 3D movements studied. Indeed it was noticed that the two principal components of motion extracted corresponded to motion across two planes. Then, a linear model with hidden states was used for modeling the relationship between the two low-dimensional variables (EMG signals and joint angles). After model training, the user could control a remote robot arm with high accuracy in 3D space.

The dimensionality reduction in the proposed method proved quite significant, since not only did it reveal some interesting aspects regarding the 3D movements studied, but it also alleviated the matching between the EMG signals and motion since signal correlations were extracted and the number of variables was drastically reduced. The latter led to the fact that a simple linear model with hidden states proved quite successful in matching EMG signals and arm motions and thus resulted in an accurate reconstruction of human motion. The second important issue presented here is that a continuous profile of 3D arm motion (including four DoFs) is extracted using only EMG signals. Furthermore, this result enables the dexterous control of robotic devices, as presented in this study.

With the use of EMG signals and robotic devices in the area of rehabilitation receiving increasing attention during the last years, our method could be proved beneficial in this area. Moreover, the proposed method can be used in a variety of applications, where an efficient human–robot interface is required. For example, prosthetic or orthotic robotic devices mainly driven by user-generated signals can be benefitted by the proposed method, resulting in a user-friendly and effective control interface.

Acknowledgments

This work is partially supported by the European Commission through contract NEUROBOTICS (FP6-IST-001917) project and by the E.U. European Social Fund (75%) and the Greek Ministry of Development GSRT (25%) through the

PENED project of GSRT. The authors also thank Michael Black and Gregory Shakhnarovich.

Appendix 11.A

11.A.1 Arm Kinematics

As shown in Figure 11.1, the arm is modeled as a four-DoF mechanism, with four rotational joints, two at the shoulder and two at the elbow. The axes of joint 1 and 2 are perpendicular to each other. Likewise, the axes of joints 3 and 4 are also perpendicular. The kinematics will be solved using the modified Denavit-Hartenberg (D-H) notation.[34] We assign frames to each rotational joint and then we can describe the relation between two consecutive frames $i - 1$ and i by using the following homogeneous transformation matrix

$$
{}_{i}^{i-1}\mathbf{T} = \begin{bmatrix} c\theta_i & -s\theta_i & 0 & a_{i-1} \\ s\theta_i c\alpha_{i-1} & c\theta_i c\alpha_{i-1} & -s\alpha_{i-1} & -s\alpha_{i-1}d_i \\ s\theta_i s\alpha_{i-1} & c\theta_i s\alpha_{i-1} & c\alpha_{i-1} & c\alpha_{i-1}d_i \\ 0 & 0 & 0 & 1 \end{bmatrix}
\tag{11.A.1}
$$

where

> c and s correspond to cos and sin, respectively
> θ, α, a, and d denote the modified D-H parameters for the arm model given in Table 11.A.1, where L_1 and L_2 are the lengths of the upper arm and forearm, respectively

These lengths are calculated using the position tracker measurements as described in (11.3) and (11.4). The frames assignment, the base reference system placed on the shoulder, as well as the modeled joints and links are all shown in Figure 11.A.1. As can be seen, the fifth frame is assigned to the wrist. Therefore, the transformation from the shoulder (frame 0) to the wrist (frame 5) is given by

Table 11.A.1   Arm Model Modified D-H Parameters

i	α_{i-1}	a_{i-1}	d_i	θ_i
1	0	0	0	q_1
2	$-90°$	0	0	q_2
3	$90°$	0	L_1	q_3
4	$-90°$	0	0	q_4
5	$90°$	0	L_2	0

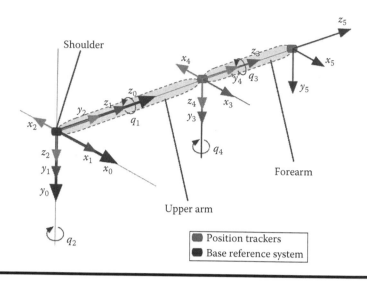

Figure 11.A.1 **Frames assignment, base reference system and position trackers along with modeled joints.**

$$\,{}^{0}_{5}\mathbf{T} = {}^{0}_{1}\mathbf{T} \cdot {}^{1}_{2}\mathbf{T} \cdot {}^{2}_{3}\mathbf{T} \cdot {}^{3}_{4}\mathbf{T} \cdot {}^{4}_{5}\mathbf{T} \tag{11.A.2}$$

where the matrices **T** are defined according to (11.A.1). The center of frame 3 is located at the elbow, where the first position tracking sensor is placed (see Figure 11.A.1). Computing the position of frame 3 using the modified D-H notation in a similar way to that used in (11.A.2), and defining it as equal to the position of the first position sensor $\mathbf{T_1} = \begin{bmatrix} x_1 & y_1 & z_1 \end{bmatrix}^T$, we have

$$x_1 = \cos\left(q_1\right) \sin\left(q_2\right) L_1 \tag{11.A.3}$$

$$y_1 = \sin\left(q_1\right) \sin\left(q_2\right) L_1 \tag{11.A.4}$$

$$z_1 = \cos\left(q_2\right) L_1 \tag{11.A.5}$$

From (11.A.3) and (11.A.4) it can easily be found that

$$q_1 = \arctan 2 \left(\pm y_1, x_1\right) \tag{11.A.6}$$

Squaring (11.A.3) and (11.A.4), adding then up, and using also (11.A.5) we have

$$q_2 = \arctan 2 \left(\pm\sqrt{x_1^2 + y_1^2}, z_1\right) \tag{11.A.7}$$

Using (11.A.2) and also the fact that the position of the second position-tracking sensor $\mathbf{T_2} = \begin{bmatrix} x_2 & y_2 & z_2 \end{bmatrix}^T$ coincides with the center of the fifth frame, the matrix ${}_5^0\mathbf{T}$ can be written as follows:

$$
{}_5^0\mathbf{T} = \begin{bmatrix} \mathbf{R}_{3\times3} & \mathbf{T_2} \\ \mathbf{0}_{1\times3} & 1 \end{bmatrix} \tag{11.A.8}
$$

where $\mathbf{R}_{3\times3}$ is the rotation matrix describing the orientation of the fifth frame and $\mathbf{0}_{3\times1}$ is a zero-element matrix with size 3×1. From (11.A.2) we have

$$
\left({}_2^1\mathbf{T}\right)^{-1} \left({}_1^0\mathbf{T}\right)^{-1} \cdot {}_5^0\mathbf{T} = {}_3^2\mathbf{T} \cdot {}_4^3\mathbf{T} \cdot {}_5^4\mathbf{T} \tag{11.A.9}
$$

Solving (11.A.9) by using (11.A.8) and equating the first three elements of the fourth column on both sides, we have

$$
x_2 c_1 c_2 + y_2 s_1 c_2 - z_2 s_2 = L_2 c_3 s_4 \tag{11.A.10}
$$

$$
x_2 c_1 s_2 + y_2 s_1 s_2 + z_2 c_2 = L_1 + L_2 c_4 \tag{11.A.11}
$$

$$
y_2 c_1 - x_2 s_1 = L_2 s_3 s_4 \tag{11.A.12}
$$

where c_i and s_i correspond to $\cos(q_i)$ and $\sin(q_i)$, where $i = 1, 2, 3, 4$ respectively. From (11.A.10) and (11.A.12) we have

$$
q_3 = \arctan 2 \left(\pm B_3, B_1\right) \tag{11.A.13}
$$

where

$$
\begin{aligned}
B_1 &= x_2 c_1 c_2 + y_2 s_1 c_2 - z_2 s_2 \\
B_3 &= y_2 c_1 - x_2 s_1
\end{aligned} \tag{11.A.14}
$$

From (11.A.10) and (11.A.11) we have

$$
q_4 = \arctan 2 \left(\pm\sqrt{B_1^2 + B_3^2}, B_2 - L_1\right) \tag{11.A.15}
$$

where

$$
B_2 = x_2 c_1 s_2 + y_2 s_1 s_2 + z_2 c_2 \tag{11.A.16}
$$

References

1. Y. Woo-Keun, T. Goshozono, H. Kawabe, M. Kinami, Y. Tsumaki, M. Uchiyama, M. Oda, and T. Doi, Model-based space robot teleoperation of ets-vii manipulator, *IEEE Transactions on Robotics and Automation* **20**(3), 602–612 (2004).
2. T. Tayh-Jong, A. Bejczy, G. Chuanfan, and X. Ning, Intelligent planning and control for telerobotic operations, *Proceedings of IEEE/RSJ International Conference on Intelligent Robots and Systems*, Munich, Germany. pp. 389–396 (1994).
3. J. Park and O. Khatib, A haptic teleoperation approach based on contact force control, *International Journal of Robotics Research* **25**(5–6), 575–591 (2006).
4. O. Fukuda, T. Tsuji, M. Kaneko, and A. Otsuka, A human-assisting manipulator teleoperated by emg signals and arm motions, *IEEE Transactions on Robotics and Automation* **19**(2), 210–222 (2003).
5. J. Zhao, Z. Xie, L. Jiang, H. Cai, H. Liu, and G. Hirzinger, Levenberg-marquardt based neural network control for a five-fingered prosthetic hand, *Proceedings of IEEE International Conference on Robotics and Automation*, Barcelona, Spain. pp. 4482–4487 (2005).
6. S. Bitzer and P. van der Smagt, Learning emg control of a robotic hand: Towards active prostheses, *Proceedings of IEEE International Conference on Robotics and Automation*, Orlando, FL. pp. 2819–2823 (2006).
7. M. Zecca, S. Micera, M. C. Carrozza, and P. Dario, Control of multifunctional prosthetic hands by processing the electromyographic signal, *Critical Review in Biomedical Engineering* **30**(4–6), 459–485 (2002).
8. F. Tenore, A. Ramos, A. Fahmy, S. Acharya, R. Etienne-Cummings, and N. Thakor, Towards the control of individual fingers of a prosthetic hand using surface emg signals, *Proceedings 29th Annual International Conference of the IEEE EMBS*, Lyon, France. pp. 6145–6148 (2007).
9. A. V. Hill, The heat of shortening and the dynamic constants of muscle, *Proceedings of the Royal Society of London: Biology*. pp. 136–195 (1938).
10. F. E. Zajac, Muscle and tendon: Properties, models, scaling, and application to biomechanics and motor control, Bourne, J. R. ed. *CRC Critical Review in Biomedical Engineering* **17**, 359–411 (1986).
11. E. Cavallaro, J. Rosen, J. C. Perry, S. Burns, and B. Hannaford, Hill-based model as a myoprocessor for a neural controlled powered exoskeleton arm- parameters optimization, *Proceedings of IEEE International Conference on Robotics and Automation*, Barcelona, Spain. pp. 4514–4519 (2005).
12. P. K. Artemiadis and K. J. Kyriakopoulos, Teleoperation of a robot manipulator using emg signals and a position tracker, *Proceedings of IEEE/RSJ International Conference Intelligent Robots and Systems*, Edmonton, Canada. pp. 1003–1008 (2005).
13. J. Potvin, R. Norman, and S. McGill, Mechanically corrected emg for the continuous estimation of erector spine muscle loading during repetitive lifting, *European Journal of Applied Physiology* **74**, 119–132 (1996).
14. D. G. Lloyd and T. F. Besier, An emg-driven musculoskeletal model to estimate muscle forces and knee joint moments in vivo, *Journal of Biomechanics* **36**, 765–776 (2003).

15. K. Manal, T. S. Buchanan, X. Shen, D. G. Lloyd, and R. V. Gonzalez, Design of a real-time emg driven virtual arm, *Computers in Biology and Medicine* **32**, 25–36 (2002).

16. Y. Koike and M. Kawato, Estimation of dynamic joint torques and trajectory formation from surface electromyography signals using a neural network model, *Biological Cybernetics* **73**, 291–300 (1995).

17. A. Hiraiwa, K. Shimohara, and Y. Tokunaga, Emg pattern analysis and classification by neural network, *Proceedings of IEEE International Conference Systems, Man, Cybernetics*. Cambridge, MA. pp. 1113–1115 (1989).

18. H. P. Huang and C. Y. Chen, Development of a myoelectric discrimination system for a multidegree prosthetic hand, *Proceedings of IEEE International Conference on Robotics and Automation*, Detroit, MI. pp. 2392–2397 (1999).

19. K. A. Farry, I. D. Walker, and R. G. Baraniuk, Myoelectric teleoperation of a complex robotic hand, *IEEE Transactions on Robotics and Automation* **12**(5), 775–788, (1996).

20. P. K. Artemiadis and K. J. Kyriakopoulos, Emg-based teleoperation of a robot arm in planar catching movements using armax model and trajectory monitoring techniques, *Proceedings of IEEE International Conference on Robotics and Automation*, Orlando, FL. pp. 3244–3249 (2006).

21. A. d'Avella, A. Portone, L. Fernandez, and F. Lacquaniti, Control of fast-reaching movements by muscle synergy combinations, *The Journal of Neuroscience* **25**(30), 7791–7810 (2006).

22. B. Lim, S. Ra, and F. Park, Movement primitives, principal component analysis, and the efficient generation of natural motions, *Proceedings of IEEE International Conference on Robotics and Automation*, Barcelon, Spain. pp. 4630–4635 (2005).

23. J. R. Cram and G. S. Kasman, *Introduction to Surface Electromyography*. (Aspen Publishers, Inc. Gaithersburg, MD, 1998).

24. R. Putz and R. Pabst, *Sobotta Atlas of Human Anatomy* (Lippincott Williams & Wilkins, Philadelphia, PA, 2001).

25. J. E. Jackson, *A User's Guide to Principal Components* (John Wiley & Sons, New York, London, Sydney, 1991).

26. I. T. Jolliffe, *Principal Component Analysis* (Springer, New York, Berlin, Heidelberg, 2002).

27. F. A. Mussa-Ivaldi and E. Bizzi, Motor learning: The combination of primitives, *Philosophical Transactions of the Royal Society of London B* **355**, 1755–1769 (2000).

28. P. M. McGinnis, *Biomechanics of Sport and Exercise* (Human Kinetics Publishers, Champaign, IL, 1989).

29. L. Ljung, *System Identification: Theory for the User* (Upper Saddle River, NJ: Prentice-Hall, 1999).

30. N. A. Mpompos, P. K. Artemiadis, A. S. Oikonomopoulos, and K. J. Kyriakopoulos, Modeling, full identification and control of the mitsubishi pa-10 robot arm, *Proceedings of IEEE/ASME International Conference on Advanced Intelligent Mechatronics*, Switzerland (2007).

31. J. M. Carmena, M. A. Lebedev, R. E. Crist, J. E. OÓDoherty, D. M. Santucci, D. F. Dimitrov, P. G. Patil, C. S. C. S. Henriquez, and M. A. L. Nicolelis, Learning to control a brain-machine interface for reaching and grasping by primates, *PLoS, Biology* **1**, 001–016 (2003).

32. P. K. Artemiadis and K. J. Kyriakopoulos, Emg-based position and force control of a robot arm: Application to teleoperation and orthosis, *Proceedings of IEEE/ASME International Conference on Advanced Intelligent Mechatronics*, Switzerland (2007).

33. P. K. Artemiadis and K. J. Kyriakopoulos, Estimating arm motion and force using EMG signals: On the control of exoskeletons, *IEEE/RSJ International Conference on Intelligent Robots and Systems (IROS)*, Nice, France, pp. 279–284, September (2008).

34. J. J. Craig, *Introduction to Robotics: Mechanisms and Control* (Addison Wesley, Reading, MA, 1989).

BIO-INSPIRED COMMUNICA-TIONS AND NETWORKS

Chapter 12

Adaptive Social Hierarchies: From Nature to Networks

Andrew Markham

Contents

Social hierarchies are a common feature of the Animal Kingdom and control access to resources according to the fitness of the individual. We use a similar concept to form an adaptive social hierarchy (ASH) among nodes in a heterogeneous wireless network so that they can discover their role in terms of their base attributes (such as energy or connectivity). Three different methods of forming the hierarchy are presented (pairwise, one way, and agent based). With Agent ASH we show that the time taken for the hierarchy to converge decreases with increasing N, leading to good scalability. The ranked attributes are used as a network underlay to enhance the behavior of existing routing protocols. We also present an example of a cross-layer protocol using ranked connectivity and energy. The ASH provides an abstraction of real-world, absolute values onto a relative framework and thus leads to a simpler and more general protocol design.

12.1 Introduction

A group of network nodes can be regarded as a society which is performing a task (such as sensing). How well the group performs this task affects information delivery and network longevity. Nodes in a network generally form a heterogeneous set, differentiated by explicit attributes (such as size of energy reserve and amount of buffer space) and implicit attributes (such as connectivity). The network must efficiently transfer information from source to sink, generally through intermediate nodes in the absence of direct connectivity between source and sink. Choosing which intermediates to transfer through is the subject of routing protocols, of which there exist a plethora.

Rather than introduce yet another routing protocol, we consider how a network could manage itself, given that nodes "know" their role within the group. For example, a node with low remaining energy level is likely to be exhausted soon and should thus adopt a less active role in the network than a node with a large amount of battery energy. What is meant by a less active role is application dependent but could range from avoiding routing other nodes' packets to disabling energy-hungry sensors or altering their duty cycle. In a network where all the nodes are initialized with identical-sized energy reserves, determining whether a node should have a low or high role is simply related to the remaining proportion of energy.

Consider now the case where nodes are not introduced with the same initial amount of energy. Such would be the case in a widely heterogeneous network. We have previously introduced the example scenario of an animal-tracking network, where the size of the host animal dictates the maximum weight of the tracking collar[1]. This results in a large degree of diversity in initial energy reserves. Assessing when a node is a low-energy node now becomes dependent on the composition of the group, and is no longer related to the absolute amount of energy. For example, in a network with nodes with energy levels of 100, 200, and 300 J, a 100 J node has the lowest energy and thus should adopt a low role. Conversely, in a network composed of nodes with energy levels of 20, 50, and 100 J, the 100 J node is the "best" in terms of energy and thus should be the most active. Although the energy of the node is the same, its behavior is dictated by the energy reserves of its peers.

This simple example demonstrates that some sort of network-wide discovery system is required in order to determine the role of nodes in the network. Nodes could send their attributes to a central server which would gather network information and from this instruct each node to adopt a certain behavior. While this is simple, it presents a single point of failure and does not scale well to large numbers of nodes in the network. A simple and lightweight localized discovery scheme is thus required so that each node can decide on its own level, based on information acquired from its peers.

We turn to the animal kingdom for inspiration, and observe that animals are able to determine their role within a society without any centralized control. Animals form collective groups, with an order of precedence, where some individuals are superior to others in their preferential access to resources (such as mates and food). We transpose this method of social organization onto our wireless network, so that nodes can determine their role and access scarce resources (such as network bandwidth and energy).

In this chapter, we first examine social hierarchies in nature, to understand how and why they form and adapt to changes. The application and construction of social hierarchies to wireless networks are discussed in Sections 12.3 and 12.3.1, respectively. A method of creating a stable social hierarchy through pairwise exchange is presented in Section 12.4. Based on the observation that pairwise exchanges place a high burden on the Media Access Control (MAC) layer, a broadcast version of social hierarchy formation is elaborated upon in Section 12.5. To remove the restriction that the network be mobile in order for the hierarchy to correctly form, pseudo mobility is introduced through the action of random network agents in Section 12.6. This section also examines how mobility models impact on the rate of convergence and adaption. The focus of the chapter then turns to suitable network attributes to rank in Section 12.7, followed by a presentation of some example scenarios showing the use of the social hierarchy approach. Lastly, our work is contrasted with related work in Section 12.9, before conclusions and future direction are posed in Section 12.10. We now examine how social hierarchies are formed in nature.

12.2 Social Hierarchies in Nature

Many animals live in societies or groups of individuals. The size and composition of these groups vary, but most have a common feature of the imposition of a social hierarchy. A social hierarchy can be regarded as the organization of individuals of a collective into roles as either leaders or followers with respect to each other. In biological terms, leading and following are referred to as dominance and submission respectively. An individual dominant over another typically has preferential access to resources such as food, water, or mating rights. Note that although one animal may be dominant over another, it itself may be dominated by yet another member of the group. Social hierarchies are such a pervasive natural formation that words describing their behavior have taken on colloquial usage, such as "pecking order," "leader of the pack," and "alpha male."

Some hierarchies are statically created through gross physiological differences, where it is impossible for certain individuals to dominate others. Such an example would be the relationship between sterile worker bees (drones) and the queen bee. It is impossible for a drone to become a queen as it lacks the required reproductive organs. We will not consider static hierarchies here, but are more interested in how hierarchies can be dynamically formed based on differences between the members in the group. This, for example, would be how the queen bee dominates her sisters for the right to lay eggs in a hive.

We examine how these hierarchies form and adapt and consider how they result in a cohesive structure.

12.2.1 Formation and Maintenance of Hierarchies

Different species form hierarchies in different ways, but there is a common feature to all: *communication* of relative dominance or submission. Based on the received communication, animals subsequently update their perception of position within the hierarchy. Communication can be implicit, such as one animal observing the physical size of another, or it can be explicit in the form of an interaction, such as an aggressive fight. Communication can take on many forms, depending on the capabilities of the particular species and can be tactile, olfactory, acoustic/vocal, or visual. Whatever its form, the communication can be regarded as a stimulus emitted by the sender which alters the receiver's behavior. We refer to the communication exchange followed by the hierarchy update as a dyadic (pairwise) *tournament*. Tournaments generally result in a winner and a loser, with the winner increasing its role in the hierarchy upward and the loser downward. An animal which frequently wins encounters with other animals is likely to have a high role in the hierarchy. We now consider some attributes which are used to construct the social hierarchy and how they are communicated.

In a hive of bees, there is only one queen. This individual is responsible for the laying of all the eggs within the hive. The queen is tended to by her sister bees, and the queen emits a hormone which suppresses the formation of ovaries in her peers.[2] This is an example of a totalitarian or despotic hierarchy.[2] If the queen bee leaves to form a new nest or dies, in the absence of the hormone, the sister bees start to regrow their ovaries. One bee will dominate the rest, becoming the new queen. A similar structure also occurs in other eusocial organisms such as wasps,[3] ants,[4] and termites.[5]

The formation of dominance structures in chickens is a well-studied area.[6,7] Hens form a linear dominance structure, with a dominant individual meting out pecks to subordinates, hence the term "pecking order."[8] When a group of hens are assembled, there are frequent fights and changes in rank initially. Over time, as the structure becomes known, there are fewer and fewer fights as the hens are aware of their role or position within the hierarchy.[2] If a new hen is introduced into the flock, the other hens will attack it as a group.[9] In a linear dominance hierarchy, an alpha or super-dominant individual dominates all others. The next individual in the hierarchy, the beta individual, dominates all bar the alpha, and so on. Linear dominance hierarchies are a common feature in many species, such as crayfish,[10] anemones,[11] goats,[12] and ibex.[13]

Hierarchies are also a common system in primates, such as baboons and rhesus monkeys.[14] Baboons have a complex culture of social interaction which cements the troop together. Within human societies, hierarchies are also a pervasive theme. Political and business structures are organized along the lines of dominance hierarchies, with presidents, vice presidents, and so forth. Even the academic world is an example of a social hierarchy, with professors, lecturers, and students being its constituents.

There are many attributes which are hypothesized to be important in the formation of social hierarchies, and these depend on the individual species. In crayfish, there is a strong correlation between claw length and social status.[10] In anemones, the length of the feelers results in a difference in ranking in individuals.[11] In hens, there was a strong correlation between rank and comb size.[9]

The exact mechanism which is behind the formation of a social hierarchy in natural systems is unclear.[15] In repeated experiments with the same fish, researchers demonstrated that the rank order formed in successive trials was not identical, as would be the case under the assumption that rank order was solely based on observed differences between individuals.[16] There is a strong correspondence between the rank of the attribute and the resulting social hierarchy ranking, but it does not appear to be the only process at work.[17] It has been hypothesized that there is a form of positive reinforcement, where an individual which has recently won a tournament is more likely to win a subsequent tournament.[16] The reasons for this are unclear, but it is assumed to be a result of increased hormone production, leading to the individual becoming more aggressive.

It must be made clear that social hierarchies are formed along the lines of comparative or relative differences between individuals, rather than absolute attributes. For example, in one group of animals, an individual may be ranked at the top, but in a fitter group of animals, it may only be ranked in the middle. Thus, an individual's role within the hierarchy is dependent not only on its own attributes but also on the composition of the group itself. Social hierarchies are not static structures and dynamically react to changes as the result of insertion or removal of other creatures. They also alter if the attributes of a creature change, such as a result of injury or age.

12.2.2 Purpose of Social Hierarchies

Why do animals and insects form social hierarchies? The strongest driver for social agglomeration and organization is thought to be the avoidance or minimization of predation. This is the "selfish herd" hypothesis of Hamilton.[18] Essentially, it states that an animal is less likely to be a target for predation if its chances of being picked by the predator are minimized. Hence, if an animal is alone, it is more likely to be chosen by a predator. However, if it is part of a large group, it is less likely to be attacked. Some social animals form a "mob" which collectively acts to drive away a predator, giving further evidence for the purpose of the formation of the group.[19]

Some predators also form societies in order to hunt more effectively. Wild dogs form packs that can bring down a large animal which one dog could not do by itself.[20] Although the prey needs to be shared among more animals, the individual effort and risk required to obtain the food is reduced. There is also a strong correlation between the size of the wild dog pack and their hunting success.[21]

In essence, the role or position of an animal within its society controls its preferential access to resources with respect to its peers. A super-dominant individual generally has first choice of mate, food, and other resources, and thus is more likely to propagate its genes onto a future generation. This comes at the cost of maintaining its position at the apex of the hierarchy, and a deposition can often result in death or serious injury. However, social structures provide for an organized access to resources, and are thought to lead to less competition over resources, as each individual is aware of its position.[2]

12.3 Using Adaptive Social Hierarchies in Wireless Networks

Wireless networks are by their very nature amenable to the imposition of a social hierarchy. A hierarchy is a relative ranking system, based on measurable differences between individual nodes. Essentially, a hierarchy provides an abstraction of the real-world resources onto a framework which indicates relative performance (such as poor, good, best). The purpose of a network is to transfer information from a source to a sink, through multiple nodes. To conserve resources (such as bandwidth and energy), the information is sent along the shortest path (as measured by some metric such as hop-count or latency) between the sink and the source, through intermediate nodes. The aim of a routing protocol is in essence to determine the "best" path for the information to take. This means that the set of all possible paths can be cast into a hierarchy, ordered by the desired metric, and the routing algorithm decides which is the best path. All routing protocols which are based on minimizing some metric are implicitly forming a hierarchy among the nodes in the network and basing routing decisions upon differences between individuals. However, explicitly constructing and adapting a hierarchy is an area which is yet to be investigated fully.

Hierarchies are a common feature of network protocols, but many are imposed at design time, based on physical differences between nodes (e.g., the Data Mule system is composed of a three-tier static structure: stationary nodes, mobile (higher power) nodes, and base-station nodes[22]). A network should be able to learn the differences between nodes, rather than have the differences specified prior to deployment. In this way, the network will be able to dynamically adapt to insertions and removals. Thus, we refer to the structure as an *adaptive social hierarchy* (ASH) to reflect the fact that the hierarchy adapts to changes in attributes and is not static.

A hierarchy can be thought of as a mapping from an absolute domain (such as 1 J of energy, three hops from sink) to a relative domain (e.g., best node in network). An example of this mapping is shown in Figure 12.1. Although the real-world values

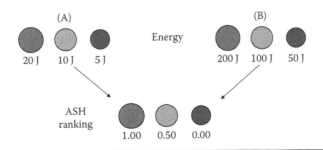

Figure 12.1   Illustration of ASH ranking procedure. Although the networks in A and B have different energy levels, they rank to the same network when viewed in the ASH sense. This resource abstraction makes network design independent of real-world values.

of the nodes in network B are 10 times as large as the metrics in network A, they both map to the same system when viewed from a relative ranking perspective. In the presented protocol, the rank is scaled to lie within [0;1], where 0 corresponds to the lowest ranked node and 1 corresponds to the highest ranked node.

Node management and routing is based on the relative parameters, rather than the absolute. This means that network decisions are based on factors which are independent of the real values, leading to scalability in the resource domain. In a relative domain, network decisions do not need to take into account real values, with the implication that a network protocol can be used in vastly different application scenarios, without having to alter network-tuning parameters. In addition, any parameter that can be measured can be placed into a social hierarchy. Thus, the design of network protocols can be separated from the real-world scenario. For example, in a static network, hop count is an important metric, whereas in a mobile network other metrics such as utility are more useful. Traditionally, each of these would require a different routing protocol, whereas if a social hierarchy is used, a single network protocol can be used for both problems. This is illustrated in Figure 12.2 for a stationary and a mobile system, showing how they both map into the same relative network, simplifying network design. In the stationary network, hop count is used as a measure of "goodness," whereas in the mobile network, transitive connectivity is used. The fixed network would require different routing rules to the mobile network, as the range of the values of the absolute parameters is different. However, if a ranking system is used, the absolute values of the parameters are mapped onto relative values which are between 0 and 1. In this way, the same routing rules can be used if a social hierarchy is constructed.

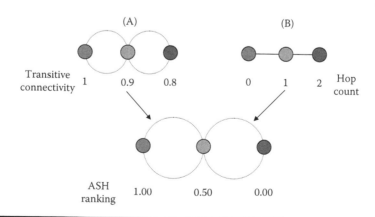

Figure 12.2 ASH working on two different connectivity metrics. In mobile network A, connectivity is measured by the transitive connectivity, whereas in the stationary network B, connectivity is measured using the traditional hop count. Note that both networks map to the same network when ranked.

Thus, there are now two different areas in network protocol design: constructing the social hierarchy and the formulation of protocol rules. The hierarchy can be regarded as an underlay which provides the resource abstraction onto a relative structure. Various methods of creating a social hierarchy are discussed in the following section.

12.3.1 Constructing an Adaptive Social Hierarchy

As previously stated, a hierarchy can be viewed as a mapping from an absolute parameter space to a relative parameter space. This mapping can be undertaken in many different ways. Ranking needs to be determined in a distributed fashion among all the nodes, as opposed to a centralized system which has a single point of failure. As nodes are typically resource-constrained, methods to determine ranking should not result in a high level of overhead in the form of ASH ranking and updating. In addition, the ranking needs to be scalable not only with the number of nodes in the network, but also with the range of parameter values. Before various ranking systems are discussed, some terminology is introduced.

Let S be the set of a particular attribute A of the N nodes in the network

$$S = \{A_1, A_2, \ldots, A_N\}. \qquad (12.1)$$

When the order of the elements in this set is taken into account, an ordered set is formed. For example, consider the four-element set of attributes $S = \{10, 4, 8, 6\}$. The ordered set $\phi = [4, 6, 8, 10]$ is the original set sorted into order of size. The rank order of the sorted set is given by $R = [4, 1, 3, 2]$ where the rank refers to the index of the original element. In our case, however, we are interested in the scaled or normalized rank, which is expressed as

$$\hat{R} = \frac{R - 1}{N - 1}. \qquad (12.2)$$

Thus, for the given example with rank order $R = [4, 1, 3, 2]$, the normalized rank is found to be $\hat{R} = [1, 0, 2/3, 1/3]$. By normalizing the rank, the rank is made independent of the number of nodes in the network, leading to scalability. This is because it is known that the maximum rank is 1 and the minimum rank is 0. Thus a node with an attribute that has a rank of 1 can be regarded as the "best" node in the network, in terms of that attribute.

The goal of our work is to present various methods of evaluating the ranking in a distributed fashion, much like animals form social hierarchies based on measurable or perceived differences. We denote the rank as estimated by a method of ranking as $\hat{E}$. First, though, we consider how to measure how well a method performs in the task of ranking a set of attributes.

There are a number of factors which we are interested in with respect to the formation and maintenance of the social hierarchies. Probably the most important factor is the rate of convergence or settling time of the social hierarchy, from initialization to a point when it is deemed to be settled. Another factor is the stability of the hierarchy—whether nodes remain in the correct order or whether there are time-varying errors in the ranking. Lastly, we are also interested in the adaptibility of the social hierarchy—how long it takes the system to react to disturbances, such as node insertion or removal or a change in a node's attribute. With all of these performance metrics, it is necessary to determine whether they are scalable in terms of increasing N, the number of nodes in the network.

To determine when the hierarchy has converged, a metric is required that reflects the degree of order of the rankings of the nodes in the system, and whether they are in concordance with the order of the absolute parameters. The simplest metric is one which indicates when the estimated ranks are perfectly ordered. In this case, we can say the system is correctly ordered when $\hat{E} = \hat{R}$.

In a network of this nature, subject to frequent changes and rank reversals, it is not necessary for the hierarchy to be perfectly sorted in order for the nodes to know their approximate role in the network. One possible measure of error is the mean squared error (MSE). This reflects how close the system is to its desired steady-state values, but does not weight incorrect orderings highly. Another metric, formulated especially for determining ordinality of the ranking, is the Kendall Correlation Coefficient, denoted by τ.[23] This indicates the normalized number of pairwise exchanges or switches required for the set of estimated ranks to be perfectly sorted. A Kendall τ of -1 corresponds to perfectly reversed ranking (e.g., [4;3;2;1]), whereas a τ of $+1$ reflects that the system is perfectly ordered.

However, we are interested in the worst case performance, and this would be expected from the node with the greatest error in ranking. For example, if a node has a small amount of energy, and it is erroneously ranked highly relative to its peers, then depending on the application, it could expire rapidly. Thus, we define another convergence time as the expected time taken for the maximum rank error in the network to drop below a certain threshold. We define the maximum (absolute) rank error as $e_{max} = \max |\hat{E} - \hat{R}|$, and we say that our system is converged when $e_{max} < \epsilon$, where ϵ is the acceptable percentage error tolerance. The time taken for this to occur is denoted as $T_{\pm\epsilon}$. Note that T_0 corresponds to the expected time to be perfectly settled. We use $T_{\pm 20\%}$, $T_{\pm 10\%}$, $T_{\pm 5\%}$, and T_0 (corresponding to a maximum error of 20%, 10%, 5%, and 0% respectively) to measure rate of convergence (ROC). Different applications will require different error thresholds depending on their requirements.

Closely related to the initial ROC is the performance of the system when nodes are subject to changes in their attributes or nodes are inserted or removed. In control theory terms, the rate of adaption (ROA) is a measure of disturbance rejection. A system which takes a long time to recover from perturbations will not be suitable for use in a wireless network, which is characterized by frequent variations in resources and connectivity. To measure the ROA, we vary 10% of the node's attributes to new, random values once the network has converged to the specified error threshold.

The ROA is defined as the time taken for the system to re-achieve $T_{\pm\epsilon}$ after the perturbation.

12.4 Pairwise ASH

Much like the way animal dominance hierarchies form through dyadic interactions, we start with a simple pairwise switch or exchange method of sorting the ranks of the nodes, in concordance to their attributes. When a pair of nodes meet, they trade their attributes and ranks. If the order of their ranks is in contradiction to the order of their attributes, their ranks are switched. If there is no contradiction, their ranks are left unchanged. The rules for this method of ranking are shown in Figure 12.3. There are four possible combinations—two correspond to ranks and attributes with the same order and the other two deal with contradictions.

A demonstration of the sorting action is shown in Figure 12.4 for a 10-node network. Initially, the nodes were assigned to have ranks in perfect reverse order. After 89 meetings, the network is perfectly ordered. Note the frequency of switches with respect to the number of meetings—initially, most meetings result in rank alteration. However, as the system becomes more ordered, fewer node meetings result in a rank exchange. This is similar to what is observed in many animal social hierarchies which are characterized by a high initial competition rate which decreases as the hierarchy becomes ordered. The corresponding maximum rank error plot for the trajectory of Figure 12.4 is shown in Figure 12.5. It can be seen that the error decreases rapidly, with the maximum error of any node being less than 30% after 30 meetings. However, to reach perfect order, it takes a further 60 meetings.

We use this simple method of sorting as a baseline to compare other methods against. The possible transitions between all the possible orderings, as well as the probability of the transition, are shown in Figure 12.6 for a three-node system. Note that the transition graph is a Directed Acyclic Graph (DAG), as it can be topologically sorted. In addition, the Kendall correlation coefficients are strictly increasing with the transitions. This has the implication that any meeting that results in a transition will result in a more ordered network. It is impossible for a meeting to result in a network which is less ordered.

	$A_j > A_k$	$A_j < A_k$
$E_j > E_k$	$E_j \leftarrow E_j$ (unchanged)	$E_j \leftarrow E_k$ (exchange)
$E_j < E_k$	$E_j \leftarrow E_k$ (exchange)	$E_j \leftarrow E_j$ (unchanged)

Figure 12.3 Rank update rules for pairwise ASH. Both nodes perform the same rules at the same time, which will lead to a simultaneous exchange of ranks in the event of a contradiction.

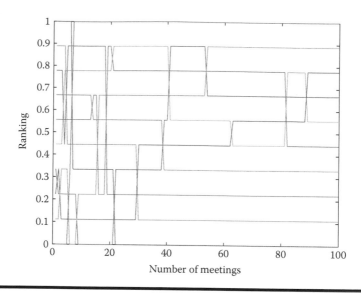

Figure 12.4 Rank trajectories for a 10-node network, starting from perfect reverse order.

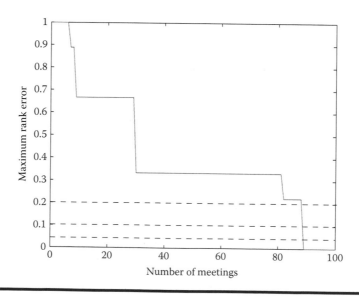

Figure 12.5 Maximum rank error for the trajectories shown in Figure 12.4. Note that the network is perfectly ordered after 89 meetings. The error thresholds corresponding to the determination of $T_{\pm 20\%}$, $T_{\pm 10\%}$, and $T_{\pm 5\%}$ are shown in the diagram as dotted lines.

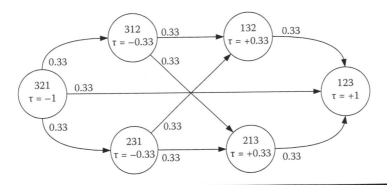

Figure 12.6   Transition diagram for the possible sequences of a three-node network. Probabilities next to each arrow are the exit probabilities from each state. The τ value shown is the Kendall rank correlation coefficient. It should be observed that τ is monotonically increasing from left to right, and that the transition graph results in strictly increasing values of τ.

We can place an upper bound on the expected number of meetings by examining how long it will take if the system traverses the maximum number of switches from perfectly disordered to perfectly ordered. The maximum number of switches is given by $N_s = N(N-1)/2$, which is the same as the number of possible combinations in which two unique nodes can be picked at random. For a three-node system, this results in the maximum number of switches being 3.

It will now be shown how to calculate the upper bound, given that the system starts from perfectly disordered, i.e., [3;2;1]. From this ordering, every possible meeting results in a transition to a new sequence. Thus the time that the system remains in this sequence before leaving is 3/3 = 1 meeting. Consider if the next ordering chosen is [3;1;2]. From this point, two out of a possible three meetings will result in a transition to a new ordering. Hence, we can expect the system to remain in this ordering for 3/2 meetings. Lastly, if the rank ordering is [1;3;2], only one meeting (one between nodes 2 and 3) will result in a transition to [1;2;3]. The system will be expected to take 3/1 meetings on average to leave. Once in the final state, the system will never leave. Thus, the total expected time to transition from completely disordered to perfectly ordered is the sum of the number of meetings required to reach this state, which is $3/3 + 3/2 + 3/1 = 5.5$ meetings. In general, the upper bound on the expected number of meetings can be written as

$$T_{\max} = \sum_{i=1}^{N_s} \frac{N_s}{i}, \tag{12.3}$$

where $N_s = N(N-1)/2$.

For large N, this can be expressed as

$$T_{\max} = N_s(\ln(N_s) + \gamma), \qquad (12.4)$$

where $\gamma = 0.57721\ldots$ is the Euler–Mascheroni constant.

Thus, in the limit, this shows that the upper bound on the settling time varies as $O(N_s \ln(N_s))$, or in terms of N as $O(N^2 \log(N))$. This has the implication that the required number of meetings for the system to converge perfectly appears to be prohibitively large as N increases. Later, in Section 12.6.2, we show that in the more realistic scenario where more than one node can meet per time interval, the effect of increasing node density counteracts the polynomial convergence time, resulting in a decrease in convergence time with increasing N.

However, while the result of Equation 12.4 is useful to place an upper bound on the performance of the switching scheme in terms of the expected number of meetings, in reality the meetings are random and thus the system is unlikely to pass through every possible combination to reach the ordered state. In addition, initial node rankings are likely to be random, not in perfect order.

To calculate the expected settling time, we can formulate this problem as a Markov Chain. The terminal state, corresponding to perfectly ordered, is called an absorbing state, and the ROC is equivalent to the expected absorption time. The possible state transitions are expressed as a transition matrix. This indicates the probability of moving from one state to another. The transition matrix for a system with N nodes is denoted by P_N. Thus, the transition matrix for the case of $N = 3$ is given by

$$P_3 = \begin{array}{c|cccccc} & 321 & 312 & 231 & 132 & 213 & 123 \\ \hline 321 & 0 & \frac{1}{3} & \frac{1}{3} & 0 & 0 & \frac{1}{3} \\ 312 & 0 & \frac{1}{3} & 0 & \frac{1}{3} & \frac{1}{3} & 0 \\ 231 & 0 & 0 & \frac{1}{3} & \frac{1}{3} & \frac{1}{3} & 0 \\ 132 & 0 & 0 & 0 & \frac{2}{3} & 0 & \frac{1}{3} \\ 213 & 0 & 0 & 0 & 0 & \frac{2}{3} & \frac{1}{3} \\ 123 & 0 & 0 & 0 & 0 & 0 & 1 \end{array} \qquad (12.5)$$

There are some things to notice about the P matrix, which we will later use to simplify our calculations. Firstly, it is of size $N! \times N!$, which has the implication that large N results in state space explosion. It is strictly upper triangular, as it is impossible for two nodes to meet and result in the system becoming more disordered. The first element on the diagonal is 0, as any meeting will result in an exit from this state. The last element on the diagonal is 1, as once the system is in this state it can never exit, corresponding to the absorbing state.

To determine the expected time to absorption, we express P in canonical form as

$$P = \left[\begin{array}{c|c} Q & R \\ \hline 0 & I \end{array}\right]. \tag{12.6}$$

Using a standard result from the theory of Markov Chains, the expected time from each state to absorption is given by

$$D = (I - Q)^{-1} = I + Q + Q^2 + Q^3 + \cdots. \tag{12.7}$$

D is a vector of dwell times, and the mean time to absorption from any state (assuming there is an equal chance of starting in any state, including starting off as perfectly ordered) can be calculated as

$$t_0 = \frac{\sum D}{N!}. \tag{12.8}$$

For P_3, the mean time to absorption is 3.167, which means that, on average, just over three pairwise meetings are required to perfectly sort the system.

To prevent state space explosion, we attempt to reduce the size of the state universe. We note that although the number of possible states varies as $N!$, the number of possible exchanges only grows as $N(N-1)/2$. We thus redefine our states as the expected number of exchanges to reach perfect order. This smaller matrix will be numerically easier to invert in order to find the expected time to absorption.

We define the reduced state transition matrix as S_N. The reduced transition matrix for a three-node network is given by

$$S_3 = \begin{array}{c|cccc} & 3 & 2 & 1 & 0 \\ \hline 3 & 0 & \frac{2}{3} & 0 & \frac{1}{3} \\ 2 & 0 & \frac{1}{3} & \frac{2}{3} & 0 \\ 1 & 0 & 0 & \frac{2}{3} & \frac{1}{3} \\ 0 & 0 & 0 & 0 & 1 \end{array} \tag{12.9}$$

The expected time to absorption still evaluates as 3.167, indicating that this approach of reducing the state universe is valid. However, even constructing a smaller matrix and inverting it is tedious. Table 12.1 shows the expected mean time to absorption as N varies for the various methods. Also shown is the convergence time as evaluated by the Monte Carlo simulation of the ranking process.

To this end, we have obtained an approximation to the expected convergence time for large N

$$T_0 = \frac{N_s}{2} + N_s(\ln(N) + \gamma). \tag{12.10}$$

Table 12.1 Convergence Times as Predicted by the Various Methods for Different N

N	T_0 (Full Markov)	T_0 (Reduced Markov)	T_0 (Simulated)	T_{max}
3	3.17	3.17	3.18	4.50
4	9.13	9.14	9.13	11.20
5	18.90	18.93	18.89	21.67
6	32.79	32.82	32.81	36.24
7	51.02	51.04	51.15	55.13

Also shown is the upper bound on the expected settling time, T_{max}.

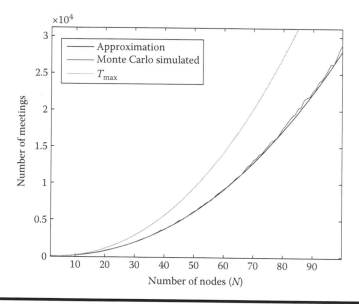

Figure 12.7 Variation in ROC against N for different error thresholds. The diagram shows that the approximation of 12.10 accurately predicts the convergence time as estimated by the Monte Carlo simulation. The upper bound on the convergence time is also shown.

A diagram showing the Monte Carlo–simulated convergence time compared with the approximation given by Equation 12.10 is shown in Figure 12.7. This shows that the approximation achieves a good correspondence to the Monte Carlo–simulated convergence times. For comparison, the upper bound of expected convergence time, T_{max}, is also shown.

Up to now, we have considered the time taken to be perfectly converged as a guideline. However, the question that must be posed is: "How close to correctly ordered must the system be to behave well?" This is because once a node has determined its approximate position in the network, the decisions made by the node are generally not fine-grained. For example, the node behavior is likely to be largely the same whether its rank is 0.16 or 0.17. Furthermore, attributes change with time, so a dynamic system is unlikely to be ever correctly ordered.

Error plots for the rank estimator are shown in Figure 12.8 for two different error thresholds ($\epsilon = 0.5$ and 0.05) for a 100-node network. Figure 12.8a shows that a maximum rank error of 0.5 leads to a wide spread of the ranks around the correct value. In most instances, this spread would make the network unusable. Figure 12.8b demonstrates that for a maximum rank error of 0.05, the correspondence between actual rank and estimated rank is very close to linear, with the majority of nodes having ranks that differ only a few percent from the correct values.

Figure 12.9 shows how $T_{\pm 20\%}$, $T_{\pm 10\%}$, and $T_{\pm 5\%}$ vary with increasing N. For comparison, T_0 is also shown. It can be seen that allowing for a looser definition of convergence results in more rapid settling times. In addition, the growth of the convergence times with increasing N slows with increasing error tolerance. This can also be seen in the plot in Figure 12.5 which shows the rapid drop in the initial error, followed by a long time to converge to perfectly settled. Based on our simulations and our prior results, it was found that the rate of convergence with N tended toward a log-linear relationship:

$$T_{\pm 20\%} = 3.32N(\ln(N)), \tag{12.11}$$

$$T_{\pm 10\%} = 7.08N(\ln(N)), \tag{12.12}$$

$$T_{\pm 5\%} = 14.2N(\ln(N)). \tag{12.13}$$

These equations show that the number of meetings required for the system to converge increases in sub-polynomial time, illustrating that this simple protocol will scale acceptably to large N. It should be noted that this is for the case when there is one pairwise meeting per unit time. In a realistic network scenario (which we encounter in Section 12.6.2), the number of meetings per unit time depends on the underlying mobility model. We deliberately decouple the mobility model from this analysis in order to provide general results that can be applied to many different application scenarios.

12.4.1 Pairwise ASH with Reinforcement

In Section 12.4, it was assumed that the initial ranks were equally spaced over the extent of the ranking interval. The restriction of equally spaced ranks makes initialization complex. Worse still, adaption to node insertions and removals is impossible, as the former will result in duplicate values and the latter in gaps in the ranking order. This makes scaling to dynamically varying numbers of nodes in the network impossible, precluding its use in most scenarios.

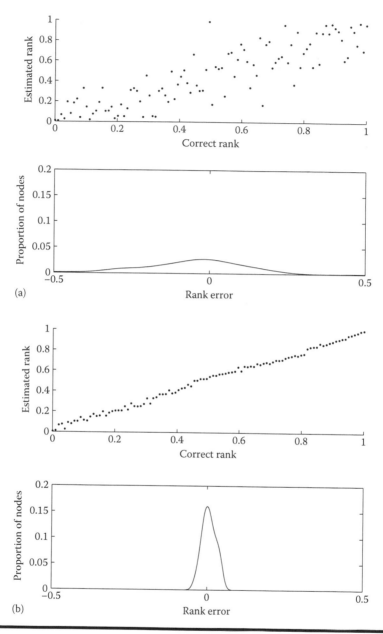

Figure 12.8 Ranks and the distribution of errors for two different error thresholds. Note how the distribution of the error becomes more peaked as the ε value is decreased, and how the distribution is approximately normal. The top plots show the scatter between estimated and correct rank and the bottom plots show the distribution of the error. (a) Error diagrams for ε = 0.5. (b) Error diagrams for ε = 0.05.

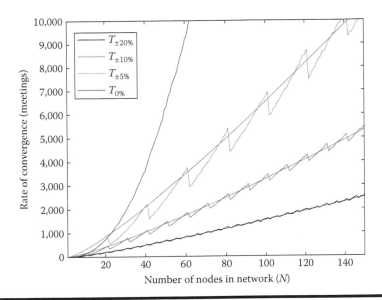

Figure 12.9 **Variation in ROC against *N* for different error thresholds. The approximations to the ROC for the differing thresholds are shown in grey. Note the slow growth of $T_{\pm 20\%}$ compared with T_0.**

To deal with this problem, nodes on startup choose a random normalized rank value between 0 and 1, with no input from their peers. Based on their meetings, nodes switch their rankings as in the prior section. However, in order that the ranks converge to be equally spaced over the interval [0;1], regardless of the number of nodes in the network, an additional update rule is introduced. If two nodes meet and their ranks are in agreement with their attributes, each node reinforces its rank. The node with the lower attribute reinforces its rank toward 0 and the node with the higher attribute reinforces its rank toward 1. Reinforcement upward is defined as

$$R = R(1 - \alpha) + \alpha \qquad (12.14)$$

and reinforcement downward as

$$R = R(1 - \alpha), \qquad (12.15)$$

both under the condition that $0 < \alpha < 1$, where α is the reinforcement parameter. The ranking update rules are shown graphically in Figure 12.10. A large α results in rapid rank updates, but leads to rank instability, whereas a very small α leads to little success in spreading the ranks equally. Note that for the case when $\alpha = 0$, this method devolves to that presented in Section 12.4. Thus, this method can be regarded as a more generalized version of pairwise ASH.

	$A_j > A_k$	$A_j < A_k$
$E_j > E_k$	$E_j \leftarrow (1 - \alpha)E_j + \alpha$ (reinforcement upward)	$E_j \leftarrow E_k$ (exchange)
$E_j < E_k$	$E_j \leftarrow E_k$ (exchange)	$E_j \leftarrow (1 - \alpha)E_j$ (reinforcement upward)

Figure 12.10 **Rank update rules for pairwise ASH with reinforcement. This allows for dynamic insertion and removal of nodes, resulting in equal spacing of ranks over time. The reinforcement parameter** α **controls the rate of adaption.**

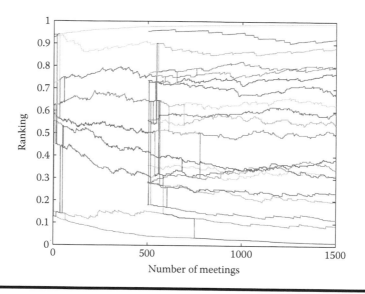

Figure 12.11 **Rank trajectories for pairwise ASH with reinforcement demonstrating node insertion. The reinforcement parameter was set to** $\alpha = 0.01$**. Initially 10 nodes with randomly assigned ranks were present in the network. After 500 meetings, another 10 nodes (also with randomly assigned ranks) were introduced into the network. Note how the trajectories spread out evenly.**

Rank trajectories highlighting how this method can handle node insertion are shown in Figure 12.11, and the corresponding maximum rank error plot in Figure 12.12. At the start of the simulation, 10 nodes with random ranks are placed into the network. The reinforcement parameter was set to be $\alpha = 0.01$. After 500 pairwise meetings, an additional 10 nodes with random ranks were injected into the network. This causes a spike in the maximum error, which is rapidly corrected through the switching (contradiction) action of the pairwise ASH protocol. The reinforcement action gradually causes the ranks to vary toward their correct values. The effect of random meetings can be seen as "noise" in the ranks of the nodes.

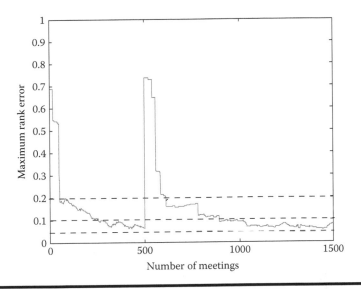

Figure 12.12 Maximum rank error for the trajectories shown in Figure 12.11. The error thresholds corresponding to the determination of $T_{\pm 20\%}$, $T_{\pm 10\%}$, and $T_{\pm 5\%}$ are shown in the diagram as dotted lines.

The value of the reinforcement parameter affects how rapidly the system is able to adapt to changes in network composition. However, if the reinforcement parameter is too large, the ranks will "chatter" and never converge to their correct value. Consider the extreme case of $\alpha = 1$. In this case, reinforcement will result in a node either having a rank of 0 or 1, depending on the direction of reinforcement. Thus, there will be no other values for the ranks in the network, leading to rapid switchings from maximum to minimum rank. To determine suitable values of α the average rate of convergence (for different error thresholds) against α for a 50-node network has been plotted in Figure 12.13.

12.5 One-Way ASH (1-ASH)

The algorithms presented so far deal with the case when a pair of nodes meet and trade their attributes and ranks. In a more typical network scenario we need to deal with the nature of the radio medium—being broadcast—from one transmitter to many receivers.

For many routing protocols, nodes emit "beacons" as network discovery packets (commonly termed "Hello" packets[24]). ASH parameters can piggyback on top of these "Hello" packets such that they do not lead to a detrimental increase in network overhead. When nodes are in receiver mode, discovering active nodes within their neighborhood, they can update their ASH ranking according to the transmitted ASH

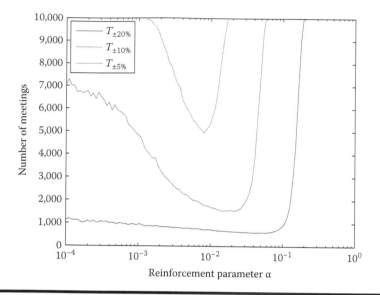

Figure 12.13 **Rate of convergence of a 50-node network with α for the various error thresholds. Note that large α results in excessively high (or infinite) convergence times.**

attribute/rank pairs. The only issue is that although the receiver nodes can update their rankings depending on the newly acquired information, the transmitting node is unable to update its ranking until it later switches to receiver mode. This is essentially an asynchronous method of updating the ranks, and it can lead to slower convergence and more churn.

12.5.1 Domination ASH

A hypothesized method of forming linear dominance hierarchies in nature is thought to be the win/loss ratio.[17] In this method, each node tracks how many nodes it dominates (i.e., the number of nodes which it exceeds in the value of its attribute) relative to the total number of nodes it meets. The rank update rules are shown in Figure 12.14. Every time a node is met, the total number of observed nodes, M, is increased by 1. If the attribute of the receiving node dominates that of the transmitter, the win counter, W, is also increased by 1. Initially, W and M are both set to zero.

As this is a ratiometric measure, it is not dependent on the absolute number of nodes in the network, leading to good scalability. It is simple to compute, but adapts slowly to changes and converges slowly. However, each node's rank converges to the correct asymptotic value with increasing M. Essentially, the domination ratio can be thought of as the probability of dominating another node chosen at random. This

	$A_j > A_k$	$A_j < A_k$
$E_j > E_k$	$E_j \leftarrow (W{+}{+})/(M{+}{+})$ (domination)	$E_j \leftarrow W/(M{+}{+})$ (submission)
$E_j < E_k$	$E_j \leftarrow (W{+}{+})/(M{+}{+})$ (domination)	$E_j \leftarrow W/(M{+}{+})$ (submission)

Figure 12.14   Rank update rules for domination ASH. As this is a one-way process, only the receivers update their ranks in relation to the transmitted attributes. *W* is a node variable which records the number of nodes dominated and *M* is the total number of nodes met.

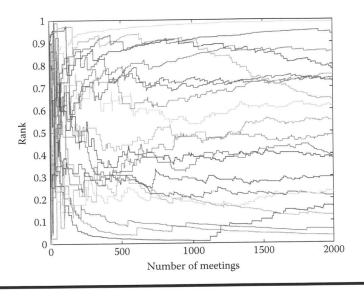

Figure 12.15   Rank trajectories for a 20-node network. After 1000 meetings, two nodes' attributes are randomly changed.

is shown in Figure 12.15 which demonstrates the long settling time coupled with the diminishing rank variation as the number of meetings increase for a 20-node network. After 1000 meetings, two nodes' (corresponding to 10% of the nodes) attribute values are randomly changed. The recovery from this disturbance is slow as *M* is large. Figure 12.16 shows the change in error with respect to time for the simulation conducted in Figure 12.15.

12.5.2 Domination Ratio with Switching

There are two main problems with the approach of Section 12.5.1. The first drawback is that it reacts very slowly to node insertions and removals, especially for

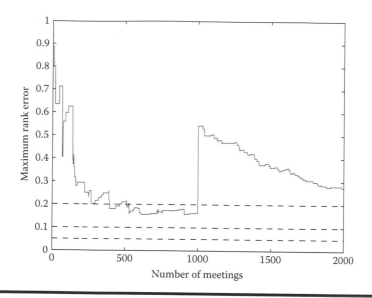

Figure 12.16   Maximum rank error for the trajectories shown in Figure 12.15. Note how the error "spikes" at 1000 meetings when two nodes randomly change their attributes and that the recovery from this error is very slow. The error thresholds corresponding to the determination of $T_{\pm 20\%}$, $T_{\pm 10\%}$, and $T_{\pm 5\%}$ are shown on the plot as dotted lines.

large M. The second issue is that it only uses the comparison between the attributes to update its rank. This is clear from Figure 12.14, where it can be seen that the rank comparison plays no part in updating the ranks of the nodes. To address these two issues, we modify the rank update rules slightly, incorporating the idea of switching from Section 12.4, and limiting the maximum value of M (and hence W).

These new rules are shown in Figure 12.17.

Before the update rules are executed by the receiving node, the total number of nodes that have been observed is compared against a limit L. If M exceeds L, both M and W are multiplied by a factor of $1 - 1/L$. This parameter controls the "memory" of the rank update. A large value of L leads to slow convergence but stable ranks, whereas a value of L which is too small leads to excessive rank oscillation. Rank trajectories for a 20-node network with $L = 100$ are shown in Figure 12.18. After 1000 meetings two nodes' attributes are randomly changed to new values. The corresponding error plot is shown in Figure 12.19. Note how the system recovers much more rapidly from the perturbation than the previous domination algorithm with no switching.

The simulation plots shown so far deal with the case when only one node is listening to the ASH broadcast. We now consider how the ROC varies when multiple nodes listen to the same transmitter in each time interval. Figure 12.20

$$if\ (M > L):$$
$$M = M(L - 1)/L$$
$$W = W(L - 1)/L$$
fi

	$A_j > A_k$	$A_j < A_k$
$E_j > E_k$	$E_j \leftarrow (W{+}{+})/(M{+}{+})$ (domination)	$W = M \times E_k$ $E_j \leftarrow (W)/(M{+}{+})$ (exchange)
$E_j < E_k$	$W = M \times E_k$ $E_j \leftarrow (W{+}{+})/(M{+}{+})$ (exchange)	$E_j \leftarrow W/(M{+}{+})$ (submission)

Figure 12.17 Rank update rules for domination ASH with switching. Before nodes update their rank, they first check the limit on M. If M exceeds the limit, both W and M are proportionally reduced. The nodes then run the update rules in the table. In the event of a contradiction, the receiver will adopt the transmitter's rank.

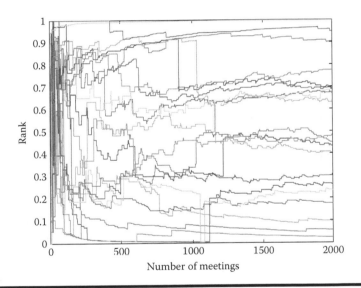

Figure 12.18 Rank trajectories for a 20-node network with a limit L = 100. After 1000 meetings, two nodes' attributes are randomly changed. Note the rapid switching action at the start of the simulation due to the rank exchange.

demonstrates how increasing the number of receivers in a time window leads to a much more rapid rate of convergence. Increasing the number of receivers results in an increase in the dissemination of information across the network.

It should be noted that if the number of receiver nodes is increased from 1 to 2, the expected rate of convergence halves, regardless of the value of N. This is an

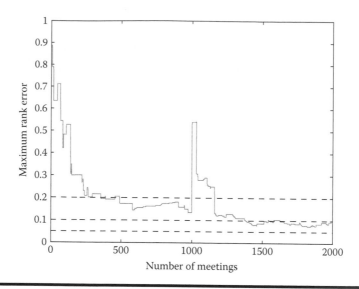

Figure 12.19   Maximum rank error for the trajectories shown in Figure 12.18. Note how the error "spikes" at 1000 meetings when two nodes randomly change their attributes. The error thresholds corresponding to the determination of $T_{\pm 20\%}$, $T_{\pm 10\%}$, and $T_{\pm 5\%}$ are shown on the plot as dotted lines.

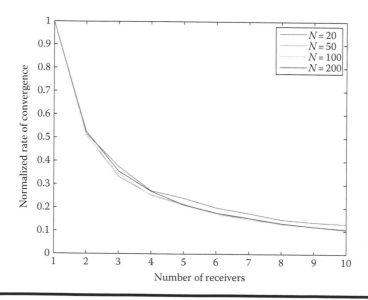

Figure 12.20   Normalized ROC ($T_{\pm 10\%}$) against number of receivers for varying number of nodes in the network. The time to converge for one receiver is taken as the base figure of 100%. Note how increasing the number of receivers leads to a rapid decrease in the normalized ROC.

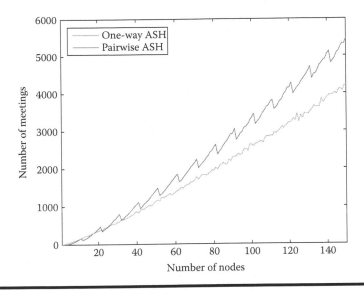

Figure 12.21 **ROC ($T_{\pm10\%}$) against number of nodes for pairwise ASH and one-way ASH. For one-way ASH, $L = 100$ and the number of receivers for each transmission was set to 3.**

important result, as it demonstrates that it is not the proportion of nodes receiving rank information, but rather the number of nodes. Thus, it would be expected that in a network with increasing N, the important factor is the average node degree, not the edge density. Lastly, in Figure 12.21, we show how one-way ASH can converge more rapidly than pairwise ASH, for the situation where there are multiple receivers (in this case three) to each node broadcast. For the case of $N = 150$, a node degree of 3 corresponds to a network density of 2%. This shows that one-way ASH can converge quickly even in sparsely connected networks.

Figure 12.21 demonstrates the performance of one-way ASH compared to pairwise ASH for varying N. The ROC for $T_{\pm10\%}$ for both approaches is shown, when each transmitter in one-way ASH transmits to three receivers. This shows that although one-way ASH suffers from asynchronous updates, it can exceed the performance of pairwise ASH while being more suited to the broadcast nature of the wireless medium.

12.6 Dealing with Mixed Mobility: An Agent-Based Approach

The prior approaches to ranking nodes discussed in Sections 12.4 and 12.5 rely on the assumption that nodes meet at random (with a uniform probability) in order to percolate the attribute/rank information throughout the whole network. If nodes are stationary, the previously presented techniques can fail as a result of the restricted neighbor horizon, leading to limited node discovery.

We can introduce pseudo-mobility by recreating the effect of randomized meetings. Nodes listen to transmissions from nodes within their immediate radio range. If they repeat these transmissions to their neighbors, two nodes which do not have a direct connection can "observe" each other and correctly update their ASH ranking. This has the effect of artificially increasing the probability of connection between any two nodes, hence increasing the network connectivity. Thus, the goal of disseminating rank/attribute pairs to a large number of nodes in the network can be achieved, leading to a more accurate representation of the social hierarchy. We refer to these rebroadcasted rank/attribute pairs as agents, as they can be viewed as independent carriers of information. The problem with the agent-based approach is that agents can be carrying outdated information. Thus churn (or rank variation with time) is expected to be higher in this scheme than in the other systems under purely random motion.

To prevent flooding and unacceptably high overhead, nodes only rebroadcast other nodes' data as part of the "Hello" packet, as in the previous sections. Nodes select at random which rank/attributes to rebroadcast from a small local buffer, and upon overhearing new data, pick a rank/attribute pair in the buffer to replace. This way, ranks are randomly rebroadcast, without detrimentally loading nodes in the network. Agent ASH uses domination with switching as presented in Section 12.5.2, with the incorporation of additional rules which control agent creation and spreading.

12.6.1 Agent Rules

Each node has a cache of length C entries. Each entry can store one rank/attribute pair. When nodes broadcast their "Hello" packets, they send their own rank/attribute pairs as before. However, they now append G "agents" or rebroadcast rank/attribute pairs from their local cache. Note that $G \leq C$. For the case where $G < C$, not all entries in the cache are re-broadcast. Each entry is thus picked without replacement at random with a uniform probability of G/C. Obviously, when $G = C$, the probability of an entry being picked from the cache is 1.

When a node overhears a broadcasted "Hello" packet, it can use this new information to update its cache. It randomly replaces entries in its local cache from the rank/attribute pairs sent in the message. There are $G + 1$ entries in the message, as each node sends its own rank/attribute pair followed by G agent entries. To populate its local cache with this new information, the receiving node places each new piece of information into a random slot in the cache. Thus, the local cache is refreshed with new information as it arrives. To prevent the cache from containing repeated information, an entry is only replaced if that node has never seen that particular rank/attribute pair.

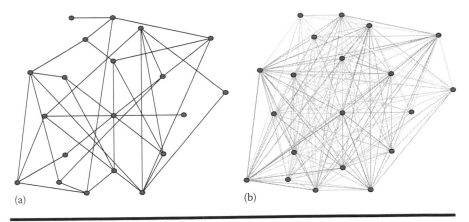

Figure 12.22 Graphs showing connectivity between nodes for a 20-node network. The graph on the right demonstrates the action of the agents which enable nodes which are not connected to each other to overhear their rank/attribute broadcasts. (a) Direct connections between nodes (indicated by solid lines). (b) Density of connectivity between nodes (line intensity denotes proportion of time edge is present).

The receiving node updates its rank using only the agent information carried in the message. It does not use the neighboring node's rank/attribute information, as this will lead to bias for frequently encountered peers, distorting the ranking process. This is not to say that a node will never use its neighbor's data. It is possible that it will observe this information indirectly through another node's agent data.

This is a very simplistic approach to using agents to spread information through the network, using no state information carried in the agents to control their spread and route. However, these random agents improve the connectivity of the underlying graph by creating "pseudo-edges." This is shown graphically in Figure 12.22, which shows the network edge density artificially increased for a 20-node stationary network.

In terms of ASH performance, we compare how Agent ASH is able to form the social hierarchy with how ordinary one-way ASH is unable to discover the nodes in the network, due to a limited horizon. This is shown in Figure 12.23, where it can be seen that the Agent approach is able to result in a lower error and faster convergence for a 100-node stationary network. This network was generated randomly to be connected with an edge density of 5%. Nodes transmitted "Hello" packets with a duty cycle of 5%. For the rest of the time, nodes were in receiving mode. The cache size, C, was set to three entries, and the number of agents transmitted per message (G) was set to one. This illustrates that Agent ASH can correctly discover the network ranking with a very modest increase in node resources,

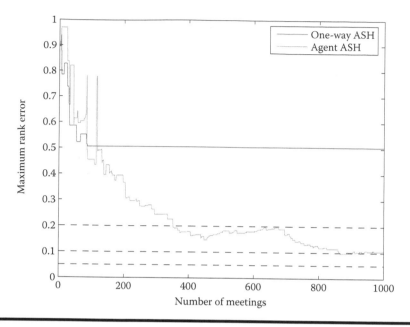

Figure 12.23 Rank errors for one-way ASH and Agent ASH applied to the same 100-node stationary network. Observe how the error for one-way ASH never drops lower than 0.5. This is as a result of the limited discovery horizon. The error thresholds corresponding to the determination of $T_{\pm 20\%}$, $T_{\pm 10\%}$, and $T_{\pm 5\%}$ are shown on the plot as dotted lines. In the simulation, $L = 500$, $G = 1$, $C = 3$.

even in sparse stationary networks. Agent ASH is not restricted to stationary networks; however, in Section 12.6.2 we demonstrate how it functions when nodes are mobile.

Ranking trajectories for a random 20-node connected network with edge density of 10% are shown in Figure 12.24. In this example, both the cache and the number of agents per message were set slightly larger, to four and two respectively. The error plot for this simulation is shown in Figure 12.25, which demonstrates the rapid recovery from the perturbation imposed after 1000 meetings. In this simulation, for fair comparison to the other protocols, one node was chosen at random to be the transmitter.

12.6.2 Realistic Meetings

In the prior sections, we have assumed that only one node is active and transmitting at any one point. In reality, the way nodes meet is dependent on the underlying mobility model. In a realistic network scenario, it is possible that, at any point in time, no nodes meet or more than two meet (possibly in geographically distinct

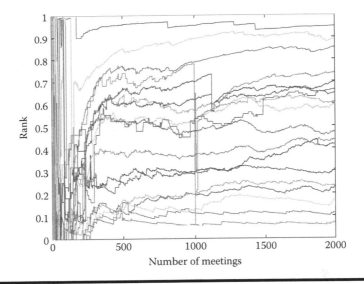

Figure 12.24 Rank trajectories for a 20-node network with $C = 4$ and $G = 2$, using Agent ASH. After 1000 meetings, two nodes' attributes were randomly changed.

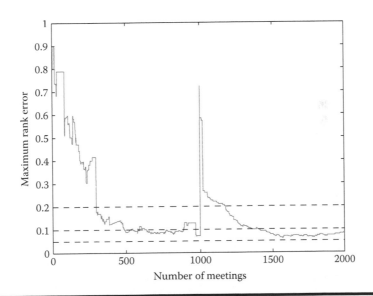

Figure 12.25 Maximum rank error for the trajectories shown in Figure 12.24. When the attributes are randomly changed after 1000 meetings, the error sharply peaks, followed by a rapid switching action to correct the erroneous ranks. The error thresholds corresponding to the determination of $T_{\pm 20\%}$, $T_{\pm 10\%}$, and $T_{\pm 5\%}$ are shown on the plot as dotted lines.

locations). To account for this, we incorporate the effect of the mobility model into the way nodes meet.

We examine how Agent ASH performs when subject to two commonly discussed mobility models: the random walk model and the random waypoint model. In the random walk model, at each point in time, a node chooses to move to another location that is one step away. In the random way-point model, nodes travel in a straight line at a randomly chosen speed until they reach their destination. Once the destination is reached, nodes choose a new destination and velocity and travel toward that. The random walk model exhibits a very slow mixing time, on the order of $O(K^2)$ where K is the length of one side of the simulation area, whereas the random waypoint has mixing times on the order of $O(K)$.[25] Thus, we would expect the performance of ASH to be worse when subject to the random walk model in comparison to the random way-point mobility model.

For the random walk model, N nodes are placed on a $K \times K$ toroidal simulation area. The radio radius, U, is varied from 5 to 15. The transmission range (along with the number of nodes) controls the connectivity of the network. The rate of convergence is examined for $N = 50$, 100, and 200. We examine the time taken for the system to converge to $\pm 20\%$ of its final value, and denote this as $t_{\pm 20\%}$, where the lower case t indicates real-time results, as opposed to the number of meetings. To introduce some realism into the MAC layer, we assume that nodes transmit "Hello" packets 10% of the time. For the remaining time, the nodes are in receiving mode. Unless otherwise stated, the parameters for the Agent ASH model were chosen to be $G = 1$, $C = 3$, and $L = 500$.

The simulation results for the random walk model are shown in Figure 12.26. The graph shows that the radio range has a large effect on the ROC. Increasing the number of nodes has the effect of reducing the simulation time. Contrast this to the previous results from Section 12.4 which showed that an increase in N resulted in an increase in the number of meetings for the system to converge. This demonstrates the scalability of the ranking system—an increase in N actually results in a faster convergence time, as it results in a more dense network.

In the simulation of the random way-point model, all common parameters are kept the same as the random walk. Nodes have a random speed uniformly chosen between 1 and 2 units/s. The results from this simulation are shown in Figure 12.27. Comparing these results to those of the random walk simulation, it can be seen how the random way-point model leads to faster convergence as it has a faster expected mixing time. Like the random walk results, it can be seen that an increase in N results in a decrease in the ROC.

The Agent ASH algorithm demonstrates that it is possible to sort or order a network according to its attributes, even if the network is purely stationary or subject to mixed degrees of mobility. In addition, only small amounts of local information are used to infer global attribute distribution. ASH piggybacks on top of existing network discovery packets and therefore does not present a large overhead burden. Now that we have described various methods of forming a social hierarchy,

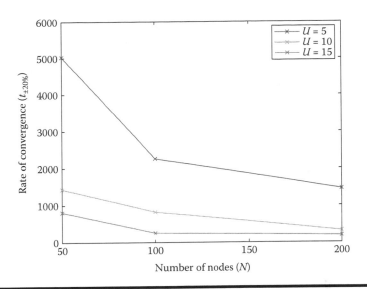

Figure 12.26   ROC to $t_{\pm20\%}$ for the random walk mobility model for different numbers of nodes. Note how the radio radius (U) has a strong impact on the ROC, as it alters the connectivity of the graph.

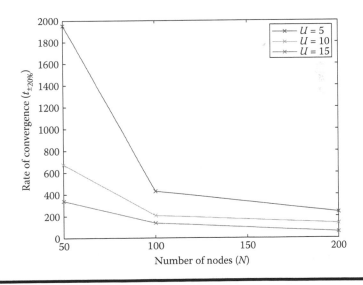

Figure 12.27   Rate of convergence to $t_{\pm20\%}$ for the Random Waypoint Mobility model for different numbers of nodes.

we consider how to use this information for network control, access, management, and routing. First, though, we examine some common network attributes.

12.7 Suitable Attributes to Rank

Any network parameter which can be measured on a node-by-node basis can be ranked in the global sense using one of the ASH ranking methods. Suitable attributes are obviously ones which have a direct and useful impact on network performance and control. We present a few useful attributes, which we will later use in Section 12.8 to demonstrate the power and flexibility of the ASH philosophy.

12.7.1 Energy/Lifetime

Of prime importance in deeply embedded and remote networks is energy, and coupled to that, the rate of use of energy. This is reflected in the vast number of energy-aware network routing protocols (refer to Jones et al.[26] and Akyildiz et al.[27] for more information). Nodes, depending on their hardware capabilities, can measure their energy reserves and usage, in an absolute sense (such as 100 J remaining). However, it is also possible to estimate energy usage in software (e.g., using the Contiki Operating System[28]), and knowing the size of the battery that the unit is equipped with, determine the remaining amount of energy. A node with a large amount of energy in relation to its peers can be expected to survive longer in the network. However, this assumption of survival is made under the pretext that all nodes consume energy at the same rate. Thus, a more useful indicator is estimated node lifetime, based on a long-term average of prior energy usage and current reserves.

A composite indicator of lifetime and connectivity can be formed from the lifetime of the path to the sink. The lifetime of the path can be regarded as the expected lifetime of the node with the lowest energy along the path, as this is the likely point of failure. To incorporate the distance from sink to source, the minimum path lifetime is decreased by a factor with each hop. The amount by which the minimum path lifetime is decreased controls whether shorter paths that travel through low energy nodes are favored over longer paths which avoid low energy nodes. This metric is similar to the Conditional Battery (MMCBR) protocol proposed by Toh.[29]

12.7.2 Connectivity

The purpose of a wireless network is to transfer information from source to sink, through intermediate nodes. The purpose of a routing protocol is to deliver the data along the "best" possible path, where the desirability of each path depends on some underlying cost function of suitability (such as delay, hop count, or redundancy).

In a network for information delivery, nodes need to send information to base station(s). At each hop, data should be sent to a node which is "closer" to a base station. Lindgren et al. present a method of evaluating transitive connectivity to a base station,[30] which is based on the observation that if A and B frequently meet, and B and C frequently meet, then A and C can be regarded as being transitively connected through B. In static networks, the conventional hop-count metric can be used as a measure of connectivity.

Another metric which impacts the delivery of information is the local traffic density. In regions of high traffic density, it would be expected that collisions will be more frequent. Nodes can assess the local error rate, possibly by monitoring how many of their transmissions fail. This information can be used to build a ranking of the expected congestion in the network. Data can be sent along paths where the expected congestion is low, thus balancing traffic across the network.

12.7.3 Buffer Space

Depending on their physical memory size and their role within the network, nodes will have varying amounts of buffer space available for messages from other nodes. This can be ranked, although the ranking would be expected to be very dynamic as buffers are cleared upon successful delivery. A long-term average of available buffer space would possibly be a better metric of delivery rate.

12.7.4 Functions of Attributes

Ranked attributes do not need to be based on a single measure of network performance, but can be a composite of multiple attributes, weighted in various ways. There are many ways of constructing a combined attribute. Context-Aware Routing (CAR) combines attributes to form a single weighting.[31] Of particular interest with the CAR approach is that the availability and predictability of the various attributes is incorporated into the weighting procedure.[32] The ranking process can be performed on the combined attributes, or the ranked attributes can be combined (and further ranked if necessary).

12.7.5 Levels and Loops

In a practical network scenario, a node needs to decide whether to send a message to another node based on the difference in their ranks. If a node sends its data to another node with a greater rank, this can lead to an explosion in traffic. In addition, there is the possibility of data forming a loop as the rank of the host varies over time.

To address these issues, we quantize the rank into L discrete levels. As all the ASH methods presented in this chapter generate linear hierarchies, it can be easily seen that each level (or bin) will contain N/L nodes. Assume that traffic is generated at a rate of λ messages per unit time, and nodes only send messages to nodes which have a greater level. The traffic density of the lowest nodes in the level is $D_1 = \lambda$ as they will only forward the packets they generate, and not route any other traffic. A level 1 node will send its messages to any node with a higher level. There are $L - 1$ levels that are higher, and they will thus each expect to receive (and subsequently be responsible for forwarding) $\lambda/(L - 1)$ messages. Thus, in general, we can express the traffic density at each level in the hierarchy using the recursive equation

$$D_k = D_{k-1}\left(1 + \frac{1}{L - k + 1}\right), \qquad (12.16)$$

where $1 \leq k \leq L$.[33]

This can be simplified to obtain an expression for the traffic volume at level k

$$D_k = \frac{\lambda L}{L - k + 1}. \qquad (12.17)$$

This important result shows that traffic density is *independent* of the number of nodes in the network, and only related to the number of levels in the quantized hierarchy. This demonstrates that the ASH framework results in good scalability by controlling traffic density.

To prevent packets from looping through the hierarchy, in the event of dynamic ranks, packets can be tagged with the rank of the node which sent the data to the current node. As packets can only ascend the hierarchy this means that loops cannot form.

12.8 Example Scenarios of ASH

We now show how ASH can be used as an underlay to enhance existing protocols and also to form a cross-layer protocol in its own right. For reasons of space, we omit any simulations, but rather discuss how ASH can be used to improve the behavior and functioning of protocols.

12.8.1 Enhancing Spray and Focus

Spray and Focus[34] is a multi-copy controlled replication routing protocol. Routing is divided into two phases. The first phase, spraying, creates multiple copies of a message around a source. There are different ways of undertaking the spraying, but the authors show that binary spraying is optimal.[34] In this situation, given that L distinct copies are to be created, the source will hand over $L/2$ copies to the first node it meets, keeping $L/2$ for itself. At each point in time, a node can hand over half the remaining copies of the message, until there are L nodes in the network, each carrying a single copy. In the original version of this protocol, Spray and Wait,[35] nodes performed direct delivery to the sink of the information. However, in the most recent version, nodes enter a Focus phase, where data are forwarded along a utility gradient toward the sink. A timer indicating the time of last contact with the destination is used as the utility parameter. To capture transitive connectivity, the timer value is updated if a node meets an intermediate node which was recently in contact with the destination. The use of the focus phase dramatically reduces delivery latency, while maintaining the benefits of controlled replication.

The details of the Focus phase are as follows. A node sends its local copy of a message to another node if and only if the utility of the receiving node exceeds the sending node by a certain threshold, U_{th}. Clearly, the choice of this threshold is critical. If it is too large, then messages will rarely be forwarded closer to the destination, leading to increased delivery delay. Conversely, for a small threshold, a message can possibly be forwarded many times before reaching its destination, leading to resource exhaustion. Tuning the threshold parameter can be done at run time by the user, but is likely to be time-consuming. Rather, by ranking the utility values, the choice of the threshold can be made independent of the underlying mobility model. Thus, using ASH as an underlay on this protocol will simplify the choice of the tuning parameter by making it scalable across wide attribute values.

12.8.2 Enhanced Context-Aware Routing

Musolesi and Mascolo presented Context-Aware Routing (CAR), which is a utility-based routing protocol for intermittently connected networks.[31] In CAR, nodes evaluate their "utility," which is a metric indicating the usefulness of a node in terms of network-specific parameters such as connectivity or energy level. These different parameters are combined to give a single utility value for each node in the network. One of the main contributions of this work was the idea of predicting future values of the utility, using Kalman filters (or the reduced form of an Exponentially Weighted Moving Average).

A more recent work considered the application of CAR to sensor networks with mixed mobility (called SCAR[36]). In this scenario, sources of information (which can be fixed or mobile) deliver information to sinks (which can also be fixed or mobile). As connectivity is intermittent, information is buffered at intermediate nodes which are more likely to be able to forward the data to a sink. Note that data can be delivered to any sink. To reduce the delivery delay, messages are replicated to multiple neighboring nodes before forwarding along the utility gradient toward the sinks, in a similar manner to Spray and Focus.[34]

Three measures of context information are used to decide on routing. The first is the change rate of connectivity (CRC), which essentially is a metric of the change in the local connectivity graph, a measure of relative mobility through the network. The degree of colocation with the sink is proposed as the second measure of utility. The last measure is the remaining battery level. The battery level is the proportion of remaining energy, relative to the initial battery level. Clearly, this protocol does not handle heterogeneity in the initial battery level. By using the ASH framework, the energy level can be ranked on the interval [0;1], regardless of the distribution of energy in the absolute sense.

The absolute level of energy is ranked using ASH, and then the energy ranking is used to construct the utility parameter. Thus, the introduction of the simple ASH protocol as an underlay vastly increases the scope of SCAR, allowing it to be used in networks with widely heterogeneous distributions in energy. It should be noted that ASH can also be used to rank the calculated utility, as SCAR too has a threshold value, ζ, which is used to control replication.

12.8.3 A Simple Cross-Layer Protocol

In this example we demonstrate how network information can be used at all levels of the traditional network stack, collapsing the strict segmentation, leading to a simpler implementation. We rank both energy and connectivity separately and use the ranked data throughout all the levels of the network stack.

12.8.3.1 Medium Access

Nodes with low ranks both in connectivity and energy are not active in network tasks such as routing and replication. They essentially act as leaf nodes, only injecting packets into the network. As a leaf node is not required to route other node's packets, there is no cause for it to ever attempt to listen for other nodes' data transmissions (it does still need to observe "Hello" packets in order to maintain its correct rank). In addition, due to its scarce resources, it should not have to compete equally with higher ranked nodes for access to the medium. Furthermore, as packets can only ascend the hierarchy, there is no point in a low-ranked node listening to a high-ranked node's transmission. We present a simple slotted MAC scheme to demonstrate how ranking can result in a sensible preferential access to the shared medium. This is shown in Figure 12.28, which shows the behavior of

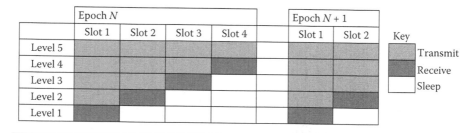

Figure 12.28 Slotted MAC organization example for a five-level network. Based on their level, nodes choose what action to take in each slot. Note that the lowest-level nodes spend the majority of their time in low power sleep mode.

each node in its assigned slot. This is only one possible arrangement of slots. A more realistic approach might be to have wider or more slots for higher-ranked nodes as they spend more time on the medium as they will have a greater amount of data to send. The scheme does not have to be slotted (relaxing the requirement of synchronization), but can also be made into a random access protocol, where the probability of listening to the medium is based on the node's level.

12.8.3.2 Routing and Replication

We now show one possible method to control data delivery in a network ranked by ASH. Low-ranked nodes perform direct delivery to higher-ranked nodes. High-ranked nodes share information amongst themselves, epidemic style. Thus, replication and delivery is controlled according to rank. Clearly, duplicating a message across low-level nodes does not achieve any useful redundancy, as these nodes are not active in disseminating information and are likely to be severely resource-constrained compared to their higher-ranked peers. Thus, messages are replicated with a probability that increases with node level. Hence, delivery amongst low-level nodes will resemble direct routing (with low traffic overhead, but high latency) and delivery between high-level nodes will resemble epidemic routing (with high traffic overhead and low latency).[37]

The probability of a level k node sending a message to a level j node is given by

$$p_{send} = \begin{array}{l} 1 : j > k, \\ p_t : j = k, \\ 0 : j < k, \end{array} \qquad (12.18)$$

where p_t is the horizontal (i.e., across the same levels in the hierarchy) transmission probability. Once the message has been sent to a higher-level node, the node can either keep the message or delete it. The probability of a level k node keeping a sent message for future replication is given by

$$p_{replicate} = r^{L-k+1}, \qquad (12.19)$$

where r is a parameter that controls the degree of the replication. This equation results in a probability of keeping a message for future replication that increases with level in the hierarchy. The value of r affects the expected number of copies present in the network.

To prevent thrashing due to rank promotion and demotion due to attribute changes, packets are not allowed to descend in rankings, even if the host's ranking has dropped. This prevents issues with packets forming loops in the hierarchy, leading to rapid network exhaustion.

12.8.3.3 Application

Information about the node's local variables, such as energy and connectivity can also be used at the application layer. A node with a low ranking in energy is likely to expire sooner than once with a high energy ranking. Thus, based on the energy ranking, the application running in the node can shut down or throttle back energy-consuming tasks. For example, in a global positioning system (GPS) based system, the GPS receiver consumes a large amount of power. To conserve energy, the sampling rate of the system can be reduced, leading to a greater node lifetime at the cost of location resolution. The application can also use lossy compression algorithms to reduce the volume of data that needs to be sent over the radio medium. The amount of compression can be controlled by the rank in the hierarchy.

12.9 Related Work

The primary contribution of this work is the presentation of methods that can be used to rank heterogeneous attributes in a network into a relative framework. As such it is complementary to many routing protocols as it can act as a network underlay, enhancing their performance.

The idea of using hierarchies in networks is not a new one, but to the best of our knowledge this is the first work that has considered the general situation of how to dynamically cast any measurable resource into a network-wide ranking system. Many networks use hierarchies that are imposed at design time. These are akin to the static hierarchies discussed in Section 12.2 where there are gross differences that result in a predefined social structure. In the Data Mule structure, there are three classes of nodes: low-energy stationary nodes, high-capability mobile nodes, and base stations.[22]

Some network hierarchies are dynamic. For example, in the Low-Energy Adaptive Clustering Hierarchy (LEACH), a node is selected from a small group to act as a "cluster head," which is responsible for relaying the combined information of the local group.[38] As nodes age, the role of cluster-head is rotated among the group. This two-level hierarchy is an example of a despotic or totalitarian structure. Other cluster-based schemes have improved on the performance of LEACH, yet still retain the two-tier structure.[39]

The work that is most closely related to ours is the role assignment algorithms of Römer et al.[40,41] In this scheme, nodes decide on their role within the network based on information acquired from their local neighborhood, by populating a cache of node properties and running node assignment rules based on their cache. Their algorithms were explicitly designed for stationary networks, as a change in the neighborhood would result in a cache update. They introduced high-level compiler directives that are used to decide on the most suitable role for a node (such as cluster head or ordinary node), while satisfying requirements such as coverage. Their algorithms require the specification of time-out factors and explicit update methods – ASH is lightweight in comparison to the role assignment algorithms as it piggybacks onto existing network control packets providing a transparent evaluation of local role. To control flooding, the role assignment algorithms use a limited hop neighborhood, whereas ASH discovers network-wide information in order to assess rank.

The ranking of nodes can be viewed as sorting them into order according to their attributes. We use a scaled domain, though, to represent the maximum value of the attribute as 1 and the minimum as 0, so this method is not entirely a simple sorting procedure. Some recent work has been performed on the theory of Random Sorting Networks.[42] Random Sorting Networks are networks for sorting information that are created at random. This is similar to our pairwise ASH scheme presented in Section 12.4 which can also be viewed as a randomized version of a Bubble Sort.

We use a very simple agent-based approach. Far more sophisticated ant-based routing and discovery methods have been presented in the literature which use stateful agents and pheremone trails.[43–45] However, it was our intention to keep ASH formation and maintenance as simple and lightweight as possible. This can be seen in our simulation results, where the error dropped to acceptable levels with a cache size of three entries and an agent repetition rate of one entry per "Hello" message.

12.10 Conclusions and Future Work

12.10.1 Future Directions

This work has discussed methods of forming a social hierarchy among nodes in various situations. However, there is scope for further exploration into some of the areas which have yet to be addressed. One such avenue is security and protecting the nodes from malicious attacks. For example, a rogue node could falsely advertise maximum rank and attribute, and so act as an attractor for packets. In this way, it could remove and use information from the network, preventing it from reaching the base station. In conventional networks, this is equivalent to a node masquerading as a base station. We are currently exploring methods of protecting the network from attacks of this sort.

All the methods for forming the ASH form a linear dominance hierarchy. In this type of hierarchy, only ordinality is relevant, not the degree of difference between nodes. In a heavily resource-partitioned network, there will be nodes with a large attribute value and nodes with low attribute values, but nothing in between. Such would be the case with small battery-powered sensors connected into a mains network, with no intermediate nodes in terms of resource size. In this case, two nodes which are close to each other in rank value could have a large difference in resource value. This could place an unfair load on a falsely ranked node. This suggests that a nonlinear hierarchy should be formed in this instance. This could be done by nodes exchanging a histogram reflecting the distribution of resource values across the network. We are currently investigating how exactly to undertake this task, while keeping network overhead low and being able to rapidly react to network changes.

In Section 12.6 we discussed how to use simple random agents to disseminate rank/attribute information across the network, to recreate the effect of randomized meetings. The agents used are extremely simple and essentially stateless. There is a large body of work on agent-based (also known as ant-inspired) algorithms for exploring networks and routing data. We are planning on extending the agent-based approach in a more refined manner, which will hopefully lead to faster convergence and lower error thresholds.

Another area which needs to be explored is how to factor in the rate of change of rank, both into the rank determination process and also the routing protocol. Thus nodes can be ranked according to their rank reliability or stability. Nodes would thus avoid using intermediates which display large rank variance. We are also looking at using some of the ideas presented in CAR,[31] in particular the prediction process to result in a more stable and useful system.

12.10.2 Conclusion

We have presented a novel biologically inspired method of forming a hierarchy based on differences between individual nodes. This hierarchy adapts to changes in node resources and provides the ability for nodes to determine their role in the network relative to their peers. Three different approaches to forming the hierarchy have been presented. The first method assumes that nodes engage in pairwise meetings and from this sort or exchange their ranks according to the relative order of their attributes. We showed that this leads to poor adaption to node insertion and removal, and to this end refined the protocol by introducing a reinforcement factor which spreads the ranks out evenly across the ranking space.

In the next method, we removed the assumption that nodes undertake pairwise exchange of information, as this imposes constraints on the MAC layer. Instead, we investigated how a single transmitter can broadcast its rank/attribute information to a number of receiving peers. Based on this newly acquired information, they update their ranks using a win/loss ratio method as used by researchers in the biological literature to measure dominance. This was found to have slow convergence and adaption to changes. To remedy this, we used the switching idea from pairwise ASH to result in rapid adaption to resource variation.

Both of the methods are designed with mobile networks in mind. In stationary networks, due to a limited discovery horizon, the ranks do not converge to their correct values. To this end, we presented our third method, Agent ASH, which replicates the effect of randomized meetings by spawning random agents which carry rank/attribute information to non-neighboring nodes, resulting in correct ASH convergence.

The effect of real-world mobility was shown to result in more rapid convergence. Agent ASH works well for both purely stationary and mixed mobility networks, whereas the other two methods work best in mobile networks. To avoid the formation of loops, levels were introduced, along with hysteresis to ensure that messages only permeate upward through the hierarchy.

Lastly we examined some possible applications for the ASH approach, showing how it can be used as a network underlay to enhance the performance of existing protocols by providing resource abstraction and also as a powerful cross-layer management and routing protocol.

In summary, ASH provides a framework for nodes to discover their role within diverse networks, by allowing resource abstraction. This leads to simpler routing and management protocols that are removed from the imposition of absolute values. This work is an example of the application of a common method of self-organization in nature to network management and control.

References

1. A. Markham and A. Wilkinson, Ecolocate: A heterogeneous wireless network system for wildlife tracking. In *International Joint Conferences on Computer, Information and Systems Sciences and Engineering (CISSE'07)* (December, 2007).
2. E. O. Wilson, *Sociobiology: The New Synthesis.* Harvard University Press, Cambridge, MA (1975).
3. K. Tsuchida, T. Saigo, N. Nagata, S. Tsujita, K. Takeuchi, and S. Miyano, Queen-worker conflicts over male production and sex allocation in a primitively eusocial wasp, *International Journal of Organic Evolution* **57**(10), 2356–2373 (2003).
4. T. Monnin and C. Peeters, Dominance hierarchy and reproductive conflicts among subordinates in a monogynous queenless ant, *Behavioral Ecology* **10**(3), 323–332 (1999).
5. B. Holldobler and E. Wilson, *The Ants.* Harvard University Press, Cambridge, MA (1990).
6. T. Schjelderup-Ebbe, Bietrage zur sozialpsychologie des haushuhns, *Zeitschrift fur Psychologie* **88**, 225–252 (1922).
7. I. Chase, Behavioral sequences during dominance hierarchy formation in chickens, *Science* **216**(4544), 439–440 (1982).
8. A. Guhl, Social behavior of the domestic fowl. In *Social Hierarchy and Dominance,* pp. 156–201. Dowden, Hutchinson, and Ross, Stroudsburg, PA (1975).
9. B. Forkman and M. J. Haskell, The maintenance of stable dominance hierarchies and the pattern of aggression: Support for the suppression hypothesis, *Ethology* **110**(9), 737–744 (2004).
10. F. A. Issa, D. J. Adamson, and D. H. Edwards, Dominance hierarchy formation in juvenile crayfish *Procambarus clarkii*, *Journal of Experimental Biology* **202**(24), 3497–3506 (1999).
11. R. C. Brace and J. Pavey, Size-dependent dominance hierarchy in the anemone actinia equina, *Nature* **273**, 752–753 (1978).
12. S. Cote, Determining social rank in ungulates: A comparison of aggressive interactions recorded at a bait site and under natural conditions, *Ethology* **106**, 945–955 (2000).

13. D. Greenberg-Cohen, P. Alkon, and Y. Yom-Tov, A linear dominance hierarchy in female nubian ibex, *Ethology* **98**(3), 210–220 (1994).

14. S. Altmann, A field study of the sociobiology of rhesus monkeys, *Annals of the New York Academy of Sciences* **102**, 338–435 (1962).

15. W. Jackson, Can individual differences in history of dominance explain the development of linear dominance hierarchies? *Ethology* **79**, 71–77 (1988).

16. I. D. Chase, C. Tovey, D. Sprangler-Martin, and M. Manfredonia, Individual differences versus social dynamics in the formation of animal dominance hierarchies, *Proceedings of the National Academy of Sciences* **99**(8), 5744–5749 (2002).

17. E. Bonabeau, G. Theraulaz, and J. Deneubourg, Dominance orders in animal societies: The self-organization hypothesis revisited, *Bulletin of Mathematical Biology* **61**(4) (1999).

18. W. Hamilton, Geometry for the selfish herd, *Journal of Theoretical Biology* **31**, 295–311 (1971).

19. H. Kruuk, Predators and anti-predator behaviour of the black-headed gull *Larus ridibundus*, *Behaviour Supplements 11* **11**, 1–129 (1964).

20. C. Carbone, J. D. Toit, and I. Gordon, Feeding success in african wild dogs: Does kleptoparasitism by spotted hyenas influence hunting group size? *The Journal of Animal Ecology* **66**(3), 318–326 (1997).

21. S. Creel and N. Creel, Communal hunting and pack size in african wild dogs, *Lycaon pictus*, *Animal Behaviour* **50**, 1325–1339 (1995).

22. D. Jea, A. Somasundara, and M. Srivastava, Multiple controlled mobile elements (data mules) for data collection in sensor networks. In *Distributed Computing in Sensor Systems*. pp. 244–257. Springer, Berlin, Germany (2005).

23. M. Kendall, *Rank Correlation Methods*. Hafner Publishing Co., New York (1955).

24. C. Perkins and E. Royer. Ad-hoc on-demand distance vector routing. In *Proceedings of IEEE Mobile Computing Systems and Applications*, pp. 90–100 (1999).

25. D. Aldous and J. Fill, *Reversible Markov Chains and Random Walks on Graphs (monograph in preparation.)*. http://stat-www.berkeley.edu/users/aldous/RWG/book.html

26. C. E. Jones, K. M. Sivalingam, P. Agrawal, and J. Chen, A survey of energy efficient network protocols for wireless networks, *Wireless Networks* **7**(4), 343–358 (2001).

27. I. F. Akyildiz, W. Su, Y. Sankarasubramaniam, and E. Cayirci, Wireless sensor networks: A survey, *Computer Networks* **38**(4), 393–422 (2002).

28. A. Dunkels, F. Osterlind, N. Tsiftes, and Z. He, Software-based on-line energy estimation for sensor nodes. In *EmNets '07: Proceedings of the Fourth Workshop on Embedded Networked Sensors*, pp. 28–32, New York (2007). ACM.

29. C. Toh, Maximum battery life routing to support ubiquitous mobile computing in wireless ad hoc networks, *IEEE Communications Magazine*, pp. 138–147, June (2001).

30. A. Lindgren, A. Doria, and O. Schelen, Probabilistic routing in intermittently connected networks. In *Service Assurance with Partial and Intermittent Resources*. Springer, Berlin (2003).

31. M. Musolesi, S. Hailes, and C. Mascolo, Adaptive routing for intermittently connected mobile ad hoc networks. In *Sixth IEEE International Symposium on a World of Wireless Mobile and Multimedia Networks (WoWMoM'05)*, Taormina, Giardini Naxos. pp. 183–189 (2005).

32. M. Musolesi and C. Mascolo, Evaluating context information predictability for autonomic communication. In *WOWMOM '06: Proceedings of the 2006 International Symposium on World of Wireless, Mobile and Multimedia Networks*, pp. 495–499, Washington, DC (2006). IEEE Computer Society.

33. A. Markham and A. Wilkinson, The adaptive social hierarchy: A self organizing network based on naturally occurring structures. In *First International Conference on Bio-Inspired mOdels of NEtwork, Information and Computing Systems (BIONETICS), Cavelese, Italy* (11–13 December, 2006).

34. T. Spyropoulos, K. Psounis, and C. Raghavendra, Spray and focus: Efficient mobility-assisted routing for heterogeneous and correlated mobility. In *Fifth IEEE International Conference on Pervasive Computing and Communications Workshops*, pp. 79–85, Los Alamitos, CA (2007). IEEE Computer Society, Washington, DC.

35. T. Spyropoulos, K. Psounis, and C. S. Raghavendra, Spray and wait: An efficient routing scheme for intermittently connected mobile networks. In *WDTN '05: Proceedings of the 2005 ACM SIGCOMM Workshop on Delay-Tolerant Networking*, pp. 252–259, New York (2005). ACM, New York.

36. C. Mascolo and M. Musolesi, SCAR: Context-aware adaptive routing in delay tolerant mobile sensor networks. In *IWCMC '06: Proceedings of the 2006 International Conference on Wireless Communications and Mobile Computing*, pp. 533–538, New York (2006). ACM, New York.

37. A. Markham and A. Wilkinson, A biomimetic ranking system for energy constrained mobile wireless sensor networks. In *Southern African Telecommunications, Networks and Applications Conference (SATNAC 2007)*, Mauritius (September, 2007).

38. W. R. Heinzelman, J. Kulik, and H. Balakrishnan, Adaptive protocols for information dissemination in wireless sensor networks. In *MobiCom '99: Proceedings of the Fifth Annual ACM/IEEE International Conference on Mobile Computing and Networking*, pp. 174–185, New York (1999). ACM, New York.

39. L. Ying and Y. Haibin, Energy adaptive cluster-head selection for wireless sensor networks. In *Sixth International Conference on Parallel and Distributed Computing Applications and Technologies (PDCAT'05)*, pp. 634–638, Los Alamitos, CA (2005). IEEE Computer Society, Los Alamitos, CA.

40. C. Frank and K. Römer, Algorithms for generic role assignment in wireless sensor networks. In *SenSys '05: Proceedings of the Third International Conference on Embedded Networked Sensor Systems*, pp. 230–242, New York (2005). ACM, New York.

41. K. Römer, C. Frank, P. J. Marrón, and C. Becker, Generic role assignment for wireless sensor networks. In *EW11: Proceedings of the 11th Workshop on ACM SIGOPS European Workshop*, p. 2, New York (2004). ACM, New York.

42. O. Angel, A. Holroyd, D. Romik, and B. Virag, Random sorting networks, *Advances in Mathematics* **215**(10), 839–868 (2007).

43. P. B. Jeon and G. Kesidis. Pheromone-aided robust multipath and multipriority routing in wireless manets. In *PE-WASUN '05: Proceedings of the Second ACM International Workshop on Performance Evaluation of Wireless Ad Hoc, Sensor, and Ubiquitous Networks*, pp. 106–113, New York (2005). ACM, New York.

44. F. Ducatelle, G. D. Caro, and L. M. Gambardella. Ant agents for hybrid multipath routing in mobile ad hoc networks. In *Second Annual Conference on Wireless On-demand Network Systems and Services (WONS'05)*, pp. 44–53, Los Alamitos, CA (2005). IEEE Computer Society. ISBN 0-7695-2290-0.
45. G. Di Caro, F. Ducatelle, and L.M.Gambardella, AntHocNet: An ant-based hybrid routing algorithm for mobile ad hoc networks. In *Parallel Problem Solving from Nature—PPSN* VIII, pp. 461–470. Springer, Berlin, Germany (2004).

Chapter 13

Chemical Relaying Protocols

Daniele Miorandi, Iacopo Carreras, Francesco De Pellegrini, Imrich Chlamtac, Vilmos Simon, and Endre Varga

Contents

13.1 Introduction

Recent advances in microelectronics and information technology are enabling the widespread diffusion of mobile computing devices equipped with short-range wireless communication capabilities. As users get more and more accustomed to such devices (smartphones, PDAs, gaming consoles, etc.) for accessing a variety of services, opportunities arise for leveraging their proximity communication capabilities in order to provide user-centric and/or location-based services. The resulting systems are variously referred to as delay-tolerant or disruption-tolerant networks (DTNs) or opportunistic communication systems, depending on the specific point of view [1,2].

Such systems are characterized by a variety of features that differentiate them from conventional networking paradigms. First of all, they are expected to experience frequent disconnections. Second, they can leverage transmission opportunities (usually referred to as "contacts" or "meetings") whenever two of the devices in the system get within mutual communication range. Third, devices in such systems are expected to be used for delivering location-based information and services to the end user.

For such kind of systems, one of the most challenging issues to be dealt with is data dissemination. Information should indeed be delivered in an effective way to the users who are potentially interested in it, and this should be accomplished in face of resource constraints and the unpredictable nature of mobility patterns. Given the aforementioned constraints, we believe that a *data-centric* architecture is the natural choice for implementing such kind of systems [3,4]. In contrast to address-based architectures (e.g., the Internet), operations in data-centric architectures are based solely on the content of the messages to be processed, and not on the identity of the hosts in the network.

At the same time, as the devices in the system may present various constraints (e.g., in terms of remaining battery or available bandwidth during a transmission opportunity), it is important—in order to ensure efficiency of the overall system—to let the users' interests drive the system behavior. This is meant to ensure that the dissemination of data is driven by the presence of users potentially interested in such kind of contents. In a recent paper, some of the authors have introduced a user-driven data-centric architecture for relaying messages in intermittently connected wireless networks [5]. Such a system is based on the use of *filtering policies* at each node which drive the requests for content based on the current interests of the user. The system is designed as static: while some adaptive mechanisms can be introduced to enhance its flexibility, the rules driving the diffusion process are in general hardwired in the system blueprint and cannot be changed at run time.

In this work, we explore the possibility of adding flexibility to the aforementioned relaying platform by means of a chemical computing model based on the use of the Fraglets language [6]. We treat messages in the systems as molecules, which can propagate from node to node. Each molecule has a "content tag" (semantic description of the message content) which defines the molecules it can react with. Each node corresponds to a chemical reactor. Policies are implemented as molecules and catalysts. Nodes periodically broadcast "query" molecules, asking for a given type of content. If such messages are received by a node hosting the requested content (in the form of a molecule with matching content tag), a reaction is triggered, leading to the delivery of a copy of the content to the requesting node. In such a way, content propagates among the nodes interested in it.

A data management scheme is also implemented in the chemical computing platform, in the form of a "dilution flux," which tries to limit the number of molecules (messages) in the reactor (node). As the policies determining the system's behavior (in terms of both relaying and data management) are expressed as fraglets, applications can dynamically rewrite them, changing the system operations at run time. The proposed solution is implemented in an extension of the OMNeT++ simulation platform [7], by including a module for interpreting the Fraglets code. Simulation results are presented in order to validate the feasibility of the proposed approach.

The remainder of this chapter is organized as follows. Section 13.2 describes related work on relaying mechanisms in opportunistic networks and presents the main features of the Fraglets chemical computing language. Section 13.3 presents the system model used, in terms of architectures and algorithms. Section 13.4 discusses the Fraglets implementation of the proposed mechanisms, whereas performance evaluation results are presented and discussed in Section 13.5. Section 13.6 concludes the paper presenting directions for extending the present work.

13.2 Related Work

13.2.1 Relaying in Intermittently Connected Wireless Networks

For a wide spectrum of wireless communication systems, connectivity is intermittent due to sudden and repeated changes of environmental conditions: nodes moving out of range, drop of link capacity, or short end-node duty cycles [1,8,9]. Such systems cannot ensure a full end-to-end path between two nodes most of the time; the emerging research field of DTNs investigates how to enforce resilience to repeated network partitions [1,10,11]. In order to overcome frequent disconnections, nodes can leverage their storage capabilities, operating according to a *store-carry-and-forward* paradigm [12]. The opportunities for pairs of nodes to communicate are rather short and, to this respect, communication among nodes is conveniently described by the set of node *contacts* [1,13].

Typically, opportunistic networks denote sparse mobile ad hoc networks [9,11–13], a domain where intermittent connectivity is due to node mobility and to limited radio coverage. In such cases, traditional end-to-end communication techniques have to be enforced or replaced with some mechanisms to reconcile disrupted paths [11].

In this chapter, we consider a combination of opportunistic and data-centric mechanisms, i.e., architectures where mobile nodes in the network can act as sources and relays, but data is potentially of interest for a wide set of nodes, a priori unknown to the source node. For opportunistic communications, a reference forwarding scheme is represented by epidemic routing [14,15].

The popularity of epidemic routing is rooted in the ease of implementation and robustness with respect to route failures, and applies naturally to the data-centric case. Epidemic routing, in fact, shares many principles with controlled flooding [16], which has been extensively described through fluid approximations and infection spreading (see for example [17]). The control of forwarding has been addressed in the ad hoc networks literature, e.g., in [18,19]. In [18], the authors describe an epidemic forwarding protocol based on the *susceptible-infected-removed* (SIR) model [17]. Authors of [18] show that it is possible to increase the message delivery probability by tuning the parameters of the underlying SIR model. In [20], the authors show that it is possible to trade off overhead for delay introducing suitable erasure coding techniques, whereas authors of [21] describe a variant known as spray-and-wait which is optimal under a uniform i.i.d. mobility model.

Notice that in the case of end-to-end opportunistic networks, the target is to deliver messages with high probability to the intended destination: all message copies stored at nodes other than the destination represent overhead. Conversely, in the case of data-centric forwarding, the trade-off involves a notion of *utility*, because not all message copies are redundant, as they can be relevant for a variety of users. In [22], this trade-off has been explored by means of a *publish–subscribe* mechanism, where each user takes into account other users' "subscriptions" for making appropriate data caching decisions. However, due to the overhead related to the regular update of users' interests, such an approach results effective only when applied to scenarios limited in both scale and dynamism.

13.2.2 Chemical Computing

Chemical computing represents an unconventional approach to computation, based on the use of analogies with chemistry. Computations are carried out as chemical reactions, in which molecules present in the system (atomic blocks of code and/or data) may get consumed and produce new ones. Two branches of chemical computing exist, dealing, respectively, with computing with real-world molecules and with artificial (digital) artifacts. In this paper, we focus on the second one.

Chemical computing system may be modelled as artificial chemistries [23]. An artificial chemistry is defined by a triple $\langle S, R, A \rangle$, where

- S represents the ensemble of all possible molecules present in the system.
- R defines the set of reactions which may possibly take place in the system. R is formally defined as a map $S^n \to S^m$.
- A defines an algorithm according to which the rules in R are applied to a collection P of molecules from S. In case P does not contain any topological constraint, it can be represented as a multi-set of elements from S. A defines the way reactions are implemented in a computing medium.

In this paper, we will focus our attention on a specific chemical computing language called Fraglets, introduced by Tschudin in 2003 [6]. Fraglets represent computation fragments, which may react based on a tag-matching process.

The Fraglets language is particularly appealing for our purposes, as (i) it has been specifically developed for implementing network protocols and (ii) it is particularly suited to a mobile environment, where the twofold nature of fraglets (code and data) can be leveraged to implement code mobility and active messages.

A fraglet is expressed as f $n[a$ b $tail]x$, where n represents the node in which the fraglet is present; a, b, and *tail* are symbols; x denotes the multiplicity of a fraglet. While we refer the interested reader to www.fraglets.net for a more detailed description of the language itself, we report here a brief description of the most important Fraglets instructions.

Instruction	Behavior	Description
Match	[match a tail1], [a tail2] → [tail1 tail2]	Two fraglets react, their tails are concatenated
Matchp	[matchp a tail1], [a tail2] → [matchp a tail1], [tail1 tail2]	"Catalytic match," i.e., the 'matchp' rule persists
Nop	[nop tail] → [tail]	Does nothing but consume the instruction tag
Nul	[nul tail] → []	Destroy a fraglet
Exch	[exch t a b tail] → [t b a tail]	Swap two tags
Fork	[fork a b tail] → [a tail], [b tail]	Copy fraglet and prepend different header symbols
Broadcast	[broadcast seg tail] → n[tail]	All neighbors 'n' attached to 'seg' receive [tail]
Send	[send seg dest tail] → dest[tail]	Send [tail] to node 'dest' attached to segment 'seg'
Split	[split seq1 * seq2] → [seq1], [seq2]	Break a fraglet into two at the first occurrence of *

13.3 System Model and Framework

In this work, we focus on systems characterized by the presence of a variety of devices, all equipped with a short-range wireless interface and mobile in nature. The density of such devices is assumed to be low, so that the network is disconnected in most time instants, and (pairwise) communications take place only whenever two devices get within mutual communication range. Such contacts are opportunistically exploited to spread information content in the system. We also assume the presence of a potentially large set of information sources, which may inject content items into the system. Content items are treated as messages, and characterized by two components. The first one is a header, which contains a semantic description of the content of the message payload. The header is used for matching purposes. The second one is the payload, which contains the actual data. Each node further maintains a list of interests. These are expressed in the form of semantic descriptions, using the same standard used for describing data items. The list is maintained in the form of a multiset, i.e., an interest may be represented by multiple instances. The concentration of a given interest represents its importance, as perceived by the user itself.

Two types of interactions are foreseen in the network. First, information sources inject information by means of periodic broadcast messages ("push" approach). Second, information exchanges are performed by means of a reactive ("pull") mechanism. Each node broadcasts periodically a beacon message, including a list of the kind of content it is interested in. Nodes receiving such broadcast messages match the request with the semantic description of the items present in its internal buffer. If a positive match is found, the corresponding message is enqueued for transmission on the wireless interface.

Information spreads in the system by means of a collaborative distributed relaying process. Nodes interested in one content item bring it around (until it gets "consumed"), making it possible for other nodes to get a copy thereof. The diffusion of information therefore depends heavily on the interests of the users in the system. The more a content is requested, the more easily and the faster it will spread in the system.

Each node implements a set of data management procedures in order to prevent buffer overflows. The node is modeled as a chemical reactor, in which different types of molecules (i.e., different types of items) are present. A molecule is associated to each content item in the data buffer. Molecules are also associated to other types of entities, which we call anti-message agents (or anti-messages for the sake of brevity). Anti-messages are characterized by a semantic tag; anti-messages may react with content items whose semantic description corresponds to their tag: the result is the erasure of the content item involved in the reaction. Two types of anti-messages are present: persistent and nonpersistent. The first are not consumed during the reaction, while the second are. The rate at which packets get erased from the buffer depends on the concentration of anti-message agents. Nonpersistent anti-messages are statically injected in the nodes; their presence ensures a (constant) "dilution flux" effect, reducing steadily the number of messages in the buffer. An adaptive mechanism is also present which ensures that the rate at which messages get erased increases with the buffer occupancy. This is done by

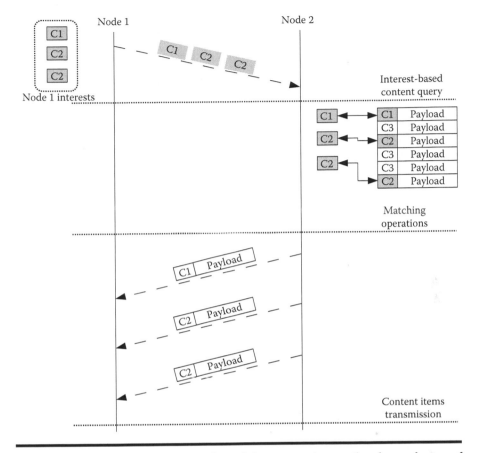

Figure 13.1 Graphical representation of the protocol operation for exchange of content items between two nodes.

ensuring that incoming messages generate (with a given probability) nonpersistent anti-messages.*

The functionality of the protocol for exchanging content items among peer nodes is reported in Figure 13.1. In such a figure, node 1 sends its interests to node 2. Upon a positive match with the content of node 2's buffer, three data items are then sent back to node 1. The process is clearly bidirectional, and may exploit piggybacking techniques to reduce overhead. A unidirectional exchange only is reported for ease of illustration.

The operations performed by the data management protocol are represented in Figure 13.2. From left to right, the arrival of an incoming packet is represented.

---

* The fact that the anti-messages generated are non-persistent is due to the need of responding with a transient increase in erasure rate to transient conditions of high message arrival rate.

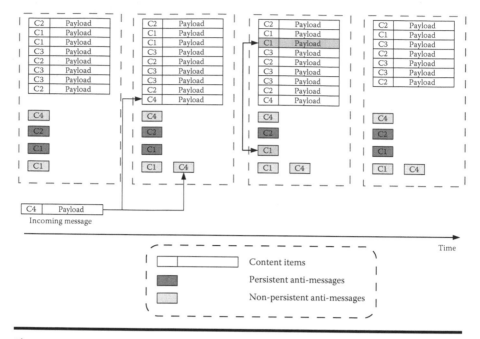

Figure 13.2 Graphical representation of the data management protocol operations.

The node is represented as a reactor, comprising both the actual data buffer and the multiset of anti-message agents. In the case represented, the incoming message gives rise to an anti-message, with the same semantic tag as the packet itself. Further, the packet gets inserted in the node's buffer. At the next execution cycle, a reaction takes place between an anti-message (corresponding to the content type C_1) and one of the messages of the buffer. The message gets erased from the buffer. As the anti-message involved in the reaction is a persistent one, it does not get erased.

Let us now focus on the dynamics of the proposed data management scheme. We focus, for the sake of simplicity, on a system having only one single type of content. Further, to simplify the analysis, we assume that there are always content items in the buffer (this is meant to represent a situation of medium-to-high load). Let us denote by F_1, F_2, and F_3 the number of message-saving agents,* the number of persistent anti-messages and nonpersistent anti-messages, respectively. Also, consider that the content items are pure passive components, and cannot initiate any reaction in the absence of active agents. Nonpersistent anti-messages are generated with probability

---

* In practice, these are agents which do not perform any task on the content items they react with. They are needed in order to limit the rate at which the items erasure process takes place.

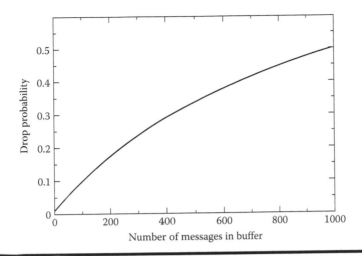

Figure 13.3 **Drop probability as a function of the number of packets in buffer for the data management scheme considered.**

p for each incoming message. Therefore, their number can be approximated as $p \cdot X$, where X represents the number of messages in the buffer. The probability of incurring a message drop event, at a given execution cycle, is therefore given by the fraction of agents leading to a message erasure:

$$P_{\text{drop}} = \frac{F_2 + F_3}{F_1 + F_2 + F_3} = \frac{F_2 + p \cdot X}{F_1 + F_2 + p \cdot X}. \qquad (13.1)$$

Plotting the drop probability as a function of the buffer occupancy, we get a concave curve, as depicted in Figure 13.3 for $F_1 = 100$, $F_2 = 1$, $p = 0.1$. Such a kind of curve does not represent a good dynamic for data management purposes. As it may be easily understood, one would prefer to have a convex function, so that few messages are dropped as the buffer is not heavily loaded, and then the drop probability rapidly increases as the buffer limit is approached. By taking F_3 to be a generic function of X, $F_3 = f(X)$, we get the following necessary and sufficient condition for the convexity of the resulting curve:

$$f''(X)[F_1 + F_2 + f(X)] \geq 2[f'(X)']^2. \qquad (13.2)$$

In our case, $f(\cdot)$ is a linear function and hence does not satisfy such a condition. In order to get more complex behavior (e.g., a quadratic one), multiple stages (implemented as sub-nodes) should be used or new match instructions, able to consume multiple molecules, could be introduced.

13.4 Fraglets Implementation

The model described in the previous section has been implemented using the Fraglets chemical computing language. The implementation includes three routines.

The first one takes place at the source nodes and is used to periodically inject data into the system. As an example, the following code creates at node *n1* every 100 execution cycles (starting from the 10th one) a fraglet that broadcasts a message of type C1:

```
t 10 every 100
{
f n1[broadcast C1 payload]
}
```

The second one takes place at each node. Periodically, a fraglet is created for broadcasting a beacon message, which includes a query for content of a given type. From the implementation standpoint, we took advantage of the twofold nature of fraglets (representing both code and data), and used *active messages*. The query message contains therefore a code which triggers the transmission, at the peer node, of one content item matching the query in terms of message content. In the following sample code, every 100 execution cycles (starting from the 10th one) *n1* broadcasts a request of content of type C1, to be sent back on channel *ch1*. The procedure does not erase the message from the peer buffer but, rather, a copy is created and then transmitted, preserving the original message. To do so, a copy of the message is created, characterized by a temporary tag *tmp1*, which is then removed when sending to node *n1*. The code looks as follows:

```
t 10 every 100
{
f n1[broadcast match C1 split match tmp1 send ch1 n1
    incoming C1 * fork tmp1 C1]
}
```

It is also possible, in an alternative implementation, to use broadcast transmission for sending back the data:

```
t 10 every 100
{
f n1[broadcast match C1 split match tmp1 broadcast
    C1 * fork tmp1 C1]
}
```

The choice between two such implementations depends on the scenario considered (unicast transmission offers better robustness to channel impairments, as it may use some media access control (MAC)-layer automatic repeat query (ARQ) mechanisms; at the same time, in case of dense scenarios, broadcast messages can present savings in terms of overhead).

In a real implementation, three additional factors have to be carefully considered:

■ As described in the previous section, interests are maintained by each node as a list of multisets. Let us assume that interests are characterized by the tag int and are maintained as active messages. Periodically, one reaction is triggered, which activates one of such active messages, broadcasting a query corresponding to one of the elements of the multiset. The probability that a query of type *Cx* is generated equals the corresponding fraction of such elements in the multiset. The resulting code for unicast transmission would look like this:

```
##profile, expressed as multiset of interests
##coded as active messages
f n1[int broadcast match C1 split match tmp1
    broadcast C1 * fork tmp1 C1]
f n1[int broadcast match C1 split match tmp1
    broadcast C1 * fork tmp1 C1]
f n1[int broadcast match C2 split match tmp2
    broadcast C2 * fork tmp1 C2]

##one element from the multiset is drawn and the
##corresponding query broadcasted
t 10 every 100
{
f n1[match int]
}
```

It is important to notice that, in such a way, an application can dynamically update the multiset of interests, by adding new fraglets in the reactor and by erasing old ones (the latter can be easily done by inserting a [match int nul] fraglet).

■ The match-copy-and-transmit routine encompasses an intermediate step, consisting in the creation of a temporary copy of the message (with the header *tmp*1) to be then sent back to the requesting node. If multiple requests arrive from different nodes, it is possible that multiple temporary messages are present in the buffer. Therefore, different temporary tags for different types of content have to be used.

■ In case a node receives a query for a content it does not have, such a query remains (dormant) in its internal buffer.

The third routine relates to the data management procedure outlined earlier. Here two types of persistent agents are encompassed (characterized by the use of `matchp` instruction). Additionally, nonpersistent anti-messages may be generated by incoming nodes. The following implementation refers to a node $n1$ able to treat three types of messages ($C1$, $C2$, and $C3$). We assume that incoming messages are characterized by the `incoming` tag. Incoming messages are processed by erasing the header; a temporary copy is made which is used to trigger the production of nonpersistent anti-messages. The code looks as follows:

```
##persistent anti-messages
f n1[matchp C1 nul]1
f n1[matchp C2 nul]1
f n1[matchp C3 nul]1

##persistent message--savers
f n1[matchp C1 nop]100
f n1[matchp C2 nop]100
f n1[matchp C3 nop]100

##non-persistent antimessages
##with probability 1% an anti-message
##is created
f n1[matchp incoming fork nop tmp]
f n1[matchp tmp exch match nul]1
f n1[matchp tmp nul]99
```

Data management procedures could also be implemented without relying on persistent matches, by using a self-reproducing code (quines), as described by Yamamoto et al. [24]. In such a case, it would be possible for the application to dynamically change the data management procedures by simply injecting new fraglets that rewrite the related code.

13.5 Performance Evaluation

In order to evaluate the performance of the proposed chemical relaying protocol we have run extensive simulations using a freely available software tool [7]. We considered 25 nodes, moving over a square playground of side $L = 600$ m, according to the random waypoint mobility model [25], with no pausing and a speed of 4 m/s. An ideal channel is assumed, and nodes can communicate if and only if their mutual distance is less than 30 m. Nodes communicate at a rate of 1 Mb/s. A carrier sense multiple access with collision avoidance (CSMA/CA) is used for resolving collisions. The following table presents a summary of the considered settings.

Parameter	Setting
Number of nodes	25
Playground size	600 × 600 m
Mobility model	Random waypoint mobility (v = 4 m/s, no pausing)
Media access	CSMA
Bit rate	1 Mb/s
Interference distance	65 m

For the simulation of the chemical reactor, a tool called OMLeTS was developed. OMLeTS is an integrated Fraglets interpreter for the popular OMNeT++ discrete event simulation library [7]. The basic building block of OMLeTS is a module called FSimpleModule (Fraglets Simple Module), which consists of three submodules called Soup, Gate, and Monitor. The Soup module is a Fraglets reaction vessel, which contains and evaluates the Fraglets code. It is also able to collect statistics and traces of the reactions taking place inside it. Every FSimpleModule uses its own Soup that executes in isolation from the other ones. Soups can be connected to each other only through their respective Gate modules. The Gate is an adapter entity that maps events in the Soup to external messages that can be sent to arbitrary OMNeT++ modules. The Gate observes the reactions inside the reaction vessel and sends out messages according to its configuration. Observable events are usually special reactions (like broadcast or send) or changes in fraglet concentrations. The Monitor is used to maintain and monitor the reaction vessel providing recovery methods if needed. This is important because the incoming fraglet code from other vessels may contain potentially damaging code that can disrupt the originally intended behavior of the Soup.

The internal architecture of a simulated node is shown in Figure 13.4. Every node in our experimental setup contains exactly one FSimpleModule that encapsulates the epidemic forwarding and buffer management routines as detailed in the previous sections. To simulate chemical relaying, the Gate was connected to the wireless network interface and it was configured to relay packets between the network and the Soup. The Monitor was used to inject payload packets periodically into source nodes, like interest profiles or content.

In order to evaluate the performance of the proposed relaying mechanism, we considered the following performance metrics:

- *Mean Buffer Occupancy* (MBO): This metric reflects the average number of messages in the nodes buffer. This is an important metric to evaluate, since in opportunistic networks nodes are expected to be resource constrained, and

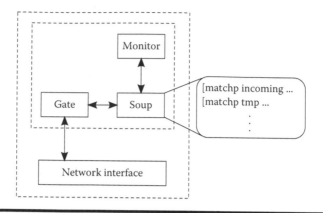

Figure 13.4   Internal architecture of a simulated node.

it is therefore important to limit the number of messages circulating in the network. The MBO is averaged over time and over different nodes, and is evaluated as a function of the traffic injected into the system.

■ *Efficiency*: The efficiency measures the fraction of nodes interested in a given type of content Cx that actually receive a message of type Cx. This metric measures how well the system behaves in delivering content to the appropriate users. As for the MBO, efficiency is computed over the set of messages generated during a simulation.

■ *Spreading Rate* (SR): The SR measures the fraction of nodes, interested in a given type of content, which get the message within a limited period of time τ from its generation. This metric reflects the ability of the system to timely deliver information to interested users.

Figure 13.5 depicts the MBO as a function of the message generation rate. The convexity of the curve is in line with the results described in Section 13.4 (concavity of the message drop probability implies indeed convexity of the buffer occupancy). Clearly, the occupancy of the internal buffer of nodes grows with the network load. Such growth is almost linear until the load reaches 1 msg/s. We also performed some tests for larger values of the traffic rate, resulting in potentially very large buffer occupancy figures. This relates to both the intrinsic limits of the "dilution flux" approach followed and the computational limits of the Fraglets interpreter itself.

Figure 13.6 reports the system efficiency as a function of the fraction of nodes interested in a given type of content. It is possible to observe that the more the users sharing a common interest, the higher is the circulation of content and therefore the efficiency of the system. This is due to the widespread diffusion of common interests that triggers a cooperative behaviour of the nodes in the system (as only content of interest is kept and diffused to other nodes). And, clearly, the more the nodes cooperate, the faster the diffusion of messages becomes. From the graph, an

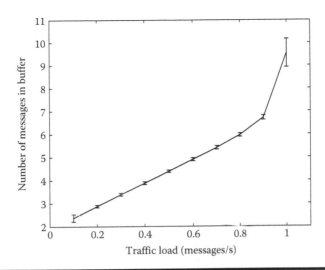

Figure 13.5   Mean buffer occupancy as a function of the traffic rate.

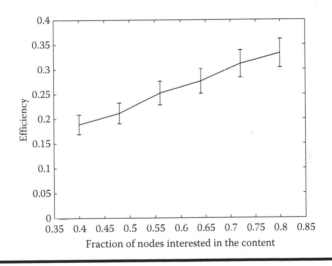

Figure 13.6   Efficiency as a function of the fraction of nodes interested in a given type of content.

approximately linear increase in efficiency with the fraction of interested nodes can be drawn.

Finally, Figure 13.7 depicts the fraction of nodes among the ones interested in a content Cx receiving a message of such type within a delay of $\tau = 500$ s from its generation. Results are again plotted as a function of the fraction of interested nodes in the system. As for Figure 13.6, the message diffusion process improves quite

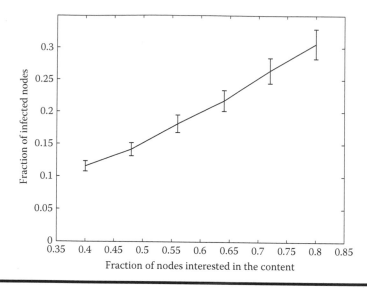

Figure 13.7 **Fraction of interested nodes infected within a delay of 500 s as a function of the fraction of nodes interested in a given type of content.**

significantly as the number of users sharing similar interests grows. It is possible to observe that the system is able to deliver a significant amount of messages already in this limited time frame.

13.6 Conclusions

In this chapter, we have introduced a chemical computing-based data-centric relaying scheme for intermittently connected wireless networks. The proposed solution has been implemented in the Fraglets language, and simulation outcomes have been used to validate the feasibility of the proposed approach. To the best of the authors' knowledge, this is the first example of a relaying scheme for DTNs implemented in chemical computing language. The integration of the Fraglets interpreter within a network-oriented event-driven simulation platform is also an important contribution of independent interest, and could provide a reference testing environment for the application of chemical computing approaches to networking problems.

With this work, our primary aim was to prove the feasibility of implementing in a chemical computing language a fully functional networking protocol for DTNs. In such highly dynamic environments, indeed, we believe that the flexibility inherently present in chemical computing models could provide advantages over more conventional static approaches. In the present work, nonetheless, such flexibility has been only partially exploited. One of the most appealing extensions would be to introduce an alternative implementation of the proposed routines, based on the use of a self-reproducing code (quines), as introduced by Yamamoto et al. [24]. This could be coupled to a feedback loop in order to provide a means for the system to autonomously adapt to varying operating environments.

References

1. K. Fall. A delay-tolerant network architecture for challenged Internets. In *Proceedings of ACM SIGCOMM*, Karlsruhe, DE (2003).
2. V. Cerf, S. Burleigh, A. Hooke, L. Torgerson, R. Durst, K. Scott, K. Fall, and H. Weiss. Delay-tolerant network architecture (2007). IETF RFC 4838. http://www.ietf.org/rfc/rfc4838.txt
3. J. Su, J. Scott, P. Hui, J. Crowcroft, E. de Lara, C. Diot, A. Goel, M. Lim, and E. Upton. Haggle: Seamless networking for mobile applications. In *Proceedings of Ubicomp*, pp. 391–408, Innsbruck, Austria (September 16–19, 2007).
4. I. Carreras, I. Chlamtac, F. D. Pellegrini, and D. Miorandi. Bionets: Bio-inspired networking for pervasive communication environments, *IEEE Transactions on Vehicular Technology* **56**(1), 218–229 (2007).
5. I. Carreras, D. Tacconi, and D. Miorandi. Data-centric information dissemination in opportunistic environments. In *Proceedings of MASS 2007—Demo Session*, Pisa, Italy (October, 2007).
6. C. Tschudin. Fraglets—A metabolistic execution model for communication protocols. In *Proceedings of 2nd Annual Symposium on Autonomous Intelligent Networks and Systems (AINS)*, Menlo Park (July, 2003).
7. AA.VV. Omnet++. URL www.omnetpp.org
8. S. Jain, K. Fall, and R. Patra. Routing in a delay tolerant network, *SIGCOMM Computer Communication Review* **34**(4), 145–158 (2004). ISSN 0146-4833. doi: http://doi.acm.org/10.1145/1030194.1015484.
9. U. Lee, E. Magistretti, B. Zhou, M. Gerla, P. Bellavista, and A. Corradi. MobEyes: Smart mobs for urban monitoring with vehicular sensor networks. Technical Report 060015, UCLA CSD (2006). URL http://netlab.cs.ucla.edu/wiki/files/mobeyestr06.pdf
10. S. Burleigh, L. Torgerson, K. Fall, V. Cerf, B. Durst, K. Scott, and H. Weiss. Delay-tolerant networking: An approach to interplanetary internet, *IEEE Communications Magazine* **41**(6), 128–136 (2003).
11. J. Burgess, B. Gallagher, D. Jensen, and B. N. Levine. Maxprop: Routing for vehicle-based disruption-tolerant networking. In *Proceedings of IEEE INFOCOM*, Barcelona, Spain (April 23–29, 2006).
12. M. M. B. Tariq, M. Ammar, and E. Zegura. Message ferry route design for sparse ad hoc networks with mobile nodes. In *Proceedings of ACM MobiHoc*, pp. 37–48, Florence, Italy (May 22–25, 2006). ISBN 1-59593-368-9. doi: http://doi.acm.org/10.1145/1132905.1132910.
13. J. Leguay, A. Lindgren, J. Scott, T. Friedman, and J. Crowcroft. Opportunistic content distribution in a urban setting. In *Proceedings of ACM Chants*, Florence, IT, pp. 205–212 (September 15, 2006).

14. A. Vahdat and D. Becker. Epidemic routing for partially connected ad hoc networks. Technical Report CS-2000-06, Duke University (2000).

15. A. Khelil, C. Becker, J. Tian, and K. Rothermel. An epidemic model for information diffusion in MANETs. In *Proceedings of ACM MSWiM*, pp. 54–60, Atlanta, GA (September 28, 2002).

16. K. Harras, K. Almeroth, and E. Belding-Royer. Delay tolerant mobile networks (DTMNs): Controlled flooding schemes in sparse mobile networks. In *Proceedings of IFIP Networking*, Waterloo, Canada, pp. 1180–1192 (May, 2005).

17. X. Zhang, G. Neglia, J. Kurose, and D. Towsley. Performance modeling of epidemic routing. Technical Report CMPSCI 05-44, University of Massachusetts (2005).

18. M. Musolesi and C. Mascolo. Controlled epidemic-style dissemination middleware for mobile ad hoc networks. In *Proceedings of ACM Mobiquitous*, San Jose, CA (July, 2006).

19. A. E. Fawal, J.-Y. L. Boudec, and K. Salamatian. Performance analysis of self limiting epidemic forwarding. Technical Report LCA-REPORT-2006-127, EPFL (2006).

20. A. E. Fawal, K. Salamatian, D. C. Y. Sasson, and J. L. Boudec. A framework for network coding in challenged wireless network. In *Proceedings of MobiSys*, Uppsala, Sweden. ACM, New York (June 19–22, 2006). http://www.sigmobile.org/mobisys/2006/demos/Fawal.pdf

21. T. Spyropoulos, K. Psounis, and C. S. Raghavendra. Spray and wait: An efficient routing scheme for intermittently connected mobile networks. In *Proceedings of SIGCOMM Workshop on Delay-Tolerant Networking (WDTN)*. pp. 252–259. ACM, New York (2005).

22. P. Costa, C. Mascolo, M. Musolesi, and G. P. Picco, Socially-aware routing for publish-subscribe in delay-tolerant mobile ad hoc networks, *IEEE JSAC* **26**(5), 748–760 (2008).

23. P. Dittrich, J. Ziegler, and W. Banzhaf. Artificial chemistries—A review, *Artificial Life* **7**(3), 225–275 (2001).

24. L. Yamamoto, D. Schreckling, and T. Meyer. Self-replicating and self-modifying programs in fraglets. In *Proceedings of BIONETICS*, Budapest, HU (2007).

25. T. Camp, J. Boleng, and V. Davies. A survey of mobility models for ad hoc network research, *Wiley Wireless Communications and Mobile Computing* **2**(5), 483–502 (2002).

Chapter 14

Attractor Selection as Self-Adaptive Control Mechanism for Communication Networks

Kenji Leibnitz, Masayuki Murata, and Tetsuya Yomo

Contents

In this chapter, we discuss a self-adaptive control mechanism inspired by cell biology and provide examples of its application to communication networks. The model originates from the dynamic behavior observed in the gene expression of operons in *Escherichia coli* cells as a reaction to the lack of nutrients in the cells' environment and it is driven by system-inherent fluctuations. We will begin this chapter by describing the influence of noise in biological systems, then illustrate the background of the model and provide an overview of alternative mathematical formulations. Based on this dynamic model, we then show two case studies on how to apply attractor selection as a robust and self-adaptive control mechanism in information and communication networks.

14.1 Introduction

Noise and fluctuations are phenomena that can be experienced almost anywhere in nature. On a microscopic level, for instance, cells may be of the same type, but the quantities describing them or their concentrations will vary and also fluctuate over time [1]. The entire process of signal transduction within a cell is driven by diffusion and fluctuations. On a macroscopic level, the dynamics of populations of species and their whole genetic evolution is influenced by noise, for example, by the random mutations of genes in the DNA. Thus, noise is a fundamental component that is inherent to many kinds of biological processes [2] and it actually is also of great importance in the adaptability and robustness of biological systems.

In this chapter, we summarize and discuss a specific class of dynamic systems that utilizes noise to adapt to changes in the environment. The basic mechanism consists of two modes of operation. In its dynamical behavior, the system is driven to a stable state (*attractor*) in spite of perturbations through system-inherent noise. The concept of attractors is a key issue for achieving *biological robustness* [3] and can be found in a wide range of nonlinear dynamical systems, for example artificial neural networks. Informally speaking, attractors can be seen as optima in a potential landscape defined by an objective function in phase space and the system state has a tendency to be "pulled" toward such an attractor. However, when an external influence occurs that impacts the dynamics of the whole system, the current system state may not be sufficient anymore and requires the transition to a different state. If the objective function is known a priori, the adaptation toward a new solution of the objective function can be done in a directed manner, for example by using gradient descent methods, to find a new stable solution. On the other hand, when the objective function cannot be defined in advance, or the influence of the system state on the dynamics of the network is unknown, no such directed adaptation can be performed. In biological systems, such an adaptation is usually determined by Brownian motion. Thus, the underlying concept that we will describe in this chapter, called *attractor selection*, is a combination of both modes described above to select a new attractor when an environmental change is encountered.

In the following, we will first outline the basic mathematical background of noise and fluctuations in dynamical systems in Section 14.2. This is followed by a summary of existing attractor models found in the literature in Section 14.3. In Section 14.4, we describe how this concept can be used as self-adaptive control mechanisms for communication networks in two specific case studies.

14.2 Noise and Fluctuations in Dynamical Systems

In this entire chapter, we deal with dynamical systems; thus, we always have a functional dependence on time. However, for the sake of simplicity, we do not always include time in the notation of the considered quantities, but only when we explicitly need to refer to it.

Before we discuss the influence of noise on the system dynamics, let us provide some preliminary definitions and assumptions. In the following, we will consider an arbitrary M-dimensional system with $M > 1$. The system state is described in a vectorized form as

$$x = \begin{pmatrix} x_1 \\ \vdots \\ x_M \end{pmatrix}$$

where without loss of generality we assume that the x_i are well defined in a real-valued, non-negative domain $\mathcal{D} \subset \mathbb{R}_0^+$ for all $i = 1, \ldots, M$.

The dynamical behavior of this system is described by its derivatives over time. We formulate this over a vectorized function $F(x)$ in a differential equations system as given by the following equation:

$$\frac{dx}{dt} = F(x) = \begin{pmatrix} f_1(x) \\ \vdots \\ f_M(x) \end{pmatrix} \tag{14.1}$$

with functions $f_i : \mathcal{D}^M \to \mathcal{D}, i = 1, \ldots, M$, which determine the dynamic behavior of the system. When we observe the system state asymptotically over time, we can formulate the steady-state vectors $x^\star$ as solutions of Equation 14.2 when there is no change over time, i.e.,

$$\frac{dx}{dt} = 0 \qquad (14.2)$$

We call a solution of Equation 14.2 an *equilibrium point* $x^\star$ of the system. Depending on the structure of the functions $f_i(x)$, one or more equilibrium points may exist in the entire state space. In general, the following formal definitions of stability of an equilibrium point $x^\star$ are used based on Strogatz [4].

Definition 14.1 (Stability) Let $x^\star$ be an equilibrium point of a system as given in Equation 14.1. Then, $x^\star$ is defined as *attracting* (cf. Figure 14.1a) if

$$\exists_{\delta > 0} : \|x(0) - x^\star\| < \delta \Rightarrow \lim_{t \to \infty} x(t) = x^\star$$

We say that $x^\star$ is *Lyapunov stable* (cf. Figure 14.1b) if

$$\forall_{\varepsilon > 0} \exists_{\delta > 0} : \|x(0) - x^\star\| < \delta \Rightarrow \forall_{t \geq 0} : \|x(t) - x^\star\| < \varepsilon$$

If $x^\star$ is both attracting and Lyapunov stable, it is called *asymptotically stable*. Finally, the point $x^\star$ is *exponentially stable*, if

$$\exists_{\alpha, \beta, \delta > 0} : \|x(0) - x^\star\| < \delta \Rightarrow \forall_{t \geq 0} : \|x(t) - x^\star\| < \alpha \|x(0) - x^\star\| e^{-\beta t}$$

We are interested in a very specific type of equilibrium points, namely the *attractors* of the system, i.e., those in which all neighboring trajectories converge. Strogatz [4] formally defines an attractor if it has the following properties.

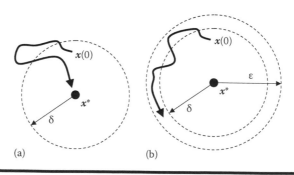

Figure 14.1 **Definitions of stability of x* for a dynamic trajectory starting at x(0): (a) attracting and (b) lyapunov stable.**

Definition 14.2 (Attractor) A closed set $\mathcal{A}$ is an *attractor* if it is the minimal set that satisfies the conditions (1) and (2).

(1) $\mathcal{A}$ is an *invariant set*, i.e.,

$$\forall_{x(0) \in \mathcal{A}} \ \forall_{t \geq 0} : \ x(t) \in \mathcal{A}$$

(2) $\mathcal{A}$ attracts all trajectories that start sufficiently near to it, i.e.,

$$\exists_{\mathcal{U} \supset \mathcal{A}} \ \forall_{x(0) \in \mathcal{U}} : \ \lim_{t \to \infty} x(t) \in \mathcal{A}$$

The largest such superset $\mathcal{U} \supset \mathcal{A}$ is called the *basin of attraction*.

Note that the formal definition of attractors only speaks of closed sets, which could lead to complex structures such as the well-known three-dimensional butterfly shaped *Lorenz attractor* [5].

The Lorenz attractor is also denoted as *strange attractor* due to its unique behavior. Its dynamics is governed by the equation system in Equation 14.3 and sample trajectories in three-dimensional phase space are illustrated in Figure 14.2.

$$\frac{dx}{dt} = \sigma(y - x) \quad \frac{dy}{dt} = x(\rho - z) - y \quad \frac{dz}{dt} = xy - \beta z \qquad (14.3)$$

Figure 14.2 also illustrates the bifurcative influence of the parameter ρ. While a small value of $\rho = 14$ in Figure 14.2a results in a stable system that deterministically evolves toward a single equilibrium point, a large value of $\rho = 28$ causes a highly complex and chaotic orbit, where the stable points repel the dynamic trajectory so that it never crosses itself (cf. Figure 14.2b). Note that although the system's

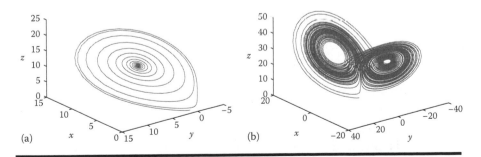

Figure 14.2 Trajectory of the Lorenz attractor in three-dimensional phase space with parameters $\sigma = 10$, $\beta = 8/3$. (a) Stable orbit for $\rho = 14$. (b) Chaotic behavior for $\rho = 28$.

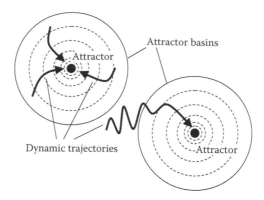

Figure 14.3   Schematic illustration of attraction of trajectories in phase space.

trajectory does not necessarily lead here to a single point as attractor, the whole structure itself forms the strange attractor in this specific case. In the following, however, we will only consider *point attractors*, while several other types of attractors may exist as well, e.g., cycle attractors, torus attractors.

Simply stated, the general dynamical properties of attractors in phase space can be illustrated as in Figure 14.3. When the initial state $x(0)$ lies within the basin of an attractor, its trajectory will bring it toward this attractor. However, if $x(0)$ initially lies outside of the attractor's basin, the system may reach the attractor if its trajectory takes it within its basin. If for any initial state $x(0)$ the system will always converge to certain attractors, we speak of *global attractors*; otherwise, like in the cases depicted in Figure 14.3, we speak of *local attractors*.

Stability analysis of an attractor is in general performed by means of the *Jacobian matrix*, which contains the first-order partial derivatives of $\partial f_i(x)/\partial x_j$ over all $i, j = 1, \ldots, M$ as given in the following equation:

$$J_{F(x)} = \begin{pmatrix} \frac{\partial f_1(x)}{\partial x_1} & \cdots & \frac{\partial f_1(x)}{\partial x_M} \\ \vdots & & \vdots \\ \frac{\partial f_M(x)}{\partial x_1} & \cdots & \frac{\partial f_M(x)}{\partial x_M} \end{pmatrix} \tag{14.4}$$

Then it can be shown that an equilibrium point $x^\star$ is a *stable attractor* if the real part of all eigenvalues λ_j of the matrix $J_{F(x)}\big|_{x=x^\star}$ is strictly negative, i.e., $\forall_j \operatorname{Re}\{\lambda_j\} < 0$.

14.2.1 Dynamic Systems under the Influence of Noise

So far we discussed some fundamental aspects of ordinary nonlinear dynamic systems. In the following, we will include noise in our formulation. In real-life systems, regardless of whether they are biological or engineered, we usually don't experience the "smooth" dynamics as described in the previous section, but are rather faced with stochastic perturbations.

In physics, randomness is expressed by *Brownian motion*, which describes the random motion of particles immersed in a fluid. This behavior is mathematically described by a *Wiener process*, a continuous-time stochastic process, and the position of the particle is approximated in its simplest form by the *Langevin equation*, see Equation 14.5:

$$\frac{dx}{dt} = -\gamma x + \eta \tag{14.5}$$

The term x describes here the velocity of the Brownian particle, γ is the friction coefficient of the underlying fluid, and η is a random noise term. A solution of the probability density function of x can be obtained through the *Fokker–Planck equation* (also known as the *Kolmogorov forward equation*).

We now extend Equation 14.1 to include noise, such that

$$\frac{d\boldsymbol{x}}{dt} = \boldsymbol{F}(\boldsymbol{x}) + \eta \tag{14.6}$$

Here, η is an M-dimensional vector of independent and identically distributed white noise terms η_i, which are zero-mean Gaussian random variables with standard deviation σ_i, $i = 1, \ldots, M$. Note that if the perturbations introduced by η are sufficiently small, the system will still converge to an attractor as defined in the previous section, as long as the state vector remains within its basin of attraction.

14.2.2 Relationship between Fluctuation and Its Response

When we consider a classical thermodynamical system in physics, the *fluctuation–dissipation theorem* [6] describes a direct relationship between the response of a system in equilibrium and when an external force is applied. However, the underlying assumption is that we have a thermodynamic quantity, and a more general discussion from the viewpoint of systems biology is discussed by Sato et al. [7]. The authors describe a measurable quantity in a biological system as a variable x, which is influenced by a parameter a. Let $\langle x \rangle_a$ be the average of x under the influence of a. Then it is shown [7] that if an external force is applied such that this parameter a becomes $a + \Delta a$, the change in the average value of x is proportional to its variance at the initial parameter value a:

$$\frac{\langle x \rangle_{a+\Delta a} - \langle x \rangle_a}{\Delta a} \propto \langle (\delta x)^2 \rangle \tag{14.7}$$

The underlying assumption is that the distribution of x is approximately Gaussian. The result in Equation 14.7 is similar to the fluctuation–dissipation theorem, however, with the slight difference that it is also valid for non-thermodynamic quantities and in cases where the fluctuation–dissipation theorem is not applicable at all. The fluctuation–response relationship can be considered as the basic model, which explains how the existence of noise enhances the adaptability of a system (see Section 14.3).

14.3 Mathematical Models of Attractor Selection

In this section we describe the basic concept of *attractor selection* that was introduced by Kashiwagi et al. [8] and its mathematical formulation. Let us first consider a very abstract view, which can be regarded as an adaptive feedback-based control mechanism that is triggered by the interaction of two dynamic equation systems.

First, there is a system of M stochastic differential equations describing the dynamics of the system state expressed by a time-dependent vector x. Secondly, there is also the influence from another dynamic function $\alpha : \mathcal{D}^M \times \mathcal{P}^N \rightarrow \mathbb{R}_0^+$ that defines the *activity* of the system or growth rate, as this value in fact represents the growth rate of biological cells as we will later see in Section 14.3.1. The term $\mathcal{P}^N$ refers to the input space of measured environmental influences and its dimension N does not necessarily need to be equal to M. Thus, basically the activity function characterizes the suitability of the current system state (output) to the environmental conditions (input), denoted in the following by vector y. Let us now define the generalized dynamics of the system state as in Equation 14.8:

$$\frac{dx}{dt} = F(x)\,\alpha(x,y) + \eta \tag{14.8}$$

Before we provide a more detailed explanation of the structure of the function F in the following sections, let us first examine the essential behavior observed in Equation 14.8. Obviously, there is a permanent noise influence due to η. The function F must be defined according to the objectives of the system and characterizes the attractors to which the system state will converge, regardless of the small perturbations introduced by η. However, as $\alpha(x,y)$ approaches 0, the influence of the first summand in Equation 14.8 diminishes, leaving the entire dynamics influenced by the noise term and essentially a random walk in the phase space is performed. In summary, attractor selection can be basically regarded as a noise-driven control loop as illustrated in Figure 14.4.

14.3.1 Mutually Inhibitory Operon Regulatory Network

Let us now describe in greater detail the background of the underlying biological model. According to Kashiwagi et al. [8], the original model was derived from experiments of a bistable switch of two mutually inhibitory operons in *E. coli* cells. An *operon* is a unit in the protein transcription process that creates messenger RNA (mRNA) and consists of an operator, a promoter, and the associated structural genes.

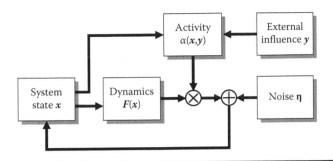

Figure 14.4 **Attractor selection as noise-driven control loop.**

In general, cells alter their gene expression through their regulatory network, which induces a change in expression in response to environmental changes or external signals. The regulatory network is essentially a network of activating or inhibiting genes and it is controlled through specialized signal transduction networks to quickly react to frequently occurring signals. Thus, a specific program in the transduction network is activated if a certain environmental event is detected. However, the number of possible environmental changes can be so large that a transduction pathway cannot exist for all possible cases, but the adaptation toward a stable expression state is nevertheless achieved.

In order to study this phenomenon, Kashiwagi et al. [8] used a synthetic gene network composed of two mutually inhibitory operons that were respectively attached with a *green fluorescence protein* (GFP) and *red fluorescence protein* (RFP) to visualize their dynamics. Exposure to a neutral medium resulted in a single monostable attractor, where the mRNA concentrations of both operons showed only weak expression. On the other hand, when the network was exposed to a medium that was lacking either one of the key nutrients, it resulted in a bistable condition with two attractors, one with high green and low red fluorescence expression and vice versa. Thus, the former single stable attractor with weak expression became unstable and fluctuations due to the noise inherent in gene expression caused the system to shift toward either one of the attractors due to their mutually inhibitory behavior.

A mathematical model of this phenomenon was derived by Kashiwagi et al. [8], describing the dynamics of the mRNA concentrations x_i of each operon $i = 1, 2$:

$$\frac{dx_1}{dt} = \frac{S(\alpha)}{1 + x_2^2} - D(\alpha) x_1 + \eta_1 \qquad \frac{dx_2}{dt} = \frac{S(\alpha)}{1 + x_1^2} - D(\alpha) x_2 + \eta_2 \qquad (14.9)$$

The functions $S(\alpha)$ and $D(\alpha)$ are the rate coefficients of the synthesis and degradation of cell growth, respectively, and can be defined over a function α, which represents the cell activity:

$$S(\alpha) = \frac{6\alpha}{2 + \alpha} \qquad D(\alpha) = \alpha \qquad (14.10)$$

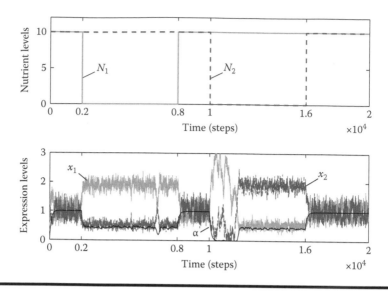

Figure 14.5 **Simulation of dynamics of network with two mutually inhibitory operons.**

The activity dynamics is explained in the supplemental material of Kashiwagi et al. [8] and defined using a classical model for cell growth rate [9]:

$$\frac{d\alpha}{dt} = \frac{P}{\prod_{i=1}^{2}\left[\left(\frac{\theta_i}{x_i + N_i}\right)^{n_i} + 1\right]} - C\alpha \qquad (14.11)$$

An example of a numerical simulation is illustrated in Figure 14.5, using the same parameter set as specified by Kashiwagi et al. [8]. Here, $P = C = 0.01$ are the production and consumption rates of activity, $N_i \in \{0, 10\}$ represent the external supplementation of the respective nutrients, $\theta_i = 2$ are the thresholds of nutrient production, and $n_i = 5$ are their sensitivities, for $i = 1, 2$. The upper figure shows the nutrient availability over time, whereas the lower figure depicts the expression levels of the two operons. Notice that roughly between time 10,000 and 12,000, the system state lies outside the basin of the proper attractor, so the activity decreases to 0 (black line marked by α) and a random walk is performed until the system finally approaches the proper attractor.

14.3.2 Sigmoid Gene Activation Model

An alternative model for the dynamics of a gene regulatory network is given by Furusawa and Kaneko [10]. Here, the dynamics of the protein concentrations x_i are described by the following differential equation system in Equation 14.12. However, in their model, the growth rate of the cell α is obtained from the metabolic flux in the cell caused by the current expression level in the on/off type of gene regulation.

$$\frac{dx_i}{dt} = \left[f\left(\sum_{j=1}^{M} w_{ij} x_j\right) - x_i \right] \alpha(\mathbf{x}, \mathbf{y}) + \eta_i \quad i = 1, \ldots, M \qquad (14.12)$$

The weights are discrete values $w_{ij} \in \{-1, 0, 1\}$ depending on whether protein i either inhibits, is neutral to, or activates protein j, and f is a sigmoid activation function. The focus of Furusawa and Kaneko [10] is on the investigation of the cell growth dynamics. In their experiments, Furusawa and Kaneko used $M = 96$ and randomly selected weights w_{ij} with positive autoregulatory weights $w_{ii} = 1$, and their results yielded an optimal range for the noise amplitude to achieve high cell growth rates without getting stuck in suboptimal attractors (see Figure 14.6). The cell growth is influenced by the dynamics of the concentration of the ith substrate in the metabolic network

$$\frac{dy_i}{dt} = \varepsilon \sum_{j=1}^{M} \sum_{k=1}^{N} Con(k, j, i)\, x_j\, y_k$$

$$- \varepsilon \sum_{j=1}^{M} \sum_{k=1}^{N} Con(i, j, k)\, x_j\, y_i + D\,(Y_i - y_i) \quad i = 1, \ldots, N \qquad (14.13)$$

where $Con(i, j, k)$ is a binary variable representing the metabolic reaction if the ith substrate to the kth substrate is catalyzed by the jth protein. The term D is the

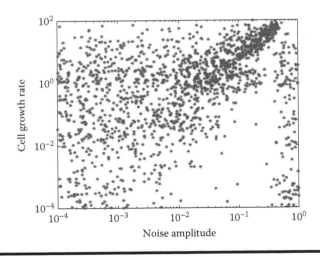

Figure 14.6 **Simulation of the sensitivity of the cell growth rate on noise amplitude.**

diffusion coefficient and Y_i is the extracellular concentration of the ith substrate. Cell growth itself is assumed to be such that it requires at least r metabolites for growth and is selected as

$$\alpha \propto \min \{y_1, \ldots, y_r\}$$

Note that in the study by Furusawa and Kaneko [10] the weights remain constant as they represent the evolved signal transduction pathway for expressing certain genes. A study on the robustness toward noise in the gene expression can be found by Kaneko [11], where mutations in the pathways reflect evolutionary changes. In the following, we can assume a generalized model inspired by Kaneko [11] as given in Equation 14.14:

$$\frac{dx_i}{dt} = \left[\tanh \left(\beta \sum_{j=1}^{M} w_{ij} x_j \right) - x_i \right] \alpha(\boldsymbol{x}) + \eta_i \quad i = 1, \ldots, M \qquad (14.14)$$

Here β is the sensitivity of activation and is a constant value. According to Kaneko [11], the weights are adapted over generations taking into account the fitness of the system is taken into account. The type of equations as given above is also often used as a neuron model in recurrent neural networks [12], where the weights are adapted over time and can take continuous values.

An example of the adaptability of the generalized model is depicted in Figure 14.7 for a dimension of $M = 5$. In this numerical simulation, at every 2000 time steps the weight values w_{ij} are randomly reset to normalized Gaussian random values with strongly self-activating values $w_{ii} = M$. The noise amplitude is constant at $\sigma = 0.01$ and activity is updated according to the dynamics in Equation 14.15, thus trying to

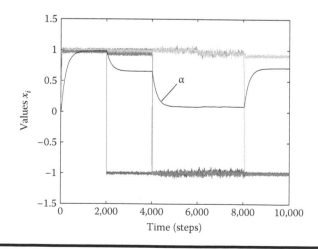

Figure 14.7 **Example of numerical simulation of generalized attractor model.**

maximize the absolute sum of x_i with an adaptation rate $\rho = 0.005$, and f is again a sigmoid function with activation threshold 0.5:

$$\frac{d\alpha}{dt} = \rho \left[f \left(\frac{1}{M} \left| \sum_{i=1}^{M} x_i \right| \right) - \alpha \right] \tag{14.15}$$

This example shows that in general the system converges almost instantly to stable attractors in the form of on/off state vectors $\boldsymbol{x}$ in spite of sudden changes of the available attractors. However, note that in the example shown in Figure 14.7 a global optimum ($\boldsymbol{x} = \pm 1$) is only reached for $t < 2000$, resulting in $\alpha \approx 1$. The reason for this behavior lies in the random selection of the weight matrix, as well as in the sensitivity β and adaptation rate ρ. In other words, the system is able to find a solution very quickly; however, the tradeoff between the quality of the solution and parameters must be carefully taken into account.

In an actual networking context, the appropriate setting of these parameters is determined by the objectives of the application and α could also include some observable (but uncontrollable) metric obtained by measurements, as we will discuss later in Section 14.4.

14.3.3 Gaussian Mixture Attractor Model

A model can also be considered where the possible target values are known in advance and attractors can be designed in an appropriate way. Such an approach is proposed by Fukuyori et al. [13] to control a 2-link robotic arm. A Gaussian mixture model is used to formulate the global potential function $F(\boldsymbol{x})$, which consists of a superposition of M-dimensional Gaussian functions around the predetermined target candidate values $\xi_1, \ldots, \xi_N$:

$$F(\boldsymbol{x}) = -\frac{1}{N v \sqrt{2\pi}} \sum_{i=1}^{N} e^{-\frac{1}{2v^2} \| \xi_i - \boldsymbol{x} \|^2} \tag{14.16}$$

and the dynamics is obtained as derivative $\partial F(\boldsymbol{x})/\partial x_i$ for $i = 1, \ldots, M$.

Thus, the target vectors ξ_i form the center of the attractors with standard deviation v, which influences the attractor basins, and activity can be formulated inversely proportional to v, which makes the potential function steep when activity is high and flat when it is low. Fukuyori et al. [13] proposed further extensions to rearranging the attractors using a clustering technique, where high activity could be maintained. Unlike the previously described methods, activity is updated by the sum of an instantaneous achievement of the robot with a decaying moving average of the previous activity values that were taken.

14.4 Application to Self-Adaptive Network Control

So far we have elaborated on the mathematical description of attractor selection and its underlying biological features. In the following, we will discuss the application and possible scenarios for the control and traffic engineering of information networks. As we have seen in the previous sections, attractor selection can be basically considered as a control mechanism utilizing the system-immanent noise. However, not only biological systems are subject to inherent noise, but also engineered systems are exposed to fluctuations in their behavior and traffic patterns, often introduced by the heterogeneity of connected users, devices, and protocols.

14.4.1 Differences between Biological Networks and Communication Networks

From the viewpoint of an engineered system, noise is generally considered as something harmful and often mechanisms are introduced to reduce variance or jitter of the observed performance metrics. For instance, traffic-shaping mechanisms in wide area networks (WAN) applying *asynchronous transfer mode* (ATM) technology have been thoroughly investigated [14,15] in the past. In more recent studies, the detrimental influence of fluctuations on the *quality of service* (QoS) in video streaming [16], *voice-over-IP* (VoIP) [17,18], or overlay networks [19] was examined. The main benefit we expect in applying the attractor selection concept to information networks is, therefore, not necessarily in designing a system that is able to perform optimally in a controlled environment, but rather a system which focuses on robustness and self-adaptation in an unknown, unpredictable, and fluctuating environment.

When applying attractor selection to information networks, we certainly cannot directly map the biological gene expression model, but must perform certain modifications in order to make its application feasible and meaningful. In this section, we will discuss some fundamental properties on applying the attractor selection scheme to communication networks before presenting some exemplary case studies.

14.4.1.1 Mapping of Growth Rate

The first point that requires to be addressed is the definition of the activity or cell growth. In the biological models we discussed so far, there is a functional relationship between the environmental input, the current state, and the growth rate. However, this is because the biological models focus on a single cell only and growth rate can be mapped directly to the cell's gene expression or metabolic flow. When we apply the attractor selection method to an information network setting, we must be aware of what the cell and ultimately the state vector x are representing. If we consider a state to consist of a certain parameter setting on an end-to-end connection oblivious to the state of the entire network, our decision may have implications that are a priori unknown, so that the activity and the influence of the current choice of attractors may be only quantified by means of measurements. In fact, growth rate could be defined as a quantity perceived from the end user's viewpoint, consisting of either objective QoS or subjective QoE (*quality of experience*) metrics.

14.4.1.2 Fluctuations: Ambient or Controllable?

Another important issue lies in the interpretation of the noise terms. In biological systems, noise is an inherent feature, and its amplitude may be considered independent of the current state of gene expression or activity. In the above examples from biological networks, a constant and independent noise was assumed, but if we design an engineered system, we have to face the following two possibilities of how to handle fluctuations:

- Noise is system-inherent and can be observed, but not influenced.
- The application can modify the noise amplitude in order to increase robustness and speed of adaptation.

At this point, we will only focus on the first point mentioned above and assume that the noise terms are not influenced by our approach. However, the analysis of noise sensitivity in a formal mathematical way is currently being investigated [20].

14.4.1.3 Centralized vs. Distributed Control

Attractors are per definition representations of the stable states of an entire system. In neural networks, attractors represent the patterns that are stored within the network for classification, for instance expressed by the minima of the energy function in *Hopfield nets* [21]. This corresponds to a centralized operation, in which information on the whole network is available. However, this is often not the case in modern communication networks, which operate in a distributed manner. If we consider overlay network topologies [22] or sensor networks [23], the operations are usually performed individually at node level, resulting in an *emergent* global behavior of the system [24]. Assuming that an attractor system may only have a limited view of the whole network, each node will have its own attractors for making its decisions based only on local information. The entire sum of individual selections results in a self-organized emergent behavior of the system as a whole. In the context of our proposal, we refer to this viewpoint as *attractor composition* (see Figure 14.8).

14.4.2 Applications to Self-Adaptive Network Control

In this section, we will briefly summarize some previous applications of attractor selection as self-adaptive control mechanism in communication networks and focus

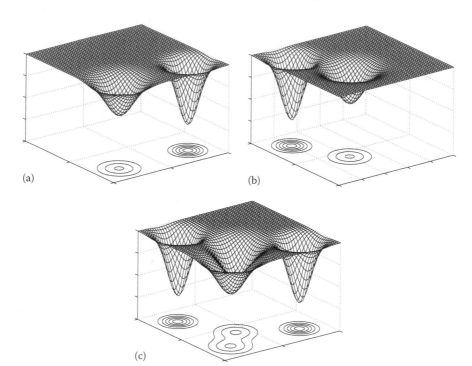

Figure 14.8 Example of composition of local attractors in phase space. (a) Local attractors of node 1. (b) Local attractors of node 2. (c) Attractor composition of nodes 1 and 2.

on two examples: path selection in overlay networks and next hop selection in ad hoc network routing.

14.4.2.1 Self-Adaptive Overlay Path Selection

The application of attractor selection as a simple, yet robust path selection method for multi-path overlay network routing was presented by Leibnitz et al. [25,26]. Overlays are logical networks formed above a physical Internet-based IP layer infrastructure (see Figure 14.9) constructed for a specific purpose, e.g., VoIP, video-on-demand (VoD), peer-to-peer (P2P) file sharing.

One of the main advantages of overlay networks is that traffic engineering can be controlled based on the requirements and the characteristics on the application layer. This facilitates the implementation in actual systems, as no modifications to the existing IP infrastructure are required. The current trends are reflected in the increased volume of Internet traffic from content delivery networks based on P2P technology [27,28], which have benefits in performance over traditional client/server architectures [29].

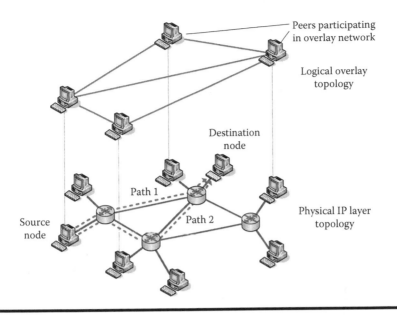

Figure 14.9 Multi-path routing in overlay network topology.

In our scenario, we consider a generic application, which connects a source node to a destination node over multiple paths. The traffic flow is split over these paths according to the current network conditions and one primary path is chosen over which the majority of traffic is transported, while secondary paths are also maintained and kept alive, transporting only a small proportion of traffic. The method described by Leibnitz et al. [26] is performed during the route maintenance phase, so it can be used in conjunction with any path discovery method during the route setup phase. Attractor selection is merely used to determine the primary path depending on the traffic conditions, and when the environment suddenly changes due to link failure or variations in traffic demand, a new primary path is automatically selected (see Figure 14.10). The advantage of our proposal is that the dynamics of each node can

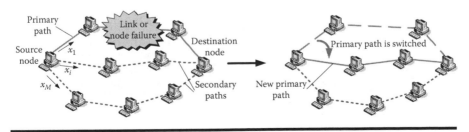

Figure 14.10 Selection of new primary path in the event of failures.

be implicitly defined using mathematical equations without giving explicit rules, which may be only limited to certain operation conditions.

The differential equation system formulating the attractors is based on that described in Section 14.3.1, but is slightly modified to yield only solutions where one of the x_i is at a high level and the others are at a low value. The selection itself is performed at the source node and the attractor solution with the high value determines the primary path over which the maximum part of the traffic is routed. Each path is chosen randomly for each packet with a probability proportional to x_i, which means that secondary paths deliver packets less frequently.

Activity is mapped by Leibnitz et al. [26] in the following way. Let l_i be the latency of packets received over path i measured at the destination node. Then, the dynamic behavior of the activity α can be defined as in Equation 14.17:

$$\frac{d\alpha}{dt} = \rho \left[\left(\prod_{i=1}^{M} \left[\left(\frac{x_i}{\max_j x_j} \frac{\min_j l_j}{l_i + \Delta} \right)^n + 1 \right] \right)^{\beta} - \alpha \right] \qquad (14.17)$$

The terms β and ρ are parameters that influence the strength and speed at which activity is adapted and Δ is a hysteresis threshold to avoid *path flapping*, i.e., alternating between two or more paths with nearly equal performance. Equation 14.17 basically tries to maximize the x_i that has minimal l_i. Comparison with other randomized mechanisms [26] reveals that the proposed mechanism can easily compensate for link failures or churn.

14.4.2.2 Next Hop Selection in Ad Hoc Network Routing

Although the path selection method in Section 14.4.2.1 assumes that the routes have been already previously determined, it still relies on a separate mechanism for finding these routes. Therefore, according to Leibnitz et al. [30], attractor selection is not applied for an end-to-end connection, but rather at each node along the path in a distributed manner for probabilistically finding the next hop node in a mobile ad hoc network (MANET) (see Figure 14.11).

In this case, activity is determined at the destination node by evaluating the received packets, but is propagated to all nodes along the reverse path. Since the main objective is now rather on finding short paths with a small number of hops, instead of those with low latency as in the overlay network scenario, the activity dynamics must be modified to take that into account, as shown in Equation 14.18:

$$\frac{d\alpha}{dt} = \rho \left[1 - \left(1 - \frac{dist(s, d)}{len} \right) \left(1 - \frac{h_{\min}}{h} \right) - \alpha \right] \qquad (14.18)$$

Here, $dist(s, d)$ denotes the physical separation between source and destination node, len is the current path length, $h_{\min}$ the minimally encountered number of

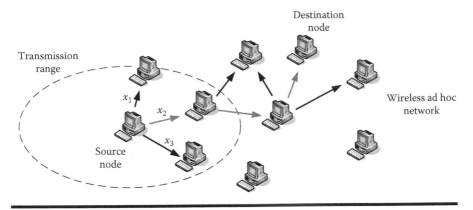

Figure 14.11 Finding the next hop node in MANET with attractor selection.

hops, and h the number of hops of the current path. Additionally, directly measured performance quantities, such as the packet delivery ratio [31] or a combination of different metrics, are also feasible.

14.5 Conclusion

In this chapter we discussed attractor selection, which is a meta-heuristic for self-adaptation of a dynamic system state to changing environmental conditions and is based on a biological process found in gene expression networks. We described its mathematical formulation and the underlying biological background, as well as alternative methods to define attractors in a way that is related to recurrent neural networks. In general, attractor selection can be considered as a robust adaptation scheme when the objective of optimization as well as the impact of the selection itself on the objective function are unknown. The adaptation itself is based on Brownian motion, which leads to a slower adaptation than if a directed method is used, but it can simply react to the response perceived from inherent random fluctuations.

Furthermore, we provided some examples of the application of attractor selection as overlay path selection and MANET routing method, and numerical evaluations [26,30] have validated that it is applicable as robust adaptation mechanism in a changing network environment. In the future, we will further investigate the influence of the parameters on the performance and convergence speed of the selection scheme applying a noise sensitivity analysis technique [32].

Acknowledgments

The authors would like to thank Naoki Wakamiya, Chikara Furusawa, and Yutaka Nakamura for their valuable contributions and discussions. This work was supported by the *"Special Coordination Funds for Promoting Science and Technology: Yuragi Project"* of the Ministry of Education, Culture, Sports, Science and Technology in Japan.

References

1. K. Kaneko, *Life: An Introduction to Complex Systems Biology* (Springer, Berlin, 2006).
2. T. Yanagida, M. Ueda, T. Murata, S. Esaki, and Y. Ishii, Brownian motion, fluctuation and life, *Biosystems.* **88**(3), 228–242 (2007).
3. H. Kitano, Biological robustness, *Nat. Rev. Genet.* **5**(11), 826–837 (2004).
4. S. H. Strogatz, *Nonlinear Dynamics and Chaos* (Westview Press, Boulder, CO, 1994).
5. E. N. Lorenz, Deterministic nonperiodic flow, *J. Atmos. Sci.* **20**, 130–141 (1963).
6. R. Kubo, The fluctuation–dissipation theorem, *Rep. Prog. Phys.* **29**(1), 255–284 (1966).
7. K. Sato, Y. Ito, T. Yomo, and K. Kaneko, On the relation between fluctuation and response in biological systems, *Proc. Natl. Acad. Sci. U.S.A.* **100**(24), 14086–14090 (2003).
8. A. Kashiwagi, I. Urabe, K. Kaneko, and T. Yomo, Adaptive response of a gene network to environmental changes by fitness-induced attractor selection, *PLoS ONE.* **1**(1), e49 (2006).
9. J. Nielsen and J. Villadsen, *Bioreaction Engineering Principles* (Plenum Press, New York, 1994).
10. C. Furusawa and K. Kaneko, A generic mechanism for adaptive growth rate regulation, *PLoS Comput. Biol.* **4**(1), e3 (2008).
11. K. Kaneko, Evolution of robustness to noise and mutation in gene expression dynamics, *PLoS ONE.* **2**(5), e434 (May, 2007).
12. M. Koda and H. Okano, A new stochastic learning algorithm for neural networks, *J. Oper. Res. Soc. Jpn.* **43**(4), 469–485 (2000).
13. I. Fukuyori, Y. Nakamura, Y. Matsumoto, and H. Ishiguro, Flexible control mechanism for multi-DOF robotic arm based on biological fluctuation. In *10th International Conference on the Simulation of Adaptive Behavior (SAB'08)*, Osaka, Japan (July, 2008).
14. J. Roberts and F. Guillemin, Jitter in ATM networks and its impact on peak rate enforcement, *Perform. Evaluation.* **16**(1–3), 35–48 (1992).
15. J. Rexford, F. Bonomi, A. Greenberg, and A. Wong, Scalable architectures for integrated traffic shaping and link scheduling in high-speed ATM switches, *IEEE J. Sel. Areas Commun.* **15**(5), 938–950 (1997).
16. M. Claypool and J. Tanner. The effects of jitter on the perceptual quality of video. In *7th International Conference on Multimedia (MULTIMEDIA'99)*, pp. 115–118, Orlando, FL (1999) (ACM, Orlando, FL).
17. L. Zheng, L. Zhang, and D. Xu. Characteristics of network delay and delay jitter and its effect on voice over IP (VoIP). In *IEEE International Conference on Communications (ICC 2001)*, Vol. 1, pp. 122–126, Helsinki, Finland (June, 2001). IEEE.
18. T. Hoßfeld and A. Binzenhöfer, Analysis of Skype VoIP traffic in UMTS: End-to-end QoS and QoE measurements, *Comput. Netw.* **52**(3), 650–666 (2008).

19. L. Subramanian, I. Stoica, H. Balakrishnan, and R. Katz, OverQoS: An overlay based architecture for enhancing internet QoS. In *1st Symposium on Networked Systems Design and Implementation (NSDI'04)*, San Francisco, CA (March, 2004) (USENIX, Berkeley, CA, 2004).

20. K. Leibnitz and M. Koda, Noise sensitivity analysis of a biologically-inspired attractor selection mechanism. In *8th Workshop on Stochastic Numerics*, Kyoto, Japan (July, 2008). Kyoto University (RIMS).

21. J. J. Hopfield, Neural networks and physical systems with emergent collective computational abilities, *Proc. Natl. Acad. Sci. U.S.A.* **79**(8), 2554–2558 (1982).

22. D. Andersen, H. Balakrishnan, M. Kaashoek, and R. Morris, Resilient overlay networks. In *18th Symposium on Operating Systems Principles (SOSP'01)*, Banff, Canada (October, 2001) (ACM, New York, 2001).

23. H. Karl and A. Willig, *Protocols and Architectures for Wireless Sensor Networks* (Wiley-Interscience, Chichester, U.K. 2007).

24. E. Bonabeau, M. Dorigo, and G. Theraulaz, *Swarm Intelligence: From Nature to Artificial Systems* (Oxford University Press, New York, 1999).

25. K. Leibnitz, N. Wakamiya, and M. Murata, Biologically inspired self-adaptive multipath routing in overlay networks, *Commun. ACM.* **49**(3), 62–67 (2006).

26. K. Leibnitz, N. Wakamiya, and M. Murata, Resilient multi-path routing based on a biological attractor selection scheme. In *2nd International Workshop on Biologically Inspired Approaches to Advanced Information Technology (BioAdit'06)*, Osaka, Japan (January, 2006) (Springer, Berlin, 2006).

27. S. Saroiu, K. P. Gummadi, R. J. Dunn, S. D. Gribble, and H. M. Levy, An analysis of internet content delivery systems. In *5th Symposium on Operating System Design and Implementation (OSDI 2002)*, Boston, MA (December, 2002). USENIX.

28. T. Hoßfeld and K. Leibnitz, A measurement survey of popular internet-based IPTV services. In *2nd International Conference on Communications and Electronics (HUT-ICCE'08)*, HoiAn, Vietnam (June, 2008).

29. K. Leibnitz, T. Hoßfeld, N. Wakamiya, and M. Murata, Peer-to-peer vs. client/server: Reliability and efficiency of a content distribution service. In *20th International Teletraffic Congress (ITC-20)*, pp. 1161–1172, Ottawa, Canada (June, 2007).

30. K. Leibnitz, N. Wakamiya, and M. Murata, Self-adaptive ad-hoc/sensor network routing with attractor-selection. In *IEEE GLOBECOM*, San Francisco, CA (November, 2006). IEEE.

31. K. Leibnitz, N. Wakamiya, and M. Murata, A bio-inspired robust routing protocol for mobile ad hoc networks. In *16th International Conference on Computer Communications and Networks (ICCCN'07)*, pp. 321–326, Honolulu, HI (August, 2007).

32. D. Dacol and H. Rabitz, Sensitivity analysis of stochastic kinetic models, *J. Math. Phys.* **25**(9), 2716–2727 (1984).

Chapter 15

Topological Robustness of Biological Systems for Information Networks—Modularity

S. Eum, S. Arakawa, and Masayuki Murata

Contents

Biological systems have evolved to withstand perturbations, and this property, called robustness, is the most commonly and ubiquitously observed fundamental system-level feature in all living organisms. Thus, understanding biological systems naturally drives us to discover the origin of robustness, and this discovery will help us to design more robust man-made systems. In this chapter, we explore topological robustness based on the insights gained from the understanding of biological systems.

15.1 Introduction

Robustness is a ubiquitously observed property of biological systems.[1] Intuitively speaking, it is not a surprising comment since life appeared on Earth about 3.7 billion years ago, and, from that time, biological systems have evolved or adapted themselves to overcome new environments, which include some expected or unexpected perturbations. This continuous survival under disturbances is conclusive proof that biological systems are robust. In addition, robust properties in biological systems have been reported in various biology papers.[2–5] Thus, the analysis of biological systems would show us the origin of robustness, and we may be able to take advantage of these lessons from the systems.

There are many different approaches to understanding robustness. Generally, robustness is reflected from its system-level behavior[6] and therefore understanding topological structure is of great importance since it represents the architecture of interactions among components that show a blueprint of system behavior. For this reason, analysis of system structure has attracted much attention among researchers. Especially, in the field of systems biology that understands biology at the system level, it has been reported that components (e.g., gene, protein, or metabolic) located in different levels of biological systems form various topological structures such as scale-free, bow-tie, hierarchy, and modularity (these will be described later in detail). Although these structures are generally believed to provide a system with a certain degree of robustness, it has not been clearly identified why and how these topological structures appear in biological systems. Thus, a thorough discovery of the detailed features of each topological structure found in biological systems is required before applying them to man-made systems such as information networks.

The rest of this chapter is organized as follows. In Section 15.2, we provide a brief description of biological systems and their topological characteristics. This section also describes what kinds of topological structures have been observed in biological systems and how these have been studied based on the theory of complex networks.

This is followed by a case study that shows how modularity structure impacts on the robustness of a system in Section 15.3. We characterize four different features of modularity structure by carrying out simulation analysis. Finally, we summarize the chapter in Section 15.4.

15.2 Topological Robustness of Biological Systems

There are various perspectives to understanding the robustness of systems. In the context of the Internet, Doyle et al.[7] described robustness of the Internet mainly from the initial military design principle of its protocol (TCP/IP); on the other hand, Albert et al.[8] explained it based on the topological structure of the Internet. Since robustness is especially appropriate for systems whose behavior results from the interplay of dynamics with a definite organizational architecture,[6] topological structure seems promising to understanding the robustness of a system.

15.2.1 Overview of Biological Systems

A system is a collection of interacting components. In the context of biology, the components can be molecule, metabolite, cell, or whole organisms which are operated in multiple levels of operation. By viewing biological systems as complex networks, these components are a set of nodes and the interactions represent directed or undirected edges between them. For instance, Xia et al.[9] expressed a cell by its network structure such as protein–protein physical interaction networks, protein–protein genetic interaction networks, expression networks, regulatory networks, metabolic networks, and signaling networks. Each subset has different types of nodes and links (we omit the detailed description of each network since it requires the background of biology). The idea here is that biological systems can be expressed or understood in terms of its topological structure, which is identified through data generated by experimental methods or computational approaches (links in biological systems represent interaction between components so it needs to be identified through various techniques[9]).

There have been two major approaches to studying biological systems. The first one is to view individual biological entities and discover their functions. This type of approach is based on reductionism, which recognizes a system by disassembling them to the interactions of each part. This approach has been a foundation of conventional biological science, such as molecular biology and biochemistry. The second view is to study the complex interaction among entities to understand the system-level behavior of a biological system. Especially, the latter has attracted huge attention among biologists since this system-level approach is the way to identify biological function (identifying all the genes and proteins in an organism is like listing all parts of an airplane[10]). In other words, the list of all the parts may provide an ambiguous impression; however, it is not sufficient for someone to understand its system-level functions.

Interestingly, this system-level understanding of biological systems is related to a deeper understanding of its robustness since robustness is determined from the behavior of systems that results from the interplay of dynamics among entities.

15.2.2 Resemblance between Biological Systems and Information Networks

Although the outward appearances of both biological systems and information networks are very different from each other, they have some common properties. Firstly, both biological systems and information networks are represented as structures consisting of entities connected by physical links or abstract interactions. For this reason, the theory of complex networks has been applied to analyze both systems. Secondly, while they remain poorly understood at the system level, it is generally known how their individual components (e.g., gene, protein, router, and switch) work. Understanding system-level behavior is of importance because a system is more than just a collection of its components. Thirdly, their evolution processes are similar. Information networks are the result of a mixture of good design principles similar to biological systems[11] in spite of very different evolution periods. These similarities enable us to apply understanding from biological systems for the designing process of information networks in the context of topological structures.

15.2.3 Topological Characteristic of Biological Networks

We briefly explore topological structures commonly observed in biological systems. All these topological structures are known to provide a certain level of robustness to systems.

15.2.3.1 Scale-Free Structure

A single node of a network can be characterized by its degree. The degree k_i of a node i is defined as the total number of links that have started from the node i. The spread of node degrees through a network is characterized as a distribution function $P(k)$ that is the probability that a randomly chosen node has k degrees. When degree distribution of a network topology follows a power law form $P(k) \sim k^{-\gamma}$, the network is called a scale-free network. It is well known that a number of large-scale complex networks, including the Internet, biological systems, and ecosystem, have a scale-free structure. This topological structure has been studied extensively from different perspectives such as the origin of this scale-free property,[12,13] the model to generate a scale-free topology,[14,15] or robustness of scale-free topology.[16]

A scale-free topology is also known to have two contradictory properties, namely, "robust yet fragile," which means it has a high degree of error tolerance, and attack vulnerability. Albert et al.[16] demonstrated the phenomenon by measuring the size of a giant cluster after the removal of a fraction of nodes. The power law degree distribution implies that a few nodes have an extremely large number of degrees while most of them have small degrees. In other words, a randomly chosen node is likely to be a small degree node—robust for random removal; on the other hand, when a large degree node is removed intentionally, it is likely to damage the system severely in terms of connectivity—fragile for an intentional attack.

15.2.3.2 Bow-Tie Structure

This structure was initially reported to describe the structure of the World Wide Web (WWW)[17] although the term "bow-tie" was introduced later. Ma and Zeng[18] also found that the macroscopic structure of the metabolic network has a bow-tie structure from the analysis of genomic data from 65 organisms. This topological structure consists of four parts: one core part whose components are tightly connected to each other, in and out parts that are connected through the core part, and one isolated part that is isolated from the body of the structure. Kitano[1] and Csete and Doyle[19] addressed the issue of the robustness of a biological system from the point of view of its bow-tie structure in two different ways. One is that the core part in a bow-tie structure of metabolic networks is the most significantly connected area of the network so that multiple routes exist between two components within the core part. The other one is that the core part plays the role of a buffer against the external stimuli to provide coordinated responses to the system.

15.2.3.3 Hierarchical Structure

This is one of the topological structures that people can understand in an intuitive manner; however, surprisingly there is no clear definition for the structure, nor a widely accepted method to quantify the strength of this structure. Gol'dshtein et al.[20] proposed a method to study this structure based on the analysis of its vulnerability. The vulnerability of a structure is measured by calculating the variation of the average shortest paths before and after a node is removed through the whole network. Thus, when removal of certain nodes causes more serious disturbance than others to the network (e.g., disconnection), the method considers the vulnerable topology to have stronger hierarchical structure. Another type of approach was shown by Costa et al.,[21–23] who focused on measuring connectivity among the identified group of nodes located in different shells which are determined by the number of hops from a particular node. In addition, Eum et al.[24] proposed an intuitive approach to discover and quantify the hierarchical structure. The method discovers nodes on the top layer of a hierarchical structure based on modularity calculations shown in the next section and then locates the rest of the nodes in different layers according to their distance from the nodes on the top layer.

15.2.3.4 Modularity Structure

Unlike hierarchical structure, this structure has been studied in many different areas under different names such as graph partitioning in graph theory and computer science, and hierarchical clustering or community structure in sociology.[25,26] In the context of complex networks, this structure is defined as groups of nodes, called modules, that are densely interconnected within modules however few the connections between the modules. This idea was formulated by Newman and Girvan[27] and used to quantify the strength of modularity structure as follows:

$$Q = \sum_i \left(e_{ii} - a_i^2 \right) \tag{15.1}$$

where
Q is the quantified modularity value
e is a symmetric matrix which represents the connectivity among modules

The dimension of e is the same as the number of modules in the network. Also, a_i represents a row or column sum of matrix $a_i = \sum_j e_{ij}$. Thus, the value Q shows the ratio between the number of links inside the modules to the number of links between the modules. In other words, a relatively small number of links between modules produces a high value of Q.

In addition, various methods have been proposed to discover the modularity structure for a given topology (a review of this modularity issue[28] can be found). For instance, Newman and Girvan[27] developed an iteration process. A most highly used link is removed iteratively to split a topology, and then the modularity value Q is calculated using Equation 15.1 whenever the removal of a link splits the topology. After all the links are removed, the maximum Q becomes the modularity value of the topology.

15.3 Modularity and Robustness of Systems

In this section, we focus on one of the topological structures described in the previous section, modularity, to investigate how this topological structure impacts on the robustness of systems. Modularity has been known to provide robustness not only for biological systems but also for many other complex systems. In ecology, it is believed that perturbations which affect some species within the ecosystem are bounded within one module and insulate the rest of the ecosystem and critical function from perturbation.[29] Also, spread of diseases, forest fire, and cascades of extinctions are normally known to arise only in non-modularized systems.[30]

In the context of biology, Ravasz et al.[31] found that 43 metabolic networks they observed were organized as highly connected structural modules. Based on the above observation, modularity seems to be a natural candidate to be investigated regarding the impact of topological structure on system robustness.

The most intuitive approach to understand robustness is to disturb a system and observe its response to the perturbation. This approach follows the general definition of robustness, which is the capacity of a system to maintain or persist with its central function in the presence of perturbation. This type of approach was initially carried out by Albert et al.,[8] who measured the diameter of a topology as the nodes were removed one by one in a random or intentional basis. In this case, the removal of the nodes becomes a perturbation, and the size of diameter of a topology represents a central function of the system.

We demonstrate four different simulation scenarios to reveal the features of modularity structures. In each scenario, an artificial perturbation is generated on modularized and non-modularized topologies, and their responses to the perturbations are observed.

15.3.1 Attack Vulnerability

Random failure and intentional attack are a well-known scenario to examine the robustness of networks. Albert et al.[8] explored the scenario and suggested that a scale-free network is robust against random attack but fragile to intentional attack. We follow the same scenario but with different topologies which have scale-free as well as modularity structures.

Figure 15.1 shows the relative size of the largest cluster of networks as a function of the number of removed nodes (%) based on the preferential removal of nodes with high betweenness centrality* (left) and randomly chosen nodes (right) respectively. By comparing the graphs in both the left and right diagrams, we observe the same result as Albert et al.[8] obtained that, by randomly removing nearly 90% of the nodes, it might not cause the system to collapse; however, if a small portion of targeted nodes (in this case around 20%) are removed, it might experience nearly the same amount of damage. This phenomenon observed in scale-free networks is known as "error tolerance and attack vulnerability."

In addition, based on the observation that the size of the largest cluster in a strong modularity structure decreases much faster than one in a weak modularity structure, a strongly modularized structure seems to be more vulnerable than one with a weak modularity structures in both direct and random node removal scenarios. Since both strong and weak modularized structures have scale-free property, we also observe "error tolerance and attack vulnerability" in this case.

---

* Shows how heavily a node is used by traffic that exists between every two nodes in a network.

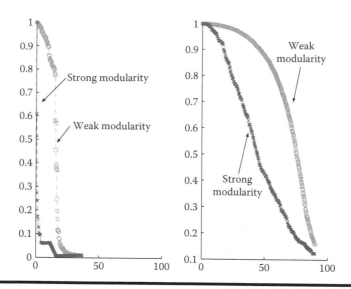

Figure 15.1   Intentional attack (left) and random attack (right).

15.3.2 *Isolation and Localization*

The most well-known hypothesis regarding why modularity structures are found in biological systems is that the topological structure localizes and isolates parts of a system against hazardous effects. However, no proof to the hypothesis has been provided so far.

In this simulation scenario, two traffic models are considered. In both models, a certain number of walkers are located in randomly chosen nodes and moved around a network. The difference between the two models is how the walkers are routed in a network. In the first model, a random routing is simulated which represents the behavior of an intruder in a system. In the second model, walkers follow the shortest path between nodes to simulate normal data exchange between nodes in real networks. In both models, to obtain intermittent fluctuation on each node, a different number of walkers are launched on different types of networks multiple times independently.

In Figure 15.2, each point in the graphs represents the variance and mean ratio (VMR) and the mean of traffic in a time series in each node. Especially, in Figure 15.2a, c, e, and g, we classify the nodes according to their location in a network using our previous work.[24] There are four different groups in these figures. Data points (upside-down triangles) at the extreme right show data observed in nodes on the top layer of the topology. Nodes on the top layer mean nodes which are used to connect among modules. Then, nodes in sub-layers are decided according to hop counts from the nodes on the top layer. Since weakly modularized topologies are well connected to each other, nodes are not classified into multiple layers as shown

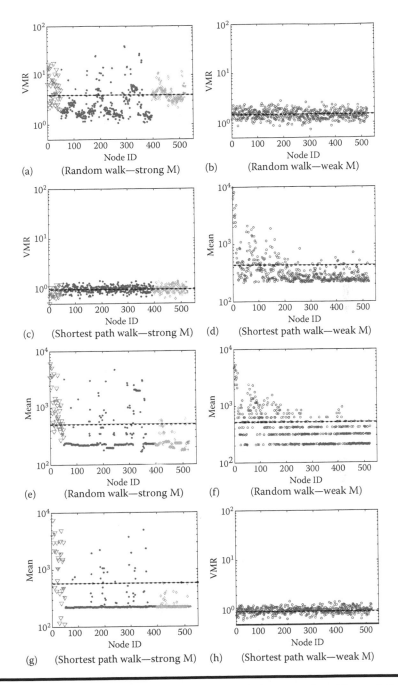

Figure 15.2 Variance and mean ratio (VMR) and mean of traffic variation in each node of strongly modularized topologies ((a), (c), (e), and (g)) and weakly modularized topology ((b), (d), (f), and (h)).

in Figure 15.2b, d, f, and h. The first thing we see is that the large amount of traffic is observed on nodes of the top layer in a modularized topology (Figure 15.2e and g). It shows how these nodes actually experience a bottleneck phenomenon.

These results show that a modularized topology experiences high fluctuation when walkers are routed randomly as shown in Figure 15.2a (nearly 70% of total nodes are under the average VMR [3.9488: dotted line]) compared to Figure 15.2b (55% of nodes are under the average VMR [1.4947: dotted line]). However, when walkers are routed along the shortest path, both modularized and non-modularized topologies experience the same fluctuation (Figure 15.2c and d).

Under the assumption that random routing represents the behavior of an intruder, the large fluctuation in Figure 15.2a can be translated as intruders who are trapped inside the modules so that they can rarely damage nodes outside the module they belong to. It shows that a modularity structure localizes or isolates damages caused by an intruder.

15.3.3 Bottleneck

This scenario involves observing the variation of critical point as traffic load increases. The critical point here is defined as the amount of traffic that makes a network begin to accumulate traffic. The simulation scenario is based on a model.[32]

In each time step, packets are generated with a probability ρ from randomly chosen nodes and forwarded along the shortest path to other randomly chosen nodes. Every node has a queue. Thus, when traffic load increases by increasing the probability ρ, packets begin to accumulate in queues and experience delay. These packets in queues are served based on the "first-in first-out" (FIFO) principle, and move to the next hop according to the probability calculated as follows:

$$\mu_{i \to j} = \frac{1}{(n_i n_j)^\gamma} \tag{15.2}$$

where

n_i and n_j represent the number of queued packets in node i and j

γ is a parameter which controls the speed of the packet forwarding process in a node

Thus, the probability $\mu_{i \to j}$ means that the probability of a packet moving from node i to node j is inversely proportional to the number of packets in node i and node j and proportional to the control parameter γ. In order to observe the saturation of a given network, we use the order parameter[32] as follows:

$$\eta(p) = \lim_{t \to \infty} \frac{1}{\rho S} \frac{\langle \Delta N \rangle}{\Delta t} \tag{15.3}$$

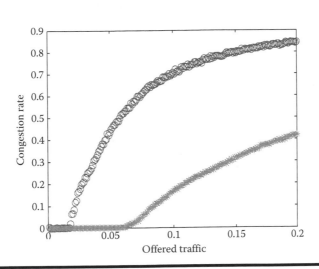

Figure 15.3 **The variation of critical points as the offered traffic increases in strongly modularized network (circles) and in weakly modularized network (asterisks).**

where

1/ρS shows the total number of packets offered to a network with the number of nodes S

$\langle \triangle N \rangle / \triangle t$ means the rate of packet increase in a network

Figure 15.3 plots the order parameters as a function of ρ with fixed $\gamma = 0.01$ in strongly modularized (circles) and weakly modularized (asterisks) topologies. The offered traffic ρ, where the order parameter η becomes non-zero, is called a critical point ρ_c at which a network begins to saturate. In the figure, modularized topology reaches the critical point much earlier than non-modularized topology. This is because some nodes in modularized topology have high betweenness centrality, so that these nodes cause strongly modularized networks to collapse much earlier than weakly modularized networks.

The early saturation experienced by modularized topology can be overcome in many different ways. Since the main reason of this early saturation is that the top-layer nodes in a modularized topology cause a bottleneck phenomenon, increasing the speed of packet processing only in these nodes releases the network from early saturation.

Figure 15.4 shows the variation of critical points as the speed of packet processing increases in a modularized topology. We decrease the values of γ in Equation 15.2 to increase the speed of packet processing in the nodes. The circles show the variation of critical points when the packet processing power of the top-layer nodes increases, and the triangles show the result when nodes located in other than the top layer are chosen to increase the packet forwarding power. It demonstrates that the top-layer

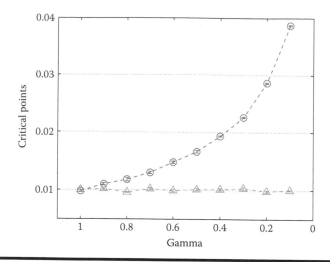

Figure 15.4 Critical points as γ in Equation 15.2 vary in nodes on the top layer of a modularized network (circles), as well as in nodes on other than the top layer (triangles).

nodes actually cause the bottleneck and this bottleneck phenomenon can be resolved by increasing the capacity of these nodes on the top layer.

15.3.4 Adaptability

As the last scenario, an attractor network model[33] is used to analyze the response of a system to different types of stimuli. Bar-Yam and Epstein[34] and Wang et al.[35] explored this dynamic attractor network model to investigate the robustness and sensitivity of scale-free and degree-correlated topologies respectively.

The basic idea of this scenario is to observe the variation in the size of the basin of attraction with different types of stimuli. A schematic illustration of the basin of attraction is provided in Figure 15.5. We can intuitively recognize that a ball will

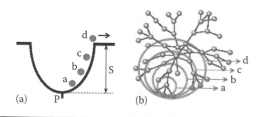

Figure 15.5 Schematic illustration of the basin of attraction.

finally converge to a predefined point P when it is dropped anywhere within the area S, which represents the size of the basin of attraction (Figure 15.5a). If the predefined point represents a certain behavior of a system which we want, any point other than the exact point P represents a system with a certain level of disturbance. When the disturbance is small (within the basin of attraction), the system will evolve to a stable state at point P; if the disturbance is too high (outside of the basin of attraction), the system will move to the other states. Thus, the size of the basin of attraction basically shows the **stability** of the system.

The method to find the size of the basin of attraction is abstractly shown in Figure 15.5a and b. The points **a** to **d** represent the levels of disturbance. Thus, the number of disturbed nodes inside the circle from **a** to **d** is increased, and then when it fails to return to the stable point P, the number of disturbed nodes becomes the size of the basin of attraction (in this example, the size of the basin of attraction is **d**). When the size of the basin of attraction of a system is large, the system is robust since it maintains its state against a large disturbance. On the other hand, a small-sized basin implies that the system is sensitive and therefore the state of the system can be changed even with a small disturbance. It is known that the topological structure of a system impacts on the size of its basin. Bar-Yam and Epstein[34] demonstrated that scale-free topology is more sensitive to direct disturbance (preferential disturbance of high-degree nodes) than random topology. Wang et al.[35] showed that the degree-correlated scale-free networks are more sensitive to directed stimuli than uncorrelated networks.

To investigate how modularized topology impacts on the size of the basin, we plot the histogram of the size of the basin in modularized scale-free topology and non-modularized scale-free topology in Figure 15.6. In each topology, we simulate two scenarios with are random and direct stimuli. For the non-modularized scale-free topology, the size of the basin of attraction is a bit smaller than 50% of the total nodes in random stimuli and 25% of total nodes in direct stimuli, which matches well with the result obtained by Bar-Yam and Epstein.[34] Modularized topology seems to create a small size of basin. In other words, a modularized topology is more sensitive than a non-modularized topology against both random and direct stimuli.

For further investigation, we artificially create the following topologies in Figure 15.7, which clearly shows the different levels of their modularity structures. Although this method of generating modularized topologies seems to be too artificial, we believe it provides at least a certain clue to how these modularity structures impact on the size of the basin of attraction. In fact, there is no systematical way to produce modularized topologies as far as we know. An interesting observation in Figure 15.8 is that the size of the basin of attraction for random stimuli decreases quite rapidly with the increase in the strength of the modularity structure while the basin of attraction for direct stimuli maintains its size through all the strength levels of modularity with small fluctuations. Robustness has two contradictory concepts, which are maintainability and adaptability. The former means a

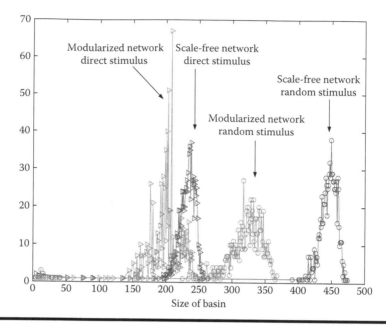

Figure 15.6 **Histograms of the size of basins in both modularized and non-modularized topologies: 1,000 nodes and 20,000 links.**

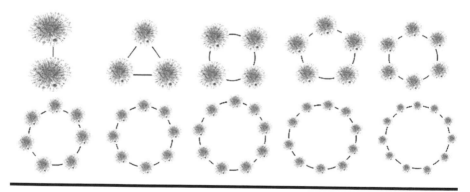

Figure 15.7 **All topologies have the same number of nodes and links: 1,000 nodes and 30000 links. The calculated modularity values for each topology from Nonmodule to 10 modules.**

robust system needs to keep its original state while the latter requires a system to change to another state. From the adaptability point of view, small-sized basins in both random and direct stimuli make modularized structures change their state even with small effort. In other words, modularized topologies have a strong adaptability property.

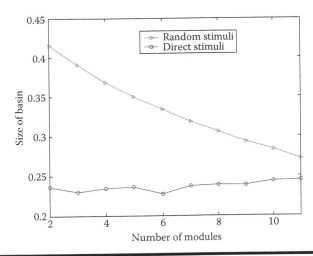

Figure 15.8   **The size of the basin of attraction as the number of modules (strength of module) increases.**

15.4 Summary

In this chapter we discussed topological robustness. We first described biological systems and their topological structures that give a comprehensive foundation for understanding biological robustness. Due to similarities between biological systems and information networks, features found in biological systems can promote man-made systems to a higher level of robustness.

We investigated one of the topological structures, modularity, to find some detailed features of this structural characteristic by carrying out four different simulation scenarios. These numerical results showed that modularized topology generally introduces a bottleneck and an attack vulnerability phenomenon to the system caused by the small number of connections between modules, and multiple modules cause parts of the system to be isolated and localized, and therefore a disturbance in the local part of a system is prohibited from spreading through the whole system. In addition, we demonstrated the adaptability of modularized topology showing that strongly modularized topology can adapt to new environments more easily than non-modularized structure.

References

1. H. Kitano, Biological robustness, *Nature Reviews Genetics* **5**(11), 826–837 (2004). ISSN 1471-0056.
2. U. Alon, M. G. Surette, N. Barkai, and S. Leibler, Robustness in bacterial chemotaxis, *Nature* **397**, 168–171 (1999).

3. N. Barkai and S. Leibler, Robustness in simple biochemical networks, *Nature* **387**, 913–917 (1997).

4. Y. M. Yi, Y. Huang, M. I. Simon, and J. Doyle, Robust perfect adaptation in bacterial chemotaxis through integral feedback control, *Proceedings of the National Academy of Sciences of the United States of America* **97**, 4649–4653 (2000).

5. G. von Dassow, E. Meir, E. M. Munro, and G. M. Odell, The segment polarity network is a robust developmental module, *Nature* **406**, 188–192 (2000).

6. E. Jen, Essays & commentaries: Stable or robust? What's the difference? *Complexity* **8**(3), 12–18 (2003). ISSN 1076-2787.

7. J. Doyle, J. Carlson, S. Low, F. Paganini, G. Vinnicombe, W. Willinger, J. Hickey, P. Parrilo, and L. Vandenberghe, Robustness and the internet: Theoretical foundations (2002). URL http://netlab.caltech.edu/internet/

8. R. Albert, H. Jeong, and A. Barabási, The Internet's Achilles' heel: Error and attack tolerance of complex networks, *Nature* **406**, 378–382 (2000).

9. Y. Xia, H. Yu, R. Jansen, M. Seringhaus, S. Baxter, D. Greenbaum, H. Zhao, and M. Gerstein, Analyzing cellular biochemistry in terms of molecular networks, *Annual Review of Biochemistry* **73**, 1051–1087 (2004).

10. H. Kitano, Systems biology: A brief overview, *Science* **295**(5560), 1662–1664 (2002). ISSN 1095-9203.

11. W. Willinger and J. Doyle, Robustness and the internet: Design and evolution (2002). URL http://netlab.caltech.edu/internet/

12. H. A. Simon, On a class of skew distribution functions, *Biometrika* **42**, 425–440 (1955).

13. M. Mitzenmacher, A brief history of generative models for power law and lognormal distributions, *Internet Mathematics* **1**(2), 226–251 (2004).

14. A. L. Barabási and R. Albert, Emergence of scaling in random networks, *Science* **286**(5439), 509–512 (1999).

15. M. Faloutsos, P. Faloutsos, and C. Faloutsos, On power-law relationships of the internet topology. In *SIGCOMM*, pp. 251–262 (1999).

16. R. Albert, H. Jeong, and A.-L. Barabási, Error and attack tolerance of complex networks, *Nature* **406**, 378 (2000).

17. A. Broder, R. Kumar, F. Maghoul, P. Raghavan, S. Rajagopalan, R. Stata, A. Tomkins, and J. Wiener, Graph structure in the web, *Computer Networks* **33**(1), 309–320 (2000). ISSN 1389-1286.

18. H.-W. Ma and A.-P. Zeng, The connectivity structure, giant strong component and centrality of metabolic networks, *Bioinformatics* **19**, 1423–1430(8) (2003).

19. M. Csete and J. Doyle, Bow ties, metabolism and disease, *Trends in Biotechnology* **22**(9), 446–450 (2004).

20. V. Gol'dshtein, G. A. Koganov, and G. I. Surdutovich, Vulnerability and hierarchy of complex networks (September, 2004). URL http://arxiv.org/pdf/cond-mat/0409298v1

21. L. da Fontoura Costa, The hierarchical backbone of complex networks, *Physical Review Letters* **93**, 098702 (2004).

22. L. da Fontoura Costa and F. Nascimento Silva, Hierarchical characterization of complex networks, *Journal of Statistical Physics* **125**, 845 (2006).

23. L. da Fontoura Costa and R. Fernandes Silva Andrade, What are the best hierarchical descriptors for complex networks? *New Journal of Physics* **9**, 311 (2007).
24. S. Eum, S. Arakawa, and M. Murata, A new approach for discovering and quantifying hierarchical structure of complex networks. In *ICAS '08: Proceedings of the Fourth International Conference on Autonomic and Autonomous Systems*, Gosier, Guadeloupe (2008).
25. M. R. Garey and D. S. Johnson, *Computers and Intractability: A Guide to the Theory of NP-Completeness* (W. H. Freeman & Co., New York, 1979).
26. J. P. Scott, *Social Network Analysis: A Handbook* (SAGE Publications, London, January 2000). ISBN 0761963391.
27. M. E. J. Newman and M. Girvan, Finding and evaluating community structure in networks, *Physical Review E* **69**, 026113 (2004).
28. S. Fortunato and C. Castellano, Community structure in graphs (2007). URL http://arxiv.org/pdf/0712.2716v1
29. S. A. Levin, *Fragile Dominion* (Perseus Publishing, Cambridge, MA, 1999).
30. R. V. Sole, S. C. Manrubia, M. Benton, and P. Bak, Self-similarity of extinction statistics in the fossil record, *Nature* **388**, 764–767, (1997).
31. E. Ravasz, A. L. Somera, D. A. Mongru, Z. N. Oltvai, and A. L. Barabási, Hierarchical organization of modularity in metabolic networks, *Science* **297**, 1551–1555 (2002).
32. A. Arenas, A. Díaz-Guilera, and R. Guimerà, Communication in networks with hierarchical branching, *Physical Review Letters* **86**(14), 3196–3199 (2001).
33. J. J. Hopfield, Neural networks and physical systems with emergent collective computational abilities, *Proceedings of the National Academy of Sciences of the United States of America* **79**(8), 2554–2558 (1982).
34. Y. Bar-Yam and I. R. Epstein, Response of complex networks to stimuli, *National Academic Science* **101**, 4341 (2004).
35. S. Wang, A. Wu, Z. Wu, X. Xu, and Y. Wang, Response of degree-correlated scale-free networks to stimuli, *Physical Review E* **75**, 046113 (2007).

Chapter 16

Biologically Inspired Dynamic Spectrum Access in Cognitive Radio Networks

Baris Atakan and Ozgur B. Akan

Contents

Cognitive radio (CR) is a promising wireless communication technology that aims to detect temporally unused spectrum bands by sensing its radio environment and communicate over these spectrum bands in order to enhance overall spectrum utilization. These objectives pose difficulties and have additional requirements such as self-organizing, self-adaptation, and scalability over conventional communication paradigms. Therefore, it is imperative to develop new communication models and algorithms. In nature, as a result of the natural evolution, biological systems have acquired great capabilities, which can be modeled and adopted for addressing the challenges in the CR domain. In this chapter, we explore the surprising similarities and mapping between cognitive radio network (CRN) architectures and natural biological systems. We introduce potential solution avenues from the biological systems toward addressing the challenges of CRN such as spectrum sensing, spectrum management, spectrum sharing, and spectrum mobility/handoff management. The objective of this chapter is to serve as a roadmap for the development of efficient scalable, adaptive, and self-organizing bio-inspired communication techniques for dynamic spectrum access in CRN architectures.

16.1 Introduction

The spectrum access of the wireless networks is generally regulated by governments via a fixed spectrum assignment policy. Thus far, the policies have satisfactorily met the spectrum needs of the wireless networks. However, in recent years, the dramatic increase in the spectrum access of pervasive wireless technologies has caused scarcity in the available spectrum bands. Therefore, the limited available spectrum and underutilization of the spectrum necessitate a new communication paradigm, referred to as dynamic spectrum access (DSA) and cognitive radio network (CRN), to exploit the entire existing wireless spectrum opportunistically.[1]

Cognitive radio (CR) has the ability to capture or sense information from its radio environment. Based on interaction with the environment, CR enables the users to communicate over the most appropriate spectrum bands which may be licensed or unlicensed. Since most of the spectrum is already assigned to licensed users by governmental policies, CR mainly strives for communication over the licensed spectrum band. In the licensed spectrum bands, CR exploits the temporally unused spectrum which is defined as the spectrum hole or white space.[2] If CR encounters the licensed user in the licensed spectrum band, it moves to another spectrum hole or stays in the same band without interfering with the licensed user by adapting its communication parameters, such as transmission power or modulation scheme. As for the unlicensed spectrum bands in which the licensed users cannot exist, all users have the same priority to access the existing unlicensed spectrum bands.

The CR technology enables the users to opportunistically access the available licensed or unlicensed spectrum bands through the following functionalities[1]:

- **Spectrum sensing**: Detecting the available spectrum holes and the presence of licensed users
- **Spectrum management**: Capturing the best available spectrum which can meet the user demand
- **Spectrum mobility**: Maintaining seamless communication requirements during the transition to better spectrum
- **Spectrum sharing**: Providing the fair spectrum-scheduling method among coexisting CRs

The above functionalities provide the CR with the ability of spectrum-aware communication while bringing new challenges that need to be addressed. Since CRN architectures may include a diverse set of communication devices and various network architectures, centralized control is not a practical solution. Instead, the CRN architectures and communication protocols must have the following capabilities:

- **Scalability**: To meet the large number of user demands over a large geographical area, the developed algorithms and protocols for the CRN must be scalable.
- **Self-adaptation**: To capture the best available channel, CR must adapt its communication parameters according to the environment conditions, such as the available spectrum band, channel parameters, existence of the licensed users, and interference level at the licensed user receiver.
- **Self-organization**: To collaboratively access the available licensed or unlicensed spectrum band, CRs must organize themselves.

In nature, biological systems intrinsically possess these capabilities. For instance, billions of blood cells that constitute the immune system can protect the organism from pathogens without any central control of the brain. Similarly, in insect colonies, insects can collaboratively allocate certain tasks according to the sensed information from the environment without any central controller. In this chapter, we explore the existing surprising similarities and mapping between communication challenges posed by CRN and the potential approaches and solutions in natural biological systems. More specifically, we first introduce a spectrum-sensing model for CRN based on the way the natural immune system distinguishes self-molecules from foreign molecules. Second, we introduce the spectrum management model for CRN based on the immune network principles through which blood cells in the immune system can adapt themselves to eliminate pathogens in the environment. Then, we introduce the spectrum-sharing model based on the adaptive task allocation model in insect colonies. Finally, we briefly discuss the spectrum mobility management model based on the biological switching mechanism. This chapter serves as a roadmap for developing scalable, adaptive, and self-organizing bio-inspired communication techniques for CRN architectures.

The rest of this chapter is organized as follows. In Section 16.2, we first give a brief introduction to the natural immune system, and introduce its relation to CRN. In Section 16.3, based on this relation, we present immune system–inspired spectrum-sensing and management models. In Section 16.4, we introduce a biological task allocation–inspired spectrum-sharing model. In Section 16.5, we explore the potential spectrum mobility management techniques inspiring by the biological switching mechanism. Finally, we discuss the concluding remarks in Section 16.6.

16.2 Immune System and Cognitive Radio Networks

Here, we first briefly introduce the immune system and its basic operation principles. Then, we discuss the similarities and the relation between the immune system and CRN.

16.2.1 Biological Immune System

The human immune system is a complex natural defense mechanism. It has the ability to learn about foreign substances (pathogens) that enter the body and to respond to them by producing antibodies that attack the antigens associated with the pathogen.[3] The adaptive immune system consists of lymphocytes, which are white blood cells, B, and T cells. Each of the B cells has a distinct molecular structure and produces antibodies from its surface. The antibody recognizes the antigen that is foreign material and eliminates it. When an antibody for a B cell binds to an antigen, the B cell becomes stimulated. The level of B cell stimulation depends not only on the success of the match to the antigen, but also on how well it matches other B cells in the immune networks.[3] As the response of the antigen, the stimulated B cells secrete the antibody to eliminate the antigen. This antibody secretion is illustrated in Figure 16.1.

To model the antibody secretion of the stimulated B cell, an analytical immune network model is proposed.[4] In this model, based on the stimulation and suppression

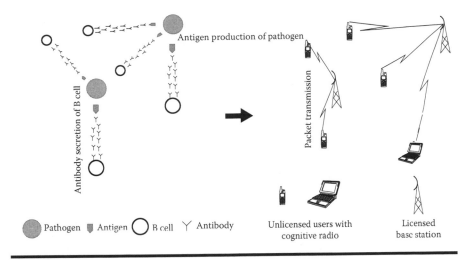

Figure 16.1 **The analogy between the immune system and cognitive radio networks.**

level of B cells and the natural extinction of antibodies, the antigen concentration is collaboratively kept at a desired level by regulating the antibody concentrations. This model is analytically given as follows:

$$\frac{dS_i(t+1)}{dt} = \left(\alpha \sum_{j=1}^{N} m_{ij}s_j(t) - \alpha \sum_{k=1}^{N} m_{ki}s_k(t) + \beta g_i - k_i \right) s_i(t) \qquad (16.1)$$

where
s_i is the concentration of antibody i
m_{ij} is the mutual coefficient of antibody i and j
N is the number of B cell types

$\sum_{j=1}^{N} m_{ij}s_j(t)$ denotes the effect of stimulated neighbors of B cell i and $\sum_{k=1}^{N} m_{ki}s_k(t)$ denotes the effect of suppressing neighbors of B cell i, g_i is the affinity between antibody i and the antigen, k_i is the natural extinction of the antibody i, and α and β are constants. S_i denotes the total stimulation of antibody i secreted by B cell i. Based on the total stimulation of antibody i (S_i), the concentration of antibody i (s_i) is given as follows:

$$s_i(t+1) = \frac{1}{1 + e^{(0.5 - S_i(t+1))}} \qquad (16.2)$$

The basic operation captured by this analytical model given in (16.1) and (16.2) can be outlined as follows:

■ When a pathogen enters the body, since antigen concentration increases, g_i increases. This results in an increase in the antibody secretion of B cell i and s_i increases.

■ If B cell i starts to be mostly suppressed by its neighbors, the effect of suppressing neighbors $\left(\sum_{k=1}^{N} m_{ki}s_k(t)\right)$ increases and s_i decreases.

■ If B cell i starts to be mostly stimulated by its neighbors and antigens, the effect of stimulating neighbors $\left(\sum_{j=1}^{N} m_{ij}s_j(t)\right)$ increases and s_i increases.

■ If the natural extinction of antibody i (k_i) increases, s_i decreases.

As will be explained in Section 16.3.2, we adopt this model given in (16.1) and (16.2) to give the efficient spectrum management model, which can enable CRs to collaboratively manage the available spectrum bands to keep the interference at a desired level.

16.2.2 Immune System–Inspired Cognitive Radio Networks

Although CR communication seems different from the natural immune system, both have a great deal of analogies when we consider their operation principles. When the immune system encounters a pathogen, B cells collaboratively *manage* their antibody densities to keep the antigen concentration at a desired level according to their stimulation and suppression levels and natural extinction of their antibodies. Similarly, when CRs encounter a spectrum channel which can be detected by means of an appropriate spectrum-sensing mechanism, CRs manage the channel by collaboratively regulating their transmission power such that the level of the interference can be kept at a desired level.

For a B cell, the level of its stimulation depends not only on how well it matches with the pathogen but also on how well it matches with its neighbor B cells such that while some of its neighbors stimulate it, some neighbors suppress it. Similarly, for a CR, the level of stimulation that results from an available channel not only depends on how well it matches with the channel but also how well it matches with its neighbors such that while some of its neighbors do not interfere with it, some neighbors interfere with it.

Antibody density of a B cell is determined according to the density of antibodies secreted by its neighbors which stimulate and suppress it, the concentration of antigen, and the natural extinction of its antibodies. Similarly, transmission power of a CR is determined according to the permissible transmission power at the selected channel, transmission power of its neighbors which use the same channel, and its path loss level to reach the channel.

Table 16.1 Relation between Immune System and Cognitive Radio networks

Immune System	Cognitive Radio Networks
B cell	Cognitive radio
Antibody density	Transmission power of cognitive radio
Pathogen	Available channel
Stimulating neighbors	Non-interfering neighbors
Suppressing neighbors	Interfering neighbors
Antigen density	Interference level in available channel
Natural extinction	Packet loss

In Table 16.1, we summarize the relation and mapping between the immune system and CRN. This relation is also illustrated in Figure 16.1. Accordingly, in Section 16.3, we introduce the immune system-inspired adaptive spectrum management approach.

Next, using the analogy between the natural immune system and CRN, we present efficient approaches that can be the roadmap to develop efficient self-organized algorithms for the CRN domain.

16.3 Immune System–Inspired Spectrum Sensing and Management in Cognitive Radio Networks

Here, we firstly discuss the spectrum-sensing problem and then introduce the immune system–inspired spectrum management model for CRN.

16.3.1 Immune System–Inspired Spectrum-Sensing Model for Cognitive Radio Networks

Spectrum sensing is one of the most important functions of CR which enables it to detect the available spectrum channels in the primary receiver by observing its local environment. In CRN, to access an available spectrum channel, CRs must ensure that the interference at the channel is within an allowable level such that CRs achieve successful transmissions to the channel without disturbing the primary transmitters using the same channel. Similarly, the immune system must detect the foreign molecules that enter the body. In the immune system, T cells take part in detecting whether a molecule entering the body is a self-molecule or a pathogen. This discrimination is achieved by T cells which have receptors on their surface that can detect foreign molecules.[5] This natural mechanism also enables T cells to learn the normal behavior of the self-molecules without requiring any prior knowledge about these molecules. This way they can detect foreign molecules through their abnormal behaviors which do not fit the normal behavior of self-molecules. Based on this natural mechanism, artificial immune system models have been developed and used for some intrusion detection problems.[5,6] As for the intrusion detection problems, this natural mechanism can provide the basis for the design of distributed harmful interference detection mechanisms for CRN.

In CRN, spectrum sensing is based on the detection of the abnormal noise level which can result in harmful interference at the primary receiver channels. However, due to the different location of CRs, each CR senses different noise power levels at the primary receiver channels. This hinders CRs to estimate the actual characteristic of the noise power at the primary receiver channels. Some CRs sense a lower power level than others and they can transmit to the channel although it does not permit a CR transmission. Therefore, the unknown power characteristic of the noise is a big drawback to reliably detect the harmful interference. Furthermore, the assumptions for noise power may lead to inaccurate detection of harmful interference. Instead of the assumptions for noise power, as with T cells in the immune system, it is possible to enable CRs to learn the power characteristics of the present noise signal without any need for a priori knowledge of the noise power. In the literature, immune system–inspired abnormal pattern detecting algorithms[5,6] have been used for detection of abnormal patterns in physical phenomena. Similarly, using the immune system–inspired abnormal pattern detection algorithms, CRs can be allowed to collaboratively learn the power characteristics of the present noise signal. Then, they can collaboratively detect the abnormal power characteristics in the received noise which does not match the power characteristics of the present noise signal. Thus, the harmful interference at the primary receiver channels can be collaboratively detected by CRs. Clearly, a bio-inspired spectrum-sensing algorithm based on the harmful interference detection can be designed and incorporated into CRs for effective spectrum sensing and dynamic spectrum access.

16.3.2 Immune System–Inspired Spectrum Management Model for Cognitive Radio Networks

As explained earlier, in the natural immune system, after a pathogen enters the body, each of the B cells distributively regulates its antibody densities to keep antigen concentration produced by the pathogen at a desired level. Similarly, in CRN, after an available channel is detected by an appropriate spectrum-sensing mechanism, each CR should distributively regulate its transmission power to keep the interference at a desired level in the channel. Based on this analogy, we adopt the model given in (16.1) and (16.2) to give an immune system–inspired spectrum management model.[7] To this end, we connect the immune system model to CR as follows:

■ We consider the transmission power of CR i at time t $(p_i(t))$ as the density of antibody i at time t which is secreted by B cell i.
■ We consider the interference level at the available channel j at time t $(I_j(t))$ as the antigen j concentration at time t produced by pathogen j.
■ When a B cell encounters an antigen, it is stimulated by the antigen and suppressed by some neighbors. Similarly, when a CR encounters an available channel, it is stimulated by its non-interfering neighbors* and it is suppressed by its interfering neighbors. Thus, we consider the total transmission power of its non-interfering neighbors $\sum_k p_k$ as the stimulation factor of neighbor B cells denoted in (16.1) by $\sum_{k=1}^{N} m_{ki} s_k(t)$. Furthermore, we consider the total transmission power of its interfering neighbors $\sum_m p_m$ as the suppression factor of neighbor B cells denoted in (16.1) by $\sum_{m=1}^{N} m_{mi} s_m(t)$.
■ We consider the packet loss rate of CR i at time t $(l_i(t))$ as the natural extinction of antibody i denoted in (16.1) by k_i.

According to the above mapping between the immune system and CR, and using (16.1), the total stimulation of CR i due to the channel j, i.e., S_{ij}, can be expressed as

$$\frac{dS_{ij}}{dt} = \left(\frac{1}{K} \sum_{k}^{K} p_k - \frac{1}{M} \sum_{m}^{M} p_m - \alpha I_j + \beta l_i \right) p_i \qquad (16.3)$$

where
α and β are the constants
K is the number of neighbors which do not interfere with CR i
M is the number of neighbors which interfere with CR i
dS_{ij}/dt denotes the deviation in the transmission power of CR i, which is imposed on CR i by its interfering and non-interfering neighbors, interference level, and its packet loss rate

To determine how this deviation affects the transmission power of CRs using channel j $(p_i, \forall j)$, we define a positive deviation threshold D such that if $dS_{ij}/dt < -D$, CR i reduces its transmission power and if $dS_{ij}/dt > D$, CR i increases

---

* We assume that if the distance between two CRs is less than r, which is called the interference radius, these CRs are the *interfering neighbors* of each other. If the distance is greater than r, they are the non-interfering neighbors of each other.

its transmission power. The model given in (16.3) is distributively employed by each CR at each time interval τ named as decision interval. The value of τ depends on the performance of the spectrum-sensing algorithm such that the smallest sensing time results in smallest τ. We assume that at each τ, each *CR i* broadcasts its transmission power (p_i) to all of the CRs in the environment and obtains interference levels in the all channels through an appropriate spectrum-sensing mechanism. Then, at each decision interval τ, using (16.3), each CR i manages the available spectrum channels as follows:

- When I_j increases due to any change in the radio environment such as location change or primary user activity, dS_{ij}/dt decreases for all CRs which use channel j ($\forall i$) such that if $dS_{ij}/dt < D$, CR i decreases its transmission power (p_i).
- When l_i increases due to any location change, p_i increases such that this can provide the acceptable packet loss rate. Since an increase in p_i results in an increase in I_j and this prevents CR i to increase p_i, CR i may not achieve the acceptable packet loss rate by increasing p_i when the channel has a higher interference level (I_j). Therefore, if the acceptable packet loss rate cannot be achieved, CR i vacates channel j and triggers the spectrum handoff.
- If the effect of interfering neighbors of CR i $\left((1/M) \sum_m^M p_m\right)$ increases due to the increase in their transmission powers, CR i decreases its transmission power to an acceptable level to minimize the interference effect of its neighbors. Therefore, if its reduced transmission power cannot meet its user requirements, it triggers spectrum handoff.

The immune system–inspired spectrum management model given above is a continuous-time model which enables CRs to collaboratively adapt to the changes in the time-varying radio environment without the need for any central controller. Furthermore, the model does not make any assumption on the available channels to optimize the channel allocation problem for an efficient and fair sharing of the available spectrum channels. In the literature, the traditional solutions for the spectrum management problem depends on some prior knowledge such as channel gains and locations of the primary receivers for all CRs. However, CRNs may not have this kind of prior information about the available channels since they do not have any infrastructure providing this kind of prior information. Thus, the immune system–inspired spectrum management approach is a potential to develop efficient spectrum management algorithms for CRNs such that these algorithms can be fully self-organized and will not need any prior knowledge about the available channels in the primary system.

16.4 Biological Task Allocation and Spectrum Sharing in Cognitive Radio Networks

In CRN, each CR senses the environment to detect the available spectrum channels, and then it must communicate over some detected spectrum channels which are the most appropriate spectrum channels meeting its quality of service (QoS) requirements and providing high spectrum utilization. *Similarly, in insect colonies, individuals sense the environment and share the available tasks such that each task is allocated to the individuals which are better equipped for that task.* Here, inspiring by this natural mechanism, we propose an effective spectrum-sharing model[8] for CRN.

16.4.1 Biological Task Allocation Model

In insect colonies, the collaborative behavior of the individuals provides a great capability to efficiently allocate and perform certain tasks. For the task allocation problem in an insect colony, every individual has a *response threshold* (θ) for each task which states the tendency of the individual to react to a task-associated stimulus. A task-associated stimulus s is defined as the intensity of an activator for a task such as a chemical concentration, or sounds. For example, if the task is larval feeding, the task-associated stimulus s can be expressed by the pheromone molecules emitted by the larvas.[9] The level of the task-associated stimulus s determines the likelihood of performing the task for the individuals. Thus, based on the respond threshold θ and the task-associate stimulus s, for an insect the probability of performing a task is given by[9]

$$T_\theta(s) = \frac{s^n}{s^n + \theta^n} \tag{16.4}$$

where n is a positive constant. For an individual having a small θ, performing the task is more likely as s increases. Performing a task having a small s is also more likely for the individuals having a higher θ.

In addition to task allocation, the natural mechanism modeled in (16.4) enables an insect colony to have self-synchronization capability, which allows the insect colony to effectively fulfill the available tasks and to increase the adaptation capability of the colony to the changes in the dynamically changing characteristics of the environment. When the stimulus level (s) in (16.4) for a task starts to increase, insects in the colony mostly perform this task because the increasing stimulus level increases the task performing probability in (16.4) of the insects for this task. Conversely, when the stimulus decreases, the insects start to perform another task having higher stimulus and providing higher task performing probabilities. Hence, all tasks can be synchronously performed by the individuals having maximum task performing probability by adapting themselves to the dynamically changing characteristics of the task-associated stimuli.

Next, inspiring by the task allocation model of the insect colony we establish an effective dynamic spectrum-sharing model[8] for CRN.

16.4.2 Biological Task Allocation–Inspired Spectrum-Sharing Model for Cognitive Radio Networks

In CRN, to share the available spectrum bands each CR senses the environment and then selects an available channel to transmits its packets. Similarly, in an insect colony, to allocate the available tasks each individual first senses the environment and then selects an available task to perform such that each task is accomplished by the individuals better equipped for that task. This allocation can be achieved by means of the task-performing probability given in (16.4). Here, we adopt the task-performing probability given in (16.4) to introduce the *channel selection probability*, which enables CRs to fairly and effectively share the available spectrum bands to increase overall spectrum utilization. For this adaptation, a CRN is modeled as an insect colony as follows:

- A CR is considered as an insect in a colony.
- An available channel is considered as a task.
- Estimated permissible power* (P_j) to a channel is considered as the task associated stimulus (s). Here, P_j denotes the estimated permissible power to channel j.
- The transmission power (p_{ij}) used by CR i to channel j is considered as the response threshold of an insect (θ).

This adaptation between CRN and an insect colony is also summarized in Table 16.2.

Using the adaptation discussed above and following (16.4), we introduce the channel selection probability T_{ij} as

$$T_{ij} = \frac{P_j^n}{P_j^n + \alpha p_{ij}^n + \beta L_{ij}^n} \qquad (16.5)$$

---

* We assume that the permissible power to a channel is estimated via noise and interference levels in this channel.[11] If a channel has higher noise and interference level such that it imposes higher bit error rate on the current communication, the permissible power to the channel is estimated to have a smaller value. If the noise and interference level is low, then the channel allows the transmission with higher power levels.

Table 16.2   Mapping between CRN and Insect Colony

Insect Colony	Cognitive Radio Networks
Insect	Cognitive radio
Task	Available channel
Task-associated stimulus (s)	Permissible power to channel (P_j)
Response threshold (θ)	Required transmission power (p_{ij})

Source: Atakan, B. and Akan, O. B., Biologically-inspired spectrum sharing in cognitive radio networks, *IEEE WCNC 2007*, Hong Kong, 2007.

where

n is a positive constant for the steepness of the channel selection probability T_{ij}

α and β are the positive constants that determine the mutual effects of L_{ij} and p_{ij}

L_{ij} is a learning factor for CR i. If channel j can meet the QoS requirements[*] of CR i

T_{ij} is increased by updating L_{ij} as $L_{ij} - \xi_0$ and channel j is learned by cognitive radio i

If channel j cannot meet the QoS requirements of CR i, T_{ij} is decreased by updating L_{ij} as $L_{ij} + \xi_1$, and channel j is forgotten by CR i. Here, ξ_0 and ξ_1 are the positive learning and forgetting coefficients, respectively, that determine the learning and forgetting rate of CRs. Using the channel selection probability for every available channel ($T_{ij}, \forall \, i,j$), CRs can fairly capture the available spectrum bands as follows:

- If $P_j \gg p_{ij}$, it is most probable that CR i transmits to channel j since T_{ij} is close to 1.
- If $P_j \ll p_{ij}$, it is almost impossible that CR i transmits to channel j since T_{ij} is close to zero. Here, $P_j \ll p_{ij}$ implies high interference level at channel j which results in small allowable power to the channel j.
- As P_j increases, T_{ij} increases, and for CRs, the likelihood of transmitting to the channel j increases.
- While p_{ij} increases, T_{ij} decreases, and the probability that CR i further uses the channel j decreases and it may vacate the channel j.
- As P_j decreases, it is highly likely that channel j is allocated by CRs having small transmission power (p_{ij}). This allows the CRs having less transmission

---

[*] Here, we characterize the QoS requirements of a CR with the desired reporting frequency rate and desired interference level at the channel. If for a CR, these requirements can be met by a channel, this channel is learned by this CR; if not, it is forgotten.

power to allocate the channels having less permissible power because for smaller P_j and p_{ij}, T_{ij} is higher than in the case for smaller P_j and higher p_{ij}. Thus, this regulation prevents the CRs from having high transmission power to transmit to the channel having small allowable transmission power in order not to exceed the interference level in the channels. Inversely, while P_j increases, channel j may be allocated by the CRs having small or high transmission power (p_{ij}). Therefore, the channels having high permissible power are allocated by CRs having less or more transmission power.

The channel selection probability (T_{ij}) given in (16.5) is periodically computed by each CR according to the changes in the time-varying environment or primary user activities, which affect the interference level at the channels and, therefore, affect permissible powers to the available channels. Thus, this enables CRs to easily adapt themselves to the changes in the time-varying radio environment.

16.5 Biological Switching–Inspired Spectrum Mobility Management in Cognitive Radio Networks

In CRN, when the current spectrum band is not available for a CR, spectrum mobility arises, which enables a CR to move another available spectrum band over which it can communicate. Spectrum mobility poses a new type of handoff referred to as *spectrum handoff*, that should be accomplished with an acceptable latency called *spectrum handoff latency*. The purpose of spectrum mobility management is to make sure that the spectrum transitions between spectrum bands are made smoothly and as soon as possible without significant performance degradation during a spectrum handoff.[1]

During some biological processes, *gene regulatory systems* might sometimes need to remember a certain state that is set by transient signals. This can be accomplished by repressor proteins in the networks of a gene. Repressor proteins are the proteins which regulate one or more genes by decreasing the rate of transcription. For example, in an environment in which both repressors can act, the system might have two stable steady states.[10]

In one state, while the gene is turned on for the first repressor, the synthesis of the second repressor is switched off. The absence of the second repressor enables the first to be synthesized. In the other steady state, the second repressor is present and the first is absent. These two steady states are controlled by the external inputs called inducer molecules. The presence of an inducer inhibits its repressor activity, and causes the production of the second repressor and the repression of the first repressor's synthesis. In this case, the concentration of the first repressor adaptively decreases with respect to the concentration of the second repressor. When the inducer is removed, the second repressor continues to reduce the level of the first repressor. In this case, the concentration of the second repressor adaptively increases with the concentration of the first repressor. Furthermore, this state transition must be realized within a certain amount of latency in order to assure proper repressor activity. A general model for this natural mechanism can be formulated as[10]

$$\frac{dx}{dt} = f(y) - \mu_1 x \tag{16.6}$$

$$\frac{dy}{dt} = g(x) - \mu_2 y \tag{16.7}$$

where
  x and y denote the concentrations of two repressor proteins
  f and g are repression functions
  μ_1 and μ_2 are positive constants indicating alleviation in the repressor concentrations

As explained earlier, the concentration of two repressor proteins can be regulated using the model given in (16.6) and (16.7). This way, a biological process can be kept in one of the steady states according to the existence of the inducer molecules.

This natural mechanism may provide an inspiration for effective spectrum mobility management, which can provide the seamless communication capabilities. Similar to the biological switching mechanism discussed earlier, when CR triggers the spectrum handoff, it should have two stable steady states without disturbing the seamless communication. The first state is the communication in the present channel which is intended to be vacated due to the spectrum handoff. The second is the communication in the next channel. Furthermore, we consider the transmission power of CR used in the present and the next channels as the first and the second repressor concentrations, respectively. Similar to the mission of inducer molecules in biological switching, these two steady states are controlled by the spectrum handoff process outlined as follows:

■ When the used channel condition provides an appropriate data rate with a small bit error rate, CR is in the first steady state and continues the communication in this channel.
■ When the used channel condition is worse or unavailable due to noise, signal loss, multi-path, and interference, CR decreases the transmission power used in this channel while increasing the transmission power used in the next channel and it moves into the second steady state.

When a CR needs a spectrum handoff, it must regulate its transmission power according to the conditions in the two channels used in the spectrum handoff process. The aim of an efficient spectrum handoff is to provide the seamless communication between the two channels. Therefore, similar to the concentration

Table 16.3 Mapping between CRNs and Biological Switching

Biological Switching	Cognitive Radio Networks
Emergence of inducer molecules	Emergence of spectrum handoff
First repressor concentration (x)	Transmission power used in ch_1 (P_1)
Second repressor concentration (y)	Transmission power used in ch_2 (P_2)
Repressor function $f(y)$	Channel condition function $f(P_2)$ in ch_2
Repressor function $f(x)$	Channel condition function $g(P_1)$ in ch_1

regulation of repressor proteins in the biological switching mechanism, in the spectrum handoff process, each CR can regulate its transmission power according to the conditions in the vacated and next channels. Thus, using the model given in (16.6) and (16.7), we present the analogy between the biological switching and spectrum handoff process as follows. We also show the analogy outlined below in Table 16.3.

■ We consider the emergence of the inducer molecules as the emergence of spectrum handoff for a channel due to bad channel condition (noise, signal loss, multi-path, interference, etc.) or unavailability of this channel.
■ We consider the concentration of the first repressor protein (x) as transmission power P_1 used in the channel ch_1 intended to be vacated.
■ We consider the concentration of the second repressor protein (y) as transmission power P_2 used in the channel ch_2 intended to be transmitted by CR.
■ We consider the repressor function $f(y)$ as the channel condition function $f(P_2)$ for the channel ch_1, determined according to the changes in P_2. If the condition in ch_1 is worse or unavailable, the channel condition function $f(P_2)$ decreases as P_2 increases. If the condition in ch_1 can provide a satisfactory data rate, $f(P_2)$ increases as P_2 decreases.
■ We consider repressor function $g(x)$ as the channel condition function $g(P_1)$ in the channel ch_2, determined according to the changes in P_1. Similar to $f(P_2)$, if the condition in ch_2 is good, the channel condition function $g(P_1)$ increases as P_1 decreases.

Based on this analogy we present a power regulation model using the model given in (16.6) and (16.7) as follows:

$$\frac{dP_1}{dt} = f(P_2) - \eta_1 P_1 \qquad (16.8)$$

$$\frac{dP_2}{dt} = g(P_1) - \eta_2 P_2 \tag{16.9}$$

where η_1 and η_2 are positive constants. The model given in (16.8) and (16.9) enables each CR to communicate among available spectrum channels by efficiently regulating the transmission power of each CR. This model does not rely on any assumption about the available channels and therefore allows each CR to sense available channels and communicate over them according to the dynamically changing radio environment.

This bio-inspired adaptive mechanism may provide seamless communication capability with efficient spectrum handoff, although the design of a detailed spectrum mobility/handoff management algorithm and its performance evaluation are yet to be proven to accurately assess its effectiveness.

16.6 Conclusion

In this chapter, we explore the existing analogy between dynamic spectrum access and CRN and natural biological systems. We have identified and discussed several potential solution approaches from biological systems to the communication challenges of CRN for dynamic spectrum sensing, sharing, management, and mobility. Intuitively, the approaches discussed here seem promising in designing adaptive, self-organizing, and scalable protocols. However, detailed design of algorithms and their performance evaluations need some additional research efforts to make the presented approaches more valuable and to prove their efficiency over the existing solutions in the CRN domain.

References

1. I. F. Akyildiz, W. Y. Lee, M. C. Vuran, S. Mohanty, NeXt generation/dynamic spectrum access/cognitive radio wireless networks: A survey, *Computer Networks Journal (Elsevier)*, 50, 2127–2159 (2006).
2. S. Haykin, Cognitive radio: Brain-empowered wireless communications, *IEEE Journal on Selected Areas in Communications*, 23, 201–220 (2005).
3. J. Timmis, M. Neal, J. Hunt, An artificial immune system for data analysis, *BIOSYSTEMS (Elsevier)*, 55, 143–150 (2000).
4. J. D. Farmer, N. H. Packard, A. S. Perelson, The immune system, adaptation, and machine learning, *Physica* 22D, 187–204 (1986).
5. D. Dasgupta, S. Forrest, Novelty detection in time series data using ideas form immunology, *International Conference on Intelligent Systems*, Nevada (1996).
6. S. Forrest, A. S. Perelson, L. Allen, R. Cherukuri, Self-nonself discrimination in a computer, *IEEE Symposium on Security and Privacy*, Oakland, CA, pp. 202–212 (1994).

7. B. Atakan, B. Gulbahar, O. B. Akan, Immune system-inspired evolutionary opportunistic spectrum access in cognitive radio ad hoc networks, *IFIP Med-Hoc-Net 2010*, France (2010).
8. B. Atakan, O. B. Akan, BIOlogically-inspired spectrum sharing in cognitive radio networks, *IEEE WCNC 2007*, Hong Kong (2007).
9. E. Bonabeau, M. Dorigo, G. Theraulaz, *Swarm Intelligence, From Natural to Artificial System*, Oxford University Press, New York (1999).
10. J. L. Cherry, F. R. Adler, How to make a biological switch, *Journal of Theoretical Biology*, 203, 117–133 (2000).
11. B. Wild, K. Ramchandran, Detecting primary receivers for cognitive radio applications, *IEEE DySPAN 2005*, 124–130 (2005).

Chapter 17

Weakly Connected Oscillatory Networks for Information Processing

Michele Bonnin, Fernando Corinto,
and Marco Gilli

Contents

Oscillatory nonlinear networks represent a circuit architecture for image and information processing. In particular, synchronous and/or entrained states can be exploited for dynamic pattern recognition and to realize associative and dynamic memories. In this chapter, we show how the equations governing the dynamical evolution of the network can be recast either to a phase equation, in which each oscillator is only described by its phase, or to an amplitude–phase equation, where each oscillator is described by its amplitude and phase. These models not only represent the ideal frameworks for investigating the emergence of collective behaviors such as synchronous or entrained oscillations but also provide a useful design tool for implementing associative and dynamic memories. The last part of the chapter is devoted to describe this application.

17.1 Introduction

Many studies in neurophysiology have revealed the existence of patterns of local synaptic circuitry in several parts of the brain. Local populations of excitatory and inhibitory neurons have extensive and strong synaptic connections between each other, so that action potentials generated by the former excite the latter, which in turn reciprocally inhibit the former. The neurons within one brain structure can be connected into a network because the excitatory (and sometimes inhibitory) neurons may have synaptic contacts with other, distant, neurons.[1]

It is widely accepted that natural and artificial biological systems can be accurately mimicked by neural networks in which the neurons, modeled as the McCulloch–Pitts neuronal units[2] or "integrate and fire" cells,[3] are coupled via adaptive synapses. Experimental observations have shown that if the neuronal activity, given by the accumulation (integration) of the spikes coming from the other neurons via the synapses, is greater than a certain threshold, then the neuron fires repetitive spikes; otherwise the neuron remains quiescent.[3] A different approach for modeling neuronal activity is founded on the replacement of neurons with periodic oscillators, where the phase of the oscillator plays the role of the spike time.[3]

There is an ongoing debate on the role of dynamics in neural computation. Collective behaviors, not intrinsic to any individual neuron, are believed to play a key role in neural information processing. The phenomenon of collective synchronization, in which an enormous system of oscillators spontaneously locks to a common frequency, is ubiquitous in physical and biological systems, and is believed to be responsible in self-organization in nature.[4] For instance, it has been observed in networks of pacemaker cells in the heart[5] and in the circadian pacemaker cells in the suprachiasmatic nucleus of the brain.[6] Partial synchrony in cortical networks is believed to generate various brain oscillations, such as the alpha and gamma electroencephalogram (EEG) rhythms, while increased synchrony may result in pathological types of activity, such as epilepsy. Current theories of visual neuroscience assume that local object features are represented by cells which are distributed across multiple visual areas in the brain. The segregation of an object requires the unique identification and integration of the pertaining cells that have to be bound into one assembly coding for the object in question.[7] Several authors have suggested that such a binding of cells could be achieved by selective synchronization.[8,9]

The theoretical understanding of the origin of collective rhythmicity may be obtained by studying its onset, that is, by treating it as a kind of phase transition or bifurcation. In many aggregate or tissue cells the constituting oscillators are assumed to be identical, despite the inevitable, small statistical fluctuation in their natural frequencies. Such randomness factors are destructive to mutual entrainment or to the formation of coherent rhythmicity. In contrast, interactions among the oscillators usually favor mutual synchronization. The balance between these opposing tendencies is common to all kinds of phase transitions.

The mathematical model of an oscillatory network consists of a large system of locally coupled nonlinear ordinary differential equations (ODEs), which may exhibit a rich spatio temporal dynamics, including the coexistence of several attractors and bifurcation phenomena.[10] For this reason the dynamics of oscillatory networks has been mainly investigated through time–domain numerical simulation. Unfortunately a global dynamic analysis, through the sole numerical simulation, would require to identify for each choice of network parameters all sets of initial conditions that converge to different attractors. This would be a formidable, and practically impossible, task. However, if the couplings among the oscillators are assumed to be weak enough, the governing equations can be recast in simpler models, which appear under the name of *phase models* and/or *amplitude–phase models*, which represent the ideal frameworks to investigate synchronization and phase-locking phenomena in large populations of oscillators.

17.2 Networks of Structurally Stable Oscillators

In this section, we consider a network composed of oscillating units, and we assume that each oscillator exhibits a self-sustained oscillation, i.e., a limit cycle in its phase space. As it corresponds to the sole neutrally stable direction, the phase, in contrast to the amplitude, can be controlled already by a weak external action. Indeed a weakly perturbed amplitude will relax to its stable value, whereas a small perturbation along the trajectory can induce a phase modulation. The difference in the relaxation time scale of perturbations of the amplitudes and the phase allows to describe the effect of a small periodic excitation with a single phase equation. This clearly demonstrates why the phase is a privileged variable,[11,12] and why we are justified in restricting our attention to its evolution only.

A weakly connected oscillatory network is described by the following set of coupled ODEs:

$$\dot{\mathbf{X}}_i(t) = \mathbf{F}_i(\mathbf{X}_i(t)) + \varepsilon \mathbf{G}_i(\mathbf{X}(t)) \quad i = 1,\ldots,N \tag{17.1}$$

where

$\mathbf{X}_i \in \mathbb{R}^n$ describes the state of the ith oscillator

$\mathbf{F}_i : \mathbb{R}^n \mapsto \mathbb{R}^n$ describes its internal dynamics

N is the total number of oscillators

$\mathbf{X} = \left(\mathbf{X}_1^T, \mathbf{X}_2^T, \ldots, \mathbf{X}_N^T\right)^T$ is the vector describing the state of the whole network

$\mathbf{G}_i : \mathbb{R}^{n \times N} \mapsto \mathbb{R}^n$ defines the coupling among the units

$\varepsilon \ll 1$ is a small parameter that guarantees a weak connection among the cells

If we denote by ω_i and $\theta_i \in S^1 = [0, 2\pi)$ the angular frequency and the phase, respectively, of each limit cycle, our goal is to reduce the state Equations 17.1 to the phase equation

$$\theta_i(t) = \omega_i t + \phi_i(\varepsilon\, t) \tag{17.2}$$

where $\phi_i(\varepsilon\, t) \in [0, 2\pi)$ is the *phase deviation* , i.e., the phase modulation induced by the couplings among the oscillators.

The possibility to reduce periodic orbits to phase models is justified by the trivial observation that any limit cycle is topologically equivalent to the unit circle. This implies the existence of a homeomorphism mapping the former to the latter (see Figure 17.1). Historically three main approaches were proposed to reduce the state equations to the phase equation. Winfree's approach[13] is based on the concept of phase response curves, which have to be determined numerically applying a proper perturbation (usually brief or weak) to the unperturbed oscillator. The method is not analytical and not suitable for high dimensional systems, since phase response curves become hyper-surfaces. The second approach is proposed by Kuramoto[4] and requires a change of coordinates that maps the solutions of the state equations to the solutions of the phase equations. While simple and elegant, it does not give any hint on the proper change of coordinates to operate. Conversely, Malkin's theorem[14–16] gives a recipe for deriving the phase deviation equation, i.e., the equation describing the phase modulation due to the coupling, provided that the limit cycle trajectory of each uncoupled oscillator is known and the oscillating frequencies are commensurable.

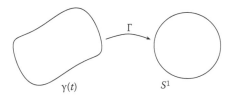

Figure 17.1 A limit cycle γ is topologically equivalent to the unit circle; thus we expect the existence of a homeomorphism Γ mapping one into the other.

THEOREM 17.1 (Malkin's Theorem[14–16]). Consider a weakly connected network described by (17.1) and assume that each uncoupled cell

$$\dot{\mathbf{X}}_i(t) = \mathbf{F}_i\left(\mathbf{X}_i(t)\right) \quad i = 1, \ldots, N \tag{17.3}$$

has a hyperbolic (either stable or unstable) periodic orbit $\gamma_i(t) \in \mathbb{R}^n$ of period T_i and angular frequency $\omega_i = 2\pi/T_i$. Let us denote by $\theta_i(t) \in [0, 2\pi)$ the phase variables, and by $\phi_i(\tau)$ the phase deviation from the natural oscillation $\gamma_i(t)$, $t \geq 0$, where $\tau = \varepsilon t$ is the slow time. Then the weakly connected oscillatory network (WCON) admits the following phase model

$$\dot{\theta}_i(t) = \omega_i + \varepsilon\, \phi_i'(\tau) \quad i = 1, \ldots, N \tag{17.4}$$

and the vector of the phase deviation $\phi = (\phi_1, \phi_2, \ldots, \phi_N)^T$ is a solution to

$$\phi_i' = H_i(\phi - \phi_i, \varepsilon) \quad i = 1, \ldots, N$$
$$\phi - \phi_i = (\phi_1 - \phi_i, \phi_2 - \phi_i, \ldots, \phi_N - \phi_i)^T \in [0, 2\pi[^N = \mathbf{T}^N \tag{17.5}$$

where $' = d/d\tau$ and

$$H_i(\phi - \phi_i, 0) = \frac{\omega}{T} \int_0^T \mathbf{Q}_i^T\, \mathbf{G}_i\left[\gamma\left(t + \frac{\phi - \phi_i}{\omega}\right)\right] dt\gamma\left(t + \frac{\phi - \phi_i}{\omega}\right)$$

$$= \left[\gamma_i^T\left(t + \frac{\phi - \phi_i}{\omega}\right), \ldots, \gamma_N^T\left(t + \frac{\phi - \phi_i}{\omega}\right)\right]^T \tag{17.6}$$

T being the minimum common multiple of $T_1, T_2, \ldots, T_n$. In the above expression (17.6) $\mathbf{Q}_i(t) \in \mathbb{R}^n$ is the unique nontrivial T_i periodic solution to the linear time-variant system:

$$\dot{\mathbf{Q}}_i(t) = -[DF_i(\gamma_i(t))]^T \mathbf{Q}_i(t) \tag{17.7}$$

satisfying the normalization condition

$$\mathbf{Q}_i^T(0)\mathbf{F}_i(\gamma_i(0)) = 1 \tag{17.8}$$

Unfortunately the periodic trajectories of almost all nontrivial oscillators can only be determined through numerical simulations. This implies that, in general, an explicit expression for the phase deviation equation is not available. It turns out that in most cases such an equation cannot be exploited for investigating the global dynamic behavior of the network and for classifying its periodic attractors. However, accurate analytical approximation of both stable and unstable limit cycles can be obtained through standard methods like the harmonic balance or the describing function technique. This suggests the idea of using an analytical approximation of the real periodic trajectory in Equations 17.6 through 17.8. Clearly, rigorous conditions which guarantee the reliability of the approach can be hardly derived. However, from a practitioner's perspective, the methodology can be a posteriori justified comparing the obtained results with those provided by numerical experiments. An approach, hereinafter referred to as the *DFMT* technique (Describing Function and Malkin's Theorem–based Technique), based on the following steps has been devised:[15,16]

(1) The periodic trajectories $\gamma_i(t)$ of the uncoupled oscillators are approximated through the *describing function technique*.
(2) Once the approximation of $\gamma_i(t)$ is known, a first harmonic approximation of $\mathbf{Q}_i(t)$ is computed by exploiting (17.7) and the normalization condition (17.8).
(3) The phase deviation Equation 17.5 is derived by analytically computing the functions (17.6) given by the *Malkin's theorem*.
(4) Finally, the phase equation is analyzed in order to determine the total number of stationary solutions (equilibrium points) and their stability properties; they correspond to the total number of limit cycles of the original weakly connected network described by (17.1).

For the sake of simplicity, let us consider a chain of identical oscillators, with space invariant connections of range r (ranging from 1, corresponding to nearest neighbors coupling, up to N for a fully connected network). It can be shown[15,16] that, for a wide class of oscillators, the network dynamics is described by the phase equation

$$\phi_i' = V(\omega) \sum_{k=-r}^{r} C_k \sin[\phi_{i+k} - \phi_i] \tag{17.9}$$

where $V(\omega)$ is a function depending on the nature of the constituent oscillators and the periodic trajectory considered, while C_k are real constants depending on the coupling functions $\mathbf{G}_i(\mathbf{X})$.

Equation 17.9 reduces a rather complex oscillatory network to a simple Kuramoto model,[4] which can be analytically dealt with. Assuming nearest neighbors' interactions and imposing continuous boundary conditions, i.e., $\phi_0 = \phi_1$, $\phi_{N+1} = \phi_N$, $\forall t \geq 0$, (17.9) reduces to

$$\begin{cases} \phi_1' = V(\omega) \, C_{+1} \sin(\phi_2 - \phi_1) \\ \phi_i' = V(\omega) \, [C_{-1} \sin(\phi_{i-1} - \phi_i) + C_{+1} \sin(\phi_{i+1} - \phi_i)] \\ \phi_N' = V(\omega) \, C_{-1} \sin(\phi_{N-1} - \phi_N) \end{cases} \tag{17.10}$$

There is a one-to-one correspondence between the equilibrium point of (17.10) and the phase-locked solution (either stable or unstable) of the weakly connected network. This greatly simplifies not only the research of entrained oscillations in the network but also their stability analysis. Imposing the equilibrium condition in (17.10) we obtain

$$\phi_{i+i} - \phi_i = \begin{cases} 0 \\ \pi \end{cases} \quad i = 1, \dots, N - 1 \tag{17.11}$$

The stability of the phase-locked states can be easily determined by looking at the eigenvalues of the jacobian matrix. According to Floquet's theory, each uncoupled limit cycle is described by a set of n characteristic (Floquet's) multipliers, one of which is unitary (the structural multiplier), while the other $n - 1$ are denoted by $\mu_1, \mu_2, \dots, \mu_{n-1}$. If the cycle is stable, the modulus of all the multipliers is less than 1; conversely, if the cycle is unstable at least one multiplier has modulus larger than 1. A limit cycle of the coupled network exhibits $n \times N$ multipliers. In the case of weak connections, $(n - 1) \times N$ of them have numerical values very close to $\mu_1, \mu_2, \dots, \mu_{n-1}$ respectively. The others (i.e., those that in the absence of coupling equal 1) will be named *coupling characteristic multipliers*, because they describe the effect of the interactions among the cells. One of them is still unitary, whereas the other $N - 1$ are in general different from 1 and determine stability and/or instability depending on their moduli. A detailed investigation can be performed,[15,16] and some interesting results are summarized in the following theorem.

THEOREM 17.2[15,16]   Consider a weakly connected oscillatory network described by the phase model (17.10), and assume that the coupling constants have the same sign, i.e., $C_{-1} C_{+1} > 0$. The jacobian matrix has one null eigenvalue, corresponding to the unitary characteristic multiplier, the number of eigenvalues with the same sign of $V(\omega) C_{+1}$ is equal to the number of phase shifts equal to π, and the number of eigenvalues with a sign opposite to $V(\omega) C_{+1}$ is equal to the number of phase shifts equal to 0.

The proof of the theorem relies upon simple arguments of linear algebra, like similarity and congruence transformations, and Sylvester's Inertia Theorem. The main information contained in the theorem is that the stability of the phase-locked states depends on both the characteristic of the uncoupled limit cycle (through the function $V(\omega)$) and the coupling constants. As an example we consider a one-dimensional array of Chua's circuits composed of seven cells with positive coupling constants. The phase portrait of each uncoupled oscillator, for a specific set of parameters values,[15,17] is shown in Figure 17.2. The circuit exhibits three unstable equilibrium points (grey dots), two stable asymmetric limit cycles (depicted with a dark grey solid lines and denoted by $A_i^{\pm}$), one stable symmetric limit cycle (depicted

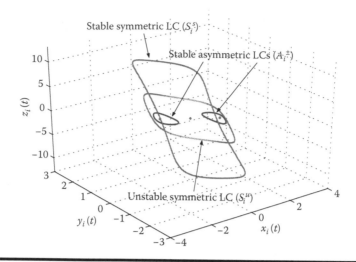

Figure 17.2 **Global dynamic behavior of a single uncoupled Chua's circuit, for a proper choice of the parameters values. The circuit exhibits three unstable equilibrium points (grey dots), two stable asymmetric limit cycles (depicted with dark grey solid lines and denoted by $A_i^{\pm}$), one stable symmetric limit cycle (depicted with a pale grey solid lines and denoted by S_i^s), and one unstable symmetric limit cycle (depicted with a pale grey dashed line and denoted by S_i^u).**

with a pale grey solid line and denoted by S_i^s), and one unstable symmetric limit cycle (depicted with a pale grey dashed line and denoted by S_i^u).

Figures 17.3 through 17.8 show the values of the characteristic multipliers computed numerically exploiting a standard numerical algorithm[17] versus the strength of the coupling coefficient C_{+1}. In all the cases, the theoretical expectations are in agreement with the numerical outcomes. In Figures 17.3 and 17.4 the oscillators are running on a stable asymmetric limit cycle. In this case $V(\omega) < 0$, so we expect as many negative eigenvalues as the number of phase shifts equal to π.

Figures 17.5 and 17.6 refer to the same network, but with the oscillators now running on a stable symmetric limit cycle. In this case we have $V(\omega) > 0$, and we expect as many negative eigenvalues as the number of null phase shifts. Figures 17.7 and 17.8 refer to the same network, but with the oscillators now running on a stable symmetric limit cycle. Now $V(\omega) < 0$, and we expect as many negative eigenvalues as the number of phase shifts equal to π.

17.3 Networks of Oscillators Close to Bifurcations

In the previous section we have shown how it is possible to obtain the phase equation for a network of structurally stable oscillators. When the oscillators are close to a

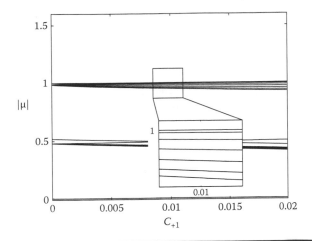

Figure 17.3 One-dimensional array of Chua's circuit, composed of seven cells. The moduli of the *characteristic multipliers* are represented as a function of the coupling parameter C_{+1}, for the asymmetric limit cycle, with all phase shifts equal to π and $C_{-1} = 0.01$. The system exhibits one unitary multiplier and six real multipliers with modulus less than 1. The moduli of all the other 14 multipliers of the system are less than 1.

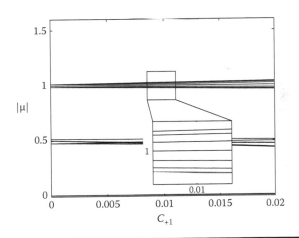

Figure 17.4 One-dimensional array of Chua's circuit, composed of seven cells. The moduli of the *characteristic multipliers* are represented as a function of the coupling parameter C_{+1} with $C_{-1} = 0.01$, for the asymmetric limit cycle. In this case three phase shifts are equal to π and three are equal to 0. The system exhibits one unitary multiplier, three multipliers with modulus less than 1, and three with modulus greater than 1. The moduli of all the other 14 multipliers of the system are less than 1.

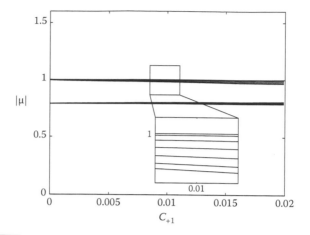

Figure 17.5 One-dimensional array of Chua's circuit, composed of seven cells. The moduli of the *characteristic multipliers* are represented as a function of the coupling parameter C_{+1} with $C_{-1} = 0.01$, for the stable symmetric limit cycle. In this case all the phase shifts are null. The system exhibits one unitary multiplier, and six with modulus less than 1. The moduli of all the other 14 multipliers of the system are less than 1.

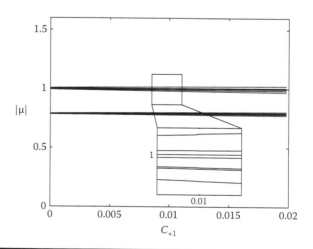

Figure 17.6 One-dimensional array of Chua's circuit, composed of seven cells. The moduli of the *characteristic multipliers* are represented as a function of the coupling parameter C_{+1} with $C_{-1} = 0.01$, for the stable symmetric limit cycle. In this case two phase shifts are equal to π and four are equal to 0. The system exhibits one unitary multiplier, four multipliers with modulus less than 1, and two with modulus greater than 1. The moduli of all the other 14 multipliers of the system are less than 1.

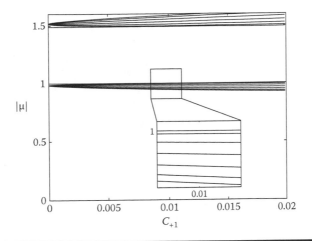

Figure 17.7 One-dimensional array of Chua's circuit, composed of seven cells. The moduli of the *characteristic multipliers* are represented as a function of the coupling parameter C_{+1} with $C_{-1} = 0.01$, for the unstable symmetric limit cycle. In this case all the phase shifts are equal to π. The system exhibits one unitary multiplier, and six with modulus less than 1. Since the limit cycles are unstable, the moduli of all the other 14 multipliers of the system are greater than 1.

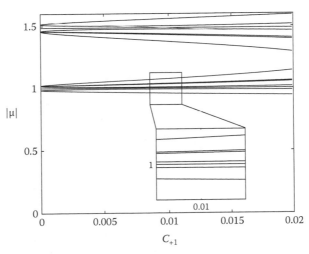

Figure 17.8 One-dimensional array of Chua's circuit, composed of seven cells. The moduli of the *characteristic multipliers* are represented as a function of the coupling parameter C_{+1} with $C_{-1} = 0.01$, for the unstable symmetric limit cycle. In this case two phase shifts are equal to π and four are equal to 0. The system exhibits one unitary multiplier, four multipliers with modulus greater than 1, and two with modulus less than 1. The moduli of all the other 14 multipliers of the system are greater than 1.

bifurcation, they do not represent structurally stable systems anymore, and also the evolution of the amplitudes must be taken into account. On the one hand, this leads to more complex equations, which, on the other hand, reveal a richer dynamical behavior.[18,19] If the trajectories are not hyperbolic, Malkin's theorem cannot be applied, but being in the neighborhood of a bifurcation, we can use center manifold reduction and normal form method.[20,21] Such a situation is of primary importance in populations of neural oscillators, where it is believed that the network local activity is not interesting from the neuro-computational point of view unless the equilibria are close to a threshold corresponding to a bifurcation.[14] If the interactions are assumed to have finite range and finite propagation speed, local events cannot be experienced by distant points without time lag, and we are induced to include delays in the governing equations. Although delay effects leading to an oscillatory behavior have been well known for a long time, especially in radio engineering sciences, only in recent times it has been emphasized that delay-induced instabilities can lead to more complex behavior.

In this section, we shall consider a network of nonlinear oscillators with delay in both the activation function and the interaction, referred to as weakly connected delayed oscillatory network (WCDON)

$$\dot{\mathbf{X}}_i(t) = \mathbf{F}_i(\mathbf{X}_i(t - \tau_i), \mu) + \varepsilon\, \mathbf{G}_i(\mathbf{X}(t - \tau), \mu) \quad i = 1, \dots, N \qquad (17.12)$$

Here $\mathbf{X}_i(t) \in \mathbb{R}^2$ describes the activity of the ith oscillator, $\mathbf{X}(t - \tau) = \left(\mathbf{X}_1^T(t - \tau_1), \dots, \mathbf{X}_N^T(t - \tau_N)\right)^T$ is the vector collecting the states of all the oscillators, and μ is a vector of parameters. Let τ_m be the largest delay, and let $C^n = C^n([-\tau_m, 0]; \mathbb{R}^n)$ denote the space of real, n-dimensional vector valued functions on the interval $[-\tau_m, 0]$. Hence $\mathbf{F}_i : C^2 \to \mathbb{R}^2$ is an activation function describing the internal dynamics of the ith oscillators and $\mathbf{G}_i : C^{2 \times N} \to \mathbb{R}^2$ is a function describing the interactions between the ith oscillators and the rest of the network. The intrinsic delays τ_i take into account the finite interaction propagation speed, while the parameter ε denotes the strength of the interaction between different oscillators. We consider the network's dynamics near a multiple Hopf bifurcation, that is, in the uncoupled limit $\varepsilon = 0$ and at an equilibrium point $(\bar{\mathbf{X}}, \bar{\mu})$, the jacobian matrix of each nonlinear oscillator $D\mathbf{F}_i(\bar{\mathbf{X}}_i, \bar{\mu})$ has a pair of complex conjugate eigenvalues $\pm i\,\omega_i$. Without loss of generality, we shall assume $(\bar{\mathbf{X}}_i, \bar{\mu}) = (0, 0)$ and $\mu(\varepsilon) = \varepsilon\mu_1 + \mathcal{O}(\varepsilon^2)$. It can be shown that the whole network admits the following canonical model, i.e., it is qualitatively equivalent to the following theorem.

THEOREM 17.3 (Canonical Model for WCDON[22])   If the weakly connected delayed oscillatory network (17.12) is close to a multiple Hopf bifurcation, and $\mu(\varepsilon) = \varepsilon\mu_1 + \mathcal{O}(\varepsilon^2)$, then its dynamics is described by the canonical model

$$z_i' = b_i z_i + d_i z_i |z_i|^2 + \sum_{j \neq i} c_{ij} z_j + \mathcal{O}\left(\sqrt{\varepsilon}\right) \tag{17.13}$$

where $\sigma = \varepsilon t$ is the slow time, $' = d/d\sigma$, and $z_i, d_i, b_i, c_{ij} \in \mathbb{C}$.

The proof of the theorem relies on center manifold reduction, normal form method, and averaging. Due to the presence of delays, Equation 17.12 is a retarded functional differential equation (RFDE).[23] As a consequence, from the formal point of view, the network represents an infinite dimensional dynamical system. The usual application of center manifold reduction allows to reduce the geometrical dimensions of the problem, eliminating all those eigenspaces above which the dynamics is trivial. In this case the number of geometrical dimensions is not reduced, but rather center manifold reduction permits to remove the delays, transforming the RFDEs into ODEs in complex variables.[22] Applying a near-to-identity transformation the ODE is transformed into its normal form, and after averaging, which removes high frequency terms, we obtain the final form (17.13).

The proof of the theorem reveals that only neurons with equal frequencies interact, at least on a scale of time $1/\varepsilon$. All neural oscillators can be divided into ensembles, according to their natural frequencies, and interactions between oscillators belonging to different ensembles are negligible even if they have nonzero synaptic contacts $c_{ij} \neq 0$. Although physiologically present and active, those synaptic connections are functionally insignificant and do not play any role in the dynamics. It is possible to speculate about a mechanism capable of regulating the natural frequency of the neurons so that the neurons can be entrained into different ensembles at different times simply by adjusting the frequencies. In this way, selective synchronization can be achieved and different ensembles can be exploited for implementing parallel computing processes.

By introducing polar coordinates

$$z_i = r_i e^{i\varphi_i} \quad b_i = \alpha_i + i\Omega_i \quad d_i = \gamma_i + i\sigma_i \quad c_{ij} = |c_{ij}| e^{i\psi_{ij}} \tag{17.14}$$

Equation 17.13 yields the *amplitude–phase equation*

$$\begin{cases} r_i' = \alpha_i r_i + \gamma_i r_i^3 + \displaystyle\sum_{j=1, j\neq i}^{N} |c_{ij}| r_j \cos\left(\varphi_j - \varphi_i + \psi_{ij}\right) \\ \varphi_i' = \Omega_i + \sigma_i r_i^2 + \dfrac{1}{r_i} \displaystyle\sum_{j=1, j\neq i}^{N} |c_{ij}| r_j \sin\left(\varphi_j - \varphi_i + \psi_{ij}\right) \end{cases} \tag{17.15}$$

Equation 17.15 describe the dynamics of coupled oscillators when the coupling strength is comparable to the attraction of the limit cycle[19] and can be exploited for the realization of neuro-computing mechanical devices.[24] With respect to the simpler phase oscillator model, amplitude–phase models exhibit a richer dynamical behavior, which includes oscillation death (Bar-Eli effect[18,19]), self-ignition,[25] phase locking, and phase drifting.

Analysis of Equation 17.15 is a daunting problem unless certain restrictive hypotheses are imposed. As an example, let us consider a chain composed of N identical oscillators, with nearest neighbors symmetric coupling and periodic boundary conditions. We focus the attention on phase-locked oscillations. By introducing the phase difference between two adjacent oscillators $\varphi_{i+1} - \varphi_i = \chi_i$, we have phase-locking if $\chi_i = \chi$, $\forall i$. Under the hypotheses, the above Equation 17.15 reduces to[22]

$$\begin{cases} r_i' = \alpha r_i + \gamma r_i^3 + |c| \sum_{j=-1}^{1} r_{i+j} \cos (\psi + j \chi) \\ \chi' = \sigma \left(r_{i+1}^2 - r_i^2 \right) + \dfrac{|c|}{r_i r_{i+1}} \sum_{j=-1}^{1} \sin (\psi + j \chi) \left(r_{i+j+1} r_i - r_{i+1} r_{i+j} \right) \end{cases}$$

(17.16)

Equilibrium points of Equation 17.16 correspond to oscillations of constant amplitudes with constant phase differences in the whole network. Now that the strength of an amplitude–phase model is manifest, the research of synchronous oscillations in the functional differential Equation 17.12 is reduced to the prospective equilibrium points of the ordinary differential Equation 17.16.

The amplitudes of the oscillations can be determined by imposing the equilibrium condition to Equation 17.16, which yields

$$r_i = \sqrt{-\frac{\alpha + |c| \sum_{j=-1}^{1} \cos (\psi + j \chi)}{\gamma}} \forall i$$

(17.17)

and turn out to depend on both the phase difference and the parameter values. Equation 17.17 also provides a condition sufficient for the existence of phase-locked states in the network. Since γ is negative (to guarantee that each oscillator exhibits a stable limit cycle[22]), the numerator must be positive or null in correspondence to the bifurcation point. Such a condition is sufficient but not necessary, since other equilibrium points might exist.

The stability of the phase-locked states can be analytically determined by looking at the eigenvalues of the jacobian matrix, and the results for the in-phase locking ($\chi = 0$) and anti-phase locking ($\chi = \pi$) are summarized in the following theorem.

THEOREM 17.4[22] Consider a chain composed of N identical delayed oscillators near a multiple Hopf bifurcation, with nearest neighbors symmetric connections and periodic boundary conditions. Let us define

$$\beta(r) = 2 \cos \left(2 \pi \frac{r-1}{N} \right) \quad r = 1, \dots, N$$

(17.18)

Then the network exhibits:

■ *In-phase, constant amplitude oscillations if* $\forall r = 1, \ldots, N$

$$\begin{cases} (\beta(r) - 4) \operatorname{Re} c - \alpha < 0 \\ (\beta(r) - 2) |c|^2 - 4 \operatorname{Re}^2 c - 2 \alpha \operatorname{Re} c - \frac{2\alpha\sigma}{\gamma} \operatorname{Im} c - \frac{4\sigma}{\gamma} \operatorname{Re} c \operatorname{Im} c < 0 \end{cases}$$
(17.19)

■ *Anti-phase, constant amplitude oscillations if* $\forall r = 1, \ldots, N$

$$\begin{cases} (\beta(r) - 4) \operatorname{Re} c + \alpha > 0 \\ (\beta(r) - 2) |c|^2 - 4 \operatorname{Re}^2 c + 2 \alpha \operatorname{Re} c + \frac{2\alpha\sigma}{\gamma} \operatorname{Im} c - \frac{4\sigma}{\gamma} \operatorname{Re} c \operatorname{Im} c < 0 \end{cases}$$
(17.20)

The assumption of periodic boundary conditions, which implies that $i + N$ coincides with i, turns out to be of capital importance. The jacobian matrix is in fact a block circulant matrix,[26] which can be reduced to a block diagonal matrix resorting to the Fourier matrices.[22] The resulting diagonal blocks are 2×2 sub-matrices, whose eigenvalues can be easily computed.

Figure 17.9 shows the regions of the parameters space satisfying these conditions for a chain composed of 10 oscillators, in the case $\sigma = 0$ (no shear condition). Notice the existence of regions in the neighborhoods of $\psi = \pi/2$ and $\psi = 3\pi/2$ where in-phase and anti-phase–locked oscillations can coexist. These regions become narrower as the number of oscillators increases, and are distorted by a nonzero value of σ, but they are not removed. The coexistence of different stable phase-locked states confirms the idea of using oscillatory networks as associative memories, where different phase-locked behaviors are associated with different stored images.

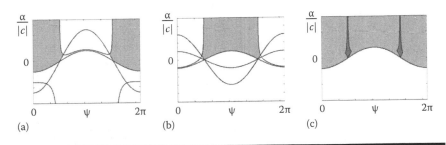

Figure 17.9 **Parameter regions corresponding to various phase-locking states. Shading indicates regions where phase-locked solutions exist. (a) In-phase locking. (b) Anti-phase locking. (c) Darker region: Coexistence of in-phase and anti-phase–locked oscillations.**

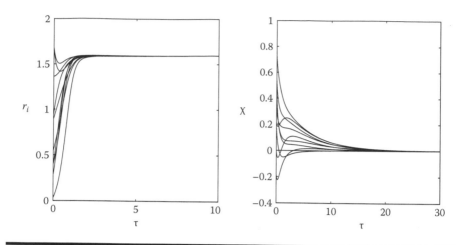

Figure 17.10 In-phase locking in a network composed of 10 oscillators. Evolution of the amplitudes (left) and phases (right) from random initial amplitudes $r_i \in (0, 2]$ and phase differences $\chi_i \in (-\pi/6, \pi/6)$.

Figure 17.10 shows the evolution of the amplitudes (left) and the phase differences (right) for a network composed of 10 oscillators for $\psi = \pi/4$. The convergence toward constant amplitude in-phase–locked oscillations is evident.

The same approach allows the analysis of more complex architectures, including two-dimensional networks, with long-range, nonsymmetric interactions.[27] Other phase-locked states, corresponding to complex collective behaviors, can also be investigated. For example, a travelling wave is a solution such that

$$\varphi_i(t) = \Omega t + \frac{2\pi}{N} i \qquad (17.21)$$

which implies

$$\chi = \frac{2\pi}{N} \qquad (17.22)$$

Figure 17.11 shows the evolution toward a phase-locked state corresponding to a traveling wave in the network.

In real-world applications the constituent oscillators are not expected to be perfectly identical, and it is natural to ask whether small statistical fluctuations in the parameter values have strong influence on the synchronization properties of the network. Let us start from the case of identical values of the parameters $\mu_i = \bar{\mu}$. Away from the bifurcation curves, the implicit function theorem guarantees the existence of a branch of equilibria for a narrow distribution of the μ_i in the neighborhood of $\bar{\mu}$, corresponding to a phase-locked state $\bar{\chi}$ close to χ. Thus, we expect the existence of phase-locked states even in the case of nearly identical oscillators, an example of which is shown in Figure 17.12.

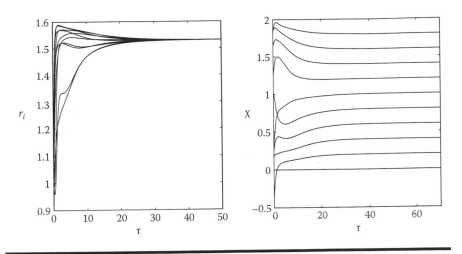

Figure 17.11 Traveling wave in a network composed of 10 oscillators. Evolution of the amplitudes (left) and the phase differences (right).

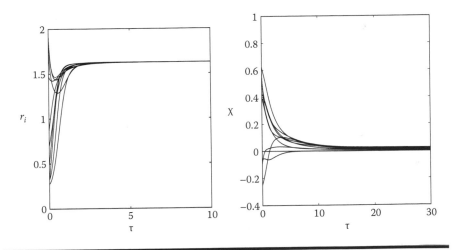

Figure 17.12 In-phase locking for a network of nearly identical oscillators. Evolution of the amplitudes (left) and of the phase differences (right). The parameter values have Gaussian distribution with 10% variance.

17.4 Pattern Recognition by Means of Weakly Connected Oscillatory Networks

This section is devoted to illustrate how pattern recognition tasks can be faced by exploiting weakly connected networks composed of structurally stable oscillators. The DFMT technique is applied to weakly connected networks having

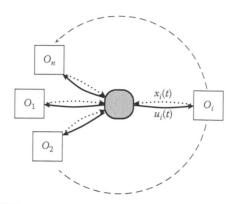

Figure 17.13 **Weakly connected oscillatory network having a star topology.**

a *star topology*[28] (see Figure 17.13). Such network architecture, mimicking the thalamus-centric organization of the brain, is functionally equivalent to a fully connected network (each cell is directly connected to all the others) but its hardware implementation is easier because a smaller number of connections are required.

The oscillators O_i ($1 \le i \le N$) interact only through the central complex cell O_0, named *master cell*, which supplies the signal $\mathbf{G}_i(\mathbf{X}_0, \mathbf{X})$ to each cell, where $\mathbf{G}_i : \mathbb{R}^{n \times (N+1)} \to \mathbb{R}^n$ and $\mathbf{X}_0 \in \mathbb{R}^n$ is the state vector of the master cell whose dynamics is described by $\dot{\mathbf{X}}_0 = \mathbf{F}_0(\mathbf{X}_0)$. It follows that Star WCONs (SWCONs) composed of N cells driven by one master cell are described by ($0 \le i \le N$):

$$\dot{\mathbf{X}}_i = \mathbf{F}_i(\mathbf{X}_i) + \varepsilon\, \mathbf{G}_i(\mathbf{X}_0, \mathbf{X}) \tag{17.23}$$

where $\mathbf{G}_0(\mathbf{X}_0, \mathbf{X}) = \mathbf{0} \in \mathbb{R}^n$.

Without losing any generality let us assume that:

a. The coupling between each cell and the master cell involves only the first scalar variables x_i of the state vector $\mathbf{X}_i$.

b. The master cell collects the signals x_i ($1 \le i \le N$) and provides to each oscillator a corresponding signal defined as ($1 \le i \le N$):

$$\mathbf{G}_i(\mathbf{X}_0, \mathbf{X}) = \left(g_i(x_0, \mathbf{x}), 0, \ldots, 0\right)^T \in \mathbb{R}^n \tag{17.24}$$

where $\mathbf{x} = (x_1, x_2, \ldots, x_N)$.

As a consequence of $\mathbf{G}_0(\mathbf{X}_0, \mathbf{X}) = \mathbf{0}$ we have $g_0(x_0, \mathbf{x}) = 0$.

As shown in Section 17.2, the global dynamical behavior of WCONs can be analytically investigated by means of the DFMT technique.[15,16] By applying steps (1)–(3) to SWCONs, whose cells belong to a wide class of oscillators, the following phase equation is obtained

$$\phi'_i = V(\omega_i) \int_0^T \cos(\omega_i t)\, g_i\left(\hat{x}_0\,(t+t_0),\dots,\hat{x}_i(t),\dots,\hat{x}_N\,(t+t_N)\right) dt \quad (17.25)$$

where $V(\omega_i)$, like in Equation 17.9, depends on the properties of the oscillators and the periodic trajectory $\gamma_i(t)$ considered, $\omega_k t_k = \phi_k - \phi_i$ for all $k = 0, 1, \dots, N$, and $\hat{x}_k(t) = A_k + B_k \sin(\omega_i t)$ is the first harmonic approximation of $x_k(t)$. The solution of the algebraic equations derived by the describing function method (step (1) of the DFMT technique) provides the bias A_k, the amplitude of the first harmonic B_k, and the angular frequency ω_k. Furthermore, the assumption $g_0(x_0, \mathbf{x}) = 0$ implies $\phi'_0 = 0$, i.e., the oscillators O_i do not influence the phase of the master cell O_0.

Hereinafter, to avoid cumbersome calculation we consider SWCONs in which the N oscillators and the master cell are identical, i.e., all the angular frequencies $\omega_0, \omega_1, \dots, \omega_N$ arc identical ($\omega_0 = \omega_1 = \cdots = \omega_N = \omega$).

By denoting $\phi_i - \phi_0 = \Phi_i = \omega t'_i$, the change of variable $t' = t - t_0$ permits to rewrite (17.25) in a compact form:

$$\Phi'_i = V(\omega) \left[\mathfrak{J}_i^A(\mathbf{\Phi})\, \cos(\Phi_i) - \mathfrak{J}_i^B(\mathbf{\Phi})\, \sin(\Phi_i) \right] \quad (17.26)$$

where $\mathbf{\Phi} = [\Phi_1, \dots, \Phi_i, \dots, \Phi_N]$ and

$$\mathfrak{J}_i^A(\mathbf{\Phi}) = \int_0^T \frac{\cos(\omega t')}{T} g_i\left(\hat{x}_0(t'),\dots,\hat{x}_i\left(t'+t'_i\right),\dots,\hat{x}_N\left(t'+t'_N\right)\right) dt'$$

$$(17.27)$$

$$\mathfrak{J}_i^B(\mathbf{\Phi}) = \int_0^T \frac{\sin(\omega t')}{T} g_i\left(\hat{x}_0(t'),\dots,\hat{x}_i\left(t'+t'_i\right),\dots,\hat{x}_N\left(t'+t'_N\right)\right) dt'$$

$$(17.28)$$

It is worth noting that functions $g_i(\cdot)$, which establish how cells O_i ($1 \le i \le N$) are interconnected, also specify the dynamical properties of the phase Equations 17.26 through the coefficients $\mathfrak{J}_i^A(\mathbf{\Phi})$ and $\mathfrak{J}_i^B(\mathbf{\Phi})$.

In pattern recognition tasks, we are mainly interested in the steady-state solutions of (17.26) that define appropriate phase relations among the synchronized oscillators. A phase pattern $\bar{\mathbf{\Phi}}$ is an equilibrium point of (17.26) if and only if the following condition is satisfied:

$$\mathfrak{J}_i^A(\bar{\mathbf{\Phi}})\, \cos(\bar{\Phi}_i) = \mathfrak{J}_i^B(\bar{\mathbf{\Phi}})\, \sin(\bar{\Phi}_i) \quad \forall\, i = 1, 2, \dots, N \quad (17.29)$$

The stability properties of $\bar{\mathbf{\Phi}}$ can be investigated by analyzing the eigenvalues of the Jacobian of (17.26) evaluated in $\bar{\mathbf{\Phi}}$. We observe that condition (17.29) is fulfilled if

$$\mathfrak{J}_i^A(\bar{\mathbf{\Phi}}^k) = 0 \quad \forall\, i = 1, 2, \dots, N \quad (17.30)$$

for all the 2^N phase patterns $\bar{\boldsymbol{\Phi}}^k = \left[\bar{\Phi}_1^k, \bar{\Phi}_2^k, \ldots \bar{\Phi}_N^k \right]$ with $k = 1, 2, \ldots, 2^N$ and $\bar{\Phi}_i^k \in \{0, \pi\}$.

It is easily shown[29] that the condition (17.30), implying that the oscillators are in phase or anti-phase, holds if, for all $i = 1, \ldots, N$, the function $g_i(\cdot)$ is odd with respect to its ith argument:

$$g_i(x_0, x_1, \ldots, -x_i, \ldots, x_N) = -g_i(x_0, x_1, \ldots, x_i, \ldots, x_N) \qquad (17.31)$$

The condition (17.31) allows us to exploit the phase of each oscillator as a single bit of binary information, e.g., 0 and π correspond to $+1$ and -1 (or 1 and 0), respectively. Hence, the SWCON (17.23) together with (17.26) works as an information storage device that represents patterns through the relative phases of the oscillators in the synchronized state.

The following subsections show how the couplings among the oscillators, i.e., functions $g_i(\cdot)$, can be designed according to the condition (17.31) in order to realize WCONs acting as associative or dynamic memories:

■ Associative memory task: Given an input phase pattern (initial condition of the phase equation), the WCON output converges to the closest stored pattern.

■ Dynamic memory task: Given an input phase pattern (initial condition of the phase equation), the WCON output can travel around stored and/or spurious patterns.

17.4.1 WCON-Based Associative Memories

It is well known that nonlinear oscillatory networks can behave as Hopfield memories, whose attractors are limit cycles instead of equilibrium points.[14,30] The SWCON given in (17.23) performs an associative memory by considering that the master cell provides a linear interconnection among cells O_i but does not influence their dynamics, i.e., $g_i(\cdot)$ does not depend on its first argument:

$$g_i(x_0(t), x_1(t), \ldots, x_N(t)) = \sum_{j=1}^{N} C_{ij} x_j(t) \qquad (17.32)$$

It follows that (17.26) can be written as

$$\Phi_i' = \sum_{j=1}^{N} \frac{1}{2} s_{ij} \sin\left(\Phi_j - \Phi_i\right) \qquad (17.33)$$

with $s_{ij} = C_{ij} V(\omega) B_j$. It is readily derived that phase configurations such that $(\Phi_j - \Phi_i) \in \{0, \pi\}$ are equilibrium points of (17.33), i.e., in-phase or anti-phase–locked states are admissible steady-state solutions. This agrees with the assumption that function (17.32) satisfies condition (17.31).

By remembering that each limit cycle (either stable or unstable) of the WCON corresponds to an equilibrium point of the phase equation, the coefficients s_{ij} can be designed according to a simple learning rule in order to store a given set of phase patterns.[14]

The parameters C_{ij} of the coupling function (17.32), which defines the interactions among the cells of the SWCONs (17.23), can be designed according to the simple hebbian learning rule[31]

$$C_{ij} = \frac{1}{V(\omega) \, B_j \, N} \sum_{k=1}^{p} \cos\left(\bar{\Phi}_i^k\right) \cos\left(\bar{\Phi}_j^k\right) \qquad (17.34)$$

In order to store a given set of p phase patterns we consider $\bar{\Phi}^k$ with $1 \leq k \leq p$. These phase patterns are in one-to-one correspondence with the following binary patterns:

$$\psi^k = \left[\psi_1^k, \psi_2^k, \ldots, \psi_n^k\right] \qquad \psi_i^k \in \{\pm 1\} \qquad 1 \leq k \leq p \qquad (17.35)$$

where $+1$ and -1 mean $\bar{\Phi}_i^k = 0$ and $\bar{\Phi}_i^k = \pi$, respectively. Thus, $\psi_i^k = \psi_j^k$ and $\psi_i^k = -\psi_j^k$ if the ith and jth oscillators are in-phase or anti-phase, respectively. It follows that ψ_i^k can be written as (for all $i = 1, 2, \ldots, N$)

$$\psi_i^k = \cos\left(\bar{\Phi}_i^k\right) \qquad \bar{\Phi}_i^k \in \{0, \pi\} \qquad (17.36)$$

and the ith and jth oscillators are:

■ In-phase if $\cos\left(\bar{\Phi}_i^k\right) \cos\left(\bar{\Phi}_j^k\right) = \cos\left(\bar{\Phi}_i^k - \bar{\Phi}_j^k\right) = +1$

■ Anti-phase if $\cos\left(\bar{\Phi}_i^k\right) \cos\left(\bar{\Phi}_j^k\right) = \cos\left(\bar{\Phi}_i^k - \bar{\Phi}_j^k\right) = -1$

Figure 17.14 shows a set of binary phase patterns for a SWCON composed of 25 (except the master cell) identical cells and cast into a regular grid with five rows and five columns. The binary value ψ_i^k is depicted as dark grey or pale grey if its value is $-1 = \cos\left(\bar{\Phi}_i^k\right)\big|_{\bar{\Phi}_i^k = \pi}$ or $+1 = \cos\left(\bar{\Phi}_i^k\right)\big|_{\bar{\Phi}_i^k = 0}$, respectively.

Once the learning process is completed, by choosing randomly the initial synaptic weights, the coefficients $s_{ij} = C_{ij} V(\omega) B_j$ so obtained are used in Equation 17.33 to solve the pattern recognition tasks.

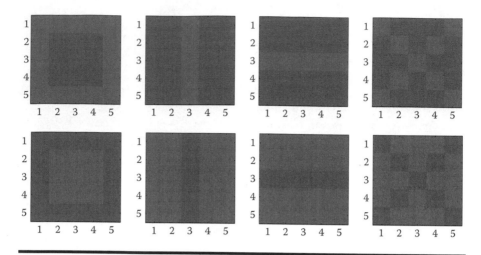

Figure 17.14 Stored phase patterns (upper row) and reversed stored phase patterns (lower row) for the Kuramoto-like model (17.33) designed with the hebbian learning rule (17.34)). From left to right, each pair of patterns (the original and the reversed one) is denoted by $\psi^{\pm 1}$, $\psi^{\pm 2}$, $\psi^{\pm 3}$, and $\psi^{\pm 4}$, respectively.

Figures 17.15 and 17.16 show the evolution, starting from a given initial phase configuration, of the phases governed by (17.33) and designed according to (17.34) for storing the binary patterns shown in Figure 17.14. In particular, Figure 17.15 presents the case in which the initial phase configuration converges toward the closest stored pattern, that is, the phase equilibrium point with the phase relations defined by ψ^2. It follows that the SWCON (17.23) presents a limit cycle whose components, corresponding to the state variables of each oscillator, are only in-phase or anti-phase in accordance with the phase relations specified by ψ^2.

Figure 17.16 shows that (17.33) can also generate and retrieve spurious phase patterns. In such a case, the SWCON exhibits a corresponding limit cycle whose components have shifted according to the phase relation given by the spurious pattern. Extensive numerical simulations have pointed out that the storage capacity of the SWCON-based associative memory is equivalent to that of the Hopfield memory.

17.4.2 WCON-Based Dynamic Memories

Despite establishing in the previous subsection that WCONs can store and retrieve oscillatory patterns, consisting of periodic limit cycle with suitable phase relations among the oscillators (due to the one-to-one correspondence between the equilibrium points of the phase equations and the limit cycles of the WCON), the state variables $x_i(t)$ are not binary. The binary information is embedded in the phase of $x_i(t)$, which can be 0 or π.

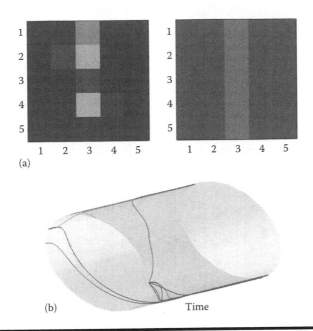

Figure 17.15 **(a) Phase pattern to be recognized given as initial condition of (17.33) designed according to (17.34) left. Recognition of a stored phase pattern right. (b) Evolution of the trajectory described by (17.33) and (17.34).**

In order to find out the phase relations by simply looking at the binary outputs of the WCON we can consider the sign of $x_i(t)$, i.e., we assume that the single oscillator's output is given by $sgn(x_i(t))$[28] ($sgn(\cdot)$ denotes the sign function). This allows us to conclude that two oscillators, described by $x_i(t)$ and $x_j(t)$, are in-phase (anti-phase) if and only if $sgn[x_i(t)\,x_j(t)] > 0$, $\forall t \geq 0$ ($sgn[x_i(t)\,x_j(t)] < 0$, $\forall t \geq 0$). If this rule is not fulfilled, then the WCON output consists of a sequence of binary patterns, thereby implying that phase relations are not 0 or π. Hence, the WCON output can travel around stored and/or spurious patterns (i.e., it acts as a dynamic memory) if there exists at least one $k \in \{1, 2, \ldots, N\}$ such that $\Phi_i^k \neq 0$ or $\bar{\Phi}_i^k \neq \pi$.

Dynamic memories can be easily obtained by considering SWCONs in which the master cell combines the output of each oscillator as follows:

$$g_i(x_0(t), x_1(t), \ldots, x_n(t)) = \left(\sum_{j=1}^{n} C_{ij}\left(x_j(t)\right) \right) |x_0(t)| - x_i(t) \qquad (17.37)$$

Some algebraic manipulations[29] lead to the following phase equation for SWCONs (17.23) with the coupling functions (17.37):

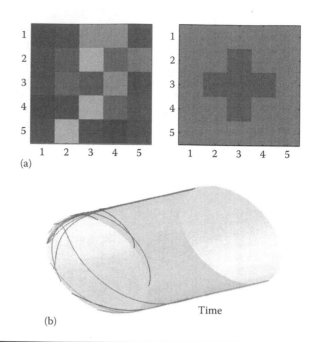

(a)

(b)

Time

Figure 17.16 (a) Phase pattern to be recognized given as initial condition of (17.33) designed according to (17.34) left. Recognition of a spurious phase pattern right. (b) Evolution of the trajectory described by (17.33) and (17.34).

$$\Phi'_i = \sum_{j=1}^{N} s_{ij} \, \mathfrak{C}^A_{ij}(\Phi_j) \cos \Phi_i - \sum_{j=1}^{N} s_{ij} \, \mathfrak{C}^B_{ij}(\Phi_j) \sin \Phi_i + \tilde{s} \, \sin(\Phi_i) \qquad (17.38)$$

with $\tilde{s} = \frac{V(\omega) \, B_i}{2}$, $s_{ij} = C_{ij} \, V(\omega) \, B_j$ and

$$\mathfrak{C}^A_{ij}(\Phi_j) = \frac{1}{2\pi} \left[1 - \cos(2\Phi_j) \right] p(\Phi_j) \qquad (17.39)$$

$$\mathfrak{C}^B_{ij}(\Phi_j) = \frac{1}{2\pi} \left[\pi - m_{2\pi}(2\Phi_j) + \sin(2\Phi_j) \right] p(\Phi_j) \qquad (17.40)$$

where $m_{2\pi}(2\eta_j)$ represents the modulus after division function, i.e., $m_{2\pi}(2\eta_j) = 2\eta_j - 2\pi \left\lfloor \frac{\eta_j}{\pi} \right\rfloor$ and $p(\eta_j)$ is a square wave with period 2π defined as:

$$p(\Phi_j) = \begin{cases} +1, & \Phi_j \in [0, \pi) \\ -1, & \Phi_j \in [\pi, 2\pi) \end{cases}$$

Figure 17.17 shows the coefficients $\mathfrak{C}^A_{ij}(\Phi_j)$ and $\mathfrak{C}^B_{ij}(\Phi_j)$ as a function of Φ_j. It is important to point out that $\mathfrak{C}^A_{ij}(\Phi_j)$ and its derivative are null for $\Phi_j = q\pi$

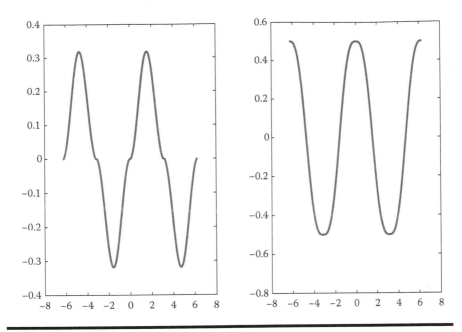

Figure 17.17 Coefficients $c_{ij}^A(\Phi_j)$ (left) and $c_{ij}^B(\Phi_j)$ (right) versus $\Phi_j \in [-2\pi, 2\pi]$.

with $q = 0, \pm 1, \pm 2, \ldots$ It follows that, as $c_{ij}^A(0) = c_{ij}^A(\pi) = 0$, all the 2^N phase patterns $\bar{\Phi}^k = \left[\bar{\Phi}_1^k, \bar{\Phi}_2^k, \ldots, \bar{\Phi}_N^k \right]$, with $k = 1, 2, \ldots, 2^N$ and $\bar{\Phi}_i^k \in \{0, \pi\}$, are equilibrium points of the phase deviation Equation 17.38. This agrees with the assumption that (17.37) satisfies the condition (17.31).

The parameters C_{ij} of the coupling function (17.37) can be designed, as in the previous subsection, according to the hebbian learning rule (17.34) for storing the binary patterns shown in Figure 17.17. Once the parameters C_{ij} are obtained, the coefficients $s_{ij} = C_{ij}V(\omega)B_j$ in (17.38) are readily derived. The calculation of the Jacobian of (17.38) in $\bar{\Phi}^k$, with $k = 1, 2, \ldots, 2^N$ and $\bar{\Phi}_i^k \in \{0, \pi\}$, shows that most of them are unstable.[29] Furthermore, it is shown[29] that the stored patterns $\psi^1, \pm\psi^2, \pm\psi^3$, and $\pm\psi^4$ are unstable, whereas $-\psi^1$ is stable.

As a consequence, if the initial conditions are chosen in a small neighborhood of ψ^3, the phase deviation Equation 17.38 converges toward an equilibrium point with phase components different from 0 or π (see Figure 17.18). In such a case the steady-state behavior of the corresponding SWCON is a periodic limit cycle, whose components do not oscillate in-phase or anti-phase. Since the output of each oscillator is given through the sign function, the SWCON output travels around a finite sequence of binary stored and spurious patterns. It turns out that the SWCON (17.23) with the connecting functions (17.37) can operate as a dynamic memory.

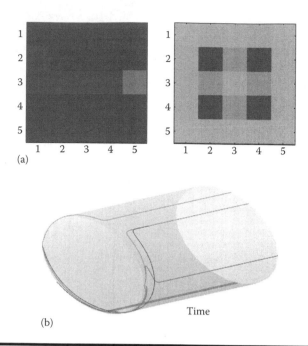

(a)

(b)

Time

Figure 17.18 **(a) Phase pattern to be recognized given as initial condition of (17.38) and designed according to (17.34) left. Recognition of a spurious phase pattern having components different from 0 or π right. (b) Evolution of the trajectory described by (17.38) and (17.34).**

Figure 17.19 shows the evolution of the binary output of a SWCON composed of 25 Chua's oscillators.[32] The simulation has been carried out by choosing the initial conditions for each Chua's cell and the single-cell parameters in such a way that each cell operates on the symmetric limit cycle (i.e., $A_i = 0$). A coupling strength on the order of $\epsilon = 0.01$ is sufficient for having synchronized binary outputs with phase shifts different from 0 and π. It is seen that the binary output of the SWCON wanders for a finite sequence of patterns as shown in Figure 17.19.

17.5 Conclusions

In this chapter we have presented some analytical techniques for investigating the emergence of synchronous or entrained oscillations in nonlinear oscillatory networks. In particular, we have shown how the equations governing the dynamical evolution of the network can be recast into either a phase equation, in which each oscillator is only described by its phase, or an amplitude–phase equation, where each oscillator is described by both its amplitude and its phase.

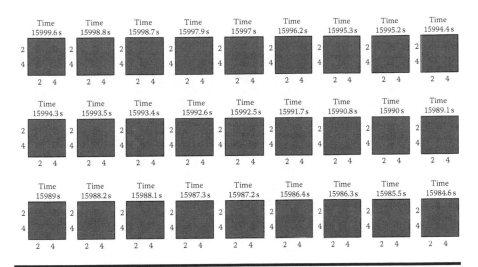

Figure 17.19 **Finite sequence of binary output patterns of a SWCON composed of 25 Chua's oscillators (operating on the symmetric limit cycle) and connected through the master cell which provides to each oscillator the signal given by (17.37).**

Finally, the one-to-one correspondence between the equilibrium points of the phase equation and the limit cycles of the WCON has allowed to store and retrieve oscillatory patterns consisting of periodic limit cycle with suitable phase relations among the oscillators. This has allowed us to realize associative and dynamic memories based on WCONs by designing the equilibria of the phase equation.

Acknowledgments

This research was partially supported by the *Ministero dell'Istruzione, dell'Università e della Ricerca*, under the FIRB project no. RBAU01LRKJ, by the *CRT Foundation*, and by *Istituto Superiore Mario Boella*.

References

1. A. Destexhe and T. J. Sejnowski, Interactions between membrane conductances underlying thalamocortical slow–wave oscillations, *Physiological Review*. **83**, 1401–1453 (2003).
2. J. J. Hopfield, Neural networks and physical systems with emergent collective computational abilities, *Proceedings of National Academy of the Sciences*. **79**, 2554–2558 (1982).

3. E. R. Kandel, J. H. Schwartz, and T. Jessel, *Principles of Neural Science* (McGraw-Hill, New York, 2000).
4. Y. Kuramoto, *Chemical Oscillations, Waves and Turbulence* (Springer, New York, 1984).
5. C. S. Peskin, *Mathematical Aspects of Heart Physiology* (Courant Institute of Mathematical Science Publication, New York, 1975).
6. C. Liu, D. R. Weaver, S. H. Strogatz, and S. M. Reppert, Cellular construction of a circadian clock: Period determination in the suprachiasmatic nuclei, *Cell.* **91**, 855–860 (1997).
7. Special issue devoted to the binding problem, *Neuron.* **24** (September 1999).
8. C. M. Gray, P. König, A. K. Engel, and W. Singer, Oscillatory responses in cat visual cortex exhibit inter–columnar synchronization which reflects global stimulus properties, *Nature.* **338**, 334–337 (1989).
9. T. B. Schillen and P. König, Binding by temporal structure in multiple features domains of an oscillatory neuronal network, *Biological Cybernetics.* **70**, 397–405 (1994).
10. Special issue on nonlinear waves, patterns and spatio-temporal chaos in dynamic arrays, *IEEE Transactions on Circuits and Systems: Part I.* **42** (1995).
11. A. Pikowsky, M. Rosenblum, and J. Kurths, Phase synchronization in regular and chaotic systems, *International Journal of Bifurcation and Chaos.* **10**(10), 2291–2305 (2000).
12. A. Pikowsky, M. Rosenblum, and J. Kurths, *Synchronization* (Cambridge University Press, Cambridge, U.K., 2001).
13. A. T. Winfree, *The Geometry of Biological Time* (Springer, New York, 1980).
14. F. C. Hoppensteadt and E. M. Izhikevich, *Weakly Connected Neural Networks* (Springer, New York, 1997).
15. M. Gilli, M. Bonnin, and F. Corinto, On global dynamic behavior of weakly connected oscillatory networks, *International Journal of Bifurcation and Chaos.* **15**(4), 1377–1393 (2005).
16. M. Bonnin, F. Corinto, and M. Gilli, Periodic oscillations in weakly connected cellular nonlinear networks, *IEEE Transactions on Circuits and Systems—I: Regular Papers.* **55**(6), 1671–1684 (July 2008).
17. M. Gilli, F. Corinto, and P. Checco, Periodic oscillations and bifurcations in cellular nonlinear networks, *IEEE Transactions on Circuit and Systems—I: Regular Papers.* **51**(5), 948–962 (May 2004).
18. K. Bar-Eli, On the stability of coupled chemical oscillators, *Physica D.* **14**, 242–252 (1985).
19. D. G. Aronson, G. B. Ermentrout, and N. Kopell, Amplitude response of coupled oscillators, *Physica D.* **41**, 403–449 (1990).
20. B. D. Hassard, N. D. Kazarinoff, and Y. H. Wan, *Theory and Applications of Hopf Bifurcation* (Cambridge University Press, Cambridge, 1981).
21. J. Guckenheimer and P. Holmes, *Nonlinear Oscillations, Dynamical Systems and Bifurcations of Vector Fields*, 5th edn. (Springer, New York, 1997).
22. M. Bonnin, F. Corinto, and M. Gilli, Bifurcations, stability and synchronization in delayed oscillatory networks, *International Journal of Bifurcation and Chaos.* **17**(11), 4033–4048 (2007).

23. J. Hale, *Theory of Functional Differential Equations* (Springer, New York, 1977).
24. F. C. Hoppensteadt and E. M. Izhikevich, Synchronization of MEMS Resonators and Mechanical Neurocomputing, *IEEE Transactions on Circuit and Systems—I: Fundamental Theory and Applications.* **48**(2), 133–138 (2001).
25. S. Smale, A mathematical model of two cells via Turing's equation, *Lectures in Applied Mathematics.* **6**, 15–26 (1974).
26. P. J. Davis, *Circulant Matrices* (John Wiley and Sons, New York, 1979).
27. M. Bonnin, F. Corinto, M. Gilli, and P. Civalleri. Waves and patterns in delayed lattices. In *Proceedings of the 2008 IEEE Symposium on Circuits and Systems*, Seattle, Washington (2008).
28. M. Itoh and L. O. Chua, Star cellular neural networks for associative and dynamic memories, *International Journal of Bifurcation and Chaos.* **14**, 1725–1772 (2004).
29. F. Corinto, M. Bonnin, and M. Gilli, Weakly connected oscillatory networks for pattern recognition via associative and dynamic memories, *International Journal of Bifurcation and Chaos.* **17**(12), 4365–4379 (2007).
30. F. C. Hoppensteadt and E. M. Izhikevich, Oscillatory neurocomputers with dynamic connectivity, *Physical Review Letters.* **82**(14), 2983–2986 (1999).
31. D. O. Hebb, *The Organization of the Behaviour* (Wiley, New York, 1949).
32. R. N. Madan, *Chua's Circuit: A Paradigm for Chaos* (World Scientific Series on Nonlinear Science, Singapore, 1993).

Chapter 18

Modeling the Dynamics of Cellular Signaling for Communication Networks

Jian-Qin Liu and Kenji Leibnitz

Contents

In this chapter, we discuss models of the dynamics of cellular signaling by characterizing their essential features in terms of dynamical network graphs. It is shown that the classical models in molecular information processing correspond to graph rewriting and graph rewiring. Inspired by the cellular signaling pathway networks, we investigate the dynamical features and illustrate how this understanding can be a step toward building more robust and resilient information networks in dynamical environments.

18.1 Introduction

In the past, a lot of research has been dedicated to studying and understanding the dynamical processes involved in the cellular signaling process [1]. Previous studies [2–11] have shown that a highly dynamical behavior is exhibited in cellular communication [12] and that the overall operation results in a high stability and robustness. Such features are also expected in the design of distributed information network infrastructures in future Internet topologies. In this chapter, we will summarize models for cellular signaling and derive our own, more tractable mathematical model of self-configuration for dynamical networks that will facilitate the dynamic control in information networks.

With the form of biological networking, cellular signaling processes carry out the biological function of cell communications at the molecular level. In this chapter, *cellular signaling* refers to the process of molecular signal transduction in biological cells, where the corresponding biochemical reactions are studied in terms of signaling pathway networks or cellular signal transduction networks, which we will unify under the term *pathway networks*. From the viewpoint of informatics, we would like to model the cellular signaling processes in order to understand how the pathway networks in cells perform robust communication to give us inspiration for designing robust information and communication networks.

Basically, the cellular communication can be distinguished into two types. Intercellular communication describes how cells interact with each other and depend

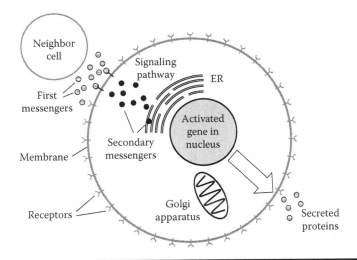

Figure 18.1   Cell communication model.

on the distance over which signaling molecules, also known as *first messengers*, are transmitted. The simplest forms are close-range signaling in which there is a direct membrane-to-membrane contact and paracrine signaling, where a local signal is diffused to neighboring cells and receptors on their membrane will either react to the incoming molecules or not. Over long distances, either synaptic signaling is performed at the axons of nerve cells or in the endocrine signaling hormones are distributed in the bloodstream to reach remotely located organs.

When an incoming first messenger molecule reaches a cell, it cannot penetrate the cell membrane, but is bound by specific receptors causing an activation of signaling proteins within the cell (see Figure 18.1). These proteins are referred to as *secondary messengers*. Once within the cell, the signals are relayed by a chemical reaction process, which results in a signaling cascade, relaying the signal to the nucleus. Once it is delivered there, a specific gene in the DNA is activated, causing the production and release of the protein it encodes. This resulting effect is referred to as *gene expression*.

Centered at the molecular information-processing mechanism of cellular signaling, the contents of this chapter are arranged as follows. In Section 18.2, the spatial and temporal dynamics of cellular signaling processes are discussed by using their quantitative descriptions. In Section 18.3, a dynamical graph representation scheme for cellular signaling processes is presented by integrating a graph representation with dynamical operations (graph rewriting). Then, this dynamical graph representation scheme is extended to the form of graph automata, where graph rewriting is embedded and the methodologies of cellular pathway networks are summarized. In Section 18.4, graph rewiring for self-configuration networking is presented as a case study of dynamic networking inspired by pathway networks. In Section 18.5, the robustness issue of cellular signaling is discussed based on two instances of cellular pathways, considering the fact that robustness is one of the major dynamical features of cellular signaling processes that are regarded as the biochemical basis of information-processing mechanisms of cellular signaling. The computational and numerical studies on robustness parameters of pathway networks discussed here can be regarded as the basis of modeling sustainable information networks in a dynamical environment.

18.2 The Dynamics of Cellular Signaling

The focus of this chapter lies on studying the dynamics of cellular signaling that takes the form of pathway networks in order to obtain a deeper understanding and inspiration from generic types of dynamical networks. Basically, the dynamics of cellular signaling can be investigated by two major aspects: *spatial dynamics* and *temporal dynamics*. In the following, we will summarize and discuss the state of the art in research of each of these approaches.

18.2.1 Spatial Dynamics

The study of spatial dynamics of cellular pathways is greatly enhanced by the advances in molecular imaging technology with high-resolution visualization of cells. Dynamical models and numerical simulations can provide us with a quantitative description of signaling molecule concentrations distributed within a cell. Thus, we can regard the spatial dynamics as the basis of the entire discussion of the cellular signal transduction using the methodologies from systems biology. Based on their background from molecular biology, cellular pathways in signal transduction are defined in terms of their biochemical functions. With parameters such as the locations of the signal molecules, we can have a close-up view of the processes of signal molecules within cells. Between two different locations, a signal molecule may travel in different ways, either by *active transportation* or *passive transportation*.

In active transportation, molecular motors actively carry so-called cargo molecules. On the other hand, diffusion is one possible way of passive transportation. It is driven by the random fluctuations following a Brownian motion and the resulting interactions among protein molecules [13]. Although these fluctuations are often perceived as inherent noise at various hierarchical levels of biological systems, they provide a fundamental function in the adaptation to external influences and the development of structures. Protein enzymes may interact through *kinase* when phospho-proteins are attached to a phosphate (*phosphorylation*) or when *phosphatase* detaches the phosphate from phosphorylated phospho-proteins (see Figure 18.2). Kholodenko [8] gives a quantitative description of the communication processes of

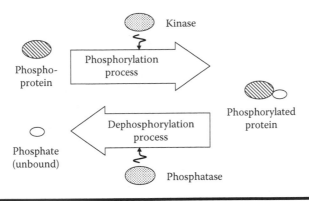

Figure 18.2 **Concept of the phosphorylation and dephosphorylation process.**

phospho-proteins where the gradients are analyzed for explanation of the communication processes regulated by kinases and phosphatases. Here, we start our discussion from a simplified representation of pathway networks based on the network's basic building blocks (referred to as *motifs*) and the pathways of kinases/phosphatases of GEFs/GAPs, which are called *universal motifs* [8]. GEF stands for *guanine exchange factor* and GAP for *GTPase activating proteins*, both playing a key role in the activation and deactivation of signaling proteins for signal transduction in cells.

For illustration purposes, let us consider a simplified example in Figure 18.3. Let X be the input and Y be the output molecules of a pathway, which could either be

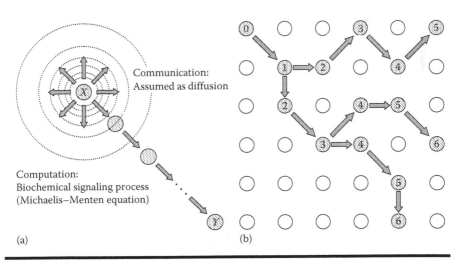

Figure 18.3 **Schematic illustration of a cellular signaling pathway. (a) Interaction among diffusion (communication) and reaction (computation). (b) Example snapshot of branching in tree traffic paths.**

indivisible or a *mitogen-activated protein kinase* (MAPK) cascade pathway. We have a cellular communication process from X to Y as illustrated in Figure 18.3a and signal molecules travel from X toward Y. A simple way to simulate this stochastic diffusion process is to use the Monte Carlo method. If the neighborhood of the signal molecules is defined by a two-dimensional lattice (e.g., Moore neighborhood in cellular automata), the correlation of the concentration values whose measures are spatially distributed provides an indication on the spreading process of molecular signaling, which exhibits a kind of *receive-and-forward* type of chemical relay [6].

The signal transmission process can thus be imagined as a broadcast algorithm under the physical constraints of a liquid medium and it may branch itself into many different traffic paths (see Figure 18.3b). Our focus lies on the features of network dynamics itself, so a network graph needs to be considered as the underlying data structure, which we will later elaborate in Section 18.3.

18.2.2 Temporal Dynamics

The temporal dynamics of signal transduction networks is usually formulated by differential equations, among which the well-known *Michaelis–Menten equation* is the most fundamental [14]. The Michaelis–Menten equation describes the reaction among molecules: a *substrate X* (input) to an *enzyme E*, resulting in a *product Y* (output) that is unbound to the enzyme (see Equation 18.1). The enzyme plays the role of catalyzer, which triggers the biochemical reaction and transforms the initial substrate into the resulting product.

$$X + E \underset{k_2}{\overset{k_1}{\rightleftarrows}} XE \overset{k_3}{\rightarrow} Y + E \tag{18.1}$$

The parameters $k_1, k_2,$ and k_3 describe the conversion rates and

$$k_m = \frac{k_2 + k_3}{k_1} \tag{18.2}$$

is defined as the *Michaelis constant*. In the following, we will denote the concentration of any chemical X as $[X]$, describing the number of molecules per unit volume. We can obtain a simple differential equation system for the kinetic dynamics of the product $[Y]$, where $[E_0]$ is the total concentration of the enzyme:

$$\frac{d[Y]}{dt} = k_3 [E_0] \frac{[X]}{k_m + [X]} \tag{18.3}$$

However, this assumption is only valid under quasi steady-state conditions, i.e., the concentration of the substrate-bound enzyme changes at a much slower rate than those of the product and substrate.

We can reformulate the Michaelis–Menten kinetics of a signal transduction pathway from Equation 18.3 in a more simplified and generalized form as

$$\frac{d[Y]}{dt} = f([X], \Phi) \tag{18.4}$$

where
the function $f(\cdot)$ corresponds to the right-hand side of Equation 18.3
Φ is the set of parameters, including $[E_0]$, k_3, and k_m

If a feedback from Y to X is embedded, we can express it by a new function $g(\cdot)$ as in Equation 18.5

$$\frac{d[X]}{dt} = f([X], \Phi) + \delta [Y] = g([X], [Y], \Phi) \tag{18.5}$$

where δ is the feedback factor from Y to X and can be positive or negative. The feedback factor could be useful for explaining the nonlinear dynamics mechanism of various types of pathway networks as instances of dynamical networks. However, we will focus here on MAPK cascades as one important instance of pathway networks.

MAPK cascades are usually considered in response to extracellular stimuli (*mitogens*), which trigger a cascade of phosphorylation finally resulting in a specific gene expression. In the case of MAPK, we have a system of differential equations as

$$\frac{d[Y_i]}{dt} = f([X_i], \Psi_i) \quad i = 1, 2, 3 \tag{18.6}$$

where
$f(\cdot)$ is again the function of the Michaelis–Menten kinetics
$[X_i]$ are the kinases for $i = 1, 2, 3$
$[Y_1]$ and $[Y_2]$ are the products (kinases) acting as an enzyme for the downstream pathway
$[Y_3]$ is the resulting phospho-protein product
Ψ_i are the set of influencing parameters and differ from Φ, previously used in Equations 18.4 and 18.5

As shown in Figure 18.4, we obtain

MAPKKK $[X_1]$

MAPKK $[Y_1] = [X_2]$

MAPK $[Y_2] = [X_3]$

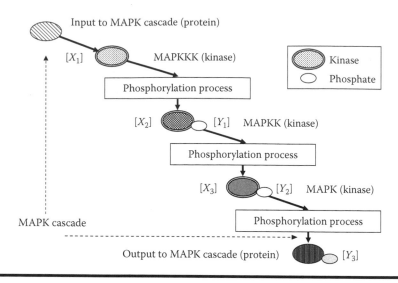

Figure 18.4   Three-layered pathway of MAPK cascade.

The feedbacks from downstream pathways to upstream pathways result in a combination of Equations 18.5 and 18.6 and can be given as follows:

$$\frac{d[X_i]}{dt} = f([X_i], \Psi_i) + \delta\,[Y_i] = g([X_i], [Y_i], \Psi_i) \quad i = 1, 2, 3 \qquad (18.7)$$

With the appropriate parameters, usually obtained from empirical observations, Equation 18.7 system permits a numerical computation of the MAPK cascade. Basically, the general behavior can be described as an amplification of the initial substrate concentration $[X_1]$, resulting in an enhanced signal with higher concentration of the resulting product $[Y_3]$.

In addition to the individual cascade pathways as expressed by the previous equations, it becomes imperative for large-scale cellular signaling networks to also take the interactions between pathways constructed by the network motifs into account, as well as crosstalks of signals from different cellular pathways, in order to model the cellular pathway network as a complex dynamic system. Therefore, Kholodenko [8] uses the diffusion equation with polar coordinates to formulate the concentration values of kinases in the MAPK cascade constrained by the distance of diffusion.

Differing from Kholodenko's formulation, we suggest a more generalized representation of pathway networks. Our proposed formulation includes the following properties:

■ An individual pathway is defined as a primitive $\langle [X], [Y], \Psi \rangle$.
■ The operators on pathway networks include feedback and crosstalk.

- The reaction and diffusion processes are modeled separately and diffusion may be a stochastic process.
- Only ordinary differential equations are employed in contrast to Kholodenko [8], who uses partial differential equations.

Another important factor in the Michaelis–Menten kinetics that we haven't discussed so far is *inhibition* [3,5]. This becomes obvious if we reveal the fact that the signaling molecules of such phospho-proteins in cells have two states: *activation* and *inactivation*. With the formulation from nonlinear dynamics, the schemes of mathematical modeling of cellular signaling could be indeed helpful to explain the nonlinear features of pathway networks as complex systems.

18.2.3 Concluding Remarks

In this section we studied various aspects of the cellular signaling mechanism and summarized some classical and well-established models. In general, we could see that the dynamics of pathway networks should take into account both the spatial and temporal dynamics, which are usually mathematically described by systems of differential equations. Additionally, the interactions among pathways in terms of feedback and crosstalk must also be considered. However, in order to provide a theoretically sound analysis, we need to first provide a suitable data representation structure to formalize the topological description of dynamical graphs on a macro level. Thus, the formalization of a data structure for dynamical graph representation will be the focus of the following section.

18.3 Dynamical Graph Representation of Cellular Signaling

Our focus in this chapter is on describing the information-processing mechanism of pathway networks based on the dynamical signaling mechanism of cellular communications with terms from computer science. Thus, our main intention is to extract abstract models from biochemical processes of pathway networks that can be applied to a broader range of dynamical networks. The methodology could be mathematical, computational, logical, etc. In recent years, the progress in *computational biology*, *bioinformatics*, and *systems biology* deserves great attention, owing to the fact that they are helpful for us to understand the complexities of biological systems in nature [15]. Here, the objective of our discussion is focused on pathway networks in biological cells. An intuitive representation could be a graph as the basic data structure for describing such pathway networks, where the nonlinear dynamical features we discussed in the previous sections are integrated within the network structure.

By formulating the dynamical operators on the corresponding network structure—normally on their topological structure—the dynamical graph representation scheme for pathway networks is given based on the spatial and temporal dynamics of the molecular mechanisms in biochemistry and biophysics. Therefore, the concept of dynamical graph representation is defined here as having an integration of graph representation as well as dynamical operators on these graphs.

Let us define a dynamical graph $\mathcal{D}$ as

$$\mathcal{D} = \langle \mathcal{G}, \mathcal{Q} \rangle$$

where
  $\mathcal{Q}$ is the set of operators
  $\mathcal{G}$ is the set of graphs defined as

$$\mathcal{G} = \langle \mathcal{V}, \mathcal{E} \rangle$$

with $\mathcal{V}$ being the set of vertices (nodes) and $\mathcal{E}$ the set of edges (links).

The graph data structure helps us to formalize the quantitative description of the changes that transform the dynamical network in a more accurate way. Consequently, we can formulate dynamical operators on the graphs that represent the pathway networks. Furthermore, when we consider the operations on sets of nodes with their connections, rather than individual nodes in a graph, we need to think about a generalized way of describing the structural variations of the dynamical network. So, the dynamical graph representation becomes the basis for formulating pathway networks using formal tools, such as *graph rewriting* and *graph automata*.

18.3.1 Formulating the Structural Variations of Pathway Networks by Graph Rewriting

From theoretical computer science, we may apply the concept of graph rewriting. Graph rewriting refers to the process in which two kinds of operators are defined:

- The insertion and/or deletion of *hyperedges*
- The insertion and/or deletion of *nodes*

A *hyperedge* is defined as a link that can connect to multiple nodes, i.e., a "macro-edge" consisting of several edges in a graph. The resulting structure is a generalized graph, called a *hypergraph*.

18.3.2 Representation of Cellular Signaling by Graph Automata

In addition to the dynamical features caused by the graph-rewriting operators on the nodes and links of networks, the topological structure and the states of dynamical networks are crucial for us to understand the graph evolution processes. Concerning the topological structure of dynamical networks, network theory provides us with the mathematical and engineering tools. However, regarding the internal states of dynamical networks, theoretical computer science can be used as the basis. Cellular automata, graph automata (in their classical definition), Petri nets, etc., are suitable to formulate the dynamical processes of updating the graph's states in a connectionist way. Recently, Sayama proposed a model called GNA [10] that integrates the graph formation processes represented in terms of state transitions with the operations on the topological structure of dynamical networks.

So, we now extend the notation for a set of dynamical networks $\mathcal{G}$ based on graph theory as follows:

$$\mathcal{G} = \langle \mathcal{V}, \mathcal{E}, \mathcal{S}, \mathcal{Q}, \mathcal{M} \rangle \tag{18.8}$$

where
$\mathcal{V}$ is the set of vertices
$\mathcal{E}$ is the set of edges
$\mathcal{S}$ is the set of states of the network
$\mathcal{Q}$ is the set of state-transition rules
$\mathcal{M}$ is the set of measures, e.g., measures given by information theory

Owing to the nature of cellular communications, the information theoretical measures of the dynamical pathway networks are important semaphores for their global and local states. A node can be regarded as source/sink and a link as channel for cellular communications. Consequently, the designed codes show us the *dynamical network information* in the sense of network informatics.

Example 18.1: (Phosphorylation/Dephosphorylation)

In the case of phosphorylation and dephosphorylation, the binary codes for signaling molecules in cells can be obtained by setting certain threshold levels of their concentrations. The *activation state* of phospho-proteins (phosphorylation) corresponds to bit "1" and the *inactivation state* (dephosphorylation) corresponds to bit "0" (see Figure 18.5a). The interacting pathways give rise to the combinatorial formation of binary codes [16]. In the case of the Rho-MBS-MLC pathway, the derived codes indicate the spatial configuration of the pathway network [17]. The three nodes are in this case MBS (Myosin-binding subunit), MLC (Myosin light chain), and a protein Rho (see Figure 18.5b). A rule set can then be defined as depicted in Figure 18.5c.

An example of state transitions and rewriting of graphs is illustrated in Figure 18.6. The dynamical changes from an initial state graph $G(t_0)$ to a state $G(t_1)$ during the dynamic transition processing can be expressed in three different ways:

- Adding the edge E_3 (graph rewriting)
- The state of node C is transformed from 1 to 0 (state transition)
- The state of graph $G(t_0)$ is changed into $G(t_1)$ (graph automata)

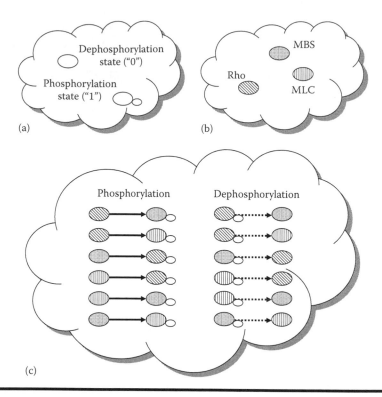

Figure 18.5 **Graph-theoretical representation of phosphorylation and dephosphorylation. (a) Set of states. (b) Set of nodes. (c) Set of rules.**

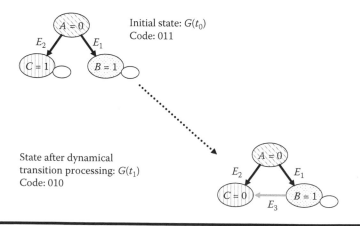

Figure 18.6 **An example of state transition and rewriting of graphs.**

18.3.3 *Methodologies of Pathway Networking*

The major factors of the computational analysis of pathway networks in cells are information, mathematics, and logic. Consequently, the methodologies of pathway networking derived from the above are summarized as

- Information-centric networks
- Formal systems in mathematics
- Logical models of network reconstruction

With the concept of network biology, it is feasible to formulate information networks based on biological evidence, i.e., pathway networks can be mapped into *information-centric networks*, where the data-centric operators and symbolic representations of the objects are formulated by links and nodes.

One of the issues we are directly facing is the computational complexity in space and time, which constrains the speed and memory of high-performance computing systems. This motivates us to study the formal system *mathematically* as the foundation for exploring the parallelism of pathway networks. In spite of the enormous amount of existing data obtained by the advancements in biological technologies, we are still only at an early stage in obtaining partial graphs of entire pathway networks for representative model organisms.

In order to obtain a systematic knowledge of the pathway networks in cells, a *logical way* to construct/reconstruct the pathway network appears to be the most promising approach. If we define "$\Rightarrow$" as the inference operator corresponding to a logical inference rule, e.g.,

IF X THEN Y

it is represented by $X \Rightarrow Y$, where X is the condition and Y its resulting conclusion. Generally speaking, we can distinguish the following three categories of logical methods:

- *Deduction* is the process of deriving a specific case C from a rule R, i.e., $R \Rightarrow C$.
- *Induction* is the process of inferring a rule R from the observation of specific cases C, i.e., $C \Rightarrow R$.
- *Abduction* allows explaining from a specific set $\{\langle R, C \rangle\}$ of cases C and rules R as a consequence of another set $\{\langle R', C' \rangle\}$ of cases C' and rules R', i.e., $\{\langle R', C' \rangle \Rightarrow \langle R, C \rangle\}$.

It is known that ontology has been applied to bioinformatics. For entire systems of pathway networks, the formal descriptions are expected to be made in terms of logical reasoning under the idea that pathway networks are formalized as logical models of network reconstruction. One of the ultimate goals of pathway informatics is to apply logical methods in the bioinformatics study on cellular communications, from which the network reconstruction technologies may lead to biologically inspired self-configuration technology toward the automatic generation of hypotheses of network functions.

18.4 Graph Rewiring Operations for Self-Configuration of Dynamical Networks

From the viewpoint of modeling entire networks for cellular signaling, an algorithmic study on dynamical networks becomes imperative. In a more general sense, it is reasonable to consider the same mathematical design tools in pathway networks as references for constructing information networks in a dynamical way. Since graph rewiring is one of the key operators in dynamical networks, in this section we will briefly discuss the graph rewiring operators for dynamical networks that are coined under the name graph rewiring operations oriented to self-configuration of dynamical networks.

Here, these operators directly change the topological configuration of the dynamical networks in a self-contained way, i.e., without interactive manipulation from an administrator. We expect to use it as a tool to explore the robustness of dynamical networks, which will be discussed in the next section. This will be helpful to develop biologically inspired networks for future communications [18]. Notice that the algorithm presented here is only limited to an abstract and generalized formulation, which not necessarily corresponds to existing biological networks.

18.4.1 Self-Configuration in Communication Networks

Self-configuration refers to the process that changes the structure of a system by using a mechanism within the system itself, where no external operators are required. It is widely applied in networking [19–22] or in the field of nano-devices [23]. One of the successful applications is in the topic of sensor and actor networks [24–26]. Bonnin et al. [27] mention that "the aspect of self-configuration is crucial—the network is expected to work without manual management or configuration."

In general, it is one of the key research goals to find self-configurable and self-adaptable mechanisms for autonomous network management in a fully distributed manner without the interaction of any centralized control entity. In some cases [18,28], the successful application of approaches inspired by biological phenomena has been reported.

In this work, we do not restrict ourselves to a specific communication network type, but rather address a more generic type of network and summarize the concept of self-configuration as follows [29].

Definition 18.1 (Self-Configuration) Self-configuration refers to the process that changes the network structure, that is, changing the nodes and links of a network (vertexes and edges of a graph that corresponds to the network) in a dynamical network according to a criterion (e.g., generating new nodes from existing nodes in a dynamical network).

18.4.2 Graph Rewiring Algorithm for Self-Configuring Dynamical Networks

One of the important aspects of self-configuration is how to dynamically update the nodes of a dynamically changing network. Here, we discuss a graph rewiring operation that generates nodes by using an information theoretic measure, the *mutual information* of codes as defined in Equation 18.9:

$$I_{i,j} = I(W_X, W_Y) = \sum_{k=0}^{K} \sum_{l=0}^{K} p(X_k, Y_l) \log \left(\frac{p(X_k, Y_l)}{p_X(X_k) \, p_Y(Y_l)} \right) \qquad (18.9)$$

where

X_k and Y_l represent the kth and lth bit of the binary codes W_X and W_k, respectively

$p(X_k, Y_l)$ is their joint probability

$p_X(X_k)$ and $p_Y(Y_l)$ are the marginal probabilities of W_X and W_Y

The self-configuration algorithm for construction of a network based on information theory is given in a pseudo-code formulation as follows.

Figure 18.7 illustrates the rewiring operator and the algorithm can be summarized by the following three major steps. First, a new node is generated from an existing node n_i out of a set of candidate neighboring nodes $\{n_j : j = 0, \ldots, J\}$. Then, the

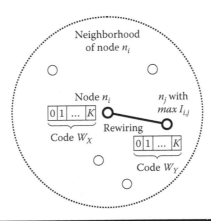

Figure 18.7 **Rewiring of nodes based on maximum mutual information of codes.**

Algorithm 18.1

---

1. $i \leftarrow 0$
2. Generate node n_i and initialize its concentration randomly using the Monte Carlo method.
3. WHILE $i < i_{max}$ DO

       generate the node set $\mathcal{N}_i = \{ n_j : j = 0, \ldots, J \}$ within the neighbor hood of node n_i

       FOR $j = 0$ TO J

           FOR $k = 0$ to K

           3.1 encode n_i by the kth bit of binary code W_X denoted as X_k

           3.2 calculate the concentration values of n_j by a nonlinear dynamics equation

           3.3 encode n_j by the kth bit of binary code W_Y denoted as Y_k, which is regarded as the output of the pathway

           3.4 calculate mutual information $I_{i,j} \leftarrow I(n_i, n_j)$ between nodes n_i and n_j

           ENDFOR k

           $j' \leftarrow \arg\max_{j} I_{i,j}$, i.e., the node is selected which maximizes the mutual information $I_{i,j}$

       ENDFOR j
4. $i \leftarrow j'$
5. ENDWHILE

---

node n_i and all the nodes in n_j from the candidate set are encoded as binary codes W_X and W_Y of length K, respectively. From the set of neighboring nodes the node j' is selected for rewiring that maximizes the mutual information $I_{i,j'}$. These steps will be iteratively repeated until the entire network is reconstructed in which mutual information is used as a measure, which we will discuss in greater detail in the following section.

18.4.3 *Information Theoretic Measures for Cellular Signaling Coding*

Revealing the dynamics of the cellular signaling processes that follow the Michaelis–Menten equation under quasi-steady-state conditions of the biochemical reactions, we can design a coding scheme for cellular signaling based on information theory as follows:

- Time is represented by the sample index within one cellular signaling process.
- The input of the biochemical reaction is the sender (Tx).

- The output of the biochemical reaction is the receiver (Rx).
- The signaling process represents the communication channel.
- The Michaelis–Menten equation is applied to calculate the input/output relation of the pathway.
- A threshold for the input and output concentration levels is set to generate binary codes according to their values.

A sample from simulations is shown in Figure 18.8a through c. We set the initial concentration value of the reactant, i.e., the input to the cellular signaling process, and the sum of enzymes as random values by using the Monte Carlo method. The total number of samples $i_{max} = 200$, which is sufficiently large to investigate the entire scenario of the corresponding signaling process. Figure 18.8a through c illustrates the simulated values of mutual information, the concentration value of the initial reactant, and the summation of enzymes, respectively. From Figure 18.8c we can observe that the communication capacity of 1 bit/pathway can be achieved by an appropriate parameter setting. This fact shows that the cellular

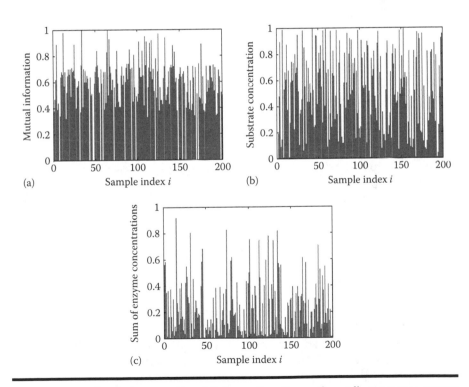

Figure 18.8 Simulation results of information theoretic coding process versus sample index. (a) Changes of mutual information. (b) Concentrations of substrate. (c) Sum of the concentrations of enzymes.

signaling process is indeed capable of transmitting 1 bit per pathway so that an information theoretical study of cellular signaling for potentially controlled cellular communication is meaningful, making an analysis of the performance of pathway networks with regard to the information flow feasible.

Furthermore, the channel capacity C can be obtained as the maximum mutual information, i.e.,

$$C = \max I(X, Y) \tag{18.10}$$

where X and Y refer to the sender and receiver, respectively.

This information theoretic view of pathway networks serves as the basis in designing dynamical operators for optimizing the channel by means of graph rewiring, graph rewriting, and even self-organizing of dynamical networks.

18.4.4 Concluding Discussion

Let us regard each node of a graph representing a pathway network as an abstract unit for information processing, regardless of the biochemical features of the pathway networks [30]. The relation between graph rewiring and pathway networks could then be unified by their self-configuration.

As we have seen above, the graph rewiring operators are performed on sets of nodes. From the data structure as graph, the next step should be the analysis of the topological structure of networks, where the effect of self-configuration can be observed. Through the graph representation of network motifs and specific measures, such as betweenness, centrality, or information theoretical metrics, they can be evaluated quantitatively [31]. Therefore, it is possible to apply graph rewiring also to explain the robustness of information networks [32] based on the graph evolution characteristics of dynamical networks found in cellular pathway networks. We will discuss robustness issues of cellular signaling in the following section.

18.5 Robustness of Pathway Networks

In the previous sections, we discussed the structural and dynamical operators on graphs in terms of network informatics. In summary, the measures from informatics for pathway networks such as graph rewriting and graph automata are necessary for modeling pathway networks. However, the parametric aspect of the cellular signaling dynamics is another important issue that needs to be taken into account. In contrast to Section 18.2, which explained the basic concepts and fundamentals of the dynamics of cellular signaling, this section discusses the quantitative form of pathway networks to study the nonlinear features of pathway networks. Robustness, in particular, could be one of such representative nonlinear features of pathway networks.

Our focus in this section is on the computational study of existing networks found in biological cells, from which we expect to find hints or inspirations to enhance our efforts on exploring unconventional approaches for new communication networks. As we have seen in the previous sections, the cellular signaling processes can be formulated in terms of bioinformatics. The information processing of *biological networks* may provide biologically inspired models for *information networking*. In particular, our focus lies on the following aspects of robustness:

- Features of biological robustness of cellular pathway networks in bioinformatics
- Quantitative descriptions of molecular mechanisms that support the biological robustness of cellular pathway models
- Theoretical models that may help us understand the mathematical principle of biological robustness of pathway networks

In this section, we will define the concept of robustness as given below.

Definition 18.2 (Robustness)   Robustness refers to a mechanism that can guarantee and realize the state transition of a (usually dynamic) system from vulnerable and unstable states to sustainable (stable) states when the system suffers from disturbance that is outside the environment and unexpected in most cases.

Although various types of examples can be found in molecular biology, two specific examples deserve our special attention. The *heat shock response* (HSR) pathway and the MAPK cascade of the phosphorylation of the Mos protein (Mos-P) are two examples to demonstrate biological robustness [7]. In both cases, the transition process can be described by the dynamic trajectory in phase space, where the system converges to attractors or fixed points. Attractors play an important role in dynamical systems and artificial neural networks, and are the underlying concept for the attractor selection mechanism, which is discussed in Chapter 14 [33].

18.5.1 The Heat Shock Response Pathway

Briefly speaking, in the case of the HSR pathway as shown in Figure 18.9, the protein folding is realized in a regular way in the cells when the temperature is normal (cf. Figure 18.9a). However, when the temperature in the environment suddenly reaches an extremely high level (shock), the HSR pathway will be activated to maintain the spatial structure of the normal protein-folding process, until the normal temperature is again restored. The biochemical process that sustains the robustness in cells shifts the state of the cellular signaling mechanism from inactive to active state of the HSR pathway. This process is formulated as a mathematical signaling process through state transitions of the dynamical system [9].

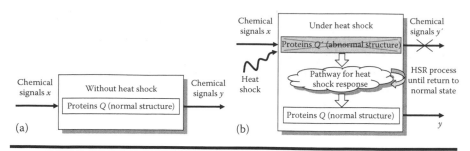

Figure 18.9 Comparison of normal pathway and heat shock response pathway. (a) Normal pathway. (b) Heat response shock pathway.

The HSR is performed as follows. Without the heat shock, the corresponding signaling proteins in the cells maintain their normal function within the three-dimensional structural folding. On the other hand, when suffering from a heat shock, the normal structure of the signaling proteins cannot be maintained, if no pathways are activated in the cells. In order to prevent any harmful effect caused by the heat shock, the pathways that carry out the HSR will be activated whenever a heat shock occurs.

18.5.2 The MAPK Pathway for Ultrasensitivity

The Mos-P pathway for transition between the stable and unstable steady states in cells is reported by Ferrell and Machleder [4]. This is another example to explain the robustness compared to the HSR pathway, however, with the difference that it includes feedback. Within the steady state of the pathway model, the quantitative relation between the input and output is described by the Michaelis–Menten equation. Outside these steady states, the formulation based on the Hill coefficient is applicable. Intuitively, it means that enzymes are changing themselves during the cellular signaling process. As Huang and Ferrell [2] reported, the nonlinear dynamics feature of different phosphorylation processes in an MAPK cascade varies, i.e., the phosphorylation concentration versus the time curve is different for each layer of an MAPK cascade. The *Hill coefficient*, which is normally defined in the *Hill equation*, is necessary for us to analyze the ultrasensitivity to explain the nonlinear phenomenon of the pathway response to the stimuli.

Intuitively, the question we would ask can be formulated as follows: *Is the steady-state point reachable in the cellular signaling system?* The system can transit to stable steady states from an unstable steady state when the system suffers from any perturbations from the outside. Here, the robustness mechanism refers to the pathway network structure (including the corresponding parameters) that realizes the aforementioned function.

The nonlinear system analysis methods for explaining the robustness mechanism of pathway networks varies depending on the studied discipline. For instance, in the theory of dynamical systems and automatic control, the research on the robust circuit design of biochemical networks by Chen et al. [11] is helpful for understanding the robustness of biochemical networks in terms of steady-state methods.

In its most general from, the concentration of Mos-P can be formulated as a fixed-point equation in the following way:

$$[\text{Mos-P}] = f([\text{Mos-P}])$$

which may give us some hint on the robustness analysis of future communication networks.

From the above brief discussion in terms of *system of systems* [34], we can see that the differential equations are necessary to formulate the robustness in a dynamical system, which is motivated from the biological robustness in pathway networks, but can be modified and extended to any abstract dynamical system where feedback is embedded.

18.6 Conclusion

In this chapter, we formalized the dynamics of pathway networks from the viewpoint of informatics to pave the path for applications to information and communication networks. In general, we can summarize that the spatial and temporal dynamics, which has been studied previously in the context of pathway networks, can be formulated in a detached way from the biomolecular signaling processes involved and the networks can be seen as abstract dynamic structures. Here, the methods from graph theory can be applied to describe the structural variations of dynamical networks by means of graph rewriting on nodes and links and graph automata for the state transition and dynamical changes of the network structure. Furthermore, graph rewiring can be considered as a technique to achieve self-configuration of dynamical networks and we provided an algorithmic approach using information theory, which proved that the dynamics of cellular signaling channels can be characterized by a channel capacity, very much like conventional communication network theory. The resulting network exhibits a high degree of robustness and we showed in the case of two example pathways how robustness is achieved in biological cells.

This study can be regarded as the starting point for a new paradigm of dynamical network formulation for self-configuration of communication networks. In the past, self-configuration and autonomous topology construction has been a key issue in information networks, with studies ranging in various fields, such as pervasive networks [35], next-generation mobile networks and autonomic computing [36], WiMAX [37], wireless sensor networks [19], and IPv6 [20]. Our goal in this chapter is to contribute to the further understanding of robustness of biological networks to reach out to new approaches, which can help in designing and operating robust and reliable communication networks.

Acknowledgments

The authors would like to thank Professor Masayuki Murata for his valuable comments and discussions. This work was partly supported by the *"Special Coordination Funds for Promoting Science and Technology: Yuragi Project"* of the Ministry of Education, Culture, Sports, Science and Technology in Japan.

References

1. B. D. Gomperts, I. M. Kramer, and P. E. Tatham, *Signal Transduction* (Academic Press, Burlington, MA, 2002).
2. C.-Y. F. Huang and J. E. Ferrell Jr., Ultrasensitivity in the mitogen-activated protein kinase cascade, *Proceedings of the National Academy of Sciences*. **93**, 10078–10083 (1996).
3. N. Sakamoto, P. de Atauri, and M. Cascante, Effects of feedback inhibition on transit time in a linear pathway of Michaelis–Menten-type reactions, *BioSystems*. **45**(3), 221–235 (1998).
4. J. E. Ferrell Jr. and E. M. Machleder, The biochemical basis of an all-or-none cell fate switch in xenopus oocytes, *Science*. **280**, 895–898 (1998).
5. C. G. Solo, M. A. Moruno, B. H. Havsteen, and R. V. Castellanos, The kinetic effect of product instability in a Michaelis–Menten mechanism with competitive inhibition, *BioSystems*. **56**(2–3), 75–82 (2000).
6. P. J. Thomas, D. J. Spencer, S. K. Hampton, P. Park, and J. P. Zurkus, The diffusion mediated biochemical signal relay channel. In eds. S. Thrun, L. Saul, and B. Schölkopf. In *Advances in Neural Information Processing Systems 16* (MIT Press, Cambridge, MA, 2004).
7. H. Kitano, Biological robustness, *Nature Reviews Genetics*. **5**(11), 826–837 (2004).
8. B. N. Kholodenko, Cell-signalling dynamics in time and space, *Nature Reviews Molecular Cell Biology*. **7**, 165–176 (2006).
9. H. Kurata, H. El-Samad, R. Iwasaki, H. Ohtake, J. C. Doyle, I. Grigorova, C. A. Gross, and M. Khammash, Module-based analysis of robustness tradeoffs in the heat shock response system, *PLoS Computational Biology*. **2**(7), e59 (2006).
10. H. Sayama. Generative network automata: A generalized framework for modeling complex dynamical systems with autonomously varying topologies. In *IEEE Symposium on Artificial Life (ALIFE '07)*, pp. 214–221, Honolulu, HI (2007).
11. B.-S. Chen, W.-S. Wu, Y.-C. Wang, and W.-H. Li, On the robust circuit design schemes of biochemical networks: Steady-state approach, *IEEE Transactions on Biomedical Circuits and Systems*. **1**(2), 91–104 (2007).
12. B. Alberts, A. Johnson, J. Lewis, M. Raff, K. Roberts, and P. Walter, *Molecular Biology of the Cell*. 4th edn. (Garland Science, New York, 2002).
13. T. Yanagida, M. Ueda, T. Murata, S. Esaki, and Y. Ishii, Brownian motion, fluctuation and life, *Biosystems*. **88**(3), 228–242 (2007).
14. L. Michaelis and M. Menten, Die Kinetik der Invertinwirkung, *Biochemische Zeitschrift* **49**, 333–369 (1913).

15. D. Bonchev and D. H. Rouvray, eds., *Complexity in Chemistry, Biology, and Ecology* (Springer, New York, 2005).

16. J.-Q. Liu and K. Shimohara. On designing error-correctable codes by biomolecular computation. In *International Symposium on Information Theory (ISIT 2005)*, pp. 2384–2388, Adelaide, Australia (September, 2005).

17. J.-Q. Liu and K. Shimohara, Partially interacted phosphorylation/dephosphorylation trees extracted from signaling pathways in cells, *Journal of Artificial Life and Robotics*. **11**(1), 123–127 (2007).

18. K. Leibnitz, N. Wakamiya, and M. Murata, Biologically inspired self-adaptive multi-path routing in overlay networks, *Communications of the ACM*. **49**(3), 62–67 (2006).

19. J.-L. Lu, F. Valois, D. Barthel, and M. Dohler, FISCO: A fully integrated scheme of self-configuration and self-organization for WSN. In *IEEE Wireless Communications and Networking Conference (WCNC 2007)*, pp. 3370–3375, Kowloon (March, 2007).

20. J. Tobella, M. Stiemerling, and M. Brunner, Towards self-configuration of IPv6 networks. In *IEEE/IFIP Network Operations and Management Symposium (NOMS 2004)*, pp. 895–896, Seoul, Korea (April, 2004).

21. H.-I. Liu, C.-P. Liu, and S.-T. Huang, A self-configuration and routing approach for proliferated sensor systems. In *IEEE Vehicular Technology Conference (VTC2004-Fall)*, pp. 2922–2924, Los Angeles, CA (September, 2004).

22. J. Wang and M. Vanninen, Self-configuration protocols for small-scale P2P networks. In *IEEE/IFIP Network Operations and Management Symposium (NOMS 2006)*, pp. 1–4, Vancouver, BC (April, 2006).

23. J. H. Collet and P. Zajac, Resilience, production yield and self-configuration in the future massively defective nanochips. In *IEEE International On-Line Testing Symposium (IOLTS 07)*, pp. 259–259, Vancouver, BC (July, 2007).

24. M. Breza, R. Anthony, and J. McCann, Scalable and efficient sensor network self-configuration in BioANS. In *First International Conference on Self-Adaptive and Self-Organizing Systems (SASO'07)*, pp. 351–354 (July, 2007).

25. A. Cerpa and D. Estrin, ASCENT: Adaptive Self-Configuring sEnsor Networks Topologies, *IEEE Transactions on Mobile Computing*. **3**(3), 272–285 (2004).

26. F. Dressler, *Self-Organization in Sensor and Actor Networks* (John Wiley & Sons, Chichester, U.K. 2007).

27. H. Karl and A. Willig, *Protocols and Architectures for Wireless Sensor Networks* (John Wiley & Sons, Chichester, U.K. 2007).

28. F. Dressler, Efficient and scalable communication in autonomous networking using bio-inspired mechanisms—An overview, *Informatica—An International Journal of Computing and Informatics*. **29**(2), 183–188 (2005).

29. J.-Q. Liu, Self-configuration of dynamical networks based on an information theoretic criterion. In *IPSJ SIG Technical Report 41*, pp. 15–18 (May, 2008).

30. B. H. Junker, Networks in biology. In eds. B. H. Junker and F. Schreiber, *Analysis of Biological Networks*, chapter 1, pp. 3–14 (Wiley, Hoboken, NJ, 2008).

31. H. Yu, P. M. Kim, E. Sprecher, V. Trifonov, and M. Gerstein, The importance of bottlenecks in protein networks: Correlation with gene essentiality and expression dynamics, *PLoS Computational Biology*. **3**(4), e59 (2007).

32. B. Parhami and M. A. Rakov, Performance, algorithmic, and robustness attributes of perfect difference networks, *IEEE Transactions on Parallel and Distributed Systems.* **16**(8), 725–736 (2005).

33. K. Leibnitz, M. Murata, and T. Yomo, Attractor selection as self-adaptive control mechanisms for communication networks. In eds. Y. Xiao and F. Hu, *Bio-Inspired Computing and Communication Networks.* Auerbach Publications, Taylor & Francis Group, CRC, Boca Raton, FL (2008).

34. V. E. Kotov, Systems of systems as communicating structures, *Hewlett Packard Computer Systems Laboratory Paper.* HPL-97-124, 1–15 (1997).

35. H. Chen, S. Hariri, B. Kim, Y. Zhang, and M. Yousif, Self-deployment and self-configuration of pervasive network services. In *IEEE/ACS International Conference on Pervasive Services (ICPS 2004)*, p. 242, Beirut, Lebanon (July, 2004).

36. R. Kühne, G. Görmer, M. Schlaeger, and G. Carle, A mechanism for charging system self-configuration in next generation mobile networks. In *3rd EuroNGI Conference on Next Generation Internet Networks*, pp. 198–204, Trondheim, Norway (May, 2007).

37. J. Li and R. Jantti, On the study of self-configuration neighbour cell list for mobile WiMAX. In *International Conference on Next Generation Mobile Applications, Services and Technologies (NGMAST'07)*, pp. 199–204, Trondheim, Norway (September, 2007).

Chapter 19

A Biologically Inspired QoS-Aware Architecture for Scalable, Adaptive, and Survivable Network Systems

Paskorn Champrasert and Junichi Suzuki

Contents

Large-scale network systems, such as grid/cloud computing systems, are increasingly expected to be autonomous, scalable, adaptive to dynamic network environments, survivable against partial system failures, and simple to implement and maintain. Based on the observation that various biological systems have overcome these requirements, the proposed architecture, SymbioticSphere, applies biological principles and mechanisms to design network systems (i.e., application services and middleware platforms). SymbioticSphere follows key biological principles such as decentralization, evolution, emergence, diversity, and symbiosis. Each application service and middleware platform is modeled as a biological entity, analogous to an individual bee in a bee colony, and implements biological mechanisms such as energy exchange, migration, replication, reproduction, and death. Each agent/platform possesses behavior policies, as genes, each of which determines when and how to invoke a particular behavior. Agents and platforms are designed to evolve and adjust their genes (behavior policies) through generations and autonomously improve their scalability, adaptability, and survivability. Through this evolution process, agents/platforms strive to satisfy given constraints for quality of service (QoS) such as response time, throughput, and workload distribution. This chapter describes the design of SymbioticSphere and evaluates how the biologically inspired mechanisms in SymbioticSphere impact the autonomy, adaptability, scalability, survivability, and simplicity of network systems.

19.1   Introduction

The scale, dynamics, heterogeneity, and complexity of network systems have been growing at an enormous rate. The Software Engineering Institute (SEI) of Carnegie Mellon University and the Department of Defence expect that the growth rate will keep increasing in the future.[1] The authors of the chapter believe that, in the very near future, the capability of network systems will exceed the capacity of humans to design, configure, monitor, and understand them. Therefore, network systems need to address new challenges that traditional systems have not considered well, for example, *autonomy*—the ability to operate with minimal human intervention; *scalability*—the ability to scale to, for example, a large number of users and a large volume of workload; *adaptability*—the ability to adapt to dynamic changes in network conditions (e.g., resource availability and network traffic); *survivability*—the ability to retain operation and performance despite partial system failures (e.g., network host failures); and *simplicity* of implementation and maintenance.

In order to meet these challenges in network systems, the authors of the chapter observe that various biological systems have already developed the mechanisms necessary to overcome those challenges.[2,3] For example, a bee colony is able to scale to a huge number of bees because all activities of the colony are carried out without centralized control. Bees act autonomously, influenced by local environmental conditions and local interactions with nearby bees. A bee colony adapts to dynamic environmental conditions. When the amount of honey in a hive is low, many bees leave the hive to gather nectar from flowers. When the hive is full of honey, most bees remain in the hive and rest. A bee colony can survive massive attacks by predators because it does not depend on any single bee, not even on the queen bee. The structure and behavior of each bee are very simple; however, a group of bees autonomously emerges with these desirable system characteristics such as adaptability and survivability through collective behaviors and interactions among ants. Based on this observation, the authors of the chapter believe that, if network systems are designed after certain biological principles and mechanisms, they may be able to meet the aforementioned challenges in network systems (i.e., autonomy, scalability, adaptability, survivability, and simplicity). Therefore, the proposed architecture, called SymbioticSphere, applies key biological principles and mechanisms to design network systems.

SymbioticSphere consists of two major system components: *application service* and *middleware platform*. Each of them is modeled as a biological entity, analogous to an individual bee in a bee colony. They are designed to follow several biological principles such as decentralization, evolution, emergence, diversity, and symbiosis. An application service is implemented as an autonomous software agent. Each agent implements a functional service (e.g., web service) and follows biological behaviors such as energy exchange, migration, replication, reproduction, death, and environment sensing. A middleware platform runs on a network host and operates agents. Each platform provides a set of runtime services that agents use to perform their services and behaviors, and implements biological behaviors such as energy exchange, replication, reproduction, death, and environment sensing.

Each agent/platform possesses *behavior policies*, each of which determines when and how to invoke a particular behavior. A behavior policy is encoded as a *gene*. In SymbioticSphere, evolution occurs on behavior policies via genetic operations such as mutation and crossover, which alter behavior policies when agents/platforms replicate themselves or reproduce their offspring. This evolution process is intended to increase the adaptability, scalability, and survivability of agents/platforms by allowing them to adjust their behavior policies to dynamic network conditions across generations. Agents/platforms evolve to satisfy given constraints for quality of service (QoS) such as response time, throughput, and workload distribution. Each constraint is defined as the upper or lower bound of a QoS measure. Evolution frees network system developers from anticipating all possible network conditions and tuning their agents and platforms to the conditions at design time. Instead, agents and platforms evolve and autonomously adjust themselves to network environments at runtime. This can be significantly simplified to implement and maintain agents/platforms.

This chapter describes the design of SymbioticSphere and evaluates the biologically inspired mechanisms in SymbioticSphere, through simulation experiments, in terms of autonomy, scalability, adaptability, and survivability. Simulation results show that agents and platforms evolve in order to autonomously scale to network size and demand volume and adapt to dynamic changes in network conditions (e.g., user locations, network traffic, and resource availability). Agents/platforms also evolve to autonomously survive partial system failures such as host failures in order to retain their availability and performance. Moreover, it is verified that agents and platforms satisfy given QoS constraints through evolution.

This chapter is organized as follows. Section 19.2 overviews key biological principles applied to SymbioticSphere. Section 19.3 describes the design of agents and platforms in SymbioticSphere. Section 19.4 discusses a series of simulation results to evaluate SymbioticSphere. Sections 19.5 and 19.6 conclude this chapter with some discussion on related and future work.

19.2 Design Principles in SymbioticSphere

SymbioticSphere applies the following biological principles to design agents and platforms.

- **Decentralization:** In various biological systems (e.g., bee colony), there are no central entities to control or coordinate individual entities for increasing scalability and survivability. Similarly, in SymbioticSphere, there are no central entities to control or coordinate agents/platforms so that they can be scalable and survivable by avoiding a single point of performance bottlenecks[4] and failures.[5]

- **Autonomy:** Inspired by biological entities (e.g., bees), agents and platforms sense their local network conditions, and based on the conditions, they behave and interact with each other without any intervention from/to other agents, platforms, and human users.

- **Emergence:** In biological systems, collective (group) behaviors emerge from local interactions of autonomous entities.[3] In SymbioticSphere, agents/platforms only interact with nearby agents/platforms. They behave according to dynamic network conditions such as user demands and resource availability. For example, an agent may invoke the migration behavior to move toward a platform that forwards a large number of request messages for its services. Also, a platform may replicate itself on a neighboring network host where resource availability is high. Through collective behaviors and interactions of individual agents and platforms, desirable system characteristics such as scalability, adaptability, and survivability emerge in a group of agents and platforms. Note that they are not present in any single agent/platform.

■ **Redundancy:** Biological entities (e.g., bees) die due to, for example, advanced age and consumption by predators. However, biological systems (e.g., bee colony) can survive because the death is compensated by the birth of new entities. In SymbioticSphere, agents/platforms replicate themselves and reproduce their offspring to retain redundancy. This redundancy enhances their survivability; they can continuously provide their services in case many agents/platforms are lost due to network failures.

■ **Lifecycle and food chain:** Biological entities strive to seek and consume food for living. In SymbioticSphere, agents store and expend *energy* for living. Each agent gains energy in exchange for performing its service to other agents or human users, and expends energy to use resources such as memory space (Figure 19.1). Each platform gains energy in exchange for providing resources to agents, and continuously evaporates energy (Figure 19.1). The abundance or scarcity of stored energy in agents/platforms affects their lifecycle. For example, an abundance of stored energy indicates high demand to an agent/platform; thus, the agent/platform may be designed to favor reproduction or replication to increase its availability and redundancy. A scarcity of stored energy indicates a lack of demand; it causes the agent/platform's death.

Also, in the ecosystem, the energy accumulated from food is transferred between different species to balance their populations. For example, producers (e.g., shrubs) convert the Sun light energy to chemical energy. The chemical energy is transferred to consumers (e.g., hares) as consumers consume producers[2] (Figure 19.2). In order to balance the populations of agents and

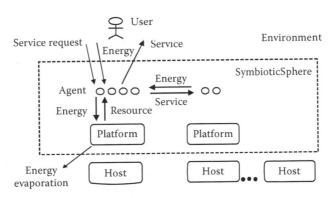

Figure 19.1 Energy exchange in SymbioticSphere.

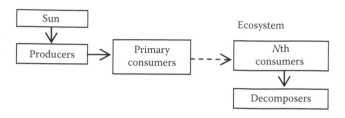

Figure 19.2   Energy flow in the ecosystem.

platforms, the energy exchange in SymbioticSphere is designed after the ecological food chain among different species. SymbioticSphere models agents and platforms as different biological species; it models human users as the Sun, which has an unlimited amount of energy, agents as producers, and platforms as consumers.

■ **Diversity:** Biological entities are slightly different with each other in each species. This can contribute to survivability against environmental changes.[6] In SymbioticSphere, agents/platforms retain *behavioral diversity*. Behavioral diversity means that different agents/platforms have different behavior policies. For example, in response to abundance of energy, an agent may migrate toward a user for reducing response time, while another agent may reproduce offspring with a mate for increasing agent availability/redundancy. Behavioral diversity is generated via genetic operations (i.e., mutation and crossover) during replication and reproduction.

■ **Evolution:** Biological entities evolve so that the entities that fit better in the environment become more abundant.[7] In SymbioticSphere, agents, and platforms evolve their genes (i.e., behavior policies) by generating behavioral diversity and executing natural selection. Natural selection is governed with agents' and platforms' energy levels. It retains the agents/platforms whose energy levels are high (i.e., the agents/platforms that have effective behavior policies, such as moving toward a user to gain more energy) and eliminates the agents/platforms whose energy levels are low (i.e., the agents/platforms that have ineffective behavior policies, such as moving too often). Through generations, effective behavior policies become abundant while ineffective ones become dormant or extinct. This allows agents/platforms to adjust their behavior policies to improve their scalability, adaptability, and survivability.

■ **Symbiosis:** Although competition for food and terrain always occurs in the biological world, several species establish symbiotic relationships to avoid excessive competition and cooperate to survive.[8] In SymbioticSphere, agents and platforms evolve to cooperate in certain circumstances in order to pursue their mutual benefits and improve their scalability, adaptability, and survivability.

19.3 SymbioticSphere

In SymbioticSphere, each agent runs (or lives) on a platform. A platform is an execution environment (or middleware) for agents. Each platform implements runtime services that agents use to perform their services and behaviors. Each platform can operate multiple agents, and each network host operates at most one platform.

19.3.1 Agents

Each agent consists of three parts: *attributes*, *body*, and *behaviors*. Attributes carry descriptive information regarding an agent, such as agent ID, energy level, description of a service the agent provides, and cost of a service (in energy unit) that the agent provides.

A body implements a service that an agent provides, and contains materials relevant to the service (e.g., application data and user profiles). For example, an agent may implement a web service and contain web pages. Another agent may implement a physical model for scientific simulations and contain parameter settings for the physical model.

Behaviors implement actions inherent to all agents.

- *Migration*: Agents may migrate from one platform to another.
- *Replication*: Agents may make a copy of themselves. A replicated (child) agent is placed on the platform that its parent agent resides on. It inherits half of the parent's energy level.
- *Reproduction*: Agents may make their offspring with their mates. A reproduced (child) agent is placed on the platform that its parent agent resides on.* It inherits half of the parent's energy level.
- *Death*: Agents may die due to energy starvation. If the energy expenditure of an agent is not balanced with its energy gain from users and other agents, it cannot pay for the resources it requires. Agents have high chances of dying from lack of energy, if they provide unwanted services and/or have wasteful behavioral policies (e.g., replicating too often). When an agent dies, an underlying platform removes it and releases the resources it consumes.

19.3.2 Platforms

Each platform consists of *attributes*, *behaviors*, and *runtime services*. Attributes carry descriptive information regarding a platform, such as platform ID, energy level, and *health level*. Health level indicates how healthy an underlying host is. It is defined as a function of resource availability on age and freshness of a host. Resource availability indicates how many resources (e.g., memory space) are available for a platform and agents on the host. Age indicates how long a host has been alive. It represents how

---

* The parent agent is an agent that invokes the reproduction behavior.

Table 19.1 Freshness and Age

Host Type	Freshness	Age
Unstable host	Low	Low
New host	High	Low
Stable host	Low	High

stable the host is. Freshness indicates how recently a host joined the network. Once a host joins the network, its freshness gradually decreases from the maximum value. When a host resumes from a failure, its freshness starts with the value that the host had when it went down. Using age and freshness can distinguish unstable hosts and new hosts. Unstable hosts tend to have low freshness and low age, and new hosts tend to have high freshness and low age (Table 19.1).

Health level affects how platforms and agents invoke behaviors. For example, higher health level indicates higher stability of and/or higher resource availability on a host that a platform resides on. Thus, the platform may replicate itself on a neighboring host if the host is healthier than the local host. This results in the adaptation of platform locations. Platforms strive to concentrate around stable and resource-rich hosts. Also, lower health level indicates that a platform runs on a host that is unstable and/or poor in resources. Thus, agents may leave the platform and migrate to a healthier (i.e., more stable and/or resource-rich) host. This results in the adaptation of agent locations. Agents strive to concentrate around stable and/or resource-rich hosts. In this case, the platforms on unstable and/or resource-poor hosts will eventually die due to energy starvation because agents do not run on the platforms and transfer energy to them. This results in the adaptation of platform population. Platforms avoid running on the hosts that are unstable and/or poor in resources.

Behaviors are the actions inherent to all platforms.

- *Replication*: Platforms may make a copy of themselves. A replicated (child) platform is placed on a neighboring host that does not run a platform. (Two or more platforms are not allowed to run on each host.) It inherits half of the parents' energy level.
- *Reproduction*: Platforms may make their offspring with their mates. A reproduced (child) platform is placed on a neighboring host that does not run a platform. It inherits half of the parents'* energy level.
- *Death*: Platforms may die due to energy starvation. A dying platform uninstalls itself and releases the resources it consumes. When a platform dies, agents running on it are killed.

---

* The parent platform is a platform that invokes the reproduction behavior.

Runtime Services are middleware services that agents and platforms use to perform their behaviors. In order to maximize decentralization and autonomy of agents/platforms, they only use the local runtime services. They are not allowed to invoke any runtime services on a remote platform.

19.3.3 Behavior Policies

Each agent/platform possesses policies for its behaviors. A behavior policy defines when and how to invoke a particular behavior. Each behavior policy consists of factors (F_i), which indicate environment conditions (e.g., network traffic) or agent/platform/host status (e.g., energy level and health level). Each factor is associated with a weight (W_i). Each agent/platform decides to invoke a behavior when the weighted sum of the behavior's factor values $\left(\sum F_i * W_i\right)$ exceeds a threshold.

19.3.3.1 Agent Behavior Policies

This chapter focuses on four agent behaviors described in Section 19.3.1 (i.e., migration, replication, reproduction, and death). The behavior policy for agent migration includes the following four factors:

1. *Energy level*: Agent energy level, which encourages agents to move in response to high energy level.
2. *Health level ratio*: The ratio of health level on a neighboring platform to the local platform, which encourages agents to move to healthier platforms. This ratio is calculated with three health level properties $(HLP_i, 1 \leq i \leq 3)$: resource availability, freshness, and age (Equation 19.1).

$$\text{Health level ratio} = \sum_{i=1}^{3} \frac{HLP_i^{\text{neighbor}} - HLP_i^{\text{local}}}{HLP_i^{\text{local}}} \qquad (19.1)$$

HLP_i^{neighbor} and HLP_i^{local} denote a health level property on a neighboring platform and local platform, respectively.
3. *Service request ratio*: The ratio of the number of incoming service requests on a neighboring platform to the local platform. This factor encourages agents to move toward human users.
4. *Migration interval*: Time interval to perform migration, which discourages agents to migrate too often.

If there are multiple neighboring platforms that an agent can migrate to, the agent calculates the weighted sum of the above factors for each of them, and moves to a platform that generates the highest weighted sum.

The behavior policy for agent reproduction and replication includes the following two factors.

1. *Energy level*: Agent energy level, which encourages agents to reproduce their offspring in response to their high energy levels.
2. *Request queue length*: The length of a queue, which the local platform stores incoming service requests to. This factor encourages agents to reproduce their offspring in response to high demands for their services.

When the weighted sum of the above factors exceeds a threshold, an agent seeks a mate from the local and neighboring platforms. If a mate is found, the agent invokes the reproduction behavior. Otherwise, the agent invokes the replication behavior. Section 19.3.5 describes how an agent seeks its mate for reproduction.

The behavior policy for agent death includes the following two factors:

1. *Energy level*: Agent energy level. Agents die when they run out of their energy.
2. *Energy loss rate*: The rate of energy loss, calculated with Equation 19.2. E_t and E_{t-1} denote the energy levels in the current and previous time instants. Agents die in response to sharp drops in demands for their services.

$$\text{Energy loss rate} = \frac{E_{t-1} - E_t}{E_{t-1}} \qquad (19.2)$$

19.3.3.2 Platform Behavior Policies

This chapter focuses on three platform behaviors described in Section 19.3.2 (i.e., reproduction, replication, and death).

The behavior policy for platform reproduction and replication includes the following three factors:

1. *Energy level*: Platform energy level, which encourages platforms to reproduce their offspring in response to their high energy levels.
2. *Health level ratio*: The ratio of health level on a neighboring host to the local host. This factor encourages platforms to reproduce their offspring on the hosts that generate higher values with Equation 19.1.
3. *The number of agents*: The number of agents working on each platform. This factor encourages platforms to reproduce their offspring in response to high agent population on them.

When the weighted sum of the above factors exceeds a threshold, a platform seeks a mate from its neighboring hosts. If a mate is found, the platform invokes the reproduction behavior. Otherwise, it invokes the replication behavior. Section 19.3.5 describes how a platform finds its mate for reproduction. If there are multiple neighboring hosts that a platform can place its child platform on, it places the child on a host whose health ratio is highest among others.

The behavior policy of platform death includes the following two factors:

1. *Energy level*: Platform energy level. Platforms die when they run out of their energy.
2. *Energy loss rate*: The rate of energy loss, calculated with Equation 19.2. Platforms die in response to sharp drops in demands for their resources.

Each agent/platform expends energy to invoke behaviors (i.e., behavior invocation cost) except the death behavior. When the energy level of an agent/platform exceeds the cost of a behavior, it decides whether it performs the behavior by calculating a weighted sum described above.

19.3.4 Energy Exchange

As described in Section 19.2, SymbioticSphere models agents and platforms as different biological species and follows ecological concepts to design energy exchange among agents, platforms, and human users. Following the energy flow in the ecosystem (Figure 19.2), SymbioticSphere models users as the Sun, agents as producers, and platforms as (primary) consumers. Similar to the Sun, users have an unlimited amount of energy. They expend energy units for services provided by agents. Agents gain energy from users and expend energy to consume resources provided by platforms. They expend 10% of the current energy level to platforms.* Platforms gain energy from agents, and periodically evaporate 10% of the current energy level.

Agents dynamically change the frequency to transfer energy to platforms, depending on the rate of incoming service requests from users. When agents receive and process more service requests from users, they consume more resources. Thus, agents transfer energy units (i.e., 10% of the current energy level) to platforms more often. On the contrary, they reduce their energy transfer rate in response to a lower rate of incoming service requests.

In order to dynamically change energy transfer rate, each agent keeps an interval time between arrivals of an incoming service request and a previous request. It records the average, shortest, and maximum intervals of previous requests (T_a, T_s, and T_m, respectively). Figure 19.3 shows how often each agent transfers energy to an underlying platform. First, an agent waits for T_s and transfers energy to an underlying platform. Then, the agent examines whether a new service request(s) has arrived during the previous T_s interval. If it has arrived, the agent updates T_a, T_s, and T_m, waits for T_a, and transfers energy. Otherwise, it waits for T_a and transfers energy. Similarly, each agent repeats energy transfers in T_a, T_s, and T_m intervals (Figure 19.3).

---

* This 10% rule is known in ecology[2] and applied to the energy exchange in SymbioticSphere.

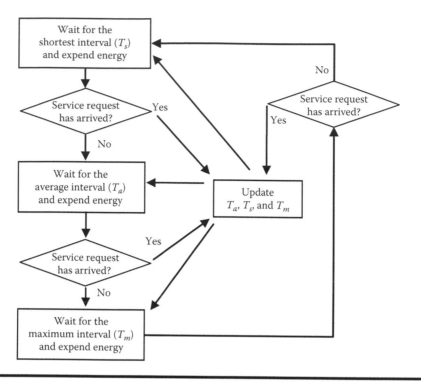

Figure 19.3   Energy exchange in SymbioticSphere.

T_a is the simple average calculated with the intervals of previous N requests. The shortest and maximum intervals play a role to adjust energy transfer rate according to dynamic changes in request rate. T_s and T_m values are periodically reset (every M service requests).

Platforms dynamically change the frequency to evaporate energy (10% of the current energy level), depending on the rate of incoming energy transfers from agents. The more often they gain energy from agents, the more often they evaporate energy. Each platform changes its energy evaporation rate in the same way as each agent changes its energy expenditure rate; each platform follows the mechanism shown in Figure 19.3.

19.3.5   Constraint-Aware Evolution

The weight and threshold values in behavior policies have significant impacts on the adaptability, scalability, and survivability of agents and platforms. However, it is hard to anticipate all possible network conditions and find an appropriate set of weight and threshold values for the conditions. As shown in Section 19.3.3, there are 18 weight and threshold variables in total (11 for agent behavior policies and

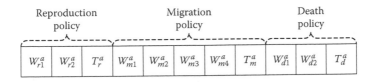

Figure 19.4   Gene structure for agent behavior policies.

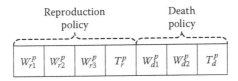

Figure 19.5   Gene structure for platform behavior policies.

7 for platform behavior policies). Assuming that 10 different values can be assigned to each variable, there are 10^{18} possible combinations of weight and threshold values.

Instead of manual assignments, SymbioticSphere allows agents and platforms to autonomously find appropriate weight and threshold values through evolution, thereby adapting themselves to dynamic network conditions. Behavior policies are encoded as genes of agents and platforms. Each gene contains one or more weight values and a threshold value for a particular behavior. Figures 19.4 and 19.5 show the gene structure for agent and platform behavior policies, respectively. For example, for the agent reproduction behavior, a gene consists of three gene elements: (1) W_{r1}^a, a weight value for the energy level factor; (2) W_{r2}^a, a weight value for the factor of request queue length; and (3) T_r^a, a threshold value (Figure 19.4).

The genes of agents and platforms are altered via genetic operations (genetic crossover and mutation) when they perform the reproduction or replication behaviors. As described in Section 19.3.3, each agent/platform selects a mate when it performs the reproduction behavior. A mate is selected by ranking agents/platforms running on the local and neighboring hosts. For this ranking process, SymbioticSphere leverages a *constraint-based domination ranking* mechanism.

Agents and platforms are ranked with the notion of *constraint domination*, which considers two factors: optimization objectives and QoS constraint violation. SymbioticSphere considers the following three objectives. For all objectives, the higher the value, the better is the result.

1. *Energy level*
2. *Total number of behavior invocations*
3. *Health level of the underlying host*

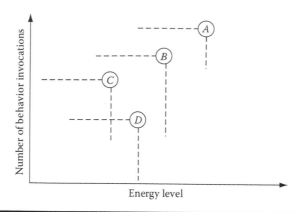

Figure 19.6   Example of agent domination.

Using these objectives, domination is determined among agents/platforms. Figure 19.6 shows an example of how to examine domination among four agents (Agents A to D). For simplicity, this figure shows only two objectives: energy level and number of behavior invocations. Agents are plotted on a two-dimensional space whose axes represent the two objectives. In this example, Agent A is the best in both objectives; it is said to *dominate* the other three agents. In other words, Agent A is *non-dominated*. Agent B is dominated by Agent A; however, it dominates the other two agents (Agent C and D). Agents C and D do not dominate with each other because one of them does not outperform the other in both objectives, and vice versa.

The second factor in the agent/platform ranking process is constraint violation on QoS such as response time, throughput, and workload distribution. Each constraint is defined as the upper or lower bound of a QoS measure. For example, a constraint may specify that response time must be lower than 1 s. When an agent/platform satisfies all of the given constraints, it is said to be *feasible*. Otherwise, it is said to be *infeasible*.

With the above two factors examined, an agent/platform i is said to *constraint-dominate* another agent/platform j if any of the following conditions are true:

1. i is feasible, and j is not.
2. Both i and j are feasible; however, i dominates j.
3. Both i and j are infeasible; however, i has a smaller degree of constraint violation than j.

Figure 19.7 shows an example of how to evaluate the degree of constraint violations by four agents. (Agents A to D are all infeasible.) The X and Y axes represent the difference between an actual QoS measure and a QoS constraint in response time and throughput, respectively. Agents A and D violate the constraints for response time and throughput. Agents B and C violate a constraint for response time, but

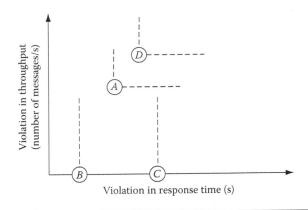

Figure 19.7   Example of constraint violation.

satisfy a constraint for throughput. In this example, Agent B constraint-dominates the other three agents because its violation is minimum in both the QoS measures. Agents A and C do not constraint-dominate each other because one of them cannot yield lower violation in both QoS measures, and vice versa. (Agent A yields lower violation in response time but higher violation in throughput than Agent B.) Agent D is constraint-dominated by Agents A and C.

Agents/platforms are ranked with constraint-domination among them. Non-constraint-dominated agents/platforms are given the lowest rank, Rank 1. When an agent/platform is constraint-dominated by a Rank N agent/platform, its rank is $N + 1$.

Figure 19.8 shows a pseudo code that shows the evolution process in SymbioticSphere. Table 19.2 shows a set of variables and functions used in the pseudo code.

In reproduction, crossover occurs. A parent and its mate contribute their genes and combine them for a child's genes. The child's gene element value is in between its parent's and the parent's mate's. It is shifted closer to its parent's and the parent's mate's depending on their fitness values. After crossover, mutation may occur on the child's genes. Each gene element is randomly altered based on a uniform distribution function ($U(0, 1)$). See Figure 19.9 for an example of crossover and mutation. In replication, a parent copies its genes to its child. Then, mutation may occur on the child's genes in the same way as the mutation in reproduction.

19.4 Evaluation

This section shows a series of simulation results to evaluate the biologically inspired mechanisms in SymbioticSphere. The objectives of simulations are to examine how the biologically inspired mechanisms impact the adaptability, scalability, and survivability of network systems. Simulations were carried out with the SymbioticSphere

main
  while not the last cycle of a simulation run
  do $\Bigg\{$
    for each $parent^a \in Agents$
    do $\Bigg\{$
      if $parent^a$ invokes the reproduction behavior
      then $\begin{cases} mate^a \leftarrow \text{FINDMATE}(parent^a) \\ \textbf{if } mate^a \neq \oslash \\ \quad \textbf{then } child^a \leftarrow \text{REPRODUCE}(parent^a, mate^a) \\ \quad \textbf{else } child^a \leftarrow \text{REPLICATE}(parent^a) \end{cases}$
    for each $parent^p \in Platforms$
    do $\Bigg\{$
      if $parent^p$ invokes the reproduction behavior
      then $\begin{cases} mate^p \leftarrow \text{FINDMATE}(parent^p) \\ \textbf{if } mate^p \neq \oslash \\ \quad \textbf{then } child^p \leftarrow \text{REPRODUCE}(parent^p, mate^p) \\ \quad \textbf{else } child^p \leftarrow \text{REPLICATE}(parent^p) \end{cases}$

procedure REPRODUCTION(*parent*, *mate*)
  $child \leftarrow \text{CROSSOVER}(parent, mate)$
  $child \leftarrow \text{MUTATE}(child)$
  return (*child*)

procedure REPLICATION(*parent*)
  $child \leftarrow \text{MUTATE}(parent)$
  return (*child*)

procedure CROSSOVER(*parent*, *mate*)
  for $i \leftarrow 1$ **to** N
  do $\begin{cases} center_i = (parent[i] + mate[i])/2 \\ offset = \frac{(\text{FITNESS}(mate) - \text{FITNESS}(parent))}{\text{FITNESS}(parent) + \text{FITNESS}(mate)} * \frac{|mate[i] - parent[i]|}{2} \\ \textbf{if } mate[i] \geq parent[i] \\ \quad \textbf{then } child[i] = center_i + offset \\ \quad \textbf{else } child[i] = center_i - offset \end{cases}$
  return (*child*)

procedure MUTATE(*parent*)
  for $i \leftarrow 1$ **to** N
  do $\begin{cases} \textbf{if } U(0,1) \leq mutationRate \\ \quad \textbf{then } child[i] = parent[i](1 + N(\mu, \sigma)) \end{cases}$
  return (*child*)

Figure 19.8 Evolution process in SymbioticSphere.

simulator, which implements the biologically inspired mechanisms described in Section 19.3.*

19.4.1 Simulation Configurations

This section describes the implementation and configuration of the SymbioticSphere simulator. The implementation and configuration are commonly used in all the simulations.

---

* The current code base of the SymbioticSphere simulator contains 15,200 lines of Java code. It is freely available at http://dssg.cs.umb.edu/projects/SymbioticSphere/

Table 19.2　Variables and Functions Used in Figure 19.8

Variable or Function	Description
Agents	A set of agents at the current simulation cycle
Platforms	A set of platforms at the current simulation cycle
parent	An agent ($parent^a$) or a platform ($parent^P$) that invokes the reproduction behavior
mate	An agent ($mate^a$) or a platform ($mate^P$) that is selected as a mate
child	An agent ($child^a$) or a platform ($child^P$) that is reproduced or replicated
child[i]	The ith gene element of a child agent/platform ($0 \leq i \leq N$)
mutationRate	The probability to perform mutation. 0.1 is currently used
FINDMATE(parent)	Returns a mate for a given parent agent/platform. Returns ⊘ if no agents/platforms exist on the local and direct neighbor (one-hop away) hosts. Returns a Rank 1 agent/platform on those hosts if the parent is feasible. Otherwise, returns a feasible agent/platform, on those hosts, which performs best in the QoS whose constraint(s) the parent violates
FITNESS(parent)	Returns the number of agents/platforms that the parent constraint-dominates on the local and direct neighbor hosts
$U(0, 1)$	Generates a random number between 0 and 1 based on the uniform distribution
$N(\mu, \sigma)$	Generates a random number based on a normal distribution whose average is μ and standard deviation is σ. Currently, 0 and 0.3 are used for μ and σ

A simulated network system is designed as a server farm. Each agent implements a web service in its body. Its behavior policies are randomly configured at the beginning of each simulation. Figure 19.10 shows a simulated network. It consists of hosts connected in an $N \times N$ grid topology, and service requests travel from users to agents via user access point. This simulation study assumes that a single (emulated) user runs on the access point and sends service requests to agents. Each host has

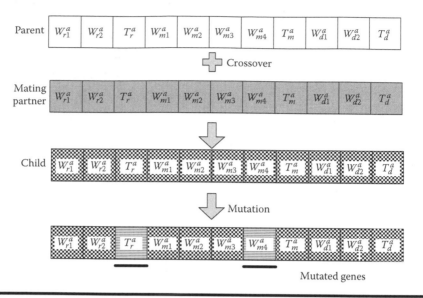

Figure 19.9   Example genetic operations.

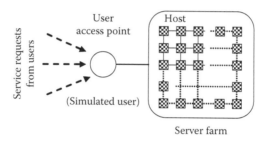

Figure 19.10   Simulated network.

256 or 320 MB of memory space.* Out of the memory space, an operating system consumes 128 MB, and Java virtual machine consumes 64 MB. The remaining space is available for a platform and agents on each host. Each agent and platform consumes 5 and 20 MB, respectively. This assumption is obtained from prior empirical experiments.[9]

Each host operates in the active or inactive state. When a platform works on a host, the host is active and consumes 60 W power. The host goes to the inactive state when a platform dies on it. An inactive host consumes 5 W power. This assumption on power consumption is obtained from Gunaratne et al.[10] A host becomes active from the inactive state using the Wake On LAN (WOL) technology.[11] When a

---

* Currently, memory availability represents resource availability on each platform/host.

```
main
  while not the last cycle of a simulation run
     ⎧ for each user
     │       ⎧ Send service requests to agents according to a
     │    do ⎨ configured service request rate.
     │       ⎩
     │   for each agent
     │       ⎧ if a service request(s) received
     │       │    then Process the request(s) and gain energy.
  do ⎨    do ⎨ Determine whether or not to invoke the reproduction,
     │       │ replication, migration and death behaviors.
     │       ⎩ Transfer energy to the local platform.
     │   for each platform
     │       ⎧ Gain energy from the local agents
     │       │ Determine whether or not to invoke the reproduction,
     │    do ⎨ replication and death behaviors.
     │       │ Update health level.
     ⎩       ⎩ Evaporate energy.
```

Figure 19.11   Pseudo code to run users, agents, and platforms in a simulation.

```
   while not the last cycle of a simulation run
      ⎧ if a discovery message(s) arrived
      │          ⎧ for each discovery message
      │          │      ⎧ if the discovery message matches one of the local agents
   do ⎨ then  do ⎨ then ⎧ Return a discovery response to the
      │          │      ⎨        ⎩ discovery originator.
      │          │      │ else ⎧ Forward the discovery message to
      ⎩          ⎩      ⎩      ⎩ neighboring platforms.
```

Figure 19.12   Pseudo code for agents' discovery process.

platform places its offspring on an inactive host, it sends a WOL packet to the host to activate it.

Figure 19.11 shows a pseudo code to run users, agents, and platforms in each simulation run. A single execution of this loop corresponds to one simulation cycle.

When a user issues a service request, the service request is passed to the local platform on which the user resides, and the platform performs a discovery process to search a target agent that can process the issued service request. The platform (discovery originator) forwards a discovery message to its neighboring platforms, asking whether they host a target agent. If a neighboring platform hosts a target agent, it returns a discovery response to the discovery originator. Otherwise, it forwards the discovery message again to its neighboring platforms. Figure 19.12 shows this decentralized agent discovery process. Note that there is no centralized directory to keep track of agent locations.

Through the above discovery process, a user finds a set of platforms hosting the target agents that can process his/her service request. The user chooses the platform closest to him/her and transmits his/her service request to the platform. When the service request arrives, the platform inserts the request into its request queue. (Each platform maintains a request queue for queuing incoming service requests and

```
while not the last cycle of a simulation run
      ⎡ if # of queued requests > # of requests that the local agents can process
      ⎢   in one simulation cycle
do ⎨    then ⎨ if there are one or more neighboring platform(s) that can process
      ⎢           queued requests
      ⎢           then Transfer queued requests to those platforms in round robin.
      ⎣ else Dispatch queued requests to the local agents.
```

Figure 19.13   Pseudo code for service request propagation.

dispatching them to agents.) Each platform inspects the number of queued service requests and the number of local agents running on it in each simulation cycle. If the number of queued requests exceeds the number of service requests that the local agents can process in one simulation cycle, the platform transfers the queued requests to neighboring platforms hosting the (idle) agents that can process the requests. This propagation continues until the number of queued requests becomes smaller than the number of service requests that the local agents can process in one simulation cycle. Figure 19.13 shows this request propagation process.

In this simulation study, SymbioticSphere is compared with the Bio-Networking Architecture (BNA).[9,12,13] In BNA, agents are modeled after biological entities and designed to evolve and adapt dynamic network conditions in a decentralized manner. See also Section 19.5.

19.4.2 Evaluation of Energy Exchange

This section evaluates SymbioticSphere's energy exchange mechanism described in Section 19.3.4. In this evaluation, an agent is deployed on a platform to accept service requests from the user. The agent does not invoke any behaviors to focus on the evaluation of energy exchange. T_s and T_m values are periodically reset every 50 service requests.

Figure 19.14 shows how the user changes service request rate over time. In order to evaluate the energy exchange in SymbioticSphere, two agents are implemented and compared. The first agent implements the energy exchange mechanism described in Section 19.3.4. It uses T_a, T_s, and T_m intervals. The second agent uses T_a only.

Figure 19.15 shows how much energy the two agents transfer to their local platforms. It demonstrates that the SymbioticSphere's energy exchange mechanism allows an agent to change its energy expenditure rate against dynamic changes in energy intake (i.e., service request rate). With T_a, T_s, and T_m, an agent's energy expenditure better follows the changes in energy intake than using T_a only.

19.4.3 Evaluation of Adaptability

This section evaluates how agents and platforms adapt to dynamic network conditions. In this evaluation, adaptability is defined as *service adaptation* and *resource*

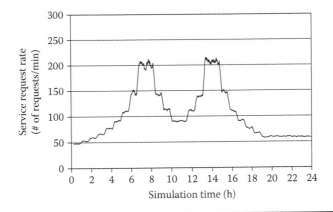

Figure 19.14   Service request rate.

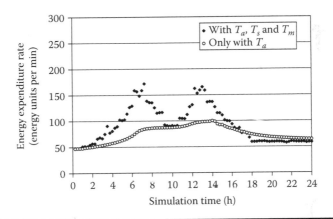

Figure 19.15   Energy expenditure rate.

adaptation. Service adaptation is a set of activities to adaptively improve the quality and availability of services provided by agents. The QoS is measured as response time and throughput of agents for processing service requests from users. Service availability is measured as the number of agents. Resource adaptation is a set of activities to adaptively improve resource availability and resource efficiency. Resource availability is measured as the number of platforms that make resources available for agents. Resource efficiency indicates how many service requests can be processed per resource utilization of agents and platforms.

A simulated network is a 7×7 network with 49 network hosts, each of which has 256 MB memory space. At the beginning of each simulation, an agent and a platform are deployed on each node (i.e., 49 platforms and 49 agents on 49 network hosts). Figure 19.16 shows how the user changes service request rate over time.

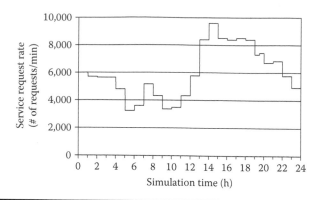

Figure 19.16   Service request rate.

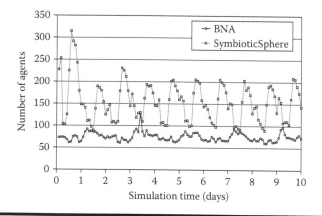

Figure 19.17   Number of agents.

This is obtained from a workload trace of the 1998 Olympic official website.[14] The peak demand is 9600 requests/min. Each simulation was carried out for 10 days in simulation time by repeating the daily workload trace 10 times.

Figure 19.17 shows how service availability (i.e., the number of agents) changes against dynamic service request rate. At the beginning of a simulation, the number of agents fluctuates in SymbioticSphere and BNA because agent behavior policies are initially random. However, as evolution continues over time, agents adapt their population to the changes in service request rate. When service request rate becomes high, agents gain more energy from a user and replicate themselves more often. In contrast, when service request rate becomes low, some agents die due to energy starvation because they cannot balance their energy gain and expenditure. This result demonstrates that the biologically inspired mechanisms in SymbioticSphere allow agents to evolve their behavior policies and adaptively adjust their availability

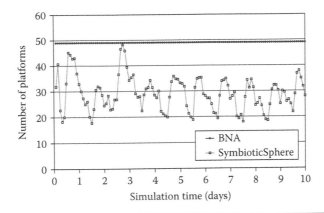

Figure 19.18 Number of platforms.

as a group. Figure 19.17 also shows that the agent population is more sensitive to workload changes in SymbioticSphere than BNA.

Figure 19.18 shows how resource availability (i.e., the number of platforms) changes against dynamic service request rate. At the beginning of a simulation, the number of platforms fluctuates in SymbioticSphere because platform behavior policies are initially random. However, as evolution continues over time, platforms adapt their population to the changes in service request rate. When service request rate becomes high, agents gain more energy and transfer more energy to platforms. In response to abundance of stored energy, platforms replicate or reproduce offspring more often. In contrast, when service request rate becomes low, some platforms die due to energy starvation because they cannot gain enough energy from agents to maintain their population. Figure 19.18 demonstrates how the biologically inspired mechanisms in SymbioticSphere allow platforms to evolve their behavior policies and adaptively adjust resource availability as a group. In BNA, platforms are not designed as biological entities; the number of platforms does not change dynamically, but remains at 49.

Figure 19.19 shows the average response time for agents to process one service request from the user. This includes the request transmission latency between the user and an agent and the processing overhead for an agent to process a service request. In SymbioticSphere, the average response time is maintained very low, at less than 5 s, because agents increase their population to process more service requests (see Figure 19.17) and migrate toward the user. Figure 19.19 demonstrates that the biologically inspired mechanisms in SymbioticSphere allow agents to autonomously improve and maintain response time. In BNA, agents fail to consistently maintain low response time on several days because they do not adapt their population well to the changes in service request rate (Figure 19.17). The agent evolution process operates better in SymbioticSphere than in BNA. See Section 19.5 for the differences in the evolution process of the two architectures.

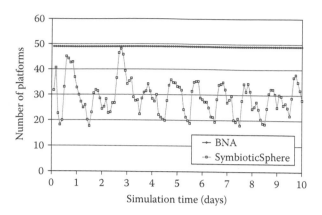

Figure 19.19   Response time.

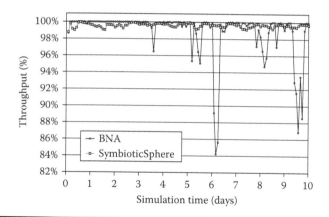

Figure 19.20   Throughput.

Figure 19.20 shows the throughput of agents. Throughput is measured as the ratio of the number of service requests agents process to the total number of service requests the user issues. In SymbioticSphere, throughput is maintained very high, at higher than 98%, because agents adapt their population to dynamic service request rate (Figure 19.17). Figure 19.20 demonstrates that the biologically inspired mechanisms in SymbioticSphere allow agents to autonomously improve and maintain throughput. In BNA, agents fail to consistently maintain high throughput on several days. For example, throughput drops to 84% on day 6. This is because agents do not adapt their population well to the changes in service request rate. The agent evolution process operates better in SymbioticSphere than in BNA. See Section 19.5 for the differences in the evolution process of the two architectures.

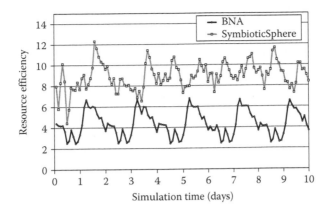

Figure 19.21   Resource efficiency.

Figure 19.21 shows how resource efficiency changes. Resource efficiency is measured with the following equation:

$$\text{Resource efficiency} = \frac{\text{the total number of service requests processed by agents}}{\text{the total amount of resources consumed by agents and platforms}}$$

$$(19.3)$$

SymbioticSphere always yields higher resource efficiency than BNA because in BNA platforms do not adapt their population to dynamic service request rate; 49 platforms are active at all times (Figure 19.18). In SymbioticSphere, both agents and platforms autonomously adapt their populations to dynamic service request rate (Figures 19.17 and 19.18); thus, resource efficiency is always higher in SymbioticSphere. Figure 19.21 demonstrates that the biologically inspired mechanisms in SymbioticSphere allow agents and platforms to retain high resource efficiency.

Figure 19.22 shows the total power consumption in SymbioticSphere and BNA. In SymbioticSphere, each host operates in an active or inactive state depending on whether a platform works on the host (Section 19.4.1). Since platforms adapt their population to dynamic service request rate (Figure 19.18), hosts change their states between active and inactive based on the changes in service request rate. More hosts become active in response to higher service request rate, and more hosts become inactive in response to lower service request rate. In BNA, all hosts are always active because platforms never die. As a result, SymbioticSphere consumes a lower amount of power(443.1 kWh) than BNA does (705.6 kWh). SymbioticSphere saves approximately 40% power consumption compared with BNA. Figure 19.22 shows that the biologically inspired mechanisms in SymbioticSphere allow platforms to improve power efficiency by adapting their population.

Figure 19.23 shows workload distribution over available platforms in SymbioticSphere. It is measured as load balancing index (LBI) with Equation 19.4. A lower

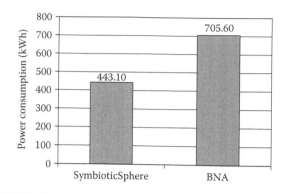

Figure 19.22   Power consumption.

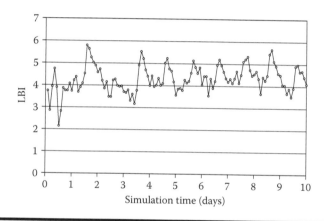

Figure 19.23   Load balancing index.

LBI indicates a higher workload distribution.

$$\text{Load balancing index} = \sqrt{\frac{\sum_{i}^{N}(X_i - \mu)^2}{N}} \qquad (19.4)$$

where

$X_i = \dfrac{\text{the number of messages processed by agents running on platform } i}{\text{the amount of resources utilized by platform } i \text{ and agents running on platform } i}$

$\mu = $ the average of X_i

$\quad = \dfrac{\text{the total number of messages processed by all agents}}{\text{the total amount of resources utilized by all platforms and all agents}}$

$N = $ the number of available platforms

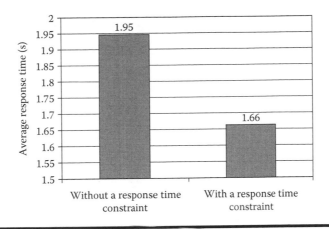

Figure 19.24   Average response time with and without a constraint.

As Figure 19.23 shows, agents and platforms always strive to improve workload distribution through evolution. This is an example of *symbiotic emergence*, a unique property that SymbioticSphere exhibits. in SymbioticSphere. Agent migration behavior policy encourages agents to move toward platforms running on healthier hosts. Platform replication behavior policy encourages platforms to replicate themselves on healthier hosts. As a result, service requests are processed by agents that are spread over the platforms running on healthy hosts. This contributes to balance workload per platform, although agent migration policy and platform replication policy do not consider platform population or workload distribution. This results in a mutual benefit for both agents and platforms. Platforms help agents decrease response time by making more resources available for them. Agents help platforms to keep their stability by distributing workload on them (i.e., by avoiding excessive resource utilization on them).

Figures 19.24 and 19.25 show the response time results with and without a QoS constraint. The constraint specifies 5 s as maximum (or worst) response time. Figure 19.24 shows that the constraint contributes to improve the average response time from 1.95 to 1.66 s. Figure 19.25 shows the number of constraint violations (i.e., the number of simulation cycles in which actual response time exceeds a given constraint). With a constraint enabled, the number of constraint violations dramatically reduces by 90% (from 849 to 78). This is because it is unlikely that agents are selected as mates when they violate a constraint. SymbioticSphere performs its evolution process to better satisfy a response time QoS constraint.

Figures 19.26 and 19.27 show the LBI results with and without a QoS constraint. The constraint specifies 30% as the maximum (or worst) difference in LBI of an agent and another running on a neighboring platform. Figure 19.26 shows that the constraint contributes to improve LBI from 4.31 to 3.96. Figure 19.27 shows that the number of constraint violations reduces by 70% (from 4841 to 1336).

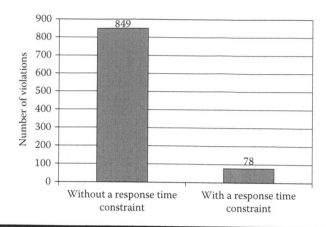

Figure 19.25   Number of constraint violations in response time.

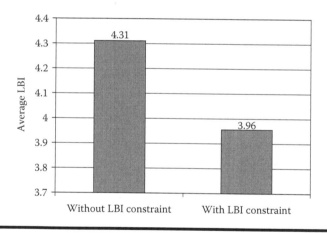

Figure 19.26   Average LBI with and without a constraint.

This is because it is unlikely that agents are selected as mates when they violate a constraint. SymbioticSphere performs its evolution process to better satisfy an LBI QoS constraint.

Figures 19.28 and 19.29 show the throughput results with and without a QoS constraint. The constraint specifies three messages/sec as the minimum (or worst) throughput. Figure 19.28 shows that the constraint contributes to improve the average throughput from 99.50% to 99.92%. Figure 19.29 shows that the number of constraint violations reduces by 70% (from 6554 to 1992). This is because it is unlikely that agents are selected as mates when they violate a constraint. SymbioticSphere performs its evolution process to better satisfy a throughput QoS constraint.

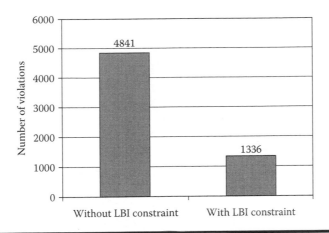

Figure 19.27 **Number of constraint violations in LBI.**

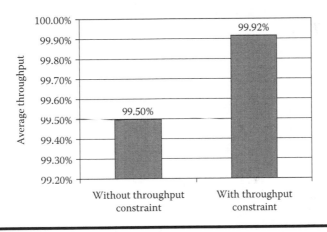

Figure 19.28 **Average throughput with and without a constraint.**

19.4.4 Evaluation of Scalability

This section evaluates how agents and platforms autonomously scale to demand volume and network size. In this simulation study, service request rate starts with 3000 requests/min, spikes to 210,000 requests/min at 8:00, and drops to 3000 requests/min at 16:30 (Figure 19.30). The peak demand and spike ratio (1:70) are obtained from a workload trace of the 1998 World Cup web site.[15] A simulated network is 7 × 7 (49 hosts) from 0:00 to 12:00 and 15 × 15 (225 hosts) from 12:00 to 24:00. This simulates that new hosts are added to a server farm at 12:00. Among the hosts (resource-rich hosts) 30% have 320 MB memory space. The other 70% are resource-poor hosts, which have 256 MB memory space. Both types of hosts are placed randomly in a server farm. At the beginning of a simulation, a single agent

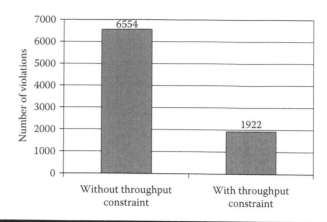

Figure 19.29   Number of constraint violations in throughput.

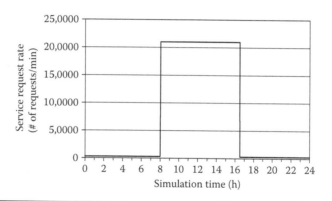

Figure 19.30   Service request rate.

and platform are deployed on each host (i.e., 49 agents and 49 platforms on 49 hosts). This simulation study uses the genes (i.e., behavior policies) obtained after a 10 day simulation in Section 19.4.3 as the initial genes for each agent and platform.

Figure 19.31 shows how the number of agents changes against the dynamic changes in service request rate and network size. At the beginning of a simulation, the number of agents is about 50 in SymbioticSphere and BNA because this number of agents is enough to process all service requests (3000 requests/min). When service request rate spikes at 8:00, agents gain more energy from the user and reproduce/replicate offspring more often. From 10:00 to 12:00, agent population does not grow due to resource limitation on available hosts. (Agents and platforms cannot reproduce/replicate any more offspring because they have consumed all the resources available on 49 hosts.) When the network size expands at 12:00, agents rapidly reproduce/replicate offspring on newly added hosts. When service

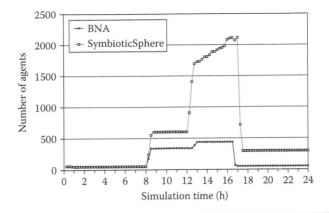

Figure 19.31   The number of agents.

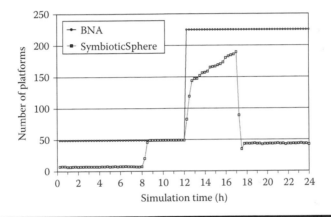

Figure 19.32   Number of platforms.

request rate drops at 16:30, many agents die due to energy starvation. This result demonstrates that the biologically inspired mechanisms in SymbioticSphere allow agents to adaptively adjust their availability to dynamic network conditions such as demand volume, spike ratio, and network size. Figure 19.31 also shows that agent population follows workload changes better in SymbioticSphere than in BNA.

Figure 19.32 shows how the number of platforms changes against the dynamic changes in service request rate and network size. In SymbioticSphere, at the beginning of a simulation, the number of platforms is about seven because this number of platforms is enough to run agents for processing service requests. When service request rate spikes at 8:00, agents gain more energy from the user and transfer more energy to underlying platforms. As a result, platforms also increase their population. From 10:00 to 12:00, platform population does not grow because platforms have

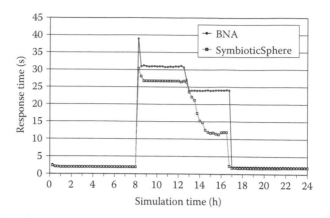

Figure 19.33   Response time.

already run on 49 hosts. When network size expands at 12:00, platforms rapidly reproduce offspring on newly added hosts. When service request rate drops at 16:30, most platforms die due to energy starvation. This result demonstrates that the biologically inspired mechanisms in SymbioticSphere allow platforms to adaptively adjust their availability to dynamic network conditions such as demand volume, spike ratio, and network size. In BNA, platforms are not designed as biological entities; they do not adaptively change their population.

Figure 19.33 show how the average response time changes. In SymbioticSphere, from 0:00 to 8:00, the average response time is maintained very low, at less than 5 s. At 8:00, response time spikes because service request rate spikes. It starts decreasing at 12:00 when agents/platforms reproduce/replicate offspring on newly added hosts. It is approximately 12 s at 16:00. At 16:30, response time drops because service request rate drops. Figure 19.33 demonstrates that the biologically inspired mechanisms in SymbioticSphere allow agents and platforms to strive to keep response time low despite demand surges. In BNA, agents do not adapt their population well to the changes in service request rate and network size (Figure 19.31); response time is rather high, at about 24 s, at 16:00.

Figure 19.34 shows how throughput changes. In SymbioticSphere, by changing their populations, agents and platforms autonomously adapt throughput to dynamic changes in demand (at 8:00 and 16:30) and network size (at 12:00). From 0:00 to 8:00, throughput is maintained very high, i.e., more than 98%. At 8:00, throughput drops (lower than 20%) because the service request rate is very high and agents cannot process all service requests in a timely manner. Then, agents and platforms improve throughput by increasing their populations. Until 12:00, throughput cannot reach 100% due to resource limitation on available hosts. After 12:00, throughput improves to 100% because agents and platforms can replicate/reproduce offspring on newly added hosts. Figure 19.34 shows that the

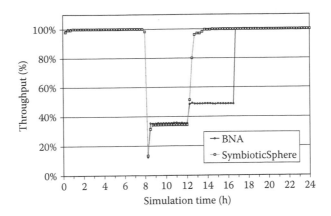

Figure 19.34   Throughput.

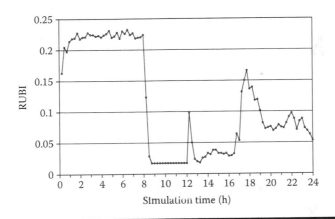

Figure 19.35   Resource utilization balancing index.

biologically inspired mechanisms in SymbioticSphere allow agents and platforms to scale well to demand volume, spike ratio, and network size. SymbioticSphere and BNA yield similar throughput from 0:00 to 12:00. From 12:00 to 16:30, BNA does not leverage newly added hosts well to improve throughput; throughput never reaches 100%.

Figure 19.35 shows how resource utilization is distributed over available hosts in SymbioticSphere. It is measured as resource utilization balancing index (RUBI) with Equation 19.5. A lower RUBI indicates a higher distribution of resource utilization.

$$\text{Resource utilization balancing index} = \sqrt{\frac{\sum_i^N (R_i - \mu)^2}{N}} \qquad (19.5)$$

where

$$R_i = \frac{\text{the amount of resources a platform and agents utilize on the host } i}{\text{the total amount of resources the host } i \text{ has}}$$

$\mu = $ the average of R_i

$$= \frac{\text{the total amount of resources utilized by all agents and platforms}}{\text{the total amount of resources available on all hosts}}$$

$N = $ the number of available platforms

From 0:00 to 8:00, RUBI is relatively constant around 0.2. When service request rate spikes at 8:00, RUBI drops and constantly remains very low, around 0.02, because available resources are fully consumed on hosts. At 12:00, RUBI spikes because resources are not consumed first on newly added nodes. Agents and platforms spread over both resource-rich and resource-poor hosts to deal with a high service request rate. Then, they seek resource-rich hosts through migration, reproduction, and replication, thereby lowering RUBI. When service request rate drops at 16:30, RUBI spikes because many agents and platforms die and resource utilization is not evenly distributed over hosts. However, they decrease RUBI again by preferentially residing on resource-rich hosts. Figure 19.35 shows that the biologically inspired mechanisms in SymbioticSphere allow agents and platforms to adaptively balance their resource utilization over a large number of heterogeneous hosts.

19.4.5 Evaluation of Survivability

This section evaluates how agents and platforms survive partial system failures due to, for example, errors by administrators and physical damages in server farm fabric. This simulation study simulates node failures. Service request rate is constantly 7200 requests/min, which is the peak in a workload trace of the IBM web site in 2001.[16] Randomly chosen 60% of hosts go down at 9:00 for 90 min. The size of a server farm is 7×7, i.e., 49 hosts, each of which has 256 MB memory space. At the beginning of a simulation, an agent and a platform are deployed on each host. This simulation study uses the genes (i.e., behavior policies) obtained after a 10 day simulation in Section 19.4.3 as the initial genes for each agent and platform.

Figures 19.36 and 19.37 show how the number of agents and the number of platforms change over time, respectively. When a host goes down, agents and platforms crash and die on the host. This is why agent and platform populations drop at 9:00. In SymbioticSphere, after hosts fail, the remaining agents and platforms increase their populations with replication and reproduction on available hosts. Around 9:45, agent and platform populations revert to the populations that they had before host failures. When failed hosts resume, platforms reproduce/replicate offspring on those hosts. Some of the agents migrate to the reproduced/replicated platforms in order to move toward the user and increase

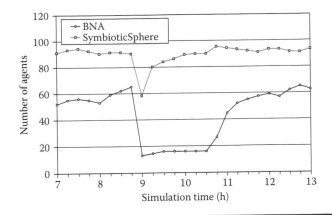

Figure 19.36   The number of agents.

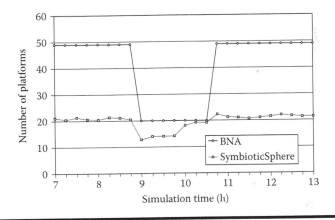

Figure 19.37   Number of platforms.

their population. The biologically inspired mechanisms in SymbioticSphere allow agents and platforms to autonomously survive host failures by adjusting their populations. Figure 19.36 also demonstrates that agent population recovers faster in SymbioticSphere than in BNA. This is because BNA does not consider workload distribution. All agents migrate toward the user; when hosts fail, most of them die. SymbioticSphere better considers survivability. In BNA, platforms are not designed as biological entities; the number of platforms is always equal to the number of hosts (Figure 19.37).

Figure 19.38 shows the average response time. At 9:00, response time increases because all agents and platforms die on failed hosts. However, response time quickly recovers in SymbioticSphere by increasing agent and platform populations immediately (Figures 19.36 and 19.37). The biologically inspired mechanisms in

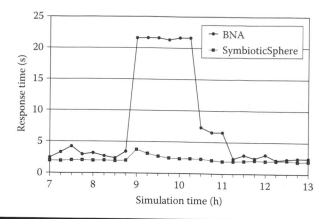

Figure 19.38   Response time.

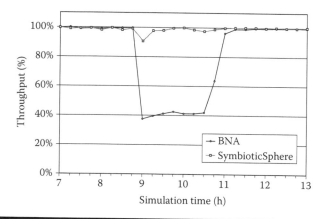

Figure 19.39   Throughput.

SymbioticSphere allow agents and platforms to autonomously survive host failures and strive to retain response time. In BNA, response time recovers when failed hosts resume at 10:30. SymbioticSphere is more resilient against host failures than BNA.

Figure 19.39 shows the throughput of agents. Throughput decreases when hosts fail at 9:00; however, it decreases only by 10% in SymbioticSphere. Upon host failures, throughput quickly recovers by increasing agent and platform populations immediately (Figures 19.36 and 19.37). Figure 19.39 demonstrates that the biologically inspired mechanisms in SymbioticSphere allow agents and platforms to autonomously retain throughput despite host failures. In BNA, throughput dramatically drops to 40% and does not recover until failed hosts resume.

19.5 Related Work

SymbioticSphere is an extension of BNA.[9,12,13] In BNA, biologically inspired agents evolve and perform service adaptation in a decentralized manner. However, as demonstrated in Section 19.4, SymbioticSphere almost always outperforms BNA. This is because the evolutionary process is designed in a more sophisticated way in SymbioticSphere. For example, mate selection, mutation, and crossover are carefully designed to better address real-value optimization to tune behavior policies. Also, in BNA, the threshold values of behavior policies are not included in genes.[12] This means that agent designers need to manually configure them through trial and error. In contrast, no manual work is necessary to configure thresholds in SymbioticSphere because they are included in genes. In addition, BNA uses a fitness function to rank agents in mate selection. It aggregates multiple objectives as a weighted sum. Agent designers need to manually configure these weight values as well. In SymbioticSphere, no parameters exist for ranking agents/platforms because of a constraint-domination ranking mechanism. As a result, SymbioticSphere incurs much less configuration tasks/costs.

Moreover, BNA does not perform resource adaptation because platforms are not designed as biological entities. In SymbioticSphere, both agents and platforms are designed as biological entities; they perform service adaptation and resource adaptation simultaneously.

Jack-in-the-Net (Ja-Net),[17,18] NetSphere,[19] and BEYOND[20] are similar to SymbioticSphere in that they extend BNA. Ja-Net focuses on spontaneous creation of network applications, each of which consists of multiple types of agents. NetSphere and BEYOND propose artificial immune systems for agents to sense network conditions and adaptively select behaviors suitable to the conditions. In Ja-Net, NetSphere, and BEYOND, platforms are not designed as biological entities; they do not address power efficiency, resource efficiency, and resource utilization balancing of network systems.

Wakamiya and Murata propose the concept of symbiosis between groups of peers (hosts) in peer-to-peer networks.[21] Peer groups symbiotically connect or disconnect with each other to improve the speed and quality of queries. A special type of peers implement the symbiotic behaviors for peer group connection/disconnection. Since the number of the symbiotic peers is statically fixed, they do not scale to network size and traffic volume. They also do not address power efficiency, resource efficiency, and survivability of network systems. In SymbioticSphere, all agents and platforms are designed to interact in a symbiotic manner. They scale well to network size and traffic volume, and achieve power efficiency, resource efficiency, and survivability.

The concept of energy in SymbioticSphere is similar to money in economy. MarketNet[22] and WALRAS[23] apply the concept of money to address market-based access control for network applications. Rather than access control, SymbioticSphere focuses on the adaptability, scalability, and survivability of network systems.

Resource Broker is designed to dynamically adjust resource allocation for server clusters via a centralized system monitor.[24] Resource Broker inspects the stability and resource availability of each host, and adjusts resource allocation for applications. In SymbioticSphere, agents and platforms adapt their populations and locations in a decentralized way. Also, Resource Broker does not consider the power efficiency of server clusters.

Shirose et al. propose a generic adaptation framework for grid computing systems.[25] It considers both service adaptation and resource adaptation. In this framework, centralized system components store the current environment conditions, and decide which adaptation strategy to execute. In contrast, SymbioticSphere does not assume any centralized system components. Each of the agents and platforms collects and stores environment conditions and autonomously decides which behaviors to invoke.

Rainbow[26] investigates the adaptability of server clusters. A centralized system monitor periodically inspects the current environment conditions (e.g., workload placed on hosts) and performs an adaptation strategy (e.g., service migration and platform replication/removal). SymbioticSphere implements a wiser set of adaptation strategies such as agent replication/reproduction (service replication/reproduction) and agent death (service removal). SymbioticSphere also addresses survivability and power efficiency as well as adaptability with the same set of agent/platform behaviors. Rainbow does not consider power efficiency and survivability.

Adam et al. propose a decentralized design for server clusters to guarantee response time to users.[27] In SymbioticSphere, agents and platforms evolve to satisfy given QoS constraints including response time; however, they do not guarantee QoS measures because the dynamic improvement of those measures is an emergent result from collective behaviors and interactions of agents and platforms. As a result, agents and platforms can adapt to unexpected environment conditions (e.g., system failures) and survive them without changing any behaviors and their policies. Moreover, Adam et al. do not consider to satisfy other QoS constraints such as throughput as SymbioticSphere does. They also do not consider consider power efficiency and survivability of network systems as SymbioticSphere does.

19.6 Concluding Remarks

This chapter describes the architectural design of SymbioticSphere and explains how it implements key biological principles, concepts, and mechanisms to design network systems. This chapter also describes how agents and platforms interact with each other to collectively exhibit the emergence of desirable system characteristics such as adaptability, scalability, and survivability. Simulation results show that agents and platforms scale well to network size and demand volume and autonomously adapt to dynamic changes in the network (e.g., user location, network traffic, and resource availability). They also survive partial system failures such as host failures to retain their availability and performance.

Several extensions to SymbioticSphere are planned. In order to further explore the impacts of symbiosis between agents and platforms on their performance, it is planned to implement and evaluate new types of behaviors, called symbiotic behaviors, for agents and platforms. It is also planned to deploy and empirically evaluate SymbioticSphere on a real network such as PlanetLab.*

References

1. L. Northrop, R. Kazman, M. Klein, D. Schmidt et al., Ultra-large scale systems: The software challenge of the future. Technical Report, Software Engineering Institute.
2. R. Alexander, *Energy for Animal Life* (Oxford University Press, New York, 1999).
3. S. Camazine, N. Franks, J. Sneyd, E. Bonabeau, J. Deneubourg, and G. Theraula, *Self-Organization in Biological Systems* (Princeton University Press, Princeton, NJ, 2003).
4. R. Albert, H. Jeong, and A. Barabasi, Error and attack tolerance of complex networks, *International Journal of Nature.* **406**(6794), 378–382 (2000).
5. N. Minar, K. Kramer, and P. Maes, *Cooperating Mobile Agents for Dynamic Network Routing.* In ed. A. Hayzelden, *Software Agents for Future Communications Systems*, Chapter 12 (Springer, Berlin, 1999).
6. S. Forrest, A. Somayaji, and D. Ackley, Building diverse computer systems. In *Proceedings of the 6th Workshop on Hot Topics in Operating System*, pp. 67–72 (1997).
7. P. Stiling, *Ecology. Theories and Applications* (Prentice-Hall, Upper Saddle River, NJ, 2002).
8. L. Margulis, *Symbiotic Planet: A New Look at Evolution* (Basic Books, Amherst, MA, 1998).
9. J. Suzuki and T. Suda, A middleware platform for a biologically inspired network architecture supporting autonomous and adaptive applications, *IEEE Journal on Selected Areas in Communications.* **23**(2), 249–260 (2005).
10. C. Gunaratne, K. Christensen, and B. Nordman, Managing energy consumption costs in desktop PCs and LAN switches with proxying, split TCP connections, and scaling of link speed, *International Journal of Network Management.* **15**(5), 297–310 (2005).
11. G. Gibson, Magic packet technology, *Advanced Micro Devices (AMD).* (1995).
12. T. Nakano and T. Suda, Self-organizing network services with evolutionary adaptation, *IEEE Transactions on Neural Networks.* **16**(5), 1269–1278 (2005).
13. T. Suda, T. Itao, and M. Matsuo, The bio-networking architecture: the biologically inspired approach to the design of scalable, adaptive, and survivable/available network applications. In ed. K. Park, *The Internet as a Large-Scale Complex System* (Princeton University Press, Oxford, UK, 2005).
14. C. Lefurgy, K. Rajamani, F. Rawson, W. Felter, M. Ki-stler, and T. Keller, Energy management for commercial servers, *IEEE Journal on Computer.* **36**(12), 39–47 (2003).

---

* http://www.planet-lab.org/

15. M. Arlitt and T. Jin, A workload characterization study of the 1998 World Cup Web Site, *IEEE Journal on Network.* **14**(3), 30–37 (2000).

16. J. Chase, D. Anderson, P. Thakar, A. Vahdat, and R. Doyle, Managing energy and server resources in hosting centers. In *Proceedings of the 18th ACM Symposium on Operating System Principles*, pp. 103–116. ACM, New York (2001).

17. T. Itao, S. Tanaka, T. Suda, and T. Aoyama, A framework for adaptive ubicomp applications based on the Jack-in-the-Net architecture, *International Journal of Wireless Networks.* **10**(3), 287–299 (2004).

18. T. Itao, T. Nakamura, M. Matsuo, T. Suda, and T. Aoyama, Adaptive creation of network applications in the Jack-in-the-Net architecture. In *Proceedings of the 2nd IFIP-TC6 Conference on Networking Technologies, Services, and Protocols; Performance of Computer and Communication Networks; and Mobile and Wireless Communications*, pp. 129–140, Pisa, Italy (2002).

19. J. Suzuki, Biologically-inspired adaptation of autonomic network applications, *International Journal of Parallel, Emergent and Distributed Systems.* **20**(2), 127–146 (2005).

20. C. Lee, H. Wada, and J. Suzuki, Towards a biologically-inspired architecture for self-regulatory and evolvable network applications. In eds. F. Dressler and I. Carreras, *Advances in Biologically Inspired Information Systems: Models, Methods and Tools*, chapter 2, pp. 21–45 (Springer, Berlin, Germany, August, 2007).

21. N. Wakamiya and M. Murata, Toward overlay network symbiosis. In *Proceedings of the 5th IEEE International Conference on Peer-to-Peer Computing*, pp. 154–155, Konstanz, Germany (2005).

22. Y. Yemini, A. Dailianas, and D. Florissi, Marketnet: A market-based architecture for survivable large-scale information systems. In *Proceedings of the 4th ISSAT International Conference on Reliability and Quality in Design*, pp. 1–6 (1998).

23. M. Wellman, A market-oriented programming environment and its application to distributed multicommodity flow problems, *Journal of Artificial Intelligence Research.* **1**, 1–23 (1993).

24. A. Othman, P. Dew, K. Djemamem, and I. Gourlay, Adaptive grid resource brokering. In *Proceedings of the 5th IEEE International Conference on Cluster Computing*, pp. 172–179 (2003).

25. K. Shirose, S. Matsuoka, H. Nakada, and H. Ogawa, Autonomous configuration of grid monitoring systems. In *Proceedings of the 4th International Symposium on Applications and the Internet Workshops*, pp. 651–657 (2004).

26. D. Garlan, S. Cheng, A. Huang, B. Schmerl, and P. Steenkiste, Rainbow: architecture-based self-adaptation with reusable infrastructure, *IEEE Journal on Computer.* **37**(10), 46–54 (2004).

27. C. Adam and R. Stadler, Adaptable server clusters with QoS objectives. In *Proceedings of the 9th IFIP/IEEE International Symposium on Integrated Network Management*, pp. 149–162, Nice, France (2005).

Index

A

ACO, *see* Ants colony optimization
Adaptive social hierarchy (ASH)
    agent-based approach (*see* Agent-based
        approach)
    1-ASH (*see* One-way ASH)
    attributes to rank
        buffer space, 339
        connectivity, 338–339
        energy/lifetime, 338
        functions, 339
        levels and loops, 339–340
    bubble sort, 345
    CAR, 341–342
    cross-layer protocol
        application, 344
        medium access, 342–343
        routing and replication, 343
    formation and maintenance
        communication, 308–309
        dominance structures, 309
        primates, 309
        tournaments, 309–310
    LEACH, 344
    network nodes, 306–307
    network-wide ranking system, 344
    nonlinear hierarchy, 345
    pairwise ASH (*see* Pairwise ASH)
    predation, 310
    random sorting networks, 345
    social agglomeration and organization,
        310
    Spray and Focus enhancement, 340–341
    wireless networks
        construction, 313–315
        Data Mule system, 311
        mapping, 311–312

        network protocol design, 312–313
        node management, 312
        routing protocol, 311
Aeschna cyanea, 9
Agent-based approach
    agent rules
        "Hello" packets, 332–333
        maximum rank error, 334–335
        network edge density, 333
        pseudo-edges, 333
        rank/attribute pairs, 332
        ranking trajectories, 334–335
    network connectivity, 332
    pseudo-mobility, 332
    realistic network meetings
        mobility model, 334, 336
        random walk model and random
            waypoint model, 336–338
Agent migration behavior policy, 507
Alouatta palliata, 12
Animal communication
    animal models, 6–7
    bio-inspired modeling approach
        agent-based modeling, 10–11
        ants task allocation, 7–8
        collective group decision making, 10
        cooperative behavior, 8
        dragonfly model, 9
        internal states, 10
        Khepera robots, 8
        learning, memory, and personality, 10
        noninsect, 9–10
        social insects, 7
        social stimuli, 8–9
    biological model
        eavesdropping, 14–15
        personality, 12–13

521

Christian Missions in Nigeria 1841–1891

Ibadan History Series
Editor Dr. K. O. Dike
Vice-Chancellor,
University of Ibadan

Christian Missions in Nigeria 1841-1891 The Making of a New Élite

J. F. ADE AJAYI
Professor of History,
University of Ibadan

Northwestern University Press
Evanston
1965

First published in the U.S.A. by
Northwestern University Press
under arrangement with Longmans,
Green and Co. Ltd., London

Library of Congress Catalog Number: 65–20800

PRINTED IN GREAT BRITAIN

To Adukẹ

Contents

List of Maps

List of Illustrations

Abbreviations

C.M.S.	Church Missionary Society
Meth. (or W.M.M.S.)	Wesleyan Methodist Missionary Society
U.P.	United Presbyterian Church of Scotland
S.B.C.	Southern Baptist Convention, U.S.A.
S.M.A.	Society of African Missionaries (Société des Missions Africaines, Lyons)
SAISC	Society for the Abolition of Inhuman and Superstitious Customs in Calabar
PP.	Parliamentary Papers
SP.	State Papers
DNB.	Dictionary of National Biography

Introduction to the Ibadan History Series

THE *Ibadan History Series* grew out of the efforts of some members of the Department of History, Ibadan University, Nigeria, to evolve a balanced and scholarly study of the history of African peoples south of the Sahara. In the years before the Second World War, the study of African history was retarded, and to some extent vitiated, by the assumption of many scholars that lack of written records in some areas of Africa meant also the absence of history. Documentary evidence had become so overwhelmingly important for the European scholar that he tended to equate written documents with history, and to take the absence of documents to mean the absence of events worthy of historical study. As a result in the nineteenth century, when Europe occupied Africa, her scholars did not attempt to understand or to build on the historical traditions in existence there; they sought instead to challenge and to supplant them. The history of European traders, missionaries, explorers, conquerors and rulers constituted, in their view, the sum total of African history.

Fortunately for the historian of today, African historical consciousness remained alive throughout the period of colonial rule: that tradition was too much a part of the African way of life to succumb to the attacks of the European scholar. Even in the heyday of white supremacy some educated Africans of the period were sufficiently dominated by their past to feel impelled to commit to writing the laws, customs, proverbs, sayings and historical traditions of their own communities. Notable among these may be mentioned James Africanus Horton of Sierra Leone, Reindorf and Sarbah of Ghana, Otomba Payne and Samuel Johnson of Nigeria, Apolo Kagwa of Uganda, to name but a few. The published works they left behind have become important sources of African history today; but they were swimming against the current of their time and made little impression on contemporaries. Historians continued to write as if Africans were not active participants in the great events that shaped their continent.

The decided change towards a new African historiography came with the movement towards independence. African nationalists rejected the European appraisal of their past. They demanded a new orientation and improved educational facilities to effect this reappraisal. With the establishment of new universities in Africa, it was inevitable that the teaching of history and the training of African historians would receive a new impetus. For obvious reasons the changeover was slow in coming. Even in the new universities the old theories for a time prevailed: besides European history, there were courses only on 'European activities in Africa' at the undergraduate level, and at the postgraduate level research was generally on British and French policy towards their African territories.

By the late 1940's, however, African research students were insisting that African history must be the history of Africans, not of Europeans *per se* in Africa; that local records and historical traditions must be used to supplement European metropolitan archives; in short, that Oral Tradition must be accepted as valid material for historical reconstruction.

x

No doubt the validity of non-written sources for historical research had been pointed out before, but it was new for university departments of history to accept it, especially in relation to African Oral Tradition. Even then not everyone was happy about it. Anthropologists replied cautiously that Oral Tradition, even when seemingly factual, was not history and could only be interpreted in terms of its functions in society and within the particular culture. But this did not destroy its validity as material for history; it only argued for a return to the link between history and sociology advocated in the fourteenth century by the famous Tunisian historian, Ibn Khaldum.

Even in studies of European impact on African societies and cultures, where European archival material still remains our major source, this source should be checked and supplemented by Oral Tradition, material artefacts and other sources of history in Africa. The achievement of the present position in the study of African history has been the result of individual and co-operative efforts of many scholars in different parts of the world, but I think it is fair to say that the Universities in Africa, and Ibadan in particular, have played and are playing their part in this pioneering work.

The History Department here has always tried to reflect the new approach to African history. It has pioneered some of the recent studies into African indigenous history and culture. These include the scheme for the Study of Benin History and Culture and two other schemes now in progress, concerned with the cultural history of the peoples of Northern and Eastern Nigeria. Our staff now include the largest concentration of trained African historians to be found anywhere in a single institution. Our postgraduate school is also expanding.

Hitherto, the fruits of our research have been published largely in articles in the *Journal of the Historical Society of Nigeria*. The aim of the Ibadan History Series is to facilitate the publication in book form of some of the major works which are beginning to emerge from the Ibadan School of History and to make available to a growing public, with the minimum of delay, the results of the latest contributions to our knowledge of the African past.

K. ONWUKA DIKE

Ibadan, Nigeria
18th January, 1965

Introduction

THE years 1841–1891 covered, roughly, the last half-century before the establishment of British rule in Nigeria. 1841, the year of the first Niger Expedition, marked the beginning of the movement to re-establish Christianity in this country, following the failure of earlier Catholic missions in Benin and Warri. 1891, the year of Bishop Crowther's death, marked the end of the first phase of this new movement, the phase when the success of the missionary enterprise was associated largely with the creation and the encouragement of a Western-educated and Christian middle class.

For the history of Christian missions in Nigeria, this first phase was only the 'seedling' time in preparation for the great expansion that came later with British rule. For the history of Nigeria, however, it was in this earlier period that the work of the missionaries has its greatest significance. After 1891 their expansion was largely incidental to the establishment of the colonial administration. Before 1891 they had a greater measure of initiative and their work had its own decisive influence. Things had not 'fallen apart'.

With the exception of Lagos, the different city-states and kingdoms, towns and villages, although increasingly under pressure from the British navy along the coast, still retained enough political authority and cultural stability to deal with missionaries more or less on a basis of equality. There was no rush to join the Churches. Conversion among men in authority was negligible, except within the Delta states. For the most part, a dialogue was still possible between missionaries and the different communities, and there was room for ideas and personalities on both sides. It was not enough for the missionaries to be Europeans to be believed. They had to use education and the technology of Europe to argue, and to convince people. Later, the missionaries as Europeans became like gods, and tended to treat their parishioners as less than men. The dialogue was virtually suspended, for gods have no need to argue. The missionaries were able to exploit the prestige and the power of the white man already won by the colonial soldiers and administrators. It was then that, in the non-Muslim areas at least, the fabric of the old society gave way and people began to flock to the missions.

In the period 1841–91 there were five principal missionary societies working in Nigeria: the Anglican Church Missionary Society (C.M.S.), many of whose missionaries at this time were Germans; the Wesleyan Methodist Missionary Society, a committee of the English Methodist Conference; the Foreign Mission Committee of the United Presbyterian Church of Scotland; the Foreign Mission Board of the Southern Baptist

Convention of the United States; and the Catholic Society of African Missions (the *Société des Missions Africaines*, S.M.A.) of France. There were some important differences between the different missionary societies, both of doctrine and approach, and some of these differences will be brought out where they are essential for understanding their work. Each mission tended to emphasize how much it differed from all the others, but it is possible to exaggerate these denominational differences. From the point of view of Nigerians, what they had in common was far more impressive. In this work, at any rate, the different missionaries, whether British, German, American or French, Catholic, Baptist, Methodist, Presbyterian or Anglican will be considered as much as possible together as 'European' missionaries forming one single factor in the history of Nigeria.

It should, however, be pointed out here that the contribution of each mission to this common factor was not equal in either men, material, length of service, or significance. The C.M.S. was the largest and the most significant in this period. Being part of the established Church, and based in London, it had the greatest influence on the British government. Its missionaries were actively connected with the Niger Expedition of 1841. They established their mission in Badagri in 1845, in Abeokuta in 1846, and they led the expansion into the Yoruba country. The Methodists beat them to it at Badagri in 1842, but were poorly represented in Nigeria in this period: Ghana was the focus of their attention, and instead of expanding from Abeokuta and Lagos into the Yoruba country, they expanded to Porto Novo and Dahomey. While the Methodists had only a short-lived mission station on the Niger at Egga, the C.M.S. had several stations all by themselves both on the Niger and in the Delta states. The Presbyterians arrived in Calabar in 1846. They concentrated their attention there almost exclusively, with only a few outstations on the Cross River. The Baptists established their first station at Ijaye in 1853, and expanded into the Yoruba country. But within ten years their work was interrupted by the American Civil War, and they had to start all over again after 1875. The Catholics came even later, in 1867. But hardly had they arrived when the Franco-Prussian War and the consequent civil disturbances in France similarly disrupted their work. Outside Lagos they were just making their initial contacts after 1875. The Holy Ghost Fathers did not arrive in Eastern Nigeria until 1884.

Thus the missionaries touched only three areas of Nigeria in this period: the coastal city states, especially Badagri and Lagos to the west, Calabar, Bonny and Brass to the east, with Warri and Benin, the old centres, conspicuously left out; secondly, the interior of the Yoruba country, especially Abeokuta, Ibadan, Ijaye, Oyo and Ogbomoso, each with a few outstations; and thirdly, the Niger Valley, especially Onitsha, Aboh, Idah, Lokoja and Egga. The missionaries laid the foundations on the coast between 1841 and 1853 and they were most concerned in that period to get the protection of the British navy for their work. Their great success came at the end of 1851 when they got the British government to order the bombardment and occupation of Lagos. This was followed by the opening in 1853–60 of several stations in the interior of the Yoruba country, and on the Niger in 1857. Wars in the interior of the Yoruba country from 1860–65, coupled with the American Civil War and the Franco-Prussian War brought this period of expansion to an end. Further, the annexation of Lagos, showing British territorial ambitions, made enemies for missionaries and led in 1867 to their expulsion from Abeokuta and restrictions on the movement of all Europeans in the Yoruba country. Expansion continued only along the coast. When the missions began to be revived after 1875, it was on the wave of the new imperialism developing in Europe.

In short, the evangelistic work of the missions was patchy in this period. But in their linguistic and educational work, in their economic policies, and above all, in the class of Western-educated élite they were seeking to create, their influence covered the whole country. They concentrated on the South, but it was the Niger-Benue waterway that first attracted them to the country, and the North remained their lodestar. They studied Hausa and Kanuri before they learnt Ibo. The significance of this educational policy was not necessarily parallel with the success or failure of their evangelistic work. In fact, the years 1867–75, when the movement of Europeans into the interior was severely limited, offered the incipient middle class such a chance for development as did not occur again till the 1950s.

This is a revised version of my Ph.D. thesis presented to London University in 1958. A look at the footnotes and the Bibliography will, I hope, reveal something of the range of my indebtedness. Perhaps it will be enough here to make this a general acknowledgement to the many friends, archivists, librarians, missionaries, government officials and colleagues who have all by their kindness and co-operation helped me in writing this book. There are, however, a few names I must mention. First, the Rev. and Mrs. Cecil Roberson, of the Baptist Mission for allowing me to use and re-visit their most valuable collection of archival material; and Dr. Akin Mabogunje who helped me with the maps.

In addition, my thanks are due to Professor Jack Simmons under whom I read history at Leicester and who suggested this subject to me; to Dr. Roland Oliver who did a pioneer study of the Missionary Factor in East Africa and has given me useful advice; to Professor G. S. Graham and the late Dr. E. Martin who introduced me to historical research, to Dr. J. E. Flint and Professor Philip Curtin who read through the earlier drafts and made valuable suggestions; to Messrs Richard Brain and Michael Crowder for editorial assistance; and to Dr. K. Onwuka Dike, Vice-Chancellor of the University of Ibadan and general editor of this series for having encouraged my work all along.

My researches were financed for two years by the University of London who awarded me the Derby Research Studentship and the Institute of Historical Research of the University who awarded me a Fellowship for a third year. A grant from the Nigerian Government enabled me to visit archives in Rome and Paris; another from the Yoruba Historical Research and the Ford Foundation enabled me to visit archives in America.

1 Christianity and Civilization

SOME people see religion as a limited set of personal beliefs about God and worship which can be isolated from a person's general culture and can be changed without necessarily upsetting that person's culture or his world-view. Others see it as an affair of the community so intimately bound up with its way of life that a change of religion necessarily involves a change of culture and the development of a new conscience.

With their emphasis on law, orthodox Muslims generally take the latter, comprehensive view of religion. But in considering conversion to Islam they think of a progression from the limited to the comprehensive view of religion as a growth from the minimal impact to a fuller realization of the faith. When Islam was introduced into Bornu and the Hausa states in the fourteenth and fifteenth centuries,[1] it spread informally at first as a set of ideas about God and worship, accommodated within the converts' monarchical and social customs. It was rather like a fashion associated with the courts and the military, mercantile and literate classes. Yet the elements of this fashion, the learned mallam as teacher, political or medical adviser; the widely-travelled Muslim trader as customer or informant; even immaterial things like charms and amulets, court music, styles of dress and architecture as symbols of status and power, all seemed to lead back to Islam. And the spread of these led to the spread of the religion, down the Niger into Nupe and Igalla, and across the Niger into Yoruba. Then at the beginning of the nineteenth century, with Usuman dan Fodio's *jihad* in the Hausa states, there was a formal attempt to convert Islam from the level of personal beliefs to one of communal law, an attempt to shake off the remnants of traditional customs and to create a theocratic empire where Islamic laws and practices would prevail. Bornu successfully resisted being incorporated into this empire only because its administration as well as its Islamic faith were reformed by El Kanemi. But even areas which had felt only

---

1 See J. S. Trimingham: *A History of Islam in West Africa*, Oxford 1961; J. Greenberg: *The Influence of Islam on a Sudanese Kingdom*, New York 1946, Monographs of the American Ethnological Society; K. O. Dike: 'Sokoto' in *Encyclopaedia Britannica*, 1956 edition.

the minimal impact of Islam were also incorporated. In Nupe there was a disputed succession; at Ilorin there was a quarrel betwoen the military governor and his sovereign at Old Oyo; and two Fulani mallams took advantage of these to convert Ilorin and Nupe into southern outposts of the Fulani empire.

Within Christianity, on the other hand, the element of theology has always been more important than that of law, and priests and kings have by no means always seen eye to eye. As long as the Church was united, Christianity tended to take the view that religion was an affair of the community, and most aspects of life were regulated by religious laws. But the Reformation, the disputes between Catholics and Protestants and between different Protestant denominations, further emphasized theology and the element of personal belief and personal commitment. For a while the communal view survived in the sixteenth century dictum of *cuius regio, eius religio,* in which the Capuchin missionaries who tried to introduce Christianity into Benin placed so much confidence. But Christianity was by then already reflecting the increasingly individualized society of Western Europe. More and more aspects of life were being regulated by ideas and beliefs outside the purview of religion. The European conception of religion became limited in the sense that it was seen as a personal and not a communal affair and that it was confined to only a special area of a man's life.

This development explains at least in part the ineffectiveness of the Christian missionaries in Benin and Warri.[1] The Oba of Benin himself had asked for Portuguese missionaries, but when they arrived in August 1515, he was away fighting the Idah War. He summoned them to join him on the battlefront, asking meanwhile that lessons on religion be postponed 'because he needed leisure for such a deep mystery'. He returned after a year and asked that one of his sons and others of the chiefs be baptized and taught to read. He was soon back at his wars and presumably the missionaries returned home. When three other missionaries arrived in 1538, they found the Oba no longer interested. Portuguese trade in Benin had declined and the missionaries were unwilling to supply ammunition as requested by the Oba. Thereafter the missionary effort in Benin was fitful and intermittent, and it failed completely to displace the traditional religion.

1 A. F. C. Ryder: 'The Benin Missions' and 'Missionary Activity in the kingdom of Warri to the Early Nineteenth Century' in *Journal of the Historical Society of Nigeria,* vol. 2, no. 2, 1961; vol. 2, no. 1, 1960. J. W. Blake: *Europeans in West Africa,* vol. I, pp. 6–12; 58–9; 78–9 (Hakluyt Society Publications, Second Series lxxxv, London 1941); Father Cuthbert O.S.F.C.: *The Capuchins,* London 1928, vol. II, pp. 306–12. Father Ralph M. Wiltgen: *Gold Coast Mission History 1471–1880,* Illinois, USA, 1956, pp. 8–13.

In the middle of the seventeenth century, Spanish and Italian Capuchins made a determined effort. But they held on to the belief that because the Oba was 'adored' by his subjects with 'fear and unbelievable reverence . . . if he were converted to the Faith, the rest of his subjects would easily be won over'.[1] They were given rooms in the palace, but denied free access to the Oba. They saw him only twice in ten months. They were denied the services of interpreters, and when in August 1651 they tried to disturb a religious festival involving human sacrifice, they were thrust out by an angry mob and subsequently deported. A number of missionaries tried to re-enter Benin through Warri, but without much success. In a moment of grave constitutional crisis in 1709–10 apparently one Oba invited the Capuchins back in the hope that their support might be useful, but he did not live long enough to give them a foothold.

In Warri, however, it did appear in the 1570s as if the Portuguese had successfully planted their religion there, since the Olu, anxious to maintain his independence from Benin, decided to enlist Portuguese support. He welcomed Augustinian missionaries from São Tomé and allowed his crown prince to be baptized as Sebastian. This prince later sent one of his sons, Domingos, to Portugal to be educated, and he came back with a Portuguese wife. The son of Domingos by this marriage, Dom Antonio Domingos, when he became the Olu, carried on this tradition of close Portuguese and Christian connections. Indeed, for about a century and a half, 1570–1733, the Warri rulers became well known as professing Christians. But the European religion did not spread beyond the court. Even at court its hold was recognized to be shaky. 'True Christianity,' said the Bishop of São Tomé in 1620, 'is almost wholly confined to the king and the prince; the rest only call themselves Christian to please the king. They take their children to baptism only with the greatest reluctance, believing that a baptized child will die immediately. The majority of them take wives without the sacrament of matrimony, they circumcise their children and practise superstitious rites and sorcery.'[2] Catholic baptism means little without the sacraments, and the Warri rulers could not even ensure a regular supply of resident clergy to administer them. Eventually the traditional religion reasserted itself at court, and from 1733 onwards the ruling family began to turn against Christianity. In the eighteenth century, the missionary enthusiasm of the Catholic countries of Europe declined; the Protestant countries showed none, and therefore few missionaries came to Warri. By the

1 Father Columbin of Nantes to the Vatican, 1641. Ryder, op. cit. JHSN, vol. 2, no. 2, p. 241.
2 Bishop Pedro da Cunha of São Tomé. Ryder, op. cit. JHSN, vol. 2, no. 1, p. 8.

beginning of the nineteenth century there was little to show for the earlier missionary endeavours: only a few relics like the huge cross in the centre of old Warri, a few church decorations surviving among the traditional shrines, a few memories reflected in oral tradition and in the ritual of traditional gods.[1]

There were, no doubt, other subsidiary factors hindering the Christian missionaries: difficulties of language, of transport, of health; inadequate numbers of missionaries and opportunities. Compared with their Muslim counterparts, the Christian missionaries laboured under great disabilities. The Muslims were usually fellow-Africans who, like the Fulani, could settle down or travel regularly with relative ease to the main centres of Islam not only in the Sudan but also in the Middle East and North Africa. The Christians were a few ailing Europeans, struggling to keep alive in the swampy creeks and depending on sailing vessels for communication with their bases in Europe. But the roots of their failure went deeper than this. After all, European traders working under similar disadvantages did not fare so badly in comparison with North African and Sudanese traders. The real problem was that to the people of Benin and Warri religion meant one thing, and to their Christian teachers quite another thing. To a people for whom religion was co-extensive with life, the Europeans presented trade and religion as two separate institutions, championed by two separate sets of people and guided by two different sets of principles. The missionaries were dependent on the traders for their transport and provisions, but they could not convincingly reconcile their teaching with the Atlantic slave-trade and slavery as practised in the New World. Ultimately, as the Dutch said, the two were incompatible. Thus the missionaries concentrated only on the aspect of personal belief and forms of worship, and consequently paid inadequate attention to education. For the same reason, they failed to understand the society they were dealing with. They saw in traditional religion no more than fetishes, idolatry and juju. The people of Warri, said a Capuchin in 1710, were 'obstinate, idolatrous and given to witchcraft and all sorts of abominable vices'.[2] As Dr. Ryder has said, not one of the Christian missionaries for all their devotion 'came near an adequate understanding of the complex religious system they were trying to displace'.

The essential point about the complex religious system was that it was not so much a matter of personal beliefs as the culture of the whole of the community. Religion, it has been said, was 'the cement of goodwill and fear that kept the family as a unit and the village as a distinctive

1 M. J. Bane: *Catholic Pioneers in West Africa*, Dublin 1956. Ryder, op. cit.
2 Father Cipriano, quoted by Ryder, op. cit. JHSN, vol. 2, no. 1.

community'.[1] The welfare of the individual, the family, village or larger community was believed to depend on the members severally and collectively maintaining the right relationships with the ancestors, gods and other unseen powers through a complex system of ritual observances. There were beliefs, of course, about the organic philosophy of the community, the proper relationships between the gods, between them and man, man and woman, the living and the dead; beliefs about the mysteries of life, sickness and death, good and ill fortune, and so on. But there was really no theology in the sense of dogmatic tenets. The traditional religion was an attitude of mind, a way of explaining the world, a way of life. It was expressed in laws and customs hallowed by time and myth as being essential for the well-being not just of the individual, but of the whole community.[2] To the problems of life to which these customs tried to provide an answer, a catechism that was no more than a set of beliefs necessary for personal salvation must have appeared irrelevant.

This is not to say that the traditional culture was rigid and unchanging. Even from our fragmentary knowledge, we know that between 1486 and 1841 there were many political, social and economic changes in Benin and Warri and all along the coast. The power and influence of Benin expanded and contracted; the nature of its monarchy and hence its religion were 'clearly subjected to profound changes'. Many people migrated to the coast to take advantage of the expanding European trade. The development of Warri as an independent state was probably part of this process. In a similar way, old fishing villages developed into the trading city-states of the Delta:[3] Calabar, or old Calabar as the Europeans called it; Kalabari, which they called New Calabar; Bonny and Brass; as well as Lagos and Badagri to the west. The European trade dominated the lives of these states. None of them could maintain its population and prosperity without it. This led to important social and political adaptations. Because of the cosmopolitan nature of the communities, the traditional social unit, the lineage or clan based on blood-relationship, was replaced by the 'House' in which economic and military organization counted for almost as much as kinship, and into which foreign slaves could be formally integrated. There were changes going on, but as long as the traditional rulers remained the arbiters of

1 C. G. Okojie: *Islam Native Laws and Customs*, Lagos 1960, p. 146.
2 Cf. R. G. Horton: The Kalabari World-view: an outline and interpretation, *Africa*, vol. xxxii, nos. pp. 197–220; E. B. Idowu: *Olodumare, God in Yoruba belief*, London 1962, pp. 62f.
3 K. O. Dike: *Trade and Politics in the Niger Delta*, Oxford 1956, chapters 1 and 2. Daryll Forde (ed.): *Efik Traders of Old Calabar*, Oxford 1957. G. I. Jones: *Trading States of the Oil Rivers*, Oxford 1963.

the destinies of their people, they saw to it that these adaptations were inspired by the traditional culture and not by the beliefs and practices of the European traders or missionaries.

They were willing to cultivate new crops from the New World, like maize and cassava. These became important not only on the coast but also over wide areas in the interior. Similarly, items of household use and luxury goods were adopted by the wealthier classes, and some of them, like umbrellas, even became symbols of traditional authority. By the beginning of the nineteenth century it was becoming fashionable in Calabar and Bonny to import pre-fabricated houses and to fill them, to excess, with imported European furniture.[1] Usually, however, their owners lived in traditional houses and displayed these only on special occasions. Above all, the leading traders along the coast were willing that their children, or more usually selected slaves, should undertake the tedious task of acquiring the skill of speaking and writing European languages and of keeping accounts. This they did either in Europe or by a system of apprenticeship on board the trading vessels. This type of education was preferred to the type which interested the missionaries, the memorizing of the catechism and the training of priests to say the mass and administer the sacraments. Secular education was in the hands of traders and they had a fair measure of success. Most European traders had no difficulty in finding interpreters locally. Portuguese was fairly widely spoken in Lagos, Warri and Benin, and English in Brass, Bonny and Calabar. In Calabar the nineteenth-century missionaries found people who were keeping not only regular accounts but also diaries in pidgin English, and had been doing so for half a century or more.

1 Cf. this description by Hope Waddell in 1846 of the state room of King Eyo II of Calabar:

'The floor was covered with oil cloth and the walls well papered with a rich, crimson-coloured paper. The windows were all glazed. There was a profusion of large and handsome mirrors. At each end was one about five feet square. On the wall facing the door were four, each about four feet long and three feet broad, between the windows, and four round, dining-room mirrors in other places. On each side of the door was a good-sized one. These were in rich gilt frames and in large letters on all *King-Eyo Honesty*. There were two good household eight-day clocks and one handsome sideboard clock in a glass frame. They were all going and keeping good time. Three well-cushioned mahogany sofas invited friendly conversation and repose. Two handsomely carved gilt and stuffed armed chairs seemed made for a king and a queen, and a superior rocking-chair, stuffed below and above with hair cloth, afforded so tempting an opportunity for getting rest and air that I wheeled it round facing the door, sat down and rocked myself to sleep. This put an end to my taking notes, of course, but there were many other excellent things, as tables, sideboards, well-covered drawers, glasses, immense blue jugs (holding) 10 or 15 gallons each, decanters (holding) 2 gallons each etc., etc., too numerous and various to mention.' Waddell: *Journals*, vol. 1, p. 56.

But there the cultural influence of Europe stopped. Not one of the coastal states so much in touch with Europeans for so long adopted the religion or the system of government, taxation, or justice, of their European customers and teachers. Rather, while continuing to trade, the coastal peoples deliberately erected barriers to shut off European cultural influence. Missionaries had been allowed to build houses in Benin and Warri, and Portuguese and Brazilian traders were allowed to build barracoons and tenements on the beach some distance from the town of Lagos. But with these exceptions it became a firm policy throughout the coast to prevent Europeans from building on land and to confine them to their trading-hulks. They were also prevented from sailing or travelling inland, with the result that they did not realize that the rivers they had long known as the Escravos, Forcados, Nun, and so on, were outlets of the legendary Niger, until the Lander brothers made the great discovery in 1830. Apart from restrictive laws, the coastal states erected cultural barriers. They tended to formalize the organization and training of the priests of the traditional cults[1] as an answer to the formal organization and training of the European missionaries. Further, they developed new integrative societies like the Ekpe of Calabar, and exploited to the full the traditional cults and religious festivals as symbols of cultural identity. Rather than accept Christianity, they were willing to offer admission into these cults and associations to a few favoured Europeans.

The leaders of the new missionary movement in the nineteenth century were much concerned about this failure of European commerce and Christianity to make a cultural impact on the coastal peoples of Nigeria. They saw it, of course, not as unique, but as part of a general failure in Africa. The experience in Warri was paralleled on a grander, more dramatic scale in the kingdom of the Kongo. And there had been the even greater failure in North Africa, where Christianity took root, flourished and flowered, and yet was nevertheless effectively displaced by Islam. The problem then was not only to re-establish Christianity in Africa, but also to ensure its permanence in the face of the tenacity of the traditional religion and the aggressiveness of Islam.

The religious impetus behind this new missionary movement was largely the evangelical revival of the late eighteenth century which owed so much to the work of John Wesley. This revival created a new and growing Methodist Church and an increasingly powerful evangelical

[1] Cf. G. Parrinder: *West African Religion*, London 1949. Paul Hazoume: 'The Priest's Revolt' in *Présence Africaine*, nos. 8–10 new series, June–November 1956.

party within the established Anglican Church. It infected all Protestants in Europe and North America with a new fervour and zeal in religious matters which resulted in the foundations of various missionary societies in the last decade of the eighteenth century. It was a Methodist leader, Dr. Coke, who in 1787 first drew up a plan for the 'Establishment of Missions to the Heathen', but the Methodist Church did not officially take this up till 1813, when they established the Wesleyan Methodist Missionary Society. In 1792, an English Baptist, William Carey, founded the Baptist Missionary Society. In 1795, attempts were made to found interdenominational missionary societies in London, Edinburgh and Glasgow, but the denominational tradition was too strong. The London society soon became Congregational, and the Scottish ones Presbyterian. In 1799 the evangelicals within the Anglican Church founded the 'Society of Missions to Africa and the East', later known as the Church Missionary Society (C.M.S.). A necessary aide to them all was the British and Foreign Bible Society, founded in 1804 to subsidize the production of the Bible in different languages and in adequate quantities.[1] It was thus the evangelical revival that inspired four of the five missionary societies we are concerned with in this study: the C.M.S., the Methodists, the Presbyterians of Edinburgh, and the Baptists from the Southern Baptist Convention, U.S.A. The fifth, the Society of African Missions from Lyons, France, was a product of the Catholic revival in Restoration France which arose partly as a reaction against the atheism of the French Revolution, and partly as the Catholic answer to the Protestant evangelical revival.

On the face of it, this evangelical Christianity seemed even less likely than sixteenth- and seventeenth-century Catholicism to make permanent impact on traditional religion in Nigeria. It abhorred ritual, dancing and finery in religious ceremonies and distrusted them in social life. Its emphasis was on theology, though of a plain and fundamentalist type, and it carried the element of personal belief and personal commitment so far that some evangelicals claimed they could tell who had achieved personal salvation and at what particular moment in time. Moreover, since denominational rivalries remained so strong in spite of the common inspiration of the evangelical faith, would-be converts inevitably had to be approached as individuals and not as communities. However, it had the advantage of a strong and indomitable faith, as well as a certain egalitarian belief that while all men were sinners until 'regenerated', all were equally capable of 'regeneration'. But probably the most signi-

---

[1] There is a good summary of 'The Missionary Awakening' in C. P. Groves: *The Planting of Christianity in Africa*, London 1948, vol. I, ch. x.

8

ficant advantage of this evangelicalism was its close association with the anti-slavery movement.

The anti-slavery movement had its roots partly in the humanitarian feeling of a small group of people whose hearts were touched when physically brought in contact with the sufferings of the slaves, and partly also in the radical philosophies of the eighteenth-century age of reason and enlightenment, with its ideas of the noble savage, the natural rights of men, the inherent values of liberty in political, social and economic relationships, and the power of environment and law to change the character of man. This humanitarian feeling was at first directed, not to abolishing the status of slavery, but improving the condition of slaves and freed slaves. John Wesley had himself felt for the suffering of the slaves during his visit to America; he had also shared the current ideas about the idyllic existence of the noble savage who must not be disturbed.[1] But as he developed his evangelical doctrine of sin and redemption, he went beyond these humanitarian ideas. He concluded that the noble savage was 'a sinner and degenerate idolater' who must be converted; slavery was not just a cruel and inhuman practice that should be improved but a sin that must be abolished. In this way Wesley brought the whole weight of the evangelical revival behind the anti-slavery movement.

This alliance was by no means inevitable. Slave-owning and extreme evangelicals continued to protest against it on the grounds that slavery was a social and economic, not a spiritual matter; and that the worst crime of the worldly Christianity of the Age of Reason against which evangelicals were reacting was in obscuring the line of division between temporal and spiritual matters. But the main body of evangelicals saw in the anti-slavery campaign a challenge with which to awaken the conscience of the Christian to do his duty to his neighbour. The deadening of this conscience, they felt, was the real guilt of the generations of Christians who had tolerated slavery and had taken little interest in the expansion of the Church. Obscuring the line between temporal and spiritual matters was dangerous, not in itself but only because it could easily lead to this deadening of the conscience. To prevent this the clergy must observe this distinction and take no risks with their own consciences so that they could always keep awake those of their parishioners. But the parishioners themselves must carry their evangelical faith into every aspect of their lives. In short, from a faith withdrawing itself from the world, evangelicalism, largely through its involvement in the anti-slavery movement, produced a pressure group seeking to exert a

1 'Thoughts on Slavery' in *The Works of the Rev. John Wesley M.A.*, London 1872, vol. XI, pp. 64–5.

9

corporate influence on many aspects of public life. It sought to do this, not directly through the clergy, who must remain withdrawn in the background, but through 'influential laymen'.

This involvement in the anti-slavery movement led the evangelicals down many an unexpected path. The movement, itself always a hetero-geneous and unorganized body, had to rely on the effective organization and the enduring devotion of the evangelicals for its propaganda and for carrying out many aspects of its programme. As this programme broadened out to cover the diverse interests of the supporters of the anti-slavery movement, the evangelical missionaries in Africa found themselves champions not only of Christianity, but also of European commerce and civilization.

The trade of Britain with West Africa up to the eighteenth century had been based on its western ports like Bristol and Liverpool. From the end of the century onwards a new group of merchants, at first principally from London, took an interest in developing new fields of trade in West Africa. In contrast to the older established traders who had adjusted themselves to trading only on the coast, they were anxious to penetrate to the interior where larger and more advanced populations were reputed to exist. As a first step, they supported exploration of the interior, especially the Niger waterway. Soon after the Lander brothers' discovery, they were arguing that 'a new hope has been opened for Africa—a new opportunity . . . of bringing into cultivation some parts at least of this vast, neglected Estate, to the great benefit of the world; that it lies with England to improve this opportunity.'[1] Among those in-terested in this commercial penetration were some of those influential laymen whose counsels carried much weight in both evangelical and anti-slavery deliberations. It was Buxton, a leading evangelical and leader of the anti-slavery movement, who took up this commercial argu-ment and in his book *The African Slave Trade and its Remedy* converted it into the anti-slavery slogan.

Buxton argued that the efforts of Britain to stop the slave trade through diplomacy in Europe and naval patrols on the Atlantic had not visibly reduced the number of slaves taken out of Africa; that the only effective remedy was to attack the slave trade at its source of supply in Africa:

> We must elevate the minds of her people and call forth the resources of her soil. . . .
> Let missionaries and schoolmasters, the plough and the spade, go

1 'The Niger Expedition', *Edinburgh Review*, January 1841, vol. 72, p. 456.

together and agriculture will flourish; the avenues to legitimate commerce will be opened; confidence between man and man will be inspired; whilst civilization will advance as the natural effect, and Christianity operate as the proximate cause, of this happy change.[1]

His plan for achieving this comprehensive programme had the attractive merit of simplicity. It was that the British government should supplement the naval patrol trying to blockade the coast with action on the mainland. He urged the government to undertake pioneer expeditions through the large waterways into the interior in order to make treaties with chiefs and to demonstrate what opportunities there were for private capital; that industrialists and merchants should follow the lead of the government and invest capital in the development of Africa; and it would be this new trade that would displace the slave trade. But in view of the fact that European trade on the coast both in slaves and recently in palm oil had had no beneficial effect, the new trade must be carefully organized to stimulate agriculture and to civilize. The old trade was controlled and restricted by the coastal chiefs, the new one must be free; it must produce both a free peasantry and also a new commercial and industrial class. For this reason, and also because of the climate, the government and the merchants supporting this programme should rely on using Africans from Sierra Leone and the Americas as their agents. These Africans, protected by Britain, guided by the missionaries, and working with capital from European merchants, would not—like European merchants—stay shyly away from the people, in hulks along the coast, but move inland and man factories at every strategic point, living together in little colonies, little cells of civilization from which the light would radiate to the regions around. As catechists and schoolmasters, they would preach Christianity; as carpenters, tailors, sawyers, masons and artisans, they would improve the standard of housing and household furniture and build the necessary roads and bridges to make a highway for legitimate trade. They would be commercial agents to encourage the cultivation of crops like cotton and indigo, which they would buy for the European market in return for European manufactures. They would teach new arts and new ideas and

1 T. F. Buxton: *The African Slave Trade and its Remedy*, London 1840, pp. 282, 511. Buxton's ideas were first printed privately in a memorandum: 'Letter to Lord Melbourne' in 1838. Then his review of the *African Slave Trade* was published alone in 1839, his *Remedy* being held over because it was feared that the French or the Arabs might try to forestall Britain on the Niger. The full book was published in 1840. Soon after, a popular abridged version was released, as well as a reprint of all the favourable reviews. A German edition was published in 1841.

in every way bring down the old society on which the slave trade was based and set up in its place a new social order.

Buxton at once set about persuading Lord Melbourne's government to try out these proposals by sending an expedition up the Niger. He gave the manuscript of his book to the government, who withheld publication of his proposals to prevent the French from forestalling British action. Buxton founded 'the Society for the Extinction of the Slave Trade and for the Civilization of Africa', with a Journal entitled *The Friend of Africa*, as well as an Agricultural Society. These, with the support of the evangelicals and the skilful manipulation of public opinion,[1] managed to convince the government. The result was the carefully prepared Niger Expedition of 1841, at a cost of some £100,000.[2] It was to symbolize the whole civilized force of Britain. There were three steam-boats; four commissioners of the government authorized to make treaties and explore the chances for a consul somewhere on the Niger; scientists of all types equipped with the latest instruments to make observations about the climate, the plants, the animals, the soil, the people themselves and their social and political institutions; commercial agents to report about the trade, the currency, the traffic on the river; a chaplain and two C.M.S. missionaries to report on the possibilities of missionary work. In addition there were agents of the Agricultural Society to acquire land at a suitable point near the confluence of the Niger and the Benue and there to establish a model farm that was to be the first cell of civilization. This was to be settled at first by twenty-four Africans who had been recruited after very wide publicity from Sierra Leone. They were to be managed by two British agriculturists, with Alfred Carr, a West Indian 'man of colour', as superintendent, and an African catechist, Thomas King, from Sierra Leone to look after their spiritual welfare before the full resident missionary party arrived.[3]

Little came of this Expedition. Treaties were signed with the Obi of Aboh and the Atta of Igalla; land was acquired at Lokoja and the model farm established. One of the ships went up the river as far as Egga. However, when forty-five of the 150 European members of the Expedition died, the model farm was wound up, the treaties were not ratified,

---

1 'Quite an epitome of the State,' wrote Buxton. 'Whig, Tory, and Radical, Dissenter, Low Church, High Church, tip-top Oxfordism, all united.' E. Stocks: *History of the Church Missionary Society*, vol. I, p. 453.
2 C. C. Ifemesia: 'The "Civilizing" Mission of 1841' in JHSN, vol. 2, no. 3, 1962.
3 Allen and Thomson: *Narrative of the Expedition to the Niger River in 1841*, London 1848; J. F. Schon and S. Crowther: *Journals of the Expedition of 1841*, C.M.S. 1842.

and 'philanthropy' was laughed to scorn as the wishy-washy dreams of old women not fit to guide the actions of governments.[1] In 1843, the African Civilization Society and the Agricultural Society were disbanded. Buxton died broken-hearted two years later. But his influence did not die with him. Between 1839 and 1842 he had given the Niger as much publicity in Europe as any other African territory not the scene of war or within the path of the charismatic David Livingstone was to receive.

It has been necessary to examine these events at some length because, in spite of immediate disillusionment, the Niger Expedition of 1841 marks the beginning of the new missionary enterprise in Nigeria. The publicity set in motion a train of events which the failure of the expedition could not hold back.[2] In particular, Buxton had achieved a far-reaching change in government policy. Hitherto generally averse to getting involved in West African local politics, the government now began to encourage the signing of slave-trade treaties to strengthen the hands of its naval officers by securing for them the support of favourably disposed African rulers on the mainland. That meant that the missionaries could now bank on more effective protection from the anti-slavery preventive squadron. Secondly, wide publicity in Sierra Leone gave new impetus to the spontaneous movement of the Liberated Africans there seeking to return home to Nigeria. In turn this led directly to the extension of the work of the Wesleyan Methodists and the C.M.S. from the Gold Coast and Sierra Leone to Badagri. Scottish missionaries in Jamaica anxious to foster a similar movement from the West Indies received inspiration directly from reading Buxton's works. They established a mission sponsored by the United Presbyterian Church[3] at Calabar in 1846. It was an account of the work of the Methodists at Badagri that led Bowen of the Southern Baptist Convention there in 1850.

Thus the failure of the Niger Expedition did not stop European advance in Nigeria. It only meant that, for a crucial period, the initiative in the matter passed from the government, and even from the anti-slavery movement as such, to the missionary societies. The missionaries in fact set out to try to accomplish the programme outlined by Buxton for the civilization of Africa. It remains to consider more carefully what they meant by civilization.

.   .   .

1 Ten years later Charles Dickens still had England laughing at the Niger Expedition, caricatured in Mrs Jellyby's philanthropic schemes for civilizing Borrioboola Gha in *Bleak House*.
2 Cf. J. Gallagher: 'Fowell Buxton and the New African Policy' in *Cambridge Historical Journal*, vol. XI, 1950.
3 I.e. the United Secession Church which on its union with the Relief Church in 1847 became the United Presbyterian Church.

Roman Catholics had always maintained that membership of the Church in any part of the world was a civilizing process both in the sense that the Church was the fountain-head of European civilization in art, music and literature, and in the old Greek sense that it was only by such membership that a man could fully justify the whole of his being. Most Protestants would have agreed with this view, although they would have emphasized that it was not so much membership of the Church that civilized as the power of the Gospel working in individual hearts. Yet in the middle of the nineteenth century it was precisely the Roman Catholic Church itself in Europe that was being attacked for its opposition to 'civilization' and 'progress'. Many Protestants also implied that there was a distinction between civilization and Christianity when they debated whether one should precede or follow after the other. Clearly, then, Buxton and the missionaries who shared his view meant by civilization more than was implicit in membership of the Church.

Civilization meant to them all that they considered best in their own way of life. In the first place, they expected conformity to their own social manners and customs. They insisted on even minor observances as necessary outward and visible signs of an inward 'civilized' state. The missionary on his first wedding anniversary at Badagri gave a tea-party and called it a token of civilization. He brought out cakes and biscuits, called his friends and assistants to the school-house, and as they sat down to tea commented:

Could our friends but behold the very interesting sight which presents itself and witness the evident token of civilization which on all sides appeared . . . they would be delighted.[1]

Similarly, the new missionary in West Africa did not mind sweating in his clerical black; rather, he was worried that an older missionary should allow himself to be served at table by a young man and a girl each in a state of 'semi-nudity . . . [having] on only the waist cloth, being from the waist upwards and from the knee downwards naked, and that too in the presence of ladies'.[2] There was a proper and an improper way of doing things in Victorian England. And many of the customs and habits were regarded not just as unimportant matters of social convenience; to the missionary, each had a religious significance. He considered it not only more 'civilized' that a boy should bow to his elders instead of prostrating; he considered also that prostration might imply a sinful element of worshipping a human being. Similarly, Victorian Christian

1 Rev. S. Annear, Journal entry for 20 Oct. 1844 (Meth.).
2 Hope Waddell, *Journals*, vol. I, p. 21, giving an account of a visit to Freeman at Cape Coast, on Waddell's first voyage to Calabar, under date 26 Mar. 1846.

virtue demanded that 'nature's secret' be kept. It was permissible perhaps that a Yoruba lady convert should wear her *buba* and *iro* and *gele*,
but it was more civilized still to wear the Victorian frock, high-necked,
long sleeved, reaching down to the ankles.[1] The Victorian frock was an
essential part of the Victorian doctrine of feminine modesty. The missionary who complained about the way houses were built in Badagri
'without any regard to anything like order or convenience'—

> Several times I followed what I supposed to be a public thoroughfare,
> but found it to terminate in a private yard—[2]

soon began to allude to the theological implications of life in the family
compounds. Quite late in the century, a missionary asked:

> Is it proper to apply the sacred name of home to a compound occupied
> by two to six or a dozen men each perhaps with a plurality of wives?[3]

He had in his mind the rows of houses hedging a straight road in his
own village, with just one man and his wife and children in each house.
The European missionary from an individualist society found the
African family system not only odd, but a negation of some of the things
he considered most vital in life, not only monogamous marriage, but
also the freedom of worship and the responsibility of each adult to God
for his own soul.

Social reform is implicit in the preaching of a new religion. Conscious-
ly, as a preacher of Christianity interpreted in the light of European
social and economic history, and unconsciously, as a man produced by
that particular environment and as a man who taught not only by word
but also by example, the missionary brought to Nigeria various aspects
of European life such as fashions in dress, architecture and town-plan-
ning, and even of eating and salutation. However, what distinguished
missionary work in the mid-nineteenth century, what made its social
and economic influence go much further than the limited number of
converted people was that the missionaries who saw civilization as allied
to Christianity attempted more than just a reform of the manners of the
converted. Early Victorian England had seen the coming of railways, of
gas lighting, of public sanitation. The products of the Industrial Revolution were by then beginning to reach down to the masses from among

1 This was less true in the case of men. Henry Townsend at least thought that
  the Yoruba costume 'is very becoming as regards the male sex'. Journal for
  quarter ending December 1847, p. 6. C.M.S. CA2/085.
2 'Journal of H. Townsend while on a Mission of Research', entry for 29 Dec.
  1842; C.M.S. CA1/0215.
3 Rev. W. T. Lumbley: *Foreign Mission Journal*, SBC, May 1897, vol. XLVII,
  no. 13, p. 405.

whom many of the missionaries arose. When the missionaries talked about civilization, it was not so much the reform of manners that they referred to—they took that more or less for granted—as the 'temporal blessings which resulted from the spread of the true religion and its inseparable companion—Civilization',[1] as a missionary put it.

Among these temporal blessings, political changes along the lines advocated by Earl Grey[2] were sometimes included, but generally politics were in the background. The emphasis was on technical and industrial changes. 1851 was the year of the Great Exhibition of Britain's industrial achievements. Prince Albert in opening a jubilee meeting of the Society for the Propagation of the Gospel in that year called the exhibition 'a festival of the Civilization of Mankind . . . This Civilization rests on Christianity, can only be maintained by Christianity.'[3] Buxton's 'Society for the Extinction of the Slave Trade and for the Civilization of Africa', after declaring it their unanimous opinion that 'the only complete cure of all the evils' that the slave trade caused in Africa was 'the introduction of Christianity into Africa', defined their programme of civilization as:

[to] adopt effectual measures for reducing the principal languages of Western and Central Africa into writing;
prevent or mitigate the prevalence of disease and suffering among the people of Africa;
encourage practical science in all its various branches;
investigate the system of drainage best calculated to succeed in a climate so humid and so hot;
assist in promoting the formation of roads and canals, the manufacture of paper and the use of the printing press;
afford essential assistance to the natives by furnishing them with useful information as to the best mode of cultivation, as to the productions which command a steady market and by introducing the most approved agricultural implements and seeds. The time may come when the knowledge of the mighty powers of steam might

1 S. Annear: Journal of a visit to the Encampment, August 1844 (Meth.).
2 In 1853, Earl Grey described Lord John Russell's policy on the Gold Coast as being:
'to keep constantly in sight the formation of a regular government on the European model, and the establishment of civilized policy as the goal ultimately to be attained. . . . The real interest of this country is gradually to train the inhabitants of this part in the arts of Civilization and government'.
Earl Grey: *The Colonial Policy of Lord John Russell's Administration*, London 1853, vol. II, p. 286.
3 Stocks, op. cit., vol. II, p. 12.

16

contribute rapidly to promote the improvement and prosperity of that country.[1]

In spite of the slogan 'The Bible and the Plough',[2] it was not so much agriculture that the missionaries considered the civilizing occupation, as the commerce that resulted from it. Agriculture was recommended to the African as a way of producing the 'legitimate' articles of the trade that would link him with Christian Europe. Agriculture in mid-nineteenth-century England was a respectable occupation of the aristocracy, but civilization was a middle-class affair. When a missionary, an Irishman, on leave from Calabar, visited Newcastle and saw for the first time the London-Edinburgh train, he declared:

> Old things are passed away, and all things are become new. The baronial and feudal age are gone never to be recalled. The railways and trains can never yield to old barbarism. Border warfare and internicene feuds fall before them. The lords of the land and the Queen of the realm must come down from their chargers and state carriages and ride in the cars of commerce made by plebeians for their own use.[3]

By no means every missionary shared the radicalism of this Irishman. But it might be said that for most of them, whether or not they encouraged the African to wear the top hat and drink tea, it was the railways, the cars of commerce, that symbolized the highest achievement of civilization. Also that they considered it the achievement of the middle classes, the type which they wished to see created in Africa.

This desire of mid-nineteenth-century missionaries to create an African middle class must be emphasized. It was reinforced by the argument that, for reasons of climate and of expense, a large part of the missionary staff had to be African. But the aim was often pursued deliberately and for its own sake. 'In the history of man,' said the American pioneer missionary, 'there has been no civilization which has not been cemented and sustained in existence by a division of the people into higher, lower and middle classes. We may affirm, indeed, that this constant attendant upon human society—gradation of classes—is indispensable to civilization in any form, however low or high.' It was to the lack of this gradation of classes that he traced African backwardness. For, he said, in Africa there was

1 'Prospectus of the Society for the Extinction of the Slave Trade and for the Civilization of Africa, instituted June 1839' at the beginning of Buxton: *The African Slave Trade and its Remedy*, pp. 8–16.
2 Buxton adopted the slogan from Read, a missionary in South Africa. Read had said: 'We take a plough with us, but let it be remembered that in Africa the Bible and the Plough go together.' Buxton, op. cit., p. 483.
3 Waddell, *Journals*, vol. X, p. 15.

no class of eminent men whose attainments may give unity, force and direction to Society; no middle class who are prepared by their attainments to receive impulses of knowledge and wisdom and power from their superiors and communicate it to the millions of the common people. With the single exception of political chiefs, themselves barbarians, the whole society of Sudan rests and stagnates on a dead level, and the people remain poor, ignorant and wretched, because they have no superiors.[1]

The emergence of such a class was perhaps the most concrete feature of the social revolution the missionaries envisaged.

There was another reason, more intimately related to their work, why missionaries wished to see an African middle class emerge. Most Catholics accepted as normal that their ordained missionaries should themselves foster the arts of civilization. On the other hand, there was always, as we have noticed, an undercurrent of opinion among Protestants, particularly evangelicals, to the effect that things temporal and spiritual did not mix well together, that ministers and ordained missionaries were essentially preachers of the Word who should not meddle with politics, trade, agriculture, or even education, except in so far as these things directly aided the work of conversion. In spite of the collaboration of evangelicals and humanitarians within the abolitionist movement, the two remained distinct. There were humanitarians like Palmerston who remained eighteenth-century sceptics and there were fervent evangelicals who did not feel the call to reform prisons or mitigate the severity of old laws or even to seek emancipation for slaves. It may be said that it was humanitarians who set out to civilize the African, the evangelicals set out to Christianize him. But as the evangelical missionary wished to see active religion influence every aspect of life at home by emphasizing the role of pious laymen like Buxton in business, government, and society as a whole, so in Africa he looked to the incipient middle class brought up in the mission to lead the civilization movement.

The missionaries placed emphasis on the development of trade because they believed that it would inevitably lead to the formation of such a class who would then themselves begin to carry out the social reforms the missionaries wished to see carried out but would rather not meddle with. Commerce, said Bowen, will aid the 'change in Society which the Gospel seeks'. For example, he said, 'commerce will erect new standards of responsibility and thus remove one of the strongest props of polygamy'.[2] The President of the C.M.S. wrote in a letter on behalf

1 T. J. Bowen: *Adventures and Missionary Labours in several Countries in the Interior of Africa*, New York 1857, pp. 339–40.
2 Ibid.

18

of the Queen to the rulers of Abeokuta in 1850: 'The commerce between nations in exchanging the fruits of the earth is blessed by God.'[1] But the C.M.S. consistently condemned the palm oil trade which was replacing the slave trade on the coast, on the grounds that it was conducted in a way that did not lead to the emergence of a middle class and so left African society unchanged. The commerce the missionaries wanted must penetrate the country, must be based on the produce of peasants, to be collected and processed by the agents of civilization. In short, they aimed more at a policy of planned economic development than just an expansion of trade.

Buxton quite naturally had looked to the Niger as the obvious highway along which to penetrate the country. Events, however, led the missionaries in the first instance not to the Niger but to Badagri and Calabar. It was Badagri in fact that became the gateway. When, in 1842, missionaries established the first station at Badagri and began to penetrate to Abeokuta, rulers everywhere else along the coast, including those at Calabar, were continuing to bar the way to European intervention in the country. It was the southern extension of the Fulani empire, referred to earlier, that had begun, as it were, to weaken the line of defence.

The Fulani conquest of Ilorin had been the last and decisive blow that led to the dissolution of the Old Oyo empire and the consequent Yoruba wars.[2] From about 1825 until they were defeated at the battle of Oshogbo in 1840, Ilorin-based Fulani armies destroyed many large towns in Northern Yoruba. Large masses of people fled southwards, increasing the population of the southern Oyo provinces and, in alliance with Ife and Ijebu adventurers, destroyed the Egba towns, settling on their land and pushing the Egba people further south still. In other words, the Yoruba replied to the Fulani threat by moving southwards and seeking among other things European arms and other means of restoration. But the supply of arms from the south did not solve the political problems in Yoruba. No single leader, town or group was able to restore order throughout the whole country. Internecine wars therefore continued to plague the Yoruba throughout the nineteenth century.

One result of this upheaval was to weaken the traditional social, political and religious ideas and institutions. The new towns and states were experimenting with new forms of government and military organizations. Turbulent characters were challenging tradition in various ways. Confidence in the old gods was shaken at many points. And as the

1 Stocks, op. cit., vol. II, p. 114.
2 J. F. A. Ajayi and R. S. Smith: *Yoruba Wars of the Nineteenth Century*, Cambridge 1964.

century progressed, there was in the air expectation of some change that would restore order and stability. Missionaries in the 1840s and 1850s often reported how they were welcomed in different places and what prophecies had preceded them about the coming of the white man who would herald in an era of peace and prosperity. Generally the missionaries interpreted this to mean that the traditional gods and religious concepts were doomed to an easy extinction. They were often surprised later to find out how stubbornly the old gods held on to life and how instinctively even genuine converts of many years' standing went back to traditional social and religious concepts. Nevertheless war-weariness and the expectation of change provided opportunities for the Christian missions, though not less for Muslim missionaries as well.[1]

One other effect of the Yoruba wars was a temporary intensification of the slave-trade. This was at a time when, on the one hand, the Spanish and Portuguese colonies were asserting their independence, and developing sugar and coffee plantations, often with English capital, and when on the other hand the British preventive squadron on the West African coast was becoming increasingly effective in capturing slave-vessels. Prisoners captured in the wars, as well as stray refugees on the way, were taken to the coast and sold into slavery. One of the earliest victims of this was the boy Ajayi, the future Bishop Crowther, aged about 15, made prisoner in the dry season of early 1821 when his town of Oshogun, a few miles south of Iseyin was sacked. He changed hands several times, but in the end was sold to Portuguese traders. He was put on board the *Esperanza Felix* on 7 April 1822; but, that same evening, while still in the Lagos roads, the Portuguese ship was captured by two ships of the British navy.[2]

When he arrived in Sierra Leone in June 1822 to embark on his extraordinary career, Ajayi was one of the earliest Yoruba there. By 1827, largely on account of the wars mentioned above, the Yoruba had

---

1 Cf. T. J. Bowen, op. cit., p. 113: 'At the time of my arrival in the country, many of the Egbas and Yorubas, looking round on their ruined country, felt sick of war and the slave trade, and sighed for a return of their former peace and prosperity.' Also D. Hinderer's conversations on his first visit to Ibadan with the Agbakin 'whose favourite topic of conversation is the peace that is to come'. Journal entries for 2, 11 and 22 June 1851 (C.M.S. CA2/049). Hinderer more than anybody else was conscious of the Christian/Muslim confrontation in the Yoruba country at this time, e.g. his Journal entry for 15 April 1855 reports an encounter with a Muslim missionary: 'He says Mohammedans must conform a little with heathen fashion because they are not yet enough in number and power to get on without.' Hinderer replied: 'Our power must be in our religion, not in our number, nor in our position.'
2 S. Crowther to Major Straith, 22 Feb. 1837; C.M.S. CA1/089. H. Macaulay: 'The Romantic Story of the life of a little Yoruba boy named Adjai' in *Nigeria Magazine*, November 1946.

become a recognizable group in the colony. They were called Akus (Ackoos, Acoos, Ockoos) because of the way they greeted, which also shows that up till then the majority of them were Oyo, since it is the Oyo-Yoruba who greet in this way. In 1830 John McCormack, an old resident in Sierra Leone, told a Parliamentary Committee that the 'Akoos' or 'Eayows' were the predominant group in the colony.[1]

During the 1830s the Egba became more numerous in Sierra Leone than the Oyo, for it was the Egba who paid the greatest price for the revolution in the Yoruba country. For this reason they were also the most ready to welcome the arrival of Europeans. Some 153 Egba towns and villages had been destroyed in the wars. In about the year 1830, under the leadership of Chief Sodeke, the Egba began to assemble on the defensible site around Olumo rock to found the town of Abeokuta. When missionaries arrived offering education, Christianity and European skills in the defence and the development of the town, Abeokuta was barely twelve years old, already with a population of thirty to forty thousand people. The task of fashioning an acceptable and workable constitution for the government of the town had scarcely been tackled yet. People from each of the old towns settled in a group with their particular ruler, behaving as if the towns were still physically separate or, as a missionary later said, 'as if all the German principalities and little kingdoms were brought together in one town, each acting, but seldom in unison'.[2]

Sodeke addressed himself primarily to the external problems of defence. He saw clearly that the greatest threat to Abeokuta came from the Ijebu, and that if the Egba were to survive, they must have direct access to ammunition from the coast, own a port of their own, and make contacts with the European world. Badagri, weak and divided, was the obvious port to choose. Sodeke therefore sought to control the Egbado and the Awori people who lived between Abeokuta and Badagri. He did not develop any machinery for the effective government of those parts, only insisting that the rulers should be people favourable to the Egba, people who would welcome them, give them land to farm, and protect their traders on their way to and from the coast.

Dahomey, freed from the rule of Old Oyo, was expanding also into the Egbado area and inevitably viewed Abeokuta as a dangerous rival. While Dahomey effectively controlled Whydah, and to some extent Porto Novo, and had acquired a monopoly of the trade and regulated

---

1 PP 1830 x, p. 69.
2 Townsend to Commander Wilmot, 5 Aug. 1851; PP 1852, LIV, *Papers relative to the Reduction of Lagos*, p. 157. Also S. O. Biobaku: *The Egba State and its Neighbours*, Oxford 1957, p. 2.

the activities of European traders on her coast, she saw to it that Abeokuta did not succeed in controlling Badagri in a similar manner. When in 1842 Sodeke sought to remove the last remaining obstacle on the Abeokuta-Badagri route by besieging the hostile town of Ado, Dahomey came to its aid. The siege was to last till 1853, when missionaries intervened to get the Egba to decamp.

Few towns can have had a more turbulent history than Badagri. It was founded soon after 1727 by a group of Egun, Hueda and Wemenu who fled from Allada and Whydah when King Agaja Trudo of Dahomey annexed those coastal states. At first the people lived on a little fishing and a small trade in salt made by evaporating sea-water, but before long they drew the attention of European slave-traders who rejoiced to see a bit of inhabited coast not under the power of a strong African monarch. Soon Badagri became an important slave market, stretching for over a mile along the coast opposite the several barracoons of Spanish, Portuguese, French, Dutch and English merchants, who lived tolerantly side by side, each doing business with one or the other of the eight wards of the town. The population increased rapidly, with traders and slaves coming, in particular, from the Porto Novo and Egbado areas. It remained politically weak because it did not develop any strong central organization; no hereditary monarch as at Bonny, no powerful council like the Ekpe at Calabar. It was an anomalous, cosmopolitan little republic, ruled over by a group of disunited chiefs. They had an ineffective council, presided over by Akran, the 'Portuguese' chief; in the event of war Possu, the 'Dutch' chief, was expected to take command of all the troops.

Divided political authority did not preclude prosperity when the slave trade was at its height, but as soon as the effects of the British preventive squadron began to be felt, Badagri fell out of favour with the slave traders. For one thing, the town might be a useful port in peaceful times, but it had not the harbour facilities necessary for eluding the vigilance of the squadron. It was on a lagoon, cut off from the ocean by a mile of sandy shore, with no outlet between Cotonou and Lagos, so that it was impossible to load a cargo and slip away to sea unobserved.

Unfortunately, the English 'legitimate' trader did not come to revive Badagri's trade because Badagri had little to offer him. The soil around Badagri was sandy and infertile,[2] and there was little palm-oil in the immediate hinterland, and no ivory, no gold. When Richard Lander

---

1 C. Newbury: *The Western Slave Coast and its Rulers*, Oxford 1961, pp. 30–2; D. N. Duckworth: 'Badagry—its place in the Pages of History' in *Nigeria Magazine*, 1952.
2 Gollmer: Journal, June 1847. 'The soil here is anything but rich.'

returned to Badagri in 1830, he reported neither slave-trader nor Englishmen. 'Everything bore an air of gloom and sadness totally different from what he had been led to expect,' said his brother. In 1840 a Captain Marmon tried to establish a factory there and failed; Captain Parsons, agent of Thomas Hutton & Co., in 1844 barely received enough oil to pay his expenses. And Captain Parsons was the one English trader the first missionaries found there in 1842.

It was because Badagri was weak, divided and poor that it was open to the Egba to build up some influence there. It was also for this reason that the emigrants, largely Egba, and not welcome in Kosoko's Lagos, found in Badagri a home or a route to the interior. In this way, Badagri became the first mission station in Nigeria. In other words, the dissolution of the Old Oyo empire and the consequent Yoruba Wars, the unsettled political situation, and the pressing demand from the Egba for European arms and ammunition had made Badagri an open door to Europeans seeking to penetrate the country. Indirectly therefore the last advance that Islam made to the south prepared the way for the advance inland of the influences of Christian Europe.

The Changing World View. A Gelede mask from Western Nigeria

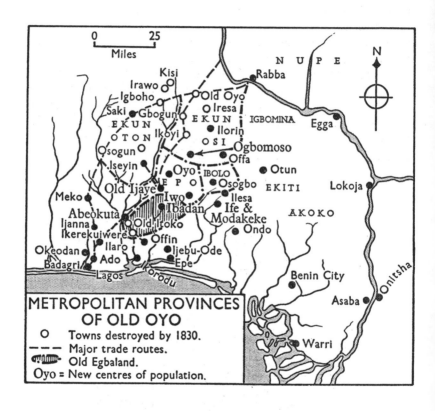

METROPOLITAN PROVINCES OF OLD OYO

○ — Towns destroyed by 1830.
- - - — Major trade routes.
⊶⊶ — Old Egbaland.
Oyo = New centres of population.

2 The Return of the Exiles

THE vision of home had a great power of attraction for the Liberated Africans in Sierra Leone, those who (unlike the Maroons or Nova Scotians)[1] had been recaptured by the British preventive squadron, often within a few days of their being shipped, and set free in Sierra Leone two or three months after that. To most officials and missionaries, they were just liberated 'Africans', but among themselves, they were Ibo or Nupe or Hausa. The Yoruba, who as a result of the wars of the early nineteenth century quickly became the most numerous group, were not even just 'Akus', but Oyo, Egba, Ijebu, Ijesa, Ife.[2] Home meant to them some remembered family homestead, father or mother, aunts, cousins, children.

On their arrival in the colony some enlisted in the West Indian regiments; some were apprenticed to artisans and traders in Freetown or settled in farming villages under superintendents; the younger ones were mostly sent to mission schools.[3] Many became Christians, others remained Muslims or continued to adhere to their traditional religion. A good number of them learnt to read and write and sought employment from the government, the missions or the commercial houses. By 1840, when few of the Liberated Africans had spent more than twenty years in the colony, some were already successful traders on their own.

---

1 Sierra Leone colonists included (a) *The Settlers*, who in 1787 founded the colony; (b) *Nova Scotians*, who were former slaves in America, emancipated on joining British troops during the American War of Independence, later located in Nova Scotia, and moved to Sierra Leone in 1791; (c) *Maroons*, who were ex-slaves of the Spaniards from whom the British seized Jamaica in 1655; they lived as free Negroes on the mountains, revolted against the British in 1795, and were removed to Sierra Leone via Halifax in 1800; (d) *Mandingoes, Kroomen,* and *Temne*, who came from the surrounding countries to look for work in the colony; (e) *Liberated Africans.*
2 Cf. also the fact that although official records were almost always silent about their ethnic origins, their own personal records, e.g. tombstones, were often specific on this point. Also, the Rev. Koelle, whose *Polyglotta Africana* (C.M.S. 1856) was evidently based on personal interviews, listed Egba, Ijesa, Oyo, Ijebu as different languages.
3 C. H. Fyfe: *A History of Sierra Leone*, Oxford 1962. R. R. Kuczynski: *Demographic Survey of the British Empire*, vol. 1. West Africa, Oxford 1958.

There was Thomas Will,[1] for example, the head of the Yoruba in Free-town, who died that year leaving behind him '£2,000 and a good corner house in Walpole Street which he had bought two years earlier for £305'. There was John Langley,[2] an Ibo, educated at the C.M.S. Regent School, who, dismissed as a village teacher, took to trading and in 1837 was appointed Superintendent of Charlotte village, the first Liberated African to be appointed to such a post. Others, like John Ezzidio, a Nupe, William Pratt, an Ibo, and Benjamin Pratt, a Yoruba, were rising young men, investing money in land, making contacts with business houses in Europe, supporting the missions fervently, seeking good education for their children, trying to live as much as possible like Victorian gentlemen.[3]

Others were rising in mission employment, the most notable of whom was Ajayi,[4] who arrived in the colony in June 1822 and was baptized by the Rev. J. C. Raban in 1825 as Samuel Crowther, after the Vicar of Christ Church, Newgate, a prominent supporter of the C.M.S. He was an industrious, intelligent, humble young man, the type beloved by missionaries. He learned to read and write; he learned some carpentry from Weeks, the Industrial Agent of the mission, who later became Bishop; he went to England with the Rev. and Mrs. Davey in 1826 and spent some months at the parish school in Islington. He returned, in 1827, in time to be the first student enrolled for the institution that was to become Fourah Bay College.[5] He taught in various mission and government schools, probably helping Raban in collecting vocabularies of the Yoruba language, studying Greek and Temne, a local language. By 1840 he had become sufficiently important to sign petitions on behalf of other Liberated Africans. In August of that year, with one John Attara, he addressed a petition to the special meeting of the local committee of the C.M.S. about higher education.[6]

1 C. H. Fyfe: 'View of the New Burial Ground' in *Sierra Leone Studies*, new series no. 2, June 1954, p. 89.   2 Ibid.
3 C. H. Fyfe: 'The Life and Times of John Ezzidio' in *Sierra Leone Studies*, new series no. 4, June 1955.   4 See p. 20.
5 Stocks, op. cit., vol. I, pp. 450–1.
6 Crowther and Attara, Petition to Special Meeting of Local Committee; C.M.S. CA1/M9, 17 Aug. 1840:
'We hear that the committee in England is very glad to do anything for our children in Europe, but is very much concerned about their health and life.
'We do not deny this: but still we have had many examples of boys who have been sent to England, several of whom have already returned to the colony after many years spent in that country, which make us hope for the best and therefore we are willing to make our children an example, on our part.
'Our chief motive in writing to you on this important step is, should it please God, after they are qualified they may be usefully employed as the servants of the Church Missionary Society in this benighted continent.'

He was at the same time advocating the establishment of a model farm and the formation of an agricultural society. In these activities he first showed his gift of leadership, his persistent advocacy of progressive measures and his unmistakable intellectual ability, as well as a firm, practical evangelical faith, expressed in a blameless life, both at home and in public office. When asked to join the Niger Expedition of 1841, he readily agreed, but when it was further suggested that he should emigrate, and be the catechist in charge of the settlers at the model farm at Lokoja, he protested, pointing out that he had a family, and he recommended instead Thomas King, who was a widower with a daughter old enough to be left with her uncle.[1]

To the majority of the Liberated Africans, however, success did not come so easily. Women were scarce; markets, farming land, opportunities in general were restricted. Many of them therefore began to look for opportunities beyond the colony. John Langley had been trading with the peoples to the north of the colony, but in 1834 the Governor had to intervene to rescue him from jail when the Alkali of Port Lokko imprisoned him for selling gunpowder to his enemies, the Mendes.[2] A little later, others, singly or in mutual aid groups, bought condemned slave-vessels and traded down the coast as far as Badagri and Lagos. It was here that some of them found people they knew and old family ties were renewed. As Crowther described it two years later:

> Some found their children, others their brothers and sisters, by whom they were entreated not to return to Sierra Leone. One of the traders had brought to Sierra Leone two of his grandchildren from Badagri to receive instruction. Several of them had gone into the interior altogether. Others in this colony have messages sent to them by their parents and relations whom the traders met at Badagry.[3]

In November 1839, while Buxton's *Remedy* was still a secret plan being urged on the government, twenty-three leading Yoruba merchants, led by Thomas Will, presented to Governor Doherty a petition which contained ideas similar to Buxton's. The humble petitioners

> feel with much thankful to Almighty God and the Queen of England, who had rescued us from being in a state of slavery, and has brought us to this colony and set us at liberty and thanks be to the God of all mercy who has sent his servants to declare unto us poor creatures the way of salvation, which illuminates our understanding so we are

1 Warburton, Sec. of Local Committee to C.M.S. Secretaries, 20 July 1841; C.M.S. CA1/o3.
2 C. H. Fyfe: 'View of the New Burial Ground', op. cit.
3 Extracts from Samuel Crowther's Journal for the term ending June 1841 in C.M.S. CA1/M9, p. 438.

brought to know we have a soul to save, and when your humble petitioners look back upon their country people who are now living in darkness, without the light of the Gospel, so we take upon ourselves to direct this our humble petition to your Excellency.

That the Queen will graciously to sympathize with her humble petitioners to establish a colony to Badagri that the same may be under the Queen's Jurisdiction and beg of her Royal Majesty to send missionary with us and by so doing the slave trade can be abolished, because the dealers can be afeared to go up to the said place so that the Gospel of Christ can be preached throughout our land. . . .

Governor Doherty recommended the proposal to the favourable consideration of Lord John Russell;

If it should consist with the designs of Her Majesty's Government for the extirpation of the slave trade and the civilization of the continent to encourage the establishment of any settlements of this description.'[1]

'We cannot send them,' the Secretary of State minuted, 'without giving them security and protection, which implies expense. But they can go if they wish.' Even in March 1841, when the Niger Expedition was about to leave England, Russell was more concerned with putting all sorts of pressure, short of actual coercion, on the Liberated Africans to emigrate to the West Indies than with assisting them to go to Nigeria. There was evidently no connection in his mind between the mass, spontaneous emigration of people who, according to his Under-Secretary, did not appear 'well-instructed in the arts of civilized life'[2] and the ordered movement of civilizing missionaries proposed by Buxton.

It must be said that the respectable petitioners, merchants with landed property, were not the people intending to emigrate. None of the twenty-three signatories of the petition is in fact known to have done so. By March 1840 Governor Doherty seems to have realized this, when he wrote another despatch on the 'pretty extensive and growing disposition' to emigrate that existed in the colony. The people concerned were smaller men and they were not waiting for the British Government to establish a colony for them. Two parties of fourteen and twenty, he said, had left.

The Governor was certainly not as enthusiastic about this mass-movement as he had been about the proposal for a British colony in

1 Doherty to Russell, 30 Nov. 1839 and enclosure dated 15 Nov. 1839; CO 265/154.
2 Minute by Vernon Smith on Doherty's dispatch of 30 Nov. 1839

Badagri. Of the two hundred people who applied, he said that he issued passports to only forty-four men and seventeen women.

The missionaries in Sierra Leone at first frowned on a movement leading their parishioners to forsake a Christian settlement for 'a land of darkness'. The same opinion was later expressed at Abeokuta by Henry Townsend. By leaving Sierra Leone, he said, the emigrants had 'left the country where God was known for this where God was not known; thus turning their backs upon them'.[1] However, as Buxton's ideas were unfolded, the missionaries soon adopted a more conciliatory attitude. The publicity given to Buxton's ideas and the arrangements for the Niger Expedition created a good deal of excitement in the colony. Governor Doherty, still reporting on the emigration movement, no longer spoke of restless, poor people leaving the colony because they could not get jobs, but of merchants who wished 'to carry back among their countrymen the arts and improvements of Europe which they had acquired here, with the fortunes which had been amassed by them'.[1] Besides the traders to whom the Expedition gave promise of opportunities, many others entered fully into the hopes and aspirations for Africa entertained by Buxton and the Evangelicals. Crowther described the excitement, and forecast that if the Expedition succeeded, many, not just Egba and Yoruba, but Kanuri, Hausa, Nupe and Ibo would emigrate.[3] And as the various churches called prayer meetings and instituted special funds to aid the project, and as friends and relatives arranged little send-off parties for the members of the Expedition, there developed some of the drama and prophecy proper to the eve of such a crusade. Perhaps the most remarkable story was that of John Langley. He had been dismissed from Charlotte, and had become a bitter critic of the government and missionaries. Then suddenly, during an illness in 1841, he was converted into a pious evangelical and fervent advocate of the taking of 'the Bible and the Plough' to the banks of the Niger.[4]

The emigration continued, however, not to the Niger but to Badagri and the Yoruba country, and the failure of the Expedition did not stop it. The Methodists, who had played a less conspicuous part than the C.M.S. in the arrangements for the Expedition, were the first to adopt the emigration wholeheartedly as an alternative way of penetrating into Nigeria. As early as June 1841, the Rev. Thomas Dove, the Superintendent of the Methodist Mission in Sierra Leone, announced that he

1 'Journal of H. Townsend while on a Mission of Research', entry for 5 Jan. 1843; C.M.S. CA1/0215.
2 Doherty to Russell, 3 Oct. 1840; CO 267/160.
3 Extracts from Samuel Crowther's Journal for the term ending June 1841, op. cit.
4 C. H. Fyfe: 'View of the New Burial Ground', op. cit.

had received two letters from the emigrants in Badagri, one anonymous, the other signed by James Fergusson, inviting missionaries to visit them urgently.[1] In recommending the letters for action by the Home Committee, Dove asserted that the desire of the emigrants for their country was

> that the Gospel of God our Saviour may be preached unto her, that schools may be established, that Bibles may be sent, that the British flag may be hoisted, and that she may rank among the civilized nations of the earth.[2]

But all this enthusiasm, much of which faded with the failure of the Niger Expedition, did not convert the movement of nostalgic exiles seeking home and opportunities into the conscious crusaders against the twin evils of idolatry and slavery for which the evangelicals were looking, though a few of them, like Crowther, did in fact become such crusaders. Many missionaries realized that the mass of emigrants[3] who in the early 1840s gave the movement its significant character of the 'return of the exiles from Babylon' and brought missionaries after them, did not exactly fit Buxton's description of the agents of civilization. Schön argued at the conclusion of his Report on the Niger Expedition of 1841 that the Africans in the West Indies were

> in many respects better qualified than the liberated Africans at Sierra Leone: they have seen more of European habits; are better acquainted with agricultural labours; and have a much greater taste for European comforts.[4]

He added, however, that Liberated Africans were preferable, since the West Indians did not fit easily into African society but were in greater and more urgent need of training. Crowther similarly was 'reluctantly led to adopt the opinion that Africa can chiefly be benefited by her own children' and they must be given the requisite training.[5] One result of this was an appeal for funds to expand the work at Fourah Bay as a training institution and to improve secondary education in Freetown. The emigrants did not wait for the training, and the improvement of

1 Published in Meth. *Missionary Notices*, new series, no. 1, pp. 801–2. December 1841.
2 Letter of Dove, 1 June 1841, ibid.
3 In a sense they were immigrants into Nigeria, emigrants from Sierra Leone, Cuba or Brazil. But since they were also emigrants from Nigeria in the first instance, I have adopted the term emigrants by which they were generally known in the nineteenth century. In Yoruba, the Sierra Leone emigrants were *Saro*, the Cuban and Brazilian, *Amaro*.
4 Schön and Crowther: *Journals of the 1841 Expedition*, pp. 62–3.
5 Ibid., p. 439.

education did not keep pace with the demand on its resources. Missionaries had to be sent after the emigrants at once.

The Methodists acted first. The Rev. Thomas Birch Freeman, the energetic Superintendent of the Methodist Mission at Cape Coast, who had shown outstanding abilities by his accounts of his two visits to Ashanti, was asked to occupy Badagri as an out-station of Cape Coast. His arrival on 24 September 1842 marked the effective beginning of missionary enterprise in Nigeria. He was accompanied by a Fanti 'assistant missionary', William de Graft, a native of Cape Coast who had been working at Winneba. He bought a small piece of ground on which he at once built a temporary bamboo chapel and, at a cost of some £300, a more elaborate mission house, the planks for which he had brought along—

> a large, airy dwelling-house, fit for an European family, raised from ten to twelve feet from the ground, on twenty-two stout coconut pillars, averaging about three-quarters of a ton each in weight . . . It appeared a thing so novel and extraordinary, that the people were often seen standing in groups at a short distance, gazing at it in astonishment.[1]

While working on it, he began holding prayer-meetings on Sundays with the Sierra Leone emigrants, settling their disputes, obtaining land for them from the authorities and sharing it out for them. He saw, however, that the majority of those who had arrived and of those who were still arriving did not stay at Badagri, but were moving out to Abeokuta, and he decided to pay them a visit. On Sunday, 11 December, Freeman entered Abeokuta and was very warmly received, both by Sodeke and the chiefs and by the emigrants. He stayed ten days, holding meetings with the emigrants, visiting the chiefs and discussing with them, looking round and making sketches of the town. He bought horses to use in his mission on the Gold Coast. On the eve of his departure, he gave a dinner party:

> Sodaka and a few members of his family and the principal men among the emigrants dined with me. We fixed a temporary table under the shed in Sodaka's yard and all things passed off well. Our party amounted, to the best of my recollection, to about twenty-five persons. Sodaka seemed much delighted; it was the first time he had ever eaten food after the manner and customs of Europeans.[2]

1 T. B. Freeman: *Journals of various visits to the Kingdoms of Ashanti, Aku and Dahomi.* Part III, published first in *Missionary Notices* and later as a book, 1844.   2 Ibid.

Freeman returned to Badagri on Christmas Eve to find Henry Townsend, a C.M.S. missionary, just arrived on a similar exploratory journey. The Local Committee of the C.M.S. in Sierra Leone, not wishing to be outdone by their Methodist friends, had decided to send a missionary to look after the interests of their own members among the emigrants. They picked on Henry Townsend, a young man of 26, frail-looking, but intelligent, determined and very ambitious. He had wished to join the Niger Expedition but had been turned down in favour of Samuel Crowther.[1] He was sent to Abeokuta only on 'a mission of research', to collect information about the country, the emigrants and the chances of a missionary establishment.[2] He was given free passage by Captain Harry Johnson, a Sierra Leonean trader, in his vessel, the *Wilberforce*, with fifty-nine emigrants on board, paying 12 dollars each, self-fed.[3] He arrived on 4 January 1843 at Abeokuta, where he received as warm a welcome as Freeman had received. He was impressed by the goodwill of Sodeke, who at once offered him land for a mission, but he refused to commit the C.M.S. in advance. Rather, he asked Sodeke to send two of his children with him to Sierra Leone to be educated, but Sodeke declined on the grounds that if missionaries were in fact going to settle in Abeokuta, there was no need.

Townsend was also impressed by the fact that the emigrants had been well received, and were treated not like ex-slaves but as honourable members of society whose skills in the arts of writing, of building houses and sawing timber and sewing clothes were being utilized. It was said that Sodeke, with characteristic warmth of heart and indifference to political calculations, suggested they could have a whole quarter of the town to themselves. The missionaries would have liked nothing better, but, as was commented in 1851, the emigrants took the suggestion lightly.[4] They had not travelled from so far, seeking friends and relatives, only to arrive and voluntarily segregate themselves; but they kept their European clothes, exploited their wide travels and made good traders.

Townsend went back to report on his mission and to prepare himself for ordination, leaving behind the emigrant who had acted as his interpreter, Andrew Wilhelm, to look after the interests of the C.M.S. Samuel Crowther, whose *Journal of the 1841 Expedition* greatly impressed the C.M.S. Committee, was called to England to spend some time at the

1 Townsend to C.M.S. Secretaries, 8 March 1841; C.M.S. CA1/079. Townsend was a little sore at the disappointment.
2 'Instructions of Local Committee' in J. Warburton to H. Townsend, 9 Nov. 1842; C.M.S. CA1/0218.
3 Townsend: Journal while on a Mission of Research, entry for 14 Nov. 1842; C.M.S. CA1/0215.
4 Townsend to Major Hector Straith, 28 Jan. 1851; C.M.S. CA2/085.

Training College, Islington. He was ordained in 1843 and was sent back to Sierra Leone to prepare for a mission to Abeokuta by beginning to conduct services in the Yoruba language.

Freeman went back to Cape Coast by way of Dahomey. Meanwhile de Graft remained in charge of the Methodist Mission in Badagri. He began regular Sunday services in the chapel for the emigrants and others who cared to come along. With his wife he opened a day school which he called the 'nursery of the infant church'. There were some forty to fifty children on the roll, belonging mostly to the emigrants and to one or two of the chiefs, Possu having sent no less than six. In the evenings de Graft preached in the open air at the market-place or instructed those who came to visit him. Within six months, in July 1843, he held his first baptismal service for five men and one woman. One of them was a refugee from the north, an elderly man of about 50, 'son of Ageza Lakunde', who had been prince of Obohon (Igboho) till he was displaced in 1835 by a wicked uncle. He compared the new teaching about Christ with 'the Moor's religion prevalent in these countries' and declared himself a convert. He was baptized Simeon, and his wife, Anna. Two of the men baptized were from his household. The other two were Fanti. To give Methodist readers in England an idea of his progress, de Graft sent extracts from his journal of which the one for Tuesday, 4 July 1843 will serve as sample. In the morning, he was engaged for the most part in

> gardening, transplanting pineapples and cocoa-roots, [i.e. coco-palm], trimming down our guinea corn and settling seeds of apples . . .

In the evening he gave a tea-party for the children of the mission school:

> The friends I invited to witness and partake of the same, made their prompt attendance, in a handsome manner, in proper time, among whom were two native chiefs with their numerous retinue. . . . The chapel was crowded and wore, on the whole, a very cheerful aspect. The children, about forty in number, both of Sierra Leoneans and of this place, were neatly dressed in the European clothes, and the members of the society who attended were all in their best; the chiefs also wore their neat country costume. The crowded meeting and the chapel, nicely arranged and well-lighted up, gave a very delightful appearance. I opened the meeting with singing and prayer, and then had the tea, cakes and bread shared out to the children, to the chiefs and to the members of our Society present; and while the children were drinking their tea, we had eight of our principal men

33

in the society by turns to improve the time by short and appropriate addresses to the children and to the meeting at large, in the English tongue as well as in the varnacular language. . . .[1]

The de Grafts were replaced in April 1844 by the Rev. Samuel Annear and his wife, who had been colleagues of Townsend at Hastings, Sierra Leone.

In January 1845 the main C.M.S. mission arrived. It consisted of the Rev. C. A. Gollmer, a Wurtemburger from the Basel Seminary who had been in Sierra Leone since 1841, as well as Townsend and Crowther. Besides these, there were two Sierra Leonean schoolmasters (Marsh and Phillips), one interpreter, four carpenters, three labourers and two servants. Their instruction[2] was to make straight for Abeokuta, but at Badagri they learnt that Sodeke had died some eight days before their arrival and they could not proceed to Abeokuta till a new ruler had been elected there. It was thus by accident that the C.M.S. came to realize the need for a coastal base, for they proceeded at once to establish a mission station at Badagri. By March they had built a church and were working on a mission house. They also built a school, and put Townsend in charge, while Gollmer and Crowther went into the streets and the outlying villages to preach.

Thus Badagri became the first missionary base in Nigeria. From it the missions hoped to penetrate into the interior. However, they found it a most difficult station. There was no open hostility or persecution, but the hold of the traditional religion on the people was very firm. They had welcomed missionaries, not because they wanted Christianity, but because they were weak and poor and they hoped that the missionaries could attract some trade back to the town.

There was no doubt that the campaign against the slave trade had brought ruin and decline to Badagri. Every missionary remarked on the poverty of the people, their pre-occupation with 'what shall I eat?' and their persistent grumbling about the absence of trade. The missionaries advised them to take up agriculture. Gollmer even tried to organize an agricultural show and offered prizes, but it had no effect. As he himself admitted, the soil was most infertile, '80 per cent sand and 20 per cent decayed vegetable matter'. They wanted not agriculture but trade. Their common saying, said Gollmer, was: 'Trade we shall, trade our fathers taught us,' even though they had nothing to sell or buy.[3] Poverty

1 De Graft to Meth. Secretary 10 July 1848, published as an appendix to Part III of Freeman's *Journal of various visits*, pp. 284 ff.
2 Dated 25 Oct. 1844, in C.M.S. CA2/11, p. 3.
3 Gollmer, Journal entry for 18 Dec. 1850; C.M.S. CA2/043.

and political instability formed part of a vicious circle, for poverty made the place weak in the face of aggression and without stability no trade could be attracted. Neither the missionary establishment nor the emigrants could flourish:

> Comparatively nothing can at present be done to promote civilization, prevent slavery, or accomplish any other good object. For if you enter into an agreement or compact with any one of the chiefs, the others are sure at once to oppose you and consider you their enemy . . .[1]

Freeman visited Abeokuta and Dahomey in order to put the Badagri mission in the care of each of those rival powers. He would, no doubt, have welcomed either of them establishing a firm and effective rule in Badagri,[2] but it was to Britain that he looked for protection. This was not as easy as it sounded. Until the Badagri mission began to make its impact on the British government, that government's attention was focused, not on the Bight of Benin, but on the Niger Delta, where the palm-oil trade was flourishing and British traders were in keen competition with other European and American traders. While the British were urging the slave trade treaties on every ruler in the Delta in the 1840s, they neglected the Bight of Benin. In 1843, when Freeman appealed to Commander Jones of the preventive squadron to sign a treaty with the chiefs of Badagri along the lines of those in the Delta,[3] no one took any notice. Freeman did the next best thing. He consulted his good friend, Governor Maclean of Cape Coast Castle, suggesting that he accept Badagri as an extension of his informal and largely illegal protectorate. Maclean sent one Fanti soldier, a Sergeant Bart of the Gold Coast Corps, with instructions dated 16 August 1843, to proceed at once to Badagri

> to hoist the English flag in the English town and afford due protection to all English subjects . . .
> to afford every protection to the Christian mission establishment there, and to all connected with it, as also to Mr. Hutton's factory and all other English traders.

1 Annear: Journal, October 1884; Meth.
2 Cf. Annear: ibid., 'If Sodeka were to send his forces here and add this place to his territories, he would confer greater benefit upon the town and all the surrounding country, and would contribute more to aid the mission cause and the diffusion of civilization than [any] other movement which he is capable of making.'
3 Freeman gives a full account of these activities in ch. XXX of his MSS book, op. cit.

35

[He] must not, however, interfere in the native palavers, but behave in all cases with moderation and forbearance.[1]

Sergeant Bart seems to have acted with much energy and zeal, for, according to Annear, the mission house became 'at the same time the fort, the jail and the temple'.[2] Annear imagined himself as having become responsible 'if any political affair go wrong demanding the interference of British law and authority'.[3] When, on the change of administration on the Gold Coast, Sergeant Bart was withdrawn in December 1844, in slightly hysterical terms Annear declared that:

> The last relic of British authority is now withdrawn. . . . Our Queen is gone: Her laws are abolished. Her invaluable protection ceases to be held out.

But that was not so, as the real basis of Sergeant Bart's influence at Badagri was not the flag but the knowledge the chiefs had that he could hoist a 'Black Peter' on the beach and attract warships. Commander Foote had issued instructions

> to all cruisers to call at Badagry now and then. To hoist a Union Jack at the main and fire a gun and wait off the place for two or three hours in order that the residents at Badagry may have an opportunity of communicating with any vessel of the squadron that may call.[4]

Thrice within the first eighteen months, this privilege was invoked. By May 1845 the chiefs of Badagri—bribed, said the missionaries, by Kosoko of Lagos—were beginning to wonder why they should be expected to harbour the English, who brought them no trade and yet exposed them to constant attacks. They decided on their expulsion. The English missionaries and traders believed that their continued existence at Badagri—indeed, in the whole Bight of Benin—had become involved in the chieftaincy dispute at Lagos.

The struggle in Lagos was between two branches of the ruling family. It was not a struggle between a slave-trading and an anti-slave-trading party.[5] Just as African chiefs had for long exploited the jealous rivalries of European traders on the coast, European traders of various

---

1 Ibid.   2 Annear: Journal, October 1844; Meth.   3 Ibid.
4 Freeman's MSS book, op. cit.: Memorandum of Captain John Foote, H.M.S. *Madagascar*, senior officer commanding the Bights division, dated 15 May 1843.
5 J. F. Ade Ajayi: 'The British occupation of Lagos 1851–61, a critical Review' in *Nigeria Magazine*, August 1961. For the early history of Lagos, see C. A. Gollmer: 'Kings of Lagos', September 1853 in *State Papers*, vol. 44, p. 1220; J. B. Wood: *History of Lagos*, London 1880.

nations had similarly exploited chieftancy disputes. At this time English traders supported Akitoye, who had been made Oba of Lagos with popular support, it was said, and who had been crowned by the Oba of Benin, after a disputed succession, in 1841. His rival, Kosoko, was supported by Brazilian and Portuguese traders who resented Akitoye's open-door policies. Kosoko succeeded in 1845 in ousting Akitoye through a general revolt in Lagos, and the latter fled to Abeokuta, where the Egba welcomed him. From there he appealed to the English, both at Cape Coast and at Badagri, for help.

The English residents at Badagri reacted at once. Led by Annear, they drew up a petition 'to any of H.M. Naval Officers on the West Coast of Africa' against Kosoko, who, they said,

> will not close hostilities until he has conquered the whole line of coast to Whydah. . . . Nothing less than the entire extirpation of the English wherever he is capable of exerting his power will satisfy him.
>
> Feeling that we have no power to resist so formidable a foe, and being sure of his intentions towards us, we feel it both our duty to the hundreds of British subjects in the interior[1] and expedient for ourselves to make you acquainted with our state and to solicit most earnestly your assistance.[2]

Captain York of H.M.S. *Albatross*, received this letter and came to Badagri a fortnight later. The missionaries were united in saying that but for him they would have been driven out.[3] He could, by the presence of the cruiser, overawe Badagri chiefs; but the restoration of Akitoye was a different matter.

In December 1845 Akitoye left Abeokuta for Badagri, where he could

---

1 I.e. the Sierra Leone emigrants. It suited missionaries to assume that they were British subjects, but in fact the legal position of those of them not born in Sierra Leone was far from clear. In 1856, after taking legal advice, the Foreign Office told the Consul at Lagos that the emigrants were not in fact British subjects.

'Liberated Africans . . . in the absence of any special legislation to that effect . . . cannot be so considered even in the Queen's Dominions, and under these circumstances they cannot, of course, be entitled to expect as a matter of Right that they should be treated as British subjects when they voluntarily return to and become Residents in the Territory of the Native Chief whose subjects they were by Birth. Nevertheless, Her Majesty's Government can never cease to take a warm interest in the Welfare and Safety of these Africans . . .' F.O. Draft to Hutchinson 19 Oct. 1856, FO 84/1001.

All that this meant was that any consul who did not wish to exert himself on their behalf had a document to cite; on other occasions, they continued to be regarded generally as British subjects.

2 Dated 20 Aug. 1845. Annear sent a copy to Methodist House with his Journal for the term ending October 1845.   3 Ibid.

the more easily plan the invasion of Lagos.[1] From Badagri, his canoemen blockaded Lagos in an attempt to cut Kosoko off from his allies in Porto Novo. Then, in March 1846, in alliance with Abeokuta, he made a combined attack on Lagos by land and lagoon. The expedition failed, but it is important for two reasons. First, it was financed by Domingo José Martinez, a Madeira trader and a great rival of the Portuguese and Brazilian traders who supported Kosoko at Lagos.[2] His motive was undoubtedly the commercial concessions that would have been his had he become king-maker in Lagos. Though he dealt in palm-oil and cowries, and sold rum and tobacco for cash, he was well known as a slave-trader and his support of Akitoye destroys the simple picture that the Lagos dispute was between slave-traders and anti-slave-traders. Second, the failure of the expedition immediately led the Egba chiefs to receive at Abeokuta the C.M.S. missionaries who had been waiting for over a year at Badagri. For, in spite of the friendliness of Sodeke, the chiefs had hesitated to allow missionaries to settle at Abeokuta before they had resolved their internal problems. They had insisted in 1845 that the missionaries should wait at Badagri till a new ruler had been appointed in place of Sodeke. But suddenly in 1846, before such a ruler had been appointed, they gave the necessary permission enabling Townsend and Crowther to move at once to Abeokuta, leaving Gollmer on the coast. The most likely explanation was that the chiefs hoped to use missionary influence to secure the support of the British navy for a second attempt on Lagos.

The arrival of missionaries at Abeokuta, was an event of considerable importance. For the first time the missionaries found themselves in a town judged to be ideal for applying Buxton's principles. Born out of upheaval, attuned to welcome new and revolutionary ideas, it offered every opportunity for agents of civilization to participate in the work of reconstruction. It was an inland town, away from the swampy coast, connected by the river Ogun to the sea, with roads leading to many large

1 'Thinking that I should have a better chance to communicate with the English and that I might be nearer Lagos to watch the movements of Kosoko and the affairs of my kingdom, I took residence at Badagry.' Petition of Akitoye to Beecroft, encl. in Beecroft to Palmerston, 24 Feb. 1851, PP 1852, LIV; *Papers relative to the Reduction of Lagos*. I find no evidence for the view sometimes taken that he was driven out of Abeokuta by the machination of Kosoko.

2 John Martin, Journal, March 1846 reported that Domingo was prepared to spend 5,000 dollars on the expedition, and commented that Domingo must have felt that the trade of Lagos was worth it. For other references to Domingo see Gollmer, Journal for March 1851, postscript; also Journal entries for 4 March 1851, 31 Aug. and 27 Nov. 1851, and Gollmer to Venn. 19 Sept. 1851; C.M.S. CA2/043.

centres of population, to Ijebu, Ife, Ibadan, Ijaye, Ketu, Porto Novo and places beyond. Here, ready to hand, was a community of two to three thousand[1] Sierra Leone emigrants, mostly Egba, once given up as lost but now returned with new ideas and skills, and all grateful to the English who had redeemed them. Abeokuta soon became a symbol of the hope of Christian missionaries in Africa—the 'Sunrise within the Tropics',[2] as it was called. Efforts were made to clear the sky, that the sun of Abeokuta might shine forth and convince Europe that the missionary plan for Africa had every chance of success if well supported, and convince Africans what good things lay in store for them if they followed Abeokuta's example.

The work of evangelization went on apace. Townsend settled at Ake, Crowther at Igbein. Each built a mission house, a church and a school. Mrs. Crowther, who was a teacher in Sierra Leone, made special efforts to establish a girls' school.[3] In addition, there were open-air sermons in the market-place, discussions in the compounds of the chiefs, and the instruction of 'enquirers' in the mission house. The first baptismal service was held within eighteen months, 5 February 1848, when three women were baptized, two of them wives of chiefs, the third Crowther's own mother, who was discovered in one of those reunions that made the return of the exiles and the coming of the missionaries such heart-warming affairs.[4] The Methodists, who had been the first to reach Abeokuta, did not wish to leave all the excitement to the C.M.S. In the absence of an ordained clergyman, Freeman sent a Fanti schoolmaster named Morgue, who had served in Badagri under Annear, to occupy the station at Ogbe, till he was relieved in 1849 by Edward Bickersteth, an Egba Sierra Leonean emigrant, who was destined to be the sole Methodist agent there till 1859. The missionary magazines proclaimed the missionary successes in conversions and gave accounts of the religious experiences of the converted. In 1849 they related how the first Christian burial provoked persecution and how persecution called forth loyalty and fervour of faith among the Christians.[5] The importance of Abeokuta in the thinking of missionary circles in England lay, however, not so much in the number of baptisms already carried out, as in the hope of

1 See below, p. 40.
2 Title of a highly publicised book by Miss Tucker, C.M.S. 1852.
3 *A Short History of the Introduction and Spread of Christianity into Egbaland under the Church Missionary Society*, compiled during the centenary celebrations of 1946 (based partly on oral tradition in the church, church registers, and perhaps some notes left by Andrew Wilhelm (p. 2), archives of St. Peter's Church, Ake, Abeokuta). By kind permission of Archdeacon Ashley Dejo.
4 Ibid.; Stocks, op. cit., vol. II, pp. 103–4.
5 Crowther: Journal entry for 13–29 Oct.; C.M.S. CA2/031b. E. Bickersteth: journal entries of the same date; Meth.

conversions still to come, and in the hope it gave for the civilization of Africa. Soon Abeokuta became a household word among the readers of missionary magazines, long before Lagos was known. Soon, it even qualified for the honour of jokes in *Punch*.[1] It became the policy of the missionaries to establish themselves firmly at Abeokuta, and foster the 'civilizing' influence of Abeokuta down to the coast, to Badagri and Lagos, and into the interior. But the true significance of this policy can only be seen against the general background of the advancing interest of Britain in Nigeria.

Meanwhile, the emigrants continued to arrive from Sierra Leone, from Brazil, from Cuba. This went on continuously throughout the century. It is very difficult to estimate the numbers of people involved. The Sierra Leone government was supposed to issue a passport to each immigrant, but the records were so notoriously defective that even in 1842 the Officer Administering the Government could only give an estimate of 500 'as near as can be ascertained'.[2] In 1844 Governor Fergusson estimated that some 600 to 800 'are now established in the Yarriba or Aku country'.[3] Since the government was anxious that the figures should not give the Colonial Office any impression of large-scale dissatisfaction with the conditions in the colony, and since many people did emigrate without passports, these figures must have been underestimates.[4] In 1851 a naval officer who reviewed the emigrants at Abeokuta said there were 3,000 of them.[5] It was the same figure that Consul Beecroft, who visited Abeokuta earlier that year, mentioned to Hope Waddell.[6] At that date, too, another naval officer referred to 'hundreds' of emigrants at Badagri.[7] There were others at Lagos, Ibadan and Ijaye, in towns and little villages as far away as Ede, Iragbiji and Ilorin. Two years later, emigration from Sierra Leone received a fresh impetus with the establishment of a British Consul at Lagos and the monthly mail packets of the African Steam Company.

One of the main arguments for securing government subsidy for the mail-boats was that it would facilitate communication between Sierra

1 *Punch* (leading British satirical magazine), 11 June 1864. Cit. below, p. 170.
2 Kuczynski, op. cit., p. 137. In December 1842 Freeman estimated 2–3,000 at Abeokuta—probably an over-estimate.
3 Kuczynski, op. cit., p. 137.
4 Kuczysnki, op. cit., pp. 132–6.
5 F. E. Forbes to Bruce, 9 Dec. 1851. *Papers relative to the Reduction of Lagos*, PP 1852, LIV, p. 180.
6 Waddell, Journal entry for 10 Feb. 1851, in U.P. *Missionary Record*, 1851, p. 120.
7 Bruce to Secretary of the Admiralty, 1 Nov. 1851. *Papers relative to the Reduction of Lagos*, p. 158.

Leone and Nigeria. McGregor Laird, who secured the contract in competition against Liverpool traders long established on the coast, argued that it would encourage traders in the development of trade in the interior of Nigeria, and he looked to small Sierra Leone traders to take an important share.[1] The mail-boats undoubtedly speeded up the rate of emigration, particularly among the mercantile class who had earlier been reluctant to emigrate. With the advent of the mail-boats, the Christian missions themselves began to encourage emigration. As the missionaries expanded inland into Yoruba country, many a missionary endeared himself to the local rulers by mentioning how many of his subjects were to be found in Sierra Leone and asking him to invite them to return home. In 1853 Townsend mentioned that within eight months of his writing such a letter for the Are of Ijaye 85 emigrants arrived in town.[2] In 1854 Hinderer and Irving, trying to overcome the reluctance of Ijebu people to welcome strangers, suggested to the Akarigbo of Ijebu Remo that they write a similar letter.[3] It was, however, to the Niger where there was no spontaneous movement of emigrants that the C.M.S. was most anxious to influence Sierra Leoneans to proceed and introduce the Gospels.

Influential Ibos in Sierra Leone were equally eager. They petitioned the local committee of the C.M.S. to take advantage of the mail-boats to extend Christianity to the Niger as it had done to the Yoruba country. In 1853 the C.M.S. asked the Rev. Edward Jones, a West Indian, Principal of Fourah Bay College, to lead an expedition of three Ibos to visit the Niger and report on the prospects awaiting emigrants there. The delegation did not reach the Niger. By the time the mail-boat had taken them to Fernando Po, Jones said he was

> fully satisfied in [his] mind from conversation with naval officers and others that it would not be possible for them to ascend the Niger and reach Aboh unless in a steamer.

He added that Beecroft was of the same opinion and that he directed them to go to Calabar instead, where there was already a sizeable colony of Ibos.[4] At Calabar King Eyo of Creek Town declared himself in favour of welcoming emigrants. From then on, emigrants began to arrive in Calabar. A few went over to Creek Town, but the majority of them settled on the mission land at Duke Town. Again it is not clear

1 PP 1852 XLIX, *Correspondence relative to the Conveyance of the Mails to the West Coast of Africa.*
2 Townsend, Journal of a visit to Ijaye, 19–21 Aug. 1852; C.M.S. CA2/085.
3 Hinderer, Journal entry for 19 Dec. 1852; C.M.S. CA2/49.
4 E. Jones, *Journal of a Mission to the Niger*, 1853; C.M.S. CA1/0129. Also U.P. *Missionary Record*, 1854, pp. 39–41.

how many of them there were in Calabar. In 1856 a missionary reported a dozen families in Duke Town;[1] in 1859, sixteen Sierra Leone men were listed as having joined the Presbyterian Church,[2] most of the others remaining either the Methodists or Anglicans that they were, though they usually did attend the Presbyterian Church.

The failure of the Jones mission was an added reason for the C.M.S. to join McGregor Laird in pressing the government for a contract (like that of carrying the mails) for a new expedition to open up the Niger. The travels of Dr. Barth in Northern Nigeria and his accounts of the resources of the country made the government ready to co-operate with Laird. The expedition went up the Niger in 1854 under Dr. William Baikie. Samuel Crowther was the C.M.S. representative on it. The expedition reported success in the prospects of trade, in the ready welcome promised to emigrants, and in the absence of any disastrous mortality among the members of the expedition.[3] McGregor Laird, with the moral support of the missionaries and of philanthropists in England, therefore redoubled the pressure on the government to grant an annual subsidy for five consecutive years to send more pioneering expeditions to trade on the Niger. 'The reasons I venture upon your Lordship to continue the exploration of Central Africa,' said Laird in a memorandum to Lord Clarendon in 1855,

> are the scientific and geographical results . . . and the advantage we possess in the colonies of the Gambia, Sierra Leone and of the Gold Coast, most efficient agents by whose means new life and energy and a higher standard of living may be introduced naturally, unobtrusively, and rapidly into the remotest regions of the interior. To succeed, this return of the civilized African to his native country, carrying English habits and language with him, must be spontaneous and self-supporting.[4]

He added that they only required regular and assured means of com-

1 William Anderson to Consul Hutchinson, 30 May 1856, encl. in Hutchinson's despatch no. 71, 24 June 1856; FO 84/1001.
2 U.P. *Missionary Record*, 1859, p. 118. There was bitter trade rivalry between the European traders in Calabar and the emigrants (see K. O. Dike, op. cit., pp. 199–26) and some of the emigrants had been obliged to return to Sierra Leone. Most notable among them was Peter Nicholl. He was Efik, liberated in Sierra Leone 1830, enlisted in West Indian Regiment, returned from the Bahamas as Sgt.-Major, became a trader in Freetown, lent his Church (Meth.) £400, interest free. Went to Calabar in 1854. Helped with translations. Was persecuted by traders. Returned to Freetown. Visited England in 1858. Died in 1880, bequeathing £50 to the Presbyterians; U.P. *Missionary Record*, 1858, p. 181; 1881, p. 208.
3 W. B. Baikie: *Narrative of an Exploring Voyage* . . ., London 1856; S. Crowther: *Journal of an Expedition up the Niger and Tshadda . . . in 1854*, London 1856.
4 Laird to Lord Clarendon, 5 March 1855; FO 2/23.

munication and they would soon be settling on the Niger. When the subsidy was granted and the 1857 Expedition was being fitted out, notices were posted up in Sierra Leone inviting emigrants who could pay their passage to take advantage of the opportunity. More than that, Crowther was again on the Expedition, this time with the Rev. J. C. Taylor, an Ibo, and 25 emigrants as schoolmasters and evangelists who opened a mission station at Onitsha and another at Igbebe at the confluence, opposite Lokoja. The period of emigration to the Niger had begun.[1] More will be heard about this later, but it must be said again that this emigration to the Niger, largely inspired by missionary and commercial expansion on the Niger, was different in character from the earlier emigration to Yoruba. As Taylor pointed out in 1866,

> There is a great difference between those who go to the Yoruba mission and the Niger. The former return to their own home, meeting their parents or surviving relatives, whilst the latter though descendants of the Ibo or Hausa are perfect strangers to the country at large. There are only two in our mission (on the Niger) who are actually sons of the country.[2]

Parallel with this movement of emigrants from Sierra Leone were various attempts to organize the return of ex-slaves from the New World. Except in a few places where first generation slaves who still remembered home secured their freedom and had a spontaneous desire to return home, these attempts were largely unsuccessful. However, even the expectation of an emigration movement from the West Indies and America led to two missionary enterprises in Nigeria.

One of the principal grounds for Buxton's optimism about the rapid development of Africa had been the reported nostalgia of free Negroes in the New World, many of whom were educated Christians and skilled artisans.

Discussing in his book the question of 'native agents', he quoted enthusiastic letters received by the Secretary of the African Civilization Society, in reply to his circular, about the many Africans in the West Indies who were longing to return to Africa.[3] In missionary circles there were stories of how, as emancipation was being celebrated in August 1838 with much laughter and many tears of joy, several Africans came forward volunteering themselves as a freedom offering to take the

1 S. Crowther and J. C. Taylor: *The Gospel on the Banks of the Niger*, C.M.S. 1859, p. 39.
2 Taylor to Henry Venn, 15 Dec. 1866; C.M.S. CA3/037.
3 Buxton, op. cit., pp. 491–7.

evangelical light which they had seen to their 'benighted' brethren in Africa. 'The conversion of Africa,' said a Baptist missionary, 'is the theme of their conversation and their prayers, and the object of their most ardent desires.'[1] A Presbyterian minister reported that:

> Our emancipated people, finding their condition so much improved by freedom, and appreciating their Christian privileges, began to commiserate their brethren in Africa. . . . All our congregations held meetings for consultation and prayer about the subject and also began to form a special fund for the benefit of Africa, which in the course of little more than a year amounted to six hundred pounds.[2]

Individual Africans, like Thomas Keith and James Keats, were reported to have embarked on pilgrimages to convert their people in Africa to Christianity.[3]

The missionaries in the West Indies encouraged this nostalgia and this enthusiasm. They made arrangements to train and to select would-be emigrants as evangelists, teachers, planters, industrialists and artisans, and urged churches in Europe and North America to sponsor them as missionaries. William Knibb, of the Baptist Mission, came to England in 1840 and persuaded his society to embark on a West African mission. The Rev. John Clarke and Dr. G. K. Prince were appointed to go on an exploratory journey.[4] They applied to join the Niger Expedition but the government, having already accepted C.M.S. missionaries, rejected their offer. One Mr. Kingdom 'in connexion with the Baptists but not sent by them as a missionary', thereupon joined the Expedition as a settler, 'with a view to make himself useful to the natives wherever he should find an opening'.[5] Prince and Clarke went to visit Fernando Po and the Cameroons. They took a favourable report to England in 1843, and Clarke, with Alfred Saker and his wife, returned to Fernando Po, accompanied by two West Indian missionaries and thirty-nine settlers.

The Scottish missionaries, who were great rivals of the English Baptists in Jamaica, similarly turned their attention to West Africa. In July 1841 the Presbytery meeting at Goschen spent two days in prayer and deliberation on the subject. They heard the Rev. Hope Masterton Waddell read extracts from Buxton's *The African Slave Trade and its Remedy* and they emerged from their retreat with resolutions declaring the time ripe for missions to Africa. They called on their congregations

1 Ibid, p. 493.
2 Rev. G. Blyth: *Reminiscences of a Missionary Life*, p. 178.
3 Groves, op. cit., vol. II, p. 28, note 2.
4 Ibid., pp. 23–30.
5 Schön and Crowther: *Journal of the 1841 Expedition*, op. cit., pp. 134–5.

at home to undertake an African mission, and eight of the ministers present, including Hope Waddell, volunteered to join it. They drew up plans 'to evangelize Africa through the means of the converted negroes of the West Indies' operating in agricultural and industrial colonies.[1] In reply, they were informed by the Board of the Scottish Missionary Society that their proposal was:

> premature, displaying more zeal than judgement, not accordant with the state of dependence in which our Jamaica Church stood, both for means and missionaries; [it was] highly presumptuous after the failure of vastly greater efforts by others than we could possibly put forth.[2] [a reference in particular to the Niger Expedition].

The missionaries were undaunted, however. In 1842 two of them, Rev. George Blyth and Rev. Peter Anderson, on leave in Scotland, canvassed the idea of a West African mission in various churches. Further, they asked Dr. Fergusson, a Liverpool merchant who had been a surgeon in West Africa, to put them in touch with supercargoes trading on the coast. From such consultations the missionaries decided that Calabar was the most eligible spot for a pioneering mission. They sent a letter through Captain Turner, a supercargo, well known in Calabar, who had been a local preacher in a Methodist church in Liverpool, to sound the views of the rulers of Calabar. Turner replied in January 1843:

> At a consultation of the chiefs held this morning in the king's house, it was settled that to sell the tract of ground required was out of the question. The land, however, will be at your service, to make such establishments as you may see proper. It will be guaranteed to its occupiers on those terms for ever. A law will be passed for its protection, and the colonists may dwell in peace and safety, none daring to make them afraid. There seems no doubt of your obtaining land sufficient for plantations for a number of families.[3]

They also consulted Beecroft, an Englishman long resident on the coast, Governor of the Spanish colony of Fernando Po and the most influential European in the whole Bight of Biafra, who reported in March 1844 that the chiefs of Calabar were favourable to the proposal

1 J. McKerrow: *History of the Foreign Missions of the Secession and United Presbyterian Church*, pp. 368–9; Hugh Goldie: *Calabar and its Mission*, Edinburgh 1901, p. 73; Hope M. Waddell: *Twenty-nine Years in the West Indies and Central Africa*, 1867.

2 Donald M. McFarlan: *Calabar, the Church of Scotland Mission, 1846–1946*, Edinburgh 1946, p. 7.

3 Goldie, op. cit., p. 75. Also Hutchinson's despatch, no. 71, 24 June 1856, enclosing copies of Capt. Turner to the missionaries, 4 Jan. 1843 and 19 Jan. 1843; and Beecroft to the same, 18 Mar. 1844; FO 84/1001.

of the missionaries. Thereupon the Presbytery of Jamaica decided to embark on the Calabar mission on their own. They obtained two years' leave of absence for Hope Waddell and asked him to lead an exploratory mission to Calabar. He was accompanied by Samuel Edgerley, an English printer, together with his wife; Andrew Chisholm, a mulatto carpenter; and Edward Miller, a Negro teacher. Hope Waddell arrived in England to gather funds for the mission, prepared if need be to form a separate missionary society to organize it.[1] However, the United Secession Church, which on its union with the Relief Church in 1847 became the United Presbyterian Church, decided to adopt the new mission. But it did so with an important modification of the original plan: Hope Waddell had to give up his idea of agricultural settlements as part of the missionary scheme.

On reaching Calabar in April 1846, Hope Waddell saw that, particularly in the Delta, trade, not agriculture, was the civilizing force, and that the chiefs were not a 'land-owning aristocracy' but middle-class traders. After visiting their houses, many of them imported from England, surveying the household furniture in them, some of which he bought for his own house, attending their weekly dinner parties and, above all, watching them conduct their trade, he saw that they would not have placed much value on a body of West Indian Negro farmers as agents of civilization. He saw King Eyo's son keep accounts, 'writing and copying into an account book the memoranda of business which his father had made on slates . . . neatly entered and all in English'. And he commented that this, in addition to other observations, convinced him that the teachers for Calabar 'must really be competent men', for neither the schoolmaster he brought along, nor the carpenter, was 'equal to this young man in writing and arithmetic'.[2]

Artisans from the West Indies would, however, have been welcome if they had been forthcoming. Waddell himself later took to recruiting from Sierra Leone the teachers, carpenters and sawyers whom he needed for the mission and for the chiefs, when he could not get them from the West Indies. The truth would appear to be that although individual West Indian missionaries continued till this century to be important in many West African missions, hopes of a large-scale emigration of nostalgic exiles from the New World, entertained in the ecstasy of the moment of emancipation, were completely false. Various attempts by different people, friendly and unfriendly to the Negro, have consistently borne this out.

1 Waddell's original appeal for the formation of a Society to take up the new mission was published in *Friend of Africa*, Journal of the African Civilization Society, III, 15–16, 1846.   2 Waddell: Journals, vol. I, pp. 94–5.

This was the experience of the Southern Baptist Convention in the United States, whose pioneer missionary, the Rev. T. J. Bowen, first appeared in Nigeria in 1850. American Baptists had followed English Baptists in undertaking foreign missions, but the American Baptist Convention was split into two by various factors that were dividing the United States into two hostile camps. The breaking-point came in 1844 when Baptists in the south asked the Mission Board, meeting at Boston and dominated by northerners, to say categorically whether they would accept as minister and missionary a slave-owner. The Board replied that they could not.[1] Baptist churches in the south therefore formed a separate convention of their own, and they were anxious to show their northern brothers that economic interest in continued slavery at home in no way interfered with evangelical faith and concern for the safety of the souls of Africans abroad. It was towards the resettlement of free Negroes in Liberia and missionary work by and among them that the Southern Baptist Convention first turned its attention. But colonization, either in Liberia or elsewhere, was unpopular among the Negroes, and the disfavour extended to missionary societies countenancing the idea, which they regarded as a device of the slave-owners to weaken the struggle for emancipation at home.[2] The S.B.C. hoped to rely to a large extent on Negro missionaries, but it constantly had to record its failure to secure suitable candidates. The Negro Baptist churches withheld their support from the Convention.[3]

Thomas Jefferson Bowen was the first outstanding missionary of the S.B.C. He volunteered for missionary work on condition that he should be sent to 'the Sudan' or 'Central Africa', that is to say, to the interior of Africa, and he relied mostly on European missionaries' training in the

1 George W. Sadler: *A Century in Nigeria*, Nashville, Tennessee 1950, pp. 29–30.
2 As the North American Convention of Coloured People said in September 1859, 'A colonization and a bitter pro-slavery man are almost convertible terms.' J. K. A. Farrell: *The History of the Negro Community in Chatham, Ontario, 1787–1865*, Ph.D. thesis, Ottawa 1955, pp. 157–8. While the Colonization Society appealed to the Northern States on the platform that emigrating Negroes would provide useful artisans, technicians and evangelists to Africa, they appealed to Southerners by saying that these would-be 'evangelists', the free Negroes, were a menace to society and their evacuation would be a blessing to America. See Jacob E. Cooke: *Frederic Bancroft, Historian, and Three Hitherto Unpublished Essays on the Colonization of American Negroes from 1801–1865*, Oklahoma 1957, p. 162. Also P. J. Staudenraus: *The African Colonization Movement, 1816–1865*, New York 1961, and Henry Wilson: *History of the Rise and Fall of the Slave Power*, pp. 208–22.
3 After the Civil War, for a while the Negro Baptist churches tried to co-operate with the S.B.C., but the experiment was short-lived and it was believed to have been unsatisfactory by both sides. H. A. Tupper: *Foreign Missions of the Southern Baptist Convention*, vol. II, p. 140; C. E. Smith in *Foreign Mission Journal*, S.B.C., vol. xxiii, no. 1, August 1891; vol. xxiv, no. 12, July 1893.

mission field itself the African staff they needed. But on his retirement in 1856, after six arduous years in Nigeria, he went whole-heartedly into a colonization project that proved once again a mirage.[1] He wished to get the United States government to sponsor a Niger Expedition like the British ones, and to get the Colonization Society (the founding fathers of Liberia) to obtain a charter from the government to set up an American colony 'in the region of Lagos and Abeokuta' where they would give 'land to free negroes and encourage their settling, free the slaves the settlers acquire after a time, give all privileges of citizenship to civilized blacks except directorships and higher offices of the colony.'[2] Bowen and the Colonization Society went so far as to get the Senate in February 1857 to pass a bill authorizing the expenditure of $250,000 on the proposed Niger Expedition, but the House of Representatives turned it down.[3] In any case, Bowen himself, the principal figure in the project, broke down in health. He began to show signs of the mental illness which but for brief interludes made his later life, until he died in 1875, a long sad epilogue to his remarkable career in Nigeria. The project was dropped. Its sponsors were succeeded by English cotton manufacturers, who formed an African Aid Society, seeking to rescue Negroes from the 'unfavourable climate' of Canada and the northern United States and the 'growing prejudices of the white population',[4] and to support their return to Africa, where their skill would be a benefit to the local populations and produce cotton for British industry. In 1859 they sent a Jamaican printer, Robert Campbell, and a Negro Canadian physician, Dr. Martin R. Delany, on a deputation to Abeokuta and other parts of Yoruba to prepare the way for the intended emigration. Their activities there will be noticed later,[5] but when Delany returned to Canada he failed to persuade the Negroes to emigrate. The Negro newspapers argued, as they had always argued, that:

the mortality among colored emigrants in Canada is no greater than among others. . . . If Africa is the real home of the Negro, so is Europe the real home of the American European.[6]

1 He had little support from his mission and opposition from his colleagues in the field. See e.g. A. D. Phillips to Poindexter, 31 Mar. 1859; S.B.C.
2 Bowen to Taylor, 18 Feb. 1818, from New York; 23 Feb. 1857, from Anap, Maryland (*Bowen Letters*). See also Bowen's 'Report on Central Africa and the Niger', presented to Congress, 4 Feb. 1857; *Senate Documents* No. 29, 37th Congress, 2nd Session, vol. iv.
3 Bowen to Lippincott and Co., 21 Nov. 1868, *Bowen Letters*.
4 *African Times*, (monthly) official organ of the African Aid Society, 23 April 1863, advertisement for funds to aid the project of settlement which had by then been diverted to Ambas Bay, near Victoria.
5 See below, pp. 191–3.
6 J. K. A. Farrell, op. cit., pp. 154–5.

To most Negroes in the New World, Africa was only vaguely their home. The slaves, with no knowledge of world geography, transported across the seas lying usually on their backs in the crowded bowels of the ships, separated quite often from everyone they could speak to in their own language, soon lost even imaginative contact with Africa. To their children, born into slavery in a strange land, home was generally either the very colony they knew and to whose development they were contributing, or some vague unattainable 'Zion' or 'Jerusalem' of the Negro spirituals. Thus the various attempts of missionary or secular bodies to convince them otherwise have always met with less than the expected success. The nostalgia of the songs had not enough practical force to attract men from the colonies on the hazardous, unromantic journey to the vast unknown of Africa which, they had often been told, offered such meagre resources for welfare that slavery in European colonies was to be preferred to liberty and freedom there.[1]

There was, however, in a few places a different kind of nostalgia. To Brazil and Cuba,[2] slaves had come continuously for centuries from the same regions of West Africa. A tradition had grown up in these countries of keeping the slaves together in linguistic groups, sometimes under appointed chiefs, and with opportunities to amuse themselves and their masters with their traditional ceremonies, dances and songs. This deliberate policy of limiting the assimilation of plantation slaves by keeping them divided in their old ethnic units and preventing them from acquiring a lingua franca they could conspire in helped to preserve their culture. It was also into these same parts of the New World that there was large scale importation of slaves at the time of the Fulani jihad and the Yoruba wars. Thus, for different reasons, the vision of home remained real among many slaves in Cuba and Brazil, and nostalgia did often mean a desire to return to some specific part of Africa. The slaves sought various ways of emancipating themselves, sometimes through the favour of a kind master, sometimes by running away from a wicked one. More usually, they tried to save up for it through the mutual aid clubs so common among African peoples. 'Often they banded together,' said Donald Pierson, 'to buy the freedom of a friend, or to work under a leader for the liberation of all. The order in which they secured their

---

1 Sociologists have shown that in fact, as is to be expected, many African customs, habits and turns of phrase have survived in the British West Indies; e.g. E. V. Goveia: *Slave Society in the British Leeward Islands*, Ph.D. thesis, London 1953. But there was little memory of Africa in concrete terms of particular places and people.
2 For accounts of Brazilian slavery, see Gilberto Freyre: *The Masters and the Slaves*, trans. Samuel Putnam, New York 1946, and Donald Pierson: *Negroes in Brazil*, Chicago 1942.

freedom was often determined by lot, the earliest liberated remaining with the rest until the last was purchased, after which they sometimes returned to Africa.'[1] With the growth of the abolitionist cause in the nineteenth century and more chances of emancipation and of repatriation, while efforts to provoke an organized return from the West Indies continued to fail, this spontaneous return of the 'exiles' from Brazil and Cuba from a trickle became one of the most important cultural streams in nineteenth-century Nigerian history.

Figures for the movement from the New World are even more difficult to get at than those for the Sierra Leonean movement. By 1853, when the Brazilian emigrants, who were mostly Portuguese-speaking Roman Catholics, gathered together and acquired a piece of land for a church in Lagos, there were perhaps only about 200–250 of them.[2] Emigration continued, sometimes direct on Brazilian merchant vessels, sometimes to England and then by the mail-boats to Nigeria. Some of the emigrants traded back to Brazil and, like the Sierra Leone emigrants before them, began to ask for a missionary of their own. However, Portuguese priests in Brazil did not share the missionary enthusiasm of English pastors in Sierra Leone.

It was a French Catholic missionary body, the Society of African Missions, with headquarters in Lyons, which set up Roman Catholic missions first in Sierra Leone, and then in Dahomey. The first missionary arrived in Dahomey in 1861, and soon began to visit Lagos, but did not assign a missionary there till 1867.[3]

However, one *emancipado* in Brazil had volunteered to answer the call on behalf of the Portuguese Fathers. He was Antonio, originally from São Tomé, brought up in a seminary in Bahia. It is not clear exactly when he arrived in Lagos, but he soon made himself a venerable figure to the emigrants. Pa Antonio, as he was called, built a little chapel and there every Sunday he tried to re-create the religious ceremonies he had known at Bahia: the catechisms, the chants, the blessings of the bread; in short,

1 Pierson, op. cit., p. 39. Cf. also Lorenzo Turner: 'African Survivals in the New World' in *Africa as seen by American Negroes*, Présence Africaine, Paris 1958: 'In Brazil manumission was comparatively easy: a child of a slave mother and a white father became free; owners who mistreated their slaves could be forced to liberate them. On 85, in some states 104, days of the year, slaves were allowed to earn money to buy their freedom at a price no higher than that at which they had been purchased.'

2 Gollmer: Journal entry for 10 April 1853; C.M.S. CA2/043. Vice-Consul Fraser reported, in December 1852, 130 families in Lagos, all Yoruba, mostly Egba; FO 2/28.

3 René F. Guilcher: *Augustin Planque*, Lyon 1928. *S.M.A. 100 Years of Missionary Achievement*, Cork 1956. J. M. Todd: *African Mission: A Historical Study of the Society of African Missions*, London 1961.

he went near saying mass. Every Saturday he gathered his flock together to recite the rosary. He baptized new-born babies, blessed marriages like the patriarchs of old, and was called to the side of the dying for the last ministrations.[1]

In 1859 Consul Campbell reported that there were 130 Brazilian families in Lagos.[2] When Father Broghero visited Lagos in 1863, there was already a Catholic church in being. At the mass he celebrated, there were 400 present.[3] When Father Bouche arrived in 1867 to take charge of the church, he reported 'about 500 Catholics'.[4] There must have been many other Brazilian and Cuban emigrants who did not go to church. There was a sizeable colony of them at Abeokuta, and like the Sierra Leone emigrants they were to be found scattered about in the Yoruba country. There was no mention of them in the Delta or on the Niger, but in 1859 Campbell issued passports to some Hausa and Nupe Brazilian emigrants.[5] A Census in Lagos in 1872 showed 1,237 Brazilians and Cubans compared with 1,533 Sierra Leonean emigrants.[6] It was after 1888 when Brazil at last abolished the status of slavery and Governor Moloney helped to establish regular communication between Lagos and Bahia for the returning 'exiles' that this other movement of emigrants reached its climax.[7]

The emigrants had an importance in Nigerian History out of all proportion to their numbers. Left to themselves, and scattered all over the whole country, they might have had no more significance than a band of ex-servicemen, people who had taken part in a nightmarish experience, with a useful stock of strange tales, and a stock of other arts for which there was not much scope at home. But the missionary movement kept most of them together in a few focal centres, gave them scope and encouragement. For the Sierra Leoneans, they offered commercial

1 Biographical Note by Father Holley in *Les Missions Catholiques*, 20 May 1881. Based on questions to Antonio himself.
2 Campbell to Malmesbury, 4 Feb. 1859; FO2/28.
3 Father Broghero in *Les Missions Catholiques*, 21 Dec. 1863. Also Journal entry for Sunday, 25 Sept. 1863; S.M.A. archives, Rome.
4 'Premiers Temps de la Mission de Lagos d'après Mère Véronique' in *Missions de la Nigéria*, S.M.A. Rome. The Lagos Census for 1872 listed 572 Catholics; Rev. J. Rhodes to Secretaries, 15 Aug. 1871, said there were about 2,000 *Amaros* in Lagos; Meth.
5 Campbell to Malmesbury, 4 Feb. 1859; FO2/28.
6 Lees to Colonial Office, 28 Feb. 1879, referring to figures collected in 1872; CO806/130. At that date, 1872, Pope Hennessy had estimated 6,000 Brazilians in Lagos (to Kimberley, 31 Dec. 1872; CO267/317), while Sir Harry Johnston estimated in 1875 4,000 Brazilians in both Lagos and Whydah; *The Negro in the New World*, New York 1916, p. 98.
7 Pierson, op. cit., p. 39. Moloney returned the figure of 2,723 for Brazilians in Lagos in 1881, and 412 arrivals between 1881 and 1886: Moloney to Holland, 20 July 1887; CO14/59.

opportunities, employment as catechists, evangelists and schoolmasters; for the Brazilians, houses to build, roads to construct, and other facilities to practise the arts they had acquired. It was the emigrants who introduced the missionaries into the country, and they were an essential and integral part of the missionary movement.

3 Missionaries, Traders and Consuls

On his arrival in Calabar in 1846, Hope Waddell found a flourishing busy trade in palm-oil[1] which had brought the town as much prosperity as in the days of the unhindered slave trade. Calabar was in some ways similar to Badagri. The Efik in Calabar, like the Egun in Badagri, had been attracted to the coast by the slave trade. In both places, settlement had brought new political divisions. In Calabar there were three separate establishments: Creek Town, Old Town and Duke Town. There was great competition especially between Creek Town and Duke Town over the control of the European trade. Further, there were internal rivalries between the various Houses in each town. But in spite of these divisions, the Efik developed in the Egbo or, more correctly, Ekpe society an organization covering the whole of Calabar and superseding the sectional interests of the various Towns or the personal ambitions of the 'gentlemen', the heads of the Houses, of whom it was largely composed.[2]

In that way Calabar not only dominated the original inhabitants of the mouth of the river (the Kwas) but also, by agreement or force, largely controlled the markets in its hinterland. Calabar was, in this, typical of the Delta city-states. What distinguished Calabar from the rest was the length and the depth of its attachment to English traders. English was the only European language spoken by Calabar traders, and Hope Waddell found 'very intelligible journals of the affairs of this country kept by its rulers, written in English, of so old a date as 1767'.[3] The predominance of English traders in Calabar encouraged and hastened the change from the slave trade to that in palm-oil, so that it became the principal source of palm-oil till it was surpassed in output by Bonny

1 Waddell: *Journals*, vol. 1, pp. 38–9, 44. For the different European traders in Calabar at this time, see the petition of the masters and supercargoes to Beecroft dated 31 Jan. 1851; State Papers, vol. 41, p. 268–9.
2 *Efik Traders of Old Calabar*, ed. Daryll Forde, Oxford 1957; Waddell: *29 Years in the West Indies and Central Africa*, London 1863; Talbot: *Southern Nigeria*, vol. I. London 1926.
3 Waddell to Sommerville, 21 Sept. 1848, in U.P. *Missionary Record*, 1849, p. 58. This was probably Antera Duke's diary which was later brought to Scotland; only some transcripts from it have survived, now published in *Efik Traders of Old Calabar*.

in the 1840s.[1] When, in 1839, the British Government, largely in response to Buxton's appeal for anti-slavery action on land, embarked on the policy of signing treaties with the coastal rulers, it was at Calabar that the readiest response was found.

The rulers were asked to make a clear choice between renunciation of the slave trade in return for some compensation as well as English trade and friendship on the one hand, and the persistent menace of the warships on the other. To them the British campaign was only another phase of the struggle that had been going on for generations between various European nations competing for the lion's share of the trade on the coast. The English had changed their article of trade; for reasons not quite clear to the Delta people, they began to refuse to deal in slaves and were calling more and more for palm-oil, for elephant's tooth, camwood and gum copal. Their intention to capture the market remained the same, however; they brought warships to fight their Portuguese and Spanish, Brazilian and French rivals. The new proposals that England began to make in 1839 were examined very critically. It appeared that England was offering not only hopes of increased trade, but also hopes of a more equitable, less one-sided trade than the merchants on the coast had hitherto engaged in.

It should be pointed out that the coastal traders had long complained of the quality of goods the Europeans brought them and their refusal to bring them the things they needed most. The coastal chiefs amassed capital from their trade but could not obtain the goods they wanted. Several chiefs imported houses they did not live in, filled like museums with European furniture and fineries. The more enterprising ones began to look for other ways of spending their money. In 1828 Captain Owen, who was making a survey of the coast for the Admiralty, reported that the chiefs of Old Calabar wished

> to be instructed in the methods of making sugar and to obtain the necessary machinery for which they say they have repeatedly applied to their friends in Liverpool without success. For these advantages they are ready to pay handsomely.[2]

The chiefs and traders wanted not just worthless finery but tools and machinery and they wanted also that their children should be taught European languages, how to keep accounts, and so on. When officers of

1 Dike, op. cit., pp. 49–51.
2 Owen to the Colonial Office, 28 April 1829; CO82/, cited in Dike, op. cit., p. 67. Cf. also Oba Adele's request in 1830 at Badagri for 'a carpenter's chest of tools, with oils, paints, and brushes', Richard and John Lander: *Journal of an Expedition to explore the course and termination of the Niger*, 3 vols., London 1832, p. 30.

the British navy came to negotiate the slave-trade treaties, it was these very things they offered. They also spoke of Buxton's 'confidence between man and man', of missionaries and teachers who would come and create a new era of things in place of the age of the slave traders and the hulk-dwelling palm-oil ruffians.

Many of the rulers were attracted by these new proposals. But even where, as in Calabar, they regarded the British as valuable and reliable customers, they still hesitated to sign the slave-trade treaties[1]. To give up their right to trade with whomsoever they wished and in whatever commodities they chose was a diminution of their sovereignty. Worse still, there was a distinct threat in the clause of the treaties which spoke about the 'severe displeasure' of the Queen of England, who claimed the right to use her navy to enforce the treaties unilaterally if she thought they were violated. The chiefs tried to negotiate cautiously. Bonny agreed in 1839 to sign a treaty, but with so many reservations and with such evident hesitation that the British officials themselves did not expect the treaty to be kept and so the government refused to ratify it. It was not till 1844, after repeated attempts, that Bonny agreed to sign another.

The Calabar chiefs were more ready to negotiate. After some hesitation, they were willing to accept the slave-trade treaty[2] provided it would in fact usher in the era of economic and industrial development promised. Asked in 1842 what they wanted for the annual 2,000 dollars compensation agreed to in the treaty, they replied:

> Now we can't sell slave again, we must have too much man for country, and want something to make work and trade; and if we can get seed for cotton and coffee, we could make trade. Plenty sugar cane live here, and if some man can teach we way for do it, we get plenty sugar too. . . .
>
> Mr. Blyth tell me England glad for send man to teach book and teach for understand God all same as whiteman. If Queen do so, I glad too much.[3]

In fact, the chiefs welcomed missionaries to Calabar principally because they wanted agricultural development, a sugar industry and

1 Capt. Blount reported in 1841 that Calabar chiefs wanted a treaty of amity and commerce but refused 'to enter into any treaty for the total suppression of the export slave trade . . . although the slave trade there is of no importance'; enclosure in Tucker to Admiralty, forwarded to the F.O., 30 July 1841.
2 Admiralty to F.O., 22 April 1843, including Raymond to Capt. Foote, 11 Dec. 1842, enclosing treaties and kings' requests, and Foote to Herbert, 12 Dec. 1842; FO 84/384.
3 Enclosure in Raymond to Capt. Foote, 11 Dec. 1842; FO 84/384.

schools. They realized that missionaries brought their own hazards.[1] Unlike traders who could be kept at arm's length, missionaries came not only to visit but to settle, to build a house and live all the year round in Calabar and to cultivate Calabar soil. King Eyo and the chiefs took all the precautions they could. They told Captain Turner that selling land to the missionaries was 'out of the question'.[2] They declared themselves willing to pay for services they received. Like shrewd businessmen, they haggled over the terms on which the missionaries were to be allowed to settle. Waddell reported that King Eyo asked him:

'Suppose man go away from his master and come to my Town [i.e. mission compound] . . . Would I keep him?' In reply I stated that my house would be no refuge for bad people; that if any servant run away from his master and beg for him and tell him to do no more bad again their master must forgive him and not flog him, neither kill him. 'That be very good,' he replied. 'You bring man back and beg for him and tell him "Go do no more bad again". That be very good.' 'You won't kill a man, king, when I beg for him?' 'No,' he promptly answered. 'We no kill man again, only for some very bad crime, same as you kill him in your country.'[3]

In 1848, the chiefs of Bonny wrote to Liverpool to ask for missionaries. They stated their terms as follows:

We agree to let them have ground for a house and garden or gardens, for a period of twenty years; we will take back the ground at the expiration of that period. And if payment for their services be required, we also agree that every gentleman sending a son to their school shall pay for the education of such son five puncheons of palm oil. They would require to bring material to build their own houses and a carpenter to put it up, as we neither have joiners nor the requisites for building a proper house for them; but we are anxious to afford them everything as well as every assistance in our power. And we further expect that those gentlemen to be sent us shall be capable of instructing our young people in the English language.[4]

Not only could missionaries not be kept at arm's length; they could not be engaged on stated terms as the chiefs of Bonny imagined. No Chris-

1 Hope Waddell records that a few days after his arrival King Eyo said his chiefs were already asking why he gave land to missionaries. They feared that 'by and by, more will come and they will take the country away from them', *Journals*, vol. i, p. 68, 21 April 1846.   2 See above, p. 45.
3 Waddell: *Journals*, vol. I, p. 61, 17 April 1846.
4 King Pepple and chiefs to Messrs C. Horsfall & Sons, 30 May 1948, passed on to the United Presbyterian Church and published in the U.P. *Missionary Record*, 1848, p. 201.

tian missionary would accept his wages from the people to whom he was sent. He must remain responsible to an outside power, to God and his conscience, as well as to the missionary society which had sent him out. He came to help; if he was not wanted in one place, he would go to another. He must remain a philanthropist, expecting not money but gratitude—which can sometimes be more exacting.

Missionaries in Nigeria did not always question the theory that Christianity, Commerce and Civilization would work together for the great benefit of Africans. But they liked nevertheless to emphasize their own philanthropy and how much it set them apart, in objectives, approach, methods and morals, from their profit-seeking fellow countrymen who came to trade. The truth of course was that traders and missionaries were interdependent. The Christian missions made a considerable impact on the trading situation. In turn, the expansion of European trade and political influence greatly facilitated the work of missionaries. Traders and missionaries often quarrelled. Many missionaries despised most traders and the compliments were fully reciprocated. Nevertheless, they had to co-operate most of the time.[1] By definition, missionaries were incurable optimists with severely limited funds and resources but unlimited hopes and aspirations, and they were for ever seeking allies in the most unlikely places, not least from traders.

It should be emphasized how limited the funds of the missionary societies were in order to appreciate the extent of their dependence on the traders. The funds at the disposal of Freeman in 1842 were £4,000 a year. With it he maintained a chain of missions along the Gold Coast, sought a footing in Ashanti, went to establish one in Badagri, and sought further openings at Abeokuta and in Dahomey.[2] The Presbyterians at Calabar in 1850 had some £2,000, and by 1860 were spending over £3,000. The C.M.S. in the Yoruba Mission started on a budget of £3,000 and gradually rose to £5,000.[3] These funds came largely from the pennies and sixpences of faithful worshippers, but often the mission

1 See below, pp. 82–3.
2 Circular of 29 Feb. 1844, imposing financial stringency on the mission. Freeman had exceeded the grant for the previous year, went on leave and raised subscriptions to clear the debt. The grant was then increased to £4,500, with the note that 'we have either to give up a large portion of the field and recall 30 or 40 missionaries, or enforce economy'. No furniture, no expensive repairs, no costly journeys without the most pressing necessity. (Meth.)
3 See annual estimates in the *Missionary Record and Proceedings of the C.M.S.* These are cited for rough comparative purposes. Miscellaneous expenses on passages, building grants, presents to African rulers, etc. will have to be added and there is the field left open for voluntary subscriptions which varied from mission to mission. The budget of the American Baptist Mission for 1855–6 stood at 3,750 dollars, which, with the rate of exchange on 1 Jan. 1856 at 4·8 dollars to the £, comes to under £850. (*Bowen Letters*, p. 74.)

ran short of funds or a new crisis in some part of the world necessitated opening new missions and special appeals had to be made. In such circumstances £100 or £50 from a Liverpool or Glasgow or London merchant made a considerable difference to the missionary budget.

The dependence of missionaries on merchants, however, went much further than mere occasional contributions. Until the establishment of the mail-boat in 1853, every missionary depended for passage, freight and correspondence on the trading vessel, as every supercargo went in his own ship and the missionary could rarely afford the expense and maintenance of a ship. Passengers and freight were taken as a favour because it was more profitable for the traders to carry trade goods than missionary houses, gift boxes and other equipment. They were quite often taken free. Even the Presbyterians who received the free loan of a brig from the well-known Liverpool trader, Robert Jamieson, with an annual £100 for its upkeep, used it in Calabar and could not always send it back to England for new recruits and new equipment needed. No missionary could afford to maintain the sort of establishment the trader maintained in his hulk at Calabar or his factory at Badagri, and it was, for example, more convenient for the missionary to hire a trader's canoe than to maintain one himself with its complement of canoemen. The traders came to expect this dependence; they objected where the missionaries tried to be self-sufficient, for fear they would import trade goods and undersell traders.[1] Usually, therefore, missionaries received from the trader's hulk or 'factory' any provision or other goods they required on bills of credit which the merchant houses in Britain or France presented for payment at the mission headquarters. The merchants thus acted also as bankers to the missions.

In return, the missionary brought some social life to the trader. While traders generally 'kept women', the missionaries brought European wives and, occasionally, children—a settled family life which an ailing trader greatly appreciated. To begin with, relations were generally cordial. The missionaries, with their faith in the virtues of commerce, made a genuine effort to co-operate with merchants in order to secure the revolution in Africa that Buxton had prophesied. At Calabar missionaries followed traders; at Badagri and Abeokuta traders were beginning to follow missionaries.

By Christmas 1846, within four months of the arrival of missionaries at Abeokuta, Thomas Hutton, Agent-General in Cape Coast of the firm of Thomas Hutton & Co. visited their factory in Badagri, still the only establishment there. He visited mission schools, contributed '120 heads

---

1 Sommerville to Goldie, 6 Sept. 1848; Letter Book, vol. I, p. 158, National Library of Scotland. Waddell: *Journals*, p. 175, 26 May 1851.

of cowries to the good cause',[1] and, together with his agent, Parsons, proceeded to Abeokuta in January 1847. There he gave the school-children a New Year feast, contributed to mission funds, and in particular looked round the town and investigated prospects for trade.

Very soon other traders followed and Badagri trade began to revive. The Huttons tried but failed to maintain a monopoly of the trade. In February 1848 two traders representing a Bristol firm arrived in Badagri and the Huttons could not dislodge them.[2] Besides Sierra Leonean emigrants trading between Freetown, Cape Coast and Badagri, there were Brazilian traders from Porto Novo and Whydah selling rum and tobacco for cowries or palm-oil.[3] About 1850, Domingo Martinez even established a factory at Ajido some 10 miles east of Badagri.[4]

By 1851 Badagri began to feel her old self as a trading centre. In May of that year Sandeman arrived with two assistants and established a factory for the firm of 'Forster & Smith', which pleased Gollmer very much. 'Now we may hope,' he said,

> the missionaries will be able to get a comparatively comfortable passage out and home in the vessels of 'F. & S.' which bears a high character on the coast, and are no longer obliged to put up with the wretched Jersey crafts.[5]

—presumably the Jersey crafts of Thomas Hutton & Co.

Missionaries and traders easily saw that they had a common interest to protect in Nigeria. They were all 'British residents'; and if ever they felt endangered they chose one of themselves, missionary or trader, to write to officers of the squadron for aid. But there were situations for which the occasional appearances of the warships on patrol were unsuitable and for which both traders and missionaries were beginning to demand the full-time services of a resident British official, that is—without going yet into the various possible conceptions of the duties of such an official—a British consul.

Already, in emergencies, traders and missionaries had been turning to the Governor of the Spanish island, Fernando Po, who happened to

1 Gollmer to C.M. Secretaries, 14 Jan. 1851; C.M.S. CA2/043.
2 Gollmer, Journal entry for 23 Feb. 1848, the Randolph Brothers; C.M.S. CA2/043.
3 Gollmer: Journal entries for 31 Aug. and 27 Nov. 1851; C.M.S. CA2/043.
4 Gollmer: Journal entry for October 1850, referring to a riot near the factory; C.M.S. CA2/043.
5 Gollmer: Journal entry for 8 May 1851; C.M.S. CA2/043. See also list of British residents who petitioned Capt. L. G. Foote for protection on 16 June 1851 in *Papers relative to the Reduction of Lagos*, p. 126. One of Sandeman's assistants was probably Thomas Tickell, later to become famous as trader, Vice-Consul and administrator in Lagos and Badagri.

be an Englishman, John Beecroft,[1] as to a British Consul; the British government also had recourse to his services. Following a visit to Calabar by two French boats in 1847, offering French trade in return for the suppression of human sacrifice at funerals and other practices (an offer which King Eyo rejected), the English missionaries and traders there asked Beecroft to send for a British warship to plant the English flag in Calabar.

Hope Waddell further persuaded the Foreign Mission Committee in Edinburgh to approach the government to set up a protectorate in Calabar, himself declaring that already 'this country is almost equivalent to an English colony'.[2] The government rejected the appeal but a British warship, accompanied by Beecroft, was sent to Calabar as a counter-demonstration to the French, and Eyo pledged that he would 'use his utmost influence and power to induce his subjects' to give up human sacrifice.[3]

It was the missionaries' rather than the traders' move that had been effective in this instance. In attempting by themselves to influence government action, British merchants tended to cancel out; in petitions to the government, they made rival demands; in evidence before Select Committees and Commissions of Inquiry, their testimony was usually divided. The missionary bodies, however, in spite of occasional differences, tended to act together; and in Britain they had country-wide organizations, and all the influence of the humanitarians. Their agents in West Africa lived among the people, wrote letters frequently and sent extracts from their daily journals, which were widely circulated in the missionary magazines, notices and records, which recent research has shown formed the bulk of the reading-matter of the Victorians.[4] When on leave, they enlarged on these in recounting their experiences to various audiences throughout the country, or in addressing the annual mammoth meeting in Exeter Hall. Some of these missionaries were beginning to penetrate into the country ahead of traders, and were therefore in possession of information not to be had elsewhere and on which even the government had to rely.

The importance of all this began to be evident during the debate

1 K. O. Dike: 'Beecroft 1835–49' in *Journal of the Nigerian Historical Society*, vol. i, no. 1, December 1956.
2 U.P. *Missionary Record*, 1848, pp. 28, 56.
3 Commander Murray to Commodore Sir Charles Hotham, 24 March 1848, in appendix to the Minutes of Evidence taken before the Select Committee of the House of Lords on the Slave Trade (1849). Also extracts from Waddell's Journal in U.P. *Missionary Record*, 1848, p. 24 ff.
4 Margaret Dalzel: *Popular Fiction 100 Years Ago*, London 1957.

provoked by the efforts of William Hutt, M.P., to secure the withdrawal from West Africa of the naval squadron which missionaries and traders alike now regarded as their lifeboat. The squadron was opposed on the grounds that it was expensive and also ineffective in its object of suppressing the slave trade. Hutt was Chairman of a Select Committee of the House of Commons which failed to reach a decision on this matter in 1848. In 1849 he used the Chairman's casting vote to recommend a withdrawal of the squadron, an increase of diplomatic measures, and the continuance of the 'instruction of the natives by missionary labours, by education and by all practical efforts'.[1]

The government was not happy at the recommendations of this committee. Lord Palmerston, then at the Foreign Office, was most conscious of the commercial interest of Britain in West Africa as well as the need for government to foster those interests. His first reaction, therefore, was to appoint John Beecroft of Fernando Po in May 1849 as Her Majesty's Consul in the Bights of Benin and Biafra. It is noteworthy, however, that while Beecroft was to remain at Fernando Po and supervise the Niger Delta which was still the focus of British interest, William Duncan, a trader was also named unpaid Vice-Consul to the kingdom of Dahomey to which British traders at Cape Coast were anxious to be admitted. In the instructions to Beecroft and Duncan no mention was made of missionaries, no mention at all of Badagri, Abeokuta and Lagos.[2] It was then that missionary influence, working through a pressure group in London, stepped into the picture.

The organizer of the group was Henry Venn, who became a member of the Parent Committee of the C.M.S. in 1835 and was soon recognized as its leading figure. He became Honorary Secretary in 1841 and remained so for 31 years (1841–72).[3] By 1849 he had grown to full stature and was able to show that in spite of the set-back caused by the failure of the Niger Expedition the Evangelical party in England, if properly organized, was still capable of exercising considerable influence on the government.

He had brought to England to give evidence before the Hutt Committee his oldest missionaries in the Yoruba Mission, Gollmer and Townsend, the one from the coast, the other from the interior, to testify

1 C. Lloyd: *The Navy and the Slave Trade*, London 1949, p. 111.
2 Palmerston to Beecroft, 30 June 1849; *Papers relative to the Reduction of Lagos*, PP 1852 LIV.
3 John Venn: 'Henry Venn of the C.M.S.' in the *Dictionary of National Biography*, and *Venn Family Annals*, London 1903. Rev. William Knight: *Memoirs of Henry Venn*, London 1881. J. F. Ade Ajayi: 'Henry Venn and the Policy of Development' in *Journal of the Historical Society of Nigeria*, vol. 1, no. 4, 1959.

to the work of the squadron at Badagri, the remarkable improvements that the decline of the slave trade and the rise of the missionary party had brought to Abeokuta, and the awful results that would come from the withdrawl of the squadron. There were other missionaries from the Gold Coast and Sierra Leone, and Hope Waddell from Calabar. Venn had also rallied naval officers like Trotter and Allen who commanded the Niger Expedition, Captain Denham, and other disciples of Buxton, not only to testify, but also to publish pamphlets, reviews, memoranda and articles in favour of the retention of the squadron.[1]

When in spite of all this, in spite of the exertions of Sir Robert Inglis, an active missionary supporter on the Hutt Committee itself, the committee began to pass hostile resolutions, Venn thought of other measures. Bishop Samuel Wilberforce, son of the more famous William, also of the missionary party, moved for a separate Committee of the House of Lords, which interviewed the same witnesses and came to a decision the exact opposite of that of the Committee of the House of Commons. The problem then was how to get the House of Commons to reject the report of a Committee of their own House and to accept that of the House of Lords.

Under the date 30 January 1850, at the opening of the parliamentary session in which the two rival reports were bound to come up, Venn wrote in his diary:

> Hastened to Sir E. Buxton's to a meeting of Abolitionists; present, Lord Monteagle, Gurney, Gurney Hoare, Captain Denham, Capt. Trotter, Capt. Beecroft. Two hours' discussions upon parliamentary tactics for the session. Agreed that the squadron must be maintained.[2]

In other words, Venn brought together missionaries, naval officers, politicians and the new Consul. They intensified their lobbying and the pamphlet warfare, and they organized more petitions including a touching one from Liberated Africans and their sons in Sierra Leone who had benefited from the services of the squadron. But there was only one parliamentary strategy that could save the squadron, namely, that the government, at that time threatened by public discontent at the handling of the 'Don Pacifico Affair', should make the issue one of confidence— in spite of the fact that many supporters of the Whig government were known to be inclined towards Hutt's views. Yet this was what another of Venn's deputations secured from Lord Palmerston on the morning of Tuesday 12 March.[3] A week later, in an unusually crowded debate in the

1 C. Lloyd, op. cit., pp. 112–13.
2 W. Knight: *Memoirs of Henry Venn*, op. cit., p. 200.
3 'To breakfast at Sir T. D. Acland's; a West African party—Lord Harrowby, Sir R. H. Inglis, Sir Edward Buxton, Capts. Denham, Pelham, Trotter, and

House of Commons, Hutt's motion was defeated by a majority of 78. The morning after the debate, *The Times* editorial lamented that in order to save the squadron the government had enslaved members of parliament by invoking the party whip:

> by a sort of vicarious bondage, the British M.P. has taken the place of the African and done task work to the music of the lash of 'Massa Russell'.

The *Spectator* thought the squadron—'this costly failure, this deadly farce'—had been forced on parliament and country. The *Morning Chronicle* added that the 'cruel, hopeless and absurd experiment' was doomed in spite of it all. But the government's paper, the *Morning Post*, replied that the editor of *The Times*

> forgets, or he never knew, that the abhorrence of the slave trade, which with him was only an expedient in political strategy, was and is with the people of England a religion.[1]

The 'Scramble for Africa' was still a very long way off. The acquisition of fresh imperial commitments was in 1850 certainly very unpopular in many circles in England. But opinion in England was not one- but many-sided. In 1846 the same Lord Palmerston supported the abolition of the sugar duties because the abolition favoured industrialists in England and encouraged the development of trade with Cuba and Brazil, whose products depended on the slave trade. In 1850 he risked office

---

Mr Evans. The consultation was to be upon parliamentary tactics in reference to Mr Hutt's motion next Tuesday for the removal of the squadron. A request was sent to Lord Palmerston at about 10 o'clock to see some of the party on the subject. The answer was that his Lordship was not up, upon which the messenger was sent back to ask for a note. The answer returned was that he would be ready to see us at 11.30 a.m., at which hour we all went in a body to his private residence, Carlton House Terrace, except Lord Harrowby. Lord Palmerston received us in his dining-room as cheerfully as if the Greek affair [Don Pacifico] existed only in Herodotus. We sat round a table. Sir T. Acland opened the business admirably, putting a few points tersely. Lord Palmerston's answers were frank and very satisfactory; the maintenance of the squadron was a government question; it was to be stated in the House that the measure had been successful to a great extent, but that our experience had taught us that it might be rendered more effectual by new arrangements without an increase of expenditure; that Lord Palmerston was to write a dispatch explaining the law respecting property employed in the slave-traffic—that it may be seized and destroyed as well as the barracoons.' Venn, Journal entry for 12 March 1850, cit. in Stocks, op. cit., pp. 107–8.

1 W. L. Mathieson: *Great Britain and the Slave Trade, 1839–65*, London 1929. Greville, the famous political commentator, hearing before the debate that the government intended to make the issue one of confidence, thought that they were 'demented at taking this violent course in reference to so unpopular a question and one so entirely fallen into disrepute'; ibid.

to keep the preventive squadron in West Africa to fight the slave trade at its source and to bolster up the expansion of British interests in Nigeria. For the real result of the great controversy over the squadron was to prompt the government to propose measures to improve its efficiency and to strengthen Palmerston's hand in extending the slave-trade-treaty system from the Delta to the Bight of Benin, with new clauses and new interpretations allowing the agents of the British government to interfere more actively in the internal affairs of the areas concerned, in pursuit of the joint policy of trade and philanthropy.

When Hope Waddell was giving evidence before the Hutt Committee, he expressed the opinion that while Britain was

> exerting itself to prevent the selling of slaves upon the coast, I think it would be equally its duty to use all its influence with the native authorities to prevent their killing their slaves, for it is certainly just as bad to kill them for nothing as to sell them, and perhaps rather worse.[1]

It is doubtful whether Waddell considered the full implications of this forceful argument. At any rate he proposed that the naval squadron should go further than the matter of the external slave trade and undertake the abolition of human sacrifice, a matter of internal reform. Even before the British government accepted the argument and included the clause in the new treaty forms, the missionaries in Calabar had taken steps to see such a measure enforced.

Hope Waddell returned to Calabar in 1849. In February of the following year, while he was on a short visit to Bonny, two notables died in Duke Town and there were rumours of human sacrifice at their funerals. William Anderson, the missionary in charge of Duke Town, feeling 'that a united moral force on the part of all white people in the neighbourhood whether missionary or trader was fully warranted if not imperatively called for',[2] convened a meeting at the mission house. Ten supercargoes, one of them a Dutchman, three surgeons, Edgerley (the missionary in Old Town), and Anderson himself met and 'resolved that all present go in a body to King Archibong and the gentlemen of Duke Town at 5 p.m. and denounce the murders committed on Tuesday and to protest against the recurrence of such barbarities'.[3] The protest meeting was duly held and King Archibong and the gentry said they

1 Minute of evidence before the Select Committee of the House of Commons on the Slave trade, 1849. Question no. 451, PP 1849 XI.
2 Anderson: Journal extracts in *Missionary Record*, 1850, pp. 105–10, entry for Wednesday, 6 Feb. 1850.
3 Ibid., Thursday, 7 Feb.

would like to meet the rulers of Creek Town and take common action. The following day the reformers met and resolved to form themselves into a 'permanent Society for the suppression of human sacrifices in Old Calabar or the destruction of human life in any way, except as the penalty of crime'.[1] They met King Eyo of Creek Town and threatened to break off all intercourse unless an Ekpe law was passed within a month to that effect. On Friday, 15 February such a law was solemnly proclaimed in both Duke Town and Creek Town.[2] The new society then met to discuss how to get such a law extended to the neighbouring towns and ensure its strict observance. By then Hope Waddell had arrived, and it was he who proposed the motion, which was carried unanimously, that the aims of the society be broadened, and the name be changed to the *Society for the Abolition of Inhuman and Superstitious Customs and for promoting Civilization in Calabar*[3] (hereinafter referred to as SAISC).

Waddell was carrying to its logical conclusion his argument before the Hutt Committee that once Britain's right to enforce prohibition of the external slave trade was accepted and extended to the enforcement of internal reform, her right to suppress anything her missionaries regarded as 'superstitious customs' must also be accepted. In November 1850 John Beecroft, Her Britannic Majesty's Consul, came to Calabar from Fernando Po and not only agreed to join the Society but also added that if Kings Eyo and Archibong had any difficulties in enforcing the laws they made under pressure from the reform society, they should appeal to him to bring warships to their aid.[4] Thus social reforms which the rulers of Calabar under pressure from missionary reformers agreed to carry out became matters which the Consul undertook to enforce by the navy of an outside power. And these reforms included not only prohibitions of human sacrifice, twin murder and trial by ordeal, but also provision for such things as the right of the mission house to be an asylum, the Presbyterian conception of Sunday observance, and Victorian dress fashions. It was, to say the least, a very questionable method of reform. SAISC as a reform society will be discussed later. What is important here is to see how both traders and the Consul took advantage of this desire of missionaries for combined intervention in

1 Ibid., Friday, 8 Feb.
2 Waddell: *Journals*, vol. VII, p. 157, entry for 15 Feb. 1850. There was thanksgiving in many Scottish churches on the second Sunday in July; Somerville to Waddell, *Letter Book*, vol. I, p. 572, in Nat. Lib. of Scotland.
3 Waddell: *Journals*, vol. VII, entry for 19 Feb. 1850.
4 Waddell: *Journals*, vol. VIII, pp. 94–5, entry for 19 Nov. 1850. Mr. George Horsfall, 'junior member of the Liverpool trading family', joined SAISC on 28 March 1851.

African society to weaken the African states without pursuing the objects the missionaries had in mind.

Even before the fate of the squadron was known, the missionary party had made new impressions on Palmerston's mind. For, unlike the traders, they wanted the squadron and the new consul for more than merely protecting coastal trade. They wished to see British influence penetrate inland to protect missionaries and support their efforts in seeking economic development as well as social reform in the country. To achieve this they began to build up a picture of Abeokuta as a land full of promise for the future evangelization and civilization of the country and as the gateway to an overland route to the north with as much potential economic value as the Niger waterway itself.

At the time when the campaign against the Hutt Committee was taken to the House of Lords, Townsend, who had come from Abeokuta bringing gifts from the chiefs to Queen Victoria, was asked by Venn to put in writing 'such considerations as appeared to him to make a deputation to Lord Palmerston desirable'. He produced a very able, clearly-expressed document, setting out the aims and objectives of the missionaries at Abeokuta, their immense success there, the 'menace' of Kosoko and of Dahomey, the huge possibilities for the economic development of the area, the friendly disposition of the chiefs and the loyalty of the emigrants, all pointing to the conclusion that Abeokuta ought to receive prior if not privileged consideration from the Government as the advance post of civilization and Christianity against slave traders and barbarism. On the basis of Townsend's paper, the C.M.S. drew up a memorandum setting out a new policy to be urged on the government. The key point was made in paragraph 2:

> Most of the advantages which were proposed by the expedition up the Niger in 1842 are now within reach of the British Government by securing the navigation of the Ogun. Traders from the banks of the Niger visit the principal markets of Abeokuta and there is little doubt that the road to Egga and Rabbah, the former of which towns was the highest point reached by the Niger Expedition, might be opened for trade through the channel.[1]

Then followed the enumeration of such advantages of Abeokuta as the favourable disposition of the Egba chiefs; the reduction of their language to writing on which the C.M.S. had been engaged; the introduction of the English language through the emigrants; and the fact that the

1 Townsend to the Secretaries of the C.M.S., Exeter, 17 Oct. 1849; *Papers relative to the Reduction of Lagos*, p. 30.

Yoruba people lived 'under a free form of constitutional government very different from the tyranny of Dahomey and Ashantee'.[1] This was all well calculated to impress Palmerston, the godfather of liberty and constitutional governments in Europe.

Both Townsend's paper and this memorandum had been presented to Palmerston on 4 December 1849. Palmerston decided to send Beecroft, who was home on leave, on a 'diplomatic' mission of enquiry to Abeokuta. His instructions quoted the C.M.S. memorandum on that place, adding that

> Lagos is therefore said to be the natural port of Abeokuta; but the slave trade being carried on at Lagos with great activity, the Yoruba people have been obliged to use the port of Badagry, between which and Abeokuta communications are carried on by a difficult road by land.[2]

He was therefore to visit Abeokuta 'to ascertain, by inquiry on the spot, the actual wants and wishes and disposition of the Yoruba people', and to report on the Lagos succession dispute. Two other points need to be noted in Beecroft's instructions of February 1850. Compared with his commission of June 1849, his functions had been extended, in fact transformed with obvious influence from the missionaries. Before going to Abeokuta, he was asked to visit all the rulers on the coast and to urge slave-trade treaties on them. He was to tell them

> that the principal object of your appointment is to encourage and promote legitimate and peaceful commerce whereby those chiefs and their people may obtain in exchange for the products of their own country those European commodities which they may want for their own use and enjoyment, so that the great natural resources of their country may be developed, their wealth and their comforts increased and the practice of stealing, buying and selling men, woman and children may be put an end to.[3]

In 1849 he was only the protector of British trade. In 1850 he was asked to be first and foremost a philanthropist. The second point to note is that it was only gradually that Palmerston, under the pressure of the missionaries, was turning to Abeokuta and Lagos. Even in February 1850 he was still waiting for the report of his envoy to convince him that the missionaries' claims for Abeokuta were justified.

Beecroft first visited Dahomey, failed to persuade the king to accept

---

1 Ibid.
2 Instructions to Beecroft, 25 Feb. 1850; *Papers relative to the Reduction of Lagos*, pp. 29–30.   3 Ibid.

a slave-trade treaty, but discovered that an attack on Abeokuta was being planned, of which he properly warned the missionaries. The missionaries appealed to the government. The government instructed the commodore of the naval squadron to send a warning to Dahomey to desist from Abeokuta because there were British subjects there and to instruct a naval officer 'to meet the wishes of the missionaries as far as practicable until the period for the Dahomian war is past'.[1] In addition, the Governor of Sierra Leone, acting on orders from home or on an appeal from the Liberated Africans in the colony, sent to Badagri four boxes of bullets for the defence of Abeokuta.[2] Further still, Palmerston was by then veering round to the missionary point of view that for Abeokuta's sake immediate action should be taken in Lagos.

Beecroft was informed that Palmerston thought Kosoko should be invited to sign a treaty or be subjected to a blockade or replaced by 'the former chief, who is understood to be now at Badagry'.[3] But Beecroft's *instructions* remained those of February, namely to enquire and report.

Beecroft arrived at Badagri on 2 January 1851. He was met on the beach by the missionaries and their schoolchildren, and the traders, including Hutton from Cape Coast. He stayed five days as the guest of Gollmer[4] and before he proceeded to Abeokuta, accompanied by Dr. van Cooten, the C.M.S. Medical Officer, Beecroft had had time to reflect on the mounting tension in Badagri. Trade was growing, but was as yet by no means enough to feed the people. In the circumstances the Badagri people resented the competition of the emigrants.[5] Above all, for four years the Badagri people had incurred the enmity of Lagos for harbouring Akitoye and a large number of his supporters. Beecroft asked Akitoye to prepare for him a petition of all his grievances against Kosoko and a declaration of his readiness to accept a slave-trade treaty.[6]

At Abeokuta, Beecroft was the guest of Townsend for twelve days. Townsend later reported how he had prepared the way for the Consul,

to give weight and influence to his visit . . . as a person of great con-

1 Commodore Farnshawe to the Admiralty, 28 Oct. 1850, reporting the mission of Lt. Boys to Badagri, 21 Oct. 1850; ibid., p. 81.
2 Gollmer to Venn, 20 March 1851; C.M.S. CA2/043.
3 F.O. Memorandum to the Admiralty, 11 Oct. 1850; *Papers relative to the Reduction of Lagos*, p. 45.
4 Gollmer to Venn, 3 Jan. 1851; Gollmer to Capt. Trotter; ibid., p. 87–8.
5 There was a clash on 3 Oct. 1850 between the emigrants and the Egun near the 'Bristol' factory; Gollmer, Journal entries for 3 and 7 Oct., C.M.S. CA2/043.
6 Beecroft's report of his visit to Abeokuta and Badagri to F.O., 21 Feb. 1851; *Papers relative to the Reduction of Lagos*, pp. 91 ff. Gollmer: Journal for January 1851; C.M.S. CA2/043.

sequence and one calculated to be the means of doing great good to their country.[1]

Townsend also wanted to impress on him the great possibilities at Abeokuta for Britain and for civilization. Among the party that went to meet him were a contingent of schoolchildren and converts and the missionaries of C.M.S. and Methodist missions. He was shown round all the mission stations. At a public meeting on the 14th, in front of Ake Ogboni House, Beecroft presented the ammunition from Sierra Leone and the chiefs agreed to sign the proposed treaty whenever the British were ready. Then Beecroft visited all the prominent chiefs individually, discussing the general political situation, Akitoye and Lagos, as well as economic development. Ogunbonna, the chief whom the missionaries found most eager to adopt European ideas, had ready for the Consul 'a load of cotton, a bag of ginger and a bag of pepper as specimens of the products of the country'. On the 18th the Baloguns (war chiefs) called as a body on the Consul and reaffirmed their willingness to accept the treaty. At the meeting the missionaries indicated what their conception of a consul was, by calling upon Beecroft to interfere in local politics and judicial procedures. Townsend raised the issue of jurisdiction among the emigrants:

> I brought before them and the Consul a grievance that I had felt, viz., that persons in the Society's employ, being also British subjects, were brought under country laws that no principle of justice could tolerate.[2]

Beecroft was impressed by what he saw at Abeokuta. By the time he returned to Badagri he had decided that Britain must intervene in the Lagos dispute on the side of Akitoye. He received the petition he wanted, duly prepared for Akitoye by Gollmer. 'My humble prayer to you,' concluded the document,

> is that you would take Lagos under your protection, that you would plant the English flag there and that you would re-establish me on my rightful throne at Lagos and protect me under my flag; and with your help I promise to enter into a treaty with England to abolish the slave trade at Lagos and to establish and carry on lawful trade, especially with the English merchants.[3]

---

1 Townsend to Major Straith, 28 Jan. 1851; C.M.S. CA2/085.　　2 Ibid.
3 Petition of Akitoye to Consul Beecroft enclosed in Beecroft's dispatch to the F.O., 24 Feb. 1851, op. cit. How much this document was of Gollmer's composition may be judged from the fact that when the arrival of Beecroft was expected in Badagri, on 9 Dec. 1850, Gollmer noted in his journal: 'Akitoye sent for me this morning to tell him what to do. I accordingly went and told

To commit the British government to intervening in the Lagos dispute, and to ensure that the dispute was not settled before the government could be persuaded to act, Beecroft decided to take Akitoye with him to Fernando Po. His argument was that Akitoye's life was in danger at Badagri. But far from seeking this protection, Akitoye pleaded that his departure would only aggravate the situation; it would be like deserting his friends in their hour of need; it would encourage his enemies and weaken the morale of his supporters.[1] Beecroft compelled him to leave all the same.

At once the tension was heightened. The two parties in the town began to consolidate: the European missionaries, traders, emigrants and other supporters of Akitoye, led in his absence by Chief Mewu on one hand; Possu and the other leading Badagri chiefs and Kosoko's supporters on the other hand. After a number of violent incidents, civil war was touched off on 11 June when a group of women traders from Lagos sang songs in the Badagri market, deriding the cowardice of the absentee Akitoye and the poverty of his supporters, in contrast to the manliness and the prosperity of Kosoko and his partisans. The British party met at once and signalled for help.[2] Five days later, Commander L. G. Heath in H.M.S. *Niger* arrived. He suggested that the Europeans should come on board for safety because, for fear of illness, he could not land his men in the rainy season to defend them. Gollmer pointed out that they could not desert their friends, and that in any case the emigrants were British subjects entitled to protection too. Commander Heath then decided to issue out arms which the emigrants could use to defend themselves. He obtained a receipt for 'One thousand pistol-ball and two thousand musket ball cartridges.'[3] Domingo Martinez also gave Mewu 'twenty guns, twenty kegs of powder, twenty iron bars for shot and a quantity of rum'.[4] In the ensuing battle, Possu and the Badagri chiefs were defeated and expelled from the town. Their attempts to force their way back,

---

him that he should clearly state his right to the Lagos throne, how he was expelled, that he desires the British government to plant the English flag there and establish him and protect him under it [did Akitoye later insist on the change from 'under it' to 'under my flag'?] at Lagos and that he would make a treaty with the British government to abolish the slave trade and carry on lawful trade, which he said he would do'; Journal, 9 Dec. 1850, C.M.S. CA2/043.

1 Beecroft's report of his visit to Abeokuta and Badagri to F.O., 24 Feb. 1851, op. cit.
2 Quite a detailed account of these events in the Journal of Gollmer for June to December 1851; C.M.S. CA2/043.
3 Commander L. G. Heath to British Residents in Badagri, 17 June 1851; *Papers relative to the Reduction of Lagos*, p. 127.
4 Gollmer to Venn, 19 Sept. 1851; C.M.S. CA2/043.

with support from Porto Novo and Lagos, subjected Badagri to a state of siege for the rest of the year. Captain Jones and Commodore Bruce saw to it that warships came frequently to aid the 'English' party in Badagri.[1]

The Commodore felt obliged to justify the policy of issuing arms to the emigrants. He referred to the right of the few British traders and missionaries to be defended, and the moral if not legal duty to defend the 'several hundred' emigrants, who were 'legitimate traders, whose freedom had been solemnly pronounced by competent British tribunals', many of them Christians. Their interference in the war had been compulsory: 'They have been obliged to fight, not that Akitoye might resume the throne', but in self-defence, because if Kosoko had conquered Badagri, 'their adopted country', they would inevitably be doomed to slavery for the remainder of their lives.[2]

In March 1851 Gezo, the King of Dahomey, invaded Abeokuta territory. He marched his troops in person, with the Amazons in the vanguard, to the very gates of the town. They failed to take it by storm and were defeated with heavy losses on both sides.[3] Gezo's action was widely reported in England, as Abeokuta was now well known and regarded with affection in many churches and in many homes. Congregations prayed for its prosperity and defence against its enemies; many an old lady contributed money, or collected gifts, or embroidered a jumper for Abeokuta schoolchildren. Exploiting this emotional upsurge, Venn exerted the utmost pressure to get the government to issue the direct order for action in Lagos and to give more arms for the defence of Abeokuta.

Palmerston, however, found it difficult to persuade the cautious Lords of the Admiralty to authorize naval action of doubtful legality in Lagos —no British subject or property was endangered or detained in Lagos. All he could do in the circumstances was to ask Beecroft to go himself to Kosoko and urge on him a treaty with a mixture of blandishment and threats—a veritable cocktail of imperialistic gin and philanthropic tonic—

Tell him that lawful commerce is more advantageous to the nations of Africa than slave trade, and that therefore the British Govern-

1 'Obba Shoron' to Capt. L. T. Jones, Badagri, 3 July 1851; *Papers relative to the Reduction of Lagos*, pp. 130–1.
2 Bruce to Secretary of the Admiralty, 1 Nov. 1851; ibid., pp. 158–9.
3 Biobaku, op. cit., pp. 43–5. H. Townsend to Major Straith, 4 March 1851; C.M.S. CA2/085. T. J. Bowen: *Missionary Labours and Adventures in Central Africa*, op. cit., p. 118 ff.; E. Dunglas: 'La première attaque des Dahoméennes contre Abéokuta' in *Etudes Dahoméennes*, 1948, Institut Français d'Afrique Noire.

ment, in putting down slave trade and in encouraging lawful commerce, is conferring a benefit upon the people and chiefs of Africa. That Great Britain is a strong power both by sea and by land; that her friendship is worth having; and that her displeasure it is well to avoid.[1]

But Venn had thought of a brilliant way to move the Lords of the Admiralty to action.

When the news of the Dahomey War reached England, Venn had taken Samuel Crowther's eldest son, of the same name, studying in London, to call on Lord Palmerston, who discussed the war and 'showed great interest in the subject, and listened with such kindness to all our remarks'. Now Venn decided to confront Palmerston and the Lords of the Admiralty with Samuel Crowther himself, once a slave-boy in Lagos, rescued by the Navy, now a clergyman of the Church of England, promoting Christianity and civilization in Abeokuta, evolving an orthography of the Yoruba language and translating the Bible into Yoruba. Crowther was the missions' greatest propaganda weapon, and Venn invoked him. He was on leave in Freetown when Venn suddenly sent for him to make haste for England. So sudden was the summons that when Townsend saw a suggestion somewhere for a black bishop for Abeokuta, he jumped to conclusions and at once forwarded a petition against Crowther's consecration.

Meetings were arranged in various places in the country for Crowther, including one at the University of Cambridge. More important still, interviews were arranged for him with ministers, in particular both Lord Palmerston and Sir F. Baring, the First Lord of the Admiralty. A memorandum on the subject-matter of these interviews repeated the old case of the Egba, 'a native tribe struggling with uncommon energy and bravery to suppress the interior traffic in slaves', and called for British aid and support. Two specific requests were made. In the first place, that

> efficient aid might be rendered by allowing a few natives of Yoruba who have been trained artillery men in Sierra Leone—and there are many such—to return to their native land with two or three light pieces of artillery to defend the walls of Abeokuta against a second attack;

And secondly,

> Mr. Crowther is able to show that if Lagos were under its lawful chief and in alliance with Great Britain, an immense extent of country,

1 Palmerston to Beecroft, 21 Feb. 1851, enclosing copy of treaty to be proposed to Kosoko; *Papers relative to the Reduction of Lagos*, p. 85.

abounding with cotton, of which he has brought specimens, would be at once thrown open to commerce, extending from the coast to the River Niger at points 200 or 300 miles from the mouth of the river.[1]

Finally Venn arranged that after Crowther had seen the ministers he should have an interview with the Queen and Prince Albert. Crowther himself wrote an account of the meeting immediately afterwards. He was led to the palace by Lord Wriothesley Russell, the Prime Minister's brother. Prince Albert received them in his study, and Crowther began with a geography lesson of West Africa, illustrated with a map from a Blue Book. Then the Queen came and joined in.

> Lord Wriothesley Russell doubted whether I was aware that the lady who took so much interest in the interview was the Queen; he made use of the words 'Your Majesty' once or twice that I might take particular notice. . . .
>
> Lagos was the particular object of inquiry as to its facility of trade, should the slave trade be abolished, which I pointed out, as I did to Lord Palmerston and Sir F. Baring. She asked what did Lord Palmerston and Sir F. Baring say? I told her that they expressed satisfaction at the information. The Prince, 'Lagos ought to be knocked down by all means.' . . .
>
> The Prince asked whether the people of Abeokuta were content at merely getting something to eat, and merely having a cloth to cover themselves. I told him that they were industrious, and were fond of finery, as well as inquisitive to get something new. He then said, 'That is right, they can easily be improved'.[2]

The Queen stayed half an hour and left; the interview with the Prince lasted an hour and a quarter in all.

Venn claimed that it was Crowther's visit that finally moved the Government to action. At any rate, when Crowther was ready to go back, Palmerston wrote to thank him 'for the important and interesting information with regard to Abeokuta and the tribes adjoining that town' which he had communicated to him at their meeting in his house in August.

> I request that you will assure your countrymen that Her Majesty's Government take a lively interest in the welfare of the Egba nation,

---

1 Straith to Palmerston, 20 Aug. 1851; *Papers relative to the reduction of Lagos*, op. cit., pp. 133–4.
2 Dated Windsor, 18 Nov. 1851, and quoted in full in Stocks, op. cit., vol. II, pp. 111–13.

and of the community settled at Abeokuta, which town seems destined to be a centre from which the lights of Christianity and civilization may spread over the neighbouring countries.[1]

In September, Palmerston drew up a new memorandum for the Admiralty, referring to Abeokuta in these same words, and arguing that the government could not any longer permit the accomplishment of the 'great purpose' of Abolition to be thwarted by Kosoko and Gezo; and that the attack of Gezo on Abeokuta in spite of the solemn warning he was given was a possible *casus belli*.

The Lords of the Admiralty were therefore asked to order a blockade of the ports of Dahomey till the king signed a treaty. Further, they should issue orders for the restoration of Akitoye, since Captain Denhan, Crowther and Beecroft had assured the government 'that there would be no great difficulty in sending into Lagos the small force which would be sufficient' for such a purpose.[2] The Admiralty then issued instructions to Commodore Bruce to institute the blockade 'according to the views of Lord Palmerston' and they left to his discretion and judgement 'the mode' of carrying out the part of the instructions relating to Lagos. They directed, however, that he was not to retain possession of the island, nor to remain there longer than was absolutely necessary.[3] He was also instructed to meet the other request of Crowther in full, to recruit volunteers from Sierra Leone and send them with an officer and ammunition to the value of £300 for the defence of Abeokuta.

Bruce obtained from the ordnance depot in Freetown

> two light field-pieces (3-pounders), with three hundred rounds of powder and shot; 159 musket flint-lock, with bayonets; twenty-eight thousand musket-ball cartridges and two barrels of flints.[4]

Though Liberated Africans had already sent thirty-two large kegs and fourteen small kegs of powder to Abeokuta for the same purpose,[5] those trained in artillery felt they could not volunteer, as no provision had been made for their remuneration. The ammunition was landed in Badagri, in November, Gollmer helping to arrange its transport to Abeokuta, the field-pieces presenting great difficulties. Commander

1 Ibid., p. 114.
2 Palmerston to Lords Commissioners of the Admiralty, 27 Sept. 1851; *Papers relative to the Reduction of Lagos*, op. cit., p. 135.
3 Lords Commissioners of the Admiralty to Commodore Bruce, 14 Oct. 1851. The letter was signed by F. T. Baring and R. D. Dundas; ibid., p. 138.
4 Bruce to the Admiralty, 6 Dec. 1851; ibid., p. 161.
5 Gollmer to Secretaries of the C.M.S., 13 Oct. 1851; C.M.S. CA2/043.

F. E. Forbes was sent to Abeokuta with a few marines. He met the emigrants there, recruited thirty of them, drilled them, supervised the mounting of the field-pieces and the repairs of Abeokuta's walls.[1]

With respect to Lagos, however, Bruce was more cautious. He went to Fernando Po to hold consultations with the Consul, met Akitoye, and on 1 November, when off São Tomé, sent to the Admiralty a despatch explaining his hesitation. 'Akitoye,' he said, 'does not appear to me to be a man likely to maintain his place by physical influence, if he could be reinstated.' And if Akitoye could not rule, what would be the future of Lagos? 'The European trade with Lagos is very considerable, particularly in Hamburg vessels.' Beecroft was of the opinion that to protect this trade effectively, 'Lagos ought to be taken under the protection of England,' but this could not be done because of the climate. He referred also to Article VI of the Convention of May 1845[2] by which England and France promised not to resort to force on the African coast without the consent of both powers.

Beecroft, however, did not hesitate to rush in where Bruce feared to tread. The evidence suggests that there was a conflict between the two men, a conflict both of personality and of policy, a conflict between a disciple of Buxton (for the Commodore was a devout evangelical) responsible to the Admiralty, and a coastal trader holding a commission from the Foreign Office. Whether there was an open quarrel we shall probably never know. But Beecroft proceeded to manoeuvre the law out of the Commodore's discretion into his own hands. He went straight to Badagri, arriving on 17 November. He sought out Commander T. G. Forbes, the naval officer in charge of that division of the coast. Together they went with four warships, and called Kosoko to a meeting on the southern tip of the Lagos island. Kosoko rejected the offer of British friendship and declined to sign a treaty, using the

---

1 F. E. Forbes to Bruce, 13 Nov. 1851, 9 Dec. 1851, enclosing the speech he declaimed to the emigrants on the 24th: 'Your religion, your duty, teach you to trust in God; but God will not save you without exertions from yourselves. . . . You are a great number and should be all men of some education; set an example to the Egbas; show them the advantage of the knowledge the white men have, through God's assistance, ingrafted in you . . .'; *Papers relative to the Reduction of Lagos*, pp. 177–8, 180–2. Townsend remarked that Forbes was a very energetic officer, a little jealous of his position *vis-à-vis* Consul Beecroft, and seemed to live only for promotion.

2 Convention between Great Britain and France for the suppression of the Traffic in Slaves, signed in London 29 May 1845, ratified June 1845. *Article VI*: 'When ever it shall be necessary to employ force, conformably to the law of nations in order to compel the due execution of any Treaty made in pursuance of the present Convention, no such force shall be resorted to either by land or sea without the consent of the Commanders both of the British and of the French Squadrons.' SP, vol. 33, pp. 8–9.

ingenious argument that Lagos was under Benin and the Oba of Benin should be persuaded to sign the treaty on his behalf. Then, wrote Beecroft later,

> It was decided to collect such a show of force as the moment could supply, with the firm belief that such force, judging from the character of African chiefs, would have the effect by simple demonstration of our power, to cause him to accede to our terms.[1]

An abortive attempt to assault Lagos took place, for which Beecroft was later severely censured,[2] but there could now be no going back, as the prestige of England had become involved in the dispute. Commander Bruce arrived off Lagos on Christmas Eve and attacked on Boxing Day. On the second day of the battle rockets from one of the boats succeeded in blowing up the royal arsenal, causing great havoc, and, with his leading supporters, Kosoko fled down the Lagoon towards Epe. Lagos was captured for the loss of only sixteen men killed and seventy-five wounded on the British side. 'Had an engineer from Woolwich been on the spot,' said Beecroft when he landed, 'the defence of Lagos could not have been better planned.'[3]

Sandeman, one of the traders at Badagri, later claimed that in signing the petitions for British intervention, he had been misled by the missionaries. 'The fact is,' he declared, 'Akitoye was made a tool to carry out the ambitious views of these two men, Messrs. Gollmer and Townsend.'[4] Akitoye was indeed made a tool. The missionaries led the way in this, but they were supported by the traders and the traders reaped much fruit from it. The British Consul also had certainly played a prominent part.

On New Year's Day, Akitoye signed a slave-trade treaty which, besides the usual clauses about renouncing the slave trade, and the British right to enforce it if violated, contained additional ones breathing

---

1 Beecroft to Palmerston, 26 Nov. 1851; *Papers relative to the Reduction of Lagos*, op. cit., p. 147.
2 Granville to Beecroft, 24 Jan. 1852: 'I have to acquaint you that Her Majesty's Government are of the opinion that you were not borne out, either by the circumstances of the case, or by your instructions from Her Majesty's Government in directing that Her Majesty's naval forces should land and attack Lagos. . . . I regret to be obliged to disapprove of your conduct in this affair.' Also 23 Feb. 1852, that this, too, was the opinion of Commodore Bruce; ibid., p. 167.
3 Beecroft to the Foreign Office, 3 Jan. 1852; ibid., pp. 188–9.
4 Sandeman to Campbell, 28 Aug. 1855, in *Correspondence relative to the Dispute between Consul Campbell and the Agents of the Church Missionary Society at Lagos*; F.O. Confidential Print 4141, 15 July 1856, p. 24.

the spirit of the times. There was a clause guaranteeing to British traders 'most-favoured nation' terms; another abolishing human sacrifice; another guaranteeing to missionaries of all nations freedom to follow 'their vocation of spreading the knowledge and doctrines of Christianity and extending the benefits of Civilization.

> Encouragement shall be given to such missionaries in the pursuits of industry, in building houses for their residence, and schools and chapels. They shall not be hindred or molested in their endeavours to teach the doctrines of Christianity to all persons willing and desirous to be taught . . .[1]

An identical treaty was signed by the chiefs of Abeokuta on 5 January, except that two additional clauses were added at the instigation of missionaries, guaranteeing freedom of movement about the country for themselves and for the emigrants. The same treaty was presented to Mewu for signature on behalf of Badagri while Possu and the other chiefs remained in exile. In March 1852 Gollmer came to Lagos and obtained from Akitoye for the C.M.S. five pieces of land on the island, 'without any condition, free of expenses, and without limit of time.'[2]

Gollmer went back to Badagri to pack as much of the mission property as had not been evacuated during the war. He reviewed the failure of the mission at that first station. For the Badagri people, as we have seen, had replied to the Gospel, with the cynical indifference of a cosmopolitan town that welcomed all comers and tried to outwit them in turn. Hardly any Egun had received baptism. But the emigrants, some refugees and traders had benefited from the missions. The C.M.S. had baptized 14 adults and 36 children, celebrated one marriage and 18 funerals in $7\frac{1}{2}$ years. The schools fared better. At one period the C.M.S. boarding school actually flourished and a boy was sent to Sierra Leone for further studies. But even the schools had been broken by war. 'The monuments of our mission in this place are the four missionary graves [in one of which lay the first Mrs. Gollmer] as witness of the devotedness of our works of faith and love among the people here.'[3] Other Europeans at Badagri and elsewhere on the coast were similarly taking stock and making up their minds to move towards the brighter prospects in Lagos. Gollmer moved faster than most: he set out for Lagos with his wife, schoolmasters, interpreters, and Scripture readers, leaving only a catechist behind. He was followed a year later by John Martin, the

1 Enclosure in Beecroft to F.O., 3 Jan. 1852; *Papers relative to the Reduction of Lagos*, pp. 188–9. Copies of these and other treaties with Badagri, Ijebu and Porto Novo will be found in *State Papers*, vol. 42.
2 'Memorandum of Agreement' dated 1 March 1852; C.M.S. CA2/043.
3 Gollmer, Journal entry for 22 June 1852; C.M.S. CA2/043.

Methodist assistant missionary who had succeeded Annear. There was an exodus also of the emigrants from Badagri, and the town gradually passes out of this story. Four months after Gollmer reached Lagos, Louis Fraser, a trader who had succeeded Duncan as vice-consul at Whydah, arrived to act as consul until he was relieved in August 1853 by Benjamin Campbell, the first Consul of Lagos.[1]

Initially Campbell stayed with the C.M.S. missionaries and seemed to concur with them in their support of Akitoye and his policy of pursuing the war against Kosoko, who had control of the principal sources of palm-oil at Ijebu.[2] But later Campbell switched to the views of the traders, who were for appeasing Kosoko in the hope that he would divert the Ijebu trade to Lagos. Missionary and trader, allies in war, became enemies in the post-war settlement. Matters came to a head between the missionaries and Campbell when the latter decided to depose Mewu, the missionaries' old ally and ruling chief at Badagri (a refugee from Porto Novo), since he was an obstacle to the trade of Porto Novo passing by Badagri on the Lagoon to Lagos. Campbell asked Townsend and the Egba to persuade Mewu to give up the British treaty he held, and offered, if this were done, to negotiate with the king of Porto Novo to allow Mewu to remain in Badagri as a private individual provided he would allow Possu and the other chiefs who had supported Kosoko to return.[3] The Egba refused to do this, and the missionaries denounced the policy of Campbell.[4] Campbell then resolved to expel Mewu by force. He brought charges of slave-trading against him but, as the Egba pointed out, Mewu could not have been more guilty in the matter than Possu or the favoured king of Porto Novo. The important point was, as Campbell said, the oil-palm trade:

> between Badagry and countries watered by a fine navigable lagoon extending through more than 100 miles of country yielding a commerce in palm oil alone of the present annual value of about a quarter of a million pounds sterling, nearly the whole of which is at present monopolized by Domingo Martins.[5]

Campbell called in a warship, restored the exiled chiefs and expelled Mewu, who took refuge in Lagos, where he died a year later. The C.M.S.

1 Bruce in recommending the appointment of consuls and vice-consuls to each of the important towns on the coast added: 'the persons best adapted for these situations would be intelligent and fairly educated Creoles of the West Indies or natives of Sierra Leone'; Bruce to the Admiralty, 17 Jan 1852, in *Papers relative to the Reduction of Lagos*, PP 1852 LIV.
2 Campbell to Gollmer, 2 Sept. 1854.
3 Campbell to Townsend, 12 May 1854; C.M.S. CA2/04.
4 Townsend to Campbell, 15 May 1854; C.M.S. CA2/04.
5 Campbell to Townsend, 30 April 1854; C.M.S. CA2/04.

sent memoranda and deputations to Lord Clarendon saying that Campbell was befriending slave-traders and antagonizing Britain's friends, particularly the Egba on behalf of whom the British had intervened in Lagos. They insisted that

> the best hope of introducing civilization into that part of Africa and of putting effectual stop to the slave trade consists in the encouragement of the Egba tribe situated at Abeokuta, which contains a thousand British subjects in the persons of the Sierra Leone emigrants.[1]

Here, then, was the crux of the disagreement between the C.M.S. and the traders between 1852 and 1855. To the C.M.S. Abeokuta counted for far more than Lagos; to the traders, Lagos was the focus of attention. The Methodists, who were not so committed to Abeokuta and who were not yet so ready to penetrate into the interior, tended to side with the traders.[2] The C.M.S. were preparing for their great advance into the Yoruba country, and under the leadership of Townsend they based their advance on an 'Abeokutan' policy.

In Henry Townsend the C.M.S. had a missionary who saw his work in Africa very largely in political terms, and he had the ambition and ability to carry it out. He was the sponsor of the policy of setting up Abeokuta rather than Lagos or anywhere else as the main centre of both missionary and British influence in Nigeria. He realized more than anybody else that the survival of the Church in Abeokuta depended on the survival of the town itself, and that the spread of the Gospel from Abeokuta depended on the outcome of the political struggle for power between Abeokuta, Ibadan, Ijaye, Oyo and Ijebu. He had the advantage of bringing the realism of a politician to reinforce the idealism of the missionary, but also the disadvantage that, like all politicians, particularly when not actually in control of affairs, he was prone to make mistakes and miscalculations. He first visited Abeokuta within twelve years of its foundation. He arrived to settle there two and a half years later and was to be a regular resident for the next twenty years. He spoke Yoruba fluently and met the rulers of Abeokuta on personal terms. He knew them and understood their aspirations. They, too, came to know

1 C.M.S. Secretaries to Lord Clarendon, encl. in Venn to Irving, 22 June 1855; C.M.S. CA2/12.
2 For details of this dispute see F.O. Confidential Print 4141, July 1856: *Correspondence relative to the Dispute between Consul Campbell and the Agents of the C.M.S. at Lagos.* The C.M.S.–Methodist dispute came to involve a doctrinal issue in Abeokuta, where C.M.S. missionaries were said to have made fun of the doctrine of the Assurance of Salvation, a fundamental tenet of Methodism, which their catechist had been labouring hard to get across to his congregation; Freeman to Secretaries, 9 Jan. 1855, Meth.

him and to repose confidence in him. It is therefore not surprising that he soon decided on the policy we have noticed of building up Abeokuta politically, militarily and economically, fostering its influence and using that to get Christianity established in the land. If the power of Abeokuta could be increased and extended along the coast, and commercial prosperity brought in through the emigrants and European traders, then the influence of such a strengthened state would carry the Gospel further inland. The supply of ammunition and the training of an emigrant artillery force for defence against Dahomey, the installation of Abeokuta's partisans, Mewu at Badagri and Akitoye at Lagos, the 'Abeokutan' port, completed the first part of this programme.

The next step was to get a British vice-consul appointed at Abeokuta. Venn approached Lord Clarendon, who had taken over the Foreign Office, but Clarendon demurred, saying that the Consul at Lagos could supervise affairs at Abeokuta, barely sixty miles away, until British trade there had grown sufficiently to justify the extra expenditure. That did not satisfy the C.M.S. The consul they expected was not Clarendon's jealous guardian of the interests of a foreign country so much as a benevolent resident helping and advising the chiefs in their daily work and a leader of the emigrants and educated Africans in their work of civilization. He was to be a secular missionary, relieving the clerical ones of the secular duties which Gollmer at Badagri said were calculated to make a clergyman sin against God and man. Venn felt a pang of conscience at having exposed Gollmer to so much political intrigue at Badagri. He consoled himself that it was in a crisis and *inter arma leges silent*, but he would avoid making the same mistake at Abeokuta. He looked round for a suitable person who would be 'receiving a salary from the society but having some recognition from the Government . . . to execute the office' of consul.[1] He picked on Dr. Edward Irving, a surgeon and commissioned officer of the Royal Navy, who had accompanied Captain Foote to Abeokuta in 1852 to continue the work begun by Captain Forbes on the defences of the town.[2] The government itself could not have picked on a better qualified consul. He was to choose an African 'Under-Secretary', who, it was hoped, would succeed him, possibly as an official consul. Dr. Irving's duties were defined as

> to co-operate with the missionaries in ameliorating the social, political and economic condition of the native tribes. . . .
> to promote their civilization and social welfare . . .
> to advise chiefs . . . respecting the principles of law and sound policy

1 Venn to Gollmer, 23 March 1853; C.M.S. CA2/L1.
2 Samuel Crowther, jun., to Venn, 28 Dec. 1852; C.M.S. CA2/032.

. . . [so that] right views of law and justice should supply a better foundation than that which is crumbling away.[1]

He was further to be 'a counsellor of the chiefs in respect of their military policy and warfare', to teach them the best way of fortifying their town and securing it from sudden attacks, and to dissuade them from aggressive warfare. Finally, he was to be a channel of communication between the mission and the consul and officers of the squadron. These functions of the 'lay agent', as he was called, were submitted in a memorandum to the Foreign Office and the Admiralty. In June 1853 Lord Clarendon 'entirely approved' of the proposal and he instructed Consul Campbell at Lagos 'to afford every assistance to Dr. Irving, in whose success Her Majesty's Government take a lively interest'.[2] When Dr. Irving was about to set out in December 1853, the C.M.S. Committee added one important detail to his duties. He was

> to make arrangements with the Vice-Consul Campbell [sic] for securing the water frontage assigned to the society as a free wharf, with special reference to native traders.

He was to study the resources of the country in marketable products like cotton, gums, indigo and dyewood, and to direct the attention of the emigrants and the Christian converts towards them, so that, in words commonly used at the time:

> these parties may rise in social position and influence while they are receiving Christian instruction and thus form themselves into a self-supporting Christian Church and give practical proof that godliness hath promise of the life that now is, as well as that which is to come.[3]

Dr. Irving's mission was a failure, largely because there were some major miscalculations in Townsend's policy which Irving was supposed to carry out. The first was that the traders and the British Consul whom the missionaries attracted to Lagos did not regard Lagos as the port of

---

1 Venn to Irving, 29 March 1853; C.M.S. CA2/L1. Irving was to secure the assistance 'of a young educated Native . . . to be himself thereby trained for some post of agency in connection with the objects before us.' In February 1855 Samuel Crowther, jun., who had had a medical training in London, was appointed Irving's 'Secretary and Assistant'. Venn to Local Committee of the Yoruba Mission, 23 Feb. 1855; C.M.S. CA2/L1.
2 Clarendon to Lord Chichester, President of the C.M.S., F.O. 11 June 1853; C.M.S. CA2/L1.
3 Final instructions to Dr. Irving, 23 Dec. 1853; C.M.S. CA2/L1. Cf. Hope Waddell, who, when King Eyo remarked that if a man was always thinking about death and his soul and his God he would become lazy, decided to preach his next sermon on 'the advantages of godliness for this life as well as for the life to come'; *Journals*, entry for 30 June 1950, vol. VIII, p. 36.

Abeokuta but as the centre where British influence was predominant and whose trade was to them an end in itself. Irving, on his arrival, wrote to Venn saying that trade was expanding in Lagos:

> The Abbeokutans are already beginning to master and have entire command over the navigation of the river [Ogun] and it would be a most desirable thing to keep infusing more and more Egban element, until Lagos becomes the Abbeokutan seaport for which it is so admirably adapted.[1]

But in fact it was the English element that was being infused into Lagos, not to supplement, but to replace existing African rule. The greatest opponents of the proposed ascendancy of Abeokuta were Kosoko and the chiefs driven away from Badagri, and Consul Campbell preferred to negotiate with them, using force to disloge Mewu from Badagri in favour of Possu and the other exiled chiefs.

Lord Clarendon approved of the negotiations with Kosoko. He disapproved of the treatment of Mewu,[2] but there could be no going back. When Irving also quarrelled with Campbell, Clarendon asked the Admiralty to send a naval officer to look into all the disputes. Before the Commission of Inquiry arrived, Dr. Irving had died.[3]

The differences between the different European groups were important and should be pointed out. However, their significance can be exaggerated. Traders and missionaries combined to get the consul installed. In 1853–55 the missionaries attacked him for being too much inclined to the views of the traders. In 1855–56 the traders wrote petitions to get him removed for meddling too much with trade affairs, antagonizing Kosoko unnecessarily, and expelling Madam Tinubu who owed them much trust.[4] From the African's point of view, these quarrels had little effect on the growing power of the consul within the African states. If anything, they tended to increase it, for the consul, in addition to powers to champion Europeans, began to acquire powers to be their umpire. The moments of disagreement could not undo what

---

1 Irving to Venn, 20 Jan. 1854; C.M.S. CA2/O52.
2 Wodehouse to Lord Chichester, President of the C.M.S., 12 June 1855, encl. in Venn to Irving, 22 June 1855; C.M.S. CA2/L2.
3 For 'The Skene Commission', see F.O. Confidential Print 4141. Also Crowther, 3 Dec. 1855; C.M.S. CA2/O31.
4 Ibid. Also Campbell to Crowther, 15 Oct. 1856, enclosing copy of the petition of the traders and asking Crowther to give him a testimonial refuting the charges. Campbell to Irving, 5 Jan. 1855, 10 Jan. 1855, on his efforts to break the Trust System in Lagos; C.M.S. CA2/O4. Crowther to Venn, 30 Sept. 1856, 3 Nov. 1856, reporting generally on the situation and the virtual revolt of the traders. A naval officer sent to inquire censured the traders; C.M.S. CA2/O31.

had been achieved in the periods of agreement, and it was the cumulative effect that mattered. The missionaries were most conscious of this interdependence.[1] That was why, in spite of recognized differences of objectives, they continued to seek the co-operation of traders and consuls for the task of social reform in the African states.

The reputation of the 'palm-oil ruffians' was not unknown to the missionaries. A good deal was written, and much of it by missionaries, about the personal degeneracy of the men, their quarrels over women, their addiction to drink, their wanton cruelty, the negation at practically every point of the morality the missionaries were preaching. Yet the missionary often stood by him. Sometimes the missionary assumed that the trader was after all inherently an honest man doing a difficult job for his company and country, far from home and in a climate he found trying. At other times he seems to have pretended that, Europe being a Christian continent, every European bearing a Christian name, however unworthy, must have something on moral grounds to offer in the reform of the African. But it was not the personal morality of the trader that separated him from the missionary so much as his trade, his very livelihood and the method in which he pursued it. Hope Waddell relates that soon after he got to Calabar, he approached a much respected trader who alone on the river had a reputation for treating his crews and workmen like human beings, and he suggested the idea of holding services on the trader's hulk every Sunday. 'With usual candour,' the trader replied that the 'mode of carrying on trade in this river was so contrary to the principles of religion that it seemed to him to savour of hypocrisy to attempt to gain both objects.'[2] A major bone of contention was the Trust system.[3]

1 In 1854, when Lagos was threatened by an attack from Kosoko, Campbell offered some ammunition to defend the C.M.S. House which he said was 'the only place in Lagos capable of making a defensive stand'. Gollmer accepted the arms, saying: 'We are entirely defenceless, having neither guns nor powder, nor people to use them and therefore cannot object to any arrangements you deem necessary'; Campbell to Gollmer, Gollmer to Campbell, 2 Sept. 1854; C.M.S. CA2/043. In April 1855, when Dr. Irving lay on his death bed, he asked for some port. His colleagues had none, and the traders had none to sell. Consul Campbell, hearing of this, sent him a bottle. Irving was for refusing it, but Crowther pressed him to take it, saying that the dispute 'was not a personal grudge but of a public nature in the cause of justice'. Irving drank the wine, murmuring, 'That was very kind; we cannot enjoy a hearty quarrel in this country being so dependent upon one another.' Crowther to Venn, 8 May 1855; C.M.S. CA2/031.
2 Waddell: *Journals*, vol. 1, entry for Sunday 21 June 1846.
3 See K. O. Dike: *Trade and Politics in the Niger Delta*, p. 60, where the Trust system has been discussed. What is here attempted is an examination of the attitude of the missionaries to the system and of the effect of their policies on it.

The philanthropists had attacked the Trust system of trade from the start. They realized that the expansion of European commerce in Africa did not of itself necessarily mean widespread economic development and social reform for the African. The promotion of 'legitimate commerce' would be a civilizing force, they said, provided it was directed to root out slavery at the source, to alter the subsistence economy on which it was based, reach down to the masses in the interior, create new wants and engage the labour of the hitherto surplus manpower. It was obvious to them that, as it was then conducted, the palm-oil trade, like the slave trade it was succeeding, was not doing this. However, they never really understood why, and quite often they acted as though they forgot the proviso that qualified the virtues of 'legitimate commerce'. To explain why the palm-oil trade was not having the desired effects on the African states, the missionaries argued that palm trees grew wild and required no cultivation, and that under the trust system a few coastal chiefs in ignorance exploited the labour of domestic slaves for the benefit of unscrupulous European traders. By taking this simple view they missed the real nature of the economic alliance between Europeans and the coastal traders. They underrated the difficulties of changing the system, and they did not realize that the cultivation of cotton, even by peasant farmers, could produce the same system if someone brought sufficient capital to develop the trade.

In the 1850s Venn embarked on an important venture to see that the production of cotton in Abeokuta was increased without the trust system, and in a way to promote the emergence of an African middle class. He went up to Manchester, met leading members of the Chamber of Commerce and finally persuaded Thomas Clegg, an industrialist, 'one of those lay men of the Church of England who form its real strength',[1] to co-operate with him. Clegg agreed to provide a few saw gins and cotton presses, train some African youths to use them and prepare cotton for the European market. However, he was to pay for cotton only when he had received it, and would not give out any credits. He promised to act 'simply to benefit the natives and to secure no more profit to himself than a bare commission upon the transaction'.[2] The boys he trained were to be independent traders on their own, transmit the cotton they gathered directly to Clegg in the name of and with the sign of each producer, taking a bare commission for their services. Needless to say, it did not work. At first there were enough farmers who

1 Venn, journal entry for Sunday 16 Nov. 1856. Also other entries for the week 11–17 Nov. for the visit to Manchester. W. Knight: *Missionary Secretariat of Henry Venn*, 1882, loc. cit., pp. 134–6.
2 Ibid., 15 Nov.

enjoyed the joke of sending their produce directly to England with their own marks, and since cotton was only a side issue they did not mind waiting six to twelve months for the returns.[1] But the young men trained in England, the first of whom arrived in February 1856, could not live on such a game when they could get much more from being agents to some palm-oil trading firm or other. By October, therefore, Venn made a concession. He approved the foundation of an Industrial Institution at Abeokuta which besides teaching brickmaking, carpentry, dyeing and so on, would act as 'a depot for receiving, preparing, and sending cotton to England'. Henry Robbin and Samuel Crowther, jun., who was also medical officer to the mission,[2] were appointed managers and therefore assured of some income. Venn, however, insisted that the Industrial Institution

> will not be a trading concern in itself, for not one shilling of the Society's must be employed in purchasing cotton or any other material. It will only transact the cotton business on commission for other parties.[3]

But within six months Venn discovered not only that Clegg had been giving out trust to the managers and was now seeking to get control of the Institution into which he had been sinking money so that he could run it in the most profitable manner, but also that the managers were supporting him in this, preferring to be his agents if he gave them enough capital to trade with. Venn wrote a stirring appeal to Robbin pointing out that the object of the Industrial Institution would be defeated if he became direct agent of a European trader. The object was to enable Africans,

> to act as Principals in the commercial transactions, to take them out of the hands of European traders who try to grind them down to the lowest mark. We hope that by God's blessing on our plans, a large body of such Native independent Growers of cotton and traders may spring up who may form an intelligent and influential class of Society and become the founders of a kingdom which shall render incalculable benefits to Africa and hold a position amongst the states of Europe.
>
> What on the other side is the object of European Traders to Africa? It is to obtain the produce of Africans for the least possible considera-

1 List of Exporters of cotton at Abeokuta in 1856 in Samuel Crowther, jun., papers; C.M.S. CA20/32.
2 See below, p. 160–1.
3 Venn to Samuel Crowther, jun., and Henry Robbin. 21 Oct. 1856; C.M.S. CA2/L2.

tion at the cheapest rate. They prefer to deal with savages whom they can cajole into parting with their goods for beads, and rum, rather than to deal with civilized and intelligent races who can compete with them in the markets of Europe. . . .

I do not speak of individuals like good Mr Clegg, but I speak of two opposite systems. Now into which of these systems will you throw your energies and those acquisitions which you have made in England through the liberality of the Friends of Africa? I am sure you cannot hesitate. You will prefer being an independent Patriot to being an agent of Europeans.[1]

The issue of the Trust system involved more than a question of patriotism and goodwill. Venn was in fact grappling with the problem familiar today as that of securing capital for 'under-developed territories' in a way that would not sacrifice the economic and political aspirations of the people of the territory to the ambitions of those who provide the capital. When the accounts of the Industrial Institution were made up in 1858 it was discovered that apart from some C.M.S. money the Institution owed trusts of up to £1,800 to Clegg. He had to be allowed to take control of the Institution and run it in the way he liked.[2] However, he made a point of using educated African agents. He retained Henry Robbin, who gradually paid off the debts by 1864. In this way Clegg was working out the system which, as we shall see, he later used on the Niger.

Large-scale capital was the essence of the Trust system. The Europeans brought capital in trade goods which they gave on credit to the chiefs, who undertook to use it to negotiate in the interior to bring down to the Europeans whatever goods they wanted. The capital enabled the chiefs to own large fleets of canoes and canoemen and other equipment necessary for the trade. By giving trust in advance, European traders secured the produce in advance and thus made it difficult for interlopers to join in the trade. By accepting trust the chiefs made profit for themselves, and were able to regulate and to limit the incursion of the European into African society. The large capital in the trade made it a highly organized business but it also imposed on it a pattern of monopoly and rigidity, particularly in the Delta, where it was most developed. The capital imprisoned both those who gave and those who received it in trust. If the European traders had been willing—which they were not— to give up their monopoly in the hope that a freer trade would open up

1 Venn to Robbin, 22 Jan. 1857; C.M.S. CA2/L2.
2 Venn to Crowther, 31 March 1857, to Townsend, 31 March 1857. In a letter to Hinderer, 21 Dec. 1859, Venn talked of 'Mr. Clegg's imprudent ventures'; C.M.S. CA2/L2.

new and increased opportunities for them, they would have had to write off a good many bad debts.[1] If the African customers wanted to regain their liberty to trade freely in cash with all comers, they would have had to expropriate forcibly, or gather in produce for a year or two to pay off past debts with the usual expenditure but no income for that period.

The Trust system partook of all the villainy and inhumanities of the trade in human beings that produced it, but normally it had a rude justice of its own. The returns for Europeans were slow; war, fire, or other accident might destroy the stock of oil for which trust had been received and expended in advance; debts were collected in difficult, often brutal, conditions; but the rate of profit was very high, particularly because the Europeans could inflate the prices and exaggerate the value of the goods they brought. On the other hand, a strong African government could by playing one set of traders against another, keep prices down and by instituting trade boycotts curb the excesses of particular traders. As long as the European traders could not expect armed support from Europe, particularly with the transition from the trade in slaves to trade in oil, which took longer to collect and required more stability on land, it was often in their interest to encourage such strong African governments on the coast. And in fact they fostered the influence of monarchs like Eyo I and Duke Ephraim of Calabar who had a reputation for fairness in adjudicating between European traders and their African customers. But when the philanthropists had a consul invested with naval power far in excess of what the African rulers possessed, and when they strengthened his hand on land by the privileges they secured for missionary intervention, they were weakening the African states and turning the Europeans from negotiators into arbiters of trade.

It should be repeated that this was far from being the intention of the missionaries. As we shall see, they continued to train and to encourage the middle class to whom they expected the powers of the chiefs to pass. They not only educated them, they put them in touch with business houses in England, and on every possible occasion urged on the merchants and government in England the policy of entrusting to them the conduct of their affairs in Nigeria.

Meanwhile, most missionaries consoled themselves with the knowledge that politics was not their main concern, that they were preachers of the Word patiently seeking to establish the Christian Church and an

1 In 1850 a supercargo on the Benin River, agent of Harrison & Co., said he was instructed to discontinue the Trust system, but first he demanded the year's output of oil at no further cost to himself than the trusts his predecessors had given out; encl. 3 and 4 in Beecroft's dispatch to Lord Palmerston, 24 Feb. 1951; *State Papers*, vol. 41, p. 268 f.

indigenous clergy and laity at whose touch all the old antiquated ways of life would wither away. Before turning to these other duties of theirs, particularly in regions where they had fewer dealings with traders and consuls, it must be added that much of the development of the power of the consul in Nigeria was due to the nature of the intervention of the missionaries themselves. While the intervention of the traders in African society was limited and intermittent, that of the missionaries was all-embracing. While traders sought to limit intervention to the coastal states, where they themselves could stay near the gunboats and expect people to bring them the goods they wanted, it was the missionaries who sought actively to penetrate into the country because they were obliged to take the Gospel to the people in every town and village.

Catholic Holy Cross Cathedral, Lagos, built 1881

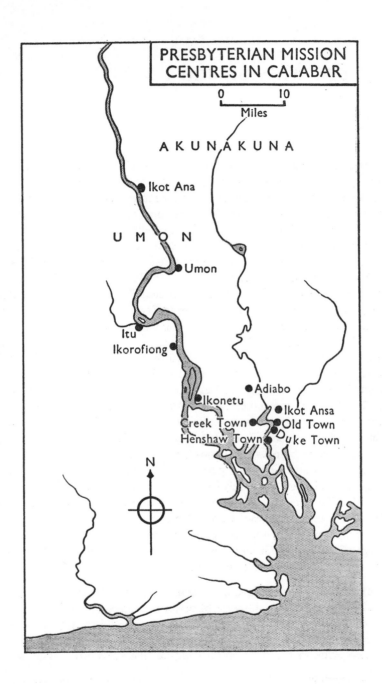

PRESBYTERIAN MISSION
CENTRES IN CALABAR

0   10
Miles

A K U N A K U N A

Ikot Ana

U M O N

Umon

Itu
Ikorofiong

Adiabo

Ikonetu

Ikot Ansa

Creek Town

Old Town

Henshaw Town

Duke Town

N

89

4 The Mission and the State

WITH the coastal bases secured at Lagos and Calabar, there followed in the years 1853–60 an expansion of missionary activities in Nigeria unequalled till the last decade of the century. Before examining the nature of these activities and discussing some of their social and political consequences it is necessary to indicate briefly the rate at which each of the missions expanded and what areas of the country they tried to cover.

The basic problem that faced each mission was the old one of how to achieve the maximum results with limited resources.[1] To take a classic example, the simple parishioners who contributed their pennies and sixpences to send David Livingstone to South Africa must soon have begun to wonder about the wisdom of his tramping north to the Zambezi.[2] In the highly emotional missionary meetings by which their enthusiasm was aroused they were told that hundreds of Africans died every day without ever having heard of Jesus and they were to make contributions to rush the Gospel to save as many of them as possible before they had to meet their Maker. Perhaps the missionary could safely spend his first year or two travelling about, and sending home exciting reports like Freeman's *Journals of Visits to Ashanti, Aku and Dahomi*. If, however, he did not settle in one place, build a church and send home accounts of conversions, the parishioners in Europe or America might begin to doubt how effectively their money was being spent. The trouble was that the same enthusiasm and sense of urgency that made the parishioners hungry for news of actual conversions tended to promote in some missionaries of the greatest ability and piety the urge to sow the seeds of the Gospel far and wide and leave to others the joy of reaping the harvest. Livingstone retorted to his critics:

> If we call the actual amount of conversions the direct result of missions, and the wide diffusion of better principles the indirect, I have no hesitation in asserting that the latter are of infinitely more importance than the former. I do not undervalue the importance of the conversion of the most abject creature that breathes; it is of over-

1 See above, pp. 57–8. 2 Groves, op. cit., vol. ii, pp. 172–3.

whelming worth to him personally, but viewing our work of wide sowing of the good seed relatively to the harvest which will be reaped when all our heads are low, there can, I think, be no comparison. . . . Time is more important than concentration.[1]

However, not every Christian could take that long-term investment view about conversion. There was no way, some people would say, of comparing the importance of the large harvest of tomorrow with the personal salvation of the individual today. To many evangelicals, Livingstone's emphasis on the 'Indirect' method, the patient planning to capture future generations by good works in place of rushing by preaching for the souls of the contemporary few, must have sounded heretical. But though evangelical theory remained unchanged, the influence of humanitarians and the consequent preoccupation with civilization and commerce as necessary adjuncts of Christianity laid down the path of virtue as being somewhere between 'time' and 'concentration'.

Freeman was on the side of Livingstone without Livingstone's unique ability or opportunity wholly to despise his critics and yet like Livingstone rise to glory by lonely, stubborn grandeur. His attempt to expand the Methodist Gold Coast Mission to the Yoruba country has already been noticed. He placed a European missionary temporarily at Kumasi, African catechists more permanently at Whydah, Badagri and Abeokuta, and whatever other missionary he could spare at Lagos. The European missionaries were young, inexperienced and by no means highly qualified. They were changed frequently and therefore rarely got down to learning the local languages.[2] To supervise them closely Freeman had to be constantly on tour. The question arose whether, with the material at his disposal, Freeman would not have done better to concentrate on a more limited area.

His view was that the wide diffusion of the pioneering missionary effort was producing 'social progress' from which his successors would reap abundant fruit. A former head gardener, he was proud of his model farms on which he cultivated cotton, coffee and vine. He was opening up the country, encouraging the people to produce better roads, houses and furniture, and he prophesied 'at no distant date . . . the general adoption

1 R. Oliver: *The Missionary Factor in East Africa*, London 1952, p. 10, loc. cit.
2 The basic problem was that the Methodists were proud of the unity of their foreign and home missions. If a young man volunteered for the unpopular climate of West Africa, after a tour or two he was offered the chance of a position at home or in some other mission field. The Gold Coast District Meeting led by Freeman in 1850 had to petition against this; Meth.

of those useful European domestic habits and usages which are capable of judicious adaptation to a tropical clime'.[1]

The Methodist Secretaries in England, because they accepted Buxton's philosophy, did not question Freeman's policy as such, especially since he did not wholly neglect the work of conversion at each of the mission stations, however isolated. Whatever vague feelings they had that Freeman's policy was wrong found outlet in their recurrent criticism of the expense it incurred. His incessant journeys made him write always 'in haste'. His accounts were ever in a muddle, and his expenditure always exceeded the estimates, sometimes by as much as a half.[2] After having been repeatedly warned, he was removed as Superintendent in August 1857 on the grounds that if every Superintendent were to act as he had done 'in the expenditure of public money, our missionary society must soon cease to exist and all its holy and benevolent operations would be paralyzed, if not entirely destroyed', and he was asked to make reparations for the amounts he had overspent by having £50 deducted from his allowance every year.[3] Freeman chose to resign from the mission so that he could try to earn more than the £250 the mission paid him and in that way be able to pay the levy more comfortably.[4] He was replaced by more orthodox and less energetic men. His policy

1 Freeman's 'Report to His Excellency Major Hill on the Social Progress of the Natives of this part', 31 Dec. 1853; Meth. Secretaries' Letter Book.
2 Secretaries to Freeman, 18 July 1844; Meth.
3 Secretaries to Freeman, 16 Aug. 1857, conveying the decision of the Wesleyan Conference meeting at Liverpool; Meth. Secretaries' Letter Book.
4 Freeman was an extraordinary man, a pious visionary, yet immensely practical in his way. He was a trusted adviser of the governors and traders of Cape Coast on political and commercial affairs; but in the financial affairs of his mission he was either superbly careless or ingeniously naïve. When, in 1849, he had exceeded his budget of £5,500 by £2,000 and the Home Secretaries began to treat his bills of credit as overdrawn cheques which they refused to honour, he argued that since they did not ask him to give up part of the mission field, he took it they were going to find money for the work. Then he called a meeting of eight European merchants of Cape Coast, on some of whom he had drawn the bills, to examine his books publicly at the mission house. When they finished, he retired and they proceeded to pass resolutions: 'That it appears to this meeting that the Management Committee [in England] has been themselves the cause of the present unsatisfactory position of the finances of the Gold Coast Mission. That this meeting is unwilling to characterize the conduct of the Management Committee for failing to give their Superintendent the checks which were necessary at the time, but it cannot avoid expressing surprise that a course which would have been considered little short of unfair dealing should have been adopted by a Religious Society. . . . That this meeting entertains the greatest respect for Mr Freeman's character.' He sent the document, dated 24 Jan. 1849, to Methodist House, marked 'Private' and continued as blissfully as before. After his resignation he farmed, traded, and worked for the government in turns. Then in 1873, after seventeen years as it were in exile, he returned to the mission as a village pastor.

was not reversed, but the removal of his dynamic supervision left the initiative to the individual missionaries, and there was only one Methodist missionary of note in this period in Nigeria. He was Thomas Champness, at Abeokuta from 1859–1863, who made an unsuccessful but noteworthy attempt to open up the Ijebu country to the Gospel.[1]

The story in Calabar was the opposite of Freeman's. There the missionaries concentrated their energies mainly on Duke and Creek Town. In 1846 Hope Waddell had made an exploratory visit to Bonny. In 1848, as we have seen, King Pepple and the gentlemen of Bonny wrote to England asking for missionaries.[2] In that year also there was in Calabar an impetuous missionary, N. B. Newhall, who rather than be an assistant in Duke Town wanted to go up the Cross River to open a new station in Umon. The two proposals were debated in Edinburgh. Waddell, who was home on leave, wrote a paper in which he opposed any premature expansion:[3] Umon, he said, was 70 to 80 miles away; the people spoke a different language from the Efik of Calabar; and, further, it was necessary to occupy the intermediate places first, for which purpose a better foundation was needed in Calabar. The Foreign Mission Committee shared some of Newhall's impatience. They declared that the missionaries should not look upon themselves

> only as the messengers of Christ to the inhabitants of Old Calabar, but as the harbingers of many to the many millions that occupy the vast regions of Central Africa drained by the Niger, the Tshadda [i.e. Benue] and the Cross River.[4]

Hope Waddell was therefore asked to expand his activities to Old Town, but before moving further into the interior, to answer the invitation from Bonny.[5] He went there in December 1849 to ask for a site and that the people should give up human sacrifice and destroy 'Juju House', the centre of traditional worship. The gentlemen who were anxious to have the school responded agreeably, but King Pepple warned Waddell that

> no man in Bonny could agree to destroy Juju House. That if anyone

1 For general accounts of Methodist expansion in Nigeria, see F. D. Walker: *Hundred Years in Nigeria*, 1942; *Thomas Birch Freeman*, 1929; and Allen Birtwhistle: *Thomas Birch Freeman*, 1949. Thomas Champness was one of the few Europeans of this period who reached Ijebu Ode, in 1860. After the death of his wife in 1862 he retired from the mission. In the 1890s he formed the *Joyful News* band of missionaries who, directly under his supervision but working in co-operation with the official Methodist missionaries on the spot, fulfilled his ambition of establishing the Methodist Church in Yoruba. He was a great friend of Townsend's and their ideas of missionary work were similar.   2 See above, p. 56.
3 Dated Edinburgh, 17 March 1849; Waddell: *Journals*, vol. VII p. 34.
4 Somerville to Anderson, 26 Oct. 1848; U.P. Letter Book, vol. I, p. 416.
5 Mission Board to King Pepple, 4 June 1849; U.P. Letter Book, vol. I, p. 306.

told me so they were deceiving me and were liars. . . . But he said that the young boys and girls now growing up who would come to me to learn book would take after my ways. I could tell them anything I pleased and they might do as I told them if they pleased.[1]

A month later, Hope Waddell returned to Bonny with his wife and little daughter and Miss Miller, another missionary, carrying printed alphabets and spelling cards in English marked 'Grand Bonny, First Lesson Book'. They spent just a fortnight there,[2] after which the Presbyterians dropped the idea of a Bonny mission. They had not the resources to work in Bonny with the same intensity as in Calabar.

Their policy was one of extreme concentration. At both Duke Town and Creek Town they maintained usually two to three European missionaries and a few West Indian teachers, besides the medical officer and other lay agents. By the standards of other missionaries they were extremely cautious in declaring converts ready for baptism: their first baptism did not come until after seven years' work, in 1853, and Waddell, who was away, thought it unduly hurried, nor did they open their first chapel until 1855.[3] As against the emphasis on open-air preaching in other missions, the schools came first with them and instruction within the compounds of the various Houses. They did not rush; but nevertheless they regarded the contemporary generation in Calabar as potentially Christian. This was why they favoured concentration and were reluctant to expand rapidly. They had of course all the suspicions of the Calabar rulers to fight against, and in 1851 three conditions were imposed on missionary travels in the interior: they must get permission, must never take traders with them, and if going beyond a day's journey, must be accompanied by an official guide.[4] These conditions did not stop

1 Waddell, Journal entries for Saturday 22 Dec. to Thursday 27 Dec. 1849; Journals, vol. VII, p. 104 ff.
2 Ibid., Monday 28 Jan. to Wednesday 13 Feb. 1850; Journals, vol. VII, p. 139 ff. Miss Miller is better known under her later name of Mrs. Sutherland.
3 Waddell: Journals, vol. X, pp. 37–43; vol. XI, p. 11.
4 Waddell objected to these conditions, arguing that in Europe people travelled wherever they pleased, being obliged to inform no one except their fathers. 'Very well,' said King Eyo, 'you be my son.' Waddell replied that on the other hand, he was the King's spiritual father. He added that he might as well join Ekpe Society so as to be free of being treated as a foreigner in Calabar, but when King Eyo appeared to be taking him seriously, he said it was only a joke; Waddell: Journals, vol. VIII, pp. 128–30. The restrictions were on the whole faithfully adhered to by the missionaries. In 1861 when Thompson threatened to call the gun-boat to assert the liberty of the British subject to wander wherever he listeth, and thus provoked the rulers of Duke Town to attempt to expel all the missionaries, both his colleagues on the spot and the Foreign Mission Board censured him severely; Somerville to Thompson, 21 June 1861; and the Resolutions adopted by the Standing Sub-Committee, 20 June 1861, both in U.P. Letter Book, vol. VI, p. 463 ff.

missionaries opening stations in the interior. In 1855 they took up Ikonetu and in 1858 Adiabo and Ikorofiong. There, however, the expansion stopped.

Some consequences of this policy of whole-hearted concentration will be discussed later. To return to the parishioners in Britain, they came to know about all the 'evils' in Calabar society and, after a while, they became bored. They wanted news of other places as well. When, in 1876, the United Presbyterian Church ordered an inquiry into the working of the various mission fields, the Secretary of the Foreign Mission Committee, fearing that they might ask the Calabar Mission to close down, advised Anderson that

> It may be well for you and all the brethren to understand that we shall need all the facts you can give us, all the evidence of progress, all the good prospects for the future you can make good, in order to convince some persons in and more out of the Board that your labour and sacrifice and the money spent and the lives surrendered might not have been better expended in some other heathen field. I hope you and the other brethren will be on the lookout for a full defence and justification. If you could open a back door into the interior!![1]

It was at that stage that Mary Slessor came in. Young and impetuous missionary, with something of the charisma of David Livingstone and following in his footsteps, she began to tramp north to Okoyong and Itu; first in spite of criticism, and then to the acclamation of admirers, she brought expansion and new life to the mission.[2]

The C.M.S. seems to have arrived more quickly at the happy medium which became a model for other missions. In 1845 the Secretaries had expressed the hope that, after Abeokuta, Rabba on the Niger would be the next place to be aimed at.[3] In January 1849 David Hinderer, a new recruit from the Basel Seminary, arrived in Badagri, with instructions that he was 'specially set apart for the study of Hausa language, for communicating with the Hausa natives and ultimately for a missionary visit to the Hausa country'.[4] But when Townsend and Gollmer went to give evidence before the Hutt Committee, Hinderer was placed tempo-

---

1 McGill to Anderson, 18 February 1876; U.P. Letter Book, vol. XVIII, p. 463 ff. In the following year a circular letter dated June 1877 was sent round to the heads of all the different missions around Calabar from the Niger to the Gaboon to express an opinion on whether the work in Calabar should be strengthened or whether the resources of the mission had better be diverted elsewhere.
2 For this account of the rate of expansion see the general histories of the mission: Goldie, op. cit., p. 7, McFarlan, op. cit., Chapters 4 and 5.
3 Secretaries to the Yoruba Mission, 25 Oct. 1845; C.M.S. CA2/I
4 Secretaries to Hinderer, 11 Jan. 1849; C.M.S. CA2/L1.

rarily at Abeokuta. It was there he became convinced that Hausaland must be approached by a chain of missions and the next link was not Rabba but Ibadan.[1] Townsend went further and, for reasons to be explained later, began to question the enthusiasm to get into Hausaland and do battle with Muslim emirates when the missionary work in Yoruba still wanted consolidation. He rejected on the one hand too much concentration on Abeokuta and on the other the idea of a chain stretching in a line. The links of the chain, he said, must not be too far apart, and the chain must stretch to all the centres of population and political power in the region already entered upon.[2]

His advice was followed in the Yoruba country. In 1853 the missionaries in Lagos and Abeokuta were reinforced, an African catechist occupied Otta, and Crowther paid a preliminary visit to Ketu. Hinderer went to settle at Ibadan; Adolphus Mann, another Basel recruit, at Ijaye. From Lagos, Gollmer stationed agents at Igbesa, Ikorodu and Offin (Sagamu) but was refused permission to go to Ijebu-Ode. From Abeokuta, between 1853 and 1858, Townsend toured incessantly,[3] west to Ibara, Isaga and Ilaro, and north to Oyo, Awaye, Iseyin, Saki and Ogbomoso, at each of which he obtained grants of land and, as soon as he could, stationed emigrant agents. In that year two European catechists on probation were stationed at Iseyin and Oyo, under the supervision of Mann. In this way Townsend's policy was carried out. Mann visited Ilorin in 1855 and Townsend himself in 1858; they received a friendly enough welcome from the Emir, but no permission was granted to open a station.[4]

1 Conclusion to Hinderer's Journal Extracts for 13 Sept. to 16 Oct. 1850: 'And now for your chains again: Two good links are already towards it—Badagri and Abeokuta, and I am sure God will graciously hear our prayers and give us Ibatan [sic] . . . Next to that may be Ilorin . . . a fourth and fifth will bring us to the Tshad where we shall shake hands with our brethren in the East'; C.M.S. CA2/049.
2 Townsend to Venn, 1 May 1850; and conclusion to his Journal Extracts for the term ending June 1851; C.M.S. CA2/085.
3 The journals of missionary travels in this period are a remarkable series of documents on the life, social, political and religious, of the Yoruba, Efik, Onitsha, Ibo, Igalla and Nupe peoples. Overland, the journey was mostly by foot along well-known trade routes, often in company of trading caravans. A catalogue of the journeys achieved by some of the missionaries reflects great credit on their zeal. Between 1852 and 1858 Townsend recorded 14 journeys: 3 return journeys to Ijaye; 2 to Lagos, 2 to Iseyin, one each to Ilorin, Badagri, Ilaro, Ado and Otta; one circular tour Abeokuta—Ijaye—Iseyin—Oyo—Ijaye—Ibadan—Abeokuta. One last one aiming at Rabba was broken off at Ijaye because of the illness of Mrs. Townsend who, as often happened, was accompanying her husband. C.M.S. CA2/085.
4 There is no general history of the Yoruba Mission, but the accounts in Stocks, op. cit., vol. ii, pp. 99–123 and Groves, op. cit., pp. 50–64 are good. See also *Church Missionary Atlas*; C.M S. 1873.

It ought to be mentioned, however, that the urge to reach Hausaland by a chain of missions remained like a legacy bequeathed to Samuel Crowther by his association with the Niger Expedition of 1841. He went up with the Expedition of 1857 and founded the Onitsha mission. He placed in charge the Rev. J. C. Taylor, born in Sierra Leone of Ibo parents, accompanied by three Christian Visitors and a Lay Agent. He went on and established a second station at Igbebe at the confluence, and put a schoolmaster in charge. When the *Dayspring*, the steamer conveying the Expedition, was wrecked near Jebba, he took the opportunity to acquire some land in the Nupe quarter at Rabba and built some huts. He said he was establishing not really a mission centre, but a missionary rest-house where passing agents of civilization could stay and where a missionary of the right sort could 'by conversation ... kind, intelligent and Christian influence ... dispel the mist of misconception and prejudice' that the Muslims had against the Christians. Meanwhile he stationed as agent there a young Kanuri, Abegga, whom the explorer Barth had taken to England in 1855 and had passed on to the C.M.S. When Crowther returned to the Niger in 1859, he was told the Emir of Bida had closed the Rabba station. Not until 1873 was Crowther able to establish a station north of the Confluence, at Kippo Hill opposite Egga. The difficulties of the Niger Mission will be discussed later,[1] but they did not diminish Crowther's rate of expansion. In 1861 he occupied Akassa at the Nun entrance of the Niger. In 1864 he was invited into Bonny by King Pepple, and that year also he established a station at Idah. In 1868 he moved into Brass. Like Freeman, Crowther believed in diffusing as widely as possible the efforts of the pioneering missionaries. 'We can act,' he said in 1869,

as rough quarry men do who hew out blocks of marbles from the quarries, which are conveyed to the workshop to be shaped and finished into perfect figures by the hands of the skilful artists. In like manner, native teachers can do, having the facility of the language in their favour, to induce their heathen countrymen to come within the reach of the means of Grace and hear the word of God. What is lacking in good training and sound Evangelical teaching,[2]

others more experienced would later supply.

Thomas J. Bowen, the American Baptist Missionary, arrived in Badagri in August 1850, having read William de Graft's account of the conversion of Old Simeon of Igboho, and it was at Igboho that he

1 See below p. 208 ff.
2 Crowther: 'A charge delivered at Lokoja at the Confluence on the 13th September 1869'; C.M.S. CA3/01.

wished to establish his first station, 'hoping in subsequent years to penetrate into Nyffe [Nupe], Hausa, etc.'[1] However, his penetration into the interior was barred for the moment by political unrest. He settled at Abeokuta as the guest of Townsend for over a year, learning the language and getting to know the people both in the town and on their farms. It was then he began to change his mind. He said he observed that the Methodists 'extend too much and act feebly'; the C.M.S., on the other hand, 'make their stations strong in men and in money and act much more efficient. A feeble attempt is a waste of means.'[2] He saw visions of a chain of Christian cities to the Niger, to Lake Chad and Abyssinia, but

> It is not the work of a day or of a generation, yet this generation may prepare the way so that our successor may do more in a year than we could accomplish by a whole life-time of toil. In my anxiety I would run forward; but this is not practical now. . . . I see the necessity of a strong foundation.[3]

In other words, he was for balancing time with concentration.

Failing to get to Igboho, he tried Ketu and Iseyin in turn but found obstacles in his way. The ruler of Ijaye barred his route to the interior, but invited him to settle in his town.[4] Bowen therefore decided to make Ijaye his first station. He went back to America and returned, in August 1853, with two other missionaries, Dennard and Lacy, and their wives. Almost immediately after, Lacy found himself going blind and had to return home with his wife. Dennard was to occupy Abeokuta, having also to maintain communication with the naval packets at Lagos. He obtained the grant of a piece of land, christened 'Alabama', and started to preach, but his wife died in January 1854 and he himself in September. That month a new missionary, William Clarke, arrived to join Bowen and his wife at Ijaye. Together they established the Baptist mission at Ijaye and Ogbomoso. Bowen had built a mission house and a little chapel at Ijaye and baptized the first convert by July 1854. An American Negro from Liberia, J. M. Harden, joined the mission and established a base for them at Lagos. In 1855 Bowen moved to Ogbomoso and with Clarke explored as far as Saki, Igboho and Ilorin. Reid, a new missionary, occupied Abeokuta in 1856 and when relieved there by Phillips, another

1 Bowen to Taylor dated Quincy, Florida, 28 Dec. 1848; *Bowen Letters*; also *Missionary Labours and Adventures in Central Africa*, op. cit., pp. 26 and 99. For Old Simeon, see above, p. 33.
2 Bowen to Taylor, 26 March 1851; *Bowen Letters*.
3 Bowen to Taylor, 7 Sept. 1851; *Bowen Letters*.
4 Bowen: *Missionary Labours and Adventures in Central Africa*, op. cit., p. 162 ff.

recruit, he moved to Ogbomoso and later founded a new mission at Oyo.[1]

Thus, on the whole, the expansion of the Christian missions in the period 1853–60, particularly in the Yoruba country, was quite impressive. Perhaps the most significant aspect of this expansion was the almost total indifference to the political institutions of the people. The missionaries began in Badagri and Calabar by trying to strengthen the hands of the traditional rulers and using their power to protect and to further the cause of the missions. But the availability of the naval power of Britain was too tempting an alternative; they began to shield behind the navy and in consequence helped to weaken the political power of the states. In the interior where the navy was not available, their indifference was all the more surprising.

The only exception, as we have seen, was Townsend's 'Abeokutan' policy.[2] The C.M.S. authorities adopted it and urged it on the British government between 1849 and 1851. The government also accepted it. Other missions began to adopt it. Bowen in 1851 said his 'present hope is in Abeokuta';[3] the Catholics up till 1864 saw in Abeokuta the key to the interior and the place that could become the centre of the mission.[4] But even Townsend soon saw how inadequate the policy was to provide the political framework for the expansion of the missions. Abeokuta remained internally divided and there was no possibility of its extending its power over Ibadan, Ijaye, Oyo or beyond. Yet no one proposed an alternative and the missions held on to the 'Abeokutan' policy till the British government rejected it in 1861.

This indifference to the political implications of missionary expansion arose from two fundamental causes. The first was the inherent distrust that most Evangelicals had for government, as if it were no more than an evil to be tolerated. The second was the assumption that as soon as missionaries established themselves in the country, the traditional way of life was doomed, the rulers would embrace Christianity and behave just as missionaries hoped they would. A good deal of the ineffectiveness of the missions in this period arose from the fact that this assumption proved false. 'I do not doubt,' wrote Townsend to Venn at the time he was expressing his lack of faith in the old Abeokutan policy,

1 For general accounts of the Baptist Mission, besides Bowen's own narratives on which most of them are based, see G. W. Sadler: *A Century in Nigeria*; Louis M. Duval: *Baptist Missions in Nigeria*; H. U. Tupper: *Foreign Missions of the Southern Baptist Convention*, Richmond 1880.
2 See above, pp. 79–80. 3 Bowen to Taylor, 7 Sept. 1851; *Bowen Letters*.
4 '*La véritable porte du Sahara du côté du Sud*' and the place '*qui puisse devenir le centre de nos éstablissements*'; Journal of Father Broghero, May 1864, p. 389; S.M.A. Rome.

I do not doubt but that the government of this country is set against the spreading of the Gospel; they see what they did not at first, that the Gospel will overturn all their system of lies which they wish to preserve as entire as possible. . . . At the same time they want us without our religion. They want us on account of the people in Sierra Leone, because they see that through us they are likely to keep open the road to the sea and obtain trade and be well supplied with guns and powder for sale or war as may be required.[1]

Townsend was beginning to observe what he did not see before: that the conversion of the rulers was not going to be easy. They sent some of their children to school, and they put few obstacles in the way of the missionaries. When representations were made to them they accepted such minor reforms as forbidding Sunday markets or putting a ban on drumming near churches on Sunday. But they did not themselves accept baptism. That was the experience of missionaries everywhere in Nigeria at this time. 'The hearts of the people,' said a missionary in Calabar, 'are wholly bent on trade and most firmly glued to their heathenish and superstitious practices.'[2] That was hardly surprising. What is important here is that although each mission had respectable statistics of conversions to show, none of them made headway in the conversion of the rulers of any of the states. It is important to consider why this was so.

Nowhere did the prospects of converting the local rulers appear brighter than in Calabar, and no missionary made greater efforts to convert them than did Hope Waddell. His failure, therefore, throws useful light on this question. He had the advantage that, while the rulers of Abeokuta were anxious mainly about defence, the rulers of Calabar were from the start eager about schools and education and even some social reform. Nowhere else on the coast, said Commander Raymond in 1842, were the people so anxious to be civilized.[3] They threw open their courts to the missionaries and, since they spoke English, themselves volunteered to act as interpreters to their people every Sunday morning. The most instructive story concerned the partnership of Waddell with King Eyo of Creek Town.

Within a few days of Waddell's arrival in Calabar they both met at breakfast on board one of the ships. Hope Waddell recorded:

I spoke to him of the value of God's word, and the love of God to us

1 Townsend to Venn, 14 Nov. 1850, marked 'Private'; C.M.S. CA2/085.
2 William Jamieson, Missionary in Calabar 1847-8, cited in Godie: *Calabar and its Mission*, p. 111.
3 Commander Raymond to Foote, 11 Dec. 1842, encl. in Foote to Herbert, 12 Dec. 1842; FO 84/495.

through his dear son, quoting the text, 'God so loved, etc.' He listened with attention, but made no remark, and soon after took his leave. He is really a fine man.

Then he reflected that there was no need to preach in Calabar that there is a God who made all things. That was known already. 'I proceed directly to the revelation of his character and will and his redeeming love in Christ Jesus.' The following Sunday he preached on a similar theme. King Eyo interpreted but later pointed out that the sermon was wasted. 'Calabar people no fit to saby all that yet—tell them about the fashions what God tells us to do or not to do.' The tendency for some missionaries was to preach with particular emphasis on the theological peculiarities of their various denominations. Now Waddell agreed with King Eyo that 'the first principles of religion must be taught first, though they may in themselves be inoperative to produce conversion: the law must be preached till it is at least understood, though it should be preached for years without making converts.'[1] Other incidents and similar exchanges of ideas deepened Waddell's respect for King Eyo. Once, the Ekpe Society passed some laws against the liberty of slaves to form associations. Waddell prepared a sermon on the duties of masters and slaves and he wondered if Eyo would agree to interpret such an embarrassing sermon. But he did so—admirably. 'He spoke loud and free and with increased energy,' interpreting and sometimes answering Waddell's rhetorical questions.[2] Once again, in 1849 Waddell felt depressed that three years' work had produced no converts in Calabar. He chose the parable of the barren fig-tree for his text. Eyo interpreted but later, like a father, consoled the missionary. 'He thought,' wrote Waddell,

> that our labours had not been quite in vain. He thought that the good seed we had sown was growing in some hearts. . . . But he added that the word of God had been preached in England for a thousand years and many there did not believe and obey it and God was patient with them and he hoped would also be patient with the Calabar people too.[3]

'There is a great deal of natural dignity about him,' said Waddell on another occasion, 'not pride, but good sense, great propriety of manners, temper and composure, with a just idea of the respect due to himself and to others.'[4]

1 Waddell, Journal entry for 14 Jan. 1856, recalling the incident; *Journals*, vol. VII, pp. 133-4.
2 Waddell, Journal entry for 31 March 1850; *Journals*, vol. VII, pp. 7-8.
3 Waddell, Journal entry for 2 Sept. 1859; *Journals*, vol. VII, p. 46.
4 Wadell, Journal entry for 5 Jan. 1850; *Journals*, vol. VII, pp. 128-9.

When Waddell and the missionaries could no longer control their impatience and they turned to the 'moral' force of the traders and the physical force of the consul and the naval squadron to achieve reform through the Society for the Abolition of Inhuman and Superstitious Customs in Calabar, Waddell noted that King Eyo was

> rather chagrined that the white people should be urging him on in matters of internal government. . . . But everything could not be done all at once [he said]. Customs and prejudices required time to subdue. He was same as a missionary, speaking to everybody same as he heard me.[1]

Differences of outlook began to emerge between king and missionary. Waddell expounded the virtues of monogamy, the evils of trial by ordeal and of substitutionary punishment, the harmlessness of twins and twins' mothers, but, shrewd and intelligent as King Eyo was, he could not see the point of view of the missionary. The Efik king regarded religion as an affair of the community, whose customs and practices could only be changed when the community became generally convinced of the need for change. To him, offence against tradition was sin, a defiling of the community that required expiation. The Christian missionary regarded sin as the responsibility of the individual, a violation of the laws of God that were absolute and independent of the traditions of the community or even the beliefs of the individual. King Eyo conceded some reforms. The Council of the Ekpe Society passed laws against human sacrifice and twin murder, restricted the use of trial by ordeal to public trials only, abolished Sunday markets, and so on. But King Eyo wished to control the rate of change. He opposed the idea of boys at school being made to sign papers committing themselves to further reform, and to their accepting baptism before the rest of the Calabar people were ready for it.

> 'Slow and sure is his motto,' said Waddell. He is not an obstructive, but a strong conservative or modern Whig wishing all reform to proceed from himself and to be conceded cautiously and sparingly and only with universal consent.[2]

Eyo went further, to argue that if he relaxed his control and allowed indiscriminate change in the name of reform there would be chaos and both the state and the mission would suffer. Waddell's answer was that the Gospel was used to finding its own way and did not depend on any one man. By 1853 he was beginning to despair of converting King

1 Waddell, Journal entry for 3 Oct. 1850; *Journals*, vol. VIII, p. 72.
2 Waddell, Journal entry for 1 Jan. 1851; *Journals*, vol. VIII, p. 105.

Eyo. He felt that though the king was intellectually convinced of the truth of much that the mission taught and did not hesitate to sweep away his *Ekpeyong* [images], his heart was unchanged; and without a change of heart he would never show that overwhelming conviction which placed personal salvation above all other considerations and signified conversion. The love of money and power he reckoned as chief among the king's besetting sins;[1] but what really symbolized his rejection of the Gospel was his refusal to give up polygamy. King Eyo added wife upon wife. By 1854 he was, in Waddell's mind, no more than 'a licentious despot . . . living a low, fleshy life', suffering from 'a miserable corruption of heart',[2] and by 1855 he had become 'the lecherous old hypocrite'.[3]

The teaching of the missions on polygamy deserve more than just a passing reference, for it was to polygamy that they attributed the failure to convert African rulers. It was to be expected that the missionaries would be opposed to polygamy; the teaching of the Church had been overwhelmingly against the practice.[4] What is surprising is the relative emphasis the missionaries placed on it in the middle of the nineteenth century in comparison with, for example, domestic slavery.

In 1849 the United Presbyterian Church broke off communion with Presbyterian churches in America who tolerated slave-owners. When, in 1853, Presbyterian missionaries in Calabar decided on having their first baptism, they asked the Foreign Mission Committee for a ruling on the question, 'Should slave-holders be received into Church fellowship?' They indicated that the answer should be 'Yes', with a proviso that converts should be made to sign a pledge declaring that since 'there is neither bond nor free in Jesus Christ', they would regard their slaves as servants, not property; that they would never sell or ill-treat them.[5] Some members of the Church in Scotland were naturally worried about this tolerance, but the missionaries argued back in strong force. They did not condone slavery, but they regarded it as a social evil to be reformed with time. They said that if they refused to receive slave-holders into the Church, 'We can form no Christian Church in Calabar'. It would mean, said Hope Waddell, that

1 Waddell, Journal entry for 30 June 1850; *Journals*, vol. VIII, p. 36.
2 Waddell, Journal entry for November 1854; *Journals*, vol. X, p. 89.
3 Waddell, Journal entry for 18 July 1855; *Journals*, vol. X, p. 181.
4 A useful historical account of the attitude of the Christian Church to polygamy will be found in the Rev. Lyndon Harries's 'Christian Marriage in African Society', Part III of the International African Institute's *Survey of African Marriage and Family Life*, edited by Arthur Phillips, 1953.
5 See Resolution of the Old Calabar Committee, 6 Dec. 1853, and Anderson to Somerville, 6 March 1854, both in U.P. *Missionary Record*, 1855, pp. 17–26.

we cannot accept as Christian brethren those whom our Lord receives and saves. . . . You treat him as a heathen after he has believed in Jesus for salvation, and that for no fault of his own.[1]

More than that—the missionaries wrote essays on the theory and practice of domestic slavery in African society which destroy the impression the missionaries give in other respects that they took little care to study the society they were trying to reform. Domestic slavery, they argued, was far different from plantation slavery. 'The native [Efik] term *Ofu*, which we render slave,' wrote Hugh Goldie, the scholar of the Calabar Mission,

> is employed to denote servant, tributary, or one who in these or similar ways acknowledges the superiority or headship of another. . . . There is but little difference in many respects between the condition of the master and the slave.[2]

Goldie added that Calabar law did not recognize manumission; that even the slaves redeemed and kept in the Missions Houses were still regarded by the Efiks as slaves. It would take time to get the ideas changed. To this essay on anthropology, Waddell added one on political science:

> Before the [state of slavery] could be changed, a new and united government for this and neighbouring countries, with a system of laws for all classes alike especially preparatory for such a change as is spoken of would be indispensable.[3]

Without such laws, and a strong central government, the heads of Houses had to exercise absolute authority over the inmates of the Houses, which meant a state of domestic slavery for all the working classes.

> Absolute property [implied by domestic slavery] is just absolute authority, and our own country, free as it is, does not disown it. The power is indeed taken out of the hands of individuals and lodged in the state under the regulation of laws—a wise and blessed change indeed—but it still exists somewhere and is exercised over some persons.

Until there was such a government to exercise such a power,

1 Waddell to Somerville, 22 Jan. 1855; copy in Waddell: *Journals*, vol. X, pp. 113–15.
2 'The nature of Calabar Slavery, the laws relating to it and the conduct of the missionaries' in U.P. *Missionary Record*, 1855, pp. 17–26.
3 Waddell to Somerville, 22 Jan., op. cit.

My plan of doing away with slavery would be as with a poison tree whose root is deep and [too] strong to be dug out. I lop its branches and prevent its growth and development continually, whereby the root will soon die. I argue not whether it is unlawful in the abstract, or in its origin, which cannot, in all cases, be traced.[1]

The Foreign Mission Committee adopted these arguments. They endorsed the solution that slave-holders could be baptized if they signed the pledge. They argued that this decision was consistent with breaking off communion with slave-owners in America, not just because domestic slavery differed from plantation slavery, but also because in Christian America it would be wrong to connive at slavery, whereas in Calabar it was necessary to tolerate slavery temporarily so that the Christian Church could be established, and when Christians became the majority and could influence laws, slavery could then be abolished.[2]

It has been necessary to detail the arguments on which the missionary attitude to domestic slavery was based, because it was in such contrast to the attitude to polygamy. In spite of the strong anti-slavery element in the missions it was decided that domestic slavery could be tolerated until the majority of the people became Christians and could see the evil and then abolish it. That was uniformly the attitude in all the missions. Freeman had argued like Goldie and Waddell in 1841–42 when Dr. Madden, a Commissioner asked to investigate, among other things, the state of slavery in the British Colonies in West Africa, recklessly proclaimed emancipation in Cape Coast.[3] In 1856 the new Bishop of Sierra Leone, Bishop Weeks, paying his first visit to Nigeria, startled C.M.S. missionaries in Yoruba by doubting the correctness of their tolerance of domestic slavery. Townsend sent anxiously to Crowther. Crowther despatched to C.M.S. House arguments like those of the Calabar missionaries:

The slaves and masters in this country live together as a family; they eat out of the same bowl, use the same dress in common and in many instances are intimate companions, so much so that, entering a family circle, a slave can scarcely be distinguished from a free man unless one is told.[4]

Bishop Weeks also thought that in baptizing wives of polygamists the

1 Waddell to James Simpson, draft dated April 1855 in Waddell: *Journals*, vol. X, pp. 158–61.
2 Resolution of the Synod of the United Presbyterian Church in 1849.
3 Freeman to Secretaries, 25 June 1843; Meth.
4 Crowther to Venn, 4 March 1857. See also Crowther's earlier letter of 2 Jan. 1857; C.M.S. CA2/031.

missionaries were showing a dangerous tolerance of polygamy. To this Crowther replied that the wife of a polygamist was an involuntary victim of a social institution and should not be denied baptism because of that. But he went further. He argued for a general policy of demanding from converts seeking baptism no more than the 'minimum qualifications necessary for salvation', leaving other refinements to time and their membership of the Church.

> That ye abstain from meats offered to idols, and from blood, and from things strangled, and from fornication; from which if ye keep yourselves, ye shall do well. (Acts 15, 29) . . .

> We use our discretion that such practices which the laws of the country allow, but not being among those in immediate requirements necessary for salvation, but which Christianity after a time will abolish, are not directly interfered with.[1]

Venn replied that whatever be the prevalent custom of a nation, the Ordinance of God could not be lowered to it; there must be one standard for the Church everywhere, as God could not condemn polygamy in an old-established Church and tolerate it in a newly established one. To clear whatever lingering doubts there might have been in Crowther's or any other person's mind that monogamy was not one of the minimum qualifications necessary for salvation, Venn proceeded to examine the Scriptural evidence for the teachings of the Church on the subject of polygamy in a paper which is still regarded as one of the most authoritative pronouncements on the subject.[2] He concluded that Christ regarded polygamy as adultery: 'It is written "Let every man have his own wife and let every woman have her own husband." ' However a polygamist's wife if not herself a believer in polygamy could be admitted. The other missions took the same view and the Lambeth Conference of 1888 confirmed it.

Perhaps more interesting than this famous paper was the covering letter issued by Venn, in which he tried to reconcile the attitude on polygamy with the contrasting attitudes to domestic slavery and the Atlantic slave trade.

1 Crowther to Venn, 3 Jan. 1857; C.M.S. CA2/031.
2 Venn's paper was issued as a Minute of the Parent Committee and sent in a circular round to the different missionaries; it was later printed as an appendix to the *Annual Report of the C.M.S. 1857* and subsequently issued as a pamphlet. It is discussed from the point of view of a High Churchman in L. Harries's 'Christian Marriage in African Society'. Crowther's reply to Venn, 6 April 1857, said the Minute was very acceptable. 'I have never at any time had a doubt in my mind as to the sinfulness of polygamy and as contrary to God's holy ordinance from the creation and confirmed in the time of the flood.' C.M.S. CA2/031.

The Committee think there'll be no hesitation to refuse baptism to a kidnapper of slaves unless he has repented and left his evil way, because the practice is directly contrary to Scripture. But the Committee would not interfere with the discretion of a missionary in admitting a slave-holder to baptism. The Word of God has not forbidden the holding of slaves, though it has forbidden the oppression and injustice of various other evils which too often, though not necessarily, cleave to the character of the slave-holder. Christianity will ameliorate the relationship between master and slave; polygamy is an offence against the law of God, and therefore is incapable of amelioration.[1]

The discretion was also left to missionaries to baptize wives of polygamists 'because this has not been decided in Scripture; it is very conceivable that a wife may have had no power to prevent the polygamy of the husband'.

One must go beyond these theological arguments to realize why the rejection of polygamy became, as it were, the most essential dogma of mid-nineteenth-century Christianity in Africa. The missionaries frequently pointed out that Islam made more rapid progress than did Christianity because Islam accepted polygamy. They implied, perhaps rather complacently, that Islam in this tolerated a lax morality and was therefore regarded by Africans as an easier option than Christianity. But polygamy raised much more than a question of standards of morality. Both for the European missionaries anxiously seeking to get converts to reject it, and for the Africans holding on tenaciously to it, polygamy was not just a plurality of wives; it was a symbol of the communal way of life in the family compounds. The acceptance of polygamy by Islam implied the acceptance of the communal way of life, and it was as a unit that Muslim missionaries sought to convert the different communities: to convert the rulers and, through them, by a new law and a new system of justice, make the people progressively Muslim. Christian missionaries from an individualistic society, where whatever folk-culture survived the Reformation and seventeenth-century Puritanism had been virtually destroyed by the Industrial Revolution[2] and the new puritanism of the Evangelical Revival, found life in the family compounds at best incomprehensible, at worst the devil's own institution. Concerned as they were not only to destroy paganism but also to

1 Secretaries to missionaries in Yoruba, 17 Feb. 1857; C.M.S. CA2/L2.
2 Nationalism has preserved some folk-culture in Scotland and Wales, and interest in anthropology has produced a twentieth century revival in parts of England, which shows that folk-culture never quite perished in the rural counties.

reform the existing social structure in Africa, they were bound, sooner or later, to attack polygamy. The crucial fact was that they refused to regard it as a social evil which could be progressively reformed, and they declared it a direct violation of the laws of God, which had to be rejected by the faithful *ab initio*. By this decision they abandoned the idea of leading the whole community as a unit gradually towards Christianity. The outward sign of his inward conviction that came to be demanded of the new convert was not so much the casting away of idols as his total rejection of life in the family compound symbolized by his adoption of monogamy.

This takes us back to the argument about time and concentration, and the dual influence in the missionary movement of humanitarians willing to work for a long-term policy of economic development and Evangelicals anxious to ensure salvation for at least a few of the contemporary generation. Besides the need to show results to the parishioners in Europe, there was also the fact of denominational competition—from which, by the way, Islam was free. In many ways it was a source of strength and impetus to missionary work; but where, as in Badagri, Lagos, Abeokuta and Ijaye, two or three different denominations existed side by side, there was no question of their regarding the community as a unit to be converted 'with time'. The community became a pool from which the various fishers of men sought in friendly—or less friendly—rivalry each to attract individual fishes into his own denominational net.[1]

Without abandoning the policy of influencing the whole community by 'indirect' methods, the main missionary effort was being shifted from the court to the Mission House, around which the Church was being built from individuals fleeing spiritually and, quite often, physically from the old society. Far from seeking to master the political situation and strengthen the hands of the faithful in the councils of the land, they advised the converts for the safety of their souls to keep aloof. They urged the converts to make a clean break with the past. They tended to regard practically everything in the old society as somehow tainted with heathenism. Though quite often the need for some adaptation was referred to, rarely was anything found that could be adapted without making concessions to heathenism. Local names, local art, local

1 One result of this was that when the Roman Catholic Church, with a traditional communal view of society, came to compete with other denominations it quickly adopted the attitude of the individualist liberalism of the Protestants. This, of course, was also what happened in Europe wherever the Roman Catholic Church was not the state church. And there were other ways in which the growing individualism of Europe had affected Roman Catholic missionaries, particularly in France, whence most of the earliest missionaries in Nigeria came.

music, local fables, were each in turn so closely bound up with the old society that objections were found to using them in the new society. An emigrant catechist was almost refused ordination because apart from being too fond of farming and hunting he was considered guilty of

that loose, superstitious way of living of many Sierra Leone Christians. . . . He did not, for instance, scruple to help to dress out some neighbours' heathen ceremonies with European fineries such as plates and other English articles which he and his wife lent. Then he made *Komo jade* feast for his heathen relations for his children.[1]

When the Rev. Thomas King died at Abeokuta in 1862, said Dr. Harrison,

there was some attempt at the native howling at the funeral, but it was checked. On the road from Igbein to the graveyard, which is a long distance off, there were a good many hymns sung.[2]

While some social segregation was inevitable, the important point here was that when the European missionaries found that the rulers were not to be easily weaned from their old ways they also began to urge political segregation on their converts.

Hope Waddell, who had been trying to secure social reforms through the Ekpe Society, bolstered up by the strong arm of the consul, began to advise his converts not to join the Society. King Eyo's eldest son, already a member, was asked to keep away from its councils, even its plays and amusements. Waddell related how Young Eyo, as he was called, insisted on joining an Ekpe parade called *Egbo Bunko*:

It is high dress Egbo which parades the town in fanciful costume and closes the day with a sort of dance and procession. It is comparatively harmless and rather fanciful, being a sort of masquerade. Young Eyo enquired what harm was in it. In being finely dressed, I told him, no harm; in being absurdly dressed with tails and other extravagances of flaunting ribbons and feathers and masks, I said, much harm to a Christian disciple. I could not recognize a brother in the Lord in such a fellow. . . . But I did not know enough even of Bunko to mention all that might be bad in it. But so far as I knew I thought the less he had to do even with it the better for himself. . . . He used to like it well, and owned that he still had an inclination that way.[3]

1 Hinderer to Crowther, November 1868, quoted in Minutes of Local Committee, 25 Feb. 1868; C.M.S. CA2/011. *Komo jade* (now called *Ikomo*), the Yoruba naming ceremony, was usually on the 8th day after birth; the child is 'brought forth', and introduced to the family and friends, who are entertained with a 'feast'.   2 Harrison to Venn, 30 Oct. 1862; C.M.S. CA2/045.
3 Waddell, Journal entry for Monday 23 Oct. 1854; *Journals*, vol. X, p. 85.

Young Eyo did not go that day, but a week later he informed Waddell that 'he had promised his father he would join [Egbo Bunko] that evening in the funeral ceremonies of the king's blacksmith, lately deceased'. Waddell said it was a 'fall' and a 'dishonour'. When they met two days later, argument was resumed. Waddell said the relevant question should be not what harm he had done but what useful purpose he had served by going. Young Eyo acquiesced on the grounds that 'however harmless it might seem to him . . . [the missionary] had more experience than he'.[1] The occasions on which Young Eyo acquiesced to the voice of missionary authority but went his own way and later expressed regret were so frequent that there can be no doubt that he was not convinced that he could give up his old culture. This became obvious when he succeeded his father in 1858.[2]

One more example from the C.M.S. must suffice. In 1859 some missionaries at Abeokuta were worried about converts and emigrants remaining in or joining the Ogboni, which was the Egba counterpart of the Ekpe Society.[3] Several of them, including some African agents who were members, prepared papers on the Ogboni, its place in the constitution and life of Abeokuta, and its initiation ceremonies. After reading these, the Parent Committee issued a minute that

> as the system of Ogboni is of the nature of a political institution and has existed from remote antiquity as a recognized power in the state, it is entitled to some degree of public respect even from those who separate themselves from it. It will be right, therefore, to avoid open opposition to the system. . . .

> That whilst there is a wide difference of opinion amongst those equally well informed respecting the connexion of the Ogboni system with idolatry, yet as all agree that it is inconsistent with the principles of the Christian religion and must fall when those principles prevail

1 Waddell, Journal entry for Friday 3 Nov. 1854; *Journals*, vol. X, p. 87. He added, 'Bunko is a town show and when I first came to the country the novelty of the scene amused myself.' Ibid., 1 Nov. 1854.

2 Young Eyo became Eyo Honesty III in December 1858. His father's house and treasure were burnt down in a fire accident in January 1859, leaving him all his father's debt as well as greatly diminished prestige. He died in 1862. The mission's epitaph on him was written by Goldie: 'Much was expected of him as a Christian ruler, nor was such expectation altogether disappointed, but taking his father's place, he was surrounded with temptation addressed to the carnality still within him and he fell'; *Calabar and its Mission*, p. 208.

3 For the place of the Ogboni in the Egba constitution, see Dr. S. O. Biobaku's 'Ogboni, the Egba Senate' in *Proceedings of the C.I.A.O.*, Ibadan 1949 (International West African Conference) Lagos 1956, pp. 257–64; 'An Historical Sketch of Egba traditional authorities' in *Africa*, vol. 22, pp. 35–49, 1952.

in the country; it is necessary that the Native Christian Church should maintain its high position of witnessing for the truth by a broad separation from this and all other questionable 'country fashions'.[1]

Because the missionaries failed to convert the rulers, because no single mission captured the 'citadel' of any one state, there was a tendency to fall back on Waddel's argument that the Gospel would find its own way in spite of local politics. They argued that in any case the pagan way of life that the rulers stood for was 'crumbling away' and must fall when the principles of Christianity prevailed in the country. In the American Baptist missionary's words,

> As Christianity advances, common sense will advance and a better form of civil government will naturally and gradually result without any political interference on the part of the missionaries.[2]

All the political activities necessary were those to ensure liberty of action in the Mission House.

The symbolic structure of the Mission House, like the cells of civilization Buxton had hoped for, began to spread in the country and to make its influence felt in many a town and village. In many of the out-stations, the influence must have been small indeed: the Mission House was hardly distinguishable from any other building, except perhaps that two or three children might sometimes be seen chanting the A B C on the verandah, or might be heard singing a few hymns with the Evangelist and his wife most evenings and mornings. The light of Christianity and civilization that many of the mission agents were capable of showing was often no more than a flicker that after a year or two might go out and not for a long time be relit. But in the larger towns the Mission House was unmistakable. Its very physical appearance was calculated to impress, to be a model by which the standard of housing in the community was to be improved. The influence of some of the earliest buildings like Freeman's at Badagri or the Presbyterians' at Calabar on

---

1 Minute of the Parent Committee on the Ogboni System, dated 23 Nov. 1861; C.M.S. CA2/L3. Gollmer discovered in 1861 that his schoolmaster, J. King, had joined Ogboni. Further enquiries revealed that 'nearly all' the schoolmasters were members and 'all the young men at Ake' had been admitted too; King to Snaith, Abeokuta, 5 Oct. 1861; C.M.S. CA2/061. King argued that 'Ogboni in this country and chiefly among the Egbas is nothing but Civil Constitution or Political Community or, in other words, African Freemason[ry]. If not a member, when Oro is out, you all have to keep in. If caught 'unwittingly' you have to join to avoid trouble.'

2 Bowen to Taylor, 7 Sept. 1851; *Bowen Letters*.

local architecture was negligible, as the material was imported.[1] Gollmer at Lagos had planks made for him locally, but he built a house like the imported ones in the 'colonial' style—a large, airy box standing on stilts, with balconies all round and a prominent flight of stairs in front.[2] But as the missionaries moved inland, their houses literally came down to earth and they built out of local materials large, commodious houses which the people could imitate.

One of the earliest duties of many missionaries in the pioneering days, besides engaging porters, was to recruit labour for their building operations. Sometimes, as at Calabar, this was given them free, but when at Abeokuta they were offered free labour on condition that in the traditional communal way they feasted the workers, the missionaries reckoned that it would be more convenient, and perhaps cheaper in the end, to pay wages.[3] Considering that by 1855 the C.M.S. had five mission stations in Abeokuta and the Methodists and Baptists one each, work on mission buildings must have become a minor occupation for many women and children. Local masons were in demand too, for they were often found to be abler craftsmen than emigrant masons where building with mud and clay was concerned. Thus in the shaping of walls the earlier Mission Houses in the interior had little that was new. It was not until the late 1850s and the 1860s, when brick-making was introduced, that the masons from Brazil and probably also Cuba began to influence architecture. It was the carpenter and the sawyer, or rather their tools, that first impressed the people. The large windows and doors, planed and fitted, in place of the traditionally carved ones, soon began to be popular. So were boxes, tables, benches, even coffins, produced in the mission carpentry sheds. At both Abeokuta and Calabar the missionaries ordered tools from England and recruited more sawyers and carpenters from Sierra Leone, not only for the service of the mission but for the local rulers as well. Besides Sierra Leone emigrants, two Liberians, J. C. Vaughan and R. Russell, followed the Baptist Mission to Ijaye in 1855 and soon had prosperous establishments[4] as carpenters.

The greatest problem in building was the roof.[5] The people roofed in

1 For a description of Freeman's house, see above, p. 31. African merchants in Calabar had of course been importing pre-fabricated houses from Liverpool before the missionaries arrived.
2 For Gollmer's house, see *Charles A. Gollmer: His Life and Missionary Labours in West Africa*, by his eldest son, London 1886, pp. 70, 144.
3 Townsend, Journal entry for 25 Sept. 1846: 'Our paying wages was a new thing in Abeokuta and many are anxious to know how we shall be able to meet what they suppose to be a wasteful expenditure of money'; C.M.S. CA2/085.
4 Fragment of Bowen's Journal for 1855 in *Bowen Letters*; R. H. Stone: *In Africa's Forest and Jungle*, New York 1899, p. 129.
5 Reference to problems of roofing are scattered, but cf. Crowther to Venn, 6 March 1855: 'I have more than four times narrowly escaped being burnt

grass or a special type of broad leaves, both of which, if kept in good repair, made excellent roofing, their greatest merit being that they kept the houses cool. But practically every dry season there were outbreaks of fire which might consume as much as a third of a town, and if a store of gunpowder was involved, wider havoc might occur. The only defence against this was to keep the most important possessions in a room with mud ceiling and no windows, likely to escape damage even if the roof was burnt. It was also necessary to smoke the roofs to keep them resistant to decay and free from vermin. The missionaries, who considered windowless rooms and smoked roofs as unhygienic, had to embark on experiments to devise alternative roofs that were as durable, cool, resistant to rain, and yet inexpensive. They failed to make tiles to stand the local weather. The iron roofs that traders in Fernando Po and Lagos were introducing were intolerably hot for living houses besides being heavy for transporting by head far into the interior and being more expensive than most missions could afford. The missionaries tried as fantastic experiments as rolls of felt laid over with several coatings of coal tar. But the problem of roofing was not settled till the coming of the corrugated 'iron' sheets made at the end of the century.

The Mission House in the towns by no means stood alone. There were out-houses for schoolmasters, interpreters, boarders, redeemed slaves and other refugees, as well as carpenters' workshops and other industrial establishments. Nearby were the church and the school. The emigrants came to build their houses near the Mission House, and converts as well were encouraged to come to build there to escape either from actual persecution or from the temptations of life in the family compounds. Thus the missionary was showing models not only of European building but also of town planning. We saw Freeman at Badagri and Gollmer at Lagos doing some town planning. The emigrants in Calabar settled on the Mission Hill at Duke Town. At Abeokuta Townsend said there was

a strong tendency for the Christians to group together. There are advantages seen and felt by themselves which impel them to it. We have not a model village nor anything like it, but Christians build houses as near to our stations as they can. I obtained land and gave it

out of both my church and house; now I think we have had a sufficient warning to prepare ourselves fire-proof roofs', and he went on to discuss some of the experiments the merchants both in Lagos and in Fernando Po were trying out. Freeman at Cape Coast was the greatest experimenter of them all. One of the earliest missionaries to adopt the iron roof was Hinderer at Ibadan in 1854. In 1866 Townsend said the cost in England of an iron roof for a house was £100, £121 for the Church. 'Tiles would be also hot and so slate: in truth there is no material so good as grass but for the fires'; to Venn, 27 July 1806; C.M.S. CA2/085.

away in building plots, obtained more and more, but we have more applications for lots than we have land. The Sierra Leone traders gravitate towards missionary stations and white men [i.e. traders] do the same.[1]

There were in fact three distinct settlements of emigrants and converts, called *Wasimi* (lit. come and rest), near the C.M.S. missions at Ake and Ikija, and the Methodist mission at Ogbe respectively. There were similar C.M.S. and Baptist settlements at Ijaye, and the same was true, though to a less extent, at Ibadan.[2]

The question then arises as to what was the relationship of these mission villages symbolized by the Mission House to the rest of the community. Townsend was anxious to insist that they were not 'Christian villages'. Christian villages were a Roman Catholic device made famous by the work of Jesuit Fathers in Paraguay, where the Fathers made mass conversions of Indians and established them in little theocratic states ruled absolutely by the Fathers. There was only one attempt to establish anything like it in Nigeria, and it deserves some notice here.

In 1863 Father Broghero, the pioneer Catholic missionary at Whydah, paid his second visit to Lagos. He referred to empty spaces along the coast and added:

If the missionaries were to gather together in these places a small Christian population and if they were to be the directors of the temporal as well as the spiritual affairs of the community, they would soon form a little Christian State which would become the example and the refuge of the scattered flock.[3]

1 Townsend, Annual Letter for 1865, dated 1 Feb. 1866; C.M.S. CA2/085.
2 See J. A. Maser: Map of Abeokuta October 1867; C.M.S. CA2/068. 'Most of the people under the care of Mr Champness live together making a little village which we hope will be the centre of light to the surrounding population'; Meth. *Missionary Notices*, vol. XVI, p. 210, 25 Nov. 1861. Mann, Journal entry 13 June 1856, 'I had some men engaged in cutting and clearing a bush on a piece of ground where I intend to build a house for converts. . . . Besides this lot, much was left to me by Are which I now parcelled out for 7 small compounds'; C.M.S. CA2/066. Hinderer to Venn, 4 Jan. 1861, that in the Missionary Party there were about 70 people including 8 African families, boarders etc.; C.M.S. CA2/049.
3 Father Broghero, dated Whydah 21 Dec. 1863, in *Annals of the Propagation of the Faith*, 1865, pp. 81–2. Besides the article referred to here, the principal sources for this account of Topo are: (A) Three contributions in a collection edited by Father L. Arial in 1921 of Reports and historical notes on the principal stations of the S.M.A. in Nigeria, called *Missions de la Nigéria*, viz: (i) '*Stations de Topo-Badagry*' by Father Ariel; (ii) '*Rapport de Topo-Badagry*' by Father L. Freyburgher with notes by Father Pages; (iii) '*Fondation de Topo d'après le P. Poirier*' (B) A collection called *La Mission de Topo* of

Owing largely to the disturbing effect on the Société des Missions Africaines of the 1870 War in France, nothing was done about it for over a decade. Then, in 1875, application was made through the Chief Justice, James Marshall, a Catholic convert from Protestantism, to the Governor of the Gold Coast (who was also responsible for the administration of Lagos) for permission to acquire a piece of land nine miles long on the land between the lagoon and the ocean just east of Badagri. The purpose was declared as

> the foundation of an agricultural establishment for raising the standard of agriculture so necessary in a colony, and so little developed in this part of Africa.[1]

The approval did not come till 1880, but an establishment was made in 1876 called St. Joseph's, Topo. It was described in an article in 1881 as an agricultural orphanage. This aspect of it will be discussed later,[2] but in spite of the letter to the Governor, Topo was more than an agricultural school. In 1888 Father Bel, the Superior of the station, said that besides the boys directly in the charge of the Fathers,

> We admit on the land of the mission families who wish to put themselves under the rules we have imposed. They cultivate the land for their own profit except for a little rent paid in kind. Their children must be baptized and brought up in the Catholic Faith. When they grow up, we see to their progress. For this reason we give them in advance a plot of land to cultivate, and when they have been sufficiently instructed to be able to live without the supervision of the mission, they get married[3].

In that year, there were still only 33 such families, 5 Catholic, 2 Protestant, 26 Pagan. They all attended mass every Sunday, and Sunday school. In addition there were, at varying distances from Topo itself,

---

documents relating to the land dispute, fuller on in-coming letters than on out-going ones. Both these are in the S.M.A. archives in Rome. (C) Articles in *Les Missions Catholiques*, cited elsewhere. See also Father M. J. Walsh: *Catholic contribution to Education in Western Nigeria (1861–1926)*, London M.A. thesis, 1953; chapter V, 'An early Agricultural Experiment', deals with Topo.

1 Father Cloud to the Governor of Gold Coast, 4 Oct. 1875, cited in Notes by Father Pages on the history of Topo, op. cit.: '*C'est la fondation d'un établissement agricole pour favoriser le développement de la culture si utile dans une colonie et si méconnue dans cette partie de l'Afrique.*'

2 See below, pp. 141–2.

3 Father Bel to Father Planque in *Les Missions Catholiques*, 21 Dec. 1888. Father Walsh, op. cit., p. 115, says that they paid three heads of each product every three months and had to clear part of the bush and keep it clean.

three villages for refugees from the surrounding country to whom the Fathers were doing their best to minister. Further away still, 30 families from Ajido on the opposite shore were allowed to settle on mission land provided they conformed to the rules—no Muslims, no idols, no dancing.

Besides being a sanatorium and a plantation, Topo also attempted to be 'a little Christian State'. But neither the local communities nor the British administration in Lagos could tolerate the Fathers wielding the powers implicit in the idea of a Christian village—much as the zeal and enterprise of the Fathers in promoting agriculture[1] was admired. There were disputes over boundaries, over the right to pick palm-fruit, which the Egun said was inalienable. Father Landais, a hard, zealous, brutal man,[2] who succeeded Father Bel in 1888 made the prosperity of Topo surpass expectation, but he began to evict the Ajido families who had come to regard the place as their home. The rights of the mission to evict them were therefore called in question. The Lagos administration stepped in. Governor Moloney, himself a Catholic, declared

> that His Excellency conceives your mission to have no higher title to that estate than the Ajidos to the farms of which they have been recently so summarily dispossessed by your mission. His Excellency has however no intention or desire to exercise to the full or indeed at all the rights of the Government as your mission's landlord, in relation to that estate.[3]

The controversy did not end with the government's dubious claim to be landlord, and it dragged on for over a decade. Long before the end of that period, the Catholics themselves had begun to ask whether the idea of a Christian village did not do the mission more harm than good, and they realized that in any case it was not politically feasible in a country not under the rule of a Catholic government. Besides Topo, there were one or two cases of persecuted Christians moving out in a body to found small villages of their own, one near Asaba in the late 1880s, and earlier on in the 1870s Egba Christians led by John Okenla colonized the villages of Shuren, Ofada and Okenla, between Abeokuta and Otta.[4] But no attempt was made to claim political independence.

1 The Fathers tried a few things including dairy farming, but soon realized that the fortune of Topo lay in coconut.
2 Father Freyburgher, in 'Rapport de Topo-Badagry', said Father Landais worked *'au milieu de beaucoup de difficultés dues pour une bonne part à la raideur de son caractère. On peut cependant se demander si le Père eut rèussi dans son travail s'il avait eu un caractére plus conciliant.'*
3 Governor Moloney's private secretary to Father Pellett, 29 Oct. 1888, cited in *Notes sur les Missions de Topo.*
4 Shuren, Ofada and Okenla were farm villages where, following the expulsion

Protestant opinion was traditionally against the idea of a Christian village, physically separate from and politically independent of the old community. With the Protestant villages in Nigeria there was not much physical separateness. There were many emigrants and converts who continued to live within the old society but were no longer part of it. It was the attitude of not belonging that characterized the Mission House, the attitude described by Bowen himself as 'our unbending foreign customs' which made the people 'feel that we are aliens'.[1] Attempts to assert independence were, however, not wanting.

When the Secretary of the Presbyterian Foreign Mission Committee detected, in 1848, a desire on the part of some of his missionaries to claim the right of asylum for the Mission House, he said it savoured of Roman Catholicism:

> The principle involved in this seems to me a dangerous one, liable to be greatly misunderstood and abused. It is the principle which in the palmy days of popery made the clergy demand exemption from the operation of the Civil Power. . . . Missionaries cannot interfere with the civil administration of a country any further than teaching what is right. Their office is instruction.[2]

In other words, the Mission House was under the civil authority of the local rulers. That put the missionary in the dilemma of obeying a civil authority he considered and declared ungodly. Waddell thought he solved the situation by saying that the local civil authority, however crude, must be regarded as rulers of their people, Christian or unconverted, 'though we should not regard them as our rulers'.[3] A few months later, Anderson, from the same mission, was writing to the same Secretary of the Foreign Mission Committee: 'Is it nothing to

of missionaries from Abeokuta, many followers of John Okenla, Townsend's protégé, settled. Rev. James Johnson said in 1877 that 'the proportion of Christians in them (130 out of 400, 120 out of 400, and 20 out of 500 respectively) is [too] small compared with the Heathen and Mohammedans to entitle them to the name of Christian villages'; Report on the State of Churches in Yorubaland 1877, C.M.S. CA2/056. In 1890 Idigo, a chief of Aguleri near Onitsha, 'seeing how difficult it would be to practise the true religion in the midst of pagans resolved to withdraw to a place three-quarters of an hour from the village, where he had a property on a lofty plateau. At the same time he offered to give land to all who would follow him provided they became Christians'; Father Lutz in a letter dated Onitsha 6 Jan. 1892, in *Annals*, 1892, pp. 252 f.

1 Bowen to Taylor, 11 April 1855; *Bowen Letters*. Bowen adds: 'Why then am I doing here just as others do? Because I cannot labour alone and have no hope of seeing men who will do as I desire to do.'
2 Somerville to Hugh Goldie, 20 Nov. 1848; U.P. *Secretaries' Letter Book*, vol. I, p. 211.
3 Waddell, Journal entry for Tuesday 4 Dec. 1849; *Journals*, vol. VII, p. 95.

encourage the hearts of the members of the United Presbyterian Church that they have erected and are maintaining "a city of refuge" for the innocent in this land of blood?'[1] In short, how far each Mission House was a 'city of refuge' depended to some extent on the theology of each missionary, but to a greater extent on the ability of the local rulers to resist European visitors tempted to take the law into their own hands.

There were missionaries who were wont to regard themselves as above local laws and the Mission House as a 'Zion' and a 'stronghold', whose independence they were willing to maintain by force, if possible. In the first year of the Methodist Mission in Badagri, in 1844, Annear prepared a military defence of the Mission House to protect a slave rescued from his owner. 'As war was now proclaimed against us,' he wrote later,

> by a people who were before professedly our warmest friends, and in the midst of whose town we lived, all hands were busily engaged preparing for the threatened attack. A large fire was kindled in the yard, around which one party were busily employed casting bullets, another sharpening their swords, while a third examined the muskets and another ransacked the store for cartridges, ten rounds of which were appointed to every man.[2]

Next morning, though the expected attack did not come, Annear decided to yield up the slave.

There were a number of other missionaries who confused rudeness to people, rulers and gods alike, with courageous zeal. In 1856 Mann was trying to get a mission established at Awaye against the opposition of Muslims and pagan priests. He quarrelled with them all, caused offence to the local rulers and priests by disparaging them, and was apt to play his harmonica and ask his boys to sing at the tops of their voices when the Sango festival was at its height, or Oro was out at night. He gained the distinction of being one of the few missionaries in the country ever to suffer personal violence.[3]

Edgerley, at Old Town, Calabar, was like Mann in his ill-controlled temper and superior attitudes. In 1849 in a fit of temper following a

---

1 Anderson's Journal entry for 31 May 1849, in U.P. *Missionary Record*, 1850, p. 25.
2 Annear, Journal entry for 1 Nov. 1844; Meth.
3 Mann, Journal entry for 22 Sept. 1856 at Awaye. He received blows from Sango worshippers at Ijaye, 17 Oct. 1856; Journal entry for that day, and Mann to Major Straith, 23 Jan. 1857; C.M.S. CA2/o66. Also at Oke-Odan in October 1863. Since he was introduced to Oke-Odan by an agent of the Lagos administration, Mann's action involved the C.M.S. in correspondence with the Colonial Office; T. F. Elliot, directed by the Duke of Newcastle, to Venn, 16 Nov. 1863, and enclosures in C.M.S. CA2/L3.

minor conflict, he broke the Ekpe drum at the local Town Hall; he later apologized after one of King Eyo's dinner parties.[1] In 1854 the ruler of Old Town died, and Edgerley reported that there was human sacrifice at his funeral. He was so eager in demanding the pursuit and punishment of the culprits that he gave the initial pretext for the destruction of Old Town, ordered later by Acting Consul Lynslager, by bombardment from H.M.S. *Antelope* in January 1855. In fact the bombardment and burning of the town were opposed by the missionaries (including Edgerley) on the grounds that it would only make the work of reform more difficult and that 'native instrumentality and co-operation are indispensable to native moral reformation.'[2] The deed was really the contrivance of the traders and supercargoes, who thought a show of British force would help them in their dealings with Duke Town and Creek Town. The traders persuaded Lynslager to report that he was acting on the invitation of the Mission, encouraged also by their own assurance that 'there is no trade whatever carried on in it . . . the total destruction of that place would be of great benefit to the other towns, to the advancement of civilization.'[3]

While the Foreign Mission Committee, faced by angry jibes in Parliament and in the Scottish Press, were trying to establish the fact that the missionaries had opposed the destruction of Old Town, and were sending deputations to the Foreign Office to have the consular ban on the rebuilding of the town lifted, another missionary, Anderson, took the cause of the independence of the Mission House in Calabar a step further. In November 1855 he gave asylum in Duke Town Mission House to two men and a woman accused of having caused the death of a little boy by *ifot* (i.e. preternatural powers) and therefore called upon to undergo trial by ordeal. In January 1856 Hutchinson, the new consul, who was sent to say that Old Town could be rebuilt if the people would observe the law against human sacrifice, supported Anderson and agreed to regard the refugees as being under his protection. The upshot of this interference with the demands of local justice[4] was that an Ekpe ban was placed on the Mission House forbidding anyone to go there or send their children there or in any way have anything to do with the missionaries or with the emigrants who lived on the mission land. At

1 Waddell, Journal entry for 4 Dec. 1849; *Journals*, vol. VII, p. 95.
2 Waddell, Journal entry for 16–19 Jan. 1855, in U.P. *Missionary Record*, 1855, pp. 206–11.
3 Consul Lynslager, Journal of Proceedings at Old Calabar, October 1855; FO 84/975.
4 As recorded by Anderson in pidgin English, the local demand was: 'When man kill man with Freemason, he must chop nut.' Journal entry for Saturday 14 June 1856, in U.P. *Missionary Record*, 1856, p. 155.

the market people refused to sell to them. On Sunday nobody came to church. When Anderson approached them, they ran away 'as if they had seen a spectre'.[1] If he knocked at the door, he was told nobody was in, or that he should please go away and not put them in trouble. But in due course a naval force (H.M.S. *Scourge*, with commodore and consul aboard) induced Duke Ephraim to lift the ban on the Mission by the threat of blowing up Duke Town, as Old Town had been blown up a year before.

A few missionaries in the interior might have shared Anderson's conception of the Mission House as a city of refuge, but, lacking Anderson's dubious advantage of a naval force to maintain it, they had to proceed on a different principle. Even Anderson grew older and wiser and came to see that personal regard for and friendship with local rulers, unconverted though they were, did the cause of the mission more good than enforcing the theory of a Mission House that was above the law. At Lagos the missionaries had little interference to fear from Dosunmu and had only consular authority and traders to contend with. At Abeokuta, Townsend initiated the tradition of personal friendship with the rulers as the way to secure the objectives of the mission; but although the central government was weak, there were always the authorities in the townships and local war-chiefs to discourage the idea of the Mission House as a sanctuary.

At Ijaye, Ibadan and Ogbomoso, the governments were more powerful still. Once or twice missionaries there smuggled out young women persecuted by their husbands to the protection of their colleagues at Abeokuta or the consul at Lagos,[2] but there was never any doubt that they acknowledged the protection and authority of the local rulers. As the Parent Committee of the C.M.S. told missionaries going into different parts of Yoruba in 1856:

> The Committee cordially approve of the wise respectful deference which has been shown by the missionaries to the authority of the Native Chiefs and they enjoin a like conduct upon all who enter the Mission. Many of the modes of exercising authority may appear at first in the eyes of an European absurd; some of their governmental institutions are associated with idolatry. Nevertheless they are the framework of society and till they are replaced by a more enlightened system, they must be respected. This respect need not involve any

1 Anderson, Journal entry for Sunday 1 June 1856, in U.P. *Missionary Record,* 1856, pp. 151–8.
2 E.g. Crowther to Venn, 7 Feb. 1856, announced the arrival in Lagos of two young Ibadan female converts escaping from the anger of their husbands; C.M.S. CA2/031.

compromise of the great principles of justice and humanity or of the personal independence of the missionary.[1]

The truth was that this acknowledgement did not destroy the large measure of independence that the missionary and the Mission House enjoyed. For where the missionary was not deliberately provocative, the areas of friction between the Mission House and the State were limited. Attempts to convert individuals and make them show contempt for tradition, or to remove them from the family compound were bound to be resented by the family. But this was not always the case. Many converts were able to fight and win their battles with their families without calling on the aid of the missionary or seeking refuge in the Mission House. In some cases, little or no conflict followed conversions. Many converts were strangers in the community.[2] Others were outcasts, various victims alike of chance and of the prejudices of men. Some, on the other hand, were wives or children of respectable families who were assigned by the head of the family to the Christian God in token of friendship to the missionary, or as one assigned some members of the family as devotees, this one to this god that to the other, one to Sango, another to Ogun, a third to Obatala, thus ensuring the goodwill of the gods at different points, in which case the new 'eccentricities' of the Christian devotees were tolerated without trouble. Persecution might follow later, or the whole family might be converted.

Those who, failing to convert the state as a whole, converted families as units are to be reckoned among the most successful missionaries. There were notable examples in Ibadan and Ogbomoso. Hinderer was the most humane of men, with a balanced, cultivated mind, and a sense of humour that Ibadan people highly appreciated. With him, religion was not an excuse for destroying human values, but for ennobling them. By his friendly disposition as a man he made friends with two families, one to whose care he was entrusted at Kudeti, the other at Aremo. These two families have been the pillars of the Church in Ibadan. They have given Christianity roots in the society and supplied in later days most of the clergymen and the two bishops Ibadan has produced.[3] William

1 'Instructions of the Parent Committee to those about to join the Yoruba Mission, 21 Oct. 1856'; C.M.S. CA2/L2.
2 Cf. Hugh Goldie: 'We have a large foreign population in Calabar [i.e. of converts] gathered in from about fourteen different tribes, to whom Efik is a foreign tongue. We have more members of the Mburutism tribe, the country of which we do not yet know, than we have of Calabar people.' In 1876 a Paper presented to the Gaboon Conference, Printed Report, op. cit., p. 12.
3 *Seventeen Years in the Yoruba Country*, memoirs of Mrs. Hinderer, C.M.S. 1872, and the romance based on it: *Swelling of Jordan* by Ellen Thorpe. 'Beginnings of missionary work in Yoruba' written by a member of Hinderer's

Clarke, the Baptist missionary at Ogbomoso, was a man like Hinderer. His great success was the conversion of the Agboola family at Oke Afo, who similarly have given the Baptist Church a sure foundation in society and many pastors and leaders of the Church.[1] Such conversion limited friction between the missionary and the community.

The other important source of friction between the Mission House and the state was the number of lawsuits involving mission agents, emigrants, or converts who sought the protection of the missionaries. These were liable to lead the missionary to repudiate the law or to criticise court procedure. Some missionaries went further, like the medieval clergy, and claimed jurisdiction not only in cases between members of the mission village but also in all cases involving any member of the mission village.[1] This was almost invariably resisted, but the rulers usually showed such deference to the wishes and convenience of the missionaries as hospitality demanded.[1] Above all, the missionaries enjoyed the widest possible discretion in the internal administration of the Mission House and village. For although they despised the family compounds and signified this by the rectangular arrangements of the houses and streets in the mission village, they were looked upon as having created family compounds of their own. Like other heads of families or Houses, they were held responsible for regulating their affairs subject to the highest tribunals of the particular community.

The missionaries jealously guarded this right of internal administration, even when they doubted their ability in terms of law and power

---

household and incorporating some of the oral tradition in the Ibadan Church; Nigerian Record Office, ECC 20/1.

1 'Notes on the Beginnings at Ogbomoso', among the collection of local histories of Baptist churches edited by Rev. Cecil Roberson for the centenary of the Baptist Church in 1953.

2 E.g., the case of Isaac Smith, a church member and a servant of the missionary, the Rev. J. J. Hoch, at Abeokuta in 1856. He was accused of having poisoned a man on his farm and was summoned for trial by the Bashorun. The C.M.S. missionaries and church elders suspected that it was a plot of the Muslims to bring the church into disrepute. They therefore declared they would not give him up unless they found him guilty. They set up a tribunal and conducted their own inquiries. They found Isaac 'innocent of poisoning, but not innocent of having imprudently provoked [the deceased] by offending words'. The Bashorun pressed to have Isaac sent for trial; the missionaries appealed to the Alake, who tried him and fined him 40 heads of cowries; Hoch, Journal entry for 31 June and 25 Aug. 1856; CA2/050. See also Gollmer, Journal for March 1851; CA2/043; though Gollmer insisted that Muslim emigrants could not enjoy the benefits of ecclesiastical jurisdiction.

3 E.g., Reid said that at Oyo the Alafin maintained the principle that 'the white man and his people must not be forced to obey country [i.e. local] customs'. He treated both the C.M.S. and the Baptist missionaries 'with great respect'; to Poindexter, 31 Aug. 1860; S.B.C.

to enforce their will. The threat of ecclesiastical sanctions was not always enough to make everybody in the mission village obey the missionary, who quite often was a young man in his twenties or thirties. The control of the missionaries, though firm enough on new converts, was never strong enough on the trading community of emigrants, some of whom were either not Christians or only just on the fringes of the Church or, like the emigrants at Calabar, did not belong to the particular denomination of the missionary,[1] but whose distinct status in society the missionary nevertheless considered essential to the progress of civilization and Christianity. The missionary, however, had other sanctions besides the ecclesiastical. He could threaten to commit the culprit to the local rulers, whose scale of punishment was likely to be harder to bear than the missionary's. He could on the other hand send him to the nearest consul or report him on the consul's next visit, since it was not generally known that the government had declared that emigrants not in a British colony were not British subjects, or that until 1872 the consuls had no judicial powers.[2]

In this way some rude sort of justice, partly ecclesiastical, partly civil, was maintained. The converts took their petty quarrels to the class leaders. The emigrants had recognized headmen who acted as magistrates, imposing fines, and, at least in Badagri and Abeokuta, had regular prisons. Serious cases involving mission agents or prominent people went to the conference of missionaries of that particular denomination, or were tried by special tribunals. Where situations cropped up that existing church regulations or known 'European Law' did not seem to have envisaged, particularly suits concerning marriage and divorce in which local law and church regulations and concern for social well-being merged into a legal jungle, the missionaries sought advice from home or just used their discretion.

By and large, the independence of the Mission House even in the interior where there was no naval force to maintain it was real and, from the point of view of the missionaries, satisfactory. So real was it that some missionaries in the Yoruba country came to consider it absolute. They began to analyse the basis of it and to see in it a vindication of the

1 The emigrants from Sierra Leone were either Anglicans or Methodists though most of them did attend the local Presbyterian Church; see Anderson to Hutchinson, 17 June 1856, encl. in Hutchinson's despatch of 24 June 1856; FO 84/1001.
2 Though the slave-trade treaties provided a legal basis for bombarding the coastal states, the consul had no legal powers 'to oblige the British supercargoes stationed in the Rivers to obey any particular code of Trading Regulations'; FO draft to Hutchinson, 12 March 1856. For this reason, many emigrants in Lagos welcomed annexation, which turned the consul into a governor who had power over Africans and Europeans alike.

policy of discouraging emigrants and converts from seeking social and political importance within the states. The states, they said, would always defer to the wishes of the European missionary when necessary, not because of the naval force on the coast, or because of the demands of hospitality to a disinterested man who had travelled from so far, but because he was a European and the African himself was the first to acknowledge the European as his superior.[1]

'I must testify,' said Townsend in June 1853,

> that the chiefs as far as I have had any intercourse with them have shown anything but a disposition to persecute. They have assisted me in every case of domestic persecution most willingly and as far as I can judge from their conduct in other respects, shown a growing attachment to Europeans and confidence in them. . . .
>
> We ought to be thankful to the Lord that we possess an influence over the natives such as causes them to submit quietly to an interference that no other than their chiefs could exercise, more especially as we have no legal right as these people are not British subjects.[2]

As long as the uneasy peace established in the interior continued, there was no occasion to test the validity of these arguments, and the missionaries settled down to their preaching and the fostering of the arts of civilization among the community of the Mission House.

1 The most explicit statement of this view at this stage was in a letter to Major Straith dated Oct. 29 1851, in Townsend's handwriting, signed by Isaac Smith, Townsend, Hinderer and Gollmer CMS CA2/016.
2 Townsend, Journal entry for June 10 1855; CMS CA2/085.

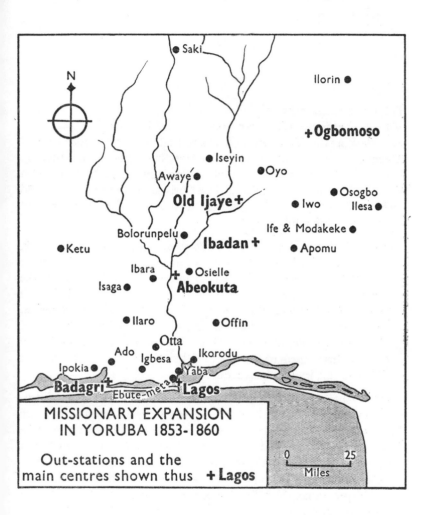

MISSIONARY EXPANSION
IN YORUBA 1853-1860

Out-stations and the
main centres shown thus **+ Lagos**

Map labels: Saki, N, Ilorin, **+Ogbomoso**, Iseyin, Oyo, Awaye, **Old Ijaye+**, Osogbo, Ilesa, Iwo, Ife & Modakeke, Bolorunpelu, **Ibadan +**, Apomu, Ketu, Ibara, + Osielle, **Abeokuta**, Isaga, Ilaro, Offin, Otta, Ado, Igbesa, Ikorodu, Ipokia, Yaba, **Badagri**, Ebute-meta, **+Lagos**

0   25
Miles

5 Civilization Around the Mission House

THE missionaries reiterated the arguments in favour of the indirect methods of evangelization from time to time as if to reassure themselves and to combat the undercurrent of evangelical distrust of mission agents engaging in activities other than preaching, baptizing and ministering. Struck by the high rate of European mortality and sickness in West Africa, and haunted by the memory that Christianity had been introduced once into West Africa and had left little or no trace behind, Bowen argued again in his book published in 1857 that

> Our designs and hopes in regard to Africa are not simply to bring as many individuals as possible to the knowledge of Christ. We desire to establish the Gospel in the hearts and minds and social life of the people, so that truth and righteousness may remain and flourish among them, without the instrumentality of foreign missionaries. This cannot be done without civilization. To establish the Gospel among any people, they must have Bibles and therefore must have the art to make them or the money to buy them. They must read the Bible and this implies instruction.[1]

Three aspects of this programme should now be noticed: that introduction of literacy, the training of missionary agents, and the fostering through technical education of a class of people 'with the art to make Bibles or the money to buy them'. The latter two were never free from controversy, but there was complete unanimity about the great importance of the first. Nothing shows the ardour of the pioneering missionaries better than the effort devoted, within the limited resources of the missions and the ability of the missionaries, to the study of the principal Nigerian languages, reducing them to writing, in most cases for the first time.

This study began in Sierra Leone where, as we have seen, several of

1 Bowen: *Missionary Labours and Adventures in Central Africa*, op. cit., pp. 321 f.

126

the languages were represented among the Liberated Africans. As early as 1830–32, the Rev. J. T. Raban of the C.M.S., obesrving that the 'Akus' were fast becoming a majority in the colony, began a study of Yoruba with a view to facilitating evangelization within the colony. When arrangements were being made for the Niger Expedition and a mission was projected for the model farm at Lokoja, the Rev. J. F. Schön was charged with the duty of training interpreters and himself acquiring the languages he considered most essential. The languages he chose were Hausa and Ibo. For the same purpose Samuel Crowther intensified his study of his own language, Yoruba. The results of these studies were published in 1843.[1]

The other missionaries in the Yoruba country were studying the language along with Crowther, comparing translations and discussing the orthography. In the winter of 1848, when Gollmer and Townsend were in England to give evidence before the Hutt Committee, the debates were carried to London. Venn secured expert advice from linguists such as Professor Lee of Cambridge, Edward Morris, Secretary of the Royal Asiatic Society, Max Müller of Oxford and, most notable of all, Professor Carl Lepsius, of Berlin. As a result of these discussions, Crowther published in 1852 a revised and enlarged edition of his *Grammar and Vocabulary*, as well as translations of four books of the New Testament.[2] It was Lepsius' orthography that continued to guide C.M.S. linguistic studies, be it in Hausa, Kanuri, Ibo or Ijaw.

The American missionary, Bowen, acquired a high degree of proficiency in Yoruba and among other things drew attention to the poetic excellence of the invocative prayers of traditional Yoruba worship, especially those of *Ifa*. But his *Grammar and Dictionary of the Yoruba Language*, containing a collection of sayings and poems, remained outside the main stream of development since it was honoured with publication along with three other unrelated works in one of the fat volumes of the *Smithsonian Contributions to Knowledge* in New York.[3] Crowther and Thomas King were the chief translators of the Bible and the Prayer Book; David Hinderer, the mission's Hebrew scholar, supervised the translation of the Old Testament and translated *The Pilgrim's Progress*, and Gollmer the Watt's Catechism and Carl Barth's *Bible Stories*.

1 J. F. Schön: (i) A *Vocabulary of the Hausa Language with Grammatical Elements pre-fixed*, 1843; (ii) *A Vocabulary of the Ibo Language*, 1843; S. A. Crowther: *Grammar and Vocabulary of the Yoruba Language*, 1843. See also J. F. Ade Ajayi: 'How Yoruba was Reduced to Writing', *Odu, Journal of Yoruba Studies*, 1961.
2 S. A. Crowther: *St Luke, Acts of the Apostles, St James and St Peter* in Yoruba, C.M.S. 1851; *Vocabulary of the Yoruba Language*, 1852.
3 T. J. Bowen: 'Grammar and Dictionary of the Yoruba Language' in the *Smithsonian Contributions to Knowledge*, vol. IX, part iv, New York 1862.

The orthography of Yoruba is today substantially that laid down by the missionaries. Their rules of grammar have been frequently criticized, but their translations are still recommended as works of high literary value.

In reducing a language to writing for the first time, Crowther said, particularly with a view to translating religious works, the most essential and most difficult thing was to get behind colloquial speech and work-a-day slang, with its preference for phrases in place of the rare meaningful words. For this reason he befriended pagans and Muslims alike, 'watched the mouth' of the elders and, while discussing theology and other serious matters with them, noted down 'suitable and significant words'. When he tried such words in common speech, he found that, like 'thrown away words', they sounded stale, but 'to the rising generation, they will sound sweet and agreeable'.[1] He went everywhere with pencil and paper—even up the Niger, at Lokoja and Rabba, where he claimed to have picked up useful Yoruba words. In December 1862 he had just returned to Lagos from the Niger and was paying a courtesy visit to the Governor when his house was burnt down. 'I had always made it a rule,' he wrote to Venn,

> that in case of a fire breaking out, not to hesitate but to snatch out the manuscripts of my translations the first thing, for security, and then I may try to save anything else if possible; but on this occasion I was not at home to put my resolution to practice. . . . Thus the manuscripts of nearly all the remaining books of the Pentateuch which I would have prepared for the press this quarter were destroyed. My collections of words and proverbs in Yoruba, of eleven years' constant observations since the publication of the last edition of my Yoruba vocabulary, were also completely destroyed. The loss of those is greater to me than anything else, in as much as it cannot be recovered with money nor can I easily recall to memory all the collections I had made during my travels at Rabba and through the Yoruba country, in which places I kept my ears open to every word to catch what I had not then secured, with which I had expected to enrich and enlarge my Yoruba vocabulary this year. Now all are gone like a dream.[2]

1 Crowther: Journal Extracts for September 1844; C.M.S. CA1/o79. He added: 'In tracing out words and their various uses, I am now and then led to search at length into some traditions or customs of the Yorubas.' The results of these researches on topics like *Egungun* and *Ifa* which he gave in some papers in 1844 have, as far as I know, never been noticed. They will be found in C.M.S. CA1/o79.
2 Crowther to Venn, 12 Dec. 1862; C.M.S. CA3/o4.

128

As a result, the revised dictionary was not published until 1870. Interest in Hausa and Ibo revived with Henry Barth's travels in Northern Nigeria and MacGregor Laird's mail contract to ascend the Niger by steamer in 1854. Attention was also paid to Kanuri. A brilliant German missionary of the C.M.S., the Rev. S. W. Koelle,[1] in 1854 published two works: *Grammar of the Bornu or Kanuri Language* and *African Native Literature in Kanuri*. Koelle worked through interpreters and it is doubtful whether he ever became fluent in Kanuri himself. But P. A. Benton, the authority on Kanuri in the British Administration, said in 1916 that Koelle's works were 'of wonderful accuracy and interest'.[2]

In 1855 Barth took to England two boys from Northern Nigeria, Abbega, a Margi, and Dorgu, a Hausa, aged about 16 or 17, each of whom spoke Hausa and Kanuri.[3] In 1856, they went to live with Schön, who had by then become a chaplain to a Naval hospital in Kent. Through them, Dorgu especially, Schön was able to revise his long-delayed works, and he published in 1857 a *Primer* and in 1862 a *Grammar of the Hausa Language*. He kept up a regular correspondence with Hausa-speaking missionaries on the Niger and he was allowed to use the extensive papers of Dr. Baikie, who led the third expedition up the Niger in 1857 and remained in Lokoja till 1864, making notes and observations and translating the Bible. From these he obtained material to enlarge his earlier vocabulary into the dictionary[4] which, published in 1876, remained the standard work till the end of the century. Owing to economic development and growing political interest, Schön began to cater for other than missionary needs. In 1877 he published a *Hausa Reading Book and Traveller's Vademecum*, one of the earliest uses of which was to train officers to command the Hausa corps that since the Ashanti War had become a principal instrument of British policy in West Africa. Besides translations from the Bible, he published in 1885, as a

1 S. W. Koelle, also author of the *Polyglotta Africana, or a Comparative Vocabulary of nearly 300 words and phrases in more than 100 distinct African Languages*, London 1854. When a naval officer discovered the only known indigenous African script it was Koelle who was sent to study it. As a result of this he published *Narrative of an Expedition into the Vy Country etc.*, 1849, *Grammar of the Vy Language*, 1854.
2 P. A. Benton: *A Bornu Almanack for the Year A.D. 1916*, p. 57, 'A note on Rev. S. W. Koelle'. Also by the same author: *Primer of Kanuri Grammar translated and revised from the German of A. von Duisburg*, Oxford 1917, Introduction.
3 For the story of Abbega and Dorgu see 'Chapter I of the Life and Travels of Dorgu dictated by himself' in appendix to Schön's *Grammar of the Hausa Language*, 1872; A. H. M. Kirk-Green: 'Abbega and Dorgu' in *West African Review*, September 1956, p. 865 ff.
4 Schön, in the introduction to *A Dictionary of the Hausa Language*, 1876.

more advanced reading book, a collection of texts called *Magana Hausa*.[1]

When Schön concentrated his attention on Hausa, the responsibility for the study of the Ibo language passed to Crowther in spite of his other commitments. He took up Schön's primer of 1843, and in 1855 recalled Schön's interpreter—Simon Jonas, who was left with the Obi of Abo in 1841 and had since been in turn Christian Instructor and Tailor to the Obi and policeman at Fernando Po—through whom he learnt some Ibo and began compiling a vocabulary.[2] In 1857 he was relieved by the Rev. J. C. Taylor, who volunteered for the Onitsha Mission. Taylor was born in Sierra Leone of Ibo parents who did not both speak the same dialect of Ibo. When he arrived in Onitsha he still had to preach through an interpreter,[3] but he quickly learnt enough Ibo for his normal duties as a missionary, though not enough to make the contribution to the study of Ibo that Crowther was making to that of Yoruba. He received little of the expert advice Crowther got, nor had he the leisure of Schön in retirement. In addition, dialect variations were a more serious problem in Ibo—by no means resolved even today— than Crowther had to face in Yoruba.

In 1861 Taylor, having missed transport up the Niger on returning from Sierra Leone, founded the new mission at Akassa. This meant adding the Ijaw language to his studies. In 1866 he completed a translation of the New Testament[4] into Ibo which he sent to England for publication. The C.M.S. Secretaries submitted it for an opinion to Schön, who was critical, and it was returned to Taylor for revision. This made him feel 'entirely disheartened and discouraged', if not angry.[5] He left the mission in 1868 and returned to Sierra Leone. For similar reasons, the works of Crowther himself on Ibo and Nupe, the Rev. C. Paul and Henry Johnson on the Nupe language, P. J. Williams on Igbirra, though useful beginnings, were far from being definitive.[6]

Besides Yoruba, Hausa and Kanuri, the missionaries had a notable success with the Efik language. There the policy of concentration yielded abundant fruit. Before Hope Waddell set out in 1846 he had collected a

1 Rev. G. P. Bargery: *A Hausa-English Dictionary and English-Hausa Dictionary*, O.U.P. 1934 for the Government of Nigeria, p. xv.
2 Crowther to Venn, 9 Dec. 1855; C.M.S. CA2/031.
3 Rev. J. C. Taylor, Journal entry for 2 Aug. 1857 in S. Crowther and J. C. Taylor: *Gospel on the Banks of the Niger*, C.M.S. 1859, p. 249.
4 Taylor to Colonel M. Dawes, 21 Nov. 1866; C.M.S. CA3/037. The other mission agents presented him with an address on the occasion.
5 Taylor to Venn, 16 April 1867; C.M.S. CA/037.
6 Crowther, S. A.: *Grammar and Vocabulary, Nupe Language*, 1864; *Vocabulary of the Ibo Language*, S.P.C.K. 1882. Johnson, Archdeacon H. A., Christaller, Rev. J. L.: *Vocabularies of the Niger and Gold Coast*, S.P.C.K. 1886.

list of Efik words from the Liverpool supercargoes and he began at once, along with the other agents, to memorize them and try to simplify and systematize the orthography. The results of these early studies were published in 1849 as the *Vocabulary of the Efik Language* in the names of Waddell and the irascible printer of the mission, Samuel Edgerley. Waddell had a clear, practical mind but he was no scholar or highly literary man. It was Hugh Goldie, who worked with him and succeeded him at Creek Town, who soon became the authority on Efik. He published in 1862 his *Principles of Efik Grammar and Specimens of the Language*, and translations from the New Testament into Efik the following year. But when the Foreign Mission Committee wished to prosecute more energetically the work of translation and Efik orthography, they sent to Calabar a younger man, a brilliant scholar, Dr Robb of Aberdeen, famous for his Hebrew. In 1866 Dr. Robb published his translations from the Old Testament.[1] But his health did not stand up well to Calabar and he was transferred to Jamaica. Goldie's ability blossomed out. In 1874 he published the *Efik Dictionary*, the *Efik Grammar in Efik* and the *Efik Grammar in English*. These works dominated the studies of other missionaries, who produced translations, primers, readers, hymns and sermons; and indeed to this day they remain the standard works on the language.

The driving force behind the work on the Nigerian languages was the anxiety to teach the converts and would-be converts to read the Bible in them. For this reason, some missionaries argued that priority should be given to teaching adults rather than children, on the grounds that

> It is the adult population that show a willingness to hear us and to receive the word we preach. . . . To the same extent that the adult population are brought under Christian instruction will children and other dependents be brought under instruction likewise.[2]

Indeed, the strong evangelical influence in the missionary movement placed great premium on the Sabbath school for teaching adult converts and catechumens who could not come to school daily during the week to read the New Testament for themselves. It was specially for their sake that so much emphasis was placed on translating the Bible into the vernaculars; for their sake, too, that throughout the work on languages the emphasis was on simplicity of orthography rather than academic perfection. As a C.M.S. Secretary said:

> All marks [of orthography] not indispensable cannot fail considerably

1 U.P. *Missionary Record*, 1866, pp. 198–200.
2 Townsend, Journal for June 1857; C.M.S. CA2/085.

to increase the difficulties in the way of a native's acquiring the art of reading, and to teach them to read is our great aim.[1]

Sometimes the Sabbath school opened before the vernacular literature was ready and a beginning had to be made none the less. The Rev. J. C. Taylor describes the first Sunday school at Akassa, at the Nun entrance of the Niger, before either he or Crowther knew many words of Ijaw. They attracted forty-seven people into church—men, women and about a dozen children.

> Mr. Crowther took the first class at the head of the table in the centre of the room, a capital place for him, with his venerable, silver-bound spectacles, a [rod] in his hand, pointing to the phonetic alphabet characters, calling out loudly the well-known letters a, b, d, e . . .[2]

The people stood mute, watching him, because they did not know what to do; but as soon as Crowther found out the Ijaw for 'repeat together' and he could say 'a, *be-be-hie*', they threw away reserve and began to imitate him

> in the pronunciation of those wonderful characters which will in due time be beneficial to them and would not fail of preparing them to read the Word of God hereafter for themselves.

Meanwhile Taylor stood at the door, enticing more of the people who stood outside to enter, shouting to them in Ijaw: '*Ebi diri ebima*', i.e. 'Good white man's book is the best' . . . '*Ebi! Ebim! Ebima! Aa, beke diri ebima!*', i.e., 'Good! Better! Best! Yes, Englishman's book is the best'.[3] Soon after that, Taylor made friends with Koko, a local trader who spoke the best English at Akassa, and together they began composing an Ijaw primer. Unfortunately Koko died a few months later and the mission fell under a cloud of suspicion.

Besides the Sunday school, one or two stations where adult literacy was taken most seriously ran evening classes during the week.[4] Some others had reading lessons during their catechumens' classes. In these different ways the majority of the early converts, who by the nature of things had to be very keen and zealous, did learn to read in the vernacular at least some portion of the Bible, usually St. Mark's Gospel, telling the story of the life of Jesus, or the Catechism setting out basic doctrine, or a few of the most popular hymns. Some of the churches have main-

---

1 W. Knight to Crowther, 23 Dec. 1852; C.M.S. CA2/085.
2 J. C. Taylor: Journal entry for Sunday 22 Dec. 1861; C.M.S. CA3/037.
3 Ibid.
4 E.g. at Abeokuta in 1858; Townsend, Journal Extracts for April to September 1858; C.M.S. CA2/085.

tained this tradition of adult literacy. The literacy was limited both in aim and achievements. Few of the adult converts were bothered about learning to write—so that they could hardly read handwritten script—in contrast with the gentlemen of Calabar and Bonny, who spoke and wrote English but as they were not familiar with the printed script had specially to learn to read the printed Bible—nor were they asked to learn to read or write English. In this lay the most important difference between the adults' and the children's schools. For in the teaching of children the missionaries were obliged to cater for the demand which prior to their arrival existed in the country for the knowledge of the English language, measurement and accounting for purposes of trade.

It was not a demand for general education as such. The trading chiefs who wanted missionaries to teach children English had their own way of bringing up their children to fit into life in the family compounds and the states. They imparted moral and religious education, with clear precepts reinforced by taboos. They gave training in the etiquette and conventions of society; they trained the minds of the children as they taught them to count yams and ears of corn, or to give answers to the conundrums, or to repeat in their own words the fables of the family history.[1] In the moonlight the children played games and told stories and learnt alliterative verses. As they grew older they were apprenticed to jobs or initiated into the further mysteries of life. There was little system, but the parents looked on it as education. What they expected from the European was not a substitute but a supplement, a system of apprenticeship by which the children acquired additional arts and skills, the art of reading and writing, gauging palm-oil or manufacturing gunpowder or sugar or building boats. As the Bonny chiefs said, when they did get a school and had to pay for it,

> They did not want religious teaching, for that the children have enough at home; they teach them that themselves; that they want them to be taught how to gauge palm-oil and the other mercantile business as soon as possible.[2]

Indeed, they wondered if three months was not enough to learn 'all book'. They would have thought little of a school that did not attempt to teach some English. At Calabar, said Goldie in 1876,

> so great is the desire to learn English that though it also is taught in

1 Cf. Crowther on the educational value of fables and proverbs in the Charge delivered to his clergy in 1869; C.M.S. CA3/04.
2 Crowther: 'Brief Statements exhibiting the characters, habits and ideas of the Natives of the Bight', 1874; C.M.S. CA3/04.

our schools and taught gratuitously in Duke Town, some of the chiefs engage at a high fee the services of any young man who may come in their way to teach their children only.[1]

The missionaries welcomed this demand, such as it was. They knew that it was one of the principal reasons why they were welcomed and allowed to settle in the city-states on the coast. They saw in schools 'the nursery of the infant Church', the principal hope for the success of their work. If most of the adults were too much wedded to the ideas of their fathers, the children, whose minds were as yet unhardened, should provide more fruitful ground for the sowing of the seed of the new religion. 'Preaching hath ever been the great ordinance,' said orthodox Evangelical doctrine. 'You must seek to convert the heart before you can instruct the mind.'[2] But the experience of many missionaries was that preaching to adults, particularly the scoffing, sceptical trading communities on the coast, was like sowing by the wayside or on the rock.[3] But let children come to school for any purpose whatever and it would be the fault of the missionary if he could not take advantage of the opportunity and make Christian converts of the children.[4] The problem was how to make the children come in sufficiently large numbers, how to make the demand for some form of European training more general on the coast, and how to create it in the interior, where it hardly existed except perhaps for the making of guns and gunpowder.[5]

1 *Report of the Conference of West African Missionaries held at Gaboon, Feb. 1876*, printed by the Mission Press, Calabar, p. 11; copy in C.M.S. CA3/013.
2 Cf. Henry Venn to William Marsh, an African catechist, 17 April 1847: 'Remember that it is to affect the heart that you must chiefly aim. It will be easy to inform the understanding when the heart is inclined to listen.' C.M.S. CA2/L1.   3 Cf. Crowther, Journal for July 1847; C.M.S. CA2/031.
4 Cf. the 'official' Baptist view at home in America: 'The only means committed to us for the conversion of men is the preaching of the Gospel with simple reliance on the power of the Holy Ghost; the Gospel is to be addressed to men and women as of old, and not merely to children, for God is able to convert the parent as well as the child . . . teaching schools may be a benevolent work, but no more the work of a missionary abroad than that of a minister at home.' Several of their missionaries arrived with this preconceived notion and, gradually becoming disillusioned at the prospects of converting adults, turned to the children, e.g. J. M. Harden, 4 May 1858, to Poindexter, *The Commission*, July 1858: 'Brethren, I tell you again that I have *no* hope of the parents; my hope is in their children.' R. H. Stone to Culpepper, 9 July 1858: 'I am fast coming to the conviction that *schools for the rising generation* must be the *basis* of all missions among *barbarous* and *savage* heathen. The Gospel should be preached regularly and steadily, faithfully and prayerfully; but through the children we get at the root of idolatry and leaven the whole lump.'
5 Cf. George Meakin, European catechist at Oyo, 1858–59, Journal entries for 15–20 June 1858. The first boy the Alafin offered for the mission school came on condition that he be taught to make 'snuffs, guns, powder, etc.'; C.M.S. CA2/069.

The response in places like Calabar, Badagri and Lagos was encouraging from the start. There were inducements, of course. The schools were free; the children received gifts from Europe—clothes, copy-books, slates, pencils and so on. At the annual public examinations, when the school was dressed up and shown off to the public, prizes were liberal. Each Christmas they had feasts, and on suitable occasions they had parades to show other children what fun they were missing. In the interior such inducements yielded some results, especially from the emigrants and their relatives and those who took quickly to European trade. The rulers began to see the advantages of having their children as clerks to write letters for them, but on the whole the response was poor. Some of the parents argued that if they were going to be deprived of the services of their children on the farms, they should be paid for it. Mann, at Ijaye, asked the C.M.S. for funds for such a purpose, on the grounds that his colleague of the rival Baptist mission was paying his pupils.[1] Other parents complained that the schools taught disrespect to elders and tradition.[2] At Abeokuta, Townsend said it was the children themselves who did not like school and he suggested that parental control was never strong enough to keep the children at school against their will. A more likely reason might be found in the inquiry which the Ake church later instituted into the causes of the deterioration in health of children used to open air being confined between benches all day long in schools which had everything in the curriculum except physical exercise.[3] Onitsha boys took a dislike to that aspect of school life from the very start. When Crowther opened the first school there in December 1858, fourteen children came regularly, all girls about six to ten years of age.

> The boys . . . like to rove about in the plantations with their bows and bamboo pointed arrows in their hands to hunt for birds, rats and lizards all day long without success; but now and then, half a dozen or more of them would rush into the [school] house and proudly gaze at the alphabet board and with an air of disdain mimick the names of the letters as pronounced by the schoolmaster and repeated by the girls, as if it were a thing only fit for females and too much confining to them as free rovers of the fields. But upon a second

1 Mann, Journal entries for 21 April 1856, 14 Aug. 1859; C.M.S. CA2/o56. A. D. Phillips to Taylor, 25 Jan. 1859, certainly asked for funds to pay school children living at home 2 to 3 cents a day; S.B.C.
2 One of the best informed and most informative documents on the attitude to education in the interior will be found in James Johnson's 'Report on Abeokuta Churches', 30 Jan. 1878, sheets 8 and 9; C.M.S. CA2/o66.
3 J. Johnson: 'Report on Abeokuta Churches', 30 Jan. 1878; C.M.S. CA2/o56.

thought, a few of them would return to the house and try to learn a letter or two.[1]

A few of them did settle down, of course, but farm work, particularly in the dry season, made their attendance very irregular. This was a difficulty the schools had to contend with everywhere. In Waddell's school at Creek Town in 1854 there were 120 names on the roll. The average attendance in July was sixty-eight, in August seventy-eight, in September eighty-one, in October seventy-five, in November fifty-four, in December forty-seven.[2] When, in November 1861, a missionary visited Isaga, an outstation to the west of Abeokuta (destroyed by Dahomey in 1862), there were nine children present, seven absent, 'among them all the best boys'. The four most advanced pupils present were all girls.[3]

The only inducement effective against irregular attendance and the premature withdrawal of pupils from school was to persuade the parents to allow them to be brought up by the missionary in his own household. For this reason the boarding-school became a regular feature of the Mission House.[4] It was hoped that from among the boarders, in close personal relationship with the missionary, the most advanced pupils, the monitors and future teachers as well as the most pious pupils, the future leaders and Pastors of the Church, would arise. The hope was often justified. Boarding out children was not new in the country. Many an indulgent father chose to send his beloved son to a trusted relative or friend for training. In the same way, missionaries who were found to like children or who made friends easily soon had children entrusted to their care. Sometimes it was calamity, not friendship, that produced this. A person in need of ready cash went to the Mission House to borrow, and in return gave his children to the missionary as a 'pawn', partly as security for the loan, partly also as interest on it. At Ijaye, where eight years of missionary endeavour had failed to bring many children to school, the war with Ibadan in 1860–62 suddenly flooded the Mission Houses of both the C.M.S. and the Baptists with children, with no conditions attached except that they should be fed and kept secure. The Baptists evacuated to Abeokuta about seventy children. The C.M.S.

1 Crowther to Venn, 2 Dec. 1858; C.M.S. CA3/04.
2 Waddell, 'Report on Creek Town Mission for 1854'; *Journals*, vol. X, pp. 130–1.
3 Dr. A. Harrison to Venn, 28 Nov. 1861; C.M.S. CA2/045.
4 Cf. Sarah M. Harden to Poindexter, 6 Aug. 1859, S.B.C. that in Lagos nobody, not even emigrants, would send children to a day school. 'Whoever wishes to instruct them must feed and clothe them too, for their idea is that it is a great favour shown to us when we are permitted to teach their children.'

at Abeokuta organized an Ijaye Relief Fund, which enabled Mann to evacuate 33 children.[1]

The boarders were in fact personal wards of the missionaries. How many each kept depended on his ability to organize private funds, as the missionary societies did not themselves allocate funds for the purpose. But in the pioneering days there were always people willing to contribute to such a cause. Personal friends and relatives of the missionaries in Europe, members of their old churches, individual humanitarians, Sunday school associations and missionary groups in different churches, perhaps as far away as Canada or Jerusalem, enabled nearly all the earlier missionaries to keep up large households. A lady in Brighton, Miss Barber, organized what was called the 'Coral Fund' specifically to enable C.M.S. missionaries to keep boarders at the rate of about £3 per child per annum. In 1851 Townsend said he had ten Coral Fund boys; in 1860, twenty-six boys and girls of various ages.[2] In 1863, Dr. Harrison, the C.M.S. Medical Officer, whose wife was an accomplished lady famous for her sewing and embroidery classes, had fourteen Coral Fund girls aged between 12 and 20.[3] In 1864 Taylor had six Coral Fund boys at Onitsha. In the 1880s the Association for the Propagation of the Faith had a similar fund to enable Roman Catholic missionaries to redeem slaves, secure 'pawns' and educate them as interns who, in the words of Father Broghero, would be

rescued from the midst of paganism, and kept safe within our fort, [to] lead a perfectly safe life, as well-regulated as any within the walls of a convent in a Christian country, to keep the Church in good order, serve at the altar, and sing the sacred canticles, assist in the religious instruction of other children and serve as interpreters.[4]

The life and duties of the boarders varied, of course, from mission to mission and even more from missionary to missionary—from the household of genial people like the Hinderers[5] to the hard school of Topo, with the reputation of an approved school—but they were similar.

1 R. H. Stone: *In Africa's Forest and Jungle*, 1899, pp. 185 ff. He had about 70 children. Mann of the C.M.S. began by criticizing the Baptist policy of taking pawns (19 Oct. 1860 in a letter to Venn), but by October 1861 he had 33 boarders where before the war he had but two; Mann to Venn, 2 Oct. 1861; C.M.S. CA2/066.
2 Townsend, conclusion to Journal Extracts for June 1851; Annual Letter for 1859 dated 30 Jan. 1860; C.M.S. CA2/085.
3 Harrison to Venn, 28 Sept. 1863; Mrs. Jane Harrison to Venn, 30 Dec. 1863; C.M.S. CA2/045.
4 Broghero to Planque, 21 Dec. 1863, in *Annals*, 1865, pp. 81–2.
5 For the household of the Hinderers see the memoirs of Anna Hinderer, *Seventeen Years in the Yoruba Country*, 1872, and the book based on it, Ellen Thorpe: *Swelling of Jordan*, 1950.

Discipline and hard work were the keynotes. The records are full of the endeavours of the missionaries to keep the morality of their wards up to standard. Harrison was not a particularly morbid man, but all the clandestine amorous dealings of his girls aroused his interest. They were not allowed to go to wash clothes at the brook 'as the company at these washing streams is so bad'. They were kept away from their mothers who were thought to be trying 'to keep their daughters down to their bad old ways'. And when a conference of missionaries had sat upon a case of alleged misconduct between one of the girls said to have been 14 years of age and a boy of 16, the rules were only further tightened up.[1] Yet in spite of such absurdities and seeming harshness many a boarder has left on record testimonies of his gratitude for spiritual and material benefits received from missionaries.

With the boarders ensuring some regular attendance at school, the missionaries gradually built up a pattern of primary education at practically every mission station. There was, of course, no system in the pattern that emerged, no common syllabus, no general inspectorate. At Calabar in the early days the Presbyterian missionaries, like the Catholic Fathers and Sisters later, were themselves teachers, assisted by the emigrants.[2] In the other missions, where the proportion of European missionaries was smaller and the few missionaries there were needed to take time off for open-air preaching, the emigrants were directly in charge of teaching and the missionaries only supervised. Efficiency in the schools varied widely. Everything depended on the ability and zeal and personal whims of the individual missionary or teacher. But the schools had the common aim of propagating the ideals of Christianity and some of the basic doctrines of the particular denomination while teaching literacy and a little arithmetic to the children. Therefore the usual curriculum consisted of the four R's: Religion, Reading, Writing and Arithmetic, with sewing for girls where there was a lady teacher.[3] At the larger mission stations, as the school progressed and in

1 Mrs. Harrison to Venn, 30 Dec. 1863; Harrison to Venn, 19 Dec. 1863; C.M.S. CA2/045.
2 Waddell said that one of his West Indian schoolmasters was unhappy at not being put in charge of the school, and he added, 'If he expected to have sole charge of the school, he knew neither me nor the importance I attached to it'; Waddell to Somerville, September 1848, in draft in Waddell's *Journals*, vol. viii, p. 30.
3 'Reading, Writing, Arithmetic, Sewing and Religious Instruction constitute the branches of education which so far we have taken up'; Hope Waddell in U.P. *Missionary Record*, 1848, p. 146, A Circular Letter to Friends of the Mission. Sarah M. Harden, Baptist, said she would teach Reading, Writing, Sewing and Knitting. She intended to begin Grammar and Geography later but 'I fear that they will not understand Geography without maps'; to Poindexter, 6 Aug. 1959, S.B.C.

place of one or two began to have four, five or more classes, this curriculum was soon elaborated by the addition of subjects like Grammar and Geography. The time-table sent out by Freeman in 1848 to the head teachers of the schools under his management may be summarized as follows:[1]

> 9.0 a.m.: Singing, Rehearsals of Scripture Passages, Reading one chapter of Scripture, Prayers.
>
> 9.15–12 noon: Grammar, Reading, Spelling, Writing, Geography, Tables [except Wednesday, when there was Catechism in place of Grammar].
>
> 2.0–4.0 p.m.: Ciphering [i.e. Arithmetic], Reading, Spelling, Meaning of Words.
>
> 4.0 p.m.: Closing Prayers.

This was more or less repeated every day except Friday, which was devoted to rehearsals of Scripture passages, revision and examinations. Girls followed a similar curriculum, but with important changes. In the afternoon session, from Monday to Thursday, they had Sewing and Embroidery, and therefore made up on Tuesday morning for the arithmetic, spelling and meaning of words they missed. On Wednesday, Bible Reading and Catechism occupied not just one period, but the whole morning session.

The first question that arose was what language were the children to be literate in. For most missionaries the demand by the people themselves for English was decisive.[2] Moreover, when the first mission schools were established at most of the large centres, the majority of the missionaries had still to learn the local languages, which were being reduced to writing. The emigrant schoolmasters had themselves been brought up on English, which they saw as 'the language of commerce and civilization', the road to success and advancement. Many, though not all, of the European missionaries themselves shared this view. The children at Creek Town, said Waddell in 1848,[3]

> are taught in English, not merely from necessity on our part, nor solely because some knew our tongue a little and all wished to learn

1 Freeman, 'Rules for Schools', 1848; Meth.
2 Among the Catholic Brazilian emigrants in Lagos, Portuguese and Spanish were the earliest languages of instruction. But English soon superseded these.
3 Waddell, in U.P. *Missionary Record*, p. 146. Cf. Dr. Harrison to Venn, 29 May 1862, that the arrival of three new Europeans to Abeokuta 'will give a considerable impetus to the progress of the English language which seems *of itself* to raise the person who is acquainted with it in the scale of civilization' ('of itself' was added as an omission, not italicised, in the original); C.M.S. CA2/045.

it, but also from a conviction of the great importance . . . of promoting among them the knowledge of our own language.

Broken English, he said, was already spoken along the coast from the Gambia to Gaboon:

> By the aid of missionaries and schools, [English] may be made the common medium of communication, yea, the literary and learned language of all Negro tribes as the Roman language was to the modern nations of Europe while yet the modern European languages were in an infantine and unwritten state.

He added that the cultivation of the native Efik language would, however, not be neglected,

> as it must ever continue to be the principal means of communicating oral instruction to the hundreds of thousands, perhaps even millions . . . who may never be able to acquire a knowledge of English.

English was the language of commerce and civilization; the vernacular, as much as possible, was the language of religious instruction. As the work on the languages progressed and vernacular literature was produced, reading and writing in the vernacular were introduced, at least in the junior classes. Although generally English remained the language of instruction, by the 1850s many missionaries were insisting that Religious Knowledge should be taught mainly, if not solely, in the vernacular, which the children most readily comprehended. Thus while on Freeman's time-table subjects like Grammar, Spelling and the Meaning of Words undoubtedly referred to the English language, and nobody thought of writing textbooks in the vernacular for subjects like Arithmetic or Geography, as soon as they were available the vernacular Bible and Catechism tended to supplant the English original. By 1854 a typical day in Waddell's school was something like this:[1]

9.0 a.m.: Prayers.

9.15–10.0 a.m.: Arithmetic for the seven different classes.

10.0–11.0 a.m.: A few verses of the scriptures taught in Efik and repeated by all. Then Efik reading and spelling lessons in the different classes—'during which period every person separately repeats to his class teacher the verse previously given out by me'.

11.0 a.m.: Roll call. Absentees enquired after. Tickets given to the most worthy of each class.

11.30 a.m.: Prayers. [Break]

3.0 p.m.: Prayers.

1 Waddell, 'Report of Creek Town Mission for 1854', op. cit.

3.15-4.0 p.m.: Writing (except Wednesday and Friday, when the two highest classes had Geography).

4.0-5.0 p.m.: Scripture verses as in the morning. Then Reading of English in the different classes—books ranging from a primer to the Bible.

5.0 p.m.: A lesson in the Calabar Catechism. Hymn, roll call, tickets.

5.30 p.m.: Prayers.

The most evident omission from this curriculum, besides physical exercise, which was hardly ever mentioned, was manual labour. There was sewing and embroidery for the girls, but nothing for the boys. The obvious choice of a manual labour subject would have been agriculture. But the parents, who were not anxious for schools except for the purpose of equipping children with new skills for trade, would have been most difficult to convince of the value of agriculture as a form of education, nor could the missionary anxiously trying—with inadequate success—to persuade children to forsake the farm and come to school make out a convincing case for putting agriculture on the curriculum. He did not even try. Many missionaries, of course, had farms on which the boarders, refugees and other residents in the Mission House worked after school hours. When the American Civil War cut off the funds of the Baptist missionaries, they considered ways and means of making their boarding-school self-supporting, largely as an alternative to closing it down. In June 1864 the older generation were 'put out on business on their own', leaving only 35 children from a total of about 70. By August 1865 classes were suspended; the girls were passed over to the C.M.S. and the boys set to work on the farm. Later the girls were taken back, and by June 1866 a regular pattern was established of boys working on the farm in the morning while the girls sewed or cooked; classes were held for 4-5 hours when it was too hot to farm, and the boys returned to the farm in the evening while the girls again cooked or sewed.[1]

But it was only at Topo that an attempt was really made to combine agricultural work with primary education. As we have seen,[2] it came later in the century, and was much more than an agricultural school. The section of it dealing with the training of children was an orphanage rather than a school. It soon acquired the reputation of an approved

1 The school was the main preoccupation of A. D. Phillips between 1860 and 1867, and the main theme of his letters. See Phillips to Taylor, 5 Oct. 1860; 5 Nov. 1860; 6 Nov. 1860; 4 Dec. 1860; 5 July 1863; 4 Feb. 1864; 4 June 1864; 29 Aug. 1865; 2 Feb. 1866; 4 May 1866; 1 June 1866; S.B.C.
2 See above, pp. 114-16.

school, with which mothers frightened naughty little children, and indeed James Marshall, Chief Justice of Lagos, used to send juvenile delinquents there as an alternative to imprisonment. Few fathers would willingly send their children to it except as punishment. Most of the children there were redeemed slaves or pawns recruited from Whydah, Porto Novo, Abeokuta and other mission stations. The daily routine began quite early in the day with mass and catechism. Then the boys went to work on the farms till midday. After lunch they had school from about 1.30 p.m. to 4.30 p.m. and they went back to work till dark.[1] In the 1890s, when girls were brought in, they had classes in the morning, and spent the afternoon gathering coconuts or making *gari*. Father Landais, the Superintendent almost continuously from 1888 to 1916, drove the boys so hard that a colleague, Father Vanleke, accused him of using them like slaves.[2] Agriculture was pursued at Topo not as part of the normal education for children, but partly for profit, partly as reformatory training, and, especially at the beginning, essentially as the only industry round which pagan families could be collected in the attempt of the Fathers to found a 'Christian village'. What was important at Topo was that the pagan recruits should be willing to obey rules once they got there—they were invited to come with their agriculture; the essence of the typical nineteenth-century mission village was that the members had already broken away from the old life, of which agriculture was the basis. In the new life of the Mission House there was to be, not idleness, but new techniques and arts to promote commerce and civilization—processing the agricultural products of the old town, for example, printing, carpentry, masonry, shoemaking and so forth. Of these only sewing and embroidery were considered suitable for primary schools; the others required basic literacy, arithmetic and maturity; they were therefore to be acquired in secondary industrial schools and training institutions.

It should be emphasized that, contrary to many assertions about missionary work in this period, the missionaries were far from fostering idleness in the mission schools, or even mere literary academic education; nor were they complacent about or indifferent to the social effects of the education they were giving the children. The mission school was conceived of as a process for drawing away children physically into the

1 'Notes sur les Missions de Topo', p. 29 f. Father M. J. Walsh: *Catholic Contribution to Education in Western Nigeria*, pp. 120–2, argues that it was an agricultural school, classified as such as an 'Industrial school' later on by the Education Department in Lagos, though he agrees that the children at Topo were mostly slaves and pawns.
2 Father L. Freyburgher: 'Rapport de Topo-Badagry'.

mission village or at least mentally and spiritually away from the family compounds. One result of this was that inevitably the children tended to regard themselves as better than their mates and elders, who did not belong to the new life of the Mission House. In a sense the missionaries thought they were, in so far as they were 'regenerated' and the others were not. But the consequent feeling their charges had of belonging to a superior caste worried the missionaries: it was dangerous for the children, dangerous also for the missionary cause, as it did not encourage parents to send more children, and, in any case, pride and indiscipline were certainly not Christian virtues. Townsend constantly inveighed against 'the pride of dress and caste'.[1] Bowen once said that 'children raised in the schools are vagabonds',[2] and Freeman once posed the question to the Gold Coast District Meeting,

> whether it is wise to educate so large a number of children as the mission has been doing with the almost certain prospect that the greater number of them will be thereby . . . rendered not only useless members of society but injurious to its well-being on account of their instrumentality in the diffusion of habits of idleness and extravagance.[3]

The important point here is that it was almost generally to the 'habits of idleness' that the unpleasant results were traced, and almost invariably manual labour was the solution offered. Only occasionally was the inefficiency of several of the schools blamed. Hardly ever was the dichotomy they created in society referred to. All the time the panacea that was urged was a practical approach to education in place of the purely literary and academic, much in the spirit of the founders and organizers of the Poor Man's and the Workmen's Institutes in Britain.[4] If anything, the doctrine of the virtue of industriousness was re-emphasized in Africa, as the new civilization that was being built could not be raised on idleness, and years of propaganda in the New World had convinced some of the missionaries that the African was a loafer who would not work unless compelled by the lashes of a slave-driver or the

---

1 Townsend, conclusion to the Journal Extracts for June 1851; C.M.S. CA2/085.
2 Bowen to Taylor, 2 Oct. 1856; *Bowen Letters*.
3 Minutes of the 1848 District Meeting, Cape Coast; Meth.
4 Cf. Mabel Tylecote: *The Mechanic's Institutes of Lancashire and Yorkshire before 1851*, Manchester 1957; R. K. Webb: *The British Working Class Reader*, London 1955. Both authors show how in fact the Institutes tended to become social and cultural—'exhibition rooms of local vanity and drowsy essay-reading' (Webb, p. 64)—rather than practical. But the emphasis of the founders like Henry Brougham and Dr. Birkbeck, and of the industrialists who financed organizations like the 'Society for the Diffusion of Useful Knowledge' was undoubtedly practical and utilitarian.

teachings of a new religion. 'Manual labour schools or none is my motto,'[1] said Bowen. 'The separation of scholastic life and manual labour,' said Henry Venn, in language more typically Victorian,

> is a refinement of advanced civilization. It may be doubted whether even in this case it is desirable; but certainly it is not desirable in a mission school or according to the example of the Apostle of the Gentiles.[2]

Finally, the continued repetition of this prescription to deal with just one of the many symptoms showing that the new education was not producing the desired effect defeated its own end. It did not cure the disease. Rather, it became a cloak under which to hide the inaction that resulted from lack of resources or lack of will.

The missionaries were moderately successful in their schemes of industrial training, where the main limiting factor was patently and admittedly the availability of money. Henry Venn, in accordance with his basic principle of training an African middle class for both Church and state, tried to draw together the available resources of humanitarians, government and traders towards a scheme of industrial training for some African youths. In 1845 he got some 'Friends of Africa'—Sir Robert Inglis, T. A. Acland, E. N. Buxton and others of the disbanded African Civilization Society—to form the 'Native Agency Committee'.

> Their object is described in the name: it is to encourage the social and religious improvement of Africa by means of her own sons.[3]

They were to send out, at their expense, European artisans to work in Africa in collaboration with the missionaries and to bring African youths to English factories and workshops to train, and to buy tools for them on their return. In 1851 they sent out a German 'mechanic', I. V. Huber, to help the missionaries in their building operations and to take charge of the industrial training of the more senior pupils in the schools. It was he who helped Captain Forbes to construct the stands on which to mount the canons at Abeokuta,[4] but he died within a year. This only confirmed the views of the Committee that the second part of their programme, of bringing a few selected Africans to train in England, was more important than sending out Europeans. In addition to the efforts

1 Bowen to Taylor, 2 Oct. 1856; *Bowen Letters.*
2 Instructions of the Parent Committee to Mr. W. Kirkham, school master, 29 Jan. 1856; C.M.S. CA/L2.
3 Venn to Robbin, 22 Dec. 1855. Also W. Knight: *Memoirs of Henry Venn,* 1880, p. 510.
4 Townsend, Journal Extracts for November 1851; C.M.S. CA2/085.

of the Native Agency Committee, Venn encouraged individual African merchants and mission agents to give their children practical industrial education in England. It should be emphasized that of all the boys known to have passed through Venn's hands in this way, only one went to England for a purely literary course—he was T. B. Macaulay, who went to the C.M.S. Training College at Islington and attended a few lectures at King's College, London. Another, Samuel Crowther, jun., after having been apprenticed to a doctor in Freetown, read Chemistry and Anatomy at the same College as part of a medical training. Two youths, Henry Robbin and Josiah Crowther, went to Thomas Clegg's factory at Tydesby, near Manchester, to learn to clean and pack cotton for the European market. Two others, Ellis and Wilson, went to Manchester to learn brick- and tile-making and building construction. Two were sent to Kew Gardens to study what new plants might be introduced into Africa; one learnt printing in London.[1] In addition, the Admiralty was prevailed upon to authorize the ships of the naval squadron to take on boys to train in navigation, so that they could become merchant ship captains. Thus at the time of the naval expeditions in Badagri and Lagos, two boys, James Davies and his brother Samuel, were on board Captain Coote's ship 'by the instruction of Commodore Bruce, for practical instruction in navigation and seamanship'. Coote felt that they did not come early enough to take easily to the sailor's life, and they stayed only fourteen months. But at the end of it, he considered them

> both capable of taking the necessary observations and obtaining the results of navigating a ship. They have done so in the ships regularly and I have placed confidence in their work. . . . I consider their abilities above the average. They have made fair progress in sail-making and rigging, but these two [arts] required an apprenticeship.[2]

The Native Agency Committee bought equipment for them, sextants and parallel rules, a spy glass and mathematical instruments. James Davies soon had charge of a Sierra Leonean schooner engaged in the coastal trade,[3] and later became Captain J. P. L. Davies of Lagos. The Commodores who succeeded Bruce were not interested in this training scheme. In March 1863, however, Commodore A. P. Eardley Wilmot

---

1 Information about this training scheme is scattered. The most useful single source is C.M.S. CA1/023. See also personal files of Capt. J. P. L. Davies, CA2/033; T. B. Macaulay, CA2/065; H. Robbin, CA2/080; S. Crowther, CA2/032; and Venn to Dr. Irving, 23 Aug. 1854, and instructions of the Parent Committee to Henry Robbin, 22 Dec. 1855, both in CA2/L2.
2 Commander R. Coote, H.M.S. *Volcano*, to Venn, 23 Oct. 1852; C.M.S. CA1/023.   3 J. P. L. Davies to Venn, 3 Sept. 1856; C.M.S. CA2/033.

revived it and took on four boys, Josiah Brown, Alfred W. Lewis, Jack T. Gibson and Francis M. Joaque, from the C.M.S. Grammar School in Freetown, and trained them at C.M.S. expense. They remained on board for over two years, at the end of which time the Commodore reported that

> their progress in navigation and mathematics generally has been very great, while in seamanship, in knowledge of the steam engine and other useful works, they have done exceedingly well.[1]

All of them soon had small vessels of their own to navigate except Joaque, who went into his father's trading business. Venn also tried to get the Admiralty to take on boys from the C.M.S. Grammar School in Freetown to send to England to qualify as surgeons for the naval squadron. Three are known to have been sent to Edinburgh; two qualified, and one of them, born in Sierra Leone of Ibo parents, Dr. James Africanus Horton, M.D., became famous on account of his published works.

The attitude of the founder of the Catholic Société des Missions Africaines was similar to Venn's. He insisted on the importance of training an African staff. Before his tragic death in 1859 at Freetown he had initiated a scheme by which one of his principal and most influential associates in Europe, the Abbé Papetard, was to open in a warm part of Spain a seminary for the training of African youths. His successor, Father Planque, did set up an S.M.A. institution for training in mechanical arts and trades, which started at an old convent outside Cadiz, moved to Puerto Real, and finally, after political changes in Spain created difficulties, settled at Bouffarick near Algiers. There, 12 Yoruba boys, brought to Wydah after their capture by Dahomey in 1862, were sent by Father Broghero and were joined by 12 more in 1865. By the end of 1867 the first of these boys were beginning to return to West Africa. They were trained in carpentry, shoemaking, masonry, tailoring, ironwork, cookery, and gardening.[3]

1 Commander A. P. Eardley Wilmot to Venn, 16 May 1865; Rev. Charles Chapman, chaplain and naval instructor, to Venn, 21 March 1864; C.M.S. CA1/023. Also *African Times*, 23 March 1865.
2 J. A. B. Horton to Venn, Dec. 1863; C.M.S. CA1/023. Horton: *West African Countries and Peoples*, 1868, *Letters on the Political Condition of the Gold Coast*, 1871. See also reference to him in two articles in *Sierra Leone Studies*: (i) J. de Hart: 'Memorial Tablets in St. George's Cathedral' (no. 11 Sgt.-Major J. A. B. Horton's tablet, no. 29 his wife's); (ii) June 1956, Dr. M. C. F. Easmon: 'Sierra Leone Doctors'.
3 This account is based largely on the work of Father Walsh, who has pieced the story together from various records, some of which, in particular those of the Jesuits at Bouffarick, I have not had the privilege of seeing. *Catholic Contribution to Education in Western Nigeria*, pp. 90 ff.

While Venn had expected many of his trainees to become independent churchmen outside the mission, the S.M.A. hoped that the Bouffarick boys would return to them and if possible become Lay Brothers. In this they were disappointed. By May 1869 it was said that only one third were giving satisfactory service to the mission. Augustin could build a wall better than any mason around, Benoit was a good gardener, Melchior baked the best bread the Fathers had tasted on the coast. Above all, Pierre was an excellent shoemaker. He had made five pairs of shoes for missionaries, one pair for the King of Dahomey, and was running a little shoe factory. Of the remaining two-thirds, some had been sent away for laziness, the rest had been discontented with conditions offered them and quitted the mission to establish on their own or join mercantile firms.[1] Even the satisfactory ones soon began to leave. Only two became Lay Brothers and apparently they soon resigned, as in 1872 two *ex-Frères* were referred to as having been engaged as teachers.

When the missionaries were so anxious to build up a wide range of African staff as teachers, catechists and clergy, it is indeed remarkable that until the late 1870s there was in Nigeria only a single Training Institution in all the five missions. It was argued not that there was not enough money for more, but that formal institutions produced academic training and were dangerous. In their place was suggested the home or family education practised by the Basel Evangelical missionaries on the Gold Coast: in effect that the best of the boarders should be trained personally by the missionary instead of sending them to a training institution, and that the training should be practical. The Basel missionaries were famous for the way in which they sought to make their missions self-supporting by training and employing carpenters, masons, sawyers and other artisans and by cultivating farms, and having a trade section— the Missionary Trade Society—to dispose of their products.[2] The number of trainees was limited, the craftsmasters were able and efficient, the standard of workmanship was high and it was based on thorough general education. A missionary who described it in 1853 called it 'a fatiguing, costly but promising method'.[3]

Until the establishment of the Theological Department in Lagos in 1879, the Methodists had no training institution, even in Sierra Leone. And until the time of the Rev. John Kilner, a disciple of Venn who became the General Secretary in 1876, the Home Committee opposed

1 Father Courdioux to Planque, May 1869; Walsh, op. cit., p. 99.
2 Groves, op. cit., vol. II, pp. 228–9.
3 Mann to Venn, October 1852; C.M.S. CA2/066.

the sending of African youths to be trained in England.[1] Academic training was not thereby avoided; rather the reverse, because, of the pious boys who became teachers, those were promoted catechists who had either been able to attend the C.M.S. Grammar School in Freetown or had been near enough to European missionaries to borrow from their stock of books to read. And before they were ordained ministers they had to pass an oral examination in Theology and Church History conducted by all the existing ministers, European and African, sitting in conference. The Home Committee helped the catechists and assistant ministers to build up their libraries by sending out books to them and periodically asking them to send lists of books they possessed, in order to ensure that they continued to study the scriptures intelligently.[2]

The Presbyterians similarly had no training institution till the Hope Waddell Institute was founded in 1895. Up to 1879 two Africans had

1 Cf. Synopsis of W. B. Boyce's letter to the Rev. William West, 23 May 1865: 'Objects to young Africans coming for education to England'; *Secretaries' Letter Book*; Meth. From West's reply to Boyce, 12 July 1865, it is possible to infer that Boyce based his objection largely on financial grounds, that the mission agents who wished to educate their children in England could not really afford the expense and would, if encouraged, be tempted to borrow money for it, or try to make money elsewhere, or appeal to the committee for aid. West disagreed with this view. Meth.

2 Cf. this list of books in the Library of the African minister, the Rev. Thomas J. Marshall, at Abeokuta, sent to Methodist House in 1874: Penny Encyclopaedia (27 vols); Chamber's Information (2 vols); Rollins Ancient History (8 vols.); Macaulay's Miscellaneous Writings. The Students' Gibbon; The Students' Hume; Africa and the West Indies; Johnson's Lives of the Poets (4 vols.); Milton's Paradise Regained; Hand of God in History; The Koran; Crowther's Vocabulary of the Yoruba Language; Yoruba Translations: Testament Titon, Testament Lailai; Walker's Pronouncing Dictionary; Graham: English Composition; Greek and English Dictionary; Dictionary of Scripture Names; [T. B. Freeman]: Missionary Enterprise no Fiction; T. J. Hutchinson]: Ten Years Wandering among the Ethiopians; Beecham: Ashanti and Gold Coast; History of Wesleyan Missions; The Student's New Testament History; The Student's Old Testament History; D'Aubigné's History of the Reformation (4 vols); Fox's Book of Martyrs; Baxter's Saint's Rest; Pentecost and the Foundation of the Church; Biddle's Scripture History; Blount on the Reformation; Bunyan's Holy War; Life of Dr. Bunting; Life of Dr. Adam Clarke; Life and Remains of Cecil; Life of Samuel Leigh; Life of Thomas Collins; Moses Right and Bishop Colenso Wrong; Thomson: The Land and the Book; Fletcher's Works (9 vols); Watson's Works (13 vols); Watson's Theological Institutes; Paley's Works; Butler's Anthology of Religions; Wesleyana; Wesley's Sermons (3 vols.); Edmondson's Short Sermons; Cassel's Family Bible; Burkill: Notes on the Old and New Testaments; Clarke's Bible; Help in the Reading of the Bible; Farrar's Bible and Theological Dictionary; Harvey's Meditations; Dick's Philosophy of Religion; Angus: the Bible Handbook; Dick's Theology; The Bible and Modern Thought; A Cyclopaedia of Illustrations of Moral and Religious Truths; Forty Days after our Lord's Ressurection; Conversion Illustrated from the Bible; Mill's Local Ministry; Gems of Piety; Jackson's Duties of Christianity;

been ordained, both outstanding men, converts made through the schools and brought up in missionary households. By that time African agents were needed in large numbers and a missionary had been set apart for training them. He described in some detail his way of avoiding a training institution. Boys from the schools who could 'read any book in Efik, write a fair hand, work in the four simple rules of Arithmetic, show a fair acquaintance with Bible history and doctrine, and write a short historical essay', if certified pious by the local missionary or agent, were admitted into a training class. But instead of putting them in an institution and thereby running the risk of making them lazy, they were made teachers, evangelists, printers in training, or appointed to any other available post. They were given a reading assignment. Once a year they assembled for a session of four to six weeks, when they received lectures in Efik on the Bible, had practices in preaching, and had their notes of the lectures examined. 'The Bible lectures,' said the missionary in charge,

are of a very heterogeneous character, being out and out expository; science, geography, history, biography, etc. being brought forward, so that the meaning of the text, in all its ramifications, so far as my small ability goes, may be laid open before them.[1]

And he went on to add that there were also lectures on anatomy, physiology, astronomy, geography and 'common things', but that he avoided the dead languages and mathematics 'though simple laws in that science have been explained to them'.

Besides two Negro missionaries from America, the first three African Baptist ministers in Nigeria came from among the boys evacuated from Ijaye in 1862 and educated at Abeokuta till 1867. The first of them, Moses Ladejo Stone, made a pastor in 1880, had been apprenticed to the carpenter J. C. Vaughan, and had been interpreter, evangelist and assistant to a European missionary for five years.[2] The Baptists had no training institution till the end of the century. Their earliest teachers were emigrants who received further training in English,[3] or had an education

---

Keysell: The Earnest Life; Rephram: Life of Faith; Tongue of Fire; Christ and the Inheritance of the Saints; Jeffrey: Eternal Sonship; Sympathy of Christ; The Exodus of Israel; Complete Duty of Man; Pierce's Principles and Polity of Wesleyan Methodists.

1 Rev. S. H. Edgerley; U.P. *Missionary Record*, 1880, pp. 35–6.
2 Rev. Cecil Roberson, 'Notes on Moses L. Stone', in a collection of local Church histories he edited for the centenary of Baptist Missions in Nigeria in 1953; Roberson Collection.
3 Bowen to Taylor, Ijaye 18 Jan. 1855, said Thomas Coker, his emigrant schoolmaster, and James Cole, the Rev. W. Clarke's interpreter, were studying Grammar; *Bowen Letters*.

similar to Ladejo's. The Roman Catholic Fathers did most of the teaching themselves, used emigrants or boys trained in their households as assistants, had no training institution till after 1900 and did not ordain the first African in Nigeria till 1918.[1]

The only institution for training African staff belonged to the C.M.S. at Abeokuta. In 1851 Townsend argued that it was not necessary. He had ten Coral Fund boys. The eldest was apprenticed to a carpenter, the next three received lessons in English grammar from one agent, and the other instructed the rest.

> We are, in fact, I am happy to think, performing the work of training native schoolmasters without an institution, and it is our aim to check that pride of dress and caste that unhappily sometimes obtains with the African so that if driving of a nail would save a door from falling off its hinges, his own hands could not drive it.[2]

However, Henry Venn was not convinced by this argument. He sent a boy, T. B. Macaulay, from the Fourah Bay Institution to the C.M.S. Training College at Islington, London, in the hope that he would be of use in a Training Institution at Abeokuta. Meanwhile he secured the services of a Cambridge graduate, grandson of Paley, the great Evangelical theologian, who went out with his wife—and an English housemaid—to found the Institution. But Venn emphasized that the Parent Committee wished to avoid the type of academic grammar school they had created in Sierra Leone. 'We are not to educate a few young gentlemen,' he said, 'but to make a model, self-supporting, educational institution, by combining industrial labour [with book learning]'.[3] Paley arrived in January 1853 and died in April. Macaulay then took over the school, but Townsend considered him too academic and in spite of the protests of some missionaries like Hinderer,[4] had him removed to do parish work at Owu (where he was similarly accused of lecturing to his congregation instead of preaching to them). In 1856 Venn engaged William Kirkham, an experienced schoolmaster in England, but he too died within a year of his arrival. A German missionary on the spot, a product of the Basle Seminary, G. F. Bühler, was then asked to take over the institution. He was there from 1857 till his death in 1864 and under him the institution flourished. But it was a constant struggle

1 S.M.A.: *100 Years of Missionary Achievement*, 1957, p. 12.
2 Townsend, Journal Extracts for June 1851; C.M.S. CA2/085.
3 Venn to Townsend, 2 Dec. 1852; C.M.S. CA2/L1.
4 Hinderer to Venn, 8 Feb. 1855, that 'Mr. Macaulay under a large hearted superintendent like the late Mr. Paley can manage such a grammar school, and it ought to be at Abeokuta, not at Lagos which is too iniquitous'; C.M.S. CA2/049. Also Bühler to Venn, 2 Nov. 1857; C.M.S. CA2/024.

between him and Townsend as to how much general education should be allowed besides theology and industrial training. In July 1861, he described the curriculum he adopted as follows:

> I lay particular stress upon Scripture history to give them a good and practical knowledge of it; in general history, they were taught the history of Rome to Constantine; in physical geography, Europe; in Bible geography, Paul's missionary journeys; in Arithmetic, fractions and application thereof; in reading, translation of verses or portions of whole chapters from English into Yoruba or vice versa.[1]

Two years later, he introduced Greek and Latin to the most advanced students. But more than once before that he had to threaten to resign because of Townsend's persistent opposition. Bühler said in March 1862:

> I entirely disagree with Mr. Townsend when he says too much instruction is given to the youths. We always differed on that point. . . .
> As, however, not all my brethren agree with my present mode of teaching and as my training is not considered to be distinctly a missionary training; further, as it is feared . . . that a superior education makes young men often useless or worse than useless . . .[2]

and as he could not but try to do his best, he had better give up the institution and go out to preach. After his death little was done in the institution till it was reorganized in Lagos in 1872.

The character of the institution in Lagos may best be judged from a resolution of the Finance Committee that they

> cannot too strongly urge on the Parent Committee their desire that nothing should be done to prevent the future Native Pastors and Evangelists obtaining an education suited to their *future work and position*. This involves their retaining a full knowledge of their native language and receiving part of their education in that language. It also involves their not contracting habits of life which would render necessary a larger stipend than the native Yoruba Church would as a rule be able to provide.[3]

1 Rev. Gottlieb Frederick Bühler's Annual Report, July 1861; C.M.S. CA2/024 He described the Training Institution to the 1865 Select Committee of the House of Commons; Questions 6086, 6092, 6098, Minutes of Evidence before the Select Committee, PP. 1865 V. See also a good discussion about the school in B. O. Rotimi: *The study of the recruitment, training and placing of teachers in the primary and secondary school levels in Nigeria, 1842–1927*, M.A. thesis, London, 1955, pp. 135–42.
2 Bühler to Chapman, C.M.S. Secretary, 30 Jan. 1863; Bühler to Venn, 3 May 1862; C.M.S. CA2/024.
3 Minutes of the Finance Committee of the C.M.S., Lagos, 26 April 1873; C.M.S. CA2/01.

They were, in short, in favour of a non-institutional education in which the mission did not have to bear the expenses of maintaining the student abroad. He was to be trained while he was carrying on active service in the Church and could be kept under observation. It required more insistence from London to get Charles Phillips and Isaac Oluwole, two future assistant bishops, to Fourah Bay where they could take advantage of the new special relationship of the College with Durham University. It is also not surprising that all the three clergymen that the C.M.S. Training Institution had produced up to 1890 passed through the hands of Bühler.

Indeed the outcome of this fear of academic training was not to avoid it, but to have more of it. Teachers and catechists and clergymen were required in large numbers. And those who could at least read and write properly had a chance of passing the necessary examinations. The result was continued dependence on the emigrants and the Sierra Leone Training Institution, and on English education. Up to 1890 both for the Yoruba and the Niger Missions forty Anglican clergymen had been ordained. Six of them had been trained in England, five at Islington, one at Highbury Training College; fourteen had been to Fourah Bay for longer or shorter periods; two to the Freetown Grammar School; six possessed no more than primary education; two went only to Sunday school. Besides the three products of the Abeokuta Training Institution, only two others had been trained wholly in Nigeria. One was Edward K. Buko, the son of Possu, from Gollmer's boarding school at Badagri; the other was Daniel Olubi, who had a few terms at the Abeokuta Training Institution but was trained mostly in the household of the Hinderers. And only these two last mentioned were not emigrants or sons of emigrants.[1]

The fact that the missionaries failed to avoid literary education in trying to avoid institutional training is even better illustrated in the development of secondary schools in Lagos which began in spite of missionaries and was moulded almost entirely by the needs of trade and the predilections of the Sierra Leonean emigrants for literary and academic education.[2] In 1859, in circumstances to be related later, Townsend had Macaulay transferred from Abeokuta and suggested that he could fit in nowhere in the Yoruba Mission and that

1 This analysis is based principally on information given about the African ministers in the Yoruba and Niger missions listed in the *C.M.S. Register of Missionaries and Native Clergy, 1804–1904*.
2 See J. F. Ade Ajayi: 'The Development of Secondary Grammar School Education in Nigeria', *Journal of the Historical Society of Nigeria*, vol. iii, no. 1, December 1963.

he belonged really to the Freetown Grammar School. Macaulay, however, pleaded that he be allowed to start a grammar school at Lagos. Crowther supported the plea, and the Parent Committee after some hesitation agreed. Macaulay arrived at Lagos, having at his disposal nothing more than his salary and four rooms in the old cotton warehouse in Lagos. It was an emigrant business man, Captain J. P. L. Davies, who advanced £50 to buy books and equipment.[1] Of the first 25 boys in the school, the parents (or guardians) of 8 were classified as merchants, 14 as traders; only one was a clergyman, one a carpenter and one a scripture reader.[2] The books ordered are also instructive. They included the usual ones in Grammar, Composition and Arithmetic; History and Geography (including atlases and globes); also Book-keeping, as well as Euclid's *Elements*, Eaton's *Latin Grammar*, Valsy's *Greek Grammar*. Lastly, there was *Plain Treatises on Natural Philosophy* with a note by the principal that 'I should like some mechanical instruments to illustrate the sciences but I am afraid there may not be money enough just now'.[3] From the fees the loans were repaid, subordinate teachers were employed, and in July 1860 the Parent Committee resolved 'that the salary of the tutor of the Grammar School be reduced by one-third and that he receives one-third of the yearly payment of £1 for each pupil which at present numbers will prove an increase of his salary'.[4]

The Lagos C.M.S. Grammar School was such a success financially as well as in raising the mission's prestige, in the increasingly prosperous Lagos community that in 1872 a Female Institution was begun by the Rev. A. Mann, formerly of Ijaye, and his wife. The prospectus, with an eye on the opportunities in Lagos as well as on probable critics, declared that the Female Institution would provide 'a good and useful education, thoroughly English, but suited as much as possible to the peculiarities of this country'.[5] The reports of the public examinations, however, indicate that little adaptation was carried out. More than that, the pattern set by the C.M.S. institutions was soon being followed by the other missions. Leading Methodists in Lagos soon began to point out that Methodism was losing ground in society because the children of Methodists tended to go to the C.M.S. schools and were in that way lost to their denomination. They therefore began to urge their mission to found a grammar school also. Tired of waiting, in 1874 they collected

1 T. B. Macaulay to Venn, 5 April 1859; C.M.S. CA2/065.
2 Macaulay, Annual Report of the Grammar School, 7 July 1859; C.M.S. CA2/065.
3 List of books encl. in Macaulay to Venn, 5 April 1859; C.M.S. CA2/065.
4 H. Straith to Townsend as Secretary of the Yoruba Mission, 10 July 1860; C.M.S. CA2/L3.
5 Printed Prospectus, 1872; C.M.S. CA2/066.

£500 and asked the mission to add a similar sum towards the cost of building and provide a suitable principal.[1] It was to this institution that a few theological students were attached when it was finally opened in 1879. The grammar school section, or the Methodist Boys' High School, as it was called, was stated explicitly to be designed to prepare young men 'for a commercial and literary life'. The subjects offered were English, Reading, Writing, Orthography, Dictation; Arithmetic and Algebra; History, secular and sacred; Geography; Grammar; Classics, prose writers and poets. These were all in the normal curriculum. There were, however, additional subjects, at extra cost, with a note that 'the principal reserves to himself in every case, on due consideration with parents and guardians, the right of deciding what additional subjects each pupil shall take up as premature attention to higher studies is often disastrous to real educational advancement'. These additional subjects were stated in two parts. Part I included Latin, Greek, Hebrew, French, and other modern languages; Geometry, Trigonometry, Book-keeping, Drawing, Rhetoric, and Logic; Moral Philosophy and Political Economy. And Part II, Roman and Grecian Histories, mythology and antiquities, Natural Philosophy in its various branches, Astronomy, Chemistry, Physiology, Geology and Botany.[2] Lest this list of subjects be taken too seriously, it should be added that the principal had but two assistants, nor was the principal himself a super-man. In 1881 the Methodist Superintendent in Lagos noted

> the great dissatisfaction that reigns among our people in consequence of the appointment of a minister as the principal of the High School instead of a student from Westminster who would give himself exclusively to the work of the school and institution.[3]

The secondary schools were in fact no more than senior primary schools, conducted wholly in English, with a bit of Latin, and mathematics and recitations of poetry, depending on the ability of the teachers, supplemented as was usual by extra lessons from leading members of the community, doctors, traders, even government officials. The important point here was that this development arose partly from the lack of adequate resources, partly also from the doctrinaire avoidance of academic training, both of which combined to make grammar school education so wholly dependent on the needs of traders. And once begun, the different missions, each criticizing the work of the others, competed

1 Petition signed by 20 members to the Rev. T. R. Picot, Acting General Superintendent, Lagos, 27 Jan. 1874; Meth.
2 Printed Prospectus, 1878; Meth.
3 John Milum, Memorandum, 26 Feb. 1881; Meth.

in the founding of similar institutions. The Catholic Fathers were extremely critical of the C.M.S. Grammar School boys. You can see them, wrote a Reverend Father,

> walk arrogantly about the streets of Lagos . . . a packet of books under their arms, believing themselves to be doctors before they are scholars, so that later when they are employed they become unbearable both to those who have to command and to those who have to obey them.[1]

But in 1881 they constituted the senior boys of their primary schools into a St. Gregory's College, and the Baptists, as soon as they could, began an Academy in 1883.

The training schemes in Europe for practical and mechanical arts, already described, were meant to supplement and to improve upon the skills the emigrants had already brought with them from Sierra Leone, Brazil or America. The missionaries always emphasized that they were not undertaking the industrial education of the country. All they needed to do as pioneers was to show the light in a few selected spheres of life and the demands of commerce and economic development would do the rest. We have seen how they were encouraging traders to follow them not only to Badagri and Lagos, but also into the interior, to Abeokuta and beyond, and up the Niger. In 1863 there were at Abeokuta, where there was none in 1851, five firms with resident agents, two of them Africans, three Europeans, besides two independent European traders and several Africans. In addition to making financial contributions, the traders sometimes helped in the schools, particularly during arithmetic lessons, when they helped to explain to the pupils—and to some of the teachers also—the mysteries of the English system of measurement and counting.[2] The important point here was that the missionaries exerted themselves to see that increased trade did not mean just more money for some European traders and a handful of African middlemen, but that it should bring in its train a wide diffusion of the knowledge of

1 '. . . passer avec fierté dans les rues . . . un paquet de livres sous les bras, se croyant docteurs avant d'être écoliers, tels ils sont dans les situations qu'ils remplissent plus tard: insupportables à ceux qui ont a leur commander ou à leur obéir'; *Les Missions Catholiques*, 27 Aug. 1880.
2 Harrison to Venn, 30 May 1863. Talking about education in C.M.S. schools at Abeokuta, said: 'Arithmetic is very hard for them to learn.' The poor children apparently knew what $\frac{9}{7}$ of £1 was, and could multiply 4/8½d. by 4½ but failed to do 'the rule of three sums' like 'If cowries are 40 strings a shilling, what is the price of 3 bags 8 heads?' which involved not only the intricacies of £ s. d. but also of 'strings', 'heads' and 'bags' of cowries. Harrison concluded by saying that Mr. Mills, one of the traders at Abeokuta, went into the Ake School to try to throw some light on the subject.

European mechanical and industrial skill. This they did by organizing a system of apprenticeship[1] through which they ensured that the knowledge of the men they had trained in England and of the best artisans among the emigrants, as well as of the traders and the lay and industrial agents of the missions, was passed on to the largest number of people, generally those connected in some way or the other with the Mission House. One of the earliest functions of the C.M.S. Industrial Institution at Abeokuta, and later of similar institutions at Onitsha and Lokoja, was to encourage the cultivation and export of cotton by giving instruction in cleaning and packing cotton. For this reason Henry Robbin and Josiah Crowther had been trained in Manchester. By 1861 there were already some three hundred gins at Abeokuta, a few at Ibadan and Ijaye, and some were beginning to ascend the Niger.[2] There were also a couple of grinding mills at Abeokuta, and in 1863 a European trader erected a steam-powered mill at Aro (near Abeokuta).

It was the missionaries who recruited apprentices for these from the primary schools, and especially from among the boarders. The missionaries who most strongly opposed academic education took the initiative in organizing the apprenticeship scheme. Townsend instituted a fund so that the mission could go on maintaining the apprentices as boarders while they went daily to their masters.[3] Besides, the missionaries maintained a panel of artisans most worthy to receive apprentices. It was Bowen who discovered the ability of Vaughan and Russell, the emigrant carpenters from Liberia resident at Ijaye.[4] Ribiero, a Brazilian emigrant, was the master tailor at Abeokuta.[5] Dr. Harrison discovered that a European trader was a watchmaker by profession and he at once apprenticed two boys to him to learn to clean clocks and repair minor disorders.[6] It is not possible to follow up in detail the results of these efforts, and in many cases they owed as much to economic development as to missionary endeavour. There were, however, three spheres in which the missions led the way, as they most intimately touched the life of the missions themselves.

The first was building and architecture, which has already been re-

1 Townsend to Venn, 17 May 1853, 15 Aug. 1854, and Annual Letter for 1859, dated 30 Jan. 1860; C.M.S. CA2/085. Harrison to Venn, 29 May and 30 June 1862; Harrison to Ford Fenn, 30 June 1864; C.M.S. CA2/045. Crowther to Venn, 6 April 1861; C.M.S. CA3/04. Mann, Journal entries for 3 July and 20 July 1855; C.M.S. CA2/066.
2 Taylor to Venn, 23 Sept. 1862; C.M.S. CA3/037.
3 Taylor to Venn, 17 May 1853; C.M.S. CA3/037.
4 Bowen, Fragment of a Journal for 1855; *Bowen Letters.*
5 Harrison to Ford Fenn, 30 June 1863; C.M.S. CA2/045.
6 Harrison to Venn, 28 Sept. 1863; C.M.S. CA2/045.

ferred to.[1] As we have seen, the carpenters and sawyers made an immediate impact on the life of the people around and there was a demand for them or their qualified apprentices. Both the C.M.S. and the Presbyterian missionaries imported sawyers and carpenters from Sierra Leone; the Baptists from Liberia; and one or two Fantis came from Cape Coast to work for the Methodists. Nevertheless there was always a long queue of people wishing to buy planks. Between 1859 and 1860 Townsend said he had ordered up to £50 worth of carpenter's tools for people at Abeokuta and received payment in full, including expenses.[2] In 1861 Crowther said that many youths who were apprenticed by the missionaries had been trained as 'house carpenters' able to make strong batten doors and windows, simple tables, chairs and coffins, and that the most advanced could make panel doors and dovetail a box.[3] This description might have been made of the majority of the village carpenters of today who, as distinct from the 'cabinet makers' of the towns, are direct descendants of the apprentices of the 1850s. They use the same tools and in fact have reproduced the simple carpenter's shed of the Mission House.

In 1862 the C.M.S. transferred to Lagos J. A. Ashcroft, the industrial agent who had been teaching brick- and tile-making in Sierra Leone, and in that year also the two boys trained at Manchester in building and construction returned. The American Civil War having cut off the funds of the Baptist Mission, their agent, J. M. Harden, at Lagos returned to his old trade of brickmaking. He established a brickworks at Iddo, and Ashcroft set up another at Ebute-Metta. Their combined influence[4] soon made itself felt on the architecture of Lagos, when not only the missionaries but also the government and the growing commercial class were able to exploit the new sources of brick. The finest examples of architecture were to be found in the Brazilian quarter and perhaps the greatest single building produced in this period was the Holy Cross Cathedral, opened in 1881.[5] Missionary influence on

1 See above, pp. 111–14.
2 Townsend, Annual Letter for 1859 dated 30 Jan. 1860.
3 Crowther to Venn, 6 April 1861; C.M.S. CA3/04.
4 For a detailed account of the organization of the brickworks, see the letters of Ashcroft to the C.M.S. Secretaries in C.M.S. CA3/020. For J. M. Harden, see Tupper, op. cit., p. 338.
5 'Le plus beau monument de la côte occidentale d'Afrique,' said Father Carambauld in an article on the opening of the cathedral in September 1881; in Les Missions Catholiques, 14 Sept. 1882. All Lagos took an interest in the building of the cathedral and a pride in the finished work. Father Carambauld was in charge of the building. The master mason was a Brazilian emigrant, Senhor Lazaro da Silva; 'Premiers Temps de la Mission de Lagos d'après Mère Véronique', in Missions de la Nigéria, ed. L. Arial.

architecture spread inland, though not without unexpected hazards. In 1864 Townsend began to build the first stone building at Abeokuta. It was in a period when relations between the Europeans in general and the local rulers were strained, and Townsend was suspected of building a fort behind which to shelter Europeans when they decided to capture Abeokuta as they had captured Lagos.[1] It required all the personal influence of Townsend to get the chiefs to allow the building to continue. The same view was taken by the Obi of Onitsha when Taylor went to Freetown and collected money for a new brick building and returned in 1867 with two of Ashcroft's boys to erect it.[2] The Obi and other elders wondered why anyone should want to make such strong foundations and build such walls if he meant to live at peace with his neighbours. When the missionaries proceeded, at a very solemn ceremony, to bury a bottle in which some curios and written documents were preserved— as was the custom when laying foundation stones—the suspicion of the elders turned to certainty that the missionaries did not mean well. They insisted at least that the bottle be dug up and destroyed.[3] However, when the initial suspicions faded, mission architecture soon passed beyond the confines of the mission village into the old town itself. In 1880 a Roman Catholic Father who visited Ado, that fortress of the old ways of life, remarked that the people were building for one of their gods a two-storey building, roofed in iron, its pillars carved with images 'les plus bizarres', and with an emigrant in charge of the interior decoration.[4]

Next to architecture, the missionaries took great interest in printing. The Presbyterians arrived with a printer and a printing press, and they began publication on the spot almost at once. In August 1849 the printer listed that he had produced eight hundred copies of the Primer, five hundred copies of Bible lessons, 150 of Arithmetical Examples, two hundred of multiplication tables, five hundred almanacs with the Commandments in Efik, three hundred copies of Elementary Arithmetic and four hundred of the Catechism in Efik and English.[5] By publishing on the spot the printer was training some apprentices. The other missions for a long time continued to rely on printing their requirements in Europe or America. However, as part of his industrial schemes, Townsend

1 Townsend to Thomas Champness, 5 Jan. 1864; Meth. This letter is important as one of the very few strictly unofficial letters by missionaries that have been available for this study.
2 Taylor to Venn, 15 Feb. 1867; C.M.S. CA3/037.
3 Crowther to Venn, 27 Nov. 1868; C.M.S. CA3/04.
4 Father Antoine Durieux, 7 June 1880; Les Missions Catholiques, 27 Aug. 1880.
5 U.P. Missionary Record, 1849, pp. 120–2.

brought out an old hand-press, probably obtained from his brother, who was a printer. He taught himself how to use it and began to teach one of his boarders. The boy did well and in 1860 Townsend said he would have liked to send him for further training in England, 'for he is the best-behaved boy I have',[1] but he was convinced it would spoil him. More boys were apprenticed and a regular printing works was established. Townsend acquired more equipment and in 1859 Robert Campbell, the Jamaican who visited Abeokuta to pave the way for the projected immigration from the New World, was able to improve on Townsend's methods.[2] The press began to publish pamphlets of hymns, catechisms and prayers, and to undertake some binding. It was in 1859, too, that Townsend began to publish the *Iwe-Irohin*, a fortnightly journal in Yoruba giving news of Church and state from near and far, and educating the growing reading public through didactic essays on history and politics.[3] In the following year an English supplement was begun, and a missionary observed that one new feature of the civilized world was introduced, an advertisement column declaring vacancies for apprentices, clerks, houseboys and others. Townsend's example was soon followed. Robert Campbell returned to Lagos in 1862 and almost at once founded the *Anglo-African*. His printers came from the C.M.S. press at Abeokuta.[4] In that way Townsend had contributed to the keen interest in journalism and the technical excellence of many of the newspapers that began to appear in Lagos in the succeeding years. It may also be claimed that the large number of one-room printing works in several large towns in Nigeria owes something to the same source.

Besides building and printing, the Christian missions took an interest in medicine. Compared with the end of the century, when the hospital almost rivalled the school as a means of evangelization, this interest was on a small scale. Crowther and Hope Waddell took an early interest in vaccination against smallpox[5] as one way of establishing good relations between the missionary and the community and as one of the marvels of civilized Europe that even the untrained missionary could exhibit; but it does not appear that the interest was kept up. Each mission tried to maintain medical agents whose attention was directed in the first instance to the health of the missionaries, but who on the one hand tried

1 Townsend to Venn, 28 Feb. 1860; C.M.S. CA2/085.
2 Townsend to Venn, 5 Nov. 1859; C.M.S. CA2/085.
3 Townsend, Annual Letter for 1860, dated 6 Feb. 1861; letter to Venn, 4 May 1860; C.M.S. CA2/085.
4 Harrison to Venn, 1 Sept. 1863; C.M.S. CA2/045.
5 Crowther to Venn, 25 March 1856; C.M.S. CA2/031. Waddell, Journal entries for 1 and 17 Sept. 1849; *Journals*, vol. I.

to train boys to help them, and on the other were willing in cases of emergency to treat all comers. For this purpose, too, some missionaries, like Bowen, not medically trained, tried to acquire some medical knowledge. It was perhaps typical of the age that except for a few men who took university degrees in medicine, several of the 'doctors' and 'surgeons' in West Africa took diploma courses as apprentices to doctors only as a stepping-stone to some political or mercantile career.[1] This part-time attitude to medicine, and the absence of hospitals as institutions which had to be carried on, made the influence of many of the missions' medical agents transient and limited. Three of them may be briefly mentioned.

One was Archibald Hewan, a Jamaican who arrived at Calabar in 1855.[2] He had been given a diploma course by the Foreign Mission Committee to take the place of an earlier Scottish student who was declared physically unfit for service in West Africa after he had taken an M.D. degree. Hewan did not appear to the Committee pious enough and he was taken on only 'on trial', but at Calabar he became a reputable doctor who did 'far beyond [the Committee's] expectations'. But he complained of the attitude of Anderson, who after Hope Waddell's departure in 1857 became the senior missionary. And it was probably to secure himself more independence of the European missionaries that he urged the Committee to build a proper hospital where the people could come for treatment. The Committee censured Anderson for his excesses in his treatment of Hewan, but they turned down the idea of a hospital. Hewan turned to the traders and tried to get them to establish a hospital with himself in charge, but nothing came of it. Hewan left Calabar in 1864 for Britain. An affable man, he made friends easily. He found people to pay his university fees, and obtained an M.D. of Edinburgh. One of his professors recommended him to the Earl of Erne, who introduced him to high class society in England and he never went back to Calabar.

Samuel Crowther, junior, did establish a dispensary at Abeokuta when he returned from England in 1852, opening three days a week between 10.0 a.m. and 4.0 p.m. He began with about twenty patients attending but the number soon rose to about 110.[3] Missionaries who visited his dispensary or whom he went to treat said he was a good doc-

1 Waddell said of the 8 ships' captains he found in Bonny in 1849. 'One was bred regularly to sea. The others had originally come out as surgeons, then became supercargoes and finally trading captains. This series of changes is quite common on this coast.' *Journals*, vol. VII, p. 120.
2 Somerville to Dr. Fergusson, Liverpool Agent of the Foreign Mission Committee, 6 Feb. 1861; Somerville to Anderson, 3 July 1863; U.P. *Secretaries' Letter Book*, vol. VIII, and Minute of the Foreign Mission Committee, 25 Feb. 1865, quoted in the *Letter Book*, vol. IX, p. 62.
3 Samuel Crowther, jun., to Venn, 26 Sept. 1852; C.M.S. CA2/032.

tor.[1] But in 1855 he was appointed Assistant and Secretary to Irving. When Irving died, though still practising medicine, he went into the cotton business; and later still he also became an architect and builder of some note, among his creations being St. Stephen's Church at Bonny.[2] Finally, Henry Venn's ambitious scheme of founding a Medical School must be noticed. In January 1861 Dr. A. A. Harrison, M.D. (Cantab.), was appointed partly to succeed Dr. Irving as political agent in 'the cherishing of confidential relations with the chiefs of important towns as far as providential opportunities may arise', but more especially to look after the health of European missionaries and for 'the training of a few promising native youths or young men in the elements of medical and surgical science. 'You will find', he was told,

> that the native doctors are not merely quacks as among many uncivilized tribes. They have already discovered many herbs and roots, as well as some mineral substances which they occasionally compound together.[3]

He was therefore to look for some such people with some of the native art of medicine who might be willing to place themselves under his instruction for a time. He was to create a fund for running this Medical School, largely from fees he charged for treating non-C.M.S. Europeans. Dr. Harrison faced his formidable tasks in a very ordinary way. As if his assignments were not numerous and onerous enough, he got himself involved in the routine administration of the mission as Secretary of the Local Committee, editor of the *Iwe-Irohin*, something of an inspector of education, with keen interest in the apprenticeship schemes and his wife's sewing institute. However, he selected four boys from the Theological Training Institution, chosen as the cleverest boys, not as having the greatest interest in medical science, African or European. They attended classes at the institution in the morning, and between the hours of 1.0 p.m. and 2.0 p.m. daily, when they could have done with some rest, they came to Dr. Harrison's lectures to cover the whole range from anatomy and physiology to botany and zoology, chemistry and mathematics:

1 E.g., Hinderer to Venn, 31 May 1858: 'Mr. Crowther Junior has kindly come up from Abeokuta to see [Mrs. Hinderer who was ill] which was a great comfort to me and his advice seems to be very good and sound'; C.M.S. CA2/049.
2 D. C. Crowther on the building of St. Clement's Church, Bonny in the appendix to Bishop Crowther's Charge to his clergy in 1874; C.M.S. CA3/04. See also his 'Sketches of Steamers in use on the Niger 1863 and 1875'; C.M.S. CA3/014.
3 Instructions of the Parent Committee to Dr. Harrison, 22 Jan. 1861; C.M.S. CA2/L3.

We began with anatomy, chiefly that of the muscles, with a little physiology. We had also some botany and *Materia Medica* and whilst waiting for a box of chemicals I expected from England, we did a little surveying with the chain and prismatic compass and made a slight attempt at Euclid but only got as far as the fifth proposition, over which the boys spent two or three lessons but did not seem able to master it. Since then we have been doing natural philosophy and chemistry . . . but I was disappointed when I went over the chemistry the last day we did it to find how little I had taught them.[1]

Their main text book was Hooper's *Vademecum* (a sort of 'Family Doctor'), which the mission bought for them and from which they had assignments for homework when they finished the afternoon session at the Training Institution. Dr. Harrison added:

I daresay they have not much time to read at home, as I fancy Mr. Bühler keeps them pretty well employed with their other lessons, which it would be a pity for them to lose.

There was no nonsense about specialization. In less than two years one was dismissed, two were sent to teach in schools in Lagos, the fourth, Nathaniel King, became personal assistant to Dr. Harrison. When Harrison died in 1865, Venn asked for King to be sent to Sierra Leone to live as a student at Fourah Bay and serve as an apprentice to Dr. Bradshaw of the Colonial Hospital there. Eventually he got to England and qualified.[2] Another boy, Obadiah[3] Johnson, who lived in Dr. Harrison's household, later went to England to qualify as a doctor. He was a brother of Samuel Johnson, the one who edited his *History of the Yorubas.*

In these and various other ways the missionaries were building up around the Mission House something of the 'civilized' community Buxton had dreamt of. In Lagos the progress was so rapid that this community of the Mission villages soon overwhelmed the community of the old town. When Lagos became a British colony in 1861, development no longer depended exclusively on missionaries: town planning and the building of roads, clearing of creeks and lagoons and sanitary

1 Harrison to Venn, 26 Feb. 1862; C.M.S. CA2/045.
2 Venn to Nicholson, 23 Dec. 1865; C.M.S. CA2/L3. Nathaniel King was son of the Rev. Theophilus King mentioned earlier, and nephew of Henry Robbin. Robbin went to England in 1871 and arranged with Venn to pay part of Nathaniel's fees at King's College, London, the C.M.S. paying the rest; Venn to Robbin, 1 May 1871. By 1878 Dr. Nathaniel King had returned to Lagos; Hutchinson to Maser, 28 June 1878; C.M.S. CA2/L4.
3 Harrison mentions Obadiah in a letter to Venn, 30 July 1861; C.M.S. CA2/045.

regulations were a responsibility of the administration—pursued energetically by governors like Glover. But the Mission House remained at the centre of the social life. In spite of the presence of Government House and occasional disagreements between Government House and Mission House, the new Lagos was made up essentially of the Christian villages of the different missions joined together. The main social events were the opening of new churches and schools, the weddings, the missionary meetings, the concerts that followed the public examinations of the schools. Nowhere else was the development so rapid; nowhere else did the community of the Christian villages come so quickly to threaten the existence of the old town. But it was in the development of places outside Lagos that the missionaries took the greatest pride.[1] For in those places the introduction of literacy and the knowledge of the industrial and mechanical arts that were the essence of European civilization depended more exlusively on their initiative. Their greatest achievement was at Abeokuta and, considering the limited ability and resources of the missionaries, it was impressive. By 1863, when the steam-powered mill was established, there was a Commercial Association—a type of Chamber of Commerce—with missionaries on the committee. For amusement, besides the usual social events mentioned above, there were public lectures and magic lantern shows. It is not clear whether Henry Robbin's photographic apparatus arrived, but there were harmoniums and harmonicas. There was a circulating library, and a bookshop was to come later. That most essential feature of Victorian society, the voluntary association, was also encouraged. Besides those with a missionary purpose, there was a Road-Building Association urging the widening of the Lagos-Otta-Abeokuta road, and undertaking the construction of the Abeokuta-Aro road so that heavy machinery could be transported on the Ogun up to Aro and wheeled by cart to Abeokuta. A Mutual Aid Society was organized. During the Ijaye war, an Ijaye Relief Committee was formed to send aid to Christians in the town and care for the children evacuated from there.

This civilization was essentially a civilization around the Mission House. That is to say the immediate beneficiaries besides European traders were those in close contact with the missionaries, the converts and their friends and relatives to some extent, the emigrants above all. Just before the session, the Consul observed that 'the progress of Lagos in civilization is much too fast to please the native chiefs' who could not compete 'either in mercantile or agricultural pursuits with the emigrants

1 For developments in Lagos see A. A. Aderibigbe: *The Expansion of the Lagos Protectorate* Ph.D. London 1959; and J. Herskovits: *Liberated Africans and the History of Lagos*, D.Phil. Oxon. 1960.

from Sierra Leone, Brazil and Cuba'.[1] At Calabar, emigrants and converts could not hope to displace the powerful chiefs and European traders, but there was scope for them as artisans and clerks. At Abeokuta, Madam Tinubu, who was an experienced coastal trader expelled from Lagos in 1856, was the only important trader independent of mission influences. The missionaries in the early days interested other chiefs like Ogunbona and Atambala in the cotton trade, but the rising trade of the town was dominated by emigrant traders.

The emigrants and converts were bound by several ties to the Mission House; clergymen, European and African, wielded great influence among them but the society being built up was far from being over-clerical. With mission agents, the training of whose children was largely dependent on opportunities in the mission, the ministry was of course the most likely goal of ambition. Though Obadiah Johnson mentioned above became a doctor, his three elder brothers who financed his training were all pastors who owed their careers to the mission. Their father, Henry Johnson, was a Liberated African in Sierra Leone sent as a young man by the Native Agency Committee to Kew Gardens to study Botany. He did not profit much by it. He returned to Freetown to farm and trade and was a leading member of the church at Hastings when Hinderer in 1859 persuaded him to emigrate to Ibadan as a Christian Visitor. He was a most effective evangelist. When he died suddenly in 1865, Hinderer mourned his loss as a 'faithful fellow-labourer and a friend . . . a sincere Christian . . . [who led] a consistent life'.[2] Hinderer and the C.M.S. gave his children every encouragement they could. The eldest, Henry, became a tutor at Fourah Bay College, was later sent to study Arabic in Palestine, and became archdeacon on the Niger, where his linguistic studies won him an honorary M.A. of Cambridge University. His brother Nathaniel was sent to the Abeokuta Training Institution under Bühler and later became a well-known pastor in Lagos. Samuel, the historian, was brought up in Hinderer's household, appointed catechist at Aremo, Ibadan, in 1875 and pastor at Oyo in 1886. However, where fathers were more directly in charge of the training of their children, the ministry was less often the goal aimed at—even when the fathers were themselves loyal members of the church. If the Rev. J. C. Taylor at Onitsha had had his way, his eldest son who went to Fourah Bay and later became a pastor would have been trained

---

1 McCoskry to Lord John Russell, 7 June 1861; Papers relating to the Occupation of Lagos, PP 1862 LXI.
2 Hinderer to Venn, 30 March 1865. See also Hinderer to Straith, 23 April 1861; C.M.S. CA2/049, and J. F. Ade Ajayi: 'Samuel Johnson, Historian of the Yoruba', *Nigeria Magazine*, June 1964.

in mercantile business or as a doctor.[1] The balance of opinion on this matter in the new society would seem to have been reflected in the family of Crowther, who took great care in the education and settlement of his children. All six of them were educated in England. Of the daughters, two married clergymen, one a trader. Of the sons, it was Dandeson, the youngest, who was set apart for the ministry. As we shall see later, he went to the C.M.S. Training Institution at Islington and later became archdeacon in the Delta. The eldest son, Samuel, as we have seen, had a scientific and medical training but later took to trade. Josiah had an industrial training in Manchester and became a business man.

The emigrants and converts were in fact trying to become what the missionaries hoped they would, a rich, inventive, powerful middle class. They were not the idle people often imagined. It was trade rather than technological development that came to dominate their lives, for such development could only flourish if it was preceded by demand stimulated by trade. It was through trade that the emigrants and converts could expect to make their money quickly, but their early training was more likely than not to have been as artisans, or in an English factory or aboard a ship. Nor was trading in palm-oil or cotton in the conditions of those days an idle enterprise. They were loud in the profession of their faith, and generally were supporters of missionary work. They owned 'pawns' and domestic slaves, usually under the guise of redeeming them, but as we have seen, many missionaries did the same. Just as missionaries recruited 'pawns' as boarders, emigrants and converts were allowed to recruit domestic slave labour for economic development, provided such recruitment was regulated and did not lead to indiscriminate slave trading. In a community dominated by trade it was not alarming that there were occasional cases of fraud. What worried the missionaries most was the frequency of cases of adultery calling for church discipline, and of polygamy symbolizing the falling away of members of the mission village back to the society of the old town. But as Townsend said in 1875, the new class was 'no better and no worse than the majority of church-goers in England'.[2] It is in the light of the fortunes of this class that the further development of the work of the Christian missions must now be examined.

1 Taylor to Venn, 21 Dec. 1864: 'I am thinking DV to send my son to Mr. Ford Fenn to complete his study in mercantile business next year. . . . I might have placed him to learn medicine for the use of the mission, but my means will not allow me to go through all the expenses.' C.M.S. CA3/037.
2 Townsend to Wright, 9 Feb. 1875; C.M.S. CA2/085.

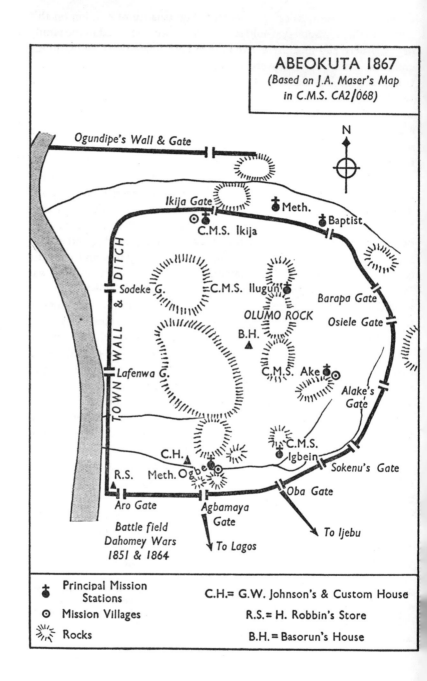

ABEOKUTA 1867
(Based on J.A. Maser's Map
in C.M.S. CA2/068)

N

Ogundipe's Wall & Gate

Ikija Gate
C.M.S. Ikija
Meth.
Baptist

DITCH
TOWN WALL &

Sodeke G.
C.M.S. Ilugun
Barapa Gate
OLUMO ROCK
Osiele Gate
B.H.
Lafenwa G.
C.M.S. Ake
Alake's
Gate

C.M.S.
Igbein
C.H.
Meth. Ogbe
Sokenu's Gate
R.S.
Oba Gate
Aro Gate
Agbamaya
Gate
To Ijebu
Battle field
Dahomey Wars
1851 & 1864
To Lagos

✝ Principal Mission
  Stations
⊙ Mission Villages
Rocks

C.H.= G.W. Johnson's & Custom House
R.S.= H. Robbin's Store
B.H.= Basorun's House

6 Towards Self-Government in Church and State

THE progress of civilization in the sense of economic development and social change raised two important political problems: one was the problem of keeping the rulers of each state reconciled to the expanding civilization around the Mission House, the other was that of securing thoroughfare on long-distance trade routes which cut across existing political boundaries and necessitated peace between the different states. Up to 1861 consuls exhibited British naval power on the coast, and missionaries pursued the 'Abeokutan policy' of aiding the Egba against Dahomey and using Abeokuta as a showpiece of British philanthropy, trade and civilization. As soon as missionaries took expansion north of Abeokuta seriously, it became doubtful whether this policy was adequate. When war was resumed in the interior, it became imperative to find an alternative policy.

By 1860 the cotton supply of Abeokuta, though small in itself,[1] was regarded as of considerable significance in the English market because of the steep rise in the price of American cotton. This importance grew as America drifted into civil war. Even before that, Clegg, who was the pioneer cotton merchant of Abeokuta, was consulting Crowther about how to expand production in other regions north of Abeokuta.[2] In 1858 Hinderer said that cotton production was growing in Ibadan. Enterprising merchants like Madam Tinubu of Abeokuta had been buying it up

1 The Consul quoted the export of cotton from Lagos (chiefly from Abeokuta) for the year ending June 1861, 'not a favourable year', as 1,303 bales (118 lbs. to a bale); Table of exports encl. in McCoskry to Lord John Russell, 7 Jan. 1862, in *Papers relating to the Occupation of Lagos*, PP 1862 LXI.
2 Venn to Clegg, 23 April 1858, referring to Clegg's offer to Samuel Crowther, jun., to join a venture to establish a cotton trading depot 'near the Niger', a scheme which Crowther, Sen., considered unrealistic. The quest for cotton from Abeokuta and the Niger began in 1850 when Venn wrote to Townsend: 'A great change has taken place since you left England [end of 1849] in the price and demand for cotton. The manufacturers have been eager for the encouragement of African cotton'; Venn to Townsend, 29 Nov. 1850; C.M.S. CA2/L1.

and taking it to the cotton gins at Abeokuta. In the following year Edward Gurney, the Quaker banker and philanthropist, gave the C.M.S. £1,000 for industrial establishments in Yoruba, particularly at Ibadan, so that the progress made at Abeokuta could be repeated there.[1] A group of English merchants and philanthropists under Lord Alfred Churchill, M.P., formed the African Aid Society with the idea of sponsoring persecuted negroes in America who wished to settle and grow cotton in West Africa. In 1859 they sent two delegates, a Canadian negro, Dr. Delany, and a Jamaican, Robert Campbell,[2] who toured Yorubaland and in December signed an agreement with the Alake by which he promised to give land and other privileges to the expected settlers. This agreement will be referred to again later in this chapter. The point here is that the missionary effort at Abeokuta succeeded well enough to prove their point that the overland route to the Niger via Abeokuta could rival the Niger waterway itself in expanding British trade.

The Niger Expedition of 1857 further drew attention to the overland route. The steamer carrying the expedition was wrecked near Jebba. During 1858 the members of the expedition made several journeys to and from Lagos and arranged a regular postal system which worked effectively in spite of the war between Ilorin and Ibadan.[3] Dr. Baikie, who led the expedition, then established a settlement—'a market' as he called it—at Lokoja and was sending despatches home about the great prospects for British trade on the Niger—so great that he considered it justifiable to stay on, in defiance of his orders to return home, until he could convince the government of the necessity to appoint a consul to succeed him at Lokoja.[4] The government seems to have wavered about which route to concentrate on for the development of the Niger trade. The river was the obvious highway, but it could be ascended by large ships during only four or five rainy and mosquito-ridden months in the year, and the Delta people showed in November 1859, when they fired on Laird's ship, that they could put up a resistance. Then Laird, the only merchant ready at that time to put capital into developing the Niger trade, died in January 1861. All this emphasized the importance for the time being of the alternative overland route. The decision to

1 Memorandum dated 14 Dec. 1859, Venn to Hinderer, 21 Dec. 1859; C.M.S. CA2/L2.
2 R. Campbell: *Pilgrimage to my Motherland*, 1860; J. K. A. Farell; *The History of the Negro Community in Chatham, Ontario, Canada, 1787–1865*, op. cit., pp. 154–5.
3 Crowther to Consul Campbell, 1 Oct. 1857: 'This is to be a regular traffic and the mail bearers are to be paid'; C.M.S. CA3/04. Also Crowther to Venn, 4 Jan. 1858; C.M.S. CA3/04.
4 Dr. Baikie to Lord John Russell, 10 Sept. 1861; Slave Trade Correspondence, Africa (Consular), PP 1862 LXI.

annex Lagos in August 1861 was, in a way, a measure of the importance attached to that route.

The great merit of the route was the fact that the entry to it was at Lagos, the next stage was at Abeokuta, and both these places, unlike Brass, were friendly towards British penetration. Provided peace was maintained along that route, the great problem was porterage. Dr. Baikie and Captain Glover solved this problem in a rough and ready manner in 1858. They declared that all slaves who volunteered for their service were *ipso facto* emancipated. Several domestic slaves, notably Hausas from Lagos, Badagri and Abeokuta, flocked to them. In retaliation some of their masters, in particular chiefs of Igbein in Abeokuta, lay in ambush for the convoy of the Niger Expedition and plundered their goods.[1] This became a frequent sequence of events. In short, the need to exploit the overland route emphasized the issue of domestic slavery. Hitherto it had been the external slave trade that was a sin, to abolish which Britain was penetrating the country. Domestic slavery, supplying labour for economic development, was a social evil which need worry no conscience unduly. Indeed, under the guise of redeeming slaves and 'pawns' it became a Christian duty. But from about 1858 onwards domestic slavery was becoming recognized as an offence against both British law and the Christian golden rule of 'do unto others as you would be done by'. The immediate consequence was to antagonize the Egba against increasing British influence at Lagos, thus endangering the very friendship on which the overland route depended. It was at that stage that war was resumed in the interior.

Latent hostility between Ibadan and Ijaye led to war in March 1860. The Alafin of Oyo supported Ibadan. The Egba, having just staved off an invasion from Dahomey, decided to join Ijaye. Soon, the Ijebu joined the Egba. Thus the war spread and came to involve practically the whole of the Yoruba country.[2]

The news of the outbreak of the war created anxiety in Lagos, for it was a threat to trade and to the overland route just when so much was expected from it. To begin with, the successive Consuls in Lagos continued to favour the Egba side. As late as April 1861, Consul Foote in his anxiety to end the war went with Crowther to Abeokuta and offered military assistance not only against Ibadan but also against Dahomey

1 Cf. Townsend, Journal entry for 12 Sept. 1859: 'visited the Alake. He was very vexed about the Niger Expedition. He did not like their taking their old slaves up the country as freemen.' C.M.S. CA2/085.
2 J. F. Ade. Ajayi and R. S. Smith: *Yoruba Wars of the Nineteenth Century*, Cambridge 1964.

which every year either invaded or threatened to invade Egbaland.[1] A military officer was in fact sent to train Egba troops and to prepare the way for 250 soldiers from Sierra Leone. However, when he went to the battle-front and saw that a speedy end of the war was unlikely, he advised against such British involvement. William McCoskry, a trader long opposed to the Egba, became Consul when Foote died suddenly in June. It was he who began the complete reversal of the 'Abeokutan policy'. That same month, Lord John Russell gave orders for the annexation of Lagos and two months later was writing to McCoskry,

> You should lose no opportunity of impressing not only on the Alake and Chiefs of Abeokuta but also on the other chiefs of the Yoruba country that Her Majesty's Government have no favour or predilection for one tribe more than another.[2]

The war had closed the overland route. It blocked further missionary expansion. In March 1862 Ijaye was destroyed completely, and with it the C.M.S. and Baptist missions there, save for several children and a few converts evacuated to Abeokuta.[3] Several refugees from Ijaye settled in Abeokuta and other places. But the war did not end until 1864, and even then the Lagos government had to send an expedition to drive the Egba away from Ikorodu in 1865.[4]

It is not easy to generalize about the reaction of missionaries to these events, for they were far from being united. The missionary at Ibadan identified himself with the Ibadan cause just as the missionaries at

---

1 In April 1864 the Egba defeated an invading army from Dahomey. The news of this was just being publicized in England by enthusiastic missionaries when Prussia invaded Schleswig-Holstein. Palmerston and Lord John Russell protested, but, to the annoyance of many people in England decided that there was nothing England could do about it. Punch published some verses about the 'savage' king of Dahomey and the 'savage' king of Prussia: the former, as was right, was foiled; the other, because he had rifles, was triumphant:

ABEOKUTA AND DYBBOL
Oh the king of Dahomey's infuriate ire
Against Abeokuta breathed slaughter and fire . . .
O, that right could at Dybbol, too, thus have prevailed,
And the savage attack upon Sonderborg failed. . . .
Great and grave is the peril wherein the world stands
From the weapons of science in savages' hands.
Let us look to our arms, that, in coming to blows
We may lick, like the Egbas, the like of their foes.
                    Punch, 11 June 1864, vol. XLVI, p. 240.

2 Russell to McCoskry, 20 Aug. 1861; Slave Trade Correspondence, Africa (Consular), PP 1862 LXI.
3 For the last days of Ijaye, see Mann to Venn, 15 May 1862; C.M.S. CA2/o66. R. H. Stone: *In Africa's Forest and Jungle*, pp. 141 ff.
4 Hinderer to Venn, 15 Nov. 1864; C.M.S. CA2/o49.

Abeokuta identified themselves with the Egba cause.[1] However, it was Townsend's reaction that has been taken as representative of missionary opinion. Placed at Abeokuta, he no doubt occupied a strategic position. Besides, the Abeokutan policy destroyed by the war was in a special sense his own policy. His reaction will therefore be examined in detail later.

The government's reaction was apparently the annexation of Lagos. This might have proved an adequate solution if, as Governor Freeman urged, the government had been willing to bring enough money, troops and munitions to make Lagos the centre of an empire, to control the coast from Whydah to Palma and to annex Abeokuta. Lagos, said Freeman in 1863,

> must become the most important point on the West African coast if we join with it the exploration of the Niger and the exploration of the lagoons—but to do this we must have the influence of superiority over the natives.[2]

That would at least have created the political framework within which the agents of civilization could work. The people could have been 'civilized' through conquest if not through conviction. But Britain was not willing. It was announced that only Lagos island was annexed, and the Lagos government had to live on the customs dues of its trade. This created more problems than it solved. Whatever were the real reasons for the annexation of Lagos, it did not stop the war or solve the political crisis in the interior, or open the overland route. Lagos island was not an economic entity. The governors of Lagos therefore embarked on unauthorized expansion. This raised the problem of what diplomatic relations there could be between a Crown Colony and indigenous African states so closely knit politically and economically. The general effect was to create the fear of impending annexation in the states around Lagos, fear and suspicion and hostility towards all European penetration, philanthropic, mercantile or consular.[3] Where missionaries had before been welcomed, and schools and trade asked for, the people began to

1 Hinderer: 'As long as Ilorin stands as a Mohammedan power in this country, it is by no means to be wished that Ibadan's war power should diminish, or the Yoruba country would be overrun with Mohammedanism and Christian missions be at an end'; 24 Sept. 1860 to Venn, also 2 Aug. 1861; C.M.S. CA2/049. Townsend: 'The Egbas are the power that represents progress and advancing civilization, and it is to be feared if they should be conquered, our cause, or rather, that of God, would suffer at least for a time immensely'; 4 Oct. 1860 to Venn; C.M.S. CA2/085.
2 Governor Freeman's Memorandum on Lagos, 10 March 1863; CA147/4.
3 Crowther, who was by no means hostile to Glover or the Lagos government in general said that at Igbessa, in November 1866, there was 'an open con-

draw back, pointing to Lagos, where schools and trade led to annexation. An old ruler summed up the fear and hostility by saying that the menaces of Europeans were like those of the gorilla in the fable. The gorilla saw a hunter and his wife who went out plucking *agidi* leaves. They left their babe on a cloth spread under a tree. The gorilla came down and took the babe and was fondling it, singing, 'Ah, child, your father will not let me play with you, your mother will not let me play with you.' The child enjoyed the fondling and laughed and laughed till the parents' attention was drawn. The father ran to his bow and took a good aim. The gorilla shielded himself with the child. Then he got up and threw down the dead child and walked away, saying, 'I told you so; your father will not let me play with you.'[1]

By 1865 an alternative policy was being canvassed—that if Britain did not intend to be an effective ruler, it was not in her interest to continue to appear like the menacing gorilla. As C. B. Adderley said in Parliament in February 1865,

> Either we must render our governments secure by sending out larger forces, the opening up of the country, the making of roads and the extending of our power, or we must do less than is being done at present and stand out of the way of the native chiefs, who, if we were not there, would have full control over their own subjects.[2]

A Select Committee of the House of Commons under his chairmanship considered the annexation of Lagos a mistake—'a strong measure, of which not only the wisdom may be questioned, but the alleged justification also', and urged a gradual return to a policy of consular influence and advice, and education through missionary schools, a policy of co-operating with African rulers by training them in 'self-government'.[3] Historians have interpreted the resolutions of this committee variously as signifying the lowest point to which British interest in West Africa sank in the middle of the nineteenth century or as mere

---

fession on the part of the elders that their indecision [about encouraging missionary work] was not from want of appreciation of our work but from fear; they were afraid that we were pioneers of the Lagos government which would follow our steps and take away their country; that they had been strongly warned against receiving us and that by persons from Lagos'; to Venn, 6 Nov. 1866; C.M.S. CA3/04.

1 Reported by Samuel Pearse, the catechist who was sent to Ado, in a letter to the Secretary of the Yoruba Mission, 7 Oct. 1863, encl. in Harrison to Venn, 21 Oct. 1863; C.M.S. CA2/045. I have substituted 'gorilla' for Pearse's 'orangoutang' as a more likely translation of the Yoruba original.
2 Hansard, 21 Feb. 1865.
3 *Report from the Select Committee on Africa (Western Coast) together with the proceedings of the Committee. Minutes of Evidence and Appendix*, PP 1865, pp. xiv–xvi.

word-mongering in the safe confines of London, signifying nothing on the spot, where British interests and the reality of events marched relentlessly on. Or else that the emphasis of the committee was purely negative, 'not so much on withdrawal as on economy'.[1] The real point was that the committee took seriously their recommendation that the Africans be trained for self-government as an alternative to intermittent British political intervention. The implication was that in colonies Britain wished to retain, Africans, who cost less to maintain than Europeans, should be advanced into posts of responsibility in government, and in those Britain did not wish to retain, the local inhabitants should be taught the European art of government. Owing to the difficulty of wars about which Britain could do nothing constructive, and to the problems of economy and climate, self-government was being urged as a positive thing good in itself, good for the interests both of Britain and of the local communities.

The 1865 Committee was not alone in this view at this period. The remarks of Earl Grey, that the interests of Britain on the Gold Coast were best to be served by training the Fanti in self-government have already been cited.[2] Earl Grey was advocating not that British interests should be abandoned, but that Africans trained more or less along Western lines were to be the channel of British policy. In India, where British rule was firmly established and Britain was far from seeking to abandon her rights, there was discernible a marked desire to train Indians to play an increasing part in the administration of the country. It was in urging the training and employment of Indians in the civil service that Lord Macaulay in a debate in the House of Commons declared that 'having become instructed in European knowledge, they may, in some future age demand European institutions. . . . Whenever it comes it will be the proudest day in English history.'[3] Education in schools and universities of the Western model was in fact intensified, and the solemn proclamation of Queen Victoria as Empress of India in 1857 contained the passage so often quoted:

> And it is our further will that, so far as may be, our subjects of whatever race or creed, be freely and impartially admitted to office in our service, the duties of which they may be qualified by their education, ability, and integrity duly to discharge.[4]

1 Flora L. Shaw (Lady Lugard): *A Tropical Dependency*, London 1905, p. 348; Dike, op. cit., pp. 167, 168.   2 See p. 16.

3 Macaulay in the House of Commons, 10 July 1833, quoted in *From Empire to Commonwealth*, ed. Jack Simmons, 1949, p. 83.

4 The Queen's Proclamation in India, 1858, ibid., p. 154. For references to this passage in West Africa, see T. F. V. Buxton: 'The Creole in West Africa', in *Journal of the African Society*, vol. XII, 1913, pp. 385–94.

This was an important principle. It was the assumption underlying a good deal of the missionary work we have been discussing, and it culminated in the doctrine of self-government as a virtue and not dictated only by the needs of economy. Economy and other necessities remained the most persuasive advocates of the doctrine, but by a few individuals who loomed large in the middle of the century, self-government was advocated for its own sake. The most important of them for this study was Henry Venn of the C.M.S. He urged the training and employment of Africans more as a virtue than as a measure of economy. He believed that the 1865 Committee had vindicated his policy:

> The positive Resolution that the natives are to be trained for ultimate self-government is an important principle on our side . . . and seems to have set at rest the senseless outcry against the [financial] support of the colonies and against the capacity of the negro.[1]

He regarded the annexation of Lagos as a mistake, not because it meant a displacement of Dosunmu, the ruling king, but, as he told Governor Freeman in 1863, because in the British administration established, Africans, whether traditional rulers or Western educated, had no place:

> many of those Europeans who had long been rivals in trade were at once exalted to places of authority and profit while no steps were taken to bring such advantages within reach of natives.[2]

And it was the fear that annexation would almost invariably lead to this that made him urge, not withdrawal, but a return to consular influence guiding an educated African middle class as the instrument of British policy.

It was easier for missionaries than for government officials to see the virtues of self-government because, though like the government they were subject to the fears of the climate, much more than the government they were pressed by the need to make every penny go as far as possible, and above all, because evangelization, much more than administration, required active response and co-operation from the people. Mere efficiency might be a sign of growth and health in an administration. 'The breath of life in a native Church,' said Henry Venn, depended on 'self-government, self-support, self-extension.'[3] That was the lesson he learnt

1 Venn to Townsend, 23 Aug. 1865, 22 Sept. 1865; C.M.S. CA2/L3.
2 Venn to Governor Freeman, 23 Oct. 1863, ibid. Freeman had called on the Parent Committee, and Venn put down for him in writing 'the main objections to your policy felt by a large body of those who have been long interested in the welfare of Africa'.
3 W. Knight: *The Missionary Secretariat of Henry Venn*, London 1882, p. 416.

from his study of the history of earlier Roman Catholic missions,[1] that the missionary who did not prepare for the day when he would no longer be in the mission by raising up an indigenous clergy and episcopacy was building on sand:

> It is expedient that the arrangements which may be made in the missions should from the first have reference to the ultimate settlement of the native Church upon the ecclesiastical basis of an indigenous episcopate, independent of foreign aid or superintendence.[2]

Venn's greatest claim to the commanding position he came to occupy in the history of the expansion of the Church was the way in which he developed these ideas into something like a Code of missions.[3]

The cardinal doctrine was that a distinction must be made between 'the office of a *missionary*, who preaches to the heathen and instructs inquirers or recent converts, and the office of a *pastor*, who ministers in holy things to a congregation of native Christians'. While the missionary, maintained by a foreign missionary society, 'should take nothing of the gentile', the pastor must be supported financially by his congregation for 'the ox that treadeth out the corn should eat of the same'.[4] The converts a missionary made should be organized as soon as possible in little bands, each under a headman, and should start at once to make contributions to a native church fund separate from the funds of the foreign missionary society. Soon the bands should come together and form a congregation under a native teacher or catechist, whom they

1 Venn published in 1862 *The Missionary Life and Labours of St Francis Xavier* in which he traced the failure of Xavier's mission in the Far East to overdependence on European political power and failure to raise a local clergy.
2 William Knight: *The Missionary Secretariat of Henry Venn* (London 1882), p. 417.
3 There are four important papers drawn up by Venn and issued by the Parent Committee as Instructions to missionaries on the subject of the Organization of Native Churches: (i) in 1851; (ii) July 1861; (iii) January 1866; (iv) 'On Nationality' June 1868. The first three are published in Appendix C of the *Missionary Secretariat of Henry Venn* and the fourth in the *Memoirs of Henry Venn*, 1880, pp. 282–7. As these papers show, Venn's ideas on this subject developed with time, especially as a result of his correspondence with the Rev. Rufus Anderson, Secretary of the American Foreign Missions Board (for whose ideas see his *Theory of Missions to the Heathen*, Boston 1845, and *Foreign Missions, their Relations and Claims*, Boston 1869). Venn and Anderson both agreed on the distinctive roles of pastor and missionary but not on the training of the pastor. Anderson would have him 'unspoilt'; Venn at first tacitly agreed, saying that the best of the pastors would be better trained and graduate into missionaries, but in the 1861 paper he explicitly revised this: the leading pastors must be well trained, remain pastors, evolve a national church and eventually become bishops to preside over it. I have quoted more from Venn's mature views.
4 July 1861; *Missionary Secretariat of Henry Venn*, op. cit., p. 416.

should endeavour to maintain. Soon the catechist or any other suitable native should be ordained pastor and the missionary should then move on to fresh ground. Thenceforth the missionary was 'to exercise his influence *ab extra*, prompting and guiding the native pastors to lead their flocks and making provision for the supply for the native Church of catechists, pastors or evangelists'.[1] If the missionary stayed behind and became the pastor, the goal, the *euthanasia* of missions and a 'self-governing, self-supporting, self-propagating Church' would never be achieved.

The pastor, said Venn, must be of the people and maintained by them. He rejected the practice in Sierra Leone, where missionaries acted as pastors and everything was done to conform to the English pattern. Similarly he rejected the other extreme that because the pastor must not be cut off from his flock, he must be stuck to their social standards and preserved 'unspoilt'. The native teacher, he said, 'should not be too highly raised above his countrymen in his habits and mode of living . . . [but] he must always be a little ahead of the civilization of the people around him and by his example and influence lead that civilization forward'.[2] At first he thought that the best-educated pastors should graduate into missionaries, but he soon changed his mind because that might be misunderstood to mean that the difference between missionary and pastor was one of an upper and a lower degree. And in any case the best-educated pastors were required to organize the different congregations into a native Church as a national institution.

> The native Church needs the most able native pastors for its fuller development. The right position of a native minister and his true independence must now be sought in the independence of the native Church and in its more capable organization under a native bishop . . .[3]
>
> Let a native Church be organized as a national institution. . . . As a native Church assumes a national character, it will ultimately supersede the denominational distinctions which are now introduced by foreign missionary societies. . . . Every national Church is at liberty to change its ceremonies and adapt itself to the national taste.[4]

But that must be the work of the native pastorate. The temptation for European missionaries to assume the role of the pastor must be resisted,

1 June 1868; *Memoirs of Henry Venn*, op. cit., p. 285.
2 Instructions to Townsend, Gollmer and Crowther, 28 Oct. 1844; C.M.S. CA2/L1.
3 January 1866; *Missionary Secretariat of Henry Venn*, op. cit., p. 425.
4 June 1868; *Memoirs of Henry Venn*, op. cit., pp. 285–6.

for 'such a scheme, even if the means were provided, would be too apt to create a feeble and dependent native Christian community'.[1]

In theory this view that all missionary effort was from the start to be directed towards the creation of a self-reliant Christian community and that this involved the training of an indigenous clergy to be raised to positions of responsibility in the Church as soon as possible was widely held in all the missions at this period. It was implicit in the church organization of the Presbyterians and Baptists, particularly the Baptists, who took the view that the pastor was directly responsible to the congregation to which he ministered. In the Methodist mission the very phrases of Venn were being echoed. In 1860 the Secretaries wrote to the Acting General Superintendent of the Methodist Mission in the Gold Coast and Yoruba:

> It is of the highest importance to the welfare of the work that wherever societies are formed they should be trained to contribute towards the support of the ministry they enjoy. . . . A missionary society may confer inestimable blessings in commencing a work in any country; but when a flock is gathered, the Divine Rule requires that it should supply the wants of its own shepherds, and not leave them to be provided for by strangers.[2]

Just when Venn was formulating his ideas into principles, Mgr de Bressilac, founder of the S.M.A., as a bishop in India, was advocating the same policy, making the same distinction between the roles of pastor and missionary, emphasizing the need to raise revenue locally, and above all, to raise an indigenous clergy who could approach the people as the foreign missionary could not. He believed passionately that

> without a numerous and indigenous clergy, nothing that has been achieved is stable, and no general movement will arise. I am so convinced of this that I would not hesitate, I am sure, to neglect more or less other activities for a few years in order to concentrate on that activity which would ensure the prosperity of everything else later on,[3]

that is, the training of a local clergy. He argued also that the training must be thorough, that it was ignorance that bred pride, not education;

1 July 1861; *Missionary Secretariat of Henry Venn,* op. cit., p. 419.
2 Secretaries to H. Wharton, Acting General Superintendent, 22 Dec. 1862; *Secretaries' Letter Book,* Meth.
3 *Le Missionaire d'après Mgr de Bressilac,* S.M.A. Lyons 1956, p. 43. See also S.M.A. *100 Years of Missionary Achievements,* p. 11; J.M. Todd: *African mission: A Historical Study of the Society of African Missions,* London 1961, Chapter II.

that the most humble, most pious and most zealous Indian priests he knew were those who had the best education.[1]

Thus, with varying degrees of emphasis, those who directed missions in Nigeria at this time were more or less agreed in theory on the important role that African pastors must play. There were, however, important denominational differences in their points of view.

For one thing, self-government must mean different things to different denominations. Venn's emphasis was on the ecclesiastical aspect, the raising of an indigenous clergy, as distinct, that is, from doctrinal independence; but the two were inseparable. For example, the mature view of Venn, probably born out of his experience of the Indian Mutiny, that for the Church in a mission field to be fully established, it must be organized as a national institution, and that this could only be done by an indigenous clergy, was in the middle of the nineteenth century hardly orthodox Anglican doctrine. Nor was it a view that Roman Catholics could share, though it must be said that Mgr de Bressilac emphasized that the universality which was the crowning glory of Catholicism did not make the Spanish Church exactly the same as the Italian and that God wanted Indians to be not Frenchmen, but Christian Indians.[2] But the essential Catholic doctrine was the universality and the pervading influence of the hierarchy of the Church, of which the local clergy were only a small part. Even Methodists and Presbyterians who accorded some measure of independence to each 'society' or presbytery had a close-knit organization with the parent Churches, and local self-government could develop only within the framework of such organization. The Baptists, taking a more 'congregational' view of church organization, accorded greater independence to each congregation.

Doctrine and church organization apart, the time-table for promoting self-government was bound to vary. For, while the Methodists and the C.M.S. began with an African staff and congregations of emigrants ready made, the Baptists and Presbyterians had practically none of such advantages. The Roman Catholics did not arrive until a generation later and they had to raise a celibate clergy. Thus while the problem of the training of African pastors and the devolution of power to them was in the C.M.S. and Methodist missions already an issue by the 1850s, it did not arise till almost a generation later in the Presbyterian mission, in the late 1870s. In the Baptist mission until the late 1880s the problem arose only in the form of recruiting 'coloured' missionaries who were supposed to be able to stand the climate.[3]

Taking such basic variations into consideration, it may still be said

1 *Le Missionnaire*, pp. 46–7.
2 Ibid., pp. 32–3.   3 Bowen to Taylor 23 Nov. 1857; *Bowen Letters.*

that just as the training of agents did not measure up to the demand, the training of Africans for self-government in church affairs did not always measure up to the challenge of Venn's theory. The greatest obstacle was probably the one referred to by Venn—the tendency of European missionaries to wish to act as pastor to the congregation they had gathered, and to regard the African pastors as their rivals.

The earliest mission to adopt some measure of self-government was that of the Presbyterians in Calabar. The departure of Hope Waddell in 1857 was taken to mark the end of the beginning of the Calabar Church. In September 1858 the 'presbytery of the Bight of Biafra' was established, giving the local Church some say in its own affairs, subject to the supervision of the Foreign Mission Board. But the missionaries there apparently did not accept Venn's distinction between missionaries and pastors, and the advancement of African pastors was as cautious as the rate of expansion. The missionaries met in presbytery as the pastors of the various congregations, European lay agents represented the congregations as elders and only where such men were not available were emigrant lay agents chosen.[1] Only gradually did indigenous church elders come forward. The reason for this extreme caution was given as the paltry ability of the Africans themselves. In 1880 the Rev. Samuel Edgerley jun., in charge of training African agents, argued in favour of limiting their education, as follows:

> That the morality and intelligence of our agents should be distinctly ahead of the morality and intelligence of the community is, I think, what we should aim at, and what we have got; but to require in our agents an educational standard in any way approaching to what would be required in missionaries and teachers at home is, I also think, requiring the unnecessary and impossible. Unnecessary, with reference to the requirements of the community; impossible, with reference to the ability of the agents. But as long as the training of the native agents is in the hands of European missionaries, the agents will be safely ahead of the community.

In the same breath, Edgerley talked of the agents'

> seeming inability to originate a scheme. They are good at imitation, but they seem as yet to be unable to do more than imitate. . . . I

---

1 Goldie, op. cit., p. 189. The members of the first Presbytery were Anderson as Moderator, Dr. Robb as Clerk, the Rev. Zerub Baillie and the Rev. William Thomson. The West Indians Archibald Hewan, medical officer, and Henry Hamilton, schoolmaster, were Elders representing the congregations of Duke Town and Creek Town respectively.

think we should not settle down to the belief that our native agents will relieve us of the pioneering duty.[1]

This attitude of the European missionaries inevitably led to conflict with some of the agents, but at Calabar, where the missionaries on the spot controlled the pace of development effectively, the conflict did not become an issue. The situation in the C.M.S. was different. Their African staff was much larger and more powerful than was the case in the Presbyterian mission, and the resistance of the missionaries on the spot could not match the encouragement the African staff received from the Secretaries in England. And there was the constant example of Sierra Leone with an even more powerful African clergy and laity with which the emigrants in Nigeria maintained close touch. Venn began to apply his principles there in 1852 when, since it was a British colony, he got the Colonial Office to appoint a Colonial Bishop and to authorize the annual appropriation of £500 of the revenue of the colony towards the endowment of the local Anglican Church. Then in the following year he drew up a Native Pastorate Scheme under which the missionaries yielded the pastoral care of each of the established C.M.S. churches to African ministers, and the C.M.S., while still retaining control over mission land and church property, shifted financial and administrative control of the churches and their clergy to a church committee and a church council on which the Bishop, the C.M.S. itself, as well as the pastors and laity, were represented. Owing to the deaths of the first three Bishops of Sierra Leone within six years, the scheme was not fully established till 1860. Meanwhile Venn had been working out a similar scheme for C.M.S. missions in Nigeria.[2]

With the same care and anxiety that Venn was developing the theory of advancing the indigenous clergy into office, and was seeking to convince not only missionaries of his own society but those of others as well of the need to put it into practice, Townsend was building up and publicizing arguments for a contrary policy—the missionary counterpart of the policy of annexation. In 1851, when Venn issued his first paper on the organization of the native Church, the only African clergyman in the mission was Samuel Crowther who, like Townsend, was ordained in London in 1843, and whom since the expedition of 1841 Townsend looked upon as a rival.[3] In that year Venn proposed that as soon as the new Bishop of Sierra Leone could get to Abeokuta, two more Africans should be ordained. One was T. B. Macaulay, who had recently returned from the C.M.S. Training Institution at Islington,

1 S. H. Edgerley, in U.P. *Missionary Record*, 1880, p. 35–6.
2 Stocks, op. cit., vol. II, pp. 100–1.   3 See p. 32.

where Townsend had been trained, and the other Theophilus King, who had distinguished himself as catechist to the ill-fated settlement at Lokoja in 1841, had gone to Fourah Bay, and was Crowther's able assistant as translator of the Bible in Yoruba. Townsend wrote to oppose their ordination.

> I have a great doubt of young black clergymen. They want years of experience to give stability to their characters; we would rather have them as schoolmasters and catechists.[1]

When a C.M.S. publication reported that at the last anniversary meeting of the society, the Rev. H. Stowell, an influential supporter, 'made a distinct proposal for the erection of an episcopal see at Abeokuta, to be occupied by a black bishop', and Townsend learnt that Crowther had been summoned to England, his doubts became a panic. He announced that he would conduct a referendum among his African staff to find out how many of them wanted to be ruled by a black bishop.[2] And since Townsend could be nothing but thorough, he drafted a petition against the proposal and obtained to it the signatures both of Hinderer and Gollmer, German Evangelicals who on doctrinal grounds were not fond of the episcopal form of government anyway, as well as of one other English missionary, Isaac Smith.

The main argument of the petition was not that an African fit to be a bishop could not be found, but that an African bishop, however worthy, would lack in the country the respect and influence necessary for his high office:

> Native teachers of whatever grade have been received and respected by the chiefs and people only as being the agents or servants of white men . . . [and] not because they are worthy. . . . Our esteemed brother Mr. Crowther was often treated as the white man's inferior and more frequently called so, notwithstanding our frequent assertions to the contrary. . . .
>
> This state of things is not the result of white men's teaching but has existed for ages past. The superiority of the white over the black man, the negro has been forward to acknowledge. The correctness of this belief no white man can deny.[3]

He went on to argue that as long as the country remained heathen,

1 Townsend to Venn, 21 Oct. 1851; C.M.S. CA2/085.
2 Townsend to Venn, 15 Oct. 1851; C.M.S. CA2/085, (55) Isaac Smith, Henry Townsend, David Hinderer and C. A. Gollmer to Major Straith, 29 Oct. 1851; C.M.S. CA2/016.
3 Isaac Smith, Henry Townsend, David Hinderer and C. A. Gollmer to Major Straith, 29 Oct. 1851; C.M.S. CA2/016.

no native who had no traditional title or rank could command respect outside the mission village, except as the agent or servant of the white man. He elaborated this by saying that the country was torn by sectional jealousies, and that the ethnic affiliation of an indigenous bishop would make his authority unacceptable even to converts of other sections. Finally, he plunged into deeper waters still :

> There is one other view that we must not lose sight of, viz., that as the negro feels a great respect for a white man, that God kindly gives a great talent to the white man in trust to be used for the negro's good. Shall we shift the responsibility? Can we do it without sin?[1]

It is not known what Venn's immediate reaction to this letter was. But though he always gave credit to Townsend's missionary zeal, ability and clarity of thought,[2] he was constantly on his guard to resist his ambition to become head of the Yoruba Mission. In a sense Townsend was the founder of the Yoruba Mission, which began with his journey to Abeokuta in 1842–3. When he was sent with Gollmer and Crowther, one a German, the other an African, to establish the mission, he thought his headship was acknowledged by his appointment as Secretary while the other two were to take turns as Chairman of the Local Committee. But even at the risk of sacrificing unity of action, Venn insisted that every ordained missionary, including newcomers like Hinderer and Mann, was directly responsible to the Parent Committee, and none was above the other. The German missionaries, particularly Gollmer and Mann, complained of Townsend trying to rule them. Townsend resigned as Secretary in 1855 and the office went into rotation. Not till 1861 did Venn appoint another permanent Secretary with headquarters at Lagos. And even then the question of one of the missionaries being head over the others did not arise. Venn insisted that the institution of a hierarchy belonged to the organization of the native Church, and that it was not the work of the missionary. The result was that Townsend was left with a feeling of frustration, for he believed, with some justification, that it was the policy that Venn was pursuing that deprived him of the office to which his achievements entitled him and in which his abilities would have full scope. In another age, or another mission, Townsend would have been a much respected, renowned, if autocratic bishop. Under Venn he remained an ambitious but frustrated leader of the opposition.

1 Ibid.
2 Cf. Venn, 25 Oct. 1864, to Colonel Ord, who was on a Commission of Inquiry to West Africa, recommending to him which missionaries were likely to give him the most useful information: 'In the Yoruba Mission, the Rev. Henry Townsend is the most intelligent and experienced of our missionaries and the most influential with the natives.' C.M.S. CA2/13.

Against Townsend's wishes, King and Macaulay were ordained when at last the Bishop of Sierra Leone reached Abeokuta in 1854. In 1857 three other Africans were ordained. But the next step Venn took towards the establishment of the native Church was more effectively thwarted. In 1855, when peace seemed established in the interior, Venn said it was time that the pastoral care of some congregations should devolve on African pastors or catechists, while the missionaries moved on to occupy fresh ground in the interior. The two most flourishing congregations were Townsend's at Ake and Crowther's at Igbein, the first two stations at Abeokuta. Venn asked Crowther to move to Lagos, with a view to joining the next expedition and establishing new stations on the Niger; Gollmer, a reputed builder and handyman of the mission, was to move to Ake and there supervise the Industrial Institution; and Townsend should move to the small congregation at Ikija, near the Northern Gate of Abeokuta, from where he was to supervise expansion into the interior. But while Crowther and Gollmer were willing to move, petitions flowed into C.M.S. House from the Ake congregation and the Alake against the proposal to transfer Townsend.[1] Townsend pointed out the political interests of the mission which would be jeopardized if he left Ake, and Venn had to yield. Gollmer went to Ikija; the Training and Industrial Institutions were removed from Ake to Igbein, and Townsend remained as Pastor of the Ake congregation. It was Townsend who opened up much of the interior of Yoruba to C.M.S. missionaries, but the supervision of the catechists there fell to the missionaries at Ijaye and Ibadan, Mann and Hinderer respectively, Townsend being tied down to pastoral work.

While Townsend was seeking office and falling just short of attaining it, Venn was calling Crowther to posts of more responsibility and power than Crowther really cared for. What impressed Venn most in Crowther was that in all the bickering and struggle for power in the mission, Crowther not only showed no great desire for office, and was for that reason a most reliable counsellor, but he was also endowed with great tact, remarkable knowledge of human psychology and a consequent ability to feel for others, understand them and, if necessary, manage them. In a letter to Crowther in 1858, Venn remarked on the

honour which you have long had of promoting harmony and brotherly love in the mission by your wise and humble spirit.[2]

Crowther was then on the Niger. He went on the expedition of 1857,

1 Gollmer to Venn, 2 Nov. 1857; C.M.S. CA2/043.
2 Venn to Crowther, 22 July 1858; C.M.S. CA3/L2.

established new stations at Onitsha and Igbebe, was shipwrecked near Jebba and stranded for almost a year, went down by the relief boat to Onitsha in October 1858, and returned to Lagos. Crowther then reported that the stations were established, that he hoped the expansion in Yoruba would push north and reach the Niger at Rabba, because he regarded the Niger Mission as an extension of the Yoruba Mission. But Venn replied:

> The Committee fully concur in your suggestion that the Niger Mission is to be regarded as an extension of the Yoruba Mission. It may ultimately be placed under the Yoruba Committee but as long as you remain in the Niger you are invested with sole authority to act and make all pecuniary and other arrangements.[1]

Again and again, Venn repeated this. Two months later he wrote:

> The Committee repose entire confidence in you as the Head and Director of the Niger Mission and commit to you all the arrangements in respect of the location of the new labourers [i.e. missionary agents].[2]

And a month later:

> The Committee still regard you as having the *direction* of the Niger Mission. You will of course consult with others but in any matters in which your judgement is decided, all must follow your directions.[3]

The repetition was due to Crowther's hesitation to assume responsibility. He returned to the Niger to visit the new stations in 1859, having to travel by canoe from Onitsha to Rabba and by horse overland through Ogbomoso, Ibadan and Abeokuta, to Lagos. Then, in January 1860, he wrote to the C.M.S. Secretaries that he had had enough of expeditions and explorations; that, besides the usual rigours of the traveller, he did not enjoy being in such close quarters with the 'mixed body of men of different characters, temper, view and aim and mostly of no right Christian principles' with whom he had had to travel and to camp for almost a year at Jebba; that he was getting on in years and his health was declining. He wished to leave the management of the Niger Mission to others. But he would not 'urge the grant of this favour' until the new mission stations at Onitsha and Igbebe were fully established, and if possible Idah be taken up as a station:

> Then I should like to spend the remainder of my days among my own people, pursuing my translations as my bequest to the nation.[4]

1 Secretaries to Crowther, 23 April 1858; C.M.S. CA3/L1.
2 Secretaries to Crowther, 22 June 1858; C.M.S. CA3/L1.
3 Secretaries to Crowther, 22 July 1858; C.M.S. CA3/LI.
4 Crowther to W. Knight, 5 Jan. 1860, also to Venn, 4 April 1860; C.M.S. CA3/04.

Venn replied at once that he hoped the period of exploration was over, but that

> When we reckon upon rest, [God] often calls us to increased exertions. . . . The Lord has honoured you by making you his instrument for opening the Niger to the Gospel. Should a native Church be established there and should He call you to preside as a missionary bishop, you would not be the person to run away like Jonah.[1]

Venn's suggestion that Crowther might become a 'missionary bishop' was a little odd. A bishop at the head of a missionary party, as distinct from the head of a native clergy, was a High Church idea that Venn specifically attacked. It confused the roles of missionary and pastor; it contradicted the Scriptural example of the apostles going out to evangelize two by two, none being placed above the other. Venn used the term 'missionary bishop' to mean an Anglican bishop in non-British territory—sometimes called a 'Jerusalem' bishop as distinct from the diocesan at home or the colonial bishop on British territory overseas.[2] But another confusion remains. Venn suggested that Crowther might be a 'missionary bishop' should a native Church be established on the Niger and he be called upon to preside over it. Now, in 1860, no Church had been formed on the Niger: it was the churches in

---

1 Venn to Crowther, 23 Feb. 1860; C.M.S. CA3/L2.
2 See the 'Bishops in Foreign Countries Act', sometimes called the Jerusalem Bishopric Act, 5 Vict c6 1841. While a license under the great seal is required for the election of the ordinary bishop, for a Bishop in Foreign Countries, the archbishop having satisfied himself about the sufficiency of the candidate in good learning, the soundness of his faith and the purity of his manners, applies for a royal license under the royal signet and sign manual authorizing the consecration. If the candidate is not a British subject (as Crowther was) the archbishop can dispense with the Oath of Allegiance. Such a bishop has no right to be called Lord Bishop, but the Rt. Rev. Bp. X, and properly speaking, is not Bishop *of* a place, but a Bishop *in* the place. (Halsbury's *Statutes of England* (3rd edition 1955), vol. xiii, pp. 19–20 and notes, p. 62. Also Minute of the Parent Committee of the C.M.S., 21 March 1871 on the controversy over the Madagascar Bishopric. The Committee distinguished between a Colonial Bishop and a Bishop appointed under the 5 Vict. c6 Act, maintaining that all Anglican missionaries automatically came under the jurisdiction of the Colonial Bishop but that in the latter case, 'the law allows a discretion as to their converts being placed under a Bishop so consecrated. . . . The Committee must add that they conceive that the proper sphere of a missionary bishop consecrated under the Jerusalem Act is the *Native Church* when it is sufficiently advanced to require a resident bishop. Under this Act, the Society took the first step in the important branch of the extension of the native episcopacy by promoting the consecration of the native minister Dr. Samuel Crowther for the Mission on the Niger'; *Missionary Secretariat of Henry Venn*, 1882, p. 443. (Crowther on the eve of his consecration was awarded an honorary D.D. by the University of Oxford.)

Yoruba that were in dire need of a bishop to conduct episcopal visitations, confirmations and ordinations. Hitherto the 'Colonial Bishop' of Sierra Leone had been performing these functions; but between 1854 and 1859 three bishops had died, each on his way from a visitation to Yoruba or shortly after his return to Sierra Leone. The Bishop of Sierra Leone was clearly too far away. It was even difficult to get a new Bishop of Sierra Leone, and Crowther's name was being canvassed. Venn said it was only with difficulty that he could prevail on the Secretary of State to leave Crowther to devote himself to the work in Nigeria.[1] He knew that the European missionaries led by Townsend would have resisted the authority of a Bishop Crowther at Lagos or Abeokuta. His mind was therefore working towards the slightly confusing compromise of a Bishop Crowther consecrated because of his outstanding position on the Niger, whose authority would with time and necessity gradually spread to the Yoruba country where it was most needed.

Even the decorum proper to the conduct of missionary affairs could not disguise the fact that Townsend and Crowther had become open rivals. With more than the usual *nolo episcopari* of bishops elect, Crowther declared himself unworthy of any other post than the one he was already 'sustaining under many disadvantages'. That, using Venn's arguments, he was a missionary, not a pastor: 'Younger persons should be trained up with the view should the native Church need a person of that capacity [of bishop] one may be chosen for the office.' Besides, he did not wish to stand in anybody's way:

> The European missionaries who have sacrificed everything to come out to Africa, taking their lives in their hands, have a greater right to this claim. . . .
>
> As a man I know something of the feelings of men. . . .[2]

1 Venn to the Rev. J. A. Lamb, Secretary of the Yoruba Mission, 23 Jan. 1864: 'Before Bishop Bowen was appointed to Sierra Leone [i.e. in 1858] the Secretary of State had determined to recommend Mr Crowther as Bishop of Sierra Leone and the late archbishop was strongly in favour of it. I objected on the ground that it was too much of an English colony and it was with difficulty that I could stop the nomination. I feel pretty sure that Bishop Beckles will resign and in that case the appointment of Crowther to Sierra Leone will be again revived and after the miserable experience of white Bishops [3 died in 6 years and the fourth apparently proved unsatisfactory] I cannot predict the result though I should rather that his services were confined to Lagos and the Niger.' C.M.S. CA2/L3.

2 Crowther to Venn, 4 April 1860. He added 'I must confess that the use that has been made of my name in the English newspapers when the Bishopric of Sierra Leone was vacant has not done me good in the mission. Though I have not heard the remarks made by individuals myself, but from those who have heard them, that feelings of great contempt have been uttered, and a threat of quitting the mission has been expressed were such the case.' C.M.S. CA3/04.

Townsend was aware of what was going on. He argued back in force in favour of his claims, though he was losing the argument all the way. Once, in October 1858, he went near producing a reasoned counter-blast against Venn's theory of the Native Church Organization:

The purely native Church is an idea, I think, not soon to be realized. The white merchant and civilization will go hand in hand with missionary work, and foreign elements must be mixed up with the native, especially in such changes as are effected by a religion intro-duced by foreigners. The change of religion in a country is a revolu-tion of the most extensive kind and the commanding minds that introduce those changes must and do become leaders. It is a law of nature and not contrary to the laws of God, and efforts to subvert such laws must produce extensive evils.[1]

Townsend knew, of course, that many of the Africans sharing in the leadership of that revolution were less concerned than himself about the mixing of foreign elements with the native, and that Venn's ideas on native Church organization implied no such exclusion of foreign ideas. Townsend's other arguments were even less impressive. Apart from the old one that an African bishop, however worthy, would command no respect, he embarked on a general denigration of Africans, using any stick whatever he could find to beat at them. Once or twice he hit the mark. He referred to Henry Robbin's and Samuel Crowther junior's mismanagement of the cotton business at Abeokuta, and recalled how two African clergymen sent from Sierra Leone to work on the Niger, 'their own country', ran back as soon as they arrived. 'And yet excuses are made for them!'[2] More often he missed the mark by a wide margin. Although European clergymen were paid £200–250 a year while the stipends of African clergymen varied between £50 and £150, by some strange arithmetic he worked it out with Dr. Harrison that the Niger Mission, using only African staff, was not only inefficient but was pro-portionately more expensive than the Yoruba Mission.[3] (This result was obtained by dividing the amount spent in each mission between 1860 and 1862 by the number of agents, making no allowance for the capital expenditure in the new mission.) Stranger still, when two German catechists who had been staying at Abeokuta were due to accompany Crowther to the Niger, Townsend not only dissuaded them from going; he wrote to Venn:

It appears you intend to work up the native clergy to notice by sending

1 Townsend to Venn, 18 Oct. 1858; C.M.S. CA2/085.
2 Townsend, Annual Letter for 1859, dated 31 Jan. 1860, ibid.
3 Harrison to Venn, 21 Nov. 1863 and encl.; C.M.S. CA2/C45.

with them inferior white men, but it won't answer. The white man must be in advance in ability, in religion, in position, to the native teachers of all kinds, or if he ceases to be so, he must leave the work.[1]

Three months later, he added

If you want young men to go to the Niger, you must give them a white man as a leader. No opinion you can form, no statement you can make, no advice you can give will make them [i.e. black men] what they are not. They are not fit to be leaders at the present time.[2]

With characteristic efficiency Townsend had exhausted all the possible arguments, including the modern one so familiar in Central and South Africa, that the superiority of one race over the other is proved if you keep the worst of the one always above the best of the other. Since Townsend felt so deeply about the matter, it is not surprising that the quarrel spread from Crowther to educated Africans as a whole. In January 1859, the Parent Committee had to inform Townsend that

such letters as were written to the Rev. J. White and Rev. J. Morgan in reply to their application for an increase of salary were not in that mild and Christian tone which ought to have characterized the letters of senior missionaries to native clergymen. The Committee have consequently felt compelled as Christian men to write to these native ministers in a different tone.[3]

The other European missionaries therefore kept studiously neutral over the controversy. Yet the controversy could not but spread. When in 1864 Venn asked the European missionaries directly whether or not they were willing to place themselves under the jurisdiction of a Bishop Crowther,[4] besides the views of Townsend and Dr. Harrison at Abeokuta, both violently opposed to the idea, Hinderer's answer was typical:

Not that I should have the slightest objection to Bishop Crowther being over myself and the congregation which God may give me. On the contrary, I can only respect and love him. . . . But . . . The country is heathen and mixed up with and held up by heathen priest-

1 Townsend to Venn, 5 Nov. 1859; C.M.S. CA2/o85.
2 Townsend to Venn, 18 Feb. 1860; C.M.S. CA2/o85.
3 Venn to Townsend, 21 Jan. 1869; C.M.S. CA2/L2.
4 Cf. Venn to Hinderer, 22 Sept. 1864: 'The Parent Committee leave both Abeokuta and Ibadan to take their choice between the jurisdictions of the one or the other [i.e. Crowther or the Bishop of Sierra Leone] though they have a strong conviction that Bishop Crowther will give the most valuable assistance.' C.M.S. CA2/L3.

craft, and we are allowed to teach and preach the Gospel not because they are tired of heathenism, but because God gives us influence as Europeans among them. This influence is very desirable and necessary to us; but if they hear that a black man is our master, they will question our respectability.[1]

By and large, until 1867, European missionaries on the spot accepted Townsend's basic thesis that the success of the Christian missions depended on the prestige and influence of missionaries as Europeans. Inevitably the controversy spread to the emigrants at Abeokuta and, during the Ijaye War, took a political turn.

The emigrants were becoming increasingly wealthy and powerful at Abeokuta and (particularly the younger ones, who were products of the missionary training schemes and a good number of whom were educated in England) they regarded Crowther as their hero. As a body they were bound to resist Townsend's campaign against the ability of the African, and in doing so to challenge the authority of Europeans in general.

They were, however, a heterogeneous company and they were not united in their reaction. Most of them were traders, and there was trade rivalry to divide them. Others were not traders and had interests which could conflict with the interests of traders. Some were Egba in origin, others were not. Some were brought up by Townsend, others had grievances against him. Most of them were prominent churchmen and leading members of the missionary villages, but while some were ardent and pious Christians, or at least conformed and accepted the cleavage between mission village and the old town, others were beginning to question the wisdom of the cleavage. Some became polygamists; others, not being traders by inclination or not having achieved success in the limited opportunities of the mission village, sought greater scope for their energies in the old town. By 1866 at least two emigrants had become important chiefs at Abeokuta;[2] several more remained in or continued to join the Ogboni. A few were beginning to argue that it was futile to expect that the old society would just disappear and that it was the duty of emigrants to leave the apron-strings of missionaries and go into the old town to guide the chiefs in the establishment of 'a civilized government'.

The first group to challenge Townsend's authority were members of Crowther's own family—his sons Samuel and Josiah, and the Rev. T. B. Macaulay, who in 1854 married one of Crowther's daughters, all of

1 Hinderer to Venn, 15 Nov. 1864; C.M.S. CA2/049.
2 Townsend to Venn, 1 Nov. 1866; C.M.S. CA2/085.

whom continued to live at Abeokuta and were of some social and political significance. Indeed Samuel, 'the doctor' or 'Johnny Africa' as the Europeans nicknamed him,[1] laid claims to being the right person to conduct the political affairs of the mission with the local rulers and to be their adviser, on the grounds that Dr. Irving had taken over those duties from Townsend in 1854 and he (Samuel) had been appointed in 1855 Irving's Secretary and Assistant with the hope expressed of being his successor. Townsend's reply was that his influence with the Alake was personal and could not be delegated.[2]

Townsend did now stop at mere claims: he tried to build up a party. He had his own partisans among the emigrants, notably Henry Robbin, Samuel's rival in the management of the cotton business; Andrew Wilhelm, who had been Townsend's interpreter, ear and mouthpiece with the chiefs since 1843 and was now a venerable old man with an unrivalled knowledge of intricate Egba politics; and David Williams, later a pastor, but still schoolmaster in Townsend's school at Ake, whom Townsend was pushing to become the Alake's regular clerk and copyist.[3] Above all, Townsend had to outbid all other claimants to being the champion of Egba interests. He threw himself wholeheartedly into the Ijaye war from the start.

Even before the war, Townsend had secured the transfer of Macaulay away from Abeokuta. It will be remembered that earlier he had Macaulay removed from the Training Institution on the grounds that he was too academic. Now he complained that Macaulay was wholly unsuited for parish work at Owu and should be transferred to teach at the grammar school in Freetown. The Local Conference considered the proposal but thought it would be a pity for a mission not over-staffed to lose his services; they therefore transferred him to open a new station at Ibadan. Crowther intervened to make a strong plea that Macaulay be allowed to

1 Champness to Methodist Secretaries, 7 Nov. 1861. Commenting on the recall of Townsend to England in 1861 following his quarrel with Samuel, Champness said, 'Mr. Townsend has worked hard and it does seem hard that in his old age he is to be cashiered because he and Johnny Africa can't agree.' Meth.

2 When in 1863 Venn suggested that Townsend might form a committee to manage the political affairs of the mission, Townsend replied: 'I don't think the mission will gain influence by it, for personal influence cannot be transferred nor absorbed by a committee.' 27 May 1863; C.M.S. CA2/085.

3 Townsend described the political activities of Andrew Wilhelm in an Obituary notice in his Annual Letter for 1861 dated March 1862. In 1865 when he first announced the formation of a 'company' of emigrants who wished to become 'chief advisers and writers of letters' for the chiefs, Townsend said, 'Robbin and others in this neighbourhood would not join them'; to Venn, 2–3 Oct. 1865, also 28 Nov. 1865. The rest of the group will be found among those who signed the Alake's declaration denouncing the Campbell and Delany treaty in February 1861; encl. in Foote to Russell, 9 March 1861, Slave Trade Correspondence, Africa (Consular), PP 1862 LXI.

found a grammar school in Lagos.[1] The Parent Committee allowed this. After only two or three months at Ibadan Macaulay moved to Lagos in 1859, where he was to make a name for himself as founder and Principal of the Grammar School till his death in 1879. Meanwhile Townsend felt easier at Abeokuta, as the American immigration scheme turned matters in his favour.

The European missionaries remarked that though the delegates of the African Aid Society, Dr. Delany and Robert Campbell, visited them and gave public lectures under their auspices, and Campbell helped to reorganize Townsend's printing works, they were not consulted about the settlement scheme.[2] The delegates probably stayed with Samuel; and his father, who was by chance in Abeokuta on his way from the Niger was the only other witness to the treaty. Though there can be no doubt that on 27 December 1859 a treaty was signed, the validity and even the terms of the treaty soon became a matter of dispute. Crowther told Venn, and it was corroborated by Bühler, that the delegates had been persuaded to modify their plans, that they had

> given up the idea of settling as a body on a land of their own purchase upon a clear explanation of the disadvantages of such a settlement, both to themselves as well as to the country at large, and they have agreed to settle and disperse among the people and mix with them in any of the towns anyone should take a liking to settle in, as one people under the protection of the native rulers, as we now live.[3]

But Article I of the version of the treaty published by the delegates in England in 1861 read:

> That the king and chiefs on their part agree to grant and assign unto the said Commissioners on behalf of the African race in America, the right and privileges of settling on any part of the territory belonging to Abeokuta not otherwise occupied.

Samuel pointed out that this was the version the delegates had drafted but that it was amended during negotiations to

> the right and privilege of farming in common with the Egba people,

1 Crowther to Venn, 30 June 1858; C.M.S. CA3/04.
2 Bühler, writing to Venn from Germany where he was on leave, 24 Feb. 1860 said: 'The Americans Mr Campbell and Dr Delany go ahead without us. They are indeed not the white man's friend though they will take his money. . . . They have greatly modified their plans but they don't ask us to assist them, they manage for themselves.' C.M.S. CA2/024.
3 Crowther to Venn, 4 April 1860, postscripts dated 9 April; C.M.S. CA3/04. Bühler to Venn, 24 Feb. 1860; C.M.S. CA2/024.

and of building their houses and residing in the town of Abeokuta, intermingling with the population.[1]

While this contention of the Crowthers must have been true to some extent, it must be regarded as an afterthought, emphasized after the treaty had been signed and was being criticized. For Samuel did not repudiate Article II of the version of the treaty published by the delegates which still read:

> That all matters requiring legal investigation among the settlers be left to themselves to be disposed of according to their own custom.[2]

The Crowthers lacked political judgement and undoubtedly they behaved in the matter with more zeal for civilization than discretion. They played into Townsend's hands. Crowther's main argument was that the technical skill that the American emigrants would bring was just what the country needed and that it was worth some risk.[3] Venn did not favour the scheme.[4] Townsend felt that settlers who from the first kept aloof from European missionaries and claimed independence of the local authorities would form an irresponsible *imperium in imperio* that might endanger the existence of the Egba state, or at least corrupt the state by overrunning it with 'a civilized heathenism under the form of Christianity'.[5] It was not difficult to convince the Alake that in signing the treaty he had put his hand to a dangerous document and that the Crowthers, who were not Egba, had been signing away Egba land. Many of the chiefs and Ogboni elders were angry that they had not been consulted. As the agreement became public knowledge, popular fury against Samuel began to mount and he had to flee from Abeokuta in February 1861. Then the Alake signed a declaration repudiating the agreement and saying that he remembered 'Dr. Delany and Mr. Campbell coming to him to ask for a lot of land for farming; which he granted them, but he had no other transaction with them' and that he would accept no one unless recommended by 'the English Consul, the Church or Wesleyan missionaries'.[6] The Foreign Office told the African Aid Society to

---

1 Two letters by Samuel Crowther, jun., to Lord Alfred Churchill dated 50 Baker St., Portman Square, London, 18 April 1861; encl. in F.O. despatch Wodehouse to Consul Foote, 23 April 1861, Slave Trade Correspondence, Africa (Consular), PP 1862 LXI.　　　2 Ibid.
3 Crowther to Venn, 6 April 1861; also the earlier letter of 4 April 1860; C.M.S. CA3/04.
4 Venn to Crowther, 23 May 1860; C.M.S. CA3/L2.
5 Townsend nowhere in his letters attacked the Settlement scheme by name, but this passage in the Annual Letter for 1860, written 6 Feb. 1861 at the height of the crisis, clearly referred to the scheme.
6 Alake's declaration of 8 Feb. 1861, signed by J. M. Turner and others, encl. to Russell, 9 March 1861, op. cit.

look elsewhere and they turned to Ambas Bay, near Victoria, in the Cameroons.

This episode really completed the political eclipse of the Crowther family at Abeokuta. In February 1861, at the time when Consul Foote was co-operating fully with the Abeokutan policy, Townsend's prestige at Abeokuta was at its height. That is, his prestige in the old town, not in missionary circles. The expulsion of Samuel brought the reaction of the emigrants in the mission against Townsend to a violent pitch. There were trials and suspensions, recriminations and indiscriminate mudslinging, enough to confuse and to cause grave anxiety to the Parent Committee.[1] The Committee asked neutral missionaries to hold an inquiry and Townsend was recalled home for consultations. Samuel Crowther junior also went to England. Venn attempted a reconciliation and Townsend promised to use his influence to get him reinstated at Abeokuta.

Townsend was thus away in England when the government reversed the 'Abeokutan policy' and annexed Lagos. The reaction of Venn to the annexation was to regret the violent way in which it was done, but to attempt to take advantage of the increased British interest and influence at Lagos for the philanthropic desires of the mission.[2] He judged it essential to send Townsend back to Abeokuta as the one man whose influence could smooth things over between the Lagos government and Abeokuta rulers besides encouraging self-support in the churches at Abeokuta with respect to the payment of class fees, the cautious introduction of school fees, the sending out of volunteer evangelists to work in outlying villages and so on.

Townsend returned in March 1862 to an impossible political situation. He arrived determined to try to establish good relations between Lagos and Abeokuta. He desired it; on it depended his claim to being the man whose influence could open all doors for missionary expansion in Yoruba, and consequently to being the right candidate for the bishopric.

1 Gollmer to Venn, 21 Jan. 1861, reporting the 'trial' of Samuel over which he presided, the procedure of which C.M.S. lawyers later condemned; C.M.S. CA2/043. James White to Venn, 15 Jan 1861, laying charges of cruelty and slaveholding against Townsend in what came to be known as the 'Lucy Talabi case'; C.M.S. CA2/087. Venn to Maser, 24 Dec. 1860, that Wood, Harrison, and Bühler have been asked to invesitgate the disputes; also Venn to Wood, 14 May 1861; C.M.S. CA3/L3.
2 Venn to Maser, 23 Sept. 1861: 'The intelligence took us by surprise. . . . We fear that king Docemo must have been put under some pressure. . . . But looking at the event as accomplished, it must prove a great benefit to the mission and give it a stability which it could not otherwise have had. We pray to God to over-rule all to his glory.' C.M.S. CA2/L3.

But it was a hopeless task. The Lagos government tended to regard Abeokuta as an ungrateful protégé and as a prospective protectorate of the Lagos Colony as soon as they could persuade the British government to take the necessary measures. The Egba, on the other, hand wished to negotiate on a basis of equality, and regarded the Lagos government as a friend who knew their secrets, deserted them and turned against them.

Although Townsend was slow in admitting it, the basis of his influence at Abeokuta was gone. Townsend refused to act for the Lagos government in persuading the Egba to accept a British vice-consul (now seen as a precursor to annexation) in Abeokuta. The Lagos government, taking Townsend's influence at his own estimation, were angry that he did not exert it on behalf of British policy. They charged him with disloyalty and with loving power too much for himself and too little for his country. Bitter polemics between Townsend and Glover ensued. The reason for this bitterness lay not in Townsend's loss of influence at Abeokuta, but in the increasing frustration resulting from his failure to achieve his real goal of the bishopric. Henry Venn, who placed implicit trust in his political judgement, censured his language but defended his politics before all England, remained suspicious of his ambitions in the mission, and pushed ahead with the scheme for making Crowther a bishop. In January 1864 Townsend wrote to his Methodist colleague and friend Thomas Champness, who was in England in retirement. He talked about the declining prestige of the white man at Abeokuta.

We are gone down so low that I am obliged to beg permission for Mr. Maser (a missionary) to come here. . . . How different to the time when a white face was a sufficient passport here.

Then he continued:

It is reported here that we are to have a black bishop, a Bishop Crowther, a bishop of the Niger to reside at Lagos and to have nothing to do with us. He will be therefore a non-resident bishop. I believe it will be done if C.M.S. can do it, but it will be a let-down.[1]

In June 1864 Crowther was consecrated in England. Venn then appealed to each of the European missionaries voluntarily to place himself under Crowther's jurisdiction. 'I do not hesitate to say,' wrote Venn,

that in all my large experience I never met with more missionary wisdom nor—I write advisedly—more of the Spirit of Christ than in him. Here I felt to him as much drawing and knitting of soul as to

1 Townsend to Champness, 5 Jan. 1864; Meth.

194

my own brother. Be you a brother to Bishop Crowther. You will be abundantly repaid. God destines him for a great work. I should rejoice to be a helper, however, to him.[1]

Townsend replied:

The appointment of a head over the Church always appeared to me desirable; the want of a head has produced weakness—a want of unity in design and a feebleness in execution.

There was indeed work for a bishop to do.

We see around us much material; how to use it is the question. I believe we have in our congregation ample material for a native ministry for the extension of the work . . . but unfortunately the person of the bishop (Crowther) is not acceptable; by acts of partisanship he has made himself obnoxious to both chiefs and people. . . .

It is now expected that we should voluntarily place ourselves under the superintendence of Bishop Crowther. If white men had been accustomed to look up to one as Superintendent it would have been easy to change one for another.[2]

In other words, a white bishop was necessary to accustom the European missionaries to episcopal obedience. When Venn pointed out that the government would not consent to consecrate a second bishop in the area, Townsend packed all his feelings into one final gesture of disappointment. He went near threatening secession:

If the British Government won't authorize the consecration of Colonial or missionary bishops, then we must get power to ordain elders in the churches elsewhere. I don't see any necessary connection between the episcopal office in a foreign country and the Crown of England. If the episcopal office be necessary for the good of the Church, then it is a positive duty to provide it by the heads of the Church.[3]

Two months later he met Crowther at Lagos and he said that once the war was over he would have no difficulty at Abeokuta.

I shall not stand in his way, I will help him rather and go to another part of the country wherever God may direct me. I don't believe in his power to become head of the Church here, notwithstanding. He is too much a native.[4]

1 Venn to Mann, 24 April 1865; C.M.S. CA3/L3.
2 Townsend to Venn, 29 Nov. 1864; C.M.S. CA2/085.
3 Townsend to Venn, June 1865; C.M.S. CA2/085.
4 Townsend to Venn, 28 July 1865; C.M.S. CA2/085.

The bishopric controversy is essential to the understanding of the situation at Abeokuta, not just because Townsend felt so deeply about it that it is impossible to understand his political role unless it is taken into account, but also because it was a major reason why in spite of his previous record, in spite of his continued hostility to the Lagos government and championship of the Egba cause, he failed to convince the Egba that all Englishmen were not agents of the Lagos government. This was largely because Townsend staked his claims to the bishopric on the fact that he was a European, too much of a European for the Egba situation of the 1860s.

In the Basorun's difficult task of building up an Egba government that would be able to stand up to the Lagos government on an equal basis, he began to take advice from Townsend's opponents among the emigrants, from the group of emigrants falling just outside the Church and seeking political advantage as advisers to the Chiefs. They were not united or highly educated; they were young and did not command much respect in society. In 1865 they acquired a leader in the person of George William Johnson. He was not impressive at first sight. He had been a tailor in Sierra Leone, and had adventured on board a merchant ship, as a footplateman and a member of the band.[1] This gave him among other things a chance to visit England. He came to Lagos in 1863. He visited Abeokuta and must have judged the situation favourable for his intervention, for he went back to Sierra Leone to pack his belongings. He had a reputation for stubbornness and doggedness. He was ambitious and a hard worker, coarse, with a bad temper; but since his was a one-track mind, he could be patient and bide his time. He later acquired the soubriquet 'Reversible Johnson'[2] but that was only because he was willing to try various means to fulfil his one unchanging ambition. In an age when most emigrants sought success in trade or industry and were advised for the safety of their souls not to get mixed up in pagan politics, he devoted himself wholeheartedly to the immense task of creating out of the chaos at Abeokuta a 'civilized form of government'.[3] This was to be an adaptation of European methods of govern-

1 Obituary Notice in the *Lagos Weekly Record*, 16 Sept. 1899.
2 There were three Johnsons very prominent among the emigrants: James Johnson, Henry Johnson, both referred to later, and G. W. Johnson. For easy identification, they were labelled 'Holy Johnson', 'Eloquent Johnson' and 'Reversible Johnson' respectively.
3 In 1872 to W. P. Richards a leading emigrant Methodist trader who advised G. W. Johnson to be moderate and allow a settlement of the Lagos Egba dispute so that trade would again be resumed, Johnson replied that he could give away nothing. He condemned traders who 'in this our world of haste to be rich' were willing to sacrifice national interests, and he added: 'I have from

ment to the situation at Abeokuta. It was to be a government of the traditional rulers with a powerful civil service of educated Africans as officials and advisers. He regarded missionaries as allies, especially those who took education seriously and placed no restrictions on the education of Africans. Missionaries like Townsend who considered that English education turned the head of the African he regarded as deceivers; missionaries, again like Townsend, who aspired to political power, he regarded as enemies.[1]

His arrival in 1865 was unnoticed. But within a few months Townsend observed, in October, that

> Some of the Sierra Leone people are jealous and ambitious of becoming chief advisers and writers of letters from the Chiefs here to Lagos. They have been forming a company to accomplish their purpose and have so far won over the Basorun to allow them to write. They have written one letter. Robbin and others in this neighbourhood would not join them.[2]

By April 1866 he announced that they were preparing to levy customs duties. Though he judged that the duties would not be paid, 'it will cause trouble amongst them for the natives are not used to them', he added:

> The Sierra Leone men are thus forcing on civilization, and English customs, teaching the people the use of writing and printing and bringing about the adoption of written laws. They are doing what we cannot, for we cannot use the means they do to accomplish their purposes. I am trying to influence them, I cannot command them.[3]

---

the beginning done all I could to get all of us united in the carrying out of the good of a civilized form of government in this our father country', and that though the endeavour had left him poor, he was undaunted. In 1868, he wrote to the Editor of the *African Times* to 'tell England that their efforts to civilize and christianize Africa by sending missionary after missionary can have but very partial success until they become convinced that something more is wanted besides sending missionaries and putting men of war on the sea, and that is to encourage the forming of self-government among educated Africans.' Johnson received much support and encouragement from the African Aid Society and the *African Times*. Fitzgerald, the Secretary of the Society and Editor of the paper, wrote to Johnson in July 1868, lecturing him on diplomacy, e.g., 'I felt convinced that there has been a good deal too much reported here about the weakness of the Abeokuta government. No government is respected unless it shows strength either real or simulated'—and suggesting that the Abeokuta government should consider appointing a London Representative, presumably himself, like the old Colonial Agents. G. W. Johnson's Papers.

1 One of the first public acts of the E.U.B.M. was to convene a public meeting, 30 Oct. 1865, which resolved that Townsend should be asked to return to England or to co-operate.
2 Townsend to Venn, 23 Oct. 1865; C.M.S. CA2/085.
3 Townsend to Venn, 3 April 1866; C.M.S. CA2/085.

Before the end of the year, he began to report rumours that there was a proposal to drive away, some said all European traders, some said all Europeans, from Abeokuta. Townsend thought that the rumours were instigated by the emigrants, who were deliberately creating an air of crisis in the town; that they were using the reforms Glover was carrying out in Lagos to play upon the fears of the people so that they could themselves get power to carry out similar reforms at Abeokuta, 'to upset the native government and make it more English than white men could if they tried . . .

> This is supposition of course. But these men are introducing a fixed duty on exports in imitation of European ways; they have a custom house, give permits. In doing this, the greatest opposition has been met but nevertheless it stands. This stir about the white men arises out of it I believe. We cannot fathom the bottom of their doings or motives. The great fact is that immense changes are taking place and the old chiefs, while providing as they think for the safety of the town and its institutions, are introducing the greatest changes.[1]

Clearly the missionaries were puzzled, but they did not all show Townsend's suspicion of the new changes. Venn told Townsend that from his experience of Indian affairs, the 'silent revolution: . . . towards a more advanced civilization' though 'urged forward by a Godless education and wordly politicians' would 'crush idolatry and make a highway for the Gospel'.

The Egba United Board of Management which resulted was conceived of as a Chancellery, with Johnson as Secretary and another emigrant as President, formulating policy and seeking to gain the backing of the traditional rulers for the execution of their policies.[2] They won over Basorun the Regent, Akodu the Seriki (an important war chief), and the Asalu, head of the Ogboni. Their Board's claims for support were based on their being essential in the struggle with Lagos, on their ability to speak the language and to understand the diplomatic trickeries of the Lagos government. Moreover they offered if properly supported to make the Basorun's government, through systematic

1 Townsend to Venn, 1 Nov. 1866; C.M.S. CA2/085.
2 Cf. Biobaku: *The Egba State and its neighbours*, p. 79. The Board was much more than an 'empty bureaucracy parading sovereign pretensions and issuing largely idle threats'. It was never intended to be a 'proper council representative of the traditional, sectional and immigrant elements in Abeokuta'. The purpose of the Board was not to replace the traditional rulers, but, in the words of Townsend quoted above, to be their 'chief advisers and writers of letters' so that 'the old chiefs while providing as they think for the safety of the town and its institutions' would be introducing great changes.

customs duties in place of arbitrary tolls, the most wealthy, the most efficient and therefore the one supreme government in Abeokuta.

The internal difficulties were formidable, since Abeokuta was so divided. When in July 1866 they tried to end the interregnum at Abeokuta by having a descendant of a former Alake of the pre-Abeokuta era recognized as king, they met stiff opposition in the Ake quarter itself, an opposition joined by Townsend and the loyal members of the Christian villages.[1] The attempt to impose customs duties alienated important emigrant merchants as well as the heads of several of the smaller townships who controlled gates of their own and were accustomed to receive tolls, Igbein most especially, who claimed the right to collect all tolls on the Ogun. In addition, customs duties involved the regulation of trade, which was the province of traditional trade chiefs, the Parakoyi, with whom a struggle was inevitable, especially as the E.U.B.M. proceeded to establish a new court to prosecute duty evaders.[2]

It was, however, the relationship with Lagos that presented the greatest difficulties. This may be illustrated by the attempt to establish a regular postal system between the two places. In December 1866 Glover sent to Abeokuta proposals for such a system. The E.U.B.M. welcomed this. In January Glover sent postal regulations, and appointed the three mail runners. He proposed to pay them for the next six months, during which time all postage on letters would be paid in Lagos. The E.U.B.M. rejected this, saying that they could not take dictation from Lagos and that the regulations ought to be negotiated.[3] Meanwhile Glover had appointed a Postmaster-General and the mails had started to run when the negotiations as well as the mails were interrupted by a more ominous quarrel.[4]

The E.U.B.M. had established a customs house at the Aro gate of Abeokuta. They soon discovered that not only was it easy to smuggle cotton and palm-oil through some other gate, but that many merchants, rather than bring their goods into Abeokuta before exporting them, sent them down the Ogun at once, thus by-passing the Aro customs

---

1 Townsend to Venn, 27 July 1866; C.M.S. CA/085. The Ake objection was based largely on memories of Oyekan's ancestor, who was a most unpopular king and was deposed. Townsend said Oyekan was his friend but 'he is of the old school. . . . I earnestly wish he may not come to that high authority. I considered the subject to be one of the greatest importance and called a meeting of the senior members of the church to urge upon them the policy of non-interference but more especially the great need of prayer to God that he may overrule the counsel of the chiefs and cause them to appoint one who may become an instrument of good for his church.' Ibid.
2 Wood to Venn, 2 May 1867, 1 June 1867, 1 Aug. 1867; C.M.S. CA2/096.
3 G. W. Johnson to Glover, 10 April 1867; G. W. Johnson Papers.
4 C. Foresythe, the Lagos P.M.G., to Johnson, 26 April 1867; Johnson to Glover, 30 April 1867; Johnson to Foresythe, 8 August 1867; ibid.

house. The only answer was to open a new customs house on the river, south of Abeokuta. This was announced in the *Iwe Irohin* in June 1867. Glover wrote to Johnson on 8 July that His Excellency the Administrator of Lagos

> considers this a fitting occasion to call the attention of the Egba Government to the undefined condition of our respective frontiers.[1]

He therefore suggested negotiations for a 'definite treaty on this subject'. The tone of the letter was calm, even flattering—

> Were there no responsible government at Abeokuta such settlement of frontier would be of little importance . . . but with the responsible government which Abeokuta at present enjoys and with which this government is in friendly intercourse, His Excellency is of the opinion that our relations with each other on our frontier should be placed on a firmer basis than the mere good or bad behaviour of petty chiefs and customs officers. . . .
>
> You will perceive, sir, that the spirit in which this communication is addressed is one of anxiety that the friendly relationship at present existing between our two governments should receive a further development.[2]

The implications, however, were grave. The Egba claimed the land down to the coast and recognized only the island of Lagos as having been annexed by the British. How, then, could Egba customs officers violate British territory? The Egba, replied Johnson, would set up 'customs houses and officers . . . in Abeokuta and its territories which are recognized to be quite free from the Island of Lagos'. That friendship may continue,

> the Bashorun and Directors of the Board of Management therefore request me to advise your Excellency to agree with them in considering the subject of our past and present communications in reference to treaty as closed.[3]

Glover did not agree. He regarded Ebute-Metta as 'Lagos Farms'. He argued that the Egba did nothing to benefit or even protect the people south of Abeokuta, from whom they wished to collect tolls. He recruited Hausa manumitted slaves, turned them into constables and placed them on the routes by land up to Otta, and by water up to Isheri. In expectation of an Egba attack, he distributed arms to some of the

1 Lt. Gerard, private secretary to Glover, 8 July 1867; C.M.S. CA2/07.
2 Ibid.
3 Johnson to Glover, 18 July 1867; G. W. Johnson Papers.

people.[1] The Egba were enraged. On 24 September Johnson wrote to Glover that it was with difficulty that the Basorun could 'keep back hundreds of their war boys who without authority were but too ready to go after the constables'.[2] It was these war boys, led by Akodu, the Seriki of the Egba, and Solanke, the Jagunna of Igbein, who started the riot to break up Mission Houses and insisted on the expulsion of all European missionaries and traders on 13 October. This was what the Egba called the *Ifole* (lit. Housebreaking).

When it came, it seemed a spontaneous uprising against Christian missionaries.[3] But it was a most unusual type of persecution. The converts were left alone, but the house of every missionary was broken up, except the one at Ikija which was protected by Ogundipe, a war chief rising into prominence and following his own policies. Libraries were torn up, harmoniums broken down; even the printing works where the *Iwe Irohin* had so often proclaimed the Egba point of view was destroyed. 'Rather than a desire to get as much as possible,' said the Methodist missionary, 'they exhibited the most wanton destruction.'[4] It was not a persecution of Christians, but a persecution of Europeans. Many missionaries realized that, but could not believe that they had been identified with the action of the Lagos government to that extent. Some suggested that it was part of a carefully laid plot hatched with the Ijebu to stop the penetration of European influences entirely.[5] The Egba themselves insisted that it was not the deliberate act of the government so much as the angry gesture by an infuriated people.[6]

Whatever be the whole truth, the *Ifole* was the culmination of the events which began in June 1861 when Consul Foote died in Lagos and McCoskry became acting consul. The reversal of the Abeokutan Policy and the annexation of Lagos created hostility to the Lagos government and gave emigrants some political power in Abeokuta. The hostility of Townsend convinced the emigrant politicians that 'all white men were the same'. The Egba hostility to the Lagos government became hostility to all white men. The Lagos government humiliated Africans;

1 Biobaku: *The Egba State and its neighbours*, pp. 82–3. Johnson to Glover, 7 Sept. 1867; G. W. Johnson Papers.
2 Johnson to Glover, 24 Sept. 1867; ibid.
3 For accounts of the *Ifole* see the Rev. J. A. Maser, 'The Second Persecution of the Abeokuta missionaries, October 1867'; C.M.S. CA2/068, printed in the *C.M. Intelligence*, January 1868. (The first persecution was in 1849.) Also Grimmer to Methodist Secretaries, 4 Nov. 1867; Meth. Crowther to Venn, 3 Dec. 1867; C.M.S. CA3/04.
4 Grimmer to Methodist Secretaries, 4 Nov. 1867; Meth.
5 Crowther to Venn, 3 Dec. 1867; C.M.S. CA3/04.
6 G. W. Johnson to Glover, 17 Oct. 1867; C.M.S. CA2/07. Crowther to Kimberley (on behalf of the Egba rulers), 6 Feb. 1872; CO147/23.

the Egba humiliated Europeans. In 1864 three elderly Quakers visiting Abeokuta were made to prostrate themselves on the banks of the Ogun before they were allowed to pass.[1] In 1866 a German catechist was beaten up at Igbein for no reason whatever.[2] Townsend noted the growing hostility towards Europeans, but could not judge that the Egba seriously contemplated driving away Europeans, even European missionaries. He left Abeokuta in April 1867 on leave to England before the land dispute with Lagos came to a head. The riots of 13 October probably got out of hand and bungled whatever action the Egba authorities were contemplating. But once it happened G. W. Johnson began to use the possible expulsion of Europeans from Abeokuta as a bargaining counter in the negotiations with Lagos. It was only when Lagos refused to yield that, almost a year later, the E.U.B.M. in September 1868 issued the proclamation prohibiting 'for the time being' the entry of all Europeans to Abeokuta except by special permit, and banning the residence of all Europeans, missionary or commercial, in Abeokuta and its territories.[3]

In that way the Egba decided to drive away the gorilla but attempt to save the baby, the Christian community in their midst, though they could not expect the baby to go unharmed. Life in the mission villages was disrupted; churches were closed. Many converts left Abeokuta with the missionaries and formed settlements in Lagos, in Ebute-Metta and on some Egba farms. But every reliable witness agreed that the converts were not the object of the persecution. They were allowed to re-open or rebuild their churches. African pastors and catechists were allowed in freely, but not Europeans. In October 1868 the pastors and catechists received a circular requiring them to wait upon the President of the Egba United Board of Management in order that they might be 'acquainted with the views of this government relative to the future course of education of the Egba children', the most important point being that only English should be taught in the schools.[4]

Crowther made it quite clear that he had no sympathy with the action or the ideas of the E.U.B.M. In a charge he delivered to his clergy in 1869, he posed the question: 'Are Africans yet able to regenerate Africa without foreign aid?' His reply was a decisive 'No'.

1 West to Methodist Secretaries, 13 Jan. 1864; Meth. Crowther to Venn, 3 Dec. 1867; C.M.S. CA3/04.
2 Venn to Lamb, 23 Aug. 1866, 23 Nov. 1866; C.M.S. CA2/L3.
3 Proclamation of the E.U.B.M. headed by Johnson's slogan, 'African shall Rise', 11 Sept. 1868; C.M.S. CA2/07.
4 Dated 13 Oct. 1868, encl. in Grimmer to Meth. Secretaries, 3 Nov. 1868; Meth.

Africa for the Africans, the rest of the world for the rest of mankind, indeed. If we have any regard for the elevation of Africa, or any real interest for evangelization of her children, our wisdom would be to cry to those Christian nations which have been so long labouring for our conversion, to redouble their Christian efforts.[1]

In January 1871 he said that he had received several invitations from the chiefs of Abeokuta but held back deliberately because

There are some half-educated, unprincipled young men about the country who would have taken advantage of any seeming countenance from me or other natives of position, as if we were backing them in their short-sighted presumption when they said they were able of themselves to civilize and evangelize their own countrymen without European aid.[2]

The effect of the *Ifole* was not to keep out European interests and influence from the country, but to emphasize that as long as Africans continued to hold sway in the country, the policy of the 1865 Committee was likely to succeed better than that of the annexation of enclaves, and that the policy of Henry Venn was infinitely wiser than that of Henry Townsend. Europeans who believed that European and Christian interests could be protected only under European rule later redoubled their energies to see such rule extended from Lagos to cover the whole country. But meanwhile, for about a dozen years after 1867, European missionaries were confined to the coast and the work in the interior of Yoruba and on the Niger was entirely under Africans led by Bishop Crowther. At the end of that period Townsend himself felt obliged to acclaim the wisdom of the policy of Henry Venn. It was a remarkable period of training in self-government. Crowther's position was the more clearly emphasized since the *Ifole* was followed by a period of acute shortage of men and means which greatly reduced the supply of European missionaries in all the missions, and in the case of one cut it off entirely.

The American Civil War had cut off the funds of the Baptists, and their missionaries had to return home; it also so depressed Lancashire (an important source of British missionary—especially Methodist—funds) that Methodist missionaries were not ready to expand into the interior till 1879, when their Lagos District was separated from the Gold Coast District.[3] Although Roman Catholic missionaries arrived in

1 Bishop Crowther's Charge, delivered to his clergy, 1869; C.M.S. CA3/04.
2 Crowther to Hutchinson, 19 Jan. 1871; C.M.S. CA3/04.
3 West to Boyce, 1 April 1863: 'With a deficiency of £1,300 in such a dreadful want, I am not surprised that the committee do not feel at liberty to strengthen

Lagos in 1867, eager to enter the interior, especially Abeokuta, the S.M.A. was severely hit by the Communes in France in 1871 and had to appeal for funds outside France before their work in Africa could continue with the occupation of Topo in 1876 and their mission at Abeokuta in 1880. The C.M.S. were diverting their attention to East Africa and India. In 1865, even before the *Ifole*, they had announced that 'our funds are sadly behind our expenditure and . . . we shall be crippled in all our missions', and Venn was talking of the 'want of men and means, the old check to expansion, and the universal check alas!' There was some expansion along the coast, and up the new route Glover was trying to open from Agbabu up to Ondo, through Ife to Ibadan. But the work was carried out by African agents and the European supervisors were resident on the coast. Some features of this period of training in self-government ought now to be examined in the light of the career of Bishop Crowther.

---

this mission.' Indeed, ten weeks later, Champness sent a draft for £10 7s. from the Methodist Church at Abeokuta to the 'Lancashire Relief Fund', 17 June 1867; Meth.

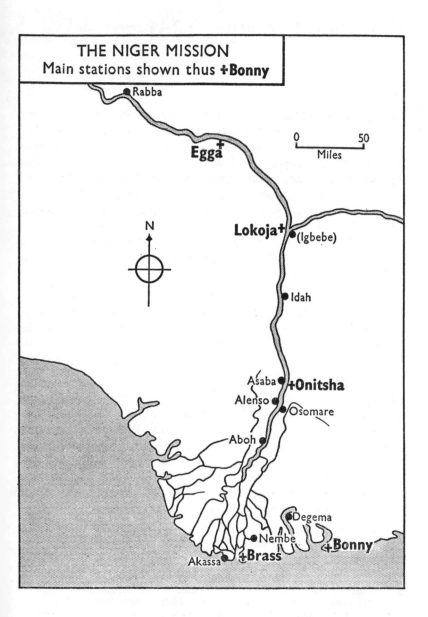

THE NIGER MISSION
Main stations shown thus **+Bonny**

Rabba

Egga +

0   50
Miles

Lokoja + (Igbebe)

Idah

Asaba +Onitsha
Alenso
Osomare

Aboh

Degema

Nembe +Bonny
Akassa +Brass

7 Bishop Crowther 1864-77

CROWTHER was consecrated not Bishop of the Niger Mission or Niger Territory, as is often asserted, but Bishop in an immense diocese described in the royal licence authorizing his consecration as 'the countries of Western Africa beyond the limits of our dominions', that is, West Africa from the Equator to the Senegal, with the exception of the British colonies of Lagos, the Gold Coast and Sierra Leone. C.M.S. lawyers drew up a 'minute on the Constitution of the Anglican native bishopric on the West Coast of Africa', approved by the Archbishop of Canterbury, further elaborating on Crowther's jurisdiction. Since the European missionaries had not declared themselves willing to be placed under his jurisdiction, the minute said:

> There are, however, existing missions of the Church Missionary Society within such limits [of Crowther's diocese] which the Bishops of Sierra Leone have been accustomed to superintend, such as the Timne Mission, near Sierra Leone, and Abeokuta near Lagos, respecting which an arrangement must be made by the two bishops as to time and circumstances of transfer.[1]

Venn added in a letter to the Yoruba Mission that the present arrangement 'can only be regarded as temporary'.[2] Meanwhile, a comic Gilbertian situation prevailed. Crowther was not bishop over Lagos; nor over Abeokuta or Ibadan, where European missionaries happened to be; but he was bishop over Otta—midway between Lagos and Abeokuta— and other places in the Yoruba country where European missionaries happened not to be. He was bishop on the Niger, later Bishop in charge of the American Episcopalian Church of Liberia.[3] He even once visited

1 'Minute on the Constitution of the Anglican Native Bishopric on the West African Coast', 1864. Copy enclosed in Crowther to Hutchinson, 4 Aug. 1873, C.M.S. CA3/04, quoted in full in appendix. Also the Parent Committee's charge to Crowther, 23 July 1864; C.M.S. CA3/L2.
2 Venn to missionaries in the Yoruba mission, 23 July 1864.
3 Crowther to Rev. J. Kimber, Secretary and General Agent of the American Episcopalian Church of Liberia, 23 Bible House, New York, 12 Feb. 1878, giving an account of his visitation January–February 1878. Also address dated 30 Jan. 1878 by the Standing Committee of the Church asking him to become their bishop; C.M.S. CA3/04.

his son-in-law, a government chaplain at Bathurst, and drew up plans for the evangelization of that area.[1] By fixing his seat at Lagos he was not being an absentee bishop, as Townsend said; he was only residing at the most central point of his diocese, to which he expected to be added before long those places in the Yoruba country over which European missionaries presided.

It must be emphasized that in 1864 Crowther controlled part of the Yoruba Mission, Otta specifically, and places not yet occupied by European missionaries; and the transfer to him of the rest was regarded as imminent. Unless this is understood, the duality of Crowther's position cannot be fully understood. Venn called his consecration the 'full development of the native African Church'.[2] This meant in theory that the mission established in Yoruba approached 'euthanasia' in two ways: that the new Church even if not wholly self-supporting, realized a measure of self-government by having an indigenous bishop; and since the bishop was directing missionary work elsewhere, the Church was also becoming self-propagating. Crowther was intended to organize the churches established into 'a national institution'. The minute referred to above went on to inform him that this did not imply a break-away from the Anglican Church; that the Archbishop of Canterbury was his Metropolitan; and that

> the Church in Western Africa over which he presides will be a branch of the United Church of England and Ireland and will be identical with the mother Church in doctrine and worship and assimilated in discipline and government as far as the same may be consistent with the peculiar circumstances of the countries in which the congregations are formed.[3]

These instructions clearly applied to the churches in Yoruba. As we shall see later, the expulsion of European missionaries from Abeokuta did hasten negotiations for the Bishop of Sierra Leone to hand over to Crowther the nominal control he exercised over Abeokuta and Ibadan. For a while Crowther undertook supervision of these missions, though his control over them was never really complete. It was not the problem of organizing the established churches in Yoruba but that of establishing new ones on the Niger that first and throughout most conspicuously engaged his attention. But his role as missionary on the Niger cannot be fully appreciated if the other perhaps less significant role in the Yoruba

1 Crowther to Wright, 18 June 1874, giving an account of his visit to Bathurst in March; C.M.S. CA3/04.
2 Venn to Lamb, 23 Jan. 1864; C.M.S. CA2/L3.
3 'Minute on the Constitution of the Anglican Native Bishopric on the West African Coast', op. cit.

Mission is completely ignored. Apart from the need to bear in mind his extra burden of Yoruba translation and orthography, unless he is seen, as Venn intended, as cut out for the headship of the churches in Yoruba, even the most obvious fact of the Niger Mission—that until the 1880s all the missionaries who worked there were Africans—does not make sense.

The Niger Mission was not originally intended to be worked in that way. Venn declared in May 1860 that the fundamental principle of the Niger Mission 'is not to be native agency and European *superintendence* or European agency and native *superintendence*, but native and European *association*'.[1] It was a series of accidents (and, as we have seen, some opposition from Townsend) that prevented any Europeans from going to work on the Niger under Crowther. But soon the Niger mission became clearly involved in the controversy about the headship of the Yoruba Mission. Venn decided in spite of the opposition of Townsend to press on with the scheme of making Crowther bishop of the 'Native Church' in Yoruba, and to bolster this up by regarding the Niger Mission as the 'self-propagation' phase of the 'full development of the Native African Church'. Hence he decided to accept Townsend's challenge of an all-African mission. It was as 'a palpable triumph of Christianity'[2] intended to show what the Church had made of Africans, and to convince those like Townsend still afflicted with doubts. It was to be a soul-warming experiment to cheer drooping missionary spirits in England[3] and to rally Africans to the cause of missions on the principle of 'self-government, self-support, self-propagation'. Since, however, Crowther never gained complete control of the churches either in Sierra Leone or Yoruba supposed to be propagating themselves as far as the Niger, the self-propagation was only partial. Thus Crowther was only nominally a 'native bishop', in practice essentially a missionary and in fact the symbol of a race on trial.

The problems of the Niger Mission were similar to those that faced the missionaries on the coast: problems of communication in days before the mail-boats, as well as the political problems of introducing foreign influences that could easily upset the existing pattern of trade and balance of power. Missionaries were dependent on traders for passage and freight, provisions and other supplies. Their salaries were

1 Venn to Crowther, 23 May 1860; C.M.S. CA3/L2.
2 Venn to Lamb, 23 Jan. 1864; C.M.S. CA2/L3.
3 Cf. when Crowther's first charge was printed and circulated, Venn wrote to him: 'It forms the substance of many a missionary meeting and will bring much gold into our treasury'; Venn to Crowther, 23 Feb. 1867; C.M.S. CA3/L2.

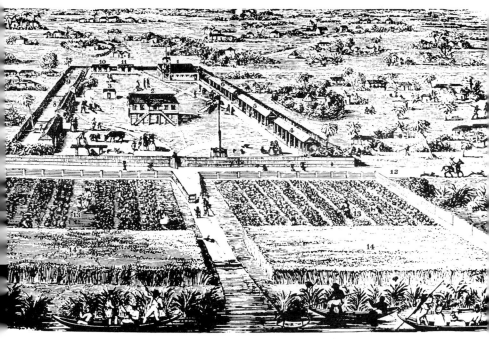

BADAGRI, 1849

The C.M.S. Mission plan-view Key: 1 The Church, 2 The
Mission House, 3 Kitchen, 4 Schoolmaster's House, 5
Boarding School, 6 Labourers' Dwelling, 7 Shed and
Carpenters' Shops, 8 Interpreter's House, 9 Watchman's
House, 10 Widows' Shelter, 11 Stable, 12 Street, 13 Vege-
table Garden, 14 Rice Garden.

The C.M.S. Mission House at Badagri, 1849.

usually paid by bills of credit issued on merchant stores from which could be obtained the beads, cloths, tobacco, salt and other goods necessary to buy whatever they needed locally. The Niger Mission could work only on the basis of Buxton's principle of opening up Africa to commerce and European civilization by a combination of European capital and African agency.

When the first mission stations were planted at Onitsha and Igbebe (at the confluence) in 1857, the difficulties of developing trade on the Niger were underestimated. Since the 1854 expedition went up without opposition,[1] and navigated the Niger to Lokoja and the Benue for another 300 miles, made a substantial profit on trade, and discovered that the use of quinine reduced European mortality, it was assumed that if enough Sierra Leoneans could be encouraged to emigrate to the river, all that McGregor Laird would require would be a government subsidy for five years, before the end of which period the Niger was expected to be swarming every rainy season with merchant vessels. However, in 1859, the lesson was driven home that until the Delta opposition had been beaten down, merchant vessels would have to be convoyed up and down the river by a warship.[2] The Foreign Office agreed to provide the convoy 'to protect all the vessels navigating the Niger for lawful commerce'.[3] But the Admiralty was at first dilatory in offering co-operation. In July 1860 Crowther gathered a large missionary force together in Laird's vessel, the *Rainbow*, at the Nun entrance of the Niger. They carried a pre-fabricated iron house for Onitsha, saw gins for cleaning cotton, provisions, and other missionary equipment. But the warship failed to arrive. They waited, sixty-five of them in the little vessel, all July, August, September and October. As Crowther said, their 'strained eyes looked towards the sea in vain'.[4] Some of the mission equipment was landed at Akassa, but the agents had to return to Sierra Leone to await the next rainy season. The evangelists at Onitsha and Igbebe, like Dr. Baikie at Lokoja, received no supplies or mail for that year. It therefore became one of the principal cares of Crowther as he returned to Lagos each dry season to organize better co-ordination between the mail boats bringing mission agents from Sierra Leone, the trading vessels from England and the naval convoy from the West African squadron.

In January 1861 Macgregor Laird, the only merchant established on the Niger, died, and his executors decided to wind up his business.

1 S. Crowther: *Journal of the Niger Expedition*, C.M.S. 1855.
2 Crowther to Venn, 23 Dec. 1859; C.M.S. CA3/04.
3 Venn to Taylor, 23 April 1860; C.M.S. CA3/L2.
4 Crowther to Venn, 6 Oct. 1860. Also previous letters from Akassa, 4 July, 9 Aug., 6 Sept 1860; C.M.S. CA3/04.

The government had also written to Dr. Baikie that the Niger Expedition was regarded as being at an end and that he should return home.[1] If the Niger Mission had not acquired a symbolic significance, it might also have been withdrawn as a mistaken, premature establishment. Fortunately Dr. Baikie insisted on staying, and his despatches soon convinced the government of the great potentialities of the Niger area for trade and for the supply of cotton. When Crowther in 1863 reported the appearance of a French ship at the Nun, the government went further. They got the rulers of Akassa, Abo and Onitsha to sign treaties prohibiting the slave trade and human sacrifice, giving protection to trade and missionaries, stipulating that 'England and Englishmen are to have first consideration in all trade transactions' and making the rulers promise to sign no other treaty or agreement 'without the full understanding and consent of the British Government'.[2] Yet in spite of this government initiative, and the desire of cotton manufacturers during the American Civil War to obtain cotton from the banks of the Niger, no merchant came forward to replace Laird. Merchants hesitated to risk their capital on a river that could only be navigated four months in the year and even then against the wishes of the Delta people and the Liverpool traders established on the coast.

In July 1861 Crowther suggested that if no new company was formed to take Laird's place, the C.M.S. should act more independently of the traders, buy a little vessel of their own which under the management of a subsidiary committee like the Native Agency Committee could take passengers up and down the Niger and export the products of the Industrial Institutions. Venn replied that the mission could not afford it: 'Our chief hope is upon a road from the Yoruba country.'[3] When, however, the Ijaye war proved to be no simple, short engagement, he turned to the idea of getting friends of the C.M.S. to form a company to replace Laird.

The West Africa Company was formed in 1863, under the direction of Thomas Clegg, whom the C.M.S. had already used in developing cotton cultivation at Abeokuta. The new company was financed largely by humanitarians at Venn's prompting; shares were even bought for Bishop Crowther and by Josiah Crowther. But one of Laird's executors in the same year formed the Company of African Merchants with con-

1 Dr. Baikie to Lord John Russell, Lokoja, 10 Sept. 1861; Russell to Baikie, F.O., 23 Dec. 1861; Slave Trade Correspondence, Africa (Consular), PP 1862 LXI.
2 Treaties, Aboh 8 Sept. 1863, Onitsha 12 Oct. 1863; SP vol. 59, p. 1187–8.
3 Crowther: 'Notices of the Delta', 29 July 1861; C.M.S. CA3/04. Venn to Bühler, 23 Aug. 1861; C.M.S. CA2/L3.

siderably greater capital.[1] Both companies applied for, but failed to obtain, a government subsidy of the kind that Laird had enjoyed.

Thus at the time when Crowther became bishop, the Niger Mission was far from being assured of regular communications. The refusal of the Treasury to grant a subsidy to tide over the inevitable losses of the years of pioneering was a great handicap. The two companies began to establish depots at the main mission centres, where they moored hulks and appointed African agents to trade for them all the year round. Each rainy season their vessels, with European agents and an Agent-General from England, convoyed through the Delta by a warship, traded up and down till convoyed out again. But the Company of African Merchants, finding the coastal trade more profitable, soon began to confine itself to the Delta. It was in fact the weaker, less well-organized West Africa Company that was the successful pioneer of the Niger trade and this was undoubtedly because of its connection with the Niger Mission.

Since Crowther regarded it as vital to the success of the Niger Mission that the company should succeed, he placed the resources of the mission at their disposal. Although the company's depots were usually physically separate from the mission stations and the management of their affairs entirely independent of the mission, and although the mission often complained that the company charged too much for their goods—sometimes 150 per cent to 200 per cent profit—or that some of their European agents were 'members of the Anthropological Society and disciples of [Richard] Burton' who considered missionary work useless if not positively harmful to Africans,[2] the company and mission realized that the growth of one depended on the growth of the other. The extent of the mutual dependence may be judged from the fact that in 1865 for freight, passage and provisions on the Niger alone, the C.M.S. paid to the company £600 5s. and in 1866 £379 18s. 3d.[3]

Thus, inevitably, the pattern that we saw earlier was repeated, of

1 *The African Times* (monthly), official organ of the African Aid Society, has much information about the Company of African Merchants launched in July 1863, not 1864 as is often said. It took over the intended Niger Chadda Company of Lyons McLeod, advertised in the same paper in March 1863.
2 Crowther to Venn, 6 June 1865, 6 Oct. 1865; C.M.S. CA3/04. The reference was to F. Burton's anthropological works, in particular to *A Mission to Gelele*, 2 vols. 1864, in which there was a chapter entitled 'The Negro's Place in Nature' dedicated to James Hunt who had just outlined the views of his Anthropological Society to the public through a paper he read to the British Association in 1863 entitled also 'The Negro's Place in Nature'. (D.N.B. article *Hunt, James 1833–69*). Burton was Consul for the Bight of Biafra (1861–64) and he visited Abeokuta in 1861. The views of the Anthropological Society are further discussed in Chapter 8.
3 Crowther to Clegg, 3 Jan. 1867; C.M.S. CA3/04.

missionaries using their local influence to pave the way for and further the interests of traders. There was, however, a difference. While it was only occasionally that a European missionary was found talking of entertaining a brother-in-law or other relation who traded to the coast, and here and there a similar name might suggest unrecorded cases of kinship between trader and missionary, the African traders and missionaries were much more close-knit. Josiah and Samuel Crowther moved to the Niger and were agents for the West Africa Company at Lokoja. James L. Thompson, their brother-in-law, was agent for Laird and became agent for the company at Onitsha. Mrs Macaulay, from Lagos, on visits to see her sister, Mrs. Thompson, also traded.[1] Once, when Mr. Thompson was ill, Isaac George, the mission's industrial agent, acted for him.[2] Indeed, following Clegg's example at Abeokuta, the West Africa Company for a while took over management of the Industrial Institution at Onitsha with all its personnel.[3] W. Romaine, a catechist and later pastor at Onitsha, had a brother working for the company, and other examples may be cited. The emigrant traders and evangelists were friends or relatives from Sierra Leone engaged on the same mission, either through self-interest or religious duty, of opening up the Niger for European trade and influence. Crowther regarded the growing success of the West Africa Company as an achievement in which he could take a pride. 'If it be acknowledged,' he wrote to Clegg in January 1867,

> both myself on behalf of the Church Missionary Society and my sons have both by labour and money contributed greatly to encourage the West Africa Company, Ltd. in their attempts to develop the trade of the Niger.[4]

There were obvious disadvantages in this degree of affinity between trade and missionary work, and Crowther was not unaware of them. African rulers might fail to distinguish between the traders and the missionaries so closely associated; missionaries might be tempted to trade at the expense of their ecclesiastical functions, and opponents of the mission, such as European traders whom the missionaries criticized or who regarded African traders as rivals, might use the connection to damn the African traders and missionaries alike. But it was not till later, when the West Africa Company had succeeded sufficiently to

1 Venn to Crowther, 23 Dec. 1867, that Mrs. Macaulay's trading had implicated Crowther's name with trading, as an observer told him that the Bishop was looking after Mrs. Macaulay's consignments; C.M.S. CA3/L2.
2 Crowther to Venn, 28 Oct. 1868; C.M.S. CA3/O4.
3 From 1867 to 1870. Crowther to Hutchinson, 12 Jan. 1875; C.M.S. CA3/O4.
4 Crowther to Clegg, 3 Jan. 1867; C.M.S. CA3/O4.

attract other competitors, that its connection with the Niger Mission became a matter of grave embarrassment to the mission.

In 1867 the West Africa Company was still running at a loss. In February of that year Venn wrote anxiously to Crowther that the mission might once again be plunged into difficulties as the losses of the company were so great that it might go bankrupt. Even as late as July 1869 Crowther still feared that the company might pull out of the Niger. However, less than four months later Crowther reported that the Agent-General found the trade so profitable that season that he changed his mind.[2] The decision of Crowther in 1870 to reconstitute the Onitsha Industrial Institution as a mission establishment, on the ground that the company had been neglecting the cotton trade, indicates an important reason for the new trading successes of the company.[3] For, since the end of the American Civil War, the demand for West African cotton had dwindled. The missionaries continued to insist that it was in the cultivation of crops like cotton by peasant farmers that there lay great civilizing forces, not in the making of palm-oil or the gathering of elephants' teeth. Nevertheless it was when the Company discovered the resources of Nupe and began to turn away from the doctrinaire search for cotton, ginger, indigo, and arrowroot to ivory and shea-butter as well as palm-oil that it began to make large profits. At once other companies from the Delta began to enter the Niger: Holland and Jaques in 1869, Alexander Miller Brothers and J. Pinnock & Co. in 1870, as well as more emigrants from Sierra Leone and Lagos, a few of them, like Captain J. P. L. Davies, with vessels of their own.[4]

Thus by 1870 the hope of more regular and assured communication with Lagos, Sierra Leone and England and between the various mission stations was greatly brightened up by the prospects of the Niger Mission. The increased trade did not bring all that Crowther and the C.M.S. had hoped and worked for. It did not destroy the trust system;[5] the peasant producer was far from exporting his produce directly to Europe. The European dealt with a middleman who was sometimes, as on the coast, a shrewd, traditional ruler who had turned to trade. More usually, however, he was an emigrant or convert educated in mission schools. Sometimes he was an independent trader; more often he was a clerk or agent

1 Venn to Crowther, 23 Feb. 1867; C.M.S. CA3/L2.
2 Crowther to Venn, 3 July 1869, 30 Oct. 1869; C.M.S. CA3/04.
3 Crowther to Hutchinson, 12 Jan. 1875; C.M.S. CA3/04.
4 Crowther to Venn, 2 Oct. 1871, 30 Oct. 1871; C.M.S. CA3/04. Also W. H. Simpson's Report to Earl Granville, 21 Nov. 1871; F.O. 84/1351, op. cit.
5 Cf. Crowther to Hutchinson, 4 Dec. 1872, that owing to the expensive visit of the Sultan of Gwandu to Bida, the Emir had been unable to pay his 'debts' to the merchants; C.M.S. CA3/04.

working directly for a European firm. Emigrants and converts were to be found in other occupations as well: they were coopers, stokers, pilots, sometimes even engineers and captains of trading vessels. And, with the All-African Mission, it can be said that Buxton's dreams had come true.

The 1870s were the golden decade for the missions' incipient middle class. The West African Company set the standard when in 1872–3 they reacted to the increasing competition on the Niger by gradually replacing all their European staff (except ships' captains and engineers) by African personnel who remained permanently on the river. In 1873 Josiah Crowther was appointed the company's Agent-General.[1] The position was further emphasized by the withdrawal of the European consul from Lokoja in 1869. The Foreign Office instead sent a Commissioner up the Niger in 1870 to sign an agreement with Masaba, Emir of Bida, whose emirate was becoming the focus of British trade on the river, committing to his care the protection of all British subjects on the river, whether traders or missionaries.[2] Crowther argued that such confidence in Africa rulers was a policy he himself advocated,[3] but that there was still need for a consul as the British representative through whom the British government could send acknowledgements of the Emir's services, as well as to give him guidance and advice.[4] The government refused to appoint a consul, and Crowther became in fact an undesignated consul on the river. Every year between 1871 and 1876 he communicated as a consul would have done with the British government and every year the British government sent to him, through the Administrator of Lagos or the Governor-General of Sierra Leone or the Gold Coast, letters and presents for Masaba, acknowledging his services. At set ceremonies in Bida or wherever the Emir was, these letters were formally read and the presents exhibited. Then Crowther, in his other capacity as missionary, presented gifts on behalf of the C.M.S., and the heads of the trading firms followed him to present their

1 Flint, op. cit., p. 26, citing Consul Hopkins to F.O., 18 Nov. 1878, discussing the organization of the Company and praising the African staff.
2 W. H. Simpson's Report, November 1871, op. cit., with Journal as encl. 3.
3 Cf. conclusion to Crowther's Journal of a visit to Bida, September 1869. Crowther described how the emir had put down a revolt and restored peace: 'Thus the gnawing worms were removed from the root of the promising tree. It may now be expected to grow and flourish so as to afford agreeable shade to all under its branches. I sincerely hope Her Majesty's Government will really see the advantages of having such an ally or allies on the banks of the Niger and in other parts of the country. . . . It is better to have to do with one ruler who keeps order and the people in subjection, although with tyranny, whether he be a heathen or Mohammedan than to have to do with a people in a state of anarchy.' C.M.S. CA3/04.
4 Crowther to Venn, 30 Oct. 1871; C.M.S. CA3/04.

gifts. In return Masaba gave his gifts of friendship to Crowther to be sent on to the Queen.[1]

It must be said that Crowther, along with MacGregor Laird and Dr. Baikie, had ensured the predominance of British trade and interests on the Niger before there was competition from the French and the Germans. They laid the foundation on which Goldie came to build. But what mattered most to Crowther in all his achievements on the Niger—the increase of trade, the increasing prosperity of emigrants, his own consular status—was not the genesis of an empire but the advancement of the missionary establishments.

We have already referred to the rate at which Crowther's missionary establishments expanded, and to the fact that he was for 'time' and not 'concentration'.[2] In 1864 he had a flourishing mission station at Onitsha and one at Igbebe, which was harassed by a disputed succession to the headship of the town. The dispute resulted in an armed conflict in 1866 and the mission was removed to the settlement at Lokoja that had grown round Dr. Baikie's 'market' and the factory of the trading company.[3] The attempt to establish a station at Rabba had failed in 1859, but one was established at Egga in 1873. An attempt at Idah in 1867 did not succeed. The Onitsha Mission gradually expanded to the surrounding villages, Obotsi in particular and to Osomare and Alenso, on the Niger, to the south.

The most remarkable expansion, however, took place in the Delta, which was first neglected in the urge to reach Hausaland. In 1861 Akassa, at the mouth of the Niger, was occupied, much as Badagri was occupied earlier when a coastal port was found necessary. But, like the people of Badagri, those of Akassa did not rush to become Christians. The death of the king as well as of his cousin, Koko, who welcomed missionaries and acted as their interpreter, within months of the establishment of the mission, was not a good omen.[4] But the mission

1 Crowther to H. Fowler, acting Administrator of Lagos, 4 Dec. 1872. Crowther to Hutchinson, 12 May 1874, enclosing copies of Crowther to George Beckley, Governor of Sierra Leone, 30 Oct. 1873; Beckley to Crowther, 25 Feb. 1874; Crowther to Beckley, 19 March 1874; Crowther to Capt. Strachan, Administrator-in-chief of the Gold Coast, 2 Oct. 1874; Crowther's Journals of visits to Bida, entries for 6 and 7 Sept. 1875, 13 Sept. 1876; C.M.S. CA3/04.
2 See p. 91. There are two valuable accounts of the Delta section of the Niger Mission: D. C. Crowther: *The Establishment of the Niger Delta Pastorate, 1864–92*, Liverpool 1907; E. M. T. Epelle: *The Church in the Niger Delta*, Niger Delta Diocese 1955. For the rest of the Niger Mission see Crowther's general review in the Charge to his Clergy, 1874; C.M.S. CA3/04.
3 Crowther, Journal of a visit to the Niger, July–September 1866; C.M.S. CA3/04.
4 J. C. Taylor's Journal Extracts, December 1861–January 1862; C.M.S. CA3/037.

established in Bonny in 1864 flourished. In 1868, the mission was invited into Brass, and they established stations at Tuwon in that year, and Nembe the following year. Here the mission had one of its most spectacular successes in the mass conversion of the Brass people between 1876 and 1879.[1] Kalabari and Okrika, the only other Delta states east of the Niger still without missionaries, were occupied in 1874 and 1879 respectively. The Kalabari mission was hampered by political difficulties between the Kalabari people and Bonny. When this culminated in war in 1882 and the Kalabari moved from the old settlement called 'Old Shipping', they carried Christianity with them to their new settlements at Abonema and Buguma.[2]

In this way, Crowther's Niger Mission came to stretch from the Nupe country to all the Delta states in the south. Between 1871 and 1875 he paid two or three visits to the Benin river but failed to get a footing there among the Tshekiri. His usual annual programme was to spend the dry season, from November or December to February or March, at Lagos, visiting Otta, writing his reports and despatches. Sometime in March he went on to Bonny and from there visited Brass and Kalabari, moved to Akassa in June and travelled up with the naval or mercantile vessel to Onitsha, Lokoja and Egga or Bida. By October he returned to Bonny, and to Lagos by November or December.

Crowther was brought up on evangelical doctrine and he subscribed to the fundamental tenets of that doctrine, but while the typical evangelical missionary was a preacher of the word, Crowther's whole inclination was to be a teacher of the word. His typical advice to a missionary at a new station was:

> Your ministerial duties will be very simple and plain: you shall have to teach more by conversation when you visit the people or they visit you at the beginning than by direct service.[3]

His missionary technique was similar to the one we have already described of using the Mission House as a nucleus of civilization and the centre of a new way of life. If anything, his mission stations tended

---

1 In 1876, King Ockiya and other leading chiefs like Spiff and Sambo in a dramatic and much publicised episode sent their idols to Salisbury Square and later began to join the Church. When King Ockiya died in 1879, the adherents of the old religion attempted to stem the tide of conversions. The Christians were by then strong enough to take up arms to establish themselves firmly as rulers of the town. Rev. John Garrick, the local pastor, to Bp. Crowther, 29 Dec. 1899; C.M.S. CA3/04.
2 Epelle, op. cit., pp. 28–9.
3 Instructions to the Rev. J. C. Taylor, encl. in Crowther to Venn, 26 Aug. 1857; C.M.S. CA3/04.

216

to be more physically separate from the old towns. At Bonny for two years boys going to school had to travel round the island by canoe before a path was made to join the mission station (at Andony) to the old town.[1] In Kalabari the site he selected was on the side of the creek opposite the town. The chiefs protested, saying that they would be greatly inconvenienced if European traders should be attracted from their side to the mission side of the creek; but Crowther insisted that the suitability of the ground he had selected outweighed such considerations.[2] When a mission was established at Egga in 1874, it was placed on a lonely eminence; in Crowther's words,

> on the slope of Kippo Hill opposite the town of Egga, about two miles across the river . . . commanding an extensive view of the river, dry and healthy, just the reverse of the town of Egga.[3]

Emigrant settlers and traders formed the nucleus of the congregations; particularly at Onitsha, Lokoja and Egga, the Mission House functioned much as at Abeokuta, Lagos or Calabar. Crowther relates in some detail how the rulers of Onitsha, at a conference with missionaries, traders and the commander of the naval escort ship in 1868, made a spirited attack on the policy of creating a dichotomy in society. They proposed not that emigrants and converts should stop going to church, but, among other things, that converts should not refuse to join their friends and relatives in performing the customary rites; that converts should be forbidden to use European clothes, and

> That an agreement should be entered into that there would be intermarriage between the children of the settlers and those of the natives of Onitsha that all may become one people, or else they could not see how we could profess to be their friends without such arrangements.[4]

Crowther, of course, rejected out of hand the suggestion that converts should continue to join the customary rites; but he insisted that the mission did not place any restrictions on dress and that marriage was a voluntary act that could not be ordered by law.

Inevitably the Mission House must maintain a separate, distinct identity. Yet it would appear that Crowther regarded the Mission House on its lonely eminence, high and dry and healthy, less as a baby that

1 Epelle, op. cit., p. 10.
2 Crowther to Hutchinson, 11 Feb. 1874, enclosing Dandeson Crowther to Crowther, 27 Jan 1874; C.M.S. CA3/04.
3 Crowther to Hutchinson, 8 Oct. and 24 Oct. 1874; C.M.S. CA3/04.
4 Crowther, Journal of a visit to the Niger Mission 1868, entry for August 12 C.M.S. CA3/04.

would grow, rival and in due course overcome the old town, than as a showpiece of the new way of life, like the example that the practical teacher held up his hand as he proceeded to educate the whole town to accept the new way of life. In this approach two important new trends should be emphasized here, as they became more generally adopted by other missionaries later, namely, a new approach to the use of education in missionary work, and an increasing emphasis on the need for missionaries and converts to exercise direct influence in changing laws and customs in the old town.

The school was Crowther's chief method of evangelization. He introduced the mission into new places by getting rulers and elders interested in the idea of having a school of their own, and usually it was to the school that he asked the senior missionary at each station to give his chief attention.[1] John Whitford, a European who traded on the Niger in the early 1870s, gives the impression that the chief feature of Sunday congregations was the preponderance of drowsy schoolchildren.[2] In later years a missionary, very critical of Crowther's methods, said he accompanied the bishop to seven meetings with chiefs in different places:

> At only one did he allude to the existence of God, when he said in one sentence at Onitsha, 'I shall conquer, God is behind me.' The existence of a future life was never once even remotely alluded to; the advantages of the mission being represented as an inexpensive way to enable their children to get good pay as clerks or engineers with the traders, to know everything but Christ and Him crucified and to set their affections on things below, not on things above.[3]

The missionary added in a footnote, however, that when Crowther went among his congregation at Lokoja, 'to our astonishment, the bishop, in his sermon, took an absolutely opposite course and preached the blood of Christ'. Crowther believed not only that civilization was an inseparable companion of Christianity, but also that the first duty of the missionary was to attract people to the mission and the doctrinal refinements would follow.[4] Education, he said in 1874, cannot but

1 Cf. Instructions to Agents given at Bonny, 27 May 1872; C.M.S. CA3/04.
2 John Whitford: *Trading Life in West and Central Africa*, Liverpoool 1877, p. 117.
3 G. W. Brooke to Robert Lang, 16 July 1889; C.M.S. G3/A3/04.
4 It was related that during a smallpox epidemic at Onitsha in 1873, a half-demented heathen priest said he saw visions that all the trouble was due to the people not accepting the missionary invitation to go to church. Crowther commented: 'We do not depend on temporary impressions of visions which after a little while waste away like fog in the air; but as man's extremity is God's opportunity, who can tell what amount of good may still be done by the people being thus frightened to the place of worship where they will receive the solid word of truth.' Annual Letter, Onitsha, 1873; C.M.S. CA3/04.

enlarge and enlighten the idea of those who are brought under its influence, especially where all the elementary school books are extracts from the Holy Scripture inculcating all virtues and condemning all vices, and vividly pointing out the folly and superstition of idolatrous worship.[1]

But besides Crowther's own predilections, there was a very practical reason why he was obliged to place emphasis on schools, namely, the old one of finance. No other missionary was as dependent as he was on securing local support and the school was the most effective organ for this.

Between 1861 and 1878 missionary establishments on the Niger more than doubled, but the C.M.S. grant to the mission hardly increased at all.[2] This state of affairs arose from the conception of Crowther as a native bishop at the head of a self-supporting, self-propagating Church, as distinct from a missionary wholly dependent on a missionary society. In 1864 Venn instituted a special endowment fund, called the 'West African Native Bishopric Fund', for which he appealed to Christians everywhere, urging them to contribute to see the native bishopric experiment succeed. Besides sums from England, there were contributions from congregations in places as remote as Madras, Quebec and Bucharest, and from European and African traders at Lagos, in the Delta and in Sierra Leone.[3] The fund was described as being at the disposal of the Bishop for the

commencement and encouragement of local missionary effort. For example, he will be able to encourage heathen kings and chiefs to receive and support native teachers and schoolmasters by grants-in-aid. . . . He can engage interpreters and copyists in reducing new languages.[4]

In fact it was recognized that apart from the mission stations at Onitsha and the confluence, for which the C.M.S. were already responsible,

1 Crowther: Paper entitled 'Brief Statements exhibiting the characters, habits and ideas of the Natives of the Bight', 1874; C.M.S. CA3/04.
2 The grant in 1861 was about £1,000 (excluding supplies in kind). In 1864, on Crowther's becoming bishop, his salary was raised from £100 to £300, and the grant therefore went up to £1,200. In 1878 two Europeans joined the mission and the grant went up again. See Estimates in the Annual Reports in *Proceedings of the C.M.S.*, 10 July 1872.
3 Crowther to Venn, 20 March 1872, mentions some subscribers to the fund. Some of the subscriptions like the £5 from Bucharest in 1868 were for stated purposes, in this case for the redemption of a slave girl to be named Sarah Bucharest and brought up in the mission. Crowther, Journal of a visit to the Niger, 1869, entry for 8 Sept.; C.M.S. CA3/04.
4 Crowther: A Charge delivered to the Clergy, 1874; C.M.S. CA3/04.

expansion elsewhere, particularly in the Delta, depended on Crowther's ability to use the Bishopric Fund to encourage local self-support. It was the keen desire for education in the Delta that Crowther had to exploit to introduce Christianity there.[1]

He began at Bonny, where the situation was particularly favourable. Crowther went to Bonny in October 1864 and negotiated with King William Dappa Pepple and the chiefs, who had asked for a missionary from the Bishop of London, an agreement by which they promised to pay 21 puncheons of oil (valued at £300–£400) as roughly 50 per cent of the capital costs of a mission school and house, the rest being met from the Bishopric Fund.[2] It was further agreed that every gentleman sending children to school would pay £2 a year for each child in the lower and £3 in the upper classes. There was a clause providing that children of poorer people would be accepted at a lower rate, but no such case was actually recorded. The fees paid the salaries of two schoolmasters, and the C.M.S. was called upon to maintain only the catechist and later the ordained missionary.

King William Pepple was succeeded by his son George, who was educated in England and was an ardent Christian, and helped to make the school a local institution in which the whole community took an interest. The teachers announced that leading heads of houses like Oko Jumbo and Ada Allison took private lessons at home. The annual public examination of the school and the Christmas feast, followed by evenings of magic-lantern shows of pictures of the Holy Land and scriptural scenes, became something of a local festival.[3] Meanwhile evangelization and social reform went on side by side, the first substantial missionary victory being the decree of the king and chiefs in 1867 that the iguana (a type of lizard), formerly held sacred to them, be allowed to be destroyed and no longer held sacred.[4]

The Bonny Mission School soon became a model which neighbouring states were anxious to emulate. And it was by agreements similar to Bonny's that Crowther extended the work of the mission to Brass in

1 When persecution broke out in Brass in 1872, and the number of children at school was reduced from 31 to 16, Crowther wrote to Venn that it 'has robbed us of the means of supporting that mission', 10 July 1872; C.M.S. CA3/04.
2 Agreement with Bonny Chiefs, 1865, referred to in Crowther to Venn, 1 May 1867. Also printed account of payment of school fees at Bonny and Brass missions, dated 31 Dec. 1869. Crowther to Venn, 10 July 1872, that in spite of the civil war in Bonny the terms of the agreement were eventually carried out; C.M.S. CA3/04.
3 Crowther to Venn, 27 Feb. 1867; Journal of Crowther's visit to Bonny 1867, entries for 12–14 Jan; C.M.S. CA3/04.
4 King George Pepple to Crowther; 22 April 1867, encl. in Crowther to Venn, 1 May 1867; C.M.S. CA3/04. Also printed in full in Appendix 3 of Epelle, op. cit.

1868 and Kalabari in 1874. In this way Crowther was pioneering the method of the village school and local self-support on which was based the very rapid expansion of missionary work in Southern Nigeria in the period 1891–1914. It worked best in the Eastern Delta and as long as trade flourished. When Crowther tried it in Warri between 1871 and 1875, wishing to take advantage of declining trade and the readiness of European traders to sell their buildings cheap, he found Chief Nana Olomu, the local ruler, too conservative and distrustful of the social effects of book-learning.[1] But as trade developed on the Niger he introduced the method of local self-support into the mission there. In 1879, for example, when the mission at Onitsha was evacuated and the buildings were destroyed by a naval bombardment of the town, Crowther went to Onitsha to convince the rulers not only that the mission had had no hand in the action, but that they had suffered a loss of up to £600–£800 and would come back only if the people helped to rebuild the mission buildings[2].

Arising out of this new approach to missionary work was the second trend referred to above, that the old dichotomy between the state and the Mission House could not be maintained. If the mission was to be supported largely by the school fees contributed by the leading citizens and rulers of the old town for the education of their children and wards, the missionary had to keep in close touch with them. Crowther himself cultivated a close personal relationship with and mutual respect for practically every ruler from Lagos to Kalabari, up the Niger to Bida and overland through Ilorin and Ibadan back to Lagos. Not only that, he came to regard the ability to work with local rulers for the reform of the old town as the most essential training for a missionary.

Crowther was emphatic on the value of literary and academic and institutional education. He could see that if it was deep and thorough enough and based on Christian principles it should not prevent a man from using his hands. He was himself a linguistic scholar of distinction who not only encouraged industrial education but was also at home in constructing buildings and planning artificial drainage. At the same time he supported Macaulay in founding the Lagos Grammar School. He took great interest in the development of the Freetown Grammar School and Fourah Bay College and, when all local resources had been exhausted, he strongly favoured education in England.

Yet, for all his belief in the value of literary and academic training,

1 Crowther to Wright, 18 June 1874; Crowther to Hutchinson, 13 April 1875; C.M.S. CA3/04.
2 Crowther, Report on the Niger Mission, 1880; C.M.S. CA3/04.

Crowther did not reckon it the most essential part of the equipment of a missionary. He made periodic analyses in 1868, in 1870 and again in 1877 of the qualifications and merits of his missionary staff, and on each occasion he came to the conclusion that he depended most on middle-aged men barely literate in English and the vernacular, farmers, carpenters, mechanics, masons, court messengers, stewards on ships and the like by profession, recommended by the Niger Mission Committee in Sierra Leone as men of proven Christian character. One of their chief merits was that they 'command more respect with chiefs than young, inexperienced, college-trained men'.[1]

Dependence on a staff of this description arose, of course, more out of necessity than choice. The frequent analyses of their qualifications was itself an indication of how Crowther was haunted by the intractable problem of recruiting an adequate staff. He had no money to found a training college until bad reports created anxiety in London and funds were made available for a Preparandi Institution in 1883, and the building was completed only in 1887. In any case, the mission was expanding faster than a new training college could have catered for. He had to rely on the existing training institutions and the men who volunteered and were recommended to him from Sierra Leone. The salaries were £36 a year for an evangelist, usually married and sometimes with as many as five children, £50–£62 for an ordained missionary. Highly qualified men, if anxious to be missionaries, chose Sierra Leone or the Yoruba Mission. For the C.M.S. insisted that all volunteers for the Niger mission must consider themselves emigrants and could not expect passages to be paid back to Sierra Leone for leave. In cases of special necessity Crowther had to make grants from the Bishopric Fund.[2]

1 (i) Crowther to Venn, 27 Nov. 1868; (ii) Paper drawn up in the summer of 1870 entitled 'Suggestions to the Parent Committee for the training of Native teachers', in which, among other things, he recommended the establishment of a 6 months' Evangelist course at Fourah Bay; (iii) Appendix to the Report of the Niger Mission, 1877; C.M.S. CA3/o4.

2 In 1870 the Rev. F. Langley, on £50 a year, had a wife and 6 children. Mr. Edward Phillips, a catechist on £36 a year, had a wife and 4 children, 2 of whom were in boarding schools in Lagos, passage alone costing £5 or £6. Mrs. Phillips undoubtedly supplemented her husband's income by trading; Appendix to the Report of the Niger Mission, 1877, op. cit. Crowther to Hutchinson, 11 Dec. 1870, said that he had to pay the passage of Phillips to Sierra Leone to enable him after 12 years' service on the Niger to visit 'his *blind aged* father, his mother having died during his absence, as well as to settle in marriage one of his grown-up daughters'. Crowther to Hutchinson, 11 Feb. 1874, recommended an increase of salaries on condition that educated wives spent at least 3½ hours a day helping at school, but nothing was done about this, presumably because there were no funds. The mission authorities emphasized that a spiritual call, not the salary, should attract people to become mission agents, but those who like Venn took a personal interest in

Within the context of that situation, Crowther evolved a system—discussed in a paper he presented to the Conference of West African Protestant Missionaries in Gaboon in 1876—by which the preliminary work in each mission station would be done by evangelists 'whose age commands respect before chiefs and elders who look to them as equals in years but as superiors in knowledge'. Younger men who had acquired at college a 'liberal education in some special branches of literature' were then placed under them to acquire '*experience* of the evil of the human heart, and how to deal with it; *experience* how to conduct themselves before old, shrewd men like their fathers, so that their youth be not despised'.[1] Then they could be ordained and placed in charge of stations. His last-born son, Dandeson, whom he dedicated to missionary work, he sent to the Lagos Grammar School. Then he kept him as his private secretary and copyist and companion on his travels for four or five years, sent him to the C.M.S. Training College at Islington, ordained him in London in 1870, and posted him to Bonny in 1872. Two years later, Crowther at Lagos received from him an account of a journey to Kalabari to argue the chiefs out of their opposition to the proposed mission on the further side of the creek. Commenting on this report, Crowther remarked on Dandeson's tact, which he said was just the type of ability the mission required.[2] Dandeson was to become Superintendent and later Archdeacon of the Delta and Lower Niger Stations.

It was, of course, impossible to try to understand the rulers of the old town and know how to deal with them without learning to respect some aspects of their way of life. Thus, as Crowther came to depend more and more on education and on bridging the gap between the Mission House and the old world, though he continued to emphasize the value of civilization and foreign ideas, he was increasingly emphasizing the value of seeking what is good in the old society and cultivating it. As a translator

---

many of the agents were not unaware that the salaries were inadequate to rear large families (and the agents did take keen interest in the education of their children, as a form of investment and insurance for the future, said Crowther). It was for this reason that Venn sought for people like Crowther who had proved themselves valuable to the mission alternative sources of income. In his case, his wife was exempted from the ban on trading, and when his position as bishop no longer warranted this, his salary was tripled.

1 Printed Report of the *Conference of West African Protestant Missionaries held at Gaboon, February 1876*, Mission Press, Old Calabar, p. 13. The Conference was called by missionaries of the American Presbyterian Foreign Missions Board. Crowther was unable to attend. His Paper was read by Dandeson Crowther, who represented the Niger Mission. For this Report and other documents about the Conference see C.M.S. CA3/013. Italics in original.
2 Crowther to Hutchinson, 11 Feb. 1874; C.M.S. CA3/04.

he had to ponder over the Yoruba equivalents of God, Devil, priest, and so on, and he consistently tried to find a term already in use. As head of a mission and in so far as he exercised authority as a bishop, this was his chief contribution to the development of the Church as a national institution.[1] 'Christianity,' he told the clergy in his charge of 1869, in which he attacked the E.U.B.M. and made quite clear his stand on the issue of nationalism—Christianity

> has come into the world to abolish and supersede all false religions, to direct mankind to the only way of obtaining peace and reconciliation with their offended God. . . . But it should be borne in mind that Christianity does not undertake to destroy national assimilation; where there are any degrading and superstititious defects, it corrects them; where they are connected with politics, such corrections should be introduced with due caution and with all meekness of wisdom, that there may be good and perfect understanding between us and the powers that be that while we render unto all their dues, we may regard it our bounden duty to stand firm in rendering to God the things that are God's.

> Their native Mutual Aid Clubs should not be despised, but where there is any with superstitious connections, it should be corrected and improved after a Christian model. Amusements are acknowledged on all hands to tend to relieve the mind and sharpen the intellect. If any such is not immoral or indecent, tending to corrupt the mind, but merely an innocent play for amusement, it should not be checked because of its being native and of a heathen origin. Of these kinds of amusements are fables, story-telling, proverbs and songs which may be regarded as stores of their national education in which they exercise their power of thinking: such will be improved upon and enriched from foreign stocks as civilization advances. Their religious terms and ceremonies should be carefully observed; the wrong use made of such terms does not depreciate their real value, but renders them more valid when we adopt them in expressing Scriptural terms in their right senses and places from which they have been misapplied for want of better knowledge.[2]

Coming from a bishop virtually on trial, the purity of whose religion was being watched, as he was accused in some quarters of baptizing too readily and lowering standards—coming from such a bishop, nothing could be clearer as a statement of policy.

1 Cf. Venn's doctrine of the Native Bishop and the Organization of the Native Church into a National Institution, pp. 176–7 above.
2 Crowther, A Charge delivered at Lokoja in 1869, op. cit.

photo by permission of Christian Missionary Society

A Missionary Group under the Wilberforce Oak, 1874. Left to right: Mr. E. Hutchinson, C.M.S. Lay Secretary, 1872-1882; The Rev. D. Hinderer, of Ibadan; Archdeacon Henry Johnson of the Niger Mission; Bishop S. Crowther; Mr. King, a medical student; The Rev., later Bishop, James Johnson; The Rev. H. Townsend, of Abeokuta. The tree and stone, just visible behind them, mark the spot where William Wilberforce resolved to introduce into Parliament his Bill for abolishing the Slave Trade in 1789.

A Chip off the Old Block. The Reverend J. Boyle and family
in Bonny, 1897.

But it is not easy to find out how this policy was interpreted in practice or how far Crowther carried his clergy with him. Marked variations were sometimes obvious. The Rev. J. C. Taylor at Onitsha in the 1860s said that he did not see anything sinful in converts taking or retaining *ozo* titles[1]—which was comparable to being a member of the Ekpe Society in Calabar or the Ogboni at Abeokuta and has for a long time since been consistently condemned by European missionaries. On the other hand, the Rev. J. Boyle at Bonny in the 1880s in a similar situation forbade members of his church to join an attempt by King George Pepple to revive an Old Bonny Secret Society, *Owu Ogbo*, as the Bonny Play Club.[2] In practice, hostility to polygamy might have varied in intensity; Crowther remained a confirmed opponent and would not baptize a polygamist, but his son went further and declared polygamy to be 'slavery for the wives'.[3] Crowther remarked in 1886 on the 'native airs' for which the church at Otta, in the Yoruba Mission, was becoming well known—songs 'of suitable Scriptural compositions of their own adapted to their native airs'[4]—but it is doubtful if anything like it was encouraged anywhere else. Such development and adaptation as was necessary to make Christianity not just a foreign religion perched on the outside of the life of the community but a way of life for the whole community, to replace the old by absorbing all that was best in it, required time and the application of several well-cultivated minds working in an atmosphere free from distrust and suspicion. It could not be the work of a generation, or of one man on trial, working with no more than the staff that Crowther had.

Yet it must be said that in the Niger Mission, more than anywhere else in this period, the effort was made consistently to convert the men

1 Taylor to Venn, 9 Oct. 1862; C.M.S. CA3/037.
2 Epelle, op. cit., p. 41 and Appendix 3, pp. 117–8. King George Pepple to Boyle, 11 Oct. 1884, argued that 'in the olden times, the play was connected with sundry sacrifices of fowls, goats, fish and the pouring of tumbo [i.e. palm wine] at the beginning as also at the end of the play. But at the present time, since the revival of the Club, by the desire of the chiefs, no sacrifice of any kind whatever has been offered, or made, or suggested, to the best of my inquiries and knowledge.' But Boyle remained unimpressed by the argument.
3 Printed Report of the Gaboon Conference, op. cit., p. 8, during a debate on polygamy.
4 Crowther to Venn, 6 Nov. 1866; C.M.S. CA3/04. It was Townsend who first remarked on the Otta Native airs in 1857. He said that the Otta converts who had gone out to meet him 'sang at my request some of their Christian hymns, words and tunes being of native composition, the first I have heard. Some of my carriers (Egba converts) were rather scandalized at this attempt, but I encouraged them to view it in a better light.' Journal entry for 29 Aug. 1857; C.M.S. CA2/085. Credit for this early attempt at evolving Christian hymns in the indigenous musical tradition goes to James White, emigrant catechist at Otta from 1854, ordained 1857, died 1890; C.M.S. CA2/087.

with influence in the old town so as to win the whole community for the new religion. The effort of Taylor has been referred to. In all the persecution the Church faced at Onitsha, the Bishop and the pastor did not have to argue alone with the chiefs. By 1868 there were already at least two titled men who were themselves being persecuted and who could in the councils of the land take up the defence of the new religion.[1] Such men, of course, were apt to argue with missionaries about how to do their work and what social customs to condemn and what to tolerate. They complicated the problem of Church discipline especially in the Delta, where the Church was most successful in converting leading members of the community. King George Pepple of Bonny was a Christian before the mission was established. Most of the chiefs in Bonny and Brass, however, rallied round the school and rivalled one another in donating money for it or for building and furnishing churches. When the time came their rivalry extended to putting away their idols, declaring themselves monogamists and accepting baptism.

The depth of some conversions may be questioned. Dandeson Crowther in 1884 referred to the prevailing sin of polygamy, particularly at Nembe (Brass), 'where the majority of the chiefs baptized have fallen into the sin of polygamy'.[2] But there can be no questioning the fact that without evoking the force of the gunboat the missionaries got the chiefs not only to pass laws against human sacrifice and twin murder, the cruelty of which, though not necessarily the sinfulness, could be easily made obvious, but also to alter national customs to the point of abolishing, destroying and desecrating objects formerly held sacred. If it is any indication, the persecutions in the Niger Mission, at Onitsha, 1863, 1868–71; in Brass, 1871–77; at Bonny, 1871–75 and 1881–86, were apt to be fiercer than elsewhere, and Bonny claimed the only martyr of the period, Joshua Hart, in November 1875.[3] The assertion may be permitted that Christianity penetrated the life of the people as communities more thoroughly in the Delta than anywhere else in Nigeria. No doubt many factors contributed to this—the long trade connections with

---

1 Taylor to Venn, 9 Oct. 1862, also 10 Jan. 1863, referring to Okosi and Mba. Okosi was steadfast for a long time; Mba soon 'backslided', but his son Isaac Mba became a mission agent. In 1868 Crowther mentioned Adam Anyabu, 'a man of good connection' who at the conference with the Obi Elders, was defending Christianity 'boldly and faithfully'; Crowther, Journal of a visit to the Niger Mission, 1868, entry for 12 Aug.; C.M.S. CA3/04.

2 Not all the chiefs were able to choose the moment of their conversion as conveniently as King Ockiya, who gave up his idols in 1876, waited another 3 years to declare himself a monogamist ready to be baptized 'Constantine' in December 1870, and died two weeks later; Garrick to Crowther, 27 Dec. 1879; C.M.S. CA3/04.

3 Stocks, op. cit., vol. ii, pp. 461–3.

Europe, and prevailing economic factors of the nineteenth century—but the effectiveness of Crowther's missionary methods must be reckoned above all factors. In 1849 King William Pepple had informed Hope Waddell that anyone who told him that Bonny people could consent to destroy 'Juju House' was a liar. In 1889, after a short devotional service, 'Juju House' was pulled down and burnt.

What contribution Crowther would have made to the development of the Church as a national institution in Nigeria if he had become what Venn hoped he would be, a bishop with the co-operation of African pastors and European and African missionaries, both on the Niger and in Yoruba, is a matter for mere conjecture. It remains to indicate briefly how near to it he came, how his achievements on the Niger convinced some of his earlier critics, and how his ideas were beginning to influence others, before the reaction against him began.

To begin with, besides holding visitations at Otta and in the surrounding area, and continuing his work of translation, there was little Crowther could do in the Yoruba country except by special licence from the Bishop of Sierra Leone. When European missionaries were expelled from Abeokuta, Venn wrote to Townsend that though residing at Lagos he was to be regarded 'in the light of the Superintendent of the Abeokuta mission to carry out the plans for the organization of the Native Church', that it was essential to ordain some more pastors to manage the Church, and that Townsend should arrange to present such people to be ordained. Venn added that their ordination 'will properly fall to the province of Bishop Crowther',[1] that is, not to the Colonial Bishop of Sierra Leone. Crowther deliberately stayed aloof lest he should appear to countenance the anti-European views of G. W. Johnson and the Egba United Board of Management, though he suggested that the churches should only be rebuilt on the basis of a scheme of local self-support,[2] like the one he was working out in the Delta. But when by 1871 the Egba had not changed their policy, the Parent Committee asked Crowther to make a visitation to Abeokuta, for which the Bishop of Sierra Leone issued a licence. Townsend himself agreed to approach the Egba to give himself permission for a short visit so that he could prepare the candidates for ordination and confirmation ready for Crowther.

He went to Abeokuta in January 1871. He found the town split into two camps over the rival candidates, Ademola and Oyekan, for the throne of the Alake. Both sides spoke fair to Townsend, and told him

1 Venn to Townsend, 21 Dec. 1869; C.M.S. CA2/L3.
2 Crowther to Hutchinson, 19 Jan. 1871; to Venn, 4 Jan. 1868; C.M.S. CA3/04.

he could return to Lagos and come back in June to stay a few weeks as he wished. But when he returned to Abeokuta in June he found the gates barred against him.[1] Crowther therefore postponed his visit. He went on to the Niger but when the vessel in which he was returning to the coast was grounded near Lokoja, he decided to return to Lagos overland, and to use the opportunity to visit the churches in the interior. He was accompanied by some of the crew and passengers, eight Europeans and about a dozen Africans. At Ogbomoso his attention was drawn to the Baptist converts, twenty-two of them, without a teacher since 1862, who had been meeting together on Sundays by themselves. A few young men who had gone to school read the Scriptures in turn; the elders engaged in prayers. Crowther had a service with them and told Baptists in Lagos of the continued existence of the little congregation. He spent a week at Ibadan, where the Rev. D. Olubi and two catechists had been looking after the three C.M.S. congregations and occasionally visiting others in the out-stations. Crowther noted, concerning the agents, that 'the Are respected and consulted them on any matter of doubt between [Ibadan] and the white man'. He visited all the congregations and at a joint service confirmed ninety-four candidates and celebrated Holy Communion for 107 people. Then he moved on to Abeokuta. He was met at the gate by envoys of the two parties in the town. From the Ademola party there were Henry Robbin and another emigrant, J. George, who had become a chief, the Lisa Oba ('an important title among the Elders'), accompanied by a messenger bearing the staff of the Alake. They wanted Crowther and his men to lodge at the C.M.S. Mission House at Ake. From the Oyekan party there was a messenger of the E.U.B.M. bearing a large envelope enclosing a permit, valid for ten days, allowing the party to enter the town but directing that they should stay at the Methodist Mission House at Ogbe near G. W. Johnson's house on the other side of the town. The Ademola party gave way. Crowther spent a week, visited all the mission stations and held a communion service for 116 people. He also held a conference with the chiefs, who were united in welcoming him and in asking him to present to the Foreign Office their grievances against the Lagos government. He returned to Lagos through Otta, where he opened a new church building.[2]

1 Crowther to Venn, 19 June 1871, suggests that this was due in part to Robbin's and Townsend's mismanagement of the situation; C.M.S. CA3/04.
2 Crowther: Journal of an overland route journey, December 1871–February 1872. Ogbomosho 16 Jan., Ibadan 20–29 Jan., Abeokuta 30 Jan.–5 Feb. See also printed memorandum, 'Acts of Liberality of King Masaba of Nupe to British subjects', Lagos February 1872, recounting and estimating the cost of the hospitality they received not only from the Emir of Bida but also all along the way; C.M.S. CA3/04. Also Crowther to Kimberley, 6 Feb. 1872; CO147/23.

The journey showed that the removal of European supervision had not shaken the Church in the interior; rather, it had encouraged it to develop self-confidence and self-support. The members had moved closer to the traditional rulers; they started praying for them in their services, took more interest in local politics and were anxious to exert their influence in the Councils of the land. More than that, they were contributing money to support their pastors and to send teachers and evangelists to outlying stations. This growing spirit of independence in the interior soon began to affect Lagos, where the wealthiest parishioners were to be found. These were now asking for some say in the affairs of the Church since they were prepared to contribute money for its support. To this end they founded in September 1872 an 'Association to further the interests of Christianity and Education in Lagos'.[1] Crowther's journey had proved that his episcopal authority was acceptable everywhere; that the Church in Yoruba which had been without episcopal visitation since 1859 needed a bishop, and that the Bishop of Sierra Leone could not conveniently fulfil the office as long as the Egba and the Ijebu continued to refuse to open the roads to Europeans. The Parent Committee and the C.M.S. therefore began to discuss the necessity for the Bishop of Sierra Leone to transfer his control of the European missionaries and Yorubaland to Crowther.

There was at the same time a movement in Sierra Leone for the formation of an independent African Church which alarmed the Parent Committee. They summoned Crowther to London together with Rev. James Johnson, a lecturer at Fourah Bay College, who was judged to have been the brain behind the movement. Venn had died in January of that year, but the Committee saw only too well from the threats of schism what was likely to happen if his principles were not fully put into practice. They therefore decided to ask Crowther to visit Sierra Leone and pacify the African pastors there;[2] to transfer James Johnson to the Yoruba Mission and find a post of responsibility for him; to press on with the introduction of the pastorate scheme to Lagos; and finally, to settle the headship of the Yoruba Mission. Meanwhile Townsend and Hinderer continued as Superintendents of their old stations, living on the coast and attempting periodic visits inland to advise and supervise the African agents.

Townsend went to Abeokuta in February. The politics were confused, but there was no marked hostility towards him. The two parties were

1 The leaders of the Association were C. Foresythe, J. Otonba Payne, and J. P. L. Davies. Payne to Maser, 1 Sept. 1873; C.M.S. CA2/011; Davies to Wright, 17 Oct. 1873, 3 Aug. 1874; C.M.S. CA2/033.
2 Crowther to Hutchinson, 24 June 1873, giving an account of his visit to Sierra Leone; C.M.S. CA3/04.

approaching the time when their internal struggle would lead them each to make overtures to the Lagos government and the missionaries. Townsend made friends with G. W. Johnson, visited him in his house and recommended that he and men like him, of the 'Civilized Party', though not church communicants, ought to be brought back into mission work, particularly on education boards and school committees.[1] Hinderer was even more welcome at Ibadan, where he spent four months. However, he remarked that it was not himself that was the centre of attraction but the Rev. D. Olubi, who enjoyed genuine respect from Christians, heathens and Muslims alike, and that, particularly since the Ashanti War, Europeans were suspect.[2] Whether Townsend and Hinderer concerted plans is not clear, but they both returned from their visits to recommend the appointment of an African bishop for the Yoruba area. 'Has not the time come,' wrote Hinderer in May 1875

> when the native bishop's jurisdiction should be further extended than the Niger, especially to his own native soil? . . . The past six or seven years surely have sufficed to show that the native teachers of Christianity are as acceptable to people and chiefs of this country (I speak here of heathen people—as for converts, that is a matter of course) as white man was some twenty years ago, without whom the native could have had no standing then and that kind of feeling was manifest to a great extent still at the Bishop's consecration, which was the chief reason as far as I know, for his not being made to preside over us. That feeling has changed entirely.

And lest it be said that he wrote that because he was an old man and that 'at my time of life I have not to care being under a native bishop,' he added,

> I could assure any young man from the knowledge I have of Bishop

1 Townsend to Wright, 9 Feb. 1875: The Civilized Party 'are men engaged in trade who have been baptized and taught and fallen into sinful habits but who attend Church regularly and are often seen at prayer meetings. They are no better and no worse than the great majority of Church goers in England. . . . It is this class I wish to get as money helpers and as having some part in managing a school system.' He repeats this in his letter of 2 March 1875 adding that 'Christianity divorced from the Western way of Life, from trade and schools etc. would wither.' He then goes on to take a more understanding view of the emigrants and converts who turned polygamist. The wife, he said, 'is educated by country maxims and rules, and these were framed ages gone by to help the husband of many wives. For instance, the wife separates from the husband until her child is weaned and that may be three or more years.' C.M.S. CA2/085.

2 Hinderer to Wright, 14 May 1875, enclosing Journal of Missionary Journey, February–April 1875, Lagos—Ondo—Ife—Ibadan. Hinderer was at Ibadan, 10 April–5 Aug.; C.M.S. CA2/049.

Crowther that he is the last person in the world to lord it over anyone. Nor would his native teachers take advantage because episcopal authority has changed colour.[1]

Townsend, consistent to the last, proposed not the extension of Crowther's jurisdiction, but the appointment of another African, James Johnson. In January 1876, the Parent Committee wrote privately and confidentially to Crowther that:

> Our good friend Townsend . . . after thinking over all the plans that might be suggested . . . has come to the opinion that the best thing to be done in his judgement in order to promote the well-being of the native Church and the progress of the work at Abeokuta and the Yoruba country would be the removal to that place of our native brother Mr. James Johnson. . . . He further suggests that with this view he should be made Bishop of Abeokuta and the Yoruba country, as you are Bishop of the Niger. . . . The Committee are disposed to regard the suggestion favourably—so that we have written to lay the matter before the Bishop of Sierra Leone.[2]

Crowther was all in favour. James Johnson, he said, would make an admirable bishop. That would leave him free to devote all his attention to improving the efficiency of the Niger Mission. Towards this end the Parent Committee launched an appeal to buy a little vessel so that the Bishop could visit the scattered stations more regularly and in greater comfort. They invited him to England to join the appeal and consult with them and Townsend in working out the details of the new arrangements.[3] That moment was the climax of his career. In the face of much suspicion and calumny, he had maintained a steady course, and now even Townsend paid him the compliment of suggesting that Venn's principles were right and that another African bishop should be created.

But Crowther's victory was short-lived. Opinions about the ability of other races are usually a matter not of fact but of crude prejudice that the careers of one or two or several men can neither justify nor disprove. The younger European missionaries Hinderer referred to were opposed to the full implementation of the Pastorate Scheme and had petitioned against Johnson's appointment. The Bishop of Sierra Leone advised caution. The committee decided to appoint Johnson for a trial

1 Ibid. He repeated this also in his letter of 10 Sept. 1875.
2 Hutchinson to Crowther, 28 Jan. 1876; C.M.S. CA3/L2.
3 Report of a Sub-committee appointed to confer with Bishop Crowther and Messrs. Townsend and Hinderer on matters connected with the Niger Mission, 20 March 1877; Committee of Correspondence, May 1877, cited in Crowther to Hutchinson, 4 Dec. 1880; C.M.S. CA3/04.

period as Superintendent of the Interior Missions and to consecrate him bishop only if he succeeded. A new age was approaching in which it was impossible for Johnson to succeed, just as Crowther's extraordinary career would have been impossible if Venn had been less persistent or if Britain had been ready in 1864 to impose political rule over the peoples of the Niger.

8 The Turning of the Tide

In 1877, just as James Johnson was setting out for Abeokuta as Super-intendent of the C.M.S. missions in the interior, and while Bishop Crowther was waiting at Lagos for the mission steamer, events were gathering force which soon altered the whole basis of missionary work in Nigeria. In that year broke out the last of the Yoruba wars. Abeokuta, Ilesha and Ekiti were menaced by Ibadan. Gradually they built up the grand alliance called Ekiti Parapo which at various times included Ilorin, Ife and Ijebu. Since Oyo aided Ibadan, the war came to affect practically every part of Yoruba. It went on intermittently, neither side gaining the advantage, and later provided an excuse for British inter-vention.

More momentous than the Yoruba war was the growing competition between the European nations to stake out claims and secure territorial possessions in Africa. The Niger and the Benue were much-coveted waterways. In 1879 the four British companies on the Niger amalga-mated into the United African Company under George Taubman Goldie. Thus the British were prepared for the competition which soon came from the French on the Niger and in Dahomey, and the Germans in the Cameroons. The English position was recognized by the Berlin Con-ference in 1885, the French Company sold out to the British Niger Company[1] which soon after, in 1886, obtained a charter to rule northern Nigeria. Britain also declared protectorates over the rest of the country. In this way, between 1877 and 1891, Nigeria, from being 'regions be-yond the Queen's dominions', became three separate British protec-torates: the Lagos Protectorate, the Oil Rivers Protectorate, and the Territories of the Royal Niger Company.

This change affected the development of the Church in Nigeria in many ways, the most notable being the impetus given to the missionary societies by the new urge of Europeans not only to trade with Africans but also to rule over them. The European missionaries confined to the

1 The United African Company (1879) became the National African Company in 1882. On obtaining a Charter in 1886 it became the Royal Niger Company. For the sake of simplicity, I refer to the two earlier companies as the Niger Company.

233

coast since 1867 began once more to move inland. The Rev. W. David came with a Negro pastor in 1875 to revive the American Baptist missions. French ambitions in Nigeria manifested themselves in the forward drive of Catholic missions. The S.M.A. moved to Abeokuta in 1880 and to Ibadan and Oyo in 1884. In 1885 they planned a new mission to the Muslim emirates and established a station at Lokoja but were soon obliged to remove to Asaba. The Methodists, under an energetic superintendent, the Rev. J. Milum, reorganized the Yoruba Mission to become in 1879 a District separate from the Gold Coast and ready to share in the general movement of expansion. Entirely new missionary societies were coming in too, the first being the Catholic Holy Ghost Fathers, who established their first station at Onitsha in 1885. They were followed by the Qua Ibo Mission in 1887 and the Primitive Methodists in 1892. Secure behind the British imperial lion, the Christian missions were ready for the great expansion that gathered pace with the British war on Ijebu in 1892, and on Benin in 1897, and was only interrupted by the First World War in 1914.[1]

This movement of expansion belonged to the future. More relevant here is the way in which the increased interests of European nations in Africa altered the basis of the missionary work we have been discussing. After being in the vanguard, dragging traders and consuls after them, missionaries were beginning to follow after the political officer.[2] They were even more than before closely allied with the national interests of their country. For a while this intensified the rivalry between French Catholics and British Protestants. But the Catholic Fathers soon adjusted themselves to the idea of working in a British colony. They brought in Irish Fathers,[3] equally Catholic and more at home in British territory, and they never lacked sympathetic Catholic officials in the Lagos administration.

It was less easy for the existing congregations and African missionaries to adjust themselves to the new situation. For, from suppliants seeking protection in the country, the European missionaries became protectors, and their attitude towards Africans changed accordingly. From fellow-

1 Groves, op. cit., vol. III, pp. 185–9 and *passim*.
2 Governor Glover in Lagos began the process. In 1863 he invited the C.M.S. to establish a mission at Oke-Odan, over which he was trying to establish a protectorate. The mission was short-lived. Again in 1871, in trying to open up the new route to the north through Agbabu and Ondo, he similarly invited the C.M.S. to consider taking up stations along the route. In that way began the work of the Rev., later Bishop, C. Phillips at Ondo. Glover to Rev. L. Nicholson, 12 June 1871, in C.M.S. CA2/04.
3 The earliest Irish Fathers began to arrive in Lagos, where the teaching of English was most essential, by the late 1870s. In 1881 an S.M.A. Seminary was opened in Ireland. S.M.A. *100 Years of Missionary Achievement*, p. 31.

men and brothers, though not without rivalry, they were becoming part of a ruling caste. Johnson was recalled from Abeokuta. The mission-educated Africans, in particular the missionaries on the Niger, were gradually discredited; European missionaries were introduced into the Niger, and in the end, Crowther was forced to resign. The 1880s were a transitional period, a decade of conflict and bitter racial feeling, of schismatic movements in all the existing missions, except, of course, the Catholic, which had been less committed to the old tradition.

At the centre of much of the controversy was James Johnson—'Holy Johnson', as he was nicknamed in Lagos.[1] He was a spare, fervent, puritanical figure, a zealous reformer with ideas very similar to those of Crowther, but with little of Crowther's deferential, diplomatic approach. A very able, conscientious man, he could not stand the pretensions of many European missionaries and he said so. Even less could he stand the supine African who continued uncritically to imitate European ways. He was a rebel from Sierra Leone and he fought all his life to see that the African of ability got his due respect and that the Church which held out so much promise to the African was made 'not an exotic but a plant become indigenous to the soil'.[2] He came to advocate a reform of the liturgy to suit local conditions.[3] When he listened to an Ifa priest converted to Christianity talking of the attributes of God as taught by his old religion, Johnson felt he had to learn from him about how to present the Christian God to the heathen.[4] When he heard the native airs at Otta, to which we have referred, he wished to see them cultivated and extended to other stations.[5] Two things he particularly wished to reform in the Church: the separateness of the Mission House from the old town, and the lack of fervour among the members. 'Christians,' he observed in Lagos in 1876,

1 Archdeacon J. Lucas: *History of St. Paul's Breadfruit Church*, Lagos 1952.
2 Report from the Rev. James Johnson, August 1877: report on Otta; C.M.S. CA2/056.
3 Cf. ibid, report on Ibadan: 'Every time I worshipped in the Church and the translation and adaptation of the prayer in the liturgy for the Queen's Majesty was read, that portion asking for victory over the enemies of the rulers of the country always grated on my ears. I felt I couldn't conscientiously say "Amen" to it.'
4 J. Johnson to Wright, 2 Aug. 1876, re Jose Meffre's address to Dosumu and his chiefs, 'an address founded upon some of the descriptive and significant names of Ifa such as the "great Almighty one", the "Child of God", the "One who came whom we have put to death with cudgels causelessly", the "One who is mightiest among the gods and prevailed to do on a certain occasion what they could not" '; C.M.S. CA2/056.
5 Report from the Rev. James Johnson, August, 1877: report on Otta; C.M.S. CA2/056.

are regarded as a people separate from [the community], as identifying themselves with a foreign people: the dress they usually assume has become a badge of distinction: the distance between them and the heathens is far greater than that between heathens and Mohammedans. Often many heathens and Mohammedans are found living together in the same house. Christians are rarely found living thus with either. All these contribute to the growth of Islamism.[1]

But he was a reformer, not a revolutionary. He sought his reforms within the existing Anglican Church in which he had been brought up. He was accustomed to believe that the African would be given a chance to make his contribution within that Church. When he felt thwarted, he spoke out passionately in a way many friends and opponents interpreted as meaning a readiness to secede from that Church. But in 1873, far from setting up an independent Church in Sierra Leone, he accepted transfer to Lagos. Again, in 1891, he disappointed seceders who hoped he would be their leader. He was an introvert, seeking the guidance of an inner light, often behaving in a way that puzzled those who thought they could predict what he would do next. He could not contemplate existence outside the Church, but others received inspiration from his words in seeking such an existence.

He arrived in Lagos in 1874 and was made pastor of the leading congregation, that of St Paul's, Breadfruit Church. His work there was devoted to the setting up of the Lagos Native Pastorate, begun in 1875, and the development of the Lagos School Board. He was far from being indulgent to his congregation. Discipline was strict over church attendance, the payment of dues, and the leading of an upright life. Even in little matters he imposed his will. It was said, for instance, when he began to preach against the unnecessary adoption of foreign names, that if any baby was presented to him for baptism, he listened as the European names were being read out and then asked if there was not one other name, a local name, that the parents wished to call their child. As soon as one was mentioned, he signed the baby with the cross and baptized him by that last name alone.[2] At first his congregation were on the verge of revolt against his high-handed manner. But such was his own manner of life that, as they grew to know him, they loved him. It was with difficulty they allowed him to go to Abeokuta in 1877.

He was sent to Abeokuta with instructions

to take the superintendence of the work in Abeokuta and Ibadan with a view on the one hand of working out the organization of the native

1 J. Johnson to Hutchinson, 6 March 1876; C.M.S. CA2/o56.
2 J. Lucas, op. cit.

Churches and, on the other, of extending the work beyond its present limits.[1]

He travelled through all the old mission stations, to Ilaro, to Ibadan and Oyo, noting the state of the churches, what parts of the liturgy needed adaptation and what reforms were most urgently called for. He referred in particular to the continued problems of polygamy, drunkenness, and domestic slavery, all of which he wished to see rooted out. Domestic slavery, he said at Ibadan, was prevalent. 'There is no Christian government to stamp out this accursed institution with the stroke of a pen. We must [work] it out through the Church and educate our people to it.'[2] Then he settled at Abeokuta, organizing church committees of each congregation, a Church council of the delegates of all the congregations, and a monthly conference of all the pastors and catechists. The council raised weekly contributions (class fees) from 1 string (about a penny) to 7½ strings of cowries, and Johnson at once began to take steps to enforce this levy.[3] Early in 1879 the Bishop of Sierra Leone visited Abeokuta and made a favourable report on Johnson's work. The only thing that worried the Bishop was the continued holding of domestic slaves and pawns by both mission agents and converts. The committee therefore wished Johnson to take steps at once to eradicate the evil. They drew up a minute on domestic slavery, a copy of which they sent to him:

> With members generally of the Christian Church in Africa, the committee do no more than appeal in loving remembrance . . . but no one in the employment of the Society shall hold man, woman or child, or have personally any connection with the practice.[4]

Johnson's high-handed methods had been causing resentment at Abeokuta. At a time of war and depression in trade he was imposing larger class fees than ever before. He went so far as to collect a list of names of defaulters and to exclude from communion those whom he judged able to afford the fees.[5] When he began to attack domestic slavery, it was not only Church members but others in the town who took alarm. There was an uproar at Abeokuta; the other mission agents, themselves under sentence of the anti-slavery edict, did not support Johnson. In October the Committee wrote to the Bishop of Sierra Leone:

1 Instructions of the Parent Committee to James Johnson, 8 Dec. 1876; C.M.S. CA2/o56.
2 Report from the Rev. James Johnson, August 1877: report on Ibadan; C.M.S. CA2/o56.
3 J. Johnson, Report on Abeokuta churches, 30 Jan. 1878.
4 Hutchinson to James Johnson, 6 Aug. 1879, enclosing 'Minutes of the Parent Committee on Domestic Slavery in the Yoruba Mission'; C.M.S. CA2/L4.
5 Secretaries to J. Johnson, 24 Oct. 1879; C.M.S. CA2/L4.

We are of opinion—and in this we are strongly supported by Mr. Townsend and Mr. Wood—that it would be injurious to the work at Abeokuta if at the present moment Mr. Johnson is removed—in case such a step can be avoided.[1]

Yet, two months later, in spite of Johnson's protest that the opposition was dying down, the Committee decided to withdraw him and replace him by a European, the Rev. V. Faulkner.

It is not clear why the Committee changed their minds. The main reason expressed was that

There seemed no hope that in the present state of things he would regain the influence for carrying out the Society's minute on domestic slavery.[2]

Domestic slavery had come to be regarded as the perennial excuse for every action, just as the external slave trade had been earlier when Britain sought footholds on the coast. But if Johnson could not enforce the minute on domestic slavery, it was unlikely that Faulkner would be able to do so. What seems clear, however, is that Johnson was not withdrawn on the merits of his case alone. Events on the Niger were beginning to loom large and to discourage any further proceeding with the idea of making Johnson a native bishop. By examining them we can trace in some detail the process whereby increased European interests in the country led to a gradual undermining of confidence in the Africans on whom the Europeans had previously depended.

By 1875 it was becoming obvious that the partnership between the Niger Mission and the West Africa Company could not be maintained much longer. The main difficulty was the 'insane' competition, largely between four British companies but also between them and emigrant traders and Brassmen from the coast. This competition forced up the prices of African produce, which meant for the European traders not only less gross profit, but also having to carry up more goods for the same amount of ivory. In the circumstances the West Africa Company was finding it irritating and burdensome to have to allocate space to missionaries who could not pay a competitive price for it. There was also the feeling among European traders that for every African missionary given passage, one or two African traders, competitors, were being surreptitiously and indirectly helped. The result was increasing hostility to African traders and missionaries alike, and much calumny was spread about them.

1 Secretaries to Bishop Chetham, 24 Oct. 1879; C.M.S. CA2/L4.
2 Secretaries to Maser, 16 Oct. 1879; C.M.S. CA2/L4.

Clegg had died. J. Edgar, the new Managing Director of the West Africa Company, wrote in April 1875 to Josiah Crowther, the Agent-General, that he had information that on one of their ships,

> there was drunken excess and extravagant waste of stores and strong liquor, hungry missionary rabble devouring everything, lazy loafers, illicit traders, dissolute black women and all implied in so naming them, smuggled on board or brought in impudently and found squatting or lying above and below among the men and much else besides.[1]

He said he wished to have full information on what the connection with the mission was costing the company. 'I am not sure we do not lose by the thing,' he said. '. . . I am afraid anything we get [for freight] for either Onitsha and especially Lokoja and Egga will never half pay us, our expenses are so heavy.'

> I shall always be ready to give the Bishop every facility for his Niger Mission that does not interfere with the requirements or the success of the trading arrangements of the company. . . .
>
> Were it not for my personal regard for the good Bishop and the deep sympathy I have with him in his Christian work, I should be unwilling to take either passenger or luggage.[2]

When Crowther saw the letter, he took the hint and began to suggest to the C.M.S. that the way to lessen the irritation of the company was to provide the mission with a little vessel of its own. But he could not let the general abuse of the missionaries go unanswered, especially since he was himself on board the *Victoria*, the vessel in question, and he saw nothing to warrant such 'sweeping, disparaging and indiscriminating charges'. In September, while travelling from Lokoja to Egga en route for Bida 'being free from pressure of business and less disturbed by constant visitors', he took the trouble to reply to Edgar like an old friend, 'as if I were at the office at Dickinson Street, Manchester, conversing with you on the subject of those communications.' He reviewed at some length the past connection between the company and the mission, the growing hostility of Europeans towards Africans on the river, and the need for friends of the Africans not to accept literally what the European traders said about them.

> My dear Sir, I am a missionary and a passenger on board your steamer; I do not lend myself to tale-telling or accusing servants to

1 Edgar to Josiah Crowther, 30 April 1875, copy in C.M.S. CA3/04.
2 Ibid.

their masters. . . . I will simply ask whether the statement [about missionary rabble] came from any crew or passenger on board the *Victoria* or from a distant observer? . . .

The fact of the case is simply this, the presence of missionaries on board the steamer or in the river is an eye-sore to some ungodly Europeans in the Niger.[1]

To illustrate this, he referred to the troubles the missionaries had over the question of Sunday observance. With the competition what it was, and the annual season of trade so short, most European traders found it highly irritating for missionaries to tell them not to work or to make others work on Sunday. Crowther quoted from the correspondence between the missionary at Onitsha and the local agent of Holland, Jacques & Co., one Cliff, who had been saying that the institution of Sunday was an invention of rascally, mischievous black missionaries unknown to the white man.[2] Such correspondence, and even the abusive tone of certain letters from traders to missionaries, were not, of course, peculiar to the Niger. In similar situations parallels can be found in the West Indies, South Africa and New Zealand. What was peculiar to the Niger was the fact that the missionaries were all Africans 'not yet proven', but, as it were, on trial, and that the C.M.S. authorities could not completely ignore the calumnies as just the result of the irritated feelings of the Europeans. The persistent reports of the Europeans were gradually undermining the confidence the C.M.S. previously had in African missionaries. This first became perceptible when Crowther did get the steamer he had asked for.

When it was known that the Bishop was about to be provided with a small steamer, the traders assured the C.M.S. that the missionaries and their trading relatives would use the steamer not only to compete against the European companies, but also to ruin the spiritual work of the mission. As a necessary precaution, therefore, Crowther agreed with the committee when he visited England in 1877 that the vessel be put in charge of a European lay agent, J. A. Ashcroft, who had been sent to the Niger in 1860 but had had to return to Sierra Leone. It was also agreed that Ashcroft was not only to take charge of the vessel, but was also to relieve Crowther of the secular affairs of the mission. He was to deal with building and repairs, payment of salaries, and management of stores, as well as with estimates and accounts, all of which Crowther had previously had to supervise alone. If Ashcroft had been sent as an official

1 Crowther to Edgar, 3 Sept. 1875; C.M.S. CA3/04.
2 Richard Cliff to the Rev. S. Perry, 31 Jan. 1875, in reply to Perry's letter of same date; encl. in Crowther to Edgar, 3 Sept. 1875; C.M.S. CA3/04. Also ibid., Cliff to Crowther, 27 Oct. 1875.

of the mission to go and take orders from the Bishop as the head of the mission, as Crowther expected, he would have been of immense help and a much needed reinforcement to the mission. But when in 1878 Ashcroft took out the *Henry Venn*, as the mission steamer was named, he was designated the 'Accountant of the Mission', associated with the Bishop, 'who is to be considered as the Secretary of the Mission'. He was to be in charge of the 'temporalities', as Crowther was in charge of the spiritual affairs of the mission.[1]

A struggle for power was inevitable between Crowther, hitherto sole Director, and the new 'Accountant', especially since Ashcroft, because of his bad temper, found it difficult to get along peaceably with anyone and, as Hutchinson, Venn's successor, told Crowther, 'Your African climate tries the temper sadly.'[2] The struggle for power was, however, only a minor aspect of the struggle for the confidence of the C.M.S., for it would appear that Hutchinson intended Ashcroft not only to share power with Crowther, but also to supersede him as the leading representative of the C.M.S. on the Niger. This was not just because the C.M.S. was a lay organization and Ashcroft was a layman, but because, owing to stories spread by European traders, Hutchinson no longer trusted the African missionaries. Ashcroft was not only given sole charge of the *Henry Venn*; he was there to keep an eye on the missionaries. It is not known what verbal instructions he had. He soon wrote to say that since the vessel could not move unless he was on board and he had to make accounts on land and keep an eye on buildings and repairs, he required a deputy, and he nominated a friend of his, James Kirk, lately returned from America. In his letter of instructions dated 6 December 1878, Hutchinson told Kirk what it was unlikely he had told the Bishop, that

> it appeared evident that the Mission was not in a satisfactory condition and that the agents, both ordained and unordained, with a few notable exceptions, were not maintaining a high tone of Christian life and conversation.[3]

It was clear that Hutchinson had lost—or probably never had—

1 'Memorandum on the financial arrangements for the Niger', encl. in Hutchinson to Ashcroft, 3 May 1878; C.M.S. CA3/L1. See also printed 'Regulations for the use of the Mission Steamer', 1878; National Archives, Ibadan; ECC 20/9. In 1883 Lang to Phillips, 26 January, said the *Henry Venn* was 'in a special sense given to the Bishop for his Mission. It was to a great extent a personal gift from those who were interested in him and in his mission'; C.M.S. C3 A3/L2.
2 Hutchinson to Crowther, 5 Sept. 1878; C.M.S. CA3/L2.
3 Hutchinson to James Kirk, 6 Dec. 1878; C.M.S. CA3/L1.

confidence in Crowther; his letters show that he put his trust in Ashcroft and Kirk, though he had yet to convince many people in the C.M.S. that Crowther was no longer worthy of their trust.

It must be said that two laymen hostile from the first to the mission were not likely to help to raise the tone of Christian life in the mission. As we shall see, they appreciated none of Crowther's problems and would not see the need for reform. In disputes between missionaries and European traders they sided with the traders. They diverted the mission vessel from facilitating episcopal visitation to trade and exploration. And all the time they harped on the theme that Africans could be trusted to work well only under European supervision.

The revival of this theme just when Townsend and Hinderer had dropped it coincided with Goldie's entry into the Niger. Out of the competing British interests he sought to create one solid bloc behind which he expected all loyal British people to range. This created a problem for the Niger Mission, as they too represented British interests but were also loyal to the African petty traders with whose interests those of Goldie's new company were hardly compatible. It was natural that he and his agents should feel that only under the supervision of Europeans could the loyalty of the African missionaries be ensured.

Josiah Crowther and other Africans like him holding positions of responsibility in the companies were dismissed when the companies were amalgamated in 1879. Systematic pressure was exerted to drive them out of the river as independent traders, unless they wished to remain as clerks and mistrusted officials. These African traders, says Dr. Flint, 'almost always caused Goldie to lose his sense of proportion'.[1] By 1886, when Goldie had acquired a charter and political power, he designated the emigrants and Brassmen as 'foreigners' in the territories of the new Royal Niger Company, who had to obtain licences if they wished to trade on the river. The effect of the licence, said Goldie himself

> will be to enforce some contribution to the revenue from—or else to exclude from the Territories—a class of men, happily now extinct, who were formerly the worst enemies of civilization in Central Africa. These were disreputable coloured men (in the past they were generally inferior clerks dismissed for peculation) who . . . lived, by surreptitious dealing in slaves . . . stirring up the natives to discontent and bloodshed . . . under a mask of ardent piety.[2]

This was hardly a fair comment on the work of Crowther and others who preceded Goldie in developing the Niger trade. Clearly the days of

1 Flint, op. cit., p. 98.
2 Goldie to Iddesleigh, 13 Dec. 1886; FO 84/1798; Flint, op. cit., p. 98.

Buxton, Laird and Clegg, of British capital investment through African agency, were over. Crowther saw this in 1879 when he refused to continue to hold the C.M.S. shares of the West Africa Company in the new company.[1] The important point here is that while in the struggle between the Niger Company and the African traders the missionaries necessarily sided with the African traders—out of sympathy, not just for helpless traders who pioneered the trade and were being hounded out, but for relatives, friends and parishioners—the authorities of the C.M.S. were not so committed to a conflict with Goldie. Some of them were suspicious of Goldie's methods and continued to sympathize with the old policy of encouraging 'African advancement', but by and large the majority of them as Englishmen felt that Goldie was in his way doing a patriotic job that needed doing and which nobody else was in a position to do. A few, in particular Hutchinson, the Lay Secretary himself, shared many of Goldie's ambitions and wished to co-operate with him and his agents. This meant that Hutchinson the more readily accepted the traders' estimate of the African missionaries, and that in cases of dispute between traders and missionaries he was inclined to believe that the missionaries were wrong.

Hutchinson, emulating the patriotic ambitions of Goldie, encouraged Ashcroft to explore up the Benue beyond Yola, and asked Crowther to consider leading an exploration from the Benue down towards the Shari River. In February 1880 Ashcroft was asked to join the search for a sanatorium for Europeans on the Cameroon mountains. In using the vessel on these speculative errands, Ashcroft was depriving the Bishop of its use. Moreover, the cost of maintaining it was far beyond the means of the mission. Hutchinson then had no choice but to give permission for Ashcroft and Kirk to trade with the vessel. Hitherto they had been allowed to take cargo, but only from the recognized agents of the European firms.[2] This was to prevent their subordinate African agents from trading surreptitiously, but it meant in effect discriminating between European and African traders in favour of Europeans. And then Ashcroft really went into trade. By collaborating with Captain McIntosh, the Agent-General of the Niger Company, he obtained potash and other produce from Nupe and carried it down to Lagos to sell to Banner Brothers & Co. and McIver & Co. Ltd.[3]

1 Crowther to Hutchinson, 16 Oct. 1879; C.M.S. CA3/04. Also Hutchinson to Crowther, 5 Sept. 1879, trying to defend the Company: 'I do hope your information as to the character of the agents is a little biased. . . . But still if you think you ought not to be a shareholder, I must ask Mr. Edgar to make other arrangements about the shares'; C.M.S. CA3/L1.
2 Hutchinson to Ashcroft, 19 Dec. 1879; C.M.S. CA3/L1.
3 Wood to Whitting, 12 Aug. 1881. Both Whitting and Wood were puzzled as to where the original capital for starting the business came from. Wood

The collaboration between McIntosh and Ashcroft inevitably went a step further. In the past most European traders, in criticizing the African missionaries, had always implied that many of their faults were attributable to Crowther's inability to supervise them adequately. Captain McIntosh, in pursuit of Goldie's policy, now condemned all African missionaries *en masse* and argued that nothing but the leadership of European missionaries would do on the Niger. But since it would not be easy to convince the C.M.S.—apart from the Lay Secretary—that they should replace Crowther by a European until he was shown to be unworthy of his post, McIntosh turned to the Methodist Superintendent at Lagos. The only Methodist station in the interior was at Abeokuta, but in November 1879 McIntosh persuaded John Milum to come with him up the Niger and establish a lonely outpost at Egga. McIntosh wrote to inform Ashcroft in January 1880:

> I have lately given passage up the river to the Rev. John Milum, Wesleyan missionary, who . . . informs me he intends to establish a station at Egga with a European at its head.[1]

In fact the Methodists had no European to take charge of the mission. At its head was William Allukura Sharpe, a Kanuri clergyman with an extraordinary career behind him.[2] He was ordained in 1878. He was longing to take the Gospel to his own people when Milum offered him the chance of going to Egga in 1879.[3] When he died in 1884, the Egga station was closed down. But for the five years he was there Methodist published records continued to give the name of the Rev. M. J. Elliott as head of the mission. The appointment of a European to compete with Crowther was the condition on which McIntosh had placed the resources of his company at the disposal of Milum, and Elliot was the man earmarked for the post, though in fact he remained in Lagos.

concluded that the capital must have been small compared with the credits of £6,000 to £7,000 which the business had with the Lagos firms; C.M.S. G3 A3/01.

1 McIntosh to Ashcroft, 5 Jan., 1880, encl. in Ashcroft to Hutchinson, 5 Jan. 1880; C.M.S. CA3/05.

2 He was born near Lake Chad and at an early age went to a Koranic school. In his early teens he was kidnapped and after various adventures which took him from Kano slave-market to Ilorin and thence to Lagos, he entered the household of the Rev. George Sharpe, the Methodist minister, in 1861, as a house boy. He learnt to read and write, graduated successively into cook, interpreter, schoolmaster and catechist. He loved books, read widely, wrote passionate, poetic prose, and must have been an eloquent preacher. W. A. Sharpe to the Rev. J. Milum, 25 March 1878, giving a short autobiography on the eve of being received as a minister.

3 Sharpe to Milum, 8 Sept. 1879, giving an account of his ambition to take the Gospel to his own people. Also his letters from Egga; Meth.

Hutchinson had gone as far as he could in introducing European supervision to the Niger without displacing Crowther. In October 1879 he said he judged that the joint control arrangement had broken down. He therefore vested management of the Niger Mission in a finance committee based in Lagos. It was to consist of five Europeans and three Africans. The Europeans were Ashcroft and four missionaries from the Yoruba mission resident in Lagos. The Africans were Crowther as Chairman, Archdeacon Dandeson Crowther, who lived at Bonny, and Archdeacon Henry Johnson, who remained in Lagos only till he was well enough to live at Lokoja. The Secretary was the Rev. J. B. Wood of the Yoruba Mission. It was in effect handing management of the Niger Mission to the European missionaries in Lagos. When that had been done, Wood was asked to go to hold an inquisition on the Niger Mission.[1]

Wood spent almost three months on the Niger with Kirk as his companion, visiting the different stations, questioning the agents and the traders as well, trying, he said, to report faithfully on the ill repute of the mission. He found the level of the Christianity of the agents and their congregations very low. The qualifications of the agents were meagre; many of their wives were illiterate and had little interest in missionary work but a good deal of interest in trade. Above all, he preferred specific charges against five ordained missionaries and ten lay agents. Four of the ordained missionaries he charged with immorality, two with trading, one with dishonesty. Charges preferred against the unordained agents ranged from the brutal flogging of a housemaid that led to her death, to immorality, immoral union with wives before marriage, dishonesty, drunkenness, building a fence without permission or just 'general unfitness' or 'downward tendency of character'.[2]

Wood said he conceived his report as a confidential document for the Parent Committee. He did not inform the agents of the charges against them; he did not call for defence; nor did he show the document to Crowther before he sent it to Hutchinson. Hutchinson, however, put it into print and circulated it to the hundred members or more of the committee. At that point someone must have asked if Crowther had seen it, for Hutchinson then wrote to ask Wood if he had shown it to Crowther. It was not until then that a copy was sent to him.[3]

1 Committee of Correspondence, Minute on the Niger Mission, 21 Oct. 1879; C.M.S. CA3/L1.
2 J. B. Wood, Report after a Visit to the Niger Mission in 1880; C.M.S. CA3/04.
3 See Wigram to Wood, 17 March 1882. Wood was angry that the Report was published, and Bishop Crowther was moved to say that Wood had gone to the Niger intending to 'break' the mission. Wigram told Wood that he considered the publication of the Report a mistake, but that 'this having been

Though the report caused disquiet in everybody's mind, it was evident that the committee was by no means united in accepting that it proved that the Niger Mission was a failure, or that it proved the incapacity of Africans to manage things on their own. Some members were in fact becoming critical of the way Crowther and the missionaries were treated, being condemned without a hearing. The committee decided to send a deputation to meet Crowther, Ashcroft, and the archdeacons, and to ask Wood to substantiate the charges he preferred. The deputation consisted of Hutchinson and the Rev. J. B. Whitting, who seemed to have represented the two sides in the committee. The conference met at Madeira in March 1881. Unfortunately Wood was unable to attend. But even on the basis of what Wood had written, Crowther was able to show that while some of the charges such as those relating to the incidence of trading or the brutal treatment of the housemaid had a factual background, much of the account was hearsay and extremely tendentious. The deputation in their report drafted by Whitting accepted this. They pointed to the case

> of the two schoolmasters, Thomas and Joseph, at Lokoja and Kippo Hill who, according to the report, were compelled by Bishop Crowther to marry their wives because they were with child by them. Bishop Crowther proved conclusively that Thomas applied for and obtained leave to visit Sierra Leone to marry a girl he had not seen for years and returned to his post a married man, while Joseph he found married on his visit to Kippo in 1878, having known nothing of the matter in any way.[1]

Crowther was therefore asked to go and investigate the charges and take necessary action.

The Madeira Conference further underlined the need for reform on the Niger. A Training Institution for Lokoja was recommended (though it was not ready till 1887), as well as boarding schools for boys and girls at Bonny and Onitsha giving slightly more advanced education and training children of agents free. An increase of salaries was granted to the agents on condition that they must regard their wives as mission employees and must sign a pledge to give up trading. The Madeira Conference did more than that. It drew the attention of Whitting and others to the way Hutchinson had been running the C.M.S. It came out

---

done, it was imperative that a copy should be sent to Bishop Crowther, especially since Kirk had got hold of a copy and, through 'indiscretion', shown it to European traders and chiefs on the Niger.

1 Printed, Report of the Deputation appointed by the Parent Committee to confer with missionaries at Madeira 1881; C.M.S. G3 A3/01.

that the mission vessel had been diverted to trade, and Hutchinson had to resign in May 1881,[1] to be replaced by a cleric, Whitting having to act as Secretary for a few months. Ashcroft was dismissed for rudeness to Bishop Crowther and to the committee, and Kirk resigned in protest when asked to wind up the trading business of the *Henry Venn*.[2] The Niger Finance Committee at Lagos was dissolved. It was replaced by a new one based at Bonny intended to show more respect to the Bishop's position. It was to consist of the Bishop as Chairman, a European as Secretary and two laymen, one African, the other European, nominated by the Bishop and approved by the committee. The new Secretary, T. Phillips, ordained deacon by an English Bishop, was admitted to priest's orders by Crowther himself.[3]

The Parent Committee had tried to redress matters, but not every effect of Hutchinson's work could be wiped out. The committee remained divided as to what conclusions to draw from the Wood Report, which Crowther after holding his investigations continued to reject.[4] Whitting remained of the opinion that much of the report 'had fallen to the ground', and he wrote to Kirk in June 1881 that 'temptations are not peculiar to the African, we find them in every parish in England';[5] whereas the Rev. F. E. Wigram, who succeeded him, wrote to Wood

1 The circumstances surrounding Hutchinson's resignation are not clear but they were undoubtedly connected with his handling of the Niger Mission. The Parent Committee considered the Report of the Deputation in April; Hutchinson resigned in May, and by June Whitting, acting as pro tem. Secretary, began to reverse Hutchinson's policy. By December both Ashcroft and Kirk were removed. Hutchinson later went to Canada, first as a lay missionary in Huron, then as an ordained missionary of the Scottish Episcopal Church. He died there in 1897. Stocks, op. cit., vol. iii, pp. 254, 261.

2 Wigram to Crowther, 16 Dec. 1881: 'Messrs. Ashcroft and Kirk are both now disconnected with the Society and the *Henry Venn* Steamer is lying idle in Lagos'; C.M.S. G3 A3/L2.

3 Instructions to the Rev. T. Phillips, 16 May 1882. He was told, 'You go out to the Niger in the double character of one of Bishop Crowther's clergy and the representative of the Parent Committee'; C.M.S. G3 A3/L2. Also, Stocks, op. cit., p. 386. Phillips was a graduate of Dublin, engaged in business when he agreed to go to the Niger.

4 Crowther's fullest answer to the Wood Report, after he had made his own investigations of the charges, was given in a Memorandum of 12 Aug. 1881; C.M.S. G3 A3/01.

5 Whitting to Kirk, 30 June 1881. Wood took strong objection to the phrase 'fell to the ground', which he heard that Whitting had used, perhaps at Madeira. Wigram later tried to explain it away by saying it referred not to Wood's Report, as to the various rumours and reports which had been reaching the Society from other sources and had provoked the commission of Wood himself. The distinction was, however, not important as Wood himself said that when he got to the Niger he found that the mission and the agents were of ill-repute and he was doing his best to give a good account of the reports current about them; C.M.S. G3 A3/L2.

nine months later: 'We recognize the absolute need of European super-vision, and that the lack of it has been the occasion of the grave offences on the Niger.'[1] There was no bridge between the two opinions in the Parent Committee, or between Africans brought up under Venn and Europeans going out in the new age when Britain was acquiring political power in Africa. It is not surprising that they could not work together.

The European Secretaries of the Niger Mission continued to follow the pattern set by Wood of one man visiting the mission, not without pre-suppositions, and making wholesale inquiries followed by confidential reports.[2] Each report was followed by charges against agents. Each charge was followed by heart-rending denials, testimonials from witnesses, investigations, trials, and heart-searching conferences of the Parent Committee trying to arrive at the right conclusions. But all the committee did was to act sometimes one way, sometimes the other; at one time to ask Crowther and his archdeacons to investigate the charges preferred, at another time, as in 1883 when the case of W. F. John,[3] the former agent guilty of the manslaughter of his housemaid became a public scandal,[4] summarily to dismiss a number of the agents accused in the report.

1 Wigram to Wood, 17 March 1882; C.M.S. G3 A3/L2.
2 T. Phillips: Report on the Agents in the Niger Mission, 1883; C.M.S. G3 A3/02. J. A. Robinson: Memorandum on the Niger, 9 July 1889; C.M.S. G3 A3/04.
3 W. F. John was a catechist in Bonny in 1869, accused of adultery with a chief's daughter. The case was plain, but instead of dismissing him, Crowther pitied his wife and infant baby and decided to give him another chance. He dismissed him from mission employment but made him a personal clerk. Subsequently John was re-engaged as an Interpreter at Onitsha. Crowther to Venn, 31 Dec. 1869; C.M.S. CA 3/04.
4 The trial was a *cause célèbre*. W. F. John and J. Williams, another C.M.S. agent, caught their housemaids, girls whom they had ransomed, trying to run away. They had them tied in the sun, flogged for a prolonged period, and their wounds peppered. One of the girls, John's, died soon after. He was acquitted on a charge of murder by the Court of Equity at Onitsha. Following Wood's Report, the C.M.S. dismissed both John and Williams. John re-turned to Sierra Leone; Williams joined the service of the Niger Company. The Governor of Sierra Leone had John arrested in 1881, and asked the British Government to make an example of him to other Sierra Leoneans. The trouble was that John had committed this crime outside British jurisdiction. The Law Officers, however, judged that he could, by a broad interpretation, be tried under the clause of an old law designed to make sailors on the high seas or deserters amenable to Law. The clause referred to those who being British subjects, 'either sailed in or belonged to and have quitted a British ship to live at the aforesaid place'. The trial had to be by a special Commission of at least four judges set up by the Lord Chancellor under the Great Seal, with a Grand Jury to ascertain the facts and a Petty Jury to determine the guilt. (Act 57 Geo. III c 53, also 46 Geo. III c 54). Williams and his wife, and Mrs.

248

Crowther insisted on investigating each charge and would himself take disciplinary action only when the charges were proved. The committee, on the other hand, insisted that charges did not have to be proved as in a law court before an agent could be declared unfit for the high and holy calling of a missionary, and that such declaration did not necessarily make him unfit for other employment.[1] Gradually Crowther acquired the reputation of an over-indulgent father shielding the wrong of his children. And although no charge, no suspicion even, attached to Crowther himself, he regarded himself as being on trial with his agents, and during his investigations he probably saw much less than there was to see. He was not unaware of faults in many of his agents. We have noted his constant anxiety about their qualifications and lack of training and adequate supervision; but he refused to accept the wholesale condemnation by Europeans. He was of a kindly, sensitive nature, feeling for the offender even when pronouncing judgement on him. Where cases were doubtful, Crowther left people to their consciences. When charges were proved, unless he felt the offender was irretrievable or utterly unsuitable for missionary employment, in which case he dismissed him, he preferred to try suspension, transfer to a new situation, and other ways of bringing the offence home to the offender in a way likely to reclaim him.[2] He told the Parent Committee on his last visit

John were also arrested. Wood and Consul Hewett helped to procure evidence from Onitsha and Lagos. Witnesses were sent to Sierra Leone. The trial lasted 22 days and Mr. and Mrs. John and Mr. Williams were sentenced to 20 years imprisonment each, Mrs. Williams, being less involved, got away with a much lighter sentence. F.O. 84/1656. T. M. Bell: *Outrage by Missionaries* . . ., Liverpool 1883, giving a full account of the trial. Sir Samuel Lewis, a Sierra Leonean, led the prosecution for the Crown—*Life of* by J. D. Hargreaves, 1958, p. 16. The trial cost about £2,000. When the vote for it came up in Parliament, critics of missions had a fine opportunity which they did not hesitate to use. The importance of the case, however, concerned the jurisdiction of the Crown; later attempts to bring Englishmen who committed outrages on the Niger to justice on the precedence of this case were not always successful.

1 Wigram to Crowther, 18 July 1883; C.M.S. G3 A3/L2.
2 Cf. the case of W. F. John above. Also the case of the Rev. W. Romaine, senior Pastor at Onitsha in 1876. Crowther said it was not surprising that in a whole year there was only one person baptized, that this was because the agents had not been living up to their profession; that Romaine was suspected of adultery and also accused of drunkenness, that adultery was always difficult to prove and therefore tended to go unpunished even in glaring cases. But that the charge of drunkenness was proved against Romaine by the testimony of both friends and opponents and Romaine's own violent acts when under the influence of drink, much to the scandal of the mission. Crowther therefore suspended him for three months, at the end of which period he judged that Romaine had shown contrition, allowed him to resume work but had him transferred to another station. Crowther, Report of a Visitation to the Niger Mission, 1876, and appendix dated Onitsha, 2 Sept. 1876; C.M.S. CA3/04.

to England in December 1889 that a newly-made fire was bound to be smoky and that his food would never be ready if the cook, instead of looking for a fan to blow the fire, began to search and pull out every stick that smoked.

> We are all weak and imperfect agents, faulty in one way or another, which need be strengthened, supported, reproved and corrected, when not beyond amendment.[1]

While we may admire the dogged way in which the Bishop stuck by this principle, 'going as though he never heard anything to the contrary',[2] as Ashcroft once remarked, it must be said that for a pioneer he was too reasonable, too soft a disciplinarian. But that is a very different thing from the Rev. J. A. Robinson's assertion, as joint Secretary of the Yoruba and Niger Missions, that 'the negro race shows almost no signs of "ruling" powers'.[3] For, where Crowther hesitated, Archdeacon Henry Johnson or Dandeson Crowther, not to talk of James Johnson, would have acted.

But if the Niger Mission agents cannot be taken at Crowther's estimation, neither can they at the estimation of the various Reports. It would take a Daniel to read through them all, and all the other documents they provoked, and sift truth from lies, hearsay and prejudice from fact. In such an atmosphere, where the merits of an African missionary were so dependent on the whim of the passing European missionary, or on whether he was willing to be 'humble and subdued' or even willing to volunteer information about other people, the truth about the mission agents was hard to discover and will not now be discovered.

The group of missionaries who came to darken Crowther's last years were able, young, zealous, impetuous, uncharitable and opinionated. The Rev. J. A. Robinson, who was appointed Secretary of the Niger Mission in 1887, was the oldest of them. He was then 29. He was a scholar at Cambridge and graduated first class in the Theological Tripos. He was ordained in 1882, but before joining the C.M.S. had been to Heidelberg, in Germany, where he had taught in a school for four years. The Rev. F. N. Eden, who succeeded him as Secretary, was of the same age, also a Cambridge man, with eleven years' pastoral experience in County Durham before joining the Niger Mission in 1890.[4] The moving

1 Crowther, Memorandum dated 9 Dec. 1889, and encl. on the Yoruba saying *Gbogbo igi l'o l'efi*; C.M.S. G3 A3/04.
2 Ashcroft to Hutchinson, 20 June 1878; C.M.S. CA3/05.
3 Robinson, Memorandum on the Niger Mission, July 1889, op. cit. He added that this was true of the Negro whether in Sierra Leone, Liberia, the West Indies or the Niger; C.M.S. G3 A3/04.
4 *Register of Missionaries and Native Clergy, 1804–1904.* C.M.S.

spirit of the group was a younger man, Graham Wilmot Brooke, aged 25 when he joined the C.M.S. in 1889 as a free-lance lay missionary, working for them but maintaining himself. He had attempted to go as a lone missionary to the Sudan, which he sought to enter first through the Nile, then through the Congo, before he asked the C.M.S. to let him go up through the Niger. For him, missionary work was a daring military adventure: he had had a military training and General Gordon was his hero.[1] When these missionaries arrived at Lokoja in 1889 and saw the building of the Preparandi Institution, a two-storey house which was completed in 1887, built by the European catechist John Burness to Archdeacon Henry Johnson's specifications, they declared it to be the grandest building in West Africa and an obstacle to the progress of Christianity:

> Its very existence is a blot on the C.M.S. . . . It tends to divert the attention of the natives; they speculate on its cost and the wealth of the white man.[2]

They said that Henry Johnson must have deceived the Parent Committee about its real size, charged him with, among other things, being extravagant and asked for him to be removed to Sierra Leone. They sold the building at once to the Royal Niger Company at the company's price, without waiting for the decision of the Parent Committee on the wisdom of the step.

They felt that material scientific progress tended to destroy faith. They re-emphasized the Evangelical distrust of the mixing of Christianity with civilization; indeed, like the anthropologists they utterly rejected the idea of 'civilization' for the African. They themselves itched to go into the villages and live like the 'natives' they despised, in the belief that by 'reasoning of the Gospel and righteousness' they would sweep them out of their old ways into a pure, simple, primitive Christianity. It did not occur to them for one single moment that to attempt to spread Christianity in any other way did not necessarily imply lack of 'spiritual powers'.

Robinson had preceded the others to the Niger as Secretary of the Mission in 1887. He returned in June 1889 to write his memorandum in which he stated these ideas and showed that the Bishop could not be made to comply with them; but that if the Bishop resigned, and Archdeacon Henry Johnson too, the group of young Europeans could work the mission better and cheaper. He would sell the expensive mission vessel; dispose of extravagant buildings; standardize salaries

1 Stocks, op. cit., vol. iii, p. 395.
2 Robinson to Lang, 21 May 1890, 8 Sept., 1890; C.M.S. G3 A3/04.

of African agents so that the attraction of money no longer drew them towards the ministry, and so on. He concluded:

> I am strongly inclined to think that if your Committee were to write a private and confidential letter to the Bishop stating that *your mind was quite made up to carry through certain reforms*, describing them in detail, and that you wished to spare him the pains of having to act in opposition to his own convictions and the policy he had so long embraced and the loss of dignity that would arise from your representative being placed in any degree in opposition to his authority, and offered him therefore this opportunity of retiring on a good pension—I am inclined to think he would be glad to accept the proposal.[1]

Once again the Parent Committee was split over Robinson's memorandum. One important factor was Brooke, who had been meeting the committee, and whom, as the Editorial Secretary said, 'to meet was to love'.[2] He was full of plans and zeal. He offered his services free, but on condition of full European control. On the other hand, nothing had been proved against the Bishop. At over 80 he was still full of life and when he heard he was being declared incompetent, he said he was willing to leave the established missions to others and break new ground elsewhere.[3] The committee called him to London, and it was his last visit. Whether they offered him the chance of retirement is not clear. If they did, he refused them that easy way out. They therefore adopted the policy of 'if in doubt, compromise'. Brooke was made leader of a European 'Sudan Party' to have their headquarters at Lokoja and to take the upper half of the Niger Mission. Archdeacon Henry Johnson was to be transferred to Sierra Leone and the Rev. C. Paul, at Egga, to be removed to the Delta. The lower Niger, with headquarters at Onitsha, was still to be under Crowther, ruled by a finance committee consisting of the Bishop, Archdeacon Dandeson Crowther, one other African pastor, and the European Secretary and two Europeans from the Sudan Party.[4] Robinson was furious at the compromise. He decided to resign and join the Sudan Party. The C.M.S. then looked round for a new

1 Robinson to Lang, 9 July 1889; C.M.S. G3 A3/o4.
2 Stocks, op. cit., vol. iii, p. 395.
3 'If others are delicate to tell me of my incompetence in the superintendence of the Niger Mission, it is my duty to relieve their minds of that delicacy. I am ready to yield place to others to act as leading managers of the Niger Mission. I am willing, as long as my health lasts, to labour as a pioneer in opening fresh grounds, while the already established stations can be worked by superior intellects and better managers.' Epelle, op. cit., p. 36.
4 Resolutions of the Parent Committee, 30 July 1889, 9 Dec. 1889; C.M.S. G3 A3/L2.

Secretary, choosing for the post the Rev. F. N. Eden. But that made little difference. Robinson and Brooke were determined young men and knew how to force the hands of the committee. In any case, they were still members of the Niger Finance Committee. Eden merely joined them. Together, they achieved the displacement of Crowther from the Niger.

Robinson left for Lokoja in January 1890, followed by Eden a month later and Brooke in May. They each indulged in the old game of visiting the missions and preferring charges so that by August, when the Finance Committee was convened at Onitsha, between them they had enough to keep the committee busy. After examining different charges against various African pastors—alleged buying of two cases of port wine for Christmas, or receiving stolen powder from sailors, and so on—in which the voting was consistently the three Europeans versus the three Africans, Eden as Secretary, overruling the Bishop as Chairman, announced the pastors suspended. He also announced that charges against the Rev. C. Paul, transferred from Egga to the Delta, had been sent to England and need not be disclosed to the committee that had been so unco-operative, but that nevertheless he was declaring Paul suspended forthwith.[1] As if it was not enough for two clergymen and one layman of the Church of England to suspend pastors of that Church in the presence of and in defiance of the Bishop who ordained them, they turned on Archdeacon Crowther himself and declared that, in his trying to defend some of the accused pastors, he had made inconsistent statements and was unworthy of his holy office and was forthwith suspended. It was unlikely that they would have gone so far if Crowther had been a European. Few scenes could have been more painful to watch than the grey-haired old Bishop of over 80 active years, tormented and insulted by the young Europeans, trembling with rage as he never trembled before, as he got up to announce his resignation from the committee.[2]

1 Minutes of the Finance Committee of the Niger Mission, meeting at Onitsha, 19–28 Aug., 1890, dated October 1890. The European missionaries, probably knowing beforehand what the issue of the meeting was likely to be, were each taking full notes of the discussions and resolutions, and the final Minutes were compiled from the notes of the three of them. Crowther testified that they were full and accurate. C.M.S. G3 A3/04.
2 Ibid. 'The Bishop: The long and short of it is that I disconnect myself from the Finance Committee if the Secretary alone is empowered to dismiss and suspend and do everything else in the Mission. . . . Will you write down, say, please, Bishop Crowther expresses surprise at the statement of the Secretary that he has power as the representative of the C.M.S. to suspend any clergyman from his duty. . . .'

Crowther's resignation caused a little stir among the members of the Parent Committee of the C.M.S., but eventually the committee's Niger Sub-committee placed the administration of what was left of the Niger Mission in the hands of a European Secretary to reside with the Bishop at Onitsha, and an Assistant Secretary to reside with the Archdeacon at Bonny:

> While the committee have no desire to interfere with the special ecclesiastical functions which the Bishop may think well to entrust to Archdeacon Crowther as Archdeacon, the Secretary personally in his section and the Assistant Secretary under his direction in the other section, as the local executives of the Finance Commitee will be, so far as the Society is concerned, responsible for the general superintendence of the work in these sections respectively.[1]

Bishop Crowther had in fact been displaced.

Eden apologized to Archdeacon Crowther,[2] but felt that the Niger Sub-committee's resolution had gone too far in censuring himself and resigned. Archdeacon Crowther said that, in the changed circumstances, he could no longer co-operate with the C.M.S. and that he could not hand over to the European Secretary the churches in the Delta, reared as they were on local support. He proceeded to declare them a self-governing Niger Delta pastorate within the Anglican communion, and left the details of organization to be worked out later.[3] Edward Blyden came to Lagos in January 1891 and delivered a public lecture urging the formation of an independent African Church. James Johnson moved the vote of thanks, but decided to remain in the Anglican Church.[4] In August a few C.M.S. and Methodist members in Lagos decided to form the United Native African Church. G. W. Johnson was present, not at the first, but at the second meeting of the founding fathers.[5] Crowther himself had a stroke in July 1891 and was removed to Lagos, where he died on 31 December. There was no question of his being

1 Resolutions of the Parent Committee dated 20 Jan. 1891. *A Memorial from Lagos to the C.M.S. on the Niger Question*, printed in Lagos, a pamphlet which assembled most of the relevant documents both in England and in Nigeria on the crisis.
2 Eden to Archdeacon Crowther, 11 Feb. 1891, cited in *A Memorial from Lagos to the C.M.S. on the Niger Question*.
3 D. C. Crowther: *The Establishment of the Niger Delta Pastorate Church, 1864–1892*.
4 E. W. Blyden: *The Return of the Exiles and the West African Church*, published text of lecture delivered in Lagos, 2 Jan. 1891.
5 Herbert Macaulay: *The History of the Development of Missionary Work in Nigeria with special reference to the United Native African Church*, printed pamphlet, Lagos 1941. First meeting 14 Aug. 1891; second meeting, 17 Aug. 1891. The U.N.A. Church was also called *Eleja*.

succeeded by an African. The European who succeeded him got the diocese Venn originally planned for Crowther, the diocese of Western Equatorial Africa, with a seat in Lagos, covering both Yoruba and the Niger and Delta. The way was then open for a fresh start under undisputed European rule.

Fundamentally it was not the merits or demerits of the African missionaries that caused the conflicts on the Niger. The disturbances in the Presbyterian Mission at Calabar, where there was a schism in 1882, in the Methodists' Church at Lagos, where one was barely averted in 1884, and in the Lagos Baptist Church, where there was a secession in 1888, all show that there was a basic conflict between the old and the new in the transitional period. It was, as Crowther once said,

like the meeting of two tides till one entirely submerges into the other when there would be an easy and regular flow in one direction.[1]

The Methodist crisis is of particular interest because, as was usually the case in Lagos, it was the congregation not the pastors who led the protest against the new European missionaries. It was also clearly a constitutional struggle into which moral issues were not dragged, and, owing to a different handling, it had a happy ending.

The Rev. John Kilner, who in 1876 became General Secretary of the Methodist Missionary Society, had been a distinguished missionary in Ceylon and India from 1847 to 1872, and he took up his appointment having evolved and put into practice a policy similar to Henry Venn's about the formation of a native ministry and a self-supporting native church.[2] But the missionaries Kilner had to use were all young men with new ideas about the ability of Africans. One of them, the Rev. Ellis Williams, on his arrival in Lagos in 1878 wrote to Kilner:

There are some fine intellects here. I have been literally astonished at the capacity of the African mind for receiving culture. We have some fine men here as preachers, leaders, members. My old notion of African incurable stupidity is, I believe, for ever gone.[3]

Or so it seemed. A struggle for power had a way of reviving such old notions.

1 Crowther to Venn, 27 Nov. 1868, referring to the upheavals in Bonny following the early successes of the mission there; C.M.S. CA3/04.
2 Obituary Notice of the Rev. John Kilner in Meth. *Minutes of Conference*, 1890, pp. 16–19. For Kilner's missionary ideas, see his article in *Methodist Magazine*, 1883–4, entitled 'An Enlightened Policy by which Missions to the Heathen may be conducted . . . Principles underlying such a policy'. Also Kilner to Milum, 15 Feb. 1881; Kilner to Coppin, 26 Sept. 1884; Meth.
3 Ellis Williams to Kilner, 3 July 1878; Meth.

The work of reorganization fell to John Milum as joint Chairman of the Gold Coast and Yoruba Missions. He was to reorganize the Yoruba Mission as a separate District, and establish the leading congregations in Lagos into self-supporting circuits under African superintendents according to the Methodist organization. He was also to distinguish between the roles of missionary and pastor so that the missionaries could be freed from pastoral work to devote attention to the expansion of the mission. Moreover, while funds for the missionaries should continue to come from abroad, Methodist funds for native pastors were to be regarded as gradually diminishing grants in aid of local funds. The real point was what the distinction would imply. Milum made it quite clear that he regarded native pastors as inferior to European missionaries:

> They should always take a subordinate position to European agents and whilst always allowed to vote on all matters pertaining to their local funds and to their own brethren, they should not be allowed (as hitherto) to vote on matters pertaining to Europeans either as to their examination or as to the distribution of funds strictly European.[1]

Thus while Europeans would meet with African pastors in a mixed district meeting, the real power was to be in the hands of a superior finance committee of the Europeans only. With that attitude distinguishing meant discriminating. Milum suggested, for example, that grants for Europeans should be increased and those for Africans reduced—so as to encourage the African pastors to economize and raise more funds locally.[2] The African pastors counter-petitioned, asking that the number

1 Milum, Notes on the Separation of the Yoruba and Popo District from the Gold Coast; recd. in Meth. House, 5 July 1878; Meth.
2 Ibid. Also 17 March 1880; Meth. An examination of salary structures will provide some indication of the changed attitude to African missionaries. Freeman, working on the principle that 'salary should not be so high that young agents will be drawn into worldliness, but it should be enough for them to maintain their respectability and influence, recommended for first-class agents, i.e. catechists, £100 p.a. for three years, then if ordained, £30 or £40 more. That, he said, was about two-thirds of what they would have got from government or mercantile establishments. Europeans got from the mission £150–£250. (Freeman: Report on the Religious State of all the Societies on the Gold Coast . . . 1845; Meth.) In 1879 Milum recommended for native ministers £42 when admitted, £49 after 2 years, £55 when in full connection, after 4 years, a maximum of £65. But if fully maintained from local funds, could after 10 years reach the maximum of £100. 'In addition to this a suitable house is allowed.' It was the status of the ministers, not the cost of living, that had gone down. For European salaries had been going up by way of allowances that did not exist in Freeman's time. Milum recommended for Europeans single £150, married £230, plus children's allowance of 16 guineas, children's education allowance 12 guineas, postage 4 guineas, medicine £10, allowance for learning the local language £12, messenger and servant allow-

256

of European missionaries be reduced so that the grants for Africans who really did the work could be increased.[1] Nothing could show more plainly the absurdity of the new attitude than the dilemma which arose over the position of Thomas Birch Freeman, who following his resignation in 1857 had rejoined the mission in 1873. It was he who had founded the mission and established the Methodist Church both on the Gold Coast and in Nigeria. He was born and brought up in England. He referred to English as 'our language'. But he was a mulatto, his father being of African origin. Now in his old age, it had to be decided not just whether he was a missionary or a pastor, but whether he was a 'European' or a 'Native'. Milum referred the matter to Kilner in these words:

> If you wish Rev. T. B. Freeman's allowance to come henceforth from *Native Grant* instead of from the *English Grant*, you had better specially refer it. I believe with our plan of working the funds we shall have sufficient native money to support him. He is an old man and the matter is delicate.[2]

Kilner was unlikely to sanction such an act. However, when Freeman died in 1890 the matter was again raised as to whether or not his wife, a Fanti lady—his two previous wives were English—was a 'Native' for the purposes of pension allowances. Milum retired in 1879. He was succeeded by men still younger than himself, with less ability, less tact, less self-assurance, but no less assertive in their claims as to the position of Europeans as Europeans. There were three of them: the Rev. W. T. Coppin, who succeeded Milum as the Chairman of the District but— so great was the shortage of staff—was transferred in 1883 to Cape Coast as Chairman of both the Gold Coast and the Yoruba Districts; M. J. Elliot, his assistant in Lagos, and Edward Tomlin, who arrived at the height of the crisis in 1884 to take charge of the High School and Theological Institution.

It was not the African pastors who led the revolt in the Methodist Church so much as the leading members of the various congregations.

---

ance £5, travelling allowance £30, also a suitable house with cost of repairs. (J. Milum, Minutes and Reports for 1879; Meth.) Kilner reduced what Milum recommended for Europeans, and improved on what he suggested for Africans. The scales approved in 1890 and 1885 respectively were *For Africans*: Lay agent max. £36; minister on probation, £50; in full connection, £80; after 10 years, £100. *For Europeans*: Minister on probation, £180; in full connection, £205 single, £230 married. Children's allowance, 8 guineas, for their education £12, allowance for family in England up to a max. of £100. (Kilner to Halligey, 4 Dec. 1890, 9 Dec. 1885 respectively. Meth.)

1 The Revs. Thos. E. Williams, W. B. George, and A. E. Franklin to Kilner, 17 March, 1882; Meth.
2 Milum to Kilner, ? 1880, a short note written on the eve of the Conference; Meth.

They were the men who in 1875 contributed £500 and asked the Methodist Missionary Society to match it and build a High School. They were angry that the principals the Society sent them were not highly qualified. Tomlin now disputed the authority of the School Board to control him. It was not surprising therefore that in April 1884 Elliot announced that the

> Lagos Circuit is in total revolt. . . . Because they are self-supporting, they question the authority of the European missionaries from intermeddling in any way with their circuit affairs.[1]

The immediate cause of the revolt concerned the authority of Coppin as Chairman to change officers of the Church, in this case circuit stewards, whose appointment was normally vested in the Quarterly Meeting of Class Leaders. In December 1883, Coppin said that the two stewards who were elected in 1879 and had continued in office for four years 'began to get officious and sometimes have been insolent to us'.[2] He decided to get them changed, though, according to Methodist practice as described in *The Polity*, he had no powers to do this.[3] He called on the stewards to resign. They not only refused; they saw to it that nobody else would accept nomination, and they refused to hand over money in their possession. Coppin called a meeting of Class Leaders to elect new stewards. According to him, 'it was a shameful failure'. The Leaders only heaped abuses on the European missionaries, saying,

> They did not want Europeans, they could do without us, there was another church they could go to. . . . When pressed to vote, they refused, shuffled their feet in contemptuous applause, made a hubbub, took up hats and sticks and walked out of the chapel.[4]

Coppin then left for Cape Coast, asking Elliot and Tomlin to act as circuit stewards, and ordering the ministers and other officials to go to them for their salaries and not to the 'ex-stewards'. In May, Coppin announced that he intended to go to Lagos to take firm disciplinary

1 Elliot to Kilner, 7 April 1884; Meth.
2 Coppin to Kilner, 9 May 1884; Meth.
3 Dr. Williams's *Polity* was at that time the recognized authority on the Constitutional law of the Methodist organization. Its oracular power even in doctrinal matters may be judged from the remark of a European missionary at Cape Coast in 1892 that 'Some people say it is very wrong to write on Sunday. I have not so found it in *Polity*, and over here, to a great extent, the *Polity* is the book by which all actions are judged'. Price to Secretaries, 28 Nov. 1892, Gold Coast Meth. The argument of Elliot, Coppin, and others was that the *Polity* referred only to Methodist circuits in England and that the Chairman of a mission district had more power than the Chairman in England; Elliot to Kilner, 7 April 1884; Meth.
4 Ibid. Coppin to Kilner, 9 May 1884; Meth.

measures, expel the trouble-mongers and discipline the ministers who continued to deal with the 'ex-stewards' knowing they should not. He wanted Kilner's advice as to whether to take legal action to recover money from the stewards. And he added:

> When I first set foot in Lagos, I found feuds. *Race feeling* is the primary cause of it. . . . Our authority and status as Europeans must be clearly laid down.[1]

It must be said that Kilner left the young men too long without advice. Coppin's letter was not replied to till Coppin had gone to Lagos and embarked on his disciplinary measures in the belief that Kilner would support him.[2] Perhaps the African ministers too thought that Kilner was behind the European missionaries, for they sent their petition not to him but direct to the President of the Methodist Conference. The conference meeting in November 1884 set up a Disciplinary Committee under Kilner to study all the papers about the Lagos dispute and take action.

The committee roundly condemned the European missionaries and upheld the contention of the circuit stewards and African ministers. They condemned Coppin's action in personally excommunicating and expelling officers of the Church when there was an established machinery for dealing with such matters.

> A Quarterly Meeting cannot be deemed factious or to have merited extinction simply because it respectfully and repeatedly declines to endorse the decision of the superintendent. . . . They cannot but think that, upon the showing, the Lagos officials and members were endeavouring to work out the organization of a Methodist circuit.

They were 'glad and thankful' to find Africans who accepted and were willing to work such an organization. And they concluded in a letter to Coppin:

> It is a far nobler thing to guide others in their efforts to govern than to act simply as an autocrat. A fair and real self-government is one of the chief objects which your predecessors have aimed at realizing, and any success of theirs is a result which we cherish with gratitude.[3]

The committee went further to pacify the Lagos Circuit. They asked

1 Coppin to Kilner, 9 May 1884; Meth.
2 'Minutes of a Minor District Meeting begun in the Wesleyan Mission House, Lagos, 11 July 1884', which also referred to copy of a Memorial by the Rev. J. B. Thomas to the President of the Methodist Conference. Coppin later complained about having been left too long without instructions; to Kilner, 6 April 1885. 3 Kilner to Coppin, 6 Nov. 1884; Meth.

John Milum to go out to Cape Coast and ask the old man of the mission, Freeman, the one man the Lagos church could not but respect, to accompany him to Lagos and talk to them. Milum was able to report that 'the decision of the Discipline Committee has given the greatest satisfaction'.[1] Kilner then looked round for an older missionary to work the circuit system in the right spirit and he picked on the Rev. J. T. F. Halligey, Chairman of the Sierra Leone District, who had to attempt to supervise the three Districts of Sierra Leone, Gold Coast and Lagos.

Coppin's guilt arose not merely from personal arrogance but even more from the ideas that were becoming current in his time. The same sort of troubles that plagued the Methodists had frustrated the working of the Lagos Pastorate, in which a European-dominated Finance Committee controlled the appointment, transfer and discipline of the pastors in charge of the 'self-governing' parishes, and the properties of the Church remained vested not in the Pastorate but in the C.M.S. Both the Church and the Lagos School Board were torn by racial feeling.[2] Kilner and Halligey alone could not alter the fact that times were changing. Nor could Whitting and his supporters in the C.M.S. Parent Committee, who continued to pursue the policy of Henry Venn.

The European missionaries entering different parts of Nigeria from the late 1870s onwards were beginning to speak a language that the African missionaries had not been accustomed to hear in times past even from their most severe critics like Anderson at Calabar or Townsend at Abeokuta. Among the petitions against James Johnson at Abeokuta in 1877 was a memorandum from the Methodist District Meeting accusing him among other things of antagonism against 'members of the ruling race'. The European missionaries were beginning to see themselves as rulers, and the word 'native' was acquiring a new and sinister meaning. In 1886 John Burness, a young Englishman in the building trade, volunteered for missionary work; because a builder was required urgently at Lokoja, instead of being sent to the Training Institution at Islington, he was sent as a probationary catechist to the Niger. He was still shy, timid and naive. About his African archdeacon and bishop he had no more than a few dark hints to report, but he was bolder

1 Milum to Kilner, 1 Jan. 1885; Meth.
2 The Lagos Pastorate was established in 1875 with only one parish. Another was added in 1879. St. Paul's Breadfruit came in only in 1881 and the Pastorate became well established. By 1889 all the parishes in Lagos except Christ's Church Cathedral had come into the scheme. Apart from St. Paul's Breadfruit, there were St. John's Aroloya; St. Peter's Faji; Holy Trinity, Ebuth Ero and St. Jude's, Ebute-Metta. Lucas: *History of St. Paul's Breadfruit Church*, op. cit.

on the character of the other agents.[1] He reported that he saw a school-master standing by a married woman at the Lokoja waterfront at 6 a.m. He said he did not question them but guessed that they had just arrived from Igbebe on the other side of the river, and that, in that case, they must have left Igbebe at 3.30 a.m., since it took at least two hours to cross at that time of the day. And he concluded in a phrase that was becoming nauseatingly common: 'To one who knows the African character, this evidence is quite sufficient to condemn him [i.e. the schoolmaster].'[2] The 'African character' was becoming for many Euro-peans synonymous with lying, hypocrisy, drunkenness and immorality. Bishop Crowther and his men were soon being caricatured as spoilt Africans, masquerading in borrowed weeds, learned perhaps, but without the heritage that even the most profane, untutored, perverse European could claim of centuries of Christian culture and civilization.[3] The African was on the way to becoming 'half devil, half child'.

This attitude, though it appeared so suddenly in the missions in Nigeria, had a more gradual growth in England and was in part the making of the publications of the earlier missionaries, both Europeans and Africans. Reference has already been made to the popular and highly emotional appeal that the missionaries made in Europe and how this coloured their work in the mission field. It also coloured the picture of the African presented to their countrymen at home. In order to excite pity and charity for Africans and maintain the flagging interest of Euro-pean Christians in the missions abroad, they tended to present the un-converted African in the worst possible light—showing the necessity of continued missionary work—while making the African around the Mission House a most docile, most teachable person—showing that the missionaries were succeeding. This dual purpose is discernible in practi-cally every missionary publication, as distinct, that is, from the letters and

1 Cf. John Burness to Lang, 12 May 1886: 'To describe the Sierra Leone men so far as I have had to do with them, I would say they are made up of *Conceit*, *Hypocrisy* and *Sensuality*. . . . I have found this spirit of hypocrisy in some of the highest dignitaries of the mission.' Also 31 March 1886; C.M.S. G3 A3/03.

2 Burness to Lang, 31 March 1886. Burness went on to say that he reported a case of misconduct by a girl, Rose Baikie, who ought to have been disciplined, but that Bishop Crowther did not take action because a schoolmaster had written to him that Burness had interfered in the matter: 'It was because of this letter under which the Bishop was smarting, and he had not the candour to tell me about it when I laid Rose Baikie's case properly before him'; C.M.S. G3 A3/03.

3 The Parent Committee of the C.M.S. itself in a circular letter to the West African Churches in 1892 spoke of the 'ripened Christianity' which twelve centuries had given Englishmen and which in West Africa 'can scarcely be looked for except in European teachers'; *C.M.S. Intelligencer*, 14 Nov. 1892, p. 61.

some of the private journals. This is not to imply that the publications were therefore deliberate distortions, but that they were most of the time arguing a case that called for a by no means objective picture of the African. The example may be given of the Secretary of the Presbyterian Foreign Mission Committee prodding his missionaries:

> You must not abate one iota of your graphic delineations; otherwise *The Record* and with it the interests felt in the mission will go down.[1]

He worried Hope Waddell to send him Calabar idols. Hope Waddell insisted that he did not think Calabar people venerated the images he saw any more than British people did their statues,[2] but the Secretary implored him:

> I am going on Sabbath and on Monday to Dundee. It is the Calabar idols, cloths, and other things from Africa that have provoked such excitements. I know that you and my excellent friend, Mrs. Waddell, do not favour such kind of things; but you must remember that I said to you that with all your wisdom I think you are mistaken. The notice that such things are to be seen packs a large house on a week-day evening and gives one the opportunity of stating to them solemn and important truths. If we cannot get the people, we cannot address them—and if we get them, it is the fault of the speaker if he does not turn the occasion to good. Oh, I do rejoice to plead the cause of the destitute millions of Central Africa before a crowded meeting. I have got Ekpeyoung, Ebok, an Abidong's rattle, cloths, and calabashes. If you can give me any other things that would help me, I shall feel very greatly obliged.[3]

That was in 1851. Fifteen years later, the same Secretary wrote to Anderson:

> I have been thinking that one reason why the old Calabar Mission is losing the hold which it once had on the Church is because of late I have got from Calabar so few details for *The Record*. . . . Unhappily, the statements are of a cruel and bloody character, but surely incidents and circumstances occur at times which it is worth while to repeat. We will not get men for the mission unless we keep up the interest.[4]

1 Somerville to Waddell, 28 Dec. 1850; U.P. *Letter Book*, vol. i, p. 682.
2 Waddell: *Journals*, vol. I, pp. 116–17.
3 Somerville to Waddell, 18 April 1851; U.P. *Letter Book*, vol. i, p. 746.
4 Somerville to Anderson, 23 Nov. 1866; U.P. *Letter Book*, vol. ix, p. 32.

The same anxiety pervaded every mission[1] and, with the best of intentions, a false picture of the African evolved in Europe.

But while the earlier missionaries had at least in theory maintained that baptism and education made the African a brother and a colleague, many who read their accounts accepted the darker side of the picture but rejected the rest. The Indian Mutiny, it is said, made people question first the wisdom of the policy of the missionaries, and then the validity of the claim that baptism and a few years in a mission school made all the difference. Then there grew up in the 1860s those whom Crowther called the 'Anthropological sort', James Hunt's 'Anthropological Society', of which R. F. Burton was a prominent member. Their emphasis was on physical anthropology; their theme, 'the place of the negro in nature',[2] and their approach, a study of the differentiation of the races, and the arrangement of the races in a hierarchical order, and the search for the missing link in the great chain between ape and European. The prevalence of these doctrines may be judged from the fact that in popular concept, Darwin's *Origin of the Species by Natural Selection* was for a long time taken to have confirmed—which it did not —the theory of the evolution of one race to another,[3] a process that could not be hurried up by evangelization and civilization. Indeed, the physical anthropologists tended to go back to the old slave trader's view of the African. Burton wrote in 1864 that the Negro had not all the 'latent capacities ascribed to him by the philanthrops' and that

> the removal of the negro from Africa is like sending a boy to school; it is his only chance of learning that there is something more in life than drumming and dancing.[4]

The picture of the African that came to prevail at the end of the century was a cross-breed between the missionary's and the physical anthropologist's, built up by those anxious to justify the new turn in the

1 Cf. Harden to Taylor, 3 March 1858: 'I will endeavour to get some curiosities ready against the time when the American vessels come here, and will try to do as you have requested.' S.B.C.

2 This was the title of the paper which James Hunt read to a meeting of the British Association at Newcastle in 1863 and marked the public launching of the Anthropological Society. See footnote 2, page 211.

3 Cf. Article on 'Anthropology' in *Encyclopaedia Britannica*, 1957, vol. ii, p. 49: 'Two important misleading research practices came into being in this period and remained in effect long after the formal recognition of a theory of evolution as expressed by Charles Darwin. First is the habit of ranking existing races in a hierarchical order, and second the practice of comparing human races with the contemporary apes for purposes of ranking them for functional or behavioural equivalents.' What Darwin said was that the different races had so much in common that it could safely be deduced that they all descended from a progenitor (*Descent of Man*, part I, Chapter 7).

4 R. F. Burton: *A Mission to Gelele*, vol. ii, pp. 200–1.

ambitions of European powers in Africa. Because of the religious scepticism of many of the 'Anthropological sort', the missionaries, African and European, for a long time refused to have anything to do with them; but their ideas could not be so easily discriminated against and before long missionaries were being affected by them. And it should be mentioned that this happened before Mary Kingsley goaded the Colonial Office into making anthropology a handmaid of Colonial administration.

Anthropology was already influencing the missionaries entering Nigeria in the late 1870s, not only towards a more contemptuous attitude to their African colleagues, but also towards a new view of the technique of evangelization. They still had faith in Victorian civilization as the highest achievement of mankind, but they no longer believed that the African had the capacity for assimilating it quickly. Consequently they were much more hesitant than before to consider civilization as an inseparable companion of Christianity. This revival of the old Evangelical attitude brought new ideas in its train. For example, while Presbyterian missionaries had been known in the past to call in the gun-boat to make it clear to the Calabar people that witchcraft and other such preternatural methods of harming people did not exist, a new C.M.S. missionary at Onitsha in 1890 not only cast doubts on this scientific disbelief, but apparently also suggested that the Church could take a hand in discovering witches and exorcizing their 'evil spirits'—much to the confusion of the converts and the scandal of the older missionaries.[1]

The point should be made that if some of the older African pastors found the newer European missionaries incomprehensible, the few European missionaries who were able, like the Africans, to remain long in the mission found them almost equally difficult to understand. The older European missionaries in the C.M.S. Yoruba Mission were beginning to retire when the newer ones were coming in. There was no conflict between them, though Hinderer, in asking in 1875 that Crowther's jurisdiction be extended, found it necessary to make a special remark about the 'younger men'. It was the Presbyterian Mission in Calabar, where older men like Anderson and Goldie continued into the 1880s, that best showed the clash between the old and new occurring between Europeans. The schism that took place there in 1882, though it came to involve local politics and the clash between two personalities, one ageing, one young, began as a conflict between older missionaries in the presbytery keeping to their accustomed method of working and resisting pressure from newer ones like Mary Slessor and Alexander

1 Onitsha Congregation to the C.M.S., 30 Aug. 1890; C.M.S. G3 A3/04.

Ross, who were keen to change the method and anxious to expand more rapidly.[1] Anderson and Ross became symbolic characters.

The Rev. Alexander Ross was a tempestuous character, much as Anderson had been twenty or thirty years earlier. But after thirty years in the mission, Anderson had mellowed into Efik society. He had come to realize that Calabar could not by the intervention of the gun-boat be turned into a copy of Scotland overnight. He was usually now at peace with the local rulers and as an old respected missionary, influential with the elders, he counted for something in local politics, as, for example, during disputed successions. Ross, on the other hand, denounced Anderson's alliance with 'cruel', 'tyrannical', unconverted chiefs, and his 'toleration' of 'barbarous' customs.[2] Anderson was convinced that Ross would break the mission. Therefore when in 1879 Ross went home on leave Anderson plotted against his return.[3]

At that stage local politics entered into the dispute. The Archibong and the Eyamba Houses in Calabar were disputing the succession to the headship of the town. Anderson supported the Archibongs.[4] The Eyambas began to support Ross in his dispute with Anderson.

The Foreign Mission Committee were so evenly divided between Anderson and Ross that more than a year elapsed before they could come to a decision. Ross did return from leave, but the disputes were resumed with petitions and counter-petitions and grievous charges. The Committee decided to send a deputation to examine the matter on the spot. The deputation censored some acts of Anderson but found against Ross and asked him to return home. He refused and started a new church on land given him by the Eyambas, carrying with him five teachers and a substantial part of the congregation.[5] Ross died in 1884. The Presbyterians thought the congregation would come back. They did not. His wife sought for a new missionary to carry on the work. As the church

1 This account is based principally on various *Minutes of the Foreign Missions Committee* between 1875 and 1882 which contain usually brief accounts of the various papers laid before the committee, discussions on them, and voting on the resolutions.

2 Ross to Secretary F.M.C., 24 Oct. 1879, 29 Oct. 1879, considered in Minute 1497 of F.M.C., 25 Nov. 1879.

3 Anderson to Secretary F.M.C., 28 Aug. 1879, considered in Minute 1497 of F.M.C., 25 Nov. 1879.

4 A correspondent of the *African Times*, 1 March 1880, alleged that the trouble started because Prince Duke of the Archibong House, whom Anderson had crowned under the confusing name of Eyamba IX, was very unpopular and many people therefore rallied round Ross in his opposition to Anderson.

5 'Ami Ndi Africanus', writing from Duke Town, 24 March 1882, in *African Times*, 1 June 1882, gives details of Ross's supporters. Also Goldie, op. cit., p. 248, that Mrs. Ross found none of the existing mission to take up her husband's church, but one of the men trained in Dr. Guiness's seminary came to her rescue.

continued under European rule, it was just as if a new missionary society had entered Calabar.

The growing interest in anthropology was affecting not only the new European missionaries; many of the mission-educated Africans, both within and on the fringes of the Church, had also been borrowing arguments from Burton and others. The borrowing began with those outside the Church in the name of nationalism. But as happened in 1867, if those like Crowther who remained in the Church did not want leadership to pass to those like G. W. Johnson on the fringes, they had to prepare an effective answer.

It was Edward Blyden who first began to popularize the idea of the anthropologists in West Africa, though with a difference. 'The despotic and overruling method,' he said in 1872,

> which had been pursued in [the African's] education by good-meaning but unphilosophical philanthropists, had so entirely mastered and warped his mind . . . All educated negroes suffer from a kind of slavery in many ways far more subversive of the real welfare of the race than the ancient physical fetters. The slavery of the mind is far worse than that of the body.[1]

Blyden was criticising the Church from without. He had been born in the Dutch West Indies and brought up in the Presbyterian Church. To be educated for the ministry, he was sent first to America and, when state laws prevented his admission to a suitable college, to Liberia College, where he distinguished himself in both classical and modern European languages. He was ordained in 1858 but his work lay outside the Church. He taught in his old college and took to politics. He became a Secretary of State in Liberia and later Ambassador in London. When he lost office he went to Freetown, where he became Government Agent for the Interior. He travelled widely, his travels including a visit to Palestine.[2] He was not a religious man. For many years he turned towards Islam, not because he was converted to it but, as he later explained, because it was more 'African', and he considered it would be better for the African to pass gradually through Islam to Christianity. His real mission was to free the African from physical and mental subservience to others. For this reason he criticized European missions and Christi-

1 Blyden to Pope-Hennessy, 11 Dec. 1872, in *Letters with Pope-Hennessy on the West African University*, Freetown 1873.
2 Biographical Notes in Blyden: *The Peoples of Africa*, p. 1. Introduction by Hon. S. Lewis in Blyden: *Christianity, Islam and the Negro Race*, London 1889; Obituary Notice in *Journal of the African Society*, vol. XLIII, no. 11, April 1912.

anity on the anthropological ground of the differentiation of races. But unlike the physical anthropologists he rejected the arrangement of the races in a hierarchical order. And he was far from rejecting either Christianity or civilization for Africans. He only urged them to develop their own. In 1872 he was advocating an Independent African Church and a West African university to recruit professors from 'Egypt, Timbuctoo and Fulah'.[1]

Blyden's pamphlets and books of essays, lectures and correspondence were widely read all over West Africa. As early as 1872, G. W. Johnson at Abeokuta was referring to one of his pamphlets and using 'the able words of our Mr. Blyden'.[2] Because of his well-known religious scepticism, however, African leaders of the Church were reluctant to follow him, though, like James Johnson in 1873, some used his arguments and were often mistaken for his disciples. James Johnson, in pressing, like Crowther, for the promotion of the indigenous culture, was always using anthropological arguments. He was emphatic about the differentiation of the races, not so much physically as culturally, arising out of the differences of geography and climate. 'It has been forgotten,' he wrote in 1883, that:

> European ideas, tastes, languages and social habits, like those of other nations, have been influenced more or less by geographical positions and climatic peculiarities, and what is esteemed by one country polite may be esteemed by another barbarous and that God does not intend to have the races confounded, but that the negro or African should be raised on its own idiosyncracies.[3]

Until the end of the century the followers of Blyden were still few, and hardly numbered any of the leading pastors. It was the congregations, particularly in Lagos, many of whose members had suffered economically from the new irruption of Europeans, who began to talk freely of secession. But even among them the idea of an African Church was for a long time far from being generally acceptable. Their ties, their loyalty even, to the European way of life were basically strong. Even when they began to take more interest in their own culture, they wished to strengthen, not weaken those ties. They tended to regard the proposed separatist African Church as a local church, a sort of tribal

1 Blyden to Hennesy, 6 Dec. 1872, in *The West African University*, op. cit.
2 G. W. Johnson to W. R. Richards, 20 Jan. 1873: 'I can behold another channel of good which—to use the able words of our Mr. Blyden—thou canst not see.' G. W. Johnson's Papers.
3 James Johnson in Report on the Lagos Native Pastorate, 1883, cited in J. O. George: *Historical Notes on the Yoruba Country*, Lecture delivered in Lagos, 1895, printed by E. Kaufmann, Lahr, Baden; p. 48.

organization cut adrift from the rest of Christendom, and they did not find that attractive.[1] The Methodist leaders in the crisis of 1884 did not threaten to found a church of their own, they threatened that 'there was another church they could go to'.

It is not surprising that the first schism in Lagos, and the first in Nigeria led by Africans, occurred within the Baptist Church with a congregational organization. When the members of the First Baptist Church found one Sunday morning in February 1888 that their African pastor, the Rev. Moses Ladejo Stone, had been dismissed by the European missionary without reference to them,[2] and the missionary did not return any satisfactory answer to them, they had no need to secede from the Church. They only needed to separate themselves from the congregation, and this they did without much ado, without even waiting to refer the matter to America.[3] In fact, by the following Sunday they had formed the new congregation called the Ebenezer Baptist Church, with the pastor at their head. It should be noticed, however, that though

1 Cf. Blyden pleading for such a Church in 1891, having to emphasize that 'while the Church should be native, we do not mean it should be local. We want to drop the conventional trammels of Europe, but we do not wish to localize religion . . . to give it any tribal colouring'; Lecture in Lagos, published as 'The Return of the Exiles and the West African Church'.

2 For the Rev. M. L. Stone see p. 149. There are various versions of the story that led to the crisis of February 1888. The story in the First Baptist Church is that Stone used to trade to supplement his salary; the missionary, Rev. W. B. David, told him to stop it. Stone then asked for an increase of salary, was refused and then resigned. So that the congregation were complaining against David's acceptance of the resignation without consulting them.— 'History of the First Baptist Church' by E. A. Alawode. The version in the Ebenezer Baptist Church is that Stone asked for an increase of salary, was refused, took to trading and was dismissed.—'History of the Ebenezer Baptist Church, Lagos' by Robertson for the centenary celebrations of the Baptist Church in Nigeria in 1958.

3 America was of course more remote from Lagos than London or Edinburgh. American Baptist missionaries had always acted more independently, and were less likely to be repudiated than their English colleagues. This was certainly true of the Rev. W. B. David, who had been 13 years in Lagos and was a hero to the Foreign Mission Board. For, despite domestic tragedy of the death of two children and two wives in succession, he had revived the missions at Lagos, Abeokuta and Ogbomoso. He had also revived interest in the work in America, having personally to collect money to support himself. Once, in 1884, he had the unusual idea of taking with him to America a Yoruba boy, Manly Ogunlana Oshodi, to parade in the masquerade dresses of *Egungun* and *Oro*. In that way, it was said, in six months of incessant touring, David excited 'more interest perhaps than was ever felt for Africa'. As a result he was able to take new European missionaries with him back to Nigeria as well as material worth 3,800 dollars with which he built the First Baptist Church in Lagos, which was completed just before the crisis. As was to be expected, when the Foreign Mission Board learnt of the crisis, they gave him a vote of confidence. Tupper: *Foreign Missions of the Southern Baptist Convention*, pp. 386, 439 and *passim*.

it was not until 1915 that the Ebenezer Baptist Church and its branches at Abeokuta and other places were brought within the same Convention as the missionary-supervised Baptist churches, the Rev. M. L. Stone had reconciled himself with the missionaries by 1894.

It is indeed remarkable how the body of African pastors remained, on the whole, in this period of stress and transition, loudly loyal to the churches in which they had been brought up. They saw in the African Church movement primarily a way of strengthening and enriching the life of the Church, and they hoped to keep it within the existing churches. When James Johnson succeeded in drawing from the C.M.S. in 1883[1] a Minute condemning the general adoption of European names at Baptism, it looked as if they would be successful. That was still two years before G. W. Johnson changed his name and became Oshokale Tejumade Johnson.[2] The movement for ceremoniously casting off European clothes hardly attracted the African pastors, but the debate and the research on the laws and customs of the people, their dancing, their elaborate court etiquette, their sayings and philosophy went on as much within as outside the churches. The most notable book of the minor renaissance that occurred, Samuel Johnson's *History of the Yorubas*, completed in 1897, was written by an African pastor.[3] The importance people in Lagos attached to the events on the Niger which culminated in the supersession of Bishop Crowther by European missionaries will be readily appreciated when it is realized that they led directly to the first attempt by Anglicans and Methodists to found a separatist African Church.

The way Bishop Crowther had been ousted showed that the transmutation of Europeans from guides to rulers was complete in the Church as it was becoming complete in the administration of the country, and that the earlier policy of encouraging the growth of an African middle class was completely overturned.

This change of policy was fundamental, though it did not lead to so complete a break with the past as men like Brooke and Robinson hoped. For they themselves died, Robinson in July 1891, Brooke in March 1892. With them died, too, much of the sensationalism of making a complete break with the work of the African missionaries. As soon as the missionaries who succeeded them got down to the not very romantic

1 Minute of the Parent Committee against the adoption of Foreign Names, in C.M.S. Intelligencer, February 1883, encl. in Lang to Hamilton, 10 Aug. 1883; C.M.S. G3 A3/L2.
2 'Notice' dated Customs House Road, 1 May 1885; G. W. Johnson's Papers.
3 J. F. Ade Ajayi: 'Nineteenth Century Origins of Nigeria Nationalism' *Journal of the Historical Society of Nigeria*, vol. ii, no. 2, 1961.

aspects of making Christianity appeal to other people, and doing so on a limited budget, they began to find many things they approved of in the policy of Crowther and his men.[1] Indeed, as roads and railways and British political power penetrated the country and created in parts of the interior as much demand for education as Crowther found in the Delta in his time, the European missionaries, not only of the C.M.S. but of other missions as well, began generally to adopt Crowther's policy of evangelization through the village school.[2] Like Crowther, they had to talk about the material advantages of having schools in the community. And just as in Crowther's time Sierra Leone had supplied teachers and evangelists, so after him, Lagos, Brass, Bonny, Calabar, Abeokuta, and to some extent Ibadan and Ogbomoso began to supply agents to the new centres of missionary work both on the coast in Benin and Ijebu and in the interior in Ekiti, Arochuku and other parts of Iboland. In this way the labours of the missionaries, African and European in the fifty years before 1891 provided abundant fruit for the Church in the fifty years after.

It is also possible to argue that much of the later harvest was made possible precisely because the earlier missionaries had placed so much emphasis on education and civilization, and because this continued to affect the work of the later missionaries who would have liked to see much less emphasis placed on these things. But for the effects of the earlier policy in the older mission centres, people in the areas of Southern Nigeria being penetrated by missionaries in the later period might have taken less readily to mission schools, and might have been less ready to pay for them. As it happened, their demand for schools was so persistent that the choice of missionaries in the matter of whether to build schools or not was severely limited. Government officials who wished to maintain the delicate balance of the Indirect Rule system and wanted only children of chiefs to go to school, did not hide their distrust of mission schools, which continued indiscriminately to accept children of chiefs and commoners. Many missionaries were inclined to agree with the government officials about the bad effects of mission schools. Some of them, as it were, smote their breasts and pleaded guilty to the sins of their predecessors in fostering an allegedly idle middle class with

1 Archdeacon Dobinson, the only survivor of the Sudan Party, later apologized openly in Freetown and Lagos for his part in the 'Great Purge'. He said he had been 'hurried along in unknown depths of fierce-flowing river . . .' He showed more appreciation of the local customs he previously despised and pleaded for more educated Africans to work on the Niger, as they did most of the work even if Europeans supervised. Dobinson to Baylis, 26 Feb., 29 March 1894; C.M.S. G3 A3/04.
2 Cf. Father J. P. Jordan: *Bishop Shanahan of Southern Nigeria*, Dublin 1949, pp. 29–31 for the new Roman Catholic attitude.

no roots in and no love for their own country, good for nothing except imitating European vices. Yet it was precisely at this period that missionaries built more purely literary schools and embarked on far fewer schemes of industrial and technical education. In spite of the new emphasis on hospitals, it was through the village school that the Church was spreading rapidly in the later period. It may also be mentioned that the political officers contributed to this. Though they declared that the installation of a Government Agent from whom henceforth the local rulers had to take orders was an unimportant change that left the people's life unaffected, it had in fact created so great a psychological revolution in the people's attitude that they tended to rush to missionaries with less hesitation and less reserve than was the case in the earlier period. It was only with the coming of the District Officer that things began to fall apart.

The ousting of Crowther from the Niger did not stop the growth of the class of educated Africans, but it meant that until they became strong enough to demand more rights and privileges, their fortunes were severely limited. They were discredited in the eyes of the new European ruling class. Little that they did received favourable comment. Even when Samuel Johnson completed his *History of the Yorubas* in 1897 the C.M.S. showed no enthusiasm to publish it.[1] Their advancement in commercial houses and the civil service was curtailed, since all the important jobs came to be regarded as specifically 'European' jobs to which only rarely favoured Africans could be admitted. Their economic opportunities declined. They took little part in the commercial expansion that resulted from the railways and the introduction of cash crops. It was the peasants who, as the philanthropists had always wished, cultivated the crops, but it was the European firms, not the educated Africans, who had the resources and the facilities to export them. Above all, their opportunities in the Church also became limited. The missions continued to rely on their African staffs, but highly-educated pastors were not encouraged, and the highest posts were reserved for Europeans.

1 The book was completed in 1897. The C.M.S. said it was too long and were interested in a short history suitable for use in schools. Apparently Johnson refused to cut it down and sent it through the C.M.S. to an English publisher. Nothing more was heard of the manuscript. The author's brother, Dr. O. Johnson, on a visit to England in 1900, was told that it had been mislaid. Within a year after that the author died and it had to be rewritten by Dr. Johnson from Samuel's notes and earlier drafts. The new version was sent to England in 1916, but owing to enemy action during the war did not reach England till 1918 and owing to shortage of paper was returned to await the end of the war. It was finally published by Routledge & Son in 1921. By 1937 the C.M.S. were anxious to publish a second edition of the book. See Samuel Johnson, op. cit. Editor's Preface.

After Crowther there were assistant bishops, but no diocesan bishop till 1953, when the constitutional changes in the country occasioned constitutional changes in the Church. It was not till 1946 that the Methodists appointed the first African Chairman of the District.

Some of the educated Africans for a while sought political careers as advisers to the local rulers in the interior, or as their agents in the capital. One effect of this may be seen in the fact that it was the centres where missionary work had been most successful in the earlier period and where the educated Africans were most influential that saw some of the most determined efforts to negotiate agreements to limit the rights of the British rulers. The Royal Niger Company had the greatest difficulty at Onitsha in its effort to obtain treaties on which to base its political privileges, and far more difficulty at Egga and Bida than at Sokoto. It was at Calabar and Bonny that the British Commissioner received the most specific conditions under which British rule would be accepted.[1] Above all, Abeokuta, where the educated Africans came nearest to political power, managed to resist British annexation till 1914, having in the meantime evolved the Egba United Government in which educated Africans continued to hold important executive posts.[2] But even that was temporary. Law and medicine, which afforded a chance of private practice and success independently of the new ruling class, became the goals of the educated Africans.

This eclipse of the educated Africans in one way delayed the full development of the Church, and in another hastened it. It delayed it because, as Venn always argued, as long as Europeans retained full control of the Church it could be no more than an exotic institution. Only the Africans themselves could make it a national institution. It is interesting to observe, for example, that little adaptation in the usages of the Church took place for many years after 1891. Venn had suggested that such adaptation should be made not by European missionaries but by the most highly accomplished and gifted of the African pastors themselves. From 1891 until quite recently, such pastors received little encouragement even in the missions sufficiently well established to produce them.

Yet this fact itself helped the development of the Church in another way by diverting the energies of many Africans towards the formation of an African Church where African usages and practices would be welcome. After some initial hesitation people began to take the United

1 Flint, op. cit., Chapter VII on Major MacDonald's Reports of 1889–90. Macdonald to Salisbury, 12 June 1889; F.O. 84/1290, and Report on the Administration of the Niger Territories; F.O. 84/2109.
2 For the Egba United Government see S. O. Biobaku: 'An Historical Sketch of Egba traditional authorities' in *Africa*, vol. 22, pp. 35–49, 1952.

Native African Church formed in 1891 a little more seriously. In 1901 a 'major' secession in the Anglican Church in Lagos led to the foundation of the African Bethel Church. Elsewhere suppressed political feelings went into Prophet movements and revivalist organizations. The African Church movement, consisting of people brought up in different denominations, was bedevilled by differences over doctrine and conflicts over leadership. In particular there was conflict between those maintaining the congregational view of the minister responsible to the congregation, and those who believed in the sacerdotal view of a priesthood with apostolic succession and a hierarchy. There were compromises and schisms. In 1907 it was agreed that the head of the new African Church would be 'Superintendent, or in other words, Ecclesiastical or Presbyterian Bishop in contradistinction to Prelatical or Historical Bishop'.[1] Some of those dissatisfied with this very obvious compromise broke away in 1908 and formed the African Salem Church.

But in spite of such schisms and of much bitterness of feeling the movement as a whole gathered strength. It showed that there were Africans who felt sufficiently deeply about the new religion that they were willing to try to express its spirit in their own way and to compete with the mission-supervised churches in spreading the Gospel to other parts of the country. It was a major outward sign that the Church had become established in Nigeria and was unlikely to die out again.

The African Church movement has another significance that should be noted in conclusion. It provides a link between the educated Africans of Crowther's age and the nationalists of our day who have re-emphasized the mid-nineteenth-century doctrines about the importance of an African middle class for the development of the country, and the distinction between the expansion of trade controlled by foreign European firms and economic development as a factor of social and economic change in the country.[2]

1 *Report and Proceedings of the African Church Organization, 1901–1908*, p. 29. 'Origin and History' Printed Pamphlet, Lagos, 1908. Also S. A. Oke: *The 'Ethiopian' National Church*, Lagos 1922; Preamble to the *Revised Constitution of the African Church* (Incorporated) ratified at Idi Ape, Abeokuta, 1941.
2 J. F. Ade Ajayi: 'Nineteenth Century Origins of Nigerian Nationalism', op. cit.; cf. Herbert Macaulay, a grandson of Bishop Crowther, regarded as Father of Nigerian Nationalism. In 1942, when the Methodists were celebrating the Centenary of Freeman's arrival in Badagri, Herbert Macaulay was using the platform of the African Church to deliver a nationalist address, welcoming the missionary movement, but condemning European imperialism. *The History of the Development of Missionary Work with special reference to the United African Church*, Printed Pamphlet, Lagos, 1942.

Appendix

Minute on the Constitution of the Anglican Native Bishopric on the West African Coast (1864)

1 The Constitution of the West African Bishopric is declared in the Acts of Parliament under the authority of which the constitution took place.

2 The Acts of Parliament are the 26 Geo. III and 5 Vict. These Acts give authority to the Bishop in these words: 'And be it further enacted that such Bishop or Bishops so consecrated may exercise within such limits as may from time to time be assigned for that purpose in such foreign countries by Her Majesty, spiritual jurisdiction over the ministers of British Congregations of the United Church of England and Ireland and over such other Protestant Congregations as may be desirous of placing themselves under his or their authority.'

3 The Queen's license for the consecration defines the locality in which the Bishop is to exercise his functions as 'Western Africa which is generally understood to comprise the countries on the West Coast lying North of the Equator as high as the River Senegal'. But as this coast contains several British colonies viz Lagos, the Gold Coast and Sierra Leone, which constitute the Diocese of Sierra Leone, the license defines the new Diocese as comprising 'the countries of Western Africa beyond the limits of our dominions'. There are however existing missions of the Church Missionary Society comprised in these limits which the Bishops of Sierra Leone have been accustomed to superintend, such as the Timneh Mission near Sierra Leone, and Abeokuta near Lagos, respecting which an arrangement must be made by the two Bishops as to the time and circumstances of transfer.

4 It follows also that within these British colonies Bishop Crowther can exercise the episcopal functions of confirmation or consecration of churches only by commission under the hand of the Bishop of Sierra Leone which commission may be either general or for specific acts, and may be at any time altered or cancelled by the Bishop of Sierra Leone.

5 Persons who are ordained by Bishop Crowther when they come to England will be in the same position as the persons ordained by

274

American Bishops or by the Anglican Bishop in Jerusalem – they can only officiate by the special permission of the Bishop of an English Diocese, which permission can only be given for two days at a time and they cannot hold a curacy or preferment in England or Ireland.

6 The Bishop must have an episcopal seal for the verification of his letters of Orders, his licenses, and other public documents.

7 Dr. Crowther has been consecrated Bishop of the United Church of England and Ireland, and the Church in Western Africa over which he presides will be a branch of the United Church of England and Ireland and will be identical with the Mother Church in doctrine and worship and assimilated in discipline and government as far as the same may be consistent with the peculiar circumstances of the countries in which the congregations are formed. In any important questions which may arise, the Bishop will have the privilege of applying for advice to the Archbishop of Canterbury as his Metropolitan, to whom he has taken the Oath of Canonical Obedience.

8 It may be well to advert to a few peculiarities in the new Bishopric which may call for some modification in the details of Episcopal Administration and in the ritual of public worship.

9 In all settled congregations of Native Christians within the new Diocese such as those which are of some years standing at Abeokuta, Ibadan, Otta, the Liturgy of the Church of England is regularly used on the Lord's day, and baptisms, marriages, and burials are performed according to the forms therein prescribed. In such cases, it has been found requisite only to make a few alterations, as in State prayers and to use the Litany as a separate service in order to reduce the length of the service. This practice must be followed in all settled congregations in the new Diocese, so that the spirit and general impress of the Church of England may be fully preserved. Translations of the Liturgy must be authorised by the Bishop.

10 But until settled Christian Congregations can be formed, arrangements will be required for purely missionary operations and for the transition of a people from heathenism to Christianity, for which the Mother Church can supply no precedents; such are the preparatory course of Instruction and the qualifications of adult candidates for Christian baptism, the times and circumstances under which that

275

sacrament is to be administered, and the duties of catechists and subordinate missionary agents. These matters have been for the most part well considered in all Church missions, and a uniformity of practice has been adopted which the Bishop will do well to establish in the missions under his direction as far as they approve themselves to his judgement, and to reduce to written Regulations issued under his authority.

11 In laying the foundations of the Native Church in the new Diocese, regard must be had to the fact that in heathen lands scattered congregations can only be held together so as to form one Church by voluntary association, and the central authority of the Bishop; all must rest upon contract or agreement. There can be no aid as in the Mother Church or the Church in the colonies from Civil power to enforce ecclesiastical authority. The main external security for the permanence and coherence of the Church and for the maintenance of episcopal authority will be the existence of a central Diocesan Fund out of which the pastors may be paid and aid contributed to the building of churches, schools, etc. The buildings, houses, lands given to the Church, should be made over to the Central Trust. For the present, the Church Missionary Society affords such a central Trust, but it will be desirable to make provision for the ultimate and normal condition of a separate Diocesan Trust when the Native Church shall have been advanced, under God's blessing, to a competent authority.

12 With a view to the establishment of a self-supporting, self-governing Native Church, it will be desirable to introduce an organisation into the Native Congregations such as the forming of the converts into classes or companies under headmen, so as to habituate the Native Church to combined action and subordination to authority. A Church Fund should be established in every congregation for receiving the weekly contributions of the people, which should be in connection with the Central Mission Trust Fund.

13 It will be desirable also in a Church which has no external aid for the enforcement of authority to hold frequent conferences or Synods in its different districts, a general Synod which delegates may attend from the District Synods. The Bishop should also be assisted by a council in the management of Funds. Every Bishop can appoint Commissaries to represent himself and communicate confidentially with him in his absence. This will be especially important in isolated

districts. Such plans will bind together the different parts of the Church. Two documents are appended which may be some guide viz a scheme for the government of the Sierra Leone Church which received the sanction of the late Archbishop of Canterbury and Bishop Bloomfield, and a Minute of the C.M.S. on the Organisation of a Native Church. These have been drawn up with much care and by the aid of lengthened missionary experiences and may furnish useful precedents in such particulars as are applicable to the new Diocese.

14 In repect of Ordination, the Apostolic injunction 'lay hands suddenly upon no man' will be especially important. It appears desirable that for some years, none should be ordained without a knowledge of the English language and English Bible. As a native literature is formed, this restriction may be relaxed.

15 It will be advisable that the Bishop keep a register of all his Episcopal and Church Acts, a copy of which should be transmitted home for preservation and information, and an annual Letter should be written to His Grace the Archbishop of Canterbury upon the progress of the Native Church under his superintendence.

(Sgd.) Approved
C. T. Cantuar.

Bibliography

PRIMARY SOURCES

1 Missionary Records

This work has been based principally on the records of the five missionary societies themselves, manuscript material as much as possible, supplemented by printed sources.

A Manuscript Materials

a CHURCH MISSIONARY SOCIETY (Archives at Salisbury Square, London, E.C.4.)

The material here is full and adequate. The correspondence from headquarters to the missionaries as individuals or as missions, and from them to the headquarters, are almost complete. In addition, up to 1880, each missionary, European or African, was obliged to keep a private journal, extracts from which were read at Local Conferences and sent to C.M.S. House every 3 or 6 months. These give useful material about social, economic and political affairs in general as well as the religious histories of converts. From 1858 onwards, the missionaries were also asked to send in Annual Letters to help the General Secretary at home to prepare his Annual Reports.

The material I have used is classified under three different missions *CA*1 'The West African Mission' i.e. Sierra Leone, for the earlier records of the emigrants; *CA*2 "The Yoruba Mission'; and *CA*3 for 'The Niger Mission'. (After 1880, these become G3 A1, G3 A2, and G3 A3 respectively. Each is subdivided into either *L* for out-going letters, and *O* for in-coming letters, journals, petitions, reports and minutes of Local Conferences, copies of correspondence with local authorities, Consuls, Governors, etc. The *L* series are serialised chronologically. Up to 1880, the *O* series are serialised according to the source, a file being for each individual missionary. After 1880, all the letters are put together and serialised chronologically, with a set of précis-books *P* giving summaries of the contents of the letters and often indicating the lines on which action over it proceeded.

The C.M.S. in Nigeria has deposited their local material in the Nigerian National Archives at Ibadan. Some of these duplicate material in London, but they also contain valuable records of events on the spot of which only abstracts or bare references will be found in the metropolitan archives. There are also important

278

personal papers of individual missionaries. Among these may be mentioned the private diary of Archdeacon Henry Johnson 1877–92, containing drafts of letters, newspaper articles and a few notes. (ECC 20/9)

b METHODIST MISSIONARY SOCIETY (Marylebone Road, London, W.1)

The records here are not as full as in the C.M.S. archives. A good deal of the letters from the missionaries on the spot, especially the General Superintendents and the 'Minutes and Reports' presented at the Local District Conferences, as well as many out-going letters (*in extenso* or in synopis), have been preserved. We miss the regular journals of the C.M.S. missionaries, but we have some useful accounts of missionary journeys. The outgoing letters are in the Secretaries' *Letter Book*. Up to 1879, the Methodist Mission in Nigeria was part of the Gold Coast District, and the incoming papers are therefore in the Gold Coast District files. After 1879, they are in the 'Yoruba and Popo District' files. At the moment, the papers are just arranged chronologically, one file for each year. In footnotes, I have indicated simply the dates. As in the case of the C.M.S., the Methodists have records in the Nigerian National Archives at Ibadan which contains interesting local material.

c UNITED PRESBYTERIAN CHURCH (in the National Library of Scotland, Edinburgh)

Surviving records consist principally of the Mission Board's *Letter Books* covering the period December 1847–August 1882, containing copies of letters to the principal missionaries. (b) Missionaries Letter Book No. 2, 31st August 1856 to 29th January 1875, being the treasurer's account book of the personal expenditure of each individual missionary (c) Five of the eleven volumes of Hope Waddell's private diaries (nos. i, vii, viii, x, xi). These are by far the most important for this study. Hope Waddell was the pioneer missionary at Calabar, an able thorough, painstaking man. He was an honest reporter, a very rare type of missionary who could record faithfully the arguments of his opponents whether fellow missionary or would-be convert. Fortunately the first volume has been preserved in which he recorded his first impressions of Calabar. The value of manuscript material for this type of study can perhaps best be illustrated by the way in which these journals, incomplete as they are, illuminate the printed material, even of other missionaries.

d SOUTHERN BAPTIST CONVENTION

i *Monument Avenue: Richmond Virginia, U.S.A.*

The archives suffered much during the American Civil War and also during removals since. They are now well located and classification was still going on when I visited them in 1959. I have therefore cited material only by the name of the writer and the date.

The outgoing letters from the Secretaries seemed to have suffered most. Fortunately, the in-coming letters from the different missionaries, which are more important for this study, are pretty full and well-preserved. There are also fragments of diaries and accounts of travels.

ii   *The Roberson Collection, Nigeria*

My first contact with Baptist Archives was through the collection of the Rev. Cecil R. Roberson, a missionary and the local authority on Baptist History in Nigeria. With the help of his wife, he has acquired or copied as much material on Baptist missionaries and their work in Nigeria as he can lay hands on. The collection is rich in early books and pamphlets. There is also a set of local histories of different Churches Commissioned in the 1950's as part of the Centenary Celebrations. The bulk of the Manuscript material is copied from the Archives in Richmond, but there is also a significant amount of personal material not available in Richmond, but acquired from relations of the missionaries and their family churches, as well as publications of these churches which give a good deal of useful background information.

e   SOCIETY OF AFRICAN MISSIONS:

i   *S.M.A. archives, Via della Nocetta, Rome.*

For the period before 1891, the archives are rather fragmentary. The most important surviving manuscript material are—
The Journal of Father Broghero; Father Planque's outgoing Letter Book vol. 1; Diaries and papers of Father Holley; Papers of Bishop J. B. Chausse.

ii   *S.M.A. Archives, Ibadan*

Again, rather fragmentary; the most important items before 1891 are: the Journal of Father Courdioux (Chronique de la Mission du Golfe du Benin 1861–67); Faculties granted to Father Broghero 11th March 1862; 5th February 1865. Father Pourets' collection of *Usages suivis à Lagos*; Courdioux's *Questions se rattachant à la conduite à tenir une mission,* and the papers of Sir James Marshall, including his correspondence with Father Chausse in 1889.

B   Printed Periodical Publications

Common to all the missions are the periodicals intended to interest the supporters of the missionary society or church in the mission fields both in Nigeria and elsewhere. C.M.S. and Methodist publications show four main types:

a   *Proceedings of the C.M.S., Minutes of the Methodist Conference*
These were annals containing the General Secretary's Report of the state of the missions, Annual Estimates of Expenditure, etc.

b   *The Church Missionary Record, The Weslyan Methodist Missionary Notices.*
These were monthly publications intended to keep the average reader in touch with the latest news from the mission fields. They contain extracts from letters and journals of missionaries and short editorial comments.

c   *The Church Missionary Intelligencer, The Methodist Magazine.*
These were more advanced publications 'for the use of educated men and women in which articles on Geography, ethnology, religions of the various mission fields could appear and what may be called the science of missions discussed, and in which important missionary letters could be published at once instead of awaiting their turn in the systematic reports and serialisation of various missions' given in type (b). (Stocks, History of the C.M.S. vol. ii, p. 51). on the *Intelligencer.*

d   The (C.M.S.) *Missionary Gleaner, The W.M.M.S. Reports.*
These were the most popular of the publications, made suitable for Sunday school use. The material in (b) was digested with more editorial comments, the front page being illustrated with a woodcut.

The principal Presbyterian publication, the U.P. *Missionary Record* is like type (b), but it also contains the Annual Estimates of Expenditure. In addition, there are the *Minutes of the Foreign Mission Committee* which gives a summary of the business conducted by the Committee, and contain resolutions passed, brief accounts of papers laid before the Committee, and, on controversial matters, some indication of the debate and the voting.

The Southern Baptist Missionary Journal of the S.B.C. was, like the Presbyterian's Missionary Record, type b, with Annual Estimates. In 1851, it was replaced by the *Home and Foreign Journal*, to combine information on both the overseas missions and the home Churches. It was interrupted by the Civil War. In 1874, the idea of a separate publication for the Overseas work was revived in the *Foreign Mission Journal.* At the most exciting periods of the missionary expansion 1849–51, 1856–61, and after 1916, there was published *The Commission*, type (d). For the work and especially the views of individual missionaries on leave, the publications of State Baptist Organisations, usually type d, are also important.

The S.M.A. relied in this period on the weekly publication *Les*

Missions Catholiques, type (b), of the Association for the Propagation of the Faith, which like the S.M.A. then had its headquarters at Lyons. There are also valuable articles on the work of the S.M.A. in the *Annales de la Propagation de la Foi* and its English version (not always identical) *Annals of the Propagation of the Faith.* These are type (c).

The greatest merit of these publications is that they have survived more uniformly in the various missions than manuscript material and they provide more continuous commentary on the work of the missionaries. However, where available, I have preferred the original manuscript material from which the information in these publications were extracted for they often give supplementary information necessary to interpret the published statements correctly. My method therefore has been to use the manuscript material exhaustively, and supplement them where lacking from the periodicals. Thus, I have used C.M.S. periodicals rarely, while the *Les Missions Catholiques* has been invaluable throughout, and the U.P. *Missionary Record* and the Methodist *Missionary Notices* have filled important gaps in the manuscript material.

2 Government Records

On a few specific issues, I have consulted documents in the Public Record Office, F.O.2 (Africa, Consular) and F.O.84 (Slave Trade) series. More usually, for reports of Consuls and Commissioners, treaties, missionary petitions, etc., I have relied on the more accessible printed sources.

i *State Papers,* published annually by the Foreign Office containing a wide selection of treaties, and important Consular despatches. There are two index volumes, no. 64 for the period before 1873, and no. 93 for the period 1873–1900.

ii *Confidential Prints of the Foreign Office,* in particular:
1856, July (4141): Correspondence relative to the Dispute between Consul Campbell and the Agents of the C.M.S. at Lagos.
1872, January: Report of W. H. Simpson, Foreign Office Commissioner, Niger Expedition 1871.

iii *Parliamentary Papers*
1840 XXXIII (57) Correspondence relating to the Niger Expedition.
1842 XI, XII (551) Report of Select Committee on British Possessions on the West Coast of Africa.
1847–8 XXII (272, 366, 536, 623) Four Reports from Select Committee on the Slave Trade.

1849 XIX (309,410) Two Reports following session.

1850 IX (53, 590) Reports of Select Committee of House of Lords.

1852 XLIX (284) Correspondence relative to the Conveyance of H.M.'s Mails to the West Coast of Africa.

1852 LIV (221) Papers relating to the reduction of Lagos by H.M.'s Forces.

1857 XXXVIII (255) Papers relating to the cultivation of cotton in Africa.

1861 LXIV (1) Slave Trade Correspondence, Africa (Consular).

1862 LXI (1) Slave Trade Correspondence, Africa (Consular).

1862 LXI (339, 365) Papers relating to the Occupation of Lagos.

1863 XXXVIII (117) Papers relating to the destruction of Epe.

1863 XXXVIII (512) Letters from the Rev. H. Venn on the conduct of missionaries at Abeokuta.

1864 XLI (571), 1865 XXXVII (907) Papers relating to the application of the Company of African Merchants for a subsidy.

1865 V (1) Report of Select Committee on State of British Settlements.

1865 XXXVII (533) Papers on War between Native Tribes in the neighbourhood of Lagos.

1887 LX (1,167) Correspondence between Native Tribes in the interior and negotiations for peace conducted by Government of Lagos.

3 Private Papers

i *Henry Venn's Family Papers* in the possession of the late Dr. J. A. Venn, President of Queen's College, Cambridge, his grandson, to whom I am obliged for permission to use them. These contain among other things a valuable diary, hitherto unpublished, relating to the period 1841–45 when Venn was learning the job of directing missions and insisting for himself, as he later did for the missionaries, on the value of keeping journals. Dr. Venn has bequeathed the papers to the C.M.S.

ii *G. W. Johnson's Papers* in the University of Ibadan Library. They contain very valuable material on the political activities of Johnson at Abeokuta between 1865 and 1875, one or two letters to his friends, drafts of articles for the *African Times*, etc. After 1872, there are only a few odd papers.

iii *Bishop Charles Phillips' Papers*, now deposited by his son, Bishop S. C. Phillips in the National Archives at Ibadan. The most relevant for this study is the diary of the elder Bishop Phillips about his

283

diplomatic missions during the negotiations of the 1880s to bring the Yoruba war to an end.

iv *Herbert Macaulay Papers*, in the University of Ibadan Library. They contain files mostly of newspaper cuttings, judgements of cases of historical interest, pamphlets of lectures, etc. all of very recent date but showing influence of the ideas of the mission educated Africans of the late 19th century on the nationalist movement in the 1930s and 40s.

SECONDARY SOURCES

1 Local Histories of Churches

I have visited most of the centres of 19th century missionary work in Nigeria. I had valuable interviews with several Church leaders, including some elderly men who knew some late 19th century missionaries. Those talks at least helped to make the work of missionaries come alive to me as of great force in moulding the lives of the particular communities. In addition, I was often shown, apart from baptismal registers and mission log-books that had survived, histories of the individual churches usually commissioned on the eve of Jubilee celebrations. Such works are always of some interest. Characters who hardly receive a mention in the official documents may turn out to have become influential and revered Church Elders in the local history.

The Rev. Cecil Roberson made a collection of such histories on the Baptist Churches. At Rome I found a collection made in 1921 of such histories written by the Fathers themselves. A select list is given below:—

(Authorship is not always clear).

A Short History of the Introduction and Spread of Christianity into Egbaland under the C.M.S. (1946) by permission of Archdeacon Ashley-Dejo, St. Peter's Church, Ake, Abeokuta.

The Beginning of Missionary Work in the Yoruba country (i.e. Ibadan) in the Nigerian Record Office (ECC 20/1).

E. A. Ojo: *History of the Ebenezer Baptist Church, Lagos*. (Ibid.).

Notes on the Beginnings at Ogobmoso (Roberson's Collection of the Histories of Baptist Churches in Nigeria).

?E. A. Alawode: *History of the First Baptist Church, Lagos* (ibid.).

Missions de la Nigeria (bound typewritten volume edited by L. Arial (S.M.A. archives, Rome) containing items like:

284

'Premier Temps de la Mission de Lagos, d'après Mère Véronique.'
'Foundation de Topo d'après le P. Poirier.'
'Stations de Topo-Badagry.'
'Rapport de Topo-Badagry'– Father L. Freyburger.
Also, a separate 'Notes sur la Mission de Topo.'

2 Missionaries' Memoirs, printed Journals, etc.

These vary in value. Some were written by the missionaries themselves, often with great care. Some, written by their friends contain valuable documents quoted in full not available elsewhere. Others are popular sentimental accounts of little historical worth. Only a select list is given here:

ANDERSON, William and Louisa	*A Record of their Life and Work in Jamaica and Old Calabar* by William Warwick.
BOWEN, T. J.	*Missionary Labours and Adventures in Central Africa* (Charleston, U.S.A. 1857).
CROWTHER, D. C.	*The Establishment of the Niger Delta Pastorate 1864 92* (Liverpool 1907).
CROWTHER, S. A.	*Journal of an Expedition up the Niger and the Tshada in 1854* (London 1855).
CROWTHER, S. A. and TAYLOR, J. C.	*The Gospel on the Banks of the Niger, Journals of the Niger Expeditions of 1857 and Missionary Notices* (London 1859).
FREEMAN, T. B.	*Journal of Various Visits to the Kingdom of Ashanti, Aku, and Dahomey* (London 1844). *Missionary Life No Fiction* (published anon. London 1871).
FREEMAN, T. B.	Unpublished book on West Africa in typescript, in the Methodist archives.
GOLDIE, Hugh	*Calabar and Its Mission* (Edinburgh 1890).
GOLLMER, C. A.	*His Life and Missionary Labours in West Africa* by his eldest son (London 1886).
HINDERER, Anna	*Seventeen Years in the Yoruba Country*, Memoirs of the wife of David Hinderer compiled by her friends (London 1872).
JOHNSON, Archdeacon Henry	*A Journey up the Niger in the Autumn of 1877* (n.d.? 1878 London).

SCHÖN, J. F. and CROWTHER, S. A.	*Journal of an Expedition up the Niger in 1841* (London 1843).
STONE, R. H.	*In Afric's Forest and Jungle* (1899).
SUTHERLAND, Mrs.	*Memorials of Mrs. Sutherland* by Agnes Waddell.
TOWNSEND, Henry	*Memoirs of Henry Townsend* by his brother, George Townsend of Exeter (London 1887).
TUCKER, Miss	*Abeokuta, or Sunrise within the Tropics* (London 1853).
VENN, Henry	*Memoirs of Henry Venn* by William Knight (London 1880).
	The Missionary Secretariat of Henry Venn by William Knight (2nd edition of the Memoirs, London, 1882).
	Our West African Colonies (London 1865).
WADDELL, Hope M.	*Twenty-nine Years in the West Indies and Central Africa* (London 1863).

3 Other Contemporary Material

ADAMS, Capt. John	*Remarks on the Country extending from Cape Palmas to the River Congo including observations on the Manners and Customs of the Inhabitants* (London 1823).
African Times	Journal of the African Aid Society (London periodical 1863–).
ALLEN, Capt. W. and THOMSON, T. R. H.	*Narrative of the Expedition to the River Niger in 1841* (2 vols. London 1848).
ANDERSON, Rufus	*Theory of Missions to the Heathen: Foreign Missions, their Relations and Claims* (Boston 1845 and 2nd edition 1869).
BAIKIE, W. B.	*Narrative of an Exploring Voyage* (London 1856).
BELL, T. M.	*Outrage by Missionaries, a Report of the whole proceedings on the Trial in Sierra Leone of W. F. John, Phoebe John, John Williams and Kezir Williams for the Murder of Amelia John at Onitsha* (Liverpool 1883).

BLYDEN, Edward   *The Western African University* (1873).
The Negro in Ancient History (1874).
Christianity, Islam and the Negro Race
(1889).
*The Return of the Exiles and the West
African Church* (1891).
West Africa Before Europe (1905).

BOWEN, T. J.   'Grammar and Vocabulary of the
Yoruba Language' in the *Smithsonian
Contributions to Knowledge*, vol. ix part iv.

BUNSEN, C. C. J.   *Christianity and Mankind*, vol iv
Appendix D. (London 1854).

BURTON, R.   *A Mission to Gelele* (2 vols, London
Abeokuta and the Cameroons (2 vols, 1864)
London 1863).
Wit and Wisdom from West Africa
(London 1863).

BUXTON, T. F.   *The African Slave Trade and its Remedy*
(London 1840).

BUXTON, T. F.   *Memoirs of* by Charles Buxton (3rd ed.
London 1851).

CAMPBELL, R.   *Pilgrimage to my Motherland:* An account
of a journey among the Egbas and
Yorubas of Central Africa (Philadelphia
1861).

CLAPPERTON,   *Journal of a Second Expedition into the*
Capt. Hugh   *Interior of Africa* (London 1829).

C.M.S.   Church Missionary Atlas (London 1873,
1896).

C.M.S.   List of Missionaries and Native Clergy
1804–1904 (London n.d.).

Conference of West African Missionaries held at Gaboon in
February, 1876, (Printed Report of,
Calabar 1876).

CROWTHER, S.A.   *Grammar and Vocabulary of the Yoruba
Language* (London 1870).
Vocabulary of the Ibo Language
(London 1882).
*Grammar and Vocabulary of the Nupe
Language* (London 1864).

CROWTHER, S. A.   *Bibeli Mimo* (London 1867) *Testamenti*
and KING, T.   *Titon* (London 1871).

287

DALZEL, A. *A History of Dahomey compiled from Authentic Memoirs* (London 1793).

DELANY, M. R. *Official Report of the Niger Valley Exploring Party* (Leeds 1861).

DESCRIBES, E. *L'Evangile au Dahomey et à la Côte des Esclaves, ou l'histoire des Missions Africaines de Lyons* (Claremont-Ferrand 1877).

Friend of Africa Journal of the African Civilization Society (1840–46).

GOLDIE, Hugh *Efik Grammar and Dictionary* (Edinburgh 1874).

HILKHAM, Mrs. Hannah *The Claims of West Africa to Christian Institutions* (London 1830).

HORTON, J. A. B. *West African Countries and Peoples* (London 1868).

HUTCHINSON, T. J. *Ten Years Wandering Among Ethiopians from Senegal to Gaboon* (London 1861).

Iwe Irohin C.M.S. Newspaper Abeokuta (1859–67).

JOHNSON, Archdeacon Henry and CHRISTALIER, J. *Vocabularies of the Niger and Gold Coast* (SPCK 1886).

KOELLE, S. W. *Polyglotta Africana or a Comparative Vocabulary of nearly 300 words and phrases in more than 100 distinct African Languages* (London 1854).
Grammar of the Bornu or Kanuri Language (London 1854).
African Native Literature in Kanuri (London 1854).

LAFITTE, P. *Le Dahomey, ou Souvenirs de Voyages et de Mission* (Tours 1873).

Lagos Weekly Record (1891–1921).

LAIRD, McGregor and OLDFIELD, R. A. K. *Narrative of an Expedition into the Interior of Africa* (2 vols. London 1837).

LANDER, Richard and John *Journal of an Expedition to Explore the Course and Termination of the Niger* (3 vols. London 1832).

McKERROW, J. *History of the Foreign Missions of the Secession and United Presbyterian Churches* (Edinburgh 1867).

288

PAYNE, O.	*A Lagos Almanack and Diary for 1878* (London 1877), *Table of Historical Events in Yorubaland* (London 1893).
SCHÖN, J. F.	*Grammar of the Hausa Language* (London 1862).
TUPPER, H.	*Foreign Missions of the Southern Baptist Convention* (Richmond, Va., U.S.A. 1880). *A Decade of Missions, 1880–90* (Richmond, Va., U.S.A. 1892)?
WESLEY, John	"Thoughts on Slavery" in *The Complete Works of John Wesley*, vol. xi (London 1872).
WHITFORD, John	*Trading Life in West and Central Africa* (Liverpool 1877).

4 Later Works

AGBEBI, Mojola	*Inaugural Sermon delivered at the celebrations of the first anniversary of the African Church* (Lagos 1902).
BANE, Father M. J.	*Catholic Pioneers in West Africa* (Dublin 1956).
BARGERY, G. P.	*A Hausa-English and English-Hausa Dictionary* (Oxford 1934).
BEIER, H. U.	*African Religion, the Story of Sacred Wood-Carvings from one small Yoruba Town* (Nigeria Magazine 1957).
BENTON, P. A.	*A Bornu Almanack for the Year A.D. 1916* (London 1916).
BIOBAKU, S. O.	*The Egba and their Neighbours 1842–72* (Oxford 1957). "*Ogboni, the Egba Senate*" in *Proceedings of C.I.A.O.* (International West African Conference) Ibadan, 1949. (Oxford 1956). "An Historical Sketch of Egba Traditional Authorities" in *Africa*, Journal of the International African Institute, vol. 22, 1952.
BIRTWHISTLE, A. W.	*Thomas Birch Freeman* (London 1949).
BLAKE, J. W.	*European Beginnings in West Africa* (Hakluyt Society, 1941).

BUXTON, T. F. V.   "The Creole in West Africa" in
                       Journal of the African Society, vol. xii,
                       1913.
COOKE, J. E.   *Frederic Bancroft, Historian, and three*
                       hitherto Unpublished Essays on the
                       Colonization of American Negroes from
                       1801–1865 (Oklahoma 1957).
DALZEL, M.   *Popular Fiction 100 Years Ago* (1957).
DARYL-FORDE (ed.)   *Efik Traders of Old Calabar* (Oxford
                       1957).
DIKE, K. O.   *Trade and Politics in the Niger Delta*
                       1830–1885 (Oxford 1956).
                       "Beecroft 1835–49" in *Journal of the*
                       Nigerian Historical Society, vol. i,
                       December 1956.
DONNAN, E.   *Documents Illustrative of the History of*
                       the Slave Trade (4 vols. Washington
                       1930–35).
DUNGLAS, E.   "Première" and "Deuxième attaque des
                       Dahoméens contre Abeokuta" in
                       Etudes Dahoméennes Nos. 1 and 2, 1949.
                       IFAN.
DUVAL, L. M.   *Baptist Missions in Nigeria* (Richmond,
                       Virginia, U.S.A. 1928).
EAST, R. M.   *A Vernacular Bibliography for the*
                       Languages of Nigeria (Zaria 1941).
EPELLE, E. M. T.   *The Church in the Niger Delta* (Port
                       Harcourt, 1955).
FINDLAY, G. and   *History of Wesleyan Missionary Society*
HOLDSWORTH, W.   (London 1921–4).
FLINT, J. E.   *Goldie and the Making of Nigeria*,
                       (Oxford 1960).
FREYRE, G.   *The Masters and the Slaves* (Trans.
                       Samuel Pitman, New York 1964).
FYFE, C. H.   *A History of Sierra Leone* (Oxford 1962).
                       "View of the New Burial Ground" in
                       Sierra Leone Studies (new series No. 2,
                       June 1954).
GALLAGHER, J.   "Fowell Buxton and the New African
                       Policy" in *Cambridge Historical Journal*
                       vol. xi, 1950.
GALLEN, M. LE.   *Vie de Mgr Marion Bresillac* (Lyons 1910).
290

GEARY, W.   "The Development of Lagos in Fifty
                           Years" in *Nigerian Pamphlets*, Colonial
                           Office Library, No. 13.
GREENBERG, J.   *The Influence of Islam on a Sudanese
                           Kingdom* (New York 1946).
GROVES, C. P.   *The Planting of Christianity in Africa*
                           (4 vols. London 1948–58).
GUILCHER, R. F.   *Augustine Planque* (Lyons 1928).
HALÉVY, E.   *A History of the English People in the
                           19th Century* (London 1949).
HARDY, Georges   *Un apôtre d'aujourd'hui: le réverend père
                           Aupiais*, (Paris 1949).
HOGDEN, M. T.   'The Negro in the Anthropology of
                           John Wesley' in *Journal of Negro
                           History*, vol. xix, no 3, July 1934).
JOHNSON, Samuel   *A History of the Yorubas from the
(Pastor of Oyo)   Yorubas from the Earliest Times to the
                           Beginning of the British Protectorate*
                           (C.M.S. 1937, first published 1921).
KIRK-GREEN, A. H. M.   'Abbega and Dorgu' in *West African
                           Review*, September 1956.
KUCZYNSKI, R. B.   *Demographic Survey of British Colonial
                           Empire*, vol. i, West Africa (Oxford 1948).
LAOTAN, A. B.   *The Torch Bearers of the Old
                           Brazilian Colony* (Lagos 1943).
LATOURETTE, S. K.   *The Expansion of Christianity* (7 vols.).
LLOYD, C.   *The Navy and the Slave Trade*
                           (London 1949).
LUCAS, J. O.   *Religion of the Yorubas* (Lagos 1948).
                           *History of St. Paul's Church Breadfruit,
                           Lagos* (Lagos 1954).
LUGARD, F. D.   *The Dual Mandate* (London 1922).
MACAULAY, Herbert   *History of the Development of
                           Missionary Work in Nigeria with Special
                           Reference to the Development of the
                           United African Church* (Lagos 1941).
MACFARLAN, Donald   *Calabar, the Church of Scotland Mission
                           1846–1946* (1946).
MATHIESON, W. L.   *Great Britain and the Slave Trade
                           1839–1865* (London 1929).
OKE, S. A.   *The Ethiopia National Church* (Lagos
                           1922).

OLIVER, R.   *The Missionary Factor in East Africa*
                            (London 1952).
PAGE, J.   *The Black Bishop, Samuel Adjai
                            Crowther* (London 1900).
PALMER, H. R.   (ed. and trans.) *Sudanese Memoirs*
                            (Lagos 1928).
PARRINDER, G.   *West African Religion* (London 1949).
                            Religion in an African City (i.e. Ibadan)
                            (London 1953).
PHILLIPS A. (ed.)   *Survey of African Marriage and
                            Family Life* (Oxford 1953).
PIERSON, Donald   *Negroes in Brazil* (Chicago 1942).
PINNOCK, S. G.   *A Romance of Missions* (Liverpool 1918).
PORTER, A. T.   *Creoldom: A Study of the Development of
                            Freetown Society* (Oxford 1963).
SADLER, G. W.   *A Century in Nigeria* (Baptist) (Nashville
                            U.S.A. 1950).
SHAW, Flora   *A Tropical Dependency* (London 1905).
(Lady Lugard)
SIMMONS, Jack   *From Empire to Commonwealth*
                            (London 1949).
S.M.A.   *100 Years of Missionary Achievement*
                            (Cork n.d. 1956).
S.M.A.   *Le Missionaire d'après Mgr de Marion
                            Bressillac* (Paris 1956).
SMITH, M. G.   'Slavery and Emancipation in two
                            Societies' (i.e. Zaria and Jamaica) in
                            Social and Economic Studies, University
                            College of the West Indies, vol. 3,
                            1954).
STOCKS, E.   *A History of the C.M.S.* (4 vols,
                            London 1899–1916).
TALBOT, P. A.   *The Peoples of Southern Nigeria*, vol. i
                            (London 1926).
THORPE, M. E.   *Swelling of Jordan* (London 1950).
TODD, J. M.   *African Mission: A Historical Study of
                            the S.M.A. since 1856* (London 1962).
TRIMINGHAM, J. S.   *A History of Islam in West Africa*
                            (Oxford 1962).
TYLECOTE, M.   *The Mechanic's Institutes of Lancashire
                            and Yorkshire 1851* (London 1957).
WALKER, F. D.   *Romance of the Black River* (London 1930).

	A Hundred Years in Nigeria (Meth.) (London 1942).
	Thomas Birch Freeman (London 1929).
WEBB, R. K.	*The British Working Class Reader* (London 1955).
WILTGEN, R. M.	*A Gold Coast Mission History 1471–1880* (Techny, Ill. U.S.A. 1965).
WISE, Colin G.	*A History of Education in British West Africa* (London 1956).

5 Unpublished Theses

FARRELL, J. K. A.	*The History of the Negro Community in Chatham, Ontario, 1787–1865.* (Ph.D. Ottawa 1955) on the attitude of Canadian Negroes to the idea of emigration back to Africa.
HERSKOVITS, J.	Liberated Africans and the History of Lagos Colony to 1886. (D.Phil. Oxon, 1960.)
MADDEN, A. F.	*The Attitude of the Evangelicals to the Empire and Imperial Problems 1820–1850* (D.Phil. Oxford).
ROTIMI, B. O.	*Recruitment and Placing of Teachers in Western Nigeria 1842–1927* (M.A. London 1955).
SMART, F. C.	*A Critical Edition of the Correspondence of Sir Thomas Fowell Buxton, Bart., with an account of his career to 1823.* (2 vols. M.A. London 1957).
WALSH, J. J.	*Catholic Contribution to Education in Western Nigeria* (M.A. London 1952).

Index

Abegga, 97, 129

Abeokuta (maps, pp. 24, 125, 166)

Politics and Commerce: origins and early wars, 19–22, 144, 170–1, 233; political rivalry in, 37–8; Hutton's visit, 58–9; Hutt Committee, 61–2, 66; Townsend's paper, 66–7; Beecroft's visit and report, 67–9; Dahomey Wars, 71–7; slave-trade treaty, 77; Dr. Irving consul, 80–1; rivalry with Lagos, 82, 193–4, 196–200, 202, 230; cotton production, 84, 167–8, 187, 190; local rulers, 100, 120; civil justice, 123; social life, 163; showpiece of British development, 167; overland route, 169; emigrants, 189–90; settlement scheme, 191–2; E.U.B.M., 198–200, 272; expulsion of Europeans, x, 117n, 201–3, 207, 227; 'Civilized Party', 230

Missions: generally, 58–9, 108–9, 156, 163, 270, and map p. 166. BAPTISTS: 48, 98, 136, 149. C.M.S.: establishment, ix, 19, 29, 32–5, 38–40, 59, 182; Crowther's ministry, 72; Townsend's 'Abeokutan Policy', 79–80, 82, 99, 167, 170–171, 193, 201; Industrial Institution, 85–6, 147, 150–2, 156, 164, 190; expansion, 95–6; the Ogboni, 110; mission houses, 112–13; case of Isaac Smith, 122 n.; education, 132 n., 135–7, 139 n., 155–7; building, 158; printing, 159; dispensary, 160; proposed See, 180–3, 186; Crowther's flock, 189–90; bishopric controversy, 193–6, 206–7, 229–32; expulsion of Europeans, x, 117 n., 201–3, 207, 227; Crowther's visitation, 227–8; Townsend barred, 227–8; then admitted, 229–30; James Johnson, 235–8, 260. *See also:* Ake; Igbein; Ikija. METHODISTS: 31, 57, 91–3, 244.

See also Ikija. ROMAN CATHOLICS (S.M.A.): 51, 142, 234

'Abeokutan Policy', 79–80, 82, 99, 167, 170–1, 201

Aboh (map, p. 205), x, 41; Obi of, 12, 130, 210

Abolition, *see* Slavery and the Slave-trade

Abonema, 216

Acland, Sir T., 62 n., 63 n.

Acland, T. A., 144

Ackoos, Acoos, *see* Akus

Adderley, C. B., M.P., 172

Adele, Oba, 54 n.

Ademola, 227–8

Adiabo (map p. 89), 95

Admiralty, 54, 71–2, 74–5, 81–2, 145, 209

Ado (maps, pp. 24, 125), 22, 96 n., 158, 172 n.

Adultery, 165, 249 n.

Africa, Scramble for, 233–4

African Aid Society, 48, 168, 191–2, 197 n., 211 n.

African Civilization Society, 12–13, 16, 17 n., 43, 144

African Merchants, Company of, 210–11, 283

African Salem Church, 273

African Steam Company, 40

African Times, 197 n., 211 n., 265 n.

Agaja Trudo, King of Dahomey, 22

Agbabu, 204, 234 n.

Agbakin, 20 n.

Agboola family, 122

Agricultural Society, 12, 13

Agriculture and farming: Buxton's plan for, 11–12, 16, 44, 92; in Missionary policy, 17–18, 27, 147, 213, 271; Waddell's policy, 45–6; and Freeman's, 91–2; in Badagri, 34; Calabar, 45–6, 55, 136, 141; Topo, 115–16, 141–2; Lokoja, 127; Abeokuta, 191–2. *See also:* Trust System; Cotton; Palm Oil, etc.

derer arrives, 95; Townsend's journeys, 96n; Gollmer's school, 152.
METHODISTS: establishment, xiv, 19,
31; Liberated Africans in, 13, 27–
30; Freeman's arrival and movements, 31–2, his Mission House, 31,
111, 113, 118; de Graft's work, 33–
34; Freeman claims British protection, 35–8; funds, 57; expansion, 91.
ROMAN CATHOLICS (S.M.A.): foundation of Topo (1876), 115
Baikie, Rose, 261n
Baikie, Dr. William B., 42, 129, 168–9,
209, 210, 215
Baillie, Rev. Zerub, 179 n.
Baloguns, 69
Banner Bros. and Co., Lagos, 243
Baptism, 126, 263; Anglican, 77;
Baptist, 98; Catholic, 3, 51; Methodist, 33; Presbyterian, 94, 103;
Crowther and, 39, 224, 279; of local
rulers, 100, 226; of polygamists, 105–
107; J. Johnson on names, 236, 269
Baptist Missionary Society (England):
founded (1782), 8; Jamaica mission,
44; followed by American Baptists,
47 (see next entry)
Baptist missions (of Southern Baptist
Convention, U.S.A., the S.B.C.),
xiii, xiv, 99n, 177–8; S.B.C. formed
from American Baptist Convention
(1844), 47; inspired by Evangelical
Revival, 8; work for emancipated
slaves, 44; Liberia, 47; finance, 57 n.;
Bowen goes to Badagri (1850), 13,
97–8; establishes missions (1854) at
Ijaye and Ogbomoso, 98; later at
Oyo, 99; mission houses, 112;
schools as aids to conversion, 134n;
boarding schools, 136, 141; mis-
sionary training, 149–50; American
Civil War, missionaries withdraw
(1862), 141, 157, 203; revival under
W. David (1875), 234; Baptist
Academy, Lagos (1883), 155; schism
(1888), Ebenezer Church formed,
268–9; then reconciliation (1894),
269; records, 279–80; journals, 281.
See also: Ijaye; Lagos; Ogbomoso
Barber, Miss, 137
Baring, Sir F., 72–3
Bart, Sergeant, 35–6
Barth, Carl, 127
Barth, Dr. Henry, 42, 97, 129
Basel: Seminary, 34, 95–6, 150; Evan-
gelical missionaries, 147

Bashorun, 122 n., 196–8, 200–1
Bathurst, 207
Beckles, Bishop, 186 n.
Beecroft, Capt. John, 40–1, 45, 60–2,
65, 67–71, 74–6
Bel, Fr., 115–16
Benin (map, p. 24). failure of early
Catholic mission, viii; later missions,
2–5; Portuguese spoken in, 6;
European houses in, 7; British war
on (1897), 234; Oba of, 2–3, 37, 76
Benin, Bight of, 35–6, 61, 64
Benin, R., 87 n., 216
Benton, P. A., 129
Benue R., x, 12, 93, 209, 233, 243
Berlin, 127; Conference (1885), 233
Biafra, Bight of, 45, 61, 179; Consul
for, 211 n.
Bible, The: supply of Bibles, 30, 126;
Bible reading, 139–40, lectures, 149;
translations: Efik, 131; Hausa, 129;
Ibo; 130, Yoruba, 72, 127, 181;
various, 8, 131–3, 140
Bickersteth, Edward, 39
Bida, 216, 239; Royal Niger Company
at, 272; Emir of, 97, 213 n., 214–15,
228 n.
Bishopric Controversy, 187, 193–6,
227
Bishopric Fund, 219–20, 222
Bishops: 'missionary', 185 and n;
'Jerusalem', 185 and n., 275; assis-
tant, 272; on the Niger, see Crow-
ther; for Yoruba area, 230–2; of
Native African Church, 273
Bloomfield, Bishop, 277
Blount, Capt., 55 n.
Blyden, Rev. Edward, 254, 266–7
Blyth, Rev. George, 45, 55
Boarders, see Schools, boarding
Bonny (map, p. 205), xiv; early history,
5, 6, 22; palm oil, 53; slave-trade
treaty, 55; chiefs ask for mission-
aries (1848), 56; Waddell's visits
(1846–50), 64, 93–4; Crowther
invited (1864), 97; literacy, 133;
Juju house, 227; and British rule,
272. Mission: C.M.S.: established by
S. Crowther (1864), 216, 220;
location, 217; school, 220, 226, 246;
Dandeson Crowther, 223, 245: Owu
Ogbo secret society, 225; new Niger
Finance Committee, 247, 254; case
of W. F. John, 248 and n; St.
Stephen's Church, 161; a source of
evangelists, 270

298

decline in missionary zeal, 3–4; Evangelical Revival and Anti-Slavery movement, 7–10; Niger Expedition of 1841, 11–13; view of civilization, 14–19; theory of church government, 174–9; and its application, 179 ff.; attitude of E.U.B.M. to, 202, 227. *See also* Discipline, Church

Churchill, Lord Alfred, M.P., 168, 192

Circumcision, 3

Civil Service, 197, 271

Civilization: and Christianity, *see* Christianity; Buxton, T. F.; around the Mission House, 126 ff., 162 ff.; and economic and social development, 167; E.U.B.M. and, 196–7; and education, 270; Traders and, 17–18, 102; Protestant view of, 14, 18, 108 n., 117; Bowen on Christianity's civilizing force, 98, 111; Evangelist's distrust of, *see* Evangelism. *See also*: African Civilization Society; Delta; Victorian civilization

'Civilized Party', 230 and n.

Clarendon, Lord, 79–82

Clarke, Rev. John, 44

Clarke, William, 98, 122

Clegg, Thomas: Manchester industrialist, 84; backs Venn's cotton-industry scheme, 84–6, 145; his business at Abeokuta, 167; and West Africa Company, 210, 212, 243; his death, 239

Clergy, 3, 9–10, 88, 147, 152, 164, 175–81

Cliff, Richard, 240

C.M.S. (Church Missionary Society): contribution to Nigeria's history, viii-x; foundation (1799), 8; Niger Expedition (1841), ix, 12, 29, 44, 61; and Liberated Africans, 13, 32; and commerce, 18–19, 210–15, 238–40, 243; Local Committee, Sierra Leone, 32; Abeokuta mission founded (1845), 32, 34, 38–9; Niger Expeditions of 1854–57, 41–2; finances, 57, 135, 219–20; Parent Committee and Venn, 61, 110, 120, 150–1, 153, 174 n., 175 n., 182, 185 n., 188, 191, 193, 222 n., 227, 229, 231, 245–9, 251–2, 254, 260, 261 n; and Hutt Committee, 61 ff., 96; memo urging government protec-

tion (1849), 66–9; occupation of Lagos, 67–9, 83 n.; and traders, 77–81, 84–6, 210, 238–40; 'Abeokutan Policy', 79–80, 99; expansion of missions, 95–9; and domestic slavery and polygamy, 105–8; and traditional politics, 110, 111 n.; role of the Mission House, 112, 114, 120, 121 n., 122 n., 162–5; linguistic studies, 127–30; building and architecture, 157–8; printing, 158–9, 191; medicine and health, 159–62; and Johnson family, 164; encourage African staff, 180; check to expansion, 204; Crowther's diocese, 206; Bishopric Fund, 219–20; Interior missions, 232–3; and Ashcroft, 240–241; Madeira conference, 246–7; on Baptismal names, 269; on evangelization through schools, 270; *Proceedings*, 281

Education: primary and boarding schools, 135–41; secondary and Grammar, 146, 148, 150, 152–4; industrial, 141, 156, 168, 210; Sunday, 131–2; adult, 13, 131–2; Higher, 26–7; Missionary training, 47, 150–2, 183, 222, 246, 251; Islington Training College, 33, 145, 150, 152, 165, 180–1, 223, 260. *See also* Schools

Coffee, 20, 55, 91

Coke, Dr. Thomas, 8

Colonial Office, 40, 180

Colonization Society, U.S.A., 47 n., 48

Coloured People, North American Convention of, 47 n.

Commerce, *see*: Trade; Traders; Economic development

Commercial Association, Abeokuta, 163

Commission, The (S.B.C.), 281

Commons, House of, 61–3; Select Committee (1865), 172–4, 203

Confirmation, 228

Congregational Missionary Society, London, 8

Conversion: and European culture, viii; appeal of Christianity, 8, 270; missionary aids to, 18–19; and hold of traditional custom, 20, 105–11, 121, 217, 223–6, 235, 264; of Liberated Africans, 25; of local rulers, 33, 39, 99–103, 226; of Crowther's mother, 39; of Africa the goal of Negro emigrants, 44; and commerce, 81, 213; Livingstone on, 90–1; Presbyterian caution in, 94; and domestic slavery, 105–8, 237; and the Mission House, 113–14, 121–4, 163–5; and education, 131 ff.; and church organization, 175; mass conversions in Brass, 216; and polygamy, 225–6; at Abeokuta, 39–40, 69, 170, 201–2; Badagri, 33; Brass, 216; Ibadan, 121; Ijaye, 98; Ogbomosa, 121–2, 228; Onitsha, 217, 264; some converts who were ordained, 150

Coote, Capt., R., 145

Coppin, Rev. W. T., 257–60

Coral Fund, 137, 150

Costume, *see* Dress

Cotonou, 22

Cotton: in Buxton's *Remedy*, 11; and British industry, 48, 145, 156, 167, 210, 285; and anti-slave trade treaties, 55, 69, 73; and Trust system, 84–5; on Freeman's model farms, 91; in Abeokuta Industrial Institution, 156; effect of American Civil War, 167, 210, 213

Trade in: 161, 187; at Abeokuta, 84, 156, 167–8, 187, 199; Ibadan, 156, 167–8; Ijaye, 156; Onitsha, 209–10, 213. *See also* Clegg, Thomas

Courdioux, Fr., 280

Cowries, 38, 59

Creek Town (map p. 89), 41, 53, 65, 93–4, 100, 119, 131, 136, 139, 179 n.

Creoles, 78 n.

Cross River, ix, 93

Crowther, Archdeacon Dandeson, youngest son of Bishop S. Crowther; educated at Lagos Grammar School, 223; enters ministry, trained at Islington, 165, 223; missionary at Bonny, Superintendent of Delta and Lower Niger stations, 223; becomes Archdeacon, 165, 223; on polygamy, 226; member of Finance Committee, Lagos, 245, 250, 252; his suspension, 253–4

Crowther, Josiah, son of Bishop S. Crowther; cotton industry training, 147, 156, 165; in cotton business, 165; challenges Townsend's authority, 189; buys shares in West Africa Company, 210; Company's agent at Lokoja, 212; Agent-General, 214, 239; his dismissal, 242

Crowther, Bishop Samuel (Ajayi): boyhood slavery, 20; liberation and baptism, 26; education and training as teacher, 26, 32–3; character and ability, 26–7, 72–4, 83 n., 183, 249; linguistic studies, Yoruba, 26, 72, 127–9, Ibo and Nupe, 130, Ijaw, 132; Niger Expedition of 1841, 27, 29, 32, 97, 180; on training of Africans, 30; is ordained, conducts services in Yoruba, 33; Abeokuta mission, 34, 38–9, 182; Igbein mission, 39, 183; called to London, Hutt Committee, 72–4; Niger Expedition 1854, 42–3; his family, 72, 165, 189–90, 193, 273 n.; visits Ketu, 96; Niger Expedition, 1857, 97, 183–184; Onitsha mission, 97, opens first school, 135; on domestic slavery, 105; polygamy, 106; 225, supports Macaulay, 153, 190; on carpenters, 157; interest in vaccination, 159; advises on cotton production, 167; and annexation of Lagos, 171 n.; rivalry with Townsend, 180–6, 195; Venn's support, 180–5, 194, 227; hero of Abeokuta emigrants, 189; and Settlement scheme, 191–2; his political eclipse at Abeokuta, 191–204; his consecration, 194–5, 206–7, 230, 275; and E.U.B.M., 202–3; his diocese defined, 206–8, 274–7; controls part of Yoruba Mission, 207–8; expansion of Niger Mission, 215–16, 231; his missionary methods, 216–227; salary, 219 n., 223 n.; founds missions at Bonny and Brass, 220, at Kalabari, 221; staff recruitment, 222; work as Bible translator, 223–4, 227; on Christianity, 203, 219–25; and Nationalism, 224–5; and Yoruba Mission, 227–32; visitation to Abeokuta, 227–8; called to London, 229; favours James Johnson for bishopric, 231; his resignation, 235; the *Victoria* affair, 239–40; gets a steamer, 239–41; visits England again (1877), 240; and Ashcroft, 240–4, 247; his replacement mooted, 244; Wood enquiry, 245–7; further enquiries, 248–50; ousted from bishopric, 250–5; last visit to London (1889), 252; resigns from Finance Committee, 253–4; his supersession, 254,

301

269–71, and death, 254; his successors, 255, 272; caricatured by Europeans, 261; and Rose Baikie, 261 n.

Crowther, Mrs. S., wife of Bishop S. Crowther, 39

Crowther, Samuel jun., son of Bishop Crowther; Medical studies in London, 72, 145, 165; secretary and assistant to Dr. Irving, 81 n., 161, 190; Medical Officer, Abeokuta, 85, 196; his dispensary, 160; manager Abeokuta Industrial Institution, 85; enters cotton business, 160, 165, 167 n., 187, 192; challenges Townsend's father's authority, 189–90; the Settlement Scheme, 191–2; visits England, 193; West Africa Company's agent at Lokoja, 212

Cuba, 63; emigrants from, 30 n., 40, 49–51, 112, 163–4

Customs: and Islam, 1; and traditional religion, 5; and civilization, 14; superstitious, 65 102; English, 197; customary rites, 217, 224–6; 'barbarous', 265; research on, 269; local generally, 270 n.

Customs duties, 197–8; Customs House at Abeokuta (map p. 166), 198–200

Da Silva, Senhor Lazaro, 157 n.

Dahomey: Methodist expansion into, ix, 33, 35, 57; Catholic mission to, 50; political expansion into Egbado area, 21–2; Wm. Duncan vice-consul to, 61; the 'menace' of, 66–7; British policy towards, 68, 168, 170; invades Abeokuta, 71, 136, 146, 169; Dahomey War, 72, 74, 80; British blockade, 74; French interests in, 233; King of, 147

Dancing, 49, 116, 263, 269

Darwin, Charles, 263 and n.

Davey, Rev. and Mrs., 26

David, Rev. W. B., 234, 268 n.

Davies, Capt. J. P. L., 145, 153, 213, 229 n.

Davies Samuel, 145

Dayspring, wreck of the, 97

de Graft, William, 31, 33–4, 97

Delany, Dr. Martin R., 48, 168, 191–192

Delta (of Niger R.): rulers' resistance to conversion, viii, 216; early history, 5; trade the civilizing force in,

302

46; absence of emigrants in, 51; English traders, 53–4, 213, 219; British interest in, 61, 64; Trust system, 86; opposition to Niger shipping, 168, 209–11; demand for education, 270

Mission: C.M.S.: ix, 220; Niger Mission's expansion into, 215, 221, 226; Archdeacon D. Crowther, 165, 223; Bishop S. Crowther's removal, 252–5. PRESBYTERIANS: Waddell on trade in 46

Denham, Capt., 62, 74

Dennard, 98

Dickens, Charles, 13 n.

Discipline, Church, 226, 236, 249, 259–60, 261 n.

Dispensary, at Abeokuta, 160

Divorce, 123

Dobinson, Archdeacon, 270 n.

Doherty, Governor, 27–9

Domingos, Prince of Warri, 3

Don Pacifico Affair, 62, 63 n.

Dorgu, 129

Dosumu, 120, 174, 235 n.

Dove, Rev. Thomas, 29–30

Dress: missionary attitude to, 14–15, 32–3, 65, 109; Africans' attitude to European, 217, 236, 269

Drunkenness, 237, 239, 249 n., 261

Duke Town (map, p. 89), Presbyterian mission, 42, 93–4, 113, 179; human sacrifice at, 64; S.A.I.S.C. formed, 65; independence of Mission House, 119–20; desire to learn English, 134; traders at, 53

Duncan, William, 61, 78

Durham, 152, 250

Dutch, see Holland

Ebenezer Baptist Church, Lagos and Abeokuta, 268–9

Ebute-Metta (map, p. 125), 157, 200, 202, 260 n.

Economic development; in missionary policy, x, 19, 66, 84, 108, 155–6, 165, 167, 169, 271, 273; in Calabar, 55; at Abeokuta, 69, 80

Ede, 40

Eden, Rev. F. N., 250–1, 253–4

Edgar, J., 238, 243 n.

Edgerley, Samuel, snr., 46, 64, 118–19, 130

Edgerley, Rev. Samuel H., jun., 179

Edinburgh, 60, 93, 146; Presbyterian Missionary Society of, 8

Education, missionary contribution to: as aid to conversion, viii, 18, 218 ff., 263, 270; influence of missionaries, x, of traders, 6; in early failures, 4; in Buxton's plans, 10; Crowther's early work for, 26, 30–1; Hutt on, 61; of middle class, 87; social effects of, 142–3, 266, 270; Lord Macaulay on, 173; and African pastors, 177–8; and ability, 179; G. W. Johnson and, 197; of Africans in England, 197, 221; E.U.B.M. and, 202; of agents' children, 246; manual labour in, 141, 143–4

In practice: general, 133–4, 147, 151, 246; primary, 133–42, 156; secondary, 30–1, 152–5; higher, 26; technical, 126, 144–7, 221, 271; moral and religious, 133, 137; teacher training, 30. *See also* Schools, etc.

Efik: people, 42 n., 53, 93, 96 n., 102, 265; language, 53, 93, 104, 130–1, 140, 149, 158

Egba people: effect of Yoruba Wars, 19, 20 n., 21–3; Liberated Africans, 25, 29, 39; welcome Akitoye, 37; receive C.M.S. missionaries at Abeokuta, 38, 66; community in Lagos, 50 n.; favourable to British, 66, 73–4, 79, 82, 169; and traffic in slaves, 72, 78; Ogboni society, 101, 111 n.; Christian colonies, 116; British aid against Dahomey, 167, 169–71; traders at Abeokuta, 189; Abeokutan politics, 190, 193–4, 196; E.U.B.M., 197 n., 198–202, 224, 227–9; *Iwe-Irohin*, 201; Seriki of the, 198, 201; converts among, 225 n.; Egba Government, 272

Egbado, 21–2

Egbaland (map, p. 24), 170

Egbo Bunko, 109–10

Egga (maps, pp. 24, 205), Methodist mission, xiv–xv; C.M.S. mission (Kippo Hill), 97, 215–17, 244, 252–253; West Africa Coy. at, 239, 272; limit of Niger Expedition (1841), 12, 66

Egun, 22, 53, 77, 116

Egungun, 128 n.

Ekiti (map, p. 24), 233, 270; Ekiti Parapo alliance, 233

Ekpe (society): symbol of cultural identity, 7, 22, 53; and human sacrifice, 65, 102; Waddell and,

94 n., 101–2, 109–10; and twin murder, 102; ban on Mission House, 119

El Kanemi, 1

Eleja, 254 n. *See also* United Native African Church

Elliott, Rev. M. J., 244, 257–8

Ellis, 145

Emigrants: their origins, 30 n.; aspirations as middle class, 165

From the Americas: 43–52, 191–2; Brazil, 40–1, 49–52, 59, 112, 139 n., 155–7, 163–4; Canada, 48; Cuba, 30 n., 40–1, 49–51, 112, 163–4; Jamaica, 13, 44–6, 48; U.S.A., 47–8; West Indies: Schön, on 30, emigration movement, 13, 28, 30, 41, 179 n.; some to become consuls 78 n.

From Sierra Leone ('Liberated Africans'): recruited for 1841 Niger expedition, 11–13; and Bishop Crowther, 20, 26–7; their dream of home, 25; occupations, 25–6; and as mission workers, 26–7; their difficulties, 27; movement to emigrate to Nigeria, 27–8, 50; first emigrants to Badagri and Yoruba country, 28–32; Buxton's *Remedy* applied, 29–30; Schön on; 30; Abeokuta, 32–4, 38–40; whether British subjects, 37 n., 79, 123; later emigration from Sierra Leone, 41–3; coastal traders, 59; petition Hutt Committee, 62; introduce English language, 66; consuls, 78 n.; call for, in Nigeria, 100; in Abeokuta, 79, 113–14, 189–90, 193, 196–8, 201; traditional customs, 109–10, 217; as artisans, 112–57; justice among, 123; as linguists, 127; as teachers, 94, 138–9; training of, 147–52; apprenticeship, 155–8; benefits of Mission House civilization, 163–4; and E.U.B.M., 196 ff.; Niger expeditions of 1854 and 1857, 209–10; Niger Mission and trade (1864–74), 212–15, 222; European attitude to, 260–1

England, Church of, *see* Anglican Church

English language: use by emigrants, 42, 66, 139, 149; by African traders, 46, 53, 133; demand for teachers of, 56, 133–4; use by local rulers, 46, 53, 100, 133; in mission schools, 139–40, 154, 202; in training

303

missionaries, 149–50; and ordination, 277

Epe (map, p. 24), 76, 283

Ephraim, Duke, of Calabar, 87, 120

Escravos, R., 7

Esperanza Felix, The, 20

E.U.B.M. (Egba United Board of Management), *see* Egba

Europe: missionary zeal in, 3; missionary movements in, 7–8, 44; trade with, 17

European; conception of religion, 2; cultural influences, 6–7; 10, 267, attitude of superiority over Africans, 179–81, 260–4; Crowther's attitude to, 186; expulsion of, from Abeokuta, 201–2; as protectors, 234–5

Evangelism: Evangelical Revival and Anti-Slavery movement, 7–10, 18, 107; Evangelicals and Humanitarians, 18, 91; and Buxton's principles, 29, 39, 75, 111; Liberated Africans and, 44–5, 47 n., 52; Evangelical Party in England, 61; Abeokuta's promise for, 66; Crowther on, 97, 202–3, 207, 216, 218; indifference to politics and distrust of 'civilization', 99, 108–11, 126, 251; local languages as aid to, 126–131; and education, 134, 218; training of African evanglists, 149; and Church government, 174–6; German evanglicals, 181; scriptural example, 185; Africans' work in field, 193; their salary, 222; supply of, 270; and anthropologists, 251, 264; technicque of, 264

Exploration, *see* Niger R. and Expeditions

Eyambas, 265

Eyo, King of Calabar: state room, 6 n.; welcome to emigrants, 41; relations with missionaries, 56, 65, 81 n., 87, 94 n., 119; French advances, 60; character, 100; his son, 46, 109–10

Ezzidio, John, 26

Faith, Association for the Propagation of the, 137, 282

Fanti people, 31, 33, 35, 39, 157, 173

Farming, *see* Agriculture

Faulkner, Rev. V., 238

Fergusson, James, 30

Fergusson, Governor of Sierra Leone, 40

Fergusson, Dr., of Liverpool, 45

Fernando Po, 41, 44–5, 59, 61, 65, 70, 75, 113, 130

Finance, Missionary, 57–8, 90, 174, 180, 219–20, 229, 256–7, 271, 275; C.M.S. Grammar School, Lagos, 153; Niger Finance Committee, 245, 247, 252–4, 260

Flint, Dr., 242

Fodio, Usuman dan, 1

Foote, Capt. John, 36, 80, 192 n., 193, 201

Forbes, Capt. F. E., 75, 80, 144

Forbes, Commander T. G., 75

Forcados, R., 7

Foreign Missions Committee, *see* Presbyterian Missions

Foreign Mission Journal (S.B.C.), 281

Foreign Office, 37 n., 61, 75, 80, 119, 192, 209, 214, 228, 282

Foresythe, C., 229 n.

Forster and Smith, 59

Fourah Bay College: Crowther first student at, 26; his interest in, 221, 222 n.; exapnsion of, 30; E. Jones principal, 41; T. B. Macaulay, 150; link with Durham, 152; Nathaniel King, 162; Henry Johnson jun. at, 164; Theophilus King, 181; James Johnson at, 229

France, French: missions, xiv, 108 n., *see also* S.M.A.; Revolution, 8; rivalry in Nigeria, 11 n., 12, 54, 60, 75, 210, 215, 233–4; merchants at Badagri, 22; effect of Communes, 115, 204; and of Franco-Prussian War, xiv–xv

Fraser, Louis, 50 n., 78

Freeman, Henry Stanhope, Governor of Lagos, 171, 174

Freeman, Rev. Thomas Birch: Superintendent, Methodist mission, Gold Coast, 31; founds Badagri mission (1842), 31–2; and returns to Cape Coast, 33; appeals for British protection for Badagri, 35–8; opens Ogbe, 39; mission funds, 57; his *Journals*, 90; attempts expansion, 91; interest in agriculture, 91–2; overspends funds, is removed (1857), 92–3; attitude to slavery, 105; his mission house, Badagri, 111, 113; his 'rules for schools', 139–40; on missionaries' salaries, 256 n.; rejoins mission as pastor (1873), 257; Methodist crisis (1884), 260; his death (1890), 257

Freetown, Sierra Leone: Liberated Africans in, 25–6, 42 n., 59, 164, 266; secondary education in, 30; ordnance depot at, 74; S. Crowther jun. at, 145; C.M.S. Grammar School, 146, 148, 152–3, 190, 221
Friend of Africa, The, 12
'Friends of Africa', 86, 144
Fulani, 4; Empire, 2, 19; *jihad*, 49
Funerals, 110, 118
Furniture, 6, 11, 46, 54, 91, 112

Gaboon, 140, 223
Gambia, 42, 140
Garrick, Rev. John, 216 n.
George, Isaac, 212
George, J. (Lisa Oba), 228
German: missionaries in C.M.S., xiii, 144, 150, 181–2, 187, 202; traders on the Niger, 215; in Cameroons, 233
Gezo, King of Dahomey, 71, 74
Ghana, *see* Gold Coast
Gibson, Jack T., 146
Ginger, 69, 213
Glasgow, 8, 58
Glover, Capt. (later Sir) John Hawley, 169, 171 n., 194, 199, 200–1, 240, 234
Gold, 23
Gold Coast (Ghana): as British colony, 16 n., 35–6, 42, 62, 115, 173, 206, 274; Methodist mission, xiv, 13, 31, 57, 91, 143, 234, 256–7, 260; C.M.S. mission, 13; Basel mission, 147. *See also* Cape Coast
Goldie, George Taubman, 233, 242–244, 264
Goldie, Rev. Hugh, 104–5, 110 n., 131, 133, 215
Gollmer, Rev. C. A.: early life and arrival (1854) with C.M.S. Badagri, 34, 183; his work there, 38, 59, 152; testifies before Hutt Committee, 61–2, 95, 127; part in Lagos occupation, 68–70, 74, 76, 80; moves to Lagos (1852), 77–8; defence of C.M.S. House (1854), 83 n.; his work at Lagos, 112–13, 124 n.; translates into Yoruba, 127; and Crowther, 181; moves to Ake, 183
Great Exhibition, of 1851, 16
Grey, Earl, 16, 173
Guinea corn, 33
Gum opal, 54
Guns, gunpowder, *see* Arms and ammunition

Gurney, Edward, 168
Gwandu, Sultan of, 213 n.

Halligey, Rev. J. T. F., 260
Hamburg, 75
Hamilton, Henry, 179 n.
Harden, J. M., 98, 157
Harden, Sarah M., 138
Harrison & Co., 87 n.
Harrison, Dr. A. A., 109, 137–8, 139 n., 156, 161–2, 187–8
Harrowby, Lord, 62 n., 63 n.
Hart, Joshua, 226
Hastings, Sierra Leone, 34, 164
Hausa: language, xv, 95, 127, 129–30; people, 25, 29, 43, 51, 95–6, 129, 169, 200
Hausaland, 1, 95–8, 215
Heath, Commander L. G., 70
Heathenism, 108, 235–6
Henry Venn (Mission steamer), 241, 243, 247, 251
Hewan, Archibald, 160, 179 n.
Hewett, Consul, 249 n.
Highbury Training College, 152
Hinderer, Rev. David: and a Muslim missionary, 20 n.; arrives in Badagri (1849), 95; at Ibadan (from 1853), 96, 167, 183, 229–30; his character, 121–2, 137 and n.; translates into Yoruba, 127; protests at Macaulay's removal, 150 and n.; and D. Olubi, 152; and Johnson family, 164; and Bp. Crowther, 181–2, 188, 231, 242, 264
Hoare, Gurney, 62
Hoch, Rev. J. J., 122 n.
Holland and the Dutch, 4, 22, 64
Holland, Jacques & Co., 213, 240
Holley, Fr., 280
Holy Ghost Fathers, xiv, 234
Holy Trinity Church, Lagos, 260 n.
Home and Foreign Journal (S.B.C.), 281
Hope Waddell Institute, 148
Horton, Dr. James Africanus, 146
Hospitals, 159–60, 271; Colonial Hospital, Sierra Leone, 162
Huber, I. V., 144
Hueda, 22
Human sacrifice, 3, 60, 64–5, 77, 93, 102, 119, 210, 226
Humanitarians: 137, 144; and evangelicals, 18, 91, 108
Hunt, James, 211 n., 263
Hutchinson, Consul, 119

Margi people, 129
Marmon, Capt., 23
Maroons, 25 and n.
Marriage, matrimony, 3, 15, 51, 77, 123, 163, 275; intermarriage, 217. *See also* Polygamy
Marsh, Mr., schoolmaster, 34
Maser, Mr., missionary, 194
Marshall, Sir James, 315, 142, 280
Martin, John, 77
Martinez, Domingo José, 38, 59, 70, 78
Mass, The, 6, 142
Massaba, 214-15. *See also* Bida, Emir of
Mba, 226 n.
Medicine, 159-62, 272
Meffre, José, 235 n.
Melbourne, Lord, 12
Mendes, the, 27
Methodist missions (Wesleyan Methodist Missionary Society): organization, xiii–xiv, 7–8, 45; expansion into Dahomey, xiv: W.M.M.S. established (1813), 8; and 1841 Niger Expedition, 13, 29; Kilner the General Secretary (1876), 255, 257–8; publications, 281–2; records, 279; Methodist families in Calabar, 42, 123 n.; influence on architecture, 157; attitude to Church government, 177–8; Methodist crisis, 255–260; Methodist Conference, xiii, 59; minutes, 281; Primitive Methodists, 234. Stations at: ABEOKUTA, 98, 112, 114; Freeman's visit, 35; sends Morgue then Bickersteth, 39; Townsend's visit, 69; Methodists support traders, 79; Thos. Champness, 93; Alake's declaration, 192; Mission House at Ogbe, 228; BADAGRI, occupied by Freeman, 31–2; de Graft, 33, Annear; 34; Sgt. Bart's protection, 35; withdrawn, 36; chieftaincy dispute, 36–8; Martin, 77–8; Freeman's house, 111–12; Annear attempts to protect a slave, 118; EGGA, 244; LAGOS, Theological dept., 147, 257; Boys' High School, 153–4, 257–8; School Board, 258; separation from Gold Coast District, 203, 234, 256, 260, 272; Methodist Crisis, 254–60; formation of United Native African Church, 254, 268–9; Disciplinary Committee, 259–60; CAPE COAST, 31, 35, 91–3, 257; SIERRA LEONE, 29–30

Merchants, *see* Traders
Merchants, Company of African, 210–11, 283
Mewu, Chief, 70, 77–8, 80, 82
Middle class, African: missionary encouragement of, xiii; opportunities of, xiii; Buxton's aim, 11; and early Missionary objective, 17–18, 87, 165, 214; Calabar chiefs as traders, 46; Venn's plan for training of, 84; later reversal of missionary policy, 269–70; modern outlook on, 273
Miller, Edward, 46
Miller, Miss, *see* Sutherland, Mrs.
Miller Brothers, Alexander, 213
Mills: grinding, 156; steam-powered, 156, 163
Mills, Mr., 155 n.
Milum, Rev. John, 244, 256–7, 260
Mission House, The; as symbol of civilization, 111–12, 114, 117–18; independence of, 119, 121–3; as refuge, 120–1; asylum in, 65, 117; Civilization round, 126 ff., 142–3, 162 ff., 167, 216–18; as boarding school, 136–41; social effects of, 142–3, 163–4, 223; carpenter's shed, 157; some destroyed, 201; house at Bonny, 220; financial support of, 221; separateness of, 235; in missionary publications, 261–2
Missionaries, Christian: difficulties of, 4, 7, 20, 56–7, 94, 174, 208–11; nineteenth-century movements, 7–10 n., 52; training of, 7, 126, 148–9, 222; and slavery, 10–13, 103–5, 165; functions of, 18, 175–9; services not for hire, 56–7; and traders, 57, 123, 154, 208–9, 211–12, 219, 238; and the state, 90 ff.; attitude to polygamy, 103, 105–8, 165, 225–6; denominational competition, 108; study of languages, 126 ff.; influence on architecture, 156–8; on printing and publishing, 158–9; on medicine, 159–62; view of adultery, 165; political ('Abeokutan') policy, 167; attitude to local wars, 170–1; European 'superiority', 179–80; expulsion of, from Abeokuta, 201–2, 227; confined to coast, 203; Crowther's advice to, 216; salaries of, 222, 246, 256; and secret societies, 225; expansion inland, 233–4. *See also under various denominations*

315

budget and source of funds, 57–8; and Hutt Committee, 61 ff.; tolerance of domestic slavery, 105; relations with local rulers, 120–1; and independence of Mission House, 123; Yoruba Church, 151, 186, 208; training of missionaries in, 152; industrial establishments, 168; Townsend v. Venn and Crowther, 182–91; Crowther's jurisdiction limited, 206–8; 264; missionaries' pay, 222; Otta Church's native airs, 225 and n.; Crowther's and Townsend's visits to Abeokuta, 227–30; separate Yoruba bishopric proposed, 231; J. B. Wood secretary, 245; J. A. Robinson joint secretary, 250; Crowther's successor given jurisdiction, 255. *Methodist mission:* 91, 93 n., 177, 234, 256–7

Yoruba language, 26, 33, 72, 79, 127–8, 151, 181, 207; literature, 159; poetry, 127

Yoruba Mission, *see under* Yoruba (country) *above*

Yoruba people, 15 and n., 20, 25–7, 29, 50 n., 67, 72, 96 n., 146; S. Johnson's *History of*, 269, 271 and n.

Yoruba Wars, xv, 19–23, 25, 49, 233, 284

Zambezi River, 90